URBAN WATER ENGINEERING AND MANAGEMENT

URBAN WATER ENGINEERING AND MANAGEMENT

MOHAMMAD KARAMOUZ ALI MORIDI SARA NAZIF

CRC Press
Taylor & Francis Group
Boca Raton London New York

CRC Press is an imprint of the
Taylor & Francis Group, an **informa** business

CRC Press
Taylor & Francis Group
6000 Broken Sound Parkway NW, Suite 300
Boca Raton, FL 33487-2742

© 2010 by Taylor and Francis Group, LLC
CRC Press is an imprint of Taylor & Francis Group, an Informa business

No claim to original U.S. Government works

Printed in the United States of America on acid-free paper
10 9 8 7 6 5 4 3 2 1

International Standard Book Number: 978-1-4398-1310-2 (Hardback)

This book contains information obtained from authentic and highly regarded sources. Reasonable efforts have been made to publish reliable data and information, but the author and publisher cannot assume responsibility for the validity of all materials or the consequences of their use. The authors and publishers have attempted to trace the copyright holders of all material reproduced in this publication and apologize to copyright holders if permission to publish in this form has not been obtained. If any copyright material has not been acknowledged please write and let us know so we may rectify in any future reprint.

Except as permitted under U.S. Copyright Law, no part of this book may be reprinted, reproduced, transmitted, or utilized in any form by any electronic, mechanical, or other means, now known or hereafter invented, including photocopying, microfilming, and recording, or in any information storage or retrieval system, without written permission from the publishers.

For permission to photocopy or use material electronically from this work, please access www.copyright.com (http://www.copyright.com/) or contact the Copyright Clearance Center, Inc. (CCC), 222 Rosewood Drive, Danvers, MA 01923, 978-750-8400. CCC is a not-for-profit organization that provides licenses and registration for a variety of users. For organizations that have been granted a photocopy license by the CCC, a separate system of payment has been arranged.

Trademark Notice: Product or corporate names may be trademarks or registered trademarks, and are used only for identification and explanation without intent to infringe.

Library of Congress Cataloging-in-Publication Data

Karamouz, Mohammad.
  Urban water engineering and management / Mohammad Karamouz, Ali Moridi, Sara Nazif.
    p. cm.
  Includes bibliographical references and index.
  ISBN 978-1-4398-1310-2 (hardcover : alk. paper)
  1. Water-supply engineering. 2. Municipal water supply--Management. 3. Water quality management. I. Moridi, Ali. II. Nazif, Sara. III. Title.

TD345.K27 2010
628.1--dc22   2009035201

Visit the Taylor & Francis Web site at
http://www.taylorandfrancis.com

and the CRC Press Web site at
http://www.crcpress.com

To my wife, Setareh, and my children, Saba Sahar and Mehrdad M. Karamouz

To my parents, Maryam and Amir, and my wife, Elahe A. Moridi

To my parents, Azar and Mostafa, and my brother, Mohsen, who taught me how to love others

Contents

Preface xix
Acknowledgments xxiii
Authors xxv

Chapter 1   Introduction 1
       1.1   Urban Water Cycle 2
             1.1.1   Components 2
             1.1.2   Impact of Urbanization 4
       1.2   Interaction of Climatic, Hydrologic Process, and Urban Components 5
             1.2.1   Climatic Effects 5
             1.2.2   Hydrologic Effects 6
             1.2.3   Qualitative Aspects 7
             1.2.4   Greenhouse Effect 8
             1.2.5   Urban Hot Islands 8
             1.2.6   Cultural Aspects 9
       1.3   Urban Water Infrastructure Management 9
             1.3.1   Life Cycle Assessment 10
             1.3.2   Life of Urban Water Infrastructure 11
             1.3.3   Environmental, Economic, and Social Performances 12
       1.4   Urban Water Cycle Management 12
       1.5   Summary and Conclusion 14
       References 14

Chapter 2   Governance and Urban Water Planning 17
       2.1   Introduction 17
       2.2   Water Sensitive Urban Design 18
       2.3   Water Governance 19
             2.3.1   Consequences of Poor Governance 20
             2.3.2   Water Governance at the Regional Level 20
       2.4   Governance Models 21
             2.4.1   Generic Governance Models for Water Supply 21
             2.4.2   Building a Governance Model 22
                     2.4.2.1   Choosing Principles 23
                     2.4.2.2   Good Governance Processes 23
       2.5   Absence of Public Participation 27
             2.5.1   Conventional Approaches 27
             2.5.2   Information Technology Approaches 27
             2.5.3   International Approaches 28
       2.6   Remaining Issues 28
       2.7   Urban Planning 29
             2.7.1   Regional and Local Strategies 30
             2.7.2   Preparing a Land and Water Strategy 30
             2.7.3   Master Plans 31

2.8	Urban Water Planning			32
	2.8.1	Paradigm Shift		32
		2.8.1.1	Shift from the Newtonian Paradigm to the Holistic Paradigm	32
		2.8.1.2	Paradigms of Urban Water Management	37
	2.8.2	Land Use Planning		38
		2.8.2.1	Dynamic Strategy Planning for Sustainable Urban Land Use Management	38
		2.8.2.2	Urban and Regional Plans and the DSR	41
		2.8.2.3	DSR Dynamic Decision Support System	42
2.9	Urban Water Assessment			44
	2.9.1	Availability and Demand		45
		2.9.1.1	Importance of Water Assessments	45
		2.9.1.2	Need for a Water Knowledge Base	45
		2.9.1.3	Objectives	45
		2.9.1.4	Demand as a Function of User Behavior and Preferences	45
		2.9.1.5	Importance of Monitoring and Gauging Systems	45
		2.9.1.6	Environmental Impact Assessments	45
		2.9.1.7	Risk Assessment Tools	46
		2.9.1.8	Risk Management	46
		2.9.1.9	Precautionary Principle	46
	2.9.2	Communication and Information Systems		46
		2.9.2.1	Communication for Enhancement of Stakeholder Involvement	46
		2.9.2.2	Information Needs for Stakeholder Involvement	46
		2.9.2.3	Stakeholder Communication Strategies	46
		2.9.2.4	Openness and Transparency	47
		2.9.2.5	Exchange of Information	47
	2.9.3	Water Allocation and Conflict Resolution		47
		2.9.3.1	Issues in Allocation	47
		2.9.3.2	Allocation by Market-Based Instruments	47
		2.9.3.3	Using Valuation to Resolve Conflicts	47
		2.9.3.4	Resolution of Upstream–Downstream Conflicts	47
		2.9.3.5	Conflict Management Techniques	47
		2.9.3.6	Valuation by Conflict Resolution Methods	47
		2.9.3.7	Valuation Research on Environmental Benefits	48
	2.9.4	Regulatory Instruments		48
		2.9.4.1	Direct Controls	48
		2.9.4.2	Economic Instruments	49
	2.9.5	Technological Instruments		50
		2.9.5.1	Technological Advances Toward Sustainability	50
		2.9.5.2	Research and Development in Technology	50
		2.9.5.3	Technology Assessment	50
		2.9.5.4	Technological Choices	50
2.10	Concluding Remarks			50
References				51

Chapter 3 Urban Water Hydrology ... 53
    3.1 Introduction ... 53
    3.2 Rainfall–Runoff Analysis in Urban Areas ... 53

		3.2.1	Drainage Area Characteristics	53
		3.2.2	Rainfall Losses	54
		3.2.3	Time of Concentration	55
	3.3	Excess Rainfall Calculation		57
		3.3.1	Interception Storage Estimation	57
		3.3.2	Estimation of Infiltration	58
			3.3.2.1 Green and Ampt Model	60
			3.3.2.2 Horton Method	62
			3.3.2.3 Modified Horton Method	67
			3.3.2.4 Holtan Method	68
			3.3.2.5 Simple Infiltration Models	70
		3.3.3	Depression Storage	72
		3.3.4	Combined Loss Models	73
			3.3.4.1 Soil Conservation Service Method	73
			3.3.4.2 Other Combined Loss Models	79
	3.4	Rainfall Measurement		79
		3.4.1	Intensity–Duration–Frequency Curves: Advantages and Disadvantages	79
	3.5	Estimation of Urban Runoff Volume		81
		3.5.1	Rational Method	81
		3.5.2	Coefficient and Regression Methods	84
		3.5.3	USGS Urban Peak Discharge Formulae	84
	3.6	Unit Hydrographs		87
		3.6.1	UH Development	88
		3.6.2	Espey 10-min UH	88
		3.6.3	SCS UH	93
		3.6.4	Gamma Function UH	96
		3.6.5	Time–Area UHs	98
		3.6.6	Application of the UH Method	102
	3.7	Revisiting a Flood Record		104
		3.7.1	Urban Effects on Peak Discharge	104
		3.7.2	Flood Record Adjusting	107
	3.8	Test of the Significance of the Urban Effect		110
		3.8.1	Spearman Test	111
		3.8.2	Spearman–Conley Test	113
	3.9	Hydrologic Modeling System (HEC-HMS)		115
		3.9.1	Modeling Basin Components	116
		3.9.2	Analysis of Meteorological Data	117
		3.9.3	Rainfall–Runoff Simulation	117
		3.9.4	Parameters Estimation	117
		3.9.5	Starting the Program	118
			3.9.5.1 Project Definition Screen	118
			3.9.5.2 Input Data in HEC-HMS Model	119
			3.9.5.3 Output Data in the HEC-HMS Model	119
	Problems			119
	References			124
Chapter 4	Urban Water Supply and Demand			127
	4.1	Introduction		127
	4.2	History of Water Supply Development		128
	4.3	Water Availability		128

	4.4	Water Development and Share of Water Users	130
	4.5	Man-Made and Natural Resources for Water Supplies	131
	4.6	Supplementary Sources of Water	131
	4.7	Water Treatment Methods	133
	4.8	Water Supply System Challenges	134
	4.9	Water Demand	134
		4.9.1 Fluctuations in Water Use	135
		4.9.2 Water Quantity Standards in Urban Areas	137
	4.10	Water Demand Forecasting	139
	4.11	Water Storage	139
	4.12	Water Distribution System	141
		4.12.1 System Components	142
	4.13	Hydraulics of Water Distribution Systems	144
		4.13.1 Energy Equation of Pipe Flow	144
		4.13.2 Evaluation of Head Loss Due to Friction	145
		4.13.2.1 Darcy–Weisbach Equation	145
		4.13.2.2 Hazen–Williams Equation for the Friction Head Loss	148
		4.13.2.3 Minor Head Loss	149
	4.14	Pipeline Analysis and Design	151
		4.14.1 Pipes in Series	154
		4.14.2 Pipes in Parallel	157
		4.14.3 Pipe Networks	159
	4.15	Water Quality Modeling in a Water Distribution Network	162
		4.15.1 Water Quality Standards	163
		4.15.2 Water Quality Model Development	163
		4.15.3 Chlorine Decay	165
		4.15.3.1 Bulk Decay	165
		4.15.3.2 Wall Decay	165
		4.15.3.3 Overall Decay Rate	166
		4.15.4 Statistic Model of Substance Concentration	168
		4.15.5 Solution Models	169
		4.15.5.1 EPANET	170
		4.15.5.2 QUALNET	170
		4.15.5.3 Event-Driven Method	171
		4.15.6 Water Quality Monitoring	171
		4.15.7 Conducting a Tracer Study	172
	4.16	Concluding Remarks	175
	Problems	175
	References	179
Chapter 5	Urban Water Demand Management	183
	5.1	Introduction	183
	5.2	Basic Definitions of Water Use	183
	5.3	Paradigm Shift in Urban Water Management: Toward Demand Management	184
		5.3.1 Supply Management	184
		5.3.2 Demand Management	185
	5.4	The "Soft Path" for Water	187
	5.5	Urban Water Demand Management: Objectives and Strategies	189
	5.6	UWDM Screening	192

	5.7	Urban Water Demand	192
		5.7.1 Urban Water Demand Estimation and Forecasting	193
		5.7.1.1 Regression Analysis	194
		5.7.1.2 Parametric Models	195
		5.7.1.3 Data Analysis: Fixed Effects—Instrumental Variables	197
		5.7.1.4 End-Use Analysis	197
		5.7.1.5 Demand Models	198
	5.8	Urban Water Demand Management Factors	199
		5.8.1 Water Loss Reduction	200
		5.8.1.1 NRW and Water Loss Indexes	200
		5.8.1.2 Reduction of NRW	203
		5.8.2 Education and Training	204
		5.8.3 Economic Incentives and Urban Water Pricing	205
		5.8.3.1 Price Elasticity	207
		5.8.3.2 Water Demand and Price Elasticity	210
		5.8.4 Institutional Measurements and Effective Legislation	211
		5.8.5 Water Reuse	211
		5.8.6 Rainwater Harvesting	214
		5.8.6.1 The Feasibility of an RRCS	215
		5.8.6.2 Design of an RRCS	216
	5.9	Demographic Considerations	224
	5.10	Evaluating Water Conservation	224
	5.11	Future Demand Supply	225
		5.11.1 Demand-Side Management	225
		5.11.2 Supply-Side Resources	226
		5.11.3 Overview of Evaluating Resource Combinations	227
	5.12	Conflict Issues in Water Demand Management	227
	5.13	Application of UWDM for Sustainable Water Development	232
		5.13.1 Provincial Action Plan	232
		5.13.2 Municipal Action Plan	233
	5.14	Concluding Remarks	233
	Problems		234
	References		237
Chapter 6	Urban Water Drainage System		241
	6.1	Introduction	241
	6.2	Urban Planning and Stormwater Drainage	241
		6.2.1 Best Management Practices	242
		6.2.1.1 Sediment Basins	242
		6.2.1.2 Bioretention Swales	243
		6.2.1.3 Bioretention Basins	244
		6.2.1.4 Sand Filters	244
		6.2.1.5 Swales and Buffer Strips	244
		6.2.1.6 Constructed Wetlands	245
		6.2.1.7 Ponds and Lakes	245
		6.2.1.8 Infiltration Systems	245
		6.2.1.9 Aquifer Storage and Recovery	246
		6.2.1.10 Porous Pavement	246
	6.3	Open-Channel Flow in Urban Watersheds	246
		6.3.1 Open-Channel Flow	247
		6.3.1.1 Open-Channel Flow Classification	249

		6.3.1.2	Hydraulic Analysis of Open-Channel Flow	249

- 6.3.2 Overland Flow ... 251
  - 6.3.2.1 Overland Flow on Impervious Surfaces ... 251
  - 6.3.2.2 Overland Flow on Pervious Surfaces ... 255
- 6.3.3 Channel Flow ... 256
  - 6.3.3.1 Muskingum Method ... 256
  - 6.3.3.2 Muskingum–Cunge Method ... 259
- 6.4 Components of Urban Stormwater Drainage Systems ... 259
  - 6.4.1 General Design Considerations ... 260
  - 6.4.2 Flow in Gutters ... 261
    - 6.4.2.1 Simple Triangular Gutters ... 261
    - 6.4.2.2 Gutters with Composite Cross Slopes ... 263
    - 6.4.2.3 Gutter Hydraulic Capacity ... 265
    - 6.4.2.4 V-Shaped Sections ... 265
  - 6.4.3 Pavement Drainage Inlets ... 265
    - 6.4.3.1 Inlet Locations ... 267
  - 6.4.4 Surface Sewer Systems ... 268
  - 6.4.5 Open Drainage Channel Design ... 269
    - 6.4.5.1 Design of Unlined Channels ... 270
    - 6.4.5.2 Grass-Lined Channel Design ... 272
- 6.5 Culverts ... 275
  - 6.5.1 Inlet Control Flow ... 277
  - 6.5.2 Outlet Control Flow ... 279
    - 6.5.2.1 Full-Flow Conditions ... 281
    - 6.5.2.2 Partly Full-Flow Conditions ... 282
  - 6.5.3 Sizing of Culverts ... 282
  - 6.5.4 Protection Downstream of Culverts ... 283
- 6.6 Design Flow of Surface Drainage Channels ... 284
  - 6.6.1 Probabilistic Description of Rainfall ... 284
    - 6.6.1.1 Return Period and Hydrological Risk ... 284
    - 6.6.1.2 Frequency Analysis ... 285
  - 6.6.2 Design Rainfall ... 287
    - 6.6.2.1 Extracting Design Rainfall from Historical Data ... 287
  - 6.6.3 Design Return Period ... 288
  - 6.6.4 Design-Storm Duration and Depth ... 289
  - 6.6.5 Spatial and Temporal Distribution of Design Rainfall ... 289
- 6.7 Stormwater Storage Facilities ... 289
  - 6.7.1 Sizing of Storage Volumes ... 291
- 6.8 Risk Issues in Urban Drainage ... 292
  - 6.8.1 Flooding of Urban Drainage Systems ... 292
  - 6.8.2 DO Depletion in Streams Due to Discharge of Combined Sewage ... 294
  - 6.8.3 Discharge of Chemicals ... 294
- 6.9 Urban Floods ... 294
  - 6.9.1 Urban Flood Control Principles ... 295
- 6.10 Sanitation ... 295
- 6.11 Wastewater Management ... 295
  - 6.11.1 Technologies for Developing Countries ... 296
  - 6.11.2 Wetlands as a Solution for Developing Countries ... 298
  - 6.11.3 Example: Caribbean Wastewater Treatment ... 298
  - 6.11.4 Example: The Lodz Combined Sewerage System ... 299

		6.11.4.1 Development and Implementation of a Sewerage System 299

Contents

	6.11.4.1	Development and Implementation of a Sewerage System ... 299

Chapter 7 — partial continuation

      6.11.4.1 Development and Implementation of a Sewerage System 299
      6.11.4.2 Upgrading the Old Sewerage System 299
  6.12 StormNET: Stormwater and Wastewater Modeling 300
    6.12.1 Hydrology Modeling Capabilities 302
    6.12.2 Hydraulic Modeling Capabilities 302
    6.12.3 Detention Pond Modeling Capabilities 302
    6.12.4 Water-Quality Modeling Capabilities 303
    6.12.5 Typical Applications of StormNET 303
    6.12.6 Program Output 303
Problems 304
References 307

Chapter 7 Environmental Impacts of Urbanization 311
  7.1 Introduction 311
  7.2 Urbanization Effects on the Hydrologic Cycle 311
  7.3 Urbanization Impacts on Water Quality 315
    7.3.1 Water Pollution Sources 316
    7.3.2 Water Quality Modeling of Urban Stormwater 319
      7.3.2.1 Unit-Area Loading 319
      7.3.2.2 Simple Empirical Method 320
      7.3.2.3 US-EPA Method 323
      7.3.2.4 Computer-Based Models 325
  7.4 Physical and Ecological Impacts of Urbanization 326
  7.5 Urbanization Impacts on Rivers and Lakes 329
  7.6 Impacts of Urbanization on Groundwater 330
    7.6.1 Unintentional Discharges into Groundwater Aquifers 331
    7.6.2 Intentional Discharges into Groundwater Aquifers 332
  7.7 Overall State of Impacts 333
Problems 334
References 335

Chapter 8 Tools and Techniques 339
  8.1 Introduction 339
  8.2 Simulation Techniques 339
    8.2.1 Stochastic Simulation 339
    8.2.2 Stochastic Processes 344
    8.2.3 Artificial Neural Networks 346
      8.2.3.1 The Multilayer Perceptron Network (Static Network) 348
      8.2.3.2 Temporal Neural Networks 352
    8.2.4 Monte Carlo Technique 359
  8.3 Optimization Techniques 359
    8.3.1 Linear Method 360
      8.3.1.1 Simplex Method 364
    8.3.2 Nonlinear Methods 366
    8.3.3 Dynamic Programming 366
    8.3.4 Evolutionary Algorithms 369
      8.3.4.1 Genetic Algorithms 369
      8.3.4.2 Simulation Annealing 372
    8.3.5 Multicriteria Optimization 373

	8.4	Fuzzy Sets and Parameter Imprecision 377
	8.5	Urban Water Systems Economics... 381
		8.5.1 Economic Analysis of Multiple Alternatives 384
		8.5.2 Economic Evaluation of Projects Using Benefit–Cost Ratio Method... 386
		8.5.3 Economic Models... 387
	Problems 388
	References 391
Chapter 9	Urban Water Infrastructure 393
	9.1	Introduction 393
	9.2	Urban Water Infrastructures.. 393
		9.2.1 Infrastructure for Water Supply (Dams and Reservoirs)........... 393
		9.2.2 Infrastructure for the Water Distribution System................... 395
		9.2.2.1 Water Mains.. 396
		9.2.2.2 Design and Construction of the Water Main System..... 396
		9.2.3 Infrastructure for Wastewater Collection and Treatment 400
		9.2.3.1 Costs of Wastewater Infrastructures 401
		9.2.3.2 Cost of Stormwater Infrastructures 401
	9.3	Interactions between the Urban Water Cycle and Urban Infrastructure Components... 402
		9.3.1 Interactions with the Wastewater Treatment System............... 403
		9.3.2 Interactions between Water and Wastewater Treatment Systems.. 404
		9.3.3 Interactions between Water Supply and Wastewater Collection Systems 404
		9.3.4 Interactions between Urban Drainage Systems and Wastewater Treatment Systems... 404
		9.3.5 Interactions between Urban Drainage Systems and Solid Waste Management... 405
		9.3.6 Interactions between Urban Water Infrastructure and Urban Transportation Infrastructures 406
	9.4	Financing Methods for Infrastructure Development..................... 407
		9.4.1 Tax-Funded System... 407
		9.4.2 Service Charge-Funded System 407
		9.4.3 Exactions and Impact Fee-Funded Systems 407
		9.4.4 Special Assessment Districts 408
	9.5	Sustainable Development of Urban Water Infrastructures........... 410
		9.5.1 Selection of Technologies 412
		9.5.1.1 Further Development of Large-Scale Centralized Systems... 412
		9.5.1.2 Separation for Recycling and Reuse 412
		9.5.1.3 Natural Treatment Systems..................................... 413
		9.5.1.4 Combining Treatment Systems............................... 413
		9.5.1.5 Changing Public Perspectives 413
		9.5.2 Assessing the Environmental Performance of Urban Water Infrastructure 413
		9.5.3 Life Cycle Assessment.. 414
		9.5.4 Developing SDI Using a Life Cycle Approach 415
	9.6	SIM of Water Mains 416
	9.7	Public Benefits from SIM Improvements 418

	9.8	Opportunities for SIM Capability Improvement	419
		9.8.1 Mobile In-Line Inspection Systems	419
		9.8.2 Mobile Nonintrusive Inspection (MNII) Systems	419
		9.8.3 Continuous Inspection Devices	421
		9.8.4 Intelligent Systems	421
	9.9	Concluding Remarks	421
	Problems		422
	References		424

Chapter 10 Urban Water System Dynamics and Conflict Resolution 427
    10.1 Introduction 427
    10.2 System Dynamics 428
        10.2.1 Modeling Dynamics of a System 429
            10.2.1.1 Define the Issue/System 430
            10.2.1.2 Test of Hypotheses 430
            10.2.1.3 Design and Test Policies 431
        10.2.2 Time Paths of a Dynamic System 431
            10.2.2.1 Linear Family 431
            10.2.2.2 Exponential Family 432
            10.2.2.3 Goal-Seeking Family 432
            10.2.2.4 Oscillation Family 433
            10.2.2.5 S-Shaped Family 434
    10.3 Conflict Resolution 440
        10.3.1 Conflict Resolution Process 441
        10.3.2 A Systematic Approach to Conflict Resolution 441
        10.3.3 Conflict Resolution Models 442
    10.4 Case Studies 447
        10.4.1 Case 1: Conflict Resolution in Water Pollution Control in Urban Areas 447
            10.4.1.1 Water Resources Characteristics in the Study Area 447
            10.4.1.2 Conflict Resolution Model 449
            10.4.1.3 Results and Discussion 451
        10.4.2 Case Study 2: Development of an Object-Oriented Programming Model for Water Transfer 452
            10.4.2.1 Area Characteristics 452
            10.4.2.2 Conflict Resolution Model for Land Resources Allocation in Each Zone 453
            10.4.2.3 Results of the Conflict Resolution Model 454
            10.4.2.4 Optimal Groundwater Withdrawal in Each Zone 455
            10.4.2.5 Sizing Channel Capacity 456
    10.5 Conclusion: Making Technologies Work 457
    Problems 457
    References 461

Chapter 11 Urban Water Disaster Management 463
    11.1 Introduction 463
    11.2 Sources and Kinds of Disasters 464
        11.2.1 Drought 464
        11.2.2 Floods 466
            11.2.2.1 Principles of Urban Flood Control Management 467

	11.2.3	Widespread Contamination	468
	11.2.4	System Failure	469
	11.2.5	Earthquakes	470
11.3	What Is UWDM?		470
	11.3.1	Policy, Legal, and Institutional Framework	470
11.4	Societal Responsibilities		471
11.5	Planning Process for UWDM		471
	11.5.1	Taking a Strategic Approach	471
	11.5.2	Scope of the Strategy Decisions	472
	11.5.3	UWDM as a Component of a Comprehensive DM	472
	11.5.4	Planning Cycle	473
11.6	Water Disaster Management Strategies		473
	11.6.1	Experience on Disaster Management	475
	11.6.2	Initiation	476
		11.6.2.1 Political and Governmental Commitment	476
		11.6.2.2 Policy Implications for Disaster Preparedness	477
		11.6.2.3 Public Participation	478
		11.6.2.4 Lessons on Community Activities	479
	11.6.3	Case Study 1: Drought Disaster Management	479
	11.6.4	Case Study 2: Drought Management in Georgia, USA	480
	11.6.5	Case Study 3: Management of Northern California Storms and Floods of January 1995	480
		11.6.5.1 Flood Characteristics	481
		11.6.5.2 Response	481
11.7	Situation Analysis		481
	11.7.1	Steps in the Development of Situation Analysis	482
		11.7.1.1 Approach	482
		11.7.1.2 Objectives	482
		11.7.1.3 Data Collection	482
	11.7.2	Urban Disasters Situation Analysis	483
11.8	Disaster Indices		483
	11.8.1	Reliability	484
		11.8.1.1 Reliability Assessment	491
		11.8.1.2 Reliability Analysis: Load-Resistance Concept	495
		11.8.1.3 Reliability Indices	496
		11.8.1.4 Direct Integration Method	497
		11.8.1.5 Margin of Safety	499
		11.8.1.6 Mean Value First-Order Second Moment (MFOSM) Method	501
		11.8.1.7 AFOSM Method	502
	11.8.2	Time-to-Failure Analysis	502
		11.8.2.1 Failure and Repair Characteristics	502
		11.8.2.2 Availability and Unavailability	502
	11.8.3	Resiliency	503
	11.8.4	Vulnerability	503
11.9	Uncertainties in Urban Water Engineering		504
	11.9.1	Implications and Analysis of Uncertainty	504
	11.9.2	Measures of Uncertainty	505
	11.9.3	Analysis of Uncertainties	505

Contents

- 11.10 Risk Analysis: Composite Hydrological and Hydraulic Risk 507
  - 11.10.1 Risk Management and Vulnerability 508
  - 11.10.2 Risk-Based Design of Water Resources Systems 509
  - 11.10.3 Creating Incentives and Constituencies for Risk Reduction 510
- 11.11 System Readiness 510
  - 11.11.1 Evaluation of WDS Readiness 511
  - 11.11.2 Hybrid Drought Index 518
  - 11.11.3 Disaster and Scale 523
  - 11.11.4 Design by Reliability 523
  - 11.11.5 Water Supply Reliability Indicators and Metrics 524
  - 11.11.6 Issues of Concern for the Public 526
- 11.12 Guidelines for Disaster Management 526
  - 11.12.1 Disaster and Technology 526
  - 11.12.2 Disaster and Training 527
  - 11.12.3 Institutional Roles in Disaster Management 527
- 11.13 Developing a Pattern for the Analysis of the System's Readiness 528
  - 11.13.1 Assessment of a Disaster Caused by Water Shortage/Drought 528
  - 11.13.2 Assessment of a Disaster Caused by System Failure 528
  - 11.13.3 Assessment of a Disaster Caused by Widespread Contamination 528
  - 11.13.4 Developing a Comprehensive Qualitative/Quantitative Monitoring System for the Water Supply, Transfer, and Distribution Network 528
  - 11.13.5 Guidelines for the Mitigation and Compensatory Activities 529
  - 11.13.6 Organization and Institutional Chart of Decision Makers in a Disaster Committee 529
- 11.14 Conclusion 532
- Problems 535
- References 537

Chapter 12 Climate Change 539
- 12.1 Introduction 539
- 12.2 Climate Change Process 539
  - 12.2.1 Increasing Temperature 540
  - 12.2.2 Variation in Evaporation and Precipitation 540
  - 12.2.3 Soil Moisture 540
  - 12.2.4 Snowfall and Snowmelt 541
  - 12.2.5 Storm Frequency and Intensity 541
  - 12.2.6 Groundwater 541
  - 12.2.7 Water Resource System Effects 542
- 12.3 Impacts of Climate Change on Urban Water Supply 542
  - 12.3.1 Direct Impacts 542
  - 12.3.2 Indirect Impacts 543
  - 12.3.3 Compound Impacts 544
- 12.4 Adaptation with Climate Change 544
  - 12.4.1 Vulnerability Analysis 545
  - 12.4.2 Integrated Resource Planning 545
  - 12.4.3 Reducing GHG Emissions 547

12.5	Climate Change Prediction		548
	12.5.1	Climate Change Scenarios	548
	12.5.2	General Circulation Models	549
	12.5.3	Spatial Variability	554
		12.5.3.1 Downscaling	554
		12.5.3.2 Regional Models	554
	12.5.4	Statistical Downscaling	556
12.6	Downscaling Models		559
	12.6.1	Statistical Downscaling Model	559
	12.6.2	LARS-WG Model	567
12.7	Case Studies		574
	12.7.1	Assessment of the Precipitation and Temperature Variations in the Aharchai River Basin under Climate Change Scenarios	574
	12.7.2	Evaluation of Urban Floods Considering Climate Change Impacts	576
12.8	Conclusion		578
Problems			578
References			580
Index			583

Preface

In the past few decades, urban water management practices have focused on optimizing the design and operation of water distribution networks, wastewater collection systems, and water and wastewater treatment plants. However, municipalities are facing hardships because of the aging urban water infrastructure whose operation must be improved and expanded to maintain and keep up with the current standards of living. As for best practice management, it has become clear that structural solutions for available urban systems' operation are not always the most efficient and economic alternative. Today's integrated urban water management focuses on engineering and planning solutions for integrating structural and nonstructural means to achieve the best operational schemes at affordable costs.

Current challenges go beyond utilizing the existing structures and the physical limitations on water availability to include technical, social, political, and economical aspects of better water and wastewater management in urban areas. The planning schemes should focus on sustainable urban water development and therefore ensure water security for communities.

In this book, a systems approach to urban water hydrology, engineering, planning, and management is taken. Urban water governance and disaster management in urban areas are some of the pressing issues that are discussed. Modeling of the urban water cycle, urban water supply and distribution systems that demand forecasting, and wastewater and stormwater collection and treatment are classical issues in urban settings that are presented in a systems perspective. Understanding the principles from simulation, optimization, economical evaluation, multiple-criterion decision making, and conflict resolution are prerequisites for integrated urban water management, which are covered in this book.

This book has been written for undergraduate and graduate courses in water resources in civil and environmental engineering, in urban planning, and in any selected urban system courses including infrastructure development. Junior and senior BS students can use Chapters 3, 4, 5, 6, and 7 of this book for engineering hydrology/water resources engineering, environmental engineering I, and water and wastewater courses. Engineers and planners, especially those who work on the design, planning, and management of urban systems and/or community development, can use this book in practice because it deals with a broad range of real-world urban water problems. Special features such as water governance and disaster management are useful to water managers, policy makers, and decision makers. The content of each chapter has been selected considering the latest developments in urban water hydrology and management. Modeling techniques are discussed with some field applications for a better understating.

Chapter 1 presents the urban water cycle as part of a hydrological cycle. Components of the urban water cycle have special characteristics that are affected by the growing population. Air, water, and soil pollution in urban areas are now getting intensified by climate change impacts. These components and the effects of urbanization, as well as physical, chemical, and biological impacts, are briefly discussed in this chapter. Urban development has caused some climatic and hydrological impacts such as greenhouse gases and heat islands, which are also discussed in this chapter.

Chapter 2 presents the institutional framework in the context of water governance. The paradigm shifts from Newtonian to holistic planning, and its impact on shifting from supply management to demand management are discussed. Water and its role in land-use planning have been a challenge for planners and decision makers. The suitability of land-use planning with respect to the availability of water resources has always been a major challenge. Municipalities that have coupled land-use planning with sustainable water resources allocation schemes have succeeded in meeting their long-term objectives while expanding environmental vitality toward sustainable development. Water resource assessment as a means to measure, evaluate, monitor, forecast, and bring consensus among engineers, water resources planners, decision makers, and stakeholders is also discussed in this chapter.

Chapter 3 presents urban runoff and drainage characteristics, excess rainfall calculation, and rainfall–runoff analyses. The design of rainfall and peak flow calculations are also discussed. Furthermore, an introduction to the effects of urbanization on peak discharge is given. The principal components of designing the urban water drainage system are also explained.

Chapter 4 presents the principles of urban water supply and distribution. In this chapter, water supply and distribution challenges are discussed. Moreover, the interdependency and interactions among the system's principal components are recognized and used for solving water supply and distribution problems, which are the essence of integrated urban water management.

Chapter 5 introduces urban water demand management and the key strategies for achieving the millennium development goals of supplying potable water and sanitation to all people in urban areas. In this chapter, the concepts of water demand forecasting and management are introduced and the methodologies that can be applied to achieve the demand management goals are described.

Chapter 6 discusses the basic principles of designing urban water drainage systems. In this chapter, an introduction to the design of surface drainage channels and the risks associated with floods is presented. The required hydraulic equations for storm channel design as well as hydrograph routing methods are explained in this chapter.

Chapter 7 explains the impacts of urbanization on various water and environmental components in urban areas. The urbanization impacts are focused on water and other elements interacting with urban water and the impacts of urban areas on the atmosphere, surface water and groundwater, soils, and wetlands. As stormwater is one of the main pollution sources of urban areas, the principles of urban stormwater quality modeling are also presented in this chapter. The material presented in this chapter can be used for stormwater pollution load estimation from different sources. This can help engineers and decision makers decide on structural and nonstructural means of dealing with pollutants and their possible impacts on human health.

Chapter 8 introduces the tools and techniques suitable for urban water engineering and management. They are classified as economic analysis, simulation, and optimization analysis. Engineering applications primarily use analysis and design techniques. The overlay between engineering and management tools in urban water is simulation tools, which can be used to test the performance of urban water systems.

Chapter 9 discusses urban water infrastructure management and maintenance of the structural integrity of its components, with particular focus on water mains, which are the most important component of urban water infrastructures. In this chapter, an introduction to the different parts of urban water infrastructures and their interactions, as well as life cycle assessment and sustainable designs of urban infrastructures are discussed. The application of standard integrity monitoring to water mains is also presented.

Chapter 10 introduces different methods for integrating of engineering and planning principles and sustainability in urban water management. In this chapter, system dynamics and conflict resolution as well as some case studies related to their applications are presented.

Chapter 11 discusses the principles of urban water disaster management. Water shortages, water pollution, flooding, aging water infrastructures as well as increasing demands for water, and other problems associated with water allocation that can devastate water systems and cause disasters are presented. The rapid oscillation of these problems has the gravest effects on developed countries. Developed countries are more vulnerable because of their high dependency on water. In developing countries, water disasters also impact the life of people because of the rapid expansion of water sectors and inadequate institutional and infrastructural setups to deal with unexpected situations. In this chapter, the planning process for urban water disaster management and methods for analyzing systems' readiness through some real-world examples are discussed.

Chapter 12 presents climate change impacts and their potentially significant implications on the urban water cycles. This chapter provides a basic understanding of climate change and consideration of the issues involved in developing suitable urban water planning and management responses to climate changes.

This book can serve water communities around the world and add significant value to engineering and applications of systems analysis techniques to urban water design, planning, and management. It can be used as a textbook by students of civil engineering, urban and regional planning, geography, and environmental science, and in courses dealing with urban water cycles. It also introduces new horizons for engineers, policy makers, and decision makers who are planning for the regional sustainability of future urban water.

We hope that this book will serve urban water communities around the world and add significant value to the application of systems analysis techniques for urban water engineering and management.

Mohammad Karamouz
Ali Moridi
Sara Nazif

Acknowledgments

Many individuals have contributed to the preparation of this book. The initial impetus for the book was provided by Professor Mark H. Houck through his class notes when he taught at Purdue University. The first author had a 2-year collaboration with Jiri Marsalek, Blanca Jimenez-Cisneros, Per-Arne Malmquist, Joel Goldenfum, and Bernard Chocat, members of UNESCO's task committee on urban water cycles. This committee published a book entitled *Urban Water Cycle Processes and Interactions* by Taylor & Francis in 2006. This collaboration was a driving force for the completion of this book. Furthermore, the *Training of Trainers Manual* and workshops developed by UNESCO Regional Center for Urban Water Management in Tehran, Iran, with input from the trainers and trainees and under the direction of Ahmad Abrishamchi and Reza Ardakanian, had an impact on the preparation of the chapters on urban water governance and disaster management. We are grateful to J. A. Tejada-Guibert at the UNESCO-IHP headquarters for encouraging and facilitating many efforts from which this book benefited.

The authors would also like to thank the PhD and MS students Azadeh Ahmadi, Masoud Taherioun, Hammed Tavakolifar, Mahdis Fallahi, Ensiyeh Ansari, Farzad Emami, Behzad Rohanizade, Saleh Ahmadinia, Navide Noori, Ana Hosseinpoor, Saleh Semsar, and Anahita Abolpour. Special thanks to Mehrdad Karamouz, Ghazale Moghadam, Elham Jazani, Sahar Karamouz, and Mona Lotfi for preparing the figures and artwork for this book. The review and constructive comments of Dr. Walter Grayman and Professor Lindell Ormsbee are greatly appreciated.

Authors

Mohammad Karamouz is a professor of civil engineering at the University of Tehran, Iran and research professor at the Polytechnic Institute of New York University. He received a BS degree in civil engineering from Shiraz University, an MS degree in water resources and environmental engineering from George Washington University, and a PhD in hydraulic and systems engineering from Purdue University. He is a licensed professional engineer in the state of New York, a fellow of the American Society of Civil Engineers (ASCE), and a diplomat of the American Academy of Water Resources Engineers.

He served as a member of the task committee on urban water cycles in UNESCO-IHP VI and was a member of the planning committee for the development of a five-year plan (2008–2013) for the UNESCO International Hydrology Program (IHP VII). Among his many professional positions and achievements, he was the former dean of engineering at Pratt Institute in Brooklyn, New York. He was also a tenured professor at Pratt Institute, Schools of Engineering and Architecture and at Polytechnic University in Tehran, Iran. Prior to that, he served Polytechnic University in Brooklyn for five years. During 2000–2003, he was a visiting professor in the department of hydrology and water resources of the University of Arizona, Tucson. He was also the founder of Arch Construction and Consulting Co., Inc., in New York City.

His areas of research and teaching interests include integrated water resources planning and management, urban water planning, water security, drought analysis and management, water-quality modeling and water pollution reduction master plans, decision-support systems, stream flow forecasting, reservoir operation, and the conjunctive use of surface water and groundwater. He has more than 200 technical publications, books, and book chapters, including a textbook titled *Water Resources System Analysis* published by Lewis Publishers in 2003, and has coauthored a book titled *Urban Water Cycle Processes and Interactions* published by Taylor & Francis in 2006. He serves nationally and internationally as a consultant to both private and governmental agencies such as UNESCO and the World Bank.

Ali Moridi is an adjunct professor at Tarbiat Modaress University. He received a BS degree in civil engineering and an MS degree in water resources engineering from Sharif University of Technology, and a PhD in water resources management from Amirkabir University. He is an international consultant on the application of systems engineering in urban water engineering and management, and has developed many state-of-the-art computer-modeling packages and decision-support systems for large-scale water resource systems. His research interests include conflict resolution, decision-support systems, multicriteria decision making (MCDM), and systems analysis and their application in urban water engineering and management.

Sara Nazif is a research associate at the Polytechnic Institute of New York University working with Professor Karamouz. She received a BS degree in civil engineering and an MS degree in water and wastewater engineering from the University of Tehran, Iran. The subject of her PhD dissertation at the University of Tehran is water management with emphasis on the impacts of climate change on urban water cycles. She is an international consultant on the application of systems engineering and computer modeling in urban water engineering and management. Her research and training interests include water distribution systems, disaster management, urban water demand management, stochastic hydrology, climate change impacts on urban water engineering and management, and systems analysis.

1 Introduction

A portrait of "A Man and His World" shows a dream town with all the technological advancement and abundance of services, utilities, and contents. With this portrait in mind, an urban population demands high quantities of energy and raw material, and removal of waste. The by-product of urban life is environmental pollution. Material and energy are the main ingredients of urbanization and all key activities of modern cities. Transportation, electric supply, water supply, waste disposal, heating, services, manufacturing, and so on are characterized by the use of energy and material. Thus, the concentration of people in urban areas dramatically alters material and energy fluxes. Drastic changes in landscape and land use have altered the fluxes of water, sediment, chemicals, and microorganisms and have increased the release of waste and heat. These changes then impact urban ecosystems, including urban waters, and result in deterioration of natural resources and degradation of supporting infrastructure. Such circumstances make the provision of water services to urban populations highly challenging, particularly in large cities. The number of these large cities keeps growing, particularly in developing countries, and this further exacerbates both human health and environmental problems on a regional scale.

For water management in time and space, the hydrologic cycle is a model of holistic nature. There are different definitions of the hydrological cycle, but it is generally defined as a conceptual model describing the storage and circulation of water between the biosphere, atmosphere, lithosphere, and hydrosphere (Karamouz and Araghinejad, 2005). Water can also be arrested in nature in the atmosphere, ice and snow packs, streams, rivers, lakes, groundwater aquifers, and oceans. Water cycle components are affected by processes such as temperature and pressure variation and condensation. The means of water movement is through precipitation, snowmelt, evapotranspiration, percolation, infiltration, and runoff.

A watershed is the best hydrological unit that can be used to carry out water studies and planning in a systematic manner. The urban setting could alter the natural movement of water. Drastic land use changes in urban areas as a subset of urban and industrial development affect natural landscapes and the hydrological response of watersheds. Although anthropogenic factors with respect to waterways, pipes, abstractions, and man-made infrastructures affect the elements of the natural environment, the main structure of the hydrological cycle remains the same in urban areas (McPherson and Schneider, 1974). But the characteristics of the hydrologic cycle are greatly altered by urbanization impacts of the services to the urban population, such as water supply, drainage, and wastewater collection and management.

As a conceptual way of looking at water balances in urban areas, the context of the urban water cycle is the total system approach. Water balances and budget studies are generally conducted on a different time scale, depending on the type of applications we are looking at in a planning horizon. For distributing water to growing populations and for coping with extreme weather and climatic variations and potential climate change, Lawrence et al. (1999) emphasized the importance of integrated urban water managements. Pressures, temperatures, and water quality buildups in urban areas have a pronounced effect on the urban water cycle, which should be attended to through best management practice (BMP) schemes.

1.1 URBAN WATER CYCLE

1.1.1 Components

Changes in the material and energy fluxes and in the amount of precipitation, evaporation, and infiltration in urban areas consequently result in changes in water cycle characteristics. The impacts of large urban areas on local microclimate have long been recognized and occurred as a result of changes in the energy regime such as air circulation patterns caused by buildings, transformation of land surfaces and land use planning as well as water transfer, waste generation, and air quality variations. These changes can be summarized as follows:

- Land use—transformation of undeveloped land into urban land, including transportation corridors
- Demand for water—increased demand because of increased concentration of people and industries in urban and nearby suburban areas
- Increased entropy on the use and redundant use of unsustainable forms of energy
- Waste production—specially solid waste and industrial hazardous wastes, and decreasing quality of different resources such as air, water, and soil
- Water and food transfer from other places to urban areas

Figure 1.1 shows the different components of the hydrologic cycle in urban areas. Each of these components is briefly explained in this section.

Temperature: The air temperature over urban areas usually exceeds that of the surrounding localities by as much as 4–7°C. These thermal variations explain the higher evaporation rates (5–20%) in urban areas. Also, convective storms can be observed in urban areas, due to higher temperatures.

Precipitation: Previous studies have shown that the total annual precipitation in large industrialized cities is generally 5–10% higher than in surrounding areas, and for individual storms, this increase in precipitation can be as high as 30%. Urban areas are more vulnerable to storms because of much higher runoff coefficients, complexity in the conveyance system as well as congestions in transportation corridors, and poor land use planning.

Evaporation: Because of the high rate of thermal and other forms of energy consumption in cities, the mean temperature and evaporation are higher in urban areas. The presence of closed conduits and storage facilities could reduce the evaporation rate compared with open and surface flow and storage in rural areas.

Transpiration: Because of land use changes and the reduction of open space and green areas, transpiration from trees and vegetation usually decreases.

Infiltration: The following factors contribute to the decrease in infiltration rate:

- Impervious areas (pavements, rooftops, parking lots, etc.)
- Man-made drainage systems

The use of compensatory devices of urbanization effects, such as infiltration trenches and pervious pavements, will increase infiltrated water flows in urbanized areas. Soil pollution caused by contaminants from surface runoff and leaking underground tanks and absorption wells as well as leaches from landfills alters the soil–water interaction. Therefore, the unsaturated zone in urban areas has different characteristics than other areas.

Runoff: The high ratio of areas covered by impervious materials results in a higher runoff rate in urban areas. Man-made drainage systems facilitate surface runoff collection and transfer; therefore, after urbanization, the peak of the unit hydrograph is greater and occurs earlier. Figure 1.2 shows the comparison between the unit hydrograph of an area before and after urbanization. It has been stated that for a drainage area of about 200 ha (1 ha = 10,000 m^2) the following change in order of magnitude could be expected after urbanization. The runoff coefficient and the peak flow increase 2–3

Introduction

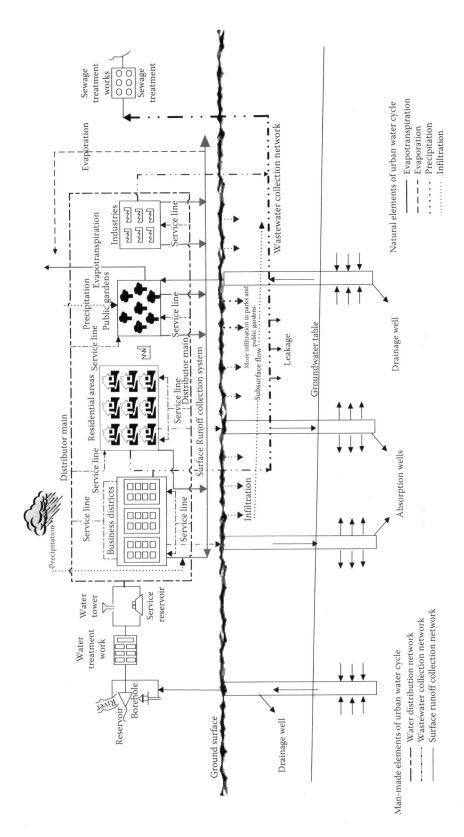

FIGURE 1.1 Different elements of the hydrologic cycle in urban areas.

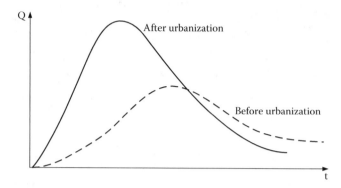

FIGURE 1.2 Unit hydrograph after and before urbanization.

times. The lag time between rainfall and runoff decreases 3 times, and the runoff volume increases 3–4 times.

Interflow: Due to very low infiltration rates in urban areas, the interflows in urban areas, as shown in Figure 1.1, are mostly supplied by leakage from water distribution and wastewater collection networks. The higher infiltration rates in parks, green areas, and areas with absorption wells are also important in increasing interflows.

Water quality: This is significantly affected by urban activities. The water quality of interflow is also adversely affected by urban life. Therefore, the effect of this important component of the hydrologic cycle is changed drastically.

1.1.2 Impact of Urbanization

Physical, chemical, and biological impacts on the water cycle have caused adverse and serious depletion of water resources in many urban areas around the world. Modifications of major drainage canals from natural to man-made structures has an impact on runoff hydrograph that affects rate of erosion and siltation. Surface runoff carries pollutants such as hydrocarbon and other organic residues, food waste, debris, and other matter. Urban drainage discharging into water bodies introduces modifications that produce different negative impacts with short- and long-term consequences on the regional ecosystem. The magnitude of the impacts depends on factors such as the condition of the water body before the discharge occurs, its carrying capacity, and also the quantity and distribution of rainfall, land use in the basin, and type and quantity of pollution transported. The problems cause esthetic changes as well as pollutions from toxic substances.

The impact can be categorized as physical, chemical, and biological. Physical impacts are those that affect the amount of surface runoff and those that affect the urban water cycle such as impact on soil consolidation. The latter impacts the specific storage of aquifers in urban areas. Air pollution from SO_2 and CO_2 is a common problem in many big cities. When mixed with precipitation, it causes acid rains that destroy trees in urban areas and have adverse effects on surface and groundwater resources.

Many infectious enteric diseases of humans are transmitted through fecal wastes to rivers or lakes. These revelations are accelerated in urban areas. In most biological reactions, oxygen is an important factor. When the rate of biological reactions increases, the dissolved oxygen in water decreases, which affects aquatic life. Industrial and municipal wastewater disposal to rivers and lakes changes water quality and causes significant changes in the hydrologic cycle in urban areas. For example, when water is polluted, its color becomes dark and it absorbs more energy from the sun. This increases the water temperature and changes the evaporation rate.

In urban areas, because of the high density of buildings and other structures, soil will be consolidated and, therefore, the porosity of soil decreases and the amount of water that can be stored and discharged from urban aquifers decreases. Another impact of urbanization on urban aquifers is the decreasing recharge to the aquifer because of decreasing infiltration. There are certain urban areas with a large number of absorption wells that are subject to a high rate of wastewater infiltration to the aquifer. Groundwater resources have been polluted by urban activities in many cities of the world. For example, landfill leakages and leakage from wastewater sewers and septic tanks are most well-known point sources that cause pollution of groundwater in urban areas.

The hydrological response of urban areas involves the determination of surface runoff hydrographs, which are then routed through the conduits and channels constituting the drainage network to produce outflow hydrographs at the urban drainage outlet. The physical characteristics of a catchment basin regarding the rainfall–runoff process can vary significantly in different parts of an urban area. The unit hydrograph is a traditional means of representing linear system response, but it suffers from the limitation that the response function is lumped over the whole catchment and does not explicitly account for spatially distributed characteristics of the catchment's properties. Schumann (1993) presented a semidistributed model where the spatial heterogeneity of the hydrological characteristics within a basin has been modeled considering each subbasin separately using geographical information system (GIS). Wang and Chen (1996) presented a linear spatially distributed model where the catchment is treated as a system that consists of a number of subcatchments, each assumed to be uniform in terms of rainfall excess and hydrological conditions. Using these approaches, different precipitation patterns and land covers, and their effects on surface runoff can be modeled. These approaches are of significant value in urban areas because of the latter's complexity and land use.

Drought and flood severity, and their impacts on urban areas are more significant. After urbanization, the peak of the unit hydrograph increases and occurs earlier and the flooding condition is more severe. Also, because of high water demand, urban areas are more vulnerable during hydrological drought events. Due to high rate of water use in urban areas, social, political, and economic issues related to water shortage in urban areas are intensified.

Water distribution and wastewater collection systems: The objectives of a municipal water distribution system are to provide safe, potable water for domestic use, adequate quantities of water at sufficient pressure for fire protection, and industrial water for manufacturing. A typical network consists of a source and a tank, treatment, pumping, and a distribution system. Domestic or sanitary wastewater refers to liquid discharges from residences, business buildings, and institutions. Industrial wastewater is discharged from manufacturing, production, power, and refinery plants. Municipal wastewater is a general term applied to the liquid collected in sanitary sewers and treated in a municipal plant. Storm runoff water in most communities is collected in a separate storm sewer system, with no known domestic or industrial connections, and is conveyed to the nearest watercourse for discharge without treatment. In many urban areas, the stormwater and wastewater collection systems are getting mixed and they operate at low efficiency and with a high ratio of system failure.

1.2 INTERACTION OF CLIMATIC, HYDROLOGIC, AND URBAN COMPONENTS

1.2.1 CLIMATIC EFFECTS

The interaction between large urban areas and local microclimate has long been recognized and occurred as a result of changes in the energy flux, air pollution, and air circulation patterns, which are caused by buildings, land transformations, and release of greenhouse gases. These factors contribute to changes in the radiation balance, and the amounts of precipitation and evaporation, and consequently changes in the hydrologic cycle.

Climate is defined as the long-term behavior of weather over a region. There are five climate categories in the world: arid and semiarid, tropical, subtropical (continental), rain shadow, and cool coastal. Hydrologic processes in urban areas under different climates are affected by hydrometeorological variables, which have different ranges in different climates. Different climates are known under the following definitions:

Arid and semiarid areas (e.g., arid areas of Asia and western USA): These areas have seasonal temperatures that range from very cold winters to very hot summers. Snow can occur; however, its effectiveness may be low. Rainfall in summer is unreliable in this climate. Thompson (1975) lists high pressure, wind direction, topography, and precipitation as four main processes that explain aridity.

Tropical areas: Interaction of climate in urban areas located in tropical and subtropical areas can be categorized by the amount of rain and rainy seasons, regional water balance over the year. According to Tucci and Porto (2001), rainfall in tropical regions usually has some of the following characteristics:

- Convective rainfall is more frequent, with a high intensity and a short duration of time, covering small areas. This type of rainfall is more critical for an urban basin that has a short time of concentration (high flow velocity due to gutter and pipe flows) and a small catchment area.
- Long periods of rainfall with high volumes of water result in water depth in the streets. This situation is critical for detention systems. Since the wet periods are concentrated in only a few months (e.g., 500 mm in 15 days has a return period of about 15 years) and there is a storage system, its critical design condition is mainly based on rainfall volumes of a few days rather than on a short period of rainfall.

Climatic conditions in the humid tropics make the urban area more prone to frequent floods, with damage aggravated by socioeconomic factors. On the other hand, the larger mean volumes of precipitation and the greater number of rainy days lead to more complex management of urban drainage. Since larger volumes must be routed somewhere, the mean transported loads of solids are larger (because runoff continues for a longer time), there is less time for urban cleaning (more sediment and refuse remain on the streets) and for maintaining drainage structures, and there is more time to develop waterborne vectors or diseases (Silveira et al., 2001).

Subtropical areas (e.g., Sahara, Arabia, Australia, and Kalahari): These areas are characterized by clear skies with high temperatures. The climate has hot summers and mild winters; hence seasonal contrasts are evident with low winter temperatures due to freezing. Convective rainfall only develops when moist air invades the region (Marsalek et al., 2006).

Rain shadow areas (e.g., mountain ranges such as the Sierra Nevada, the Great Dividing Range in Australia, and the Andes in South America): A rain shadow, or, more accurately, a precipitation shadow, is a dry region of land that is leeward of a mountain range or other geographic features, with respect to the prevailing wind direction. Mountains block the passage of rain-producing weather systems, casting a "shadow" of dryness behind them. The land gets little precipitation because all the moisture is lost on the mountains (Whiteman, 2000).

Cool coastal areas (e.g., Namib and the Pacific coast of Mexico): These areas have reasonably constant conditions with a cool humid environment. When temperate inversions are weakened by moist air aloft, thunderstorms can develop. Cold ocean currents are onshore winds blowing across a cold ocean current close to the shore, which will be rapidly cooled in the lower layers (up to 500 m). Mist and fog may be resulted, as found along the coasts of Oman, Peru, and Namibia, but the warm air aloft creates an inversion preventing the ascent of air, and hence there is little or no precipitation.

1.2.2 Hydrologic Effects

Urbanization increases surface runoff volumes and peak flows, and such excess rainfall may lead to flooding, sediment erosion and deposition, habitat washout (Borchardt and Statzner, 1990), geomorphologic changes (Schueler, 1987), and reduced recharge of groundwater aquifers. These effects may be divided into two categories: acute and cumulative. Flooding and stream channel incision belong to the first category and lowering of groundwater tables and changing the morphology belong to the second category.

In arid climates (as defined in the previous section), the urban surface collects heat, which causes the heating of stormwater runoff and increasing of temperatures in receiving water bodies by up to 10°C (Schueler, 1987). In highly developed watersheds, these processes may lead to algal succession from cold-water species, impacts on invertebrates, and cold-water fishery succession by a warm-water fishery (Galli, 1991).

Urban stormwater ponds, or small lakes, become chemically stratified during winter months (Marsalek et al., 2006), mostly by chlorides originating from road salting. The resulting environmental effects include high levels of dissolved solids and densimetric stratification, which inhibit vertical mixing and transport of oxygenated water to the bottom layers, and may also enhance the release of metal bounds in sediments.

In arid areas, the physical process of soil formation is active, resulting in heterogeneous soil types, with properties that do not differ greatly from the parent material and with soil profiles that retain their heterogeneous characteristics. Soils in arid lands may contain hardened or cemented horizons known as pans, and are classified according to cementing agents such as iron and so on. The extent to which the horizons affect infiltration and salinization depends on their thickness and depth of formation as they constitute an obstacle to water and root penetration. In salt media, salt crusts can form at the surface under specific conditions. Salt-affected and sodic soils have a very loose surface structure, making them susceptible to wind erosion and water erosion.

1.2.3 Qualitative Aspects

Urbanization is immediately associated with the pollution of water bodies due to untreated domestic sewage and industrial discharges. Recently, however, it was perceived that part of this pollution generated in urban areas also comes from surface runoff. Surface runoff carries pollutants such as organic matter, poisons, bacteria, and others. Thus, urban drainage discharge into water bodies introduces modifications that produce different negative impacts with short- and long-term consequences on the aquatic ecosystem.

Pollution caused by the surface runoff of water in urban zones is called diffuse, since it comes from pollution-depositing activities, sparsely, over the river basin contribution area. The source of diffuse pollution is very diverse, and abrasion and wear of roads by vehicles, accumulated refuse on streets and sidewalks, organic wastes from birds and domestic animals, building activities, fuel, oil, and grease residues emitted by vehicles, air pollution, and so on contribute to it. The main pollutants carried in this manner are sediments, organic matter, bacteria, metals such as copper, zinc, magnesium, iron, and lead, hydrocarbons from petrol, and toxic substances such as pesticides and air pollutants that are deposited on surfaces. Storm events may raise the toxic metal concentrations in the receiving body to acute levels (Ellis, 1986).

Stormwater runoff becomes polluted when it washes off concentrated and diffused pollution sources spread across the catchments. In addition to soil erosion caused by raindrop impacts and shear stress action, two major sources contribute to stormwater pollution in temperate climate zones:

a. Diffused sources originating primarily from atmospheric fallout and vehicle emission
b. Concentrated sources originating mostly from human activities, such as poor housekeeping practices, industrial wastes, and chemicals spread washoff by urban storm runoff

Both the processes generate soluble and suspended material. Throughout the process of transport, depending on hydraulic conditions, settling and resuspension take place on the surface and in pipes, as well as biological and chemical reactions. These processes are often considered to be more intense in the initial phase of the storm (first flush effect); however, due to the temporal and spatial variability of rainfall and flowing water, first flush effects are more pronounced in pipes than on surfaces.

1.2.4 GREENHOUSE EFFECT

The heat generated by the sun is partly absorbed by the earth, but a substantial part of it is radiated back into space. The heat absorbed is radiated by the earth as infrared radiation. Greenhouse gases such as water vapor, carbon dioxide, methane, nitrous oxides, and others absorb this infrared radiation and in turn reradiate it in the form of heat. The amount of greenhouse gases in the atmosphere is increasing, and has been scientifically established. Recent research enables us to state that it is human interference that is changing the global climate. Owing to the emission of greenhouse gases, humans are contributing actively to global warming. The most well-known effect of the greenhouse effect is the increasing average atmosphere heat by 2–5°C, which is expected to occur by the year 2050. As the concentration of greenhouse gases increases, further climate change can be expected. In urban areas, some scientists suggest that the magnitude of the effects of changing climate on water supplies is more than in rural areas due to the added heat in flux in urban areas, but may be much less important than changes in population, technologies, economics, or environmental regulations (Lins and Stakhiv, 1998).

1.2.5 URBAN HEAT ISLANDS

Urban heat islands (UHI) have been forming over a period of time around the world. The UHI phenomenon occurs when air in the urban city is 1–5°C (2–8°F) hotter than the surrounding rural area. Scientific data have shown that the maximum temperature of July during the last 30–80 years has been steadily increasing at a rate of 1.5°F every 10 years. Each city's UHI varies based on city layout, structure, infrastructure, and the range of temperature variations within the island. The urban area will have a higher temperature than the rural area as a result of the absorption and storage of solar energy by the urban environment and the heat released into the atmosphere from industrial and communal processes (Ytuarte, 2005). The UHI effect can adversely affect a city's public health, air quality, energy demand, and infrastructure costs (ICLEI, 2005). Attention should be paid to the following issues in UHI:

- *Poor air quality*: Hotter air in cities increases both the frequency and the intensity of ground-level ozone (the main ingredient in smog) and can push metropolitan areas out of compliance. Smog is formed when air pollutants such as nitrogen oxides (NO_x) and volatile organic compounds (VOCs) are mixed with sunlight and heat. The rate of this chemical reaction increases when temperature exceeds 5°C.
- *Risks to public health*: The UHI effect prolongs and intensifies heat waves in cities, making residents and workers uncomfortable and putting them at increased risk for heat exhaustion and heat stroke. In addition, high concentrations of ground-level ozone aggravate respiratory problems such as asthma, putting children and the elderly at particular risk.
- *High energy use*: Higher temperatures increase the demand for air conditioning, thus increasing energy use when demand is already high. This in turn results in power shortages and raises energy expenditures at a time when energy costs are at their highest.
- *Global warming*: Global warming is in large part caused by the burning of fossil fuels to produce electricity for heating and cooling buildings. UHI contribute to global warming by increasing the demand for electricity to cool buildings. Depending on the fuel mix used in producing electricity in the region, each kWh of electricity consumed can produce up to 1.0 kg of carbon dioxide (CO_2), the main greenhouse gas contributing to global warming.

Mitigating UHI is a simple way of decreasing the risk to public health during heat waves, while also reducing energy use, the emissions that contribute to global warming, and the conditions that cause smog.

Cities in cold climates may actually benefit from the wintertime warming effect of heat islands. Warmer temperatures can reduce heating energy needs and may help melt ice and snow on roads.

In the summertime, however, the same city may experience the negative effects of heat islands. Fortunately, there are a number of steps that communities can take to lessen the impacts of heat islands. These "heat island reduction strategies" include the following:

- Reducing the high emission from transportation through traffic zoning and well-managed public transportation
- Installing ventilated roofs and utilizing passive sources of energy in buildings
- Planting trees and other vegetation

1.2.6 Cultural Aspects

Sustainable solutions to water-related problems must reflect the cultural (emotional, intellectual, and moral) dimensions of people's interactions with water. Culture is a powerful aspect of water resources management. Water is known to be a valuable blessing in most arid and semiarid countries and by most religions. There are two cultural characteristics that cause direct impacts on water resources management in urban areas: urban architecture and people's lifestyle.

The practice of architecture in urban areas is often reflected by the climate characteristics of the area. However, the traditional architecture in many large cities is going to be replaced by modern architecture because of population increase and globalization. This may also cause many changes in urban hydrology. The density of population and buildings, rainwater collection systems, material used in construction, and wastewater collection systems are major factors, among others, that alter the urban hydrologic cycle. The change in design paradigm has made major changes in architecture and moves it toward ecological-based design.

Lifestyles in urban areas affect the hydrologic cycle through changes in domestic water demands. Water use per capita and water used in public centers such as parks and green areas are the main characteristics that define the lifestyle in large cities. Even though economic factors are important in determining these characteristics, the patterns of water use, tradition, and culture have more significant effects on the lifestyle in urban areas.

1.3 URBAN WATER INFRASTRUCTURE MANAGEMENT

There are three main urban water infrastructures: water supply, sewerage, and stormwater drainage. Managing urban water infrastructures is an extremely complex issue. While values and technology have changed, the nature of water infrastructure prevents the system from keeping pace with changes. Generally, in the last two decades, increased emphasis has been placed on environmental outcomes of water infrastructure. Previously, social and economic outcomes dominated decisions on water infrastructure.

This change in focus has led to a perception that centralized, large-scale systems ought to be converted to alternative systems. This perception is not always correct, as the best results usually come from using a mix of systems, and this mix will change with location and time. Centralized, large-scale systems will still dominate water infrastructure in many regions for the next few decades, partly because they are there and because it is difficult to change them due to engineering, environmental, economic, and social reasons. For example, headwater dams and other facilities will be too difficult and costly to alter to any significant degree. The smaller alternative systems will have an increasingly important role, but their role will be limited by the source of supply in most cases. The more congested the site and the more the property interests involved, the harder it will be to replace centralized systems with other alternatives.

The challenges of introducing alternative systems include the following:

- Ensuring public health protection
- Increasing the environmentally sensitive water infrastructures such as natural treatment facilities

- Introducing water pricing that reflects its true values
- Publicly accepting demand management measures

The challenges of improving water infrastructure include the following:

- Allocating sufficient money for infrastructure construction, repair, and replacement—financial planning and appropriation
- Allocating resources on the basis of priorities—highest property to the basic needs in the most affected areas
- Developing a standardized data set of water infrastructure

The increasing population of urban cities around the world will eventually outstrip the available sources of water if the current supply and demand conditions do not seek equilibrium. To ensure that water demand does not exceed supply, efficiency improvements, demand and supply management initiatives, and alternative technologies must all be employed. Improvements are also needed for sewerage systems. This may be easier to implement as these systems are more frequently modified because they require more maintenance. People's perception is that urban water authorities cannot keep up with new technologies as these are advancing faster compared to the actual water supply systems enhancements (Howells, 2002).

1.3.1 Life Cycle Assessment

An ecological way of thinking emphasizes on the urban water system of a city as a complex system characterized by continuous processes of change and development. Aspects such as energy, natural resources, transportation, and waste that are directly or indirectly included in an urban water cycle can be regarded as flows or processes in this system. Acts of maintaining, restoring, stimulating, and closing cycles contribute to sustainable development of an urban area. The measurement of urban water system performance raises specific methodological problems, in particular concerns about physical and operational limits, time horizons, and uncertainties associated with functional units operation. Conceptually, it is necessary to know the full environmental consequences of each decision or action made on water resources performances and usage in order to evaluate different performances or compare options. One of the basic methods for this is life cycle assessment (LCA). This constitutes a basis for estimation of the medium- and long-term outcomes of urban water resources planning and management, particularly those related to economic consequences of these actions. Measures of urban water system performance provide an improved basis for decision makers to evaluate their decisions efficiently.

LCA can be defined as a systematic inventory and analysis of the environmental effect that is caused by a product or process starting from the extraction of raw materials, production, and use up to waste treatment. For each of these steps, there will be an inventory of the use of material and energy, and emissions to the environment. With this inventory, an environmental profile will be set up, which makes it possible to identify weak points in the life cycle of the urban water system, including resources, water supply, treatment, distribution, wastewater collection and treatment, and drainage systems. These weak points are the focal points for improving the performance of the urban water system from an environmental point of view and in the movement toward more sustainability.

LCA establishes the relation between objectives and indicators, where one objective can relate to several indicators and one indicator can be used to assess the fulfillment of several objectives. Furthermore, LCA can enlarge traditional system limits in space, in time, and in the number of concerned aspects. LCA can be directly and structurally related to life cycle cost as well as to other types of social and cultural impact assessment methods. LCA proceeds in four steps: (1) goal and scope definition, (2) inventory of extraction and emissions, (3) impact assessment, and (4) evaluation and interpretation.

The application of LCA to the urban water system is only relevant if situated within the larger conceptual framework of sustainable urban development. Furthermore, the traditional focus on environmental impacts has to be completed by taking into account aspects of the long-term water resource conservation of urban areas.

The extension of LCA to water resources must respect and take into account cultural limitations in order to be effectively utilized. The sustainable development of available urban water resources is certainly at the beginning of a longer development. Its objective cannot be to close the debate by simplified procedures before a larger discussion has taken place. The urban water system has unique physical and social characteristics. It furthermore has a continuity that is both spatial and temporal. This continuity constitutes a basic value; it is a fundamental urban resource and must be protected. Protection is required to ensure sustainability through the continuity of development and the embedded social and physical values for urban residents (Hassler et al., 2004).

1.3.2 Life of Urban Water Infrastructure

The life of urban water infrastructure is related to its original design and construction, its maintenance, and its ability to meet water related demands of increasing population. Most urban water infrastructures have been designed to last for a specific time, determined at the time of construction by predicting its economic life (the time at which replacement is more cost effective than repair) and its structural life (the time at which material failure will occur).

In nearly all cases, the life of urban water infrastructure extends beyond that originally planned. This has been achieved by the careful maintenance, preservation, renovation, and appropriate replacement of structural systems. The pressure on the life of water infrastructure caused by ever-increasing water demand can be slowed down by employing demand management strategies. This involves tools such as water efficiency devices, publicity campaigns, and introducing water use and wastewater charges.

Issues of concern with the life of water infrastructure include the following:

- There is considerable variability in the quality of maintenance and replacement programs for water supply and sewerage systems in water authorities and local councils.
- Maintenance attention may focus disproportionately on areas where failures are readily noticeable and consequently attract considerable negative comment. This can result in the other areas taking substantially longer time to resolve.
- Stormwater systems, notably those designed to cope with fairly rare flood events, are maintained less rigorously. Management of these assets is hampered by inadequate funding and current administrative arrangements that occur on local government boundaries rather than on catchment boundaries. Standards vary considerably between the local councils. Most asset management systems are inadequate. The current administrative arrangements also promote a reactive rather than a proactive approach to maintenance.
- There are problems in maintaining environmental assets such as waterways and facilities designed to protect them, including gross pollutant traps and bank protection works.
- The difference in the expected life and the economic life of infrastructure results in systems being designed in a manner whereby future changes are not excluded, but may not be the most economical at the initial stage of planning.
- Most of the net present value or benefit to cost assessments are irrelevant once the planning horizon exceeds 10 years.
- Local councils are establishing asset management systems for stormwater drains, flood protection works, and stormwater treatment facilities, but few are allocating adequate funds for anticipated replacements and renewals.

1.3.3 Environmental, Economic, and Social Performances

The environmental, economic, and social performances of urban water systems vary considerably in time and space, and their performances have been judged differently over time as community values change. Previously, water infrastructure development was seen to increase economic prosperity and social production of water resources and environmental consequences were not seen as important. Such views have changed with the increase in environmental awareness. In addition, the necessity of greater efficiency in water usages due to social and environmental aspects, has changed the perspective of what economic benefit of water infrastructure development delivers.

Given that values have changed since most water infrastructures were built, the performances of water infrastructures are now viewed by many as unacceptable. While some problems such as the discharge of partly treated sewage on the shoreline have been fixed, others such as the failure of some sewerage systems to meet their original technical specifications have not. Given that perspectives have changed over time, it is not productive to linger over the past and to fix blame. Rather we should see the urban water systems in place and seek to preserve their desirable features and transform them into more appropriate systems where possible.

1.4 URBAN WATER CYCLE MANAGEMENT

The urban water cycle starts with water extracted from streams and aquifers, stored in reservoirs, and then processed to potable quality before delivery through an extensive pipe system to consumers. Some of this water is then used to transport wastes through a network of sewers to treatment plants that discharge effluent into receiving waters such as rivers, lakes, and oceans. Rainfall on the consumer's allotment contributes to the urban catchment's stormwater, which is collected by an extensive drainage system for disposal into receiving waters (Figure 1.3) (UNU-INWEH, 2006).

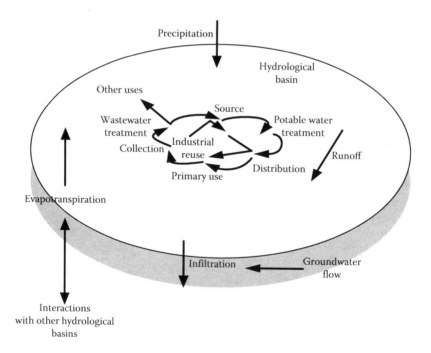

FIGURE 1.3 The urban water cycle consists of (1) source, (2) water treatment and distribution, (3) use and reuse, and (4) wastewater treatment and deposition, as well as the connection of the cycle to surrounding and adjacent hydrological basins. (Adapted from United Nations University, International Network on Water, Environment and Health [UNU-INWEH]. 2006. Four pillars. Available at http://www.inweh.unu.edu/inweh/).

Introduction

The cycle concept is a useful tool for stakeholder and community education and for consensus building around an action plan. The goal, that of restoring and maintaining the balance between the current demands of the community for water and the need to preserve the aquatic ecosystem for the benefit of future generations, becomes understandable. Table 1.1 lists some of the enabling systems and practices that can be developed within an urban community.

The starting point for wise water stewardship is the understanding that urban water, from source to final disposition, flows through a series of four interrelated stages, listed in Table 1.1, in a continuous cycle. The cycle concept illustrates some important facts about urban water:

- Waste and contamination at any stage negatively affect the sustainability of the cycle as a whole and the health and safety of the community using that water.
- Urban planning, without considering the water cycle, results in water supply shortages, deteriorating aquifer water quality, groundwater infiltration into the distribution system, endemic health problems, and other symptoms of an unsustainable situation.

TABLE 1.1
Examples of Community-Based Enabling Systems for a Sustainable Urban Water Cycle

1. Source
   - A long-term urban and watershed management master plan
   - A source water quality and quantity monitoring system
   - A geographic information and decision support system
   - An inspection and enforcement system to protect source water
   - A community education program
2. Use/reuse
   - A metering and billing system
   - An industrial discharge control program
   - Regulations and bylaws
   - An industrial incentive program
   - A community education program on water conservation
   - A network of supporting laboratories
   - A monitoring and control system
   - An emergency spill response system
3. Treatment/distribution
   - A potable water quantity and quality monitoring and control system
   - A utility operation and maintenance system, including training and accreditation of operators
   - A financial, administrative, and technical management structure
   - A flexible water treatment process
   - An operation, maintenance, leak detection, and repair system
   - Continuous pressurization
4. Treatment/disposition
   - An effluent quality monitoring and control system
   - A utility operation and maintenance system, including training and accreditation of operators
   - An environmentally sustainable biosolids management program
   - A financial, administrative, and technical management structure
   - A flexible treatment system
   - An end-user market

Source: United Nations University, International Network on Water, Environment and Health (UNU-INWEH). 2006. Four pillars. Available at http://www.inweh.unu.edu/inweh/ (visited in March).

- Every citizen, institution, agency, and enterprise in the community has a contribution to make toward the goal of sustainability.

It should also be noted that all components of the urban water cycle meet with water consumed and stormwater and wastewater discharge. Source control through management of the cycle at this level offers the opportunity to provide benefits for the consumer and the environment. The philosophy of source control is to minimize the cost of providing water and collection of stormwater and wastewater. Source control can be implemented through retention of roof rainwater (rainwater tanks), stormwater detention, on-site treatment of gray water (laundry, bathroom, and kitchen) and black water (toilet), use of water-efficient appliances and practices, and on-site infiltration.

1.5 SUMMARY AND CONCLUSION

In this chapter, the urban water cycle has been introduced as part of the hydrological cycle. Components of the urban water cycle are the same as those of the hydrological cycle, but they have special characteristics that must be attended to for successful management and design of urban water. These components and effects of urbanization as well as the physical, chemical, and biological impacts have been discussed. Urbanization disturbs and changes different natural processes. The effects of urban area developments in different aspects such as climatic, hydrologic, cultural, and other special effects such as greenhouse and hot islands have been discussed in this chapter. These effects must be considered for future designs and decision making about urban development, especially regarding water supply, because these effects result in change of water need, availability, and accessibility.

Urban water infrastructures are a major part of the urban water cycle and have three main categories: water supply, sewerage, and stormwater drainage. Managing these widespread systems is complex and there are many challenges and obstacles. The management of the urban water cycle and urban water infrastructures have been briefly introduced in this chapter. In the following chapters, different aspects of water governance, the urban water cycle and hydrology, urban storm drainage concepts, and water quality issues are described. Then planning concepts, the application of system dynamics, and conflict resolution as well as urban infrastructure are presented. Finally, stormwater pollution and disaster management and the impact of climate change on the urban water cycle are discussed.

REFERENCES

Borchardt, D. and Statzner, B. 1990. Ecological impact of urban stormwater runoff studied in experimental flumes: Population loss by drift and availability of refugial space. *Aquatic Sciences* 52(4): 299–314.
Ellis, J. B. 1986. Pollutional respects of urban runoff. In: H. C. Trono, J. Marsalek, and M. Desbordes, Eds, *Urban Runoff Pollution*, pp. 1–38. Springer, Berlin.
Galli, F. J. 1991. *Thermal Impacts Associated with Urbanization and BMPs in Maryland*. Metropolitan Washington Council of Governments, Washington, DC.
Hassler, U., Algreen-Ussing, G., and Kohler, N. 2004. Urban life cycle analysis and the conservation of the urban fabric. SUIT Position Paper (6).
Howells, L. 2002. Submission to the Inquiry into Urban Water Infrastructure. NSW Legislative Assembly. Sydney Panel of the National Committee on Water Engineering Sydney Division. Available at www.ieaust.org.au.
International Council for Local Environmental Initiatives (ICLEI). 2005. Why Should Cities and Counties Care About Urban Heat Islands? Available at http://www.iclei.org/ (visited in December).
Karamouz, M. and Araghinejad, Sh. 2005. *Advanced Hydrology*. Amirkabir University Press, Tehran, Iran.
Lawrence, A. I., Ellis, J. B., Marsalek, J., Urbonas, B., and Phillips, B. C. 1999. Total urban water cycle based management. In: I. B. Joliffe and J. E. Ball, Eds, *Urban Storm Drainage, Proc. 8th Int. Conf. on Urban Drainage* 3: 1142–1149.
Lins, H. F. and Stakhiv, E. Z. 1998. Managing a nation's water in a changing climate. *Journal of the American Water Resources Association* 34: 1255–1264.

Marsalek, J., Jimenez-Cisneros, M., Karamouz, P. A., Malmquist, Goldenfum, J., and Choat, B. 2006. *Urban Water Cycle Processes and Interactions*. Urban Water Series—UNESCO-IHP, Taylor & Francis, Paris (ISBN: 978-0-415-45346-2).

McPherson, M. B. and Schneider, W. J. 1974. Problems in modeling urban watersheds. *Water Resources Research* 10(3): 434–440.

Schueler, T. R. 1987. *Controlling Urban Runoff: A Practical Manual for Planning and Designing Urban BMPs*. Metropolitan Washington Water Resources Planning Board, Washington, DC.

Schumann, A. H. 1993. Development of conceptual semi-distributed hydrological models and estimation of their parameters with the aid of GIS. *Journal of Hydrological Sciences* 38(6): 519–528.

Silveira, A. L. L., Goldenfum, J. A., and Fendrich, R. 2001. Urban drainage control measures. In: C. E. M. Tucci, Ed., *Urban Drainage in Humid Tropics*. Urban Drainage in Specific Climates, C. Maksimovic (ch. ed.). *UNESCO Technical Documents in Hydrology*, UNESCO, Paris, France, 40(I): 125–154.

Thompson, R. D. 1975. The climatology of the arid world. Geography Papers—GP#35, University of Reading, UK.

Tucci, C. E. M. and Porto, R. L. 2001. Storm hydrology and urban drainage. In: C. E. M. Tucci (Org.) *Urban Drainage in Humid Tropics*, 1st edn, UNESCO, Paris, 1: 69–102.

United Nations University, International Network on Water, Environment and Health (UNU-INWEH). 2006. Four pillars. Available at http://www.inweh.unu.edu/inweh/ (visited in March).

Wang, S. L. and Chen, C. T. A. 1996. Comparison of seawater carbonate parameters in the East China Sea and the Sea of Japan. *Lamer* 34: 59–64.

Whiteman, C. D. 2000. *Mountain Meteorology: Fundamentals and Applications*. Oxford University Press, New York, NY (ISBN 0-19-513271-8).

Ytuarte, S. L. 2005. Urban hot island phenomenon of San Antonio Texas using time series of temperature from Modis images. *Geological Society of America Abstracts with Programs* 37(3), 4–5.

2 Governance and Urban Water Planning

2.1 INTRODUCTION

In the past decades, management of the urban water cycle has been largely based on large-scale, centralized systems, which have been successful in improving the quality of life, particularly through the reliable provision of clean water acts, and reducing the risk of infectious diseases. Nevertheless, the change of paradigm from Newtonian to holistic has changed urban water governance globally. New changes in community values, technology, and economic circumstances have resulted in many challenges and changes in the course of action in urban water management. A consensus is emerging that the long-term continuation of traditional urban water management practices is "unsustainable" on a variety of social, economic, and environmental grounds. This viewpoint seeks the development and implementation of alternative approaches to urban water management that are part of a broader framework of "ecologically sustainable development" (McAlister, 2007). The challenges of traditional urban water management are as follows:

- Building new dams to meet the growth in water demand in urban areas is no longer possible, especially in developing countries.
- Increasing the importance of environmental issues and significant regional interest in the reinstatement of "environmental flows" in streams and into estuaries is a must. Traditional water infrastructure systems have significant hidden costs, including the degradation of rivers, wetlands, oceans, and fisheries.
- Increasing public life standards leads to the need for higher standards of treatment for stormwater and sewage discharges.
- Capacity building or extending conventional water infrastructure systems to expanding outer suburbs is increasingly problematic.
- Water infrastructure aging and the increasing cost of maintaining and rebuilding existing water infrastructure systems is problematic.
- Looking at urban stormwater and wastewater as potentially useful resources rather than undesirable waste products and usage of potable-standard water for low-quality purposes such as laundry, toilet, and outdoor use is a highly wasteful use of water resources and treatment infrastructure.
- Efficient drainage systems are also effective in transporting pollutants from impermeable urban surfaces to receiving waters.
- Urbanization affects groundwater processes, contributing to rising groundwater and urban salinity problems.
- Advances in technology give the potential for greater localization of water infrastructure.

As a new vision for urban water management, using stormwater and wastewater for different water usage in urban areas has become a new concept in urban planning and management, which is called Water Sensitive Urban Design (WSUD). Since this new approach in urban planning and management needs communication between different departments in many countries, this chapter discusses the institutional change in water governance. The change in the direct result of new planning concepts and their evolution has also been discussed. A paradigm shift from Newtonian to holistic has been explored and its impact on shifting from supply management to demand management has been addressed. Water and its role in land use planning has been a challenge for planners and decision makers.

The suitability of land use based on availability of water resources has always been a major obstacle, and only those societies and municipalities that have coupled land use planning with water resources sustainable allocation schemes have succeeded in meeting their long-term objectives and have maintained environmental integrity and vitality toward sustainable development. Water resources assessment as a means of measuring, evaluating, monitoring, forecasting, and bringing consensus into water resources planning is also discussed. Technology has always been a scapegoat for planners. In the Newtonian paradigm, decision makers and planners look at technology as a way of salvation of many performance shortcomings. The holistic view of planning requires developing tools and methods that are adaptive and sensitive to the nontechnical aspects of planning such as social, cultural, and environmental factors, and that can be used with the current state of local knowledge and water infrastructure.

2.2 WATER SENSITIVE URBAN DESIGN

WSUD was originally coined in Western Australia (Whelans and Halpern Glick Maunsell, 1994) to describe a new Australian approach to urban planning and design, and was first referred to in various publications in the early 1990s (as summarized in Lloyd, 2001). The emergence of WSUD in Australia has paralleled a wider international movement toward the concept of integrated land and water planning and management, which is discussed in this chapter. The main goal of this shift is the need to provide more economical, and less environmentally damaging, ways of providing urban water, wastewater, and stormwater solutions.

WSUD is part of the contemporary trend toward more "sustainable" solutions that protect the environment. It seeks to ensure that development is carefully designed, constructed, and maintained so as to minimize impacts on the natural water cycle.

Utilizing WSUD can help counteract many of the negative impacts of urban development on the natural water cycle. By utilizing appropriate measures in the design and operation of development, it is possible to maintain and restore the natural water balance, reduce flood risk in urban areas, reduce the erosion of waterways, slopes, and banks, improve water quality in streams and groundwater, make more efficient use of water resources, reduce the cost of providing and maintaining water infrastructure, protect and restore aquatic and riparian ecosystems and habitats, and protect the scenic, landscape, and recreational values of streams.

As mentioned above, traditional water supply, stormwater, and wastewater practices are largely based on centralized collection, conveyance, treatment, and disposal of water flows. But WSUD is more attuned to natural hydrological and ecological processes and promotes a more local and decentralized approach. It gives greater emphasis to on-site collection, treatment, and utilization of water flows as part of an integrated "treatment train" that may be applied in addition to or in lieu of conventional stormwater measures. Elements in the treatment may include the following:

- Rainwater harvesting and the use of roof water for toilet flushing, laundry use, hot water systems, or irrigation
- Surface runoff and gray water reuse for irrigation purposes
- Groundwater recharge by infiltration of stormwater to underground aquifers
- Cleansing runoff and conserving water by specially designed landscaping
- Protection of native vegetation to minimize site disturbance and conserve habitat
- Protection of stream corridors for their environmental, recreational, and cultural values

Designers should respond to the constraints and opportunities of each individual site based on WSUD concepts. Consequently, careful consideration must be given to site characteristics such as soil type, slope, groundwater conditions, rainfall, and the scale and density of development.

Fundamental to the philosophy of WSUD is the integrated adoption of appropriate best planning practices (BPPs) and BMPs and its objectives are more than simply constructing a lake or wetland system. WSUD calls for an enhanced, or more considered, approach to the integration of land and water planning at all levels in the urban development process, for example strategic planning and concept planning to detailed design. A BPP refers to a site assessment, planning, and design component of WSUD and it is defined as the best practical planning approach for achieving defined management objectives in an urban situation. A BPP includes the site assessment of the physical and natural attributes of the site and capability assessment. The next step is integrating water and related environmental management objectives into site planning and design (Dahlenburg, 2003).

BPPs may be implemented at the strategic level or at the design level. At the strategic level, BPPs may be the decision to make provision for arterial infrastructure or to include water-sensitive policy provisions or design guidelines in town planning schemes. At the design level, BPPs refer to specific design approaches.

In the following, some topics for BPPs are described (McAlister, 2007):

- Planning for an integrated stormwater system, incorporating storage locations, drainage and overflow lines, and discharge points
- Land use planning and identification of developable and nondevelopable areas
- Planning for public open space networks, including natural drainage lines, and recreational, cultural, and environmental features
- Planning for the use of water-conserving measures at the design level for road layouts, housing layouts, and streetscapes

This chapter presents an introduction about land use management as an important component of this new approach of urban planning and design, which has a pronounced effect on urban water governance. The structural and nonstructural elements of a design that perform the prevention, collection, treatment, conveyance, storage, and reuse functions of WSUD are referred to as BMP.

2.3 WATER GOVERNANCE

The term "governance" generally refers to the relationship between a society and its government or between an organization and its governing entity. It is often referred to as the "art of steering societies and organizations." Other definitions of governance could be stated, depending on its context.

Ottawa's institute definition of governance* is used here, as follows:

> Governance is the process by which stakeholders articulate their interests, their input is absorbed, decisions are taken and implemented, and decision makers are held accountable.

It is known that water is essential for our survival, and yet more than 1 billion people today cannot meet their basic human needs because of the lack of access to clean water (UNESCO-WWAP, 2003). Water scarcity plagued 27 nations, and an additional 16 nations are considered to be water stressed (WRI, 2003). Water scarcity is one of four major factors to threaten human and ecological health over the next generation. Since public health, development, economy, and nature are suffering, the importance of ensuring access to clean water has become a high priority for governments.

All governments throughout the world are facing the common problem of how to address the growing water crisis. It is necessary to manage water in ways that are efficient, equitable, and environmentally sound. A great deal of capital investment and legal and economic reforms are needed to be able to improve water efficiency, which is often beyond the capacity of members of the public

---

*The Institute on Governance, based in Ottawa, is a nonprofit organization founded in 1990 to promote effective governance. For more information, see http://www.iog.ca/.

who are directly impacted by the lack of clean water. A detailed understanding of interrelated hydrodynamic, socioeconomic, and ecological systems is also required for the equitable allocation and stewardship of water resources. Often those responsible for such decisions at the local, provincial, and national scales lack the necessary knowledge.

Although critical knowledge about water management has been distributed across different groups, organizations, and the water users themselves, a broad understanding for each initiative is required by regional water managers. Governments need to gather early participation in order to effectively realize such water initiatives. The next step is to actively involve other segments of the society, including those with the most vulnerability to water limitations. Tools to support such efforts are becoming readily available. However, the lack of awareness about these tools has severely limited their application. It is for these reasons that the world has become unified in integrating public participation in the implementation and decision making of urban water management.

The current goal is to further global understanding, techniques, and approaches for the integration of public participation and improvement of water management. Sound water management is the main requirement of policies and activities that serve to provide clean water to meet human needs and sustain the water-related ecological systems that we depend on. Water management generally aims to address interests and integrate usage across hydrological units, such as watersheds. A broader geographical scope may be required in some management aspects, such as transboundary flow across multiple basins and interbasin water transfers via channels or virtual water.

Public participation is described by the International Association for Public Participation as "any process that involves the public in problem solving or decision making and uses public input to make better decisions." It should be noted that this term is slightly different from "stakeholder involvement," which involves those affected by a decision as well as those able to effect its intended outcome. Public participation should also include the interests of those who are usually marginalized.

It was emphasized in the 1992 Rio Declaration on Environment and Development (UNCED, 1992) that environmental issues such as water management "are best handled with the participation of all concerned citizens." Nations are urged to facilitate public participation by increasing transparency, participatory decision making, and accountability. These elements are respectively described as (a) informing people of water management issues that may affect them, (b) involving the public in the decision making, and (c) attempting to provide compensation to those adversely affected by these decisions and activities.

2.3.1 Consequences of Poor Governance

The consequences of poor governance in water supply can be life threatening. Communities are now realizing the extent of problems facing the water sector. These problems have mostly arisen from poor governance, although some of the issues are clearly technical. Low investment, low quality of service, and low revenue are examples of the problems created by poor governance in municipal water supply systems. Some water utilities have become trapped in a cycle of such problems. Low levels of revenue relative to costs result in a low quality of service, which makes it politically difficult, in many cases, to justify raising the water rates. When water companies do not adjust water rates to recover all of their costs, these problems become particularly critical. In addition, it may take a decade to plan, design, and build water and wastewater infrastructure, and even longer to measure the impacts of water abstraction or effluent release on the environment and on water quality; therefore, political cycles, which are often much shorter than the infrastructure life span, may sometimes work against long-term planning and development.

2.3.2 Water Governance at the Regional Level

Water management on an operational level will now be discussed. On behalf of the public, ownership of water is vested in provincial and territorial governments. Governments have created licensing regimes under which water use licenses are issued to individuals and corporations, although legal frameworks vary. The basis on which these permits are issued varies in some parts. For example, rights to use water are based on property ownership in eastern Canada, whereas in western Canada, water rights are allocated on the basis of first come, first served. The allocation of water rights is based on a hierarchy of public purposes established by statute in different territories. As with water rights, laws governing municipal water supply, including the range of business and governance models that are legally permitted, vary from one province or state to the next.

Given these differences, municipal business structure and governance models vary from place to place. The following sections provide general information that outlines good principles of governance as well as good governance models for restructuring water supply systems.

2.4 GOVERNANCE MODELS

A workable description of the principles of governance along with responsibilities and relationships among stakeholders is given through a governance model. Governance models are usually considered when organizations wish to improve governance. A set of structures, functions, and practices are usually determined, which defines who does what, and how. For example, a governance model in the case of municipal water supply would specify the distribution of decision-making authority between the community and operational managers.

There are many governance models, for both the profit and nonprofit sectors, that can be applied to internal governance of organizations as well as the provision of having business plans. Even though information on governance has increased, there is still no single model that can be applied to all circumstances.

2.4.1 Generic Governance Models for Water Supply

Governance and business models are closely interrelated. Sometimes good governance can guide business models and sometimes it may constrain it, and vice versa. The distribution of risks and responsibility for all aspects of water supply management can be determined with governance and business models together.

The three main models of resource management that are at the center for debate over water supply governance and control sharing are the planning model, the market model, and the community model (Table 2.1). These three stakeholder governance models also apply to public services more generally. Many public services have elements of more than one model; however, Table 2.1 simplifies this more complex picture.

Some hybrid models are as follows: municipal services boards or commissions, delegated management contracts, and corporative utilities that adopt elements of both the planning and market models. In all cases, balanced interests of the public and private sectors are needed.

In France, for example, private sector management of municipally owned water supply infrastructure via long-term management contracts is common practice. Municipal governments are, however, forbidden by law to sell their infrastructure, but in many cases they retain control over long-term strategic planning.

In England, where the water supply and wastewater industry were fully privatized in 1989, the market model was chosen. However, like most jurisdictions that employ the market model, England has created extensive regulatory frameworks and regulatory agencies for the water sector to protect consumers and public health. Companies are forbidden from disconnecting domestic consumers (even for nonpayment of bills) and are required to create special, low tariffs for vulnerable consumers.

Strategic financial plans as well as resource development and water supply management plans should be submitted by companies to an economic regulator and an environmental regulator for review. Despite being privatized, the water industry in England has been reregulated rather than deregulated, and many high-level decisions are characterized by planning governance attributes (Bakker, 2003).

TABLE 2.1
Example of Governance Models for Local Public Utility Services

Public Service Elements	Planning Model	Market Model	Community Model
Asset owner	Government	Private corporation	Users
Asset manager	Government	Private corporation	Users
Consumer role	Citizens	Customers	Community members
Organizational structure	Civil service	Corporation	Association/network
Accountability mechanism	Hierarchy	Contract	Community norms
Primary decision makers	Administrators, experts, public officials	Individual households, experts, companies	Leaders and members of community organizations
Primary goals of decision makers	Minimize risk Meet legal/policy requirements	Maximize profit Efficient performance	Serve community interest Effective performance
Key incentives for good performance	Expert/managerial feedback in public policy process Voter/ratepayer opinion	Price signals (share movements or bond ratings) Customer opinion	Community norms and shared goals Community opinion/sanctions
Key sanctions for failure to maintain services	State authority backed by coercion Political process via elections Litigation	Financial loss Takeover Litigation	Livelihood needs Social pressure Litigation (in some cases)
Participation of consumers	Collective, top-down	Individualistic	Collective, bottom-up
Associated business venture	Municipally owned utility	Private corporate utility	Community cooperative

Source: Adapted from McGranahan, G. et al. 2001. *The Citizens at Risk: From Urban Sanitation to Sustainable Cities.* Earthscan, London. With permission.

There are major differences between the planning, market, and community governance models, from the standpoint of consumer representation; accountability and how it is structured in each model; and the goals under each model and how goals will lead to distinct policy and management outcomes. Different models also incorporate stakeholder preferences.

When making a transition and creating a hybrid between two models, governments must carefully consider the implications of changes in the role of consumers, incentives, penalties, and accountability structures. This is particularly important when considering a hybrid model, as problems are likely to arise when the function and component of a governance model are incoherent. For example, a disjuncture between (shorter) political time cycles (2–4 years) and (longer) infrastructure life cycles (10–50 years) can compromise the sustainability of financing.

2.4.2 Building a Governance Model

Organizations will find it useful to define principles of good governance and to articulate responsibilities and relationships between stakeholders. This is particularly important during periods of rapid urban and industrial development and significant social change, when expectations may not be clear.

Although there is a widespread acknowledgment of the importance of governance, definitions and models of good governance vary considerably. In this section, general criteria for good governance are discussed. The detailed structure of a governance model will vary, depending on the type and charter of different organizations.

2.4.2.1 Choosing Principles

Good governance is characterized by a set of principles that guide decision-making processes and management practices. For example, in making recommendations on the role of municipal governments in water supply management, it could be argued that public safety is paramount, and four principles can be considered (O'Connor, 2002):

- Public accountability for decisions relating to the water system
- Effective exercise of owners' oversight responsibilities
- Competence and effectiveness in the management and operation of the system
- Full transparency in decision making

Principles of good governance and prioritization principles vary between organizations and jurisdictions due to a broader framework of political governance. In Great Britain, regulation is relatively nonlegalistic and discretionary compared to North American governance models.

Good governance implicitly makes assumptions about the legitimacy of different stakeholders and decision makers and about accepted processes of decision making. Making robust decisions is a challenge. Good governance is thus to some degree dependent on how a society interprets the democratic practice. There is, accordingly, no one receipt for good governance options; Tables 2.2 and 2.3 list several examples of governance principles for water supply management.

2.4.2.2 Good Governance Processes

Some of the most frequently occurring good governance principles in water management, according to O'Connor (2002) (in no particular order), include

- Protection of public health and safety
- Environmental protection
- Accountability for stewardship and performance
- Transparency
- User participation
- Balancing equity, efficiency, and effectiveness in performance
- Financial sustainability

Some of the organizations surveyed included only some of these principles. Some chose to rank the principles in order of priority, whereas others chose to balance them.

2.4.2.2.1 Coherency and Prioritization

Internal consistency between the different principles creates coherency. Corporatization as an example of hybrid models should ensure coherence between different aspects of the governance model. However, there is conflict among operational management situations, policies, and objectives that flow from governance principles. For this reason, it is important to prioritize governance principles.

For example, the first principle in choosing any management or operational structure for water and wastewater could always be safety (i.e., public health) (O'Connor, 2002).

2.4.2.2.2 Create Objectives and Policies

Governance principles, by themselves, are insufficient for good governance. Concrete objectives or goals must be specified in order to enable the practice of good governance. Chapter 18 of Agenda 21

TABLE 2.2
Selected Examples of Governance Principles for Water Management

Source	Principles
O'Brien et al. (2002)	• Accountability • Responsiveness • Effectiveness • Efficiency • Transparency • Participation • Respect for the rule of law
Dublin Principles (1992) International Conference on Water and the Environment in Dublin set out a statement on Water and Sustainable Development, which is known as the "Dublin Principles"	• Freshwater is a finite and vulnerable resource, essential to sustain life, development, and the environment • Water development and management should be based on a participatory approach, involving users, planners, and policy makers at all levels • Women play a central part in the provision, management, and safeguarding of water • Water has an economic value in all its competing uses, and should be recognized as an economic good
Agenda 21 (2002)	• Full public participation • Multisectoral approach to water management • Sustainable water use
Federation of Canadian Municipalities (2002)	• Quality of life • Shared responsibility (between governments) • Municipal government leadership • Adaptability • User pay • Maintenance and rehabilitation • Continuous improvement • Partnerships

on water resources is an example of a detailed set of objectives flowing from a set of governance principles.* For example, a good governance principle of public participation might lead to an objective of community involvement in standard setting. In turn, this might lead to specific policies, such as multistakeholder participation in a drinking water advisory role.

Similarly, a good governance principle of accountability might lead to the following objectives: clear lines of accountability, good communication, and trust. In turn, this might lead to specific policies, for example: a consumers' right to know (access to information) policy; holding service providers to a statutory standard of care. O'Connor (2002) recommended in a report that "since the safety of drinking water is essential for public health, those who discharge the oversight responsibilities of the municipal government should be held to a statutory standard of care."

---

*Chapter 18 of Agenda 21 deals with the Protection of the quality and supply of freshwater resources: application of integrated approaches to the development, management, and use of water resources. Agenda 21 is a plan of action adopted by more than 178 governments at the United Nations Conference on Environment and Development (UNCED) held in Rio in 1992. Agenda 21 was reaffirmed at the World Summit on Sustainable Development held in Johannesburg in 2002.

2.4.2.2.3 Responsiveness

Responsiveness implies a capacity for and commitment to self-reflection, in which stakeholders learn and feed lessons learned back into an evolving vision. This might imply, for example, the need to carry out periodic reviews of management and operating structures for water supply systems. This is recommended by O'Connor (2002): "Municipal governments should review the management and operating structure for their water system to ensure that it is capable of providing safe drinking water on a reliable basis." Table 2.3 shows a summary of the characteristics of a good model.

2.4.2.2.4 Sound Information

Governance bodies in all sectors must have access to sound information. Decision makers need it to provide accountable governance, while those who are governed need to hold decision makers accountable. Decision makers must use information to analyze the need for change and then act when necessary. Of particular importance when restructuring governance, is the considerations of reporting requirements, and timely flows of information to and from decision makers. In assessing the restructuring options open to municipal governments, the cost of regulation and oversight should be factored into the assessment. O'Connor (2002) notes that "the cost to a municipal government of due diligence before entering a contract, and of compliance monitoring over the term of a contract, is an important part of its oversight responsibilities and, as such, the full cost of water services." Information is required by the municipal government not only on the restructuring options, but also on the broader cost and quality implications for municipal governments.

2.4.2.2.5 Transparent Decision-Making Process

A good governance process is one that enables stakeholders collectively to design and implement policies and management strategies that meet their goals effectively and acceptably. Given the importance of water supply for public health, the Walkerton report emphasis should be placed on the need for openness in water supply governance. The following three recommendations of particular relevance can be made:

1. Municipal contracts with external operating agencies should be made public.
2. All service contracts between the corporative agencies should be publicly available.

TABLE 2.3
Applying Good Governance Principles to Water Management

Principle	Example of Application
Accountability	Demonstrating adherence to capital plans for water and sewage infrastructure through publicly available audited financial statements
Responsiveness	Developing a long-term plan to ensure water and sewage system capacity to accommodate future growth
Effectiveness and efficiency	Scheduling water main repairs at the same time as road repairs
Transparency	Making results of raw and treated water quality testing publicly available
Participation	Soliciting public comments about restructuring options
Financial sustainability	Full life cycle investment needs are the basis for program spending
Respect for the rule of law	Ensuring that minimum chlorine residuals are maintained in the water distribution system

Source: Adapted from O'Brien, J. J. et al. 2002. Governance and methods of service delivery for water and sewage systems. Commissioned Paper 17, The Walkerton Inquiry. Queen's Printer for Ontario, Toronto.

3. The provincial government should require all owners of municipal water systems to have an accredited operating agency, whether internal or external to the municipal government.

In O'Connor's (2002) view, this accredited operating agency should operate in accordance with provincially recognized management standards, should be periodically evaluated or independently audited, and the results of such review or audit should be made public.

2.4.2.2.6 Participation of Stakeholders

The participation of stakeholders in decision-making processes—including users—is a critical factor in good governance. This does not imply that participation is always necessary or that more participation is better; some authors even question whether participation might be the "new tyranny." Participation can be structured along a "ladder" of options, from public information, through consultation, to full-fledged representation. Good governance processes will incorporate different levels of participation when and as appropriate.

Participation is important for three reasons. It can help make decisions more effective; it may increase the political acceptability of decisions; and it fosters accountability. After a decade of experimentation, the British and Welsh water industry have evolved a multistakeholder model of participation in water policy making that seeks broad representation and formalizes public participation through customer committees known as Water Voice (formerly Customer Service Committees). Figure 2.1 shows the role of stakeholders in British policy-making process. A summary of the characteristics of a good governance model is as follows:

- The model articulates a set of governance principles or expresses a vision.
- The governance principles are coherent and are ranked in order of priority.

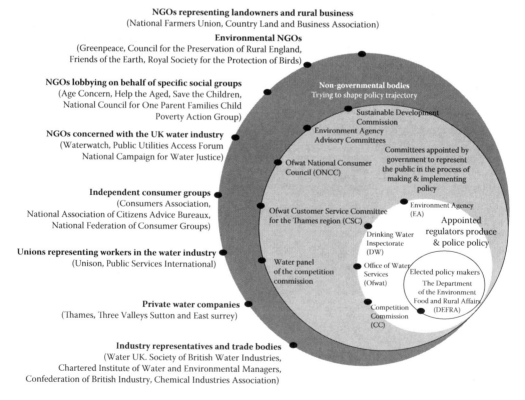

FIGURE 2.1 Stakeholders in the British water policy-making process. (Adapted from Bakker, K. 2004. *Good Governance in Restructuring Water Supply: A Handbook.* Federation of Canadian Municipalities, Canada.)

Governance and Urban Water Planning

- The model builds on the governance principles to create objectives and policies.
- The model is responsive; learning and reviewing options will inform restructuring.
- The model enables the production and dissemination of high-quality information.
- The model includes an open, transparent decision-making process.
- The model facilitates the participation of stakeholders.

In many cases, the result of this process will be a detailed description of the governance of the water supply utility, including routine decision making and higher-level policy process. Governance attributes will vary, depending on the business model chosen.

2.5 ABSENCE OF PUBLIC PARTICIPATION

Quite often people are denied the right to participate in water management decisions and policies that concern them. For instance, large dams for water supply and irrigation have forcibly displaced tens of thousands, even millions of intended beneficiaries, across India, Mauritania, Brazil, and many other places around the world. Numerous news reports also highlight how inadequate governance continues to allow industries to degrade the environment and damage their neighbors through water supplies pollution in China, Indonesia, and elsewhere. Meanwhile, some governments have even intentionally used water policy to harm the disenfranchised, such as Iraq's years of draining the wetlands upon which its Marsh Arabs depended for millennia.

These examples illustrate how inadequate public participation in governments' water management can result in tremendous social upheaval and the violation of the basic human rights of their citizens. How can public participation in water management provide for our basic human need for water and ensure no thirst? Three approaches explored by participants in the United Nations symposium are summarized in the following sections (UNCED, 1992).

2.5.1 Conventional Approaches

Approaches that utilize processes to inform, consult, involve, collaborate with, and empower the public are considered conventional approaches to public participation. Public participation will achieve by informing the public about the actions that will be decided in a participatory manner. Sometimes, even with the best intentions, governments face challenges of including participatory management of water resources under conditions of limited institutional capacity and substantial spatiotemporal variability in water quantity. The timing and manner in which participatory mechanisms are incorporated into the Environmental Impact Assessment (EIA) process can have a tremendous impact on the ultimate utility of the resulting water management regime.

To date, there has been little transfer of public participation practices or lessons between water management initiatives. As a result, there remain many potentially useful, though underutilized, tools [such as Transboundary Environmental Impact Assessment (TEIA), consensus building, and joint fact findings]. Furthermore, it is often unclear how to measure and determine the success of conventional public participation initiatives (UNCED, 1992).

2.5.2 Information Technology Approaches

The main question is how information technology (IT) can be utilized to promote public participation in the management of water. The Internet and its various applications as well as nonnetworked decision support and geographical information systems are kinds of tools and technologies. IT should be followed as an end instead of a means to improve participation and effective water management. Other pertinent externalities and misconceptions are also presented. Public involvement initiatives should also be based on a set of culturally and politically relevant principles. With respect to public participation in international waters management, in particular, the International Waters Learning

Exchange and Resource Network (IW: LEARN) has established a collaborative platform for the international waters community, which is accessible on-line. GEF IW (Global Environmental Facility International Waters): LEARN and its partners invite international water managers, interested members of the public and private sectors, and civil society at large to participate in the workshop series design, development, and evaluation.

There are developments in decision support systems (DSS) for use in water resources management recently in the literature. As DSS expands from desktop to the Internet, along with concurrent increases in IT processing and storage, there is great potential for applying DSS in real time to both long-term structural decisions (Planning DSS) and short-term operational decisions (Operation DSS). One of several challenges considered is how to meld intuitive user interfaces effectively with powerful databases and valid computational models. These IT narratives collectively indicate an emerging "toolbox" for increasing public awareness and participation in urban water management. A common challenge will be to adapt and apply appropriate IT or ICT (information and communication technologies) tools to the specific needs of urban water managers in real time (UNCED, 1992).

2.5.3 INTERNATIONAL APPROACHES

International organizations play a vital role in enhancing public participation around transboundary water resources management. The goal is to extend public participation across political boundaries and to empower the broader public to participate in decision making and monitoring relating to projects and actions that concern two or more countries. International financing of water management projects may catalyze the inclusion of participatory processes, as is the case with policies of the Global Environment Facility (GEF). According to GEF's policy, public involvement consists of three related, and often overlapping, processes: information dissemination, consultation, and stakeholder participation. Permanent international basin organizations also play a long-term role in promoting public participation to improve watershed management. Conflicting interests between civil society and modern development and between local and national priorities and contexts have escalated through both participatory and extraparticipatory processes.

Ending global thirst depends on providing the public with a voice in water-related decisions that directly affect them. Where the public are not included in decisions that affect their welfare, they may resist change, protest, or otherwise obstruct the implementation of such decisions (UNCED, 1992). Those living along international watercourses, near international borders, and far removed from central governments are particularly difficult to include in such decision making. Yet, for the stakeholders, transboundary participation is also critical. However, improving public participation across international boundaries also requires addressing difficult transboundary challenges such as sovereign water rights, migratory populations, linguistic and cultural differences, and distinct political, economic, and legal frameworks among riparian nations. Nonetheless, public involvement, associated with ongoing reform in governance, holds the promise of improving the management of watercourses and reducing the potential for national and international conflict over water issues.

To realize this potential will require a more comprehensive understanding and systematic application of public participation processes across all boundaries. This should begin, first and foremost, with a review of existing approaches and available tools. Other tools should be tailor-made for specific application based on technical and nontechnical suitability of the tools in a regional urban or rural setting or across an international network of cities and communications with water dependencies (UNCED, 1992).

2.6 REMAINING ISSUES

It is clear that much progress has been achieved in water management, as new approaches and methodologies have been developed to promote participation across local and regional and transboundary political boundaries and watersheds. Participatory processes are recognized as important in water management from project identification through design and implementation to monitoring and evaluation.

Capacity building for participatory water management is one of the most needed parts of development projects. In this respect, it is being recognized that there is a need for improved understanding and identification of the institutional and organizational arrangements required for effective decision making and operational process. There is a need to define more clearly and adapt key terms to promote public participation in water governance. The governing factor could also include the following five attributes: (1) level of public participation, (2) process of public participation, including who initiated the process and who participated at each stage, (3) the communication platform for public participation, (4) role of facilitation and consultation, and (5) role of science and technology (UNCED, 1992).

Each of the above five spheres has its own speed of functioning, and these different levels are connected vertically and not horizontally. This way of thinking is needed when dealing with resources management so that adaptive management can be used for implementing ideas in the real world. These various characterizations should be scrutinized in adapting a definition of public participation that is pertinent to water management, in particular. The result should clearly describe how public participation fits into adaptive water management regimes at local city-wide, national, and international natural and political boundaries (watershed) scales.

Although water management has historically used the river or lake basin as its unit of management, the growth in population and urbanization has placed increasing pressure on water resources across multiple basins in a region. It is equally important to learn from local people to respect different attitudes and experiences and to seek out win–win situations based on such learning. Even in the twenty-first century, there continues to be a strong need for respected experts and decision makers to solicit and accept the different viewpoints of the affected public.

Monitoring and evaluation indicators are frequently used to measure the progress and impact of water resources management activities. Such indicators of success are also needed to track the success of public participation as it contributes both to water management and to broader societal goals, such as good governance. Specific public participation indicators should measure both progress (e.g., the development of and timely adherence to a stakeholder involvement plan, broad acceptance of a collective "watershed vision," the creation of basin-wide citizen advisory committees, etc.) and its impact (e.g., the public are generally satisfied with the result, or indicate being better off thereafter). As mentioned earlier, transparent, participatory, and accountable governances are essential foundations for sustainable development. When linked to a clear vision or description of success, ongoing measures of such indicators will be key to determining the overall success of the public participation process.

Finally, public participation may provide the means for making the transition from dependence to empowerment. Local people and government officials may cynically consider "public participation" as a key to getting more funds from donor agencies. Experience also shows that foreign aid from donors can lead to a culture of dependence on the part of recipient countries. Fortunately, developing formal processes for public awareness and participation can be an effective means to increase local ownership, thereby laying the basis for locally sustained stewardship of water resources. Moreover, active oversight by donors and civil society may also ensure that water resources management projects and basin organizations are not spoiled or co-opted by corruption.

2.7 URBAN PLANNING

One of the greatest opportunities for promoting sustainable outcomes for the urban water cycle occurs in the "plan making" or "strategic" phases of the planning process. The planning applies to both the redevelopment of existing urban areas and to the development of "greenfield" land on the urban margin (McAlister, 2007). Because of the variety of local conditions within different council areas, and the variety of relevant planning documents, the following matters are considered:

- Regional and local strategies
- Local environmental plans
- Urban investigation and release process
- Master plans

2.7.1 Regional and Local Strategies

In order to outline a policy approach to planning at the regional or local scale, regional and local strategies should be planned. These documents seek to influence development generally by establishing broad principles and they do not directly "control" individual development proposals. Furthermore, they set targets or create coordination mechanisms. These plans are flexible documents that help to form the shape of statutory plans such as local environmental plans, but they are not legally binding.

Regional strategies prepared by the agencies involved in infrastructure, planning and natural resources, and water boards and authorities address a comprehensive range of planning issues affecting a region. Other strategies are prepared in relation to specific issues at the regional or local level, often drawing upon similar strategies prepared at the state or national levels such as biodiversity strategies, catchment blueprints, water cycle strategies, stormwater management plans, housing and infrastructure strategies, economic development strategies, and tourism strategies. Such documents provide a broad direction for urban structure within a policy context.

2.7.2 Preparing a Land and Water Strategy

In order to provide an overall policy direction to manage land and water resources considering their interactions, a land and water strategy should be provided by councils. This kind of strategy should be prepared in conjunction with an extensive community engagement process to elicit community views, involvement, and partnerships in urban water policy making.

The strategy should contain the visions, objectives, and targets and provide important directional statements that guide the content and implementation of more detailed policies and programs. They should be reflected in important corporate and statutory planning documents, such as the watershed council's management plan and the local environmental plan. Purpose, issues, visions, objectives, targets, and action plans are important, would need to be addressed by the strategy using the following questions (Dahlenburg, 2003):

- Why is the strategy necessary?
- Which matters are of public importance, concern, or contention, and why?
- What are the desired future outcomes that the community needs and wants for land and water resources?
- Which values are to be pursued and which attainments need to be achieved in the future?
- Are there any detailed or quantitative statements of outcomes against which the success of the strategy should be measured and evaluated?
- Which programs and actions are needed to implement the strategy?

Water utilization and efficiency, flood risk management, stream protection and restoration, water balance and stream flow, salinity and other groundwater-related issues, water quality, water-dependent ecosystems, erosion and sedimentation, and contaminated land management are examples of objectives and targets that should be considered in the strategic plans of urban areas.

The visions, objectives, and targets are developed through appropriate community engagement and scoping processes. Suitable visions, objectives, and principles can generally be drawn from existing policy documents, such as catchment blueprints, stormwater management plans, flood risk management plans, and so on.

The principles of ecologically sustainable development and other compatible principles should provide a cornerstone for watershed council decisions affecting the water cycle. As part of a watershed council's charter to properly manage, develop, protect, restore, enhance, and conserve the environment, such principles should be considered when

- Planning new or existing urban areas
- Undertaking regulatory functions, such as approving development applications
- Undertaking service functions, such as roads and stormwater drainage
- Setting rates and charges
- Managing community land, such as parks and reserves

A local water cycle strategy should be implemented through more detailed strategies and plans relating to specific issues such as stormwater management plans, flood risk management plans, coast and estuary management plans, environmental improvement programs, and stream and riparian restoration programs.

2.7.3 Master Plans

In order to implement large-scale development projects such as greenfield residential estates, high-density residential precincts, and commercial and industrial areas, a guideline for each area is necessary, which is called a "master plan." Such projects occur incrementally over long periods of time, and thus need to be strategically planned and coordinated.

Master plans serve to (Dahlenburg, 2003)

- Determine the outline of the desired location, layout, form, and staging of development.
- Provide a long-term planning framework, particularly where staged development and approvals are involved.
- Determine the way in which development should respond to relevant planning policies and strategies and the way in which the site's constraints, opportunities, and context should be expressed.
- Help the public understand the project's future outcomes.
- Assist councils when considering development applications.

Master plans provide a critical, one-off opportunity to ensure that the structure and configuration of new development support sustainable water cycle outcomes. For example, they can be used to ensure that land allocations for water management purposes are both sufficient in area and suitably located. It is generally not possible to allocate land for such purposes at the later subdivision and building phases since the overall configuration for the area has already been established. Achievement of desired water management outcomes may be severely constrained if the necessary land requirements are not met early in the planning process.

Water cycle management is one of many issues addressed by master plans. In the master plan, the proposed management responses will need to be compatible with other important issues, including the layout and configuration of infrastructures, streetscape, and local character.

In order to provide a sound basis for water-sensitive design at the subdivision and street and lot levels, master plans should cover the following statements, as mentioned by Dahlenburg (2003):

1. Determine the strategies at regional and local scales and their implications for water management.
2. Determine a vision for how the water cycle will be managed on the site, and how this will determine, in broad terms, the structure, layout, and design of development.

3. Incorporate and respond to a site analysis of natural and built conditions.
4. Water-sensitive design principles for landscape structure and open space system, street layout and design, major stormwater systems and building design and sitting.
5. Allocate sufficient land for water infrastructure in suitable locations by a conceptual urban structure model.
6. Resolve conflicts and tensions that may exist between water management and other planning imperatives that influence layout design.
7. Incorporate with an integrated urban water plan.

2.8 URBAN WATER PLANNING

2.8.1 Paradigm Shift

In the last three decades, the "ecological" paradigm comes to the forefront as a new alternative to solve man-made problems. This new way of thinking began from new discoveries and theories in science, such as Einstein's theory of relativity, Heisenberg's uncertainty principle, and chaos theory. Contrary to Newton's mechanistic universe, these new theories have taken away the foundation of the opposing paradigm, the Cartesian model. Because our way of thinking has been based so much on the Cartesian paradigm, this paradigm shift demands to change the way in which we perceive and understand our world.

Water resources management is not an exception in this dramatic shift of the paradigm. Throughout the history of water resources management, many methodologies have reflected some aspects of ecological thinking and have been shaping the water version of the new paradigm.

2.8.1.1 Shift from the Newtonian Paradigm to the Holistic Paradigm

In the ongoing theoretical debate on environmental planning, it seems to be far-reaching consensus concerning one thing: There has been a great shift in planning thoughts during the last 50 years. Some researchers refer to this as a paradigm shift from rational (Newtonian) to communicative (holistic) planning. This shift is often combined with a reference to the change of modernity, with a statement that the communicative planning paradigm is adapted to new postmodern times. In water resources planning, this shift has occurred, which is the basis for integrated approaches as well as environmental consciousness and the notion of sustainable development in water-related activities. It has also triggered the shift from supply management to demand management. In the following sections, Newtonian and holistic paradigms are discussed in a broad fashion.

2.8.1.1.1 Newtonian Paradigm
The worldview that laid the basis for modern culture was created in the early stages of the Renaissance. Before this change, the primary worldview was holistic, organic, and ecological. People lived their lives in small communities and had a holistic, spiritual relationship with nature. Even the nature of science at that time was quite different from that of modern science. Rather than using means of prediction and control, the main goal was to understand the existential meaning of things.

The medieval approach was transformed radically in the sixteenth and seventeenth centuries. The idea of the holistic, organic, and spiritual universe was replaced by a world-machine metaphor. This metaphor became a major part of the modern way of thinking. This transformation came with the Scientific Revolution involving Galileo and Newton, and a new method of reasoning by Bacon and Descartes.

With his scientific observations of celestial phenomena, Galileo was the first to combine scientific discoveries with mathematical explanations to formulate the laws of nature. These two aspects of Galileo's work, his experimental approach and his mathematical description of nature, became the dominant character of science in that time and have remained the same to date. However, because

Galileo's strategy was so successful in a quantifiable domain, nonquantifiable properties, such as subjective and mental projection of human beings, were excluded from the domain of science. This trend has become a major hurdle to integrating all human properties as a whole.

In England, Francis Bacon introduced the inductive method that used experiments and drew general conclusions from them to make usable knowledge. The new reasoning method heavily changed the character and objective of science. From ancient times, the main goal of science had been to gain wisdom and to understand nature in harmony with science. Since Bacon, the goal of science has been to gain knowledge to control and exploit nature. Nature, in his view, had to be "hounded in her wanderings," "bound into service," and made a "slave." This kind of attitude represents the prevalence of patriarchal attitudes in science fields.

The shift of worldviews was completed by Descartes and Newton. Descartes doubted everything he could until he reached an indubitable conclusion, the being of himself as a thinker—*Cogito, ergo sum*. From this he deduced that nothing but thought was the essence of human nature. This deduction led him to the conclusion that mind and matter are separate and totally different entities. This Cartesian division has made us think of ourselves as isolated entities inside material bodies.

Descartes saw the material world as a machine that had no life or spirit. The natural world functioned according to mechanical laws and everything could be explained in terms of the mechanical movements of their parts. This mechanical image of nature became the dominant model of science since Descartes. Even Descartes applied his mechanical worldview to living organisms. Plants, animals, and even human beings belonged to a machine category. The human body became a container activated by a soul connected through the pineal gland in the brain.

The man who undertook Descartes' task was Isaac Newton. With his new mathematical method, differential calculus, Newton came out with a mathematical formulation that fully completed the mechanical worldview. This deterministic character gave birth to a belief that if you know the state of a system, such as time and location, with respect to all details you could predict the future of the system with absolute certainty.

The Newtonian universe produced the image of absolute space and time—both independent of each other, and indivisible units, atoms. Absolute space was the space of Euclidean geometry—an unchangeable three-dimensional container. Absolute time flows evenly with no respect to external phenomena. The drastic change from organism to machine in the image of nature greatly affected the way in which people viewed the natural world.

The Newtonian paradigm has governed the way in which water resources planners and managers viewed water management for many years. However, it suffers from a number of serious shortcomings. Its applicability as a framework for recent theories of water resources management is increasingly questionable. As our understanding of the behaviors of complex adaptive systems increases, the complex framework becomes more relevant to studies of water resources management than the Newtonian paradigm. The principal reason may be found in the concepts of linearity and nonlinearity: Water resources management is an inherently nonlinear phenomenon. The Newtonian paradigm rests firmly on linear principles, whereas the holistic paradigm that follows complexity theory embraces nonlinearity.

Linear systems played an important role in the development of science and engineering, as their behaviors are easily modeled, analyzed, and simulated. A linear system has two defining mathematical characteristics. First, it displays proportionality. If some input X to the system gives an output of Y, then multiplying the input by a constant factor A yields an output of AY. The second characteristic of linear systems is superposition. That is, if inputs $X1$ and $X2$ give outputs $Y1$ and $Y2$, respectively, then an input equal to $X1 + X2$ gives an output of $Y1 + Y2$. Systems that do not display these characteristics are called nonlinear systems. Importantly, linear systems of equations can be solved analytically or numerically. Given a set of linear equations and initial conditions, we can calculate the future values of the variables. Consequently, if we can describe a system by a linear mathematical model, we can determine its future states exactly from its given initial state. A large body of mathematics has grown up around linear systems and techniques for their solution.

Nevertheless, the vast majority of systems and phenomena in the real world are nonlinear. As the name implies, nonlinear systems do not display linear characteristics of proportionality and superposition. Analytical solutions to nonlinear equations are generally the exception rather than the rule. Thus, the future states of nonlinear systems can often only be approximated. One method of approximating the behavior of nonlinear systems involves linearizing them, and then employing linear systems analysis on the approximated system. Unfortunately, such techniques suppress or even eliminate many of the important dynamical characteristics of nonlinear systems; for example, chaos cannot exist without nonlinearities. However, the advent of modern digital computers has brought about a revolution in the study of nonlinear systems. Computers have made it possible to simulate their rich dynamical behaviors such as chaos that might otherwise not exist in linearized approximations.

Linearity is the cornerstone of the Newtonian paradigm. This has several important ramifications for water resources management. First, the condition of water resources under the Newtonian paradigm is deterministically predictable, as effects are in principle calculable from their underlying causes. Given enough information about the current state of water resources and demand with "laws" of planning, a commander should be able to precisely determine the outcome of the management. Determining the outcome of water management becomes a simple exercise if a sufficient amount of precise information is available, much as the future states of a linear system of equations can be exactly computed. With intelligence and situational awareness approaching perfection, the Newtonian paradigm reduces fog and friction to a bare minimum just as chaos is banished from linear systems.

Reductionism is a second important consequence of the Newtonian paradigm. This is a methodology for solving problems. The analyst breaks the problem into its constituent pieces, solves each piece separately, and then sums the results from the pieces to obtain the overall solution to the problem. This is a natural consequence of superposition. The history of water resources management covers a range of examples of reductionism. For example, managers generally break the water systems into a series of separate systems with different goals, analyze each system independently of all others to determine aim points, and then sum the results to generate the overall plan. Historical analyses of water resources are frequently reductionism—what is the isolated, independent cause (or causes) that led to the outcome of the conflict? As linearity allows and even encourages this mindset, reductionism is a principal characteristic of Newtonian management.

A third consequence of the Newtonian paradigm is the view of systems as closed entities, isolated from their environments. Outside events do not influence such a system; the only dynamics are those arising from its internal workings. The analyst thus has an inward focus, with a concentration on efficiency. The emphasis on efficiency is especially noteworthy for water resources operations. How can the planner obtain the desired objectives with the least cost? What targets must the planner select to most efficiently and economically accomplish the objectives? Isolated, closed systems are perhaps easier to analyze, as outside forces and influences are of no consequence. However, isolated systems form only a small fraction of the physical universe.

The Newtonian paradigm creates a simplified, idealized view of water resources management. It is an appealing, comfortable framework as it offers simple means for analysis, methodical rules for planning and executing operations, and the illusion of predicting the future given enough information about the present. However, water resources management is intrinsically more complicated than this simplistic framework allows.

2.8.1.1.2 Holistic–Ecological Paradigm
Ecological thinking is in heavy debt to the feats of modern physics. The advent of electromagnetics, Einstein's theory of relativity, quantum physics, and the newly born chaos theory are major contributors to this new paradigm. The discovery of electric and magnetic phenomena by Michael Faraday and James Clerk Maxwell was the first step to undermining the mechanistic world model. The real collapse of the Newtonian worldview came with relativity theory and quantum theory.

In atomic physics, several discoveries unaccountable in terms of classical physics appeared, such as the discovery of the subatomic world and the duality of the subatomic entity. One of the most fascinating discoveries was the fact that "depending on how we look at subatomic units, they appear sometimes as particles, sometimes as waves; and this dual nature is also exhibited by light which can take the form of electromagnetic waves or of particles."

Quantum theory showed the holistic nature of the universe. In a different context, the discovery of evolution in biology challenged the illusion of a deterministic Cartesian world. Instead of thinking that every single element is created in the beginning of this universe and run by a system of laws, evolutionary concepts opened the possibility of development from simple forms to complex structures.

However, the evolutionary idea contrasted with the laws of thermodynamics in physics, which means that every phenomenon moves from the orderly state to the disorderly. Thermodynamics was a double-edged sword, since it at once denied the reversible time flow of the Newtonian universe and contrasted with evolution theory. This problem remained until the advent of "nonlinear revolution" with Chaos and Self-organization Theory.

Self-organization Theory demonstrated that "far from equilibrium," unlike "near equilibrium," a system of matter tends to self-organize itself through positive feedback. Once fed by this constructive loop, this system starts to evolve to a higher level of organization. In other words, the "far from equilibrium" state has a high sensitivity to even a tiny change in the system. This means that on the earth, a typical example of the "far from equilibrium" state, every entity is interdependent and interconnected to others and even a tiny transformation has the possibility of changing the whole world. Because this leap to another level of system is unpredictable, the deterministic predictions of the Cartesian worldview lost its validity.

One of the interesting discoveries in the "far from equilibrium" system is a unique structure of the system, fractal. "Fractals have an internal 'microstructure' that exhibits the phenomenon of scaled self-similar layering. Finer and finer magnification of a fractal reveals smaller and smaller versions of the same structure at all levels. Therefore, fractals are infinitely complex. No matter how small a piece you take, it is an equally complex microcosm of the whole. Fractal self-similarity and scaling are particularly important because they repudiate two Newtonian assumptions." One is that reducing and segmenting methods make problems simple, and the other is that we can measure everything objectively with an absolute scale.

The "nonlinear revolution" not only stopped the illusive universality of Cartesian worldview but solved the problematic coexistence of Newtonian and new physics in different domains. Because it shows the dynamic and diverse nature of the universe from microcosms to macrocosms, even the Newtonian method can survive in a specific range, the middle range of the ecological world.

All these developments in science revived the original image of the world, which is holistic, organic, ecological, and spiritual. The Cartesian–Newtonian paradigms caused a dramatic scientific revolution that gave us at once enormous prosperity and problems. The new ecological thinking will change again the way in which we perceive the world and will give us a new alternative to cure the fallacy of the old paradigm.

The holistic paradigm offers a broader, far more useful framework for water resources management. This paradigm is based on open, nonlinear systems in far-from-equilibrium conditions. A complex adaptive system has several defining characteristics. First, it is composed of a large number of interacting parts or "agents." The interactions between the agents are nonlinear. The interactions and behaviors of the agents influence the environment in which the system exists. Changes in the environment in turn influence the agents and their interactions. The agents and environment thus continuously affect and are affected by each other. Second, the agents characteristically organize into hierarchies. Agents at one level of the hierarchy cluster to form a "superagent" at the next higher level. Third, there are intercommunicating layers within the hierarchy. Agents exchange information in given levels of the hierarchy, and different levels pass information between themselves as well. Finally, the complex system has a number of disparate time and space scales. For example, water resources operations at the low level are highly localized and may occur very rapidly compared to events at the high level. Complex adaptive systems in widely varying disciplines appear to share these four characteristics (Figure 2.2).

Complex adaptive systems exhibit a number of common behaviors. The first is emergence: The interactions of agents may lead to emerging global properties that are strikingly different from the behaviors of individual agents. These properties cannot be predicted from a prior knowledge of the agents. The global properties in turn affect the environment that each agent "sees," influencing the agents' behaviors. A synergistic feedback loop has thus created interactions between agents to determine emerging global properties, which in turn influence the agents. A key ramification of emergence is that reductionism does not apply to complex systems. Since emergent behaviors do not arise from simple superpositions of inputs and outputs, reductionism cannot be used to analyze the behaviors of complex systems. The emergence of coherent, global behavior in a large collection of agents is one of the hallmarks of a complex system.

A second fundamental behavior of complex systems is adaptive self-organization. This appears to be an innate property of complex systems. Self-organization arises as the system reacts and adapts to its externally imposed environment. Such order occurs in a wide variety of systems, including, for example, convective fluids, chemical reactions, certain animal species, and societies. In particular, economic systems are subject to self-organization.

A third important behavior of complex systems is evolution at the edge of chaos. Dynamical systems occupy a "universe" composed of three regions. The first is an ordered, stable region. Perturbations to the system tend to die out rapidly, creating only local damage or changes to the system. Information does not flow readily between the agents. In the second region, chaotic behavior is the rule. Disturbances propagate rapidly throughout the system, often leading to destructive effects. The final region is the boundary between the stable and chaotic zones. Known as the complex region or the "edge of chaos," it is a phase transition zone between the stable and chaotic regions. Systems poised in this boundary zone are optimized to evolve, adapt, and process information about their environments. As complex systems evolve, they appear to move toward this boundary between stability and chaos, and become increasingly more complex. Evolution toward the edge of chaos appears to be a natural property of complex systems.

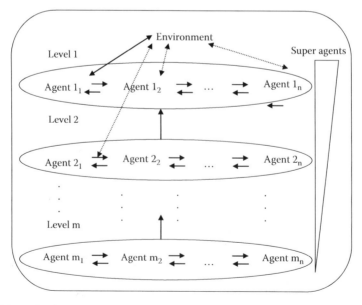

FIGURE 2.2 Schematic of a complex adaptive system.

The final key behavior of complex systems is their ability to process information. The systems sense their environments and collect information about surrounding conditions. They then respond to this information via a set of internal models that guide their actions. Systems may also encode data about new situations for use at a later date. This characteristic is closely related to adaptation near the edge of chaos. Information processing is a common characteristic of complex systems and enables them to adapt to changing environments.

The holistic paradigm provides a powerful framework for analyzing water resources systems. Indeed, water resources management is a nonlinear, complex, adaptive phenomenon with two or more coevolving competitors. The actions of every concept of system that influence and shape of the environment should be determined, and adaptations should be made in all levels of managing environment. As both suitable and nonsuitable phenomena influence the environment, the water resources system involves the coevolution of all involved parties. When viewed from this vantage point, it is clear that the Newtonian paradigm is too limited to adequately cover the many aspects of water resources management. Water resources systems are more appropriately analyzed under the holistic framework than the Newtonian framework. It is time to shift paradigms.

When described by those leading the debate today, planning theory in the 1960s and 1970s is depicted as "modernist planning." The structure of this approach is often seen as a "dominating paradigm" of time, constituted by normative arguments about what planning is or ought to be. By separating the expertise practice of planning from political policy making, the planner is characterized as the one who is able to use instrumental rationality to improve policy making by means of scientific knowledge. Planning is articulated as a tool for social progress. With its origin in "enlightenment epistemology" or "positivist epistemology," it could make the world work better. In this way, the planner is depicted as "the expert" or "the knower" and as the one who could decide what is best for "the public." Planning is science or a way of creating good decisions by using science, and therefore reason, in the decision-making process. Today's theorists in planning often relate the identified paradigm to "the modern project," but it goes under different names as "systematic planning," "the system and rational approach to planning," or "the rational comprehensive model."

In comparison, in the 1990s, a new emerging paradigm has come to dominate the theoretical debate. The "communicative" or "collaborative planning" approach is fully understood as the heir to modernist planning, and hence as a sign of a paradigm shift in planning thought.

In the communicative turn, planning is described as an interactive, communicative activity. A normative theory is formulated on this observation. Based on Habermas theory of communicative action, planning ought to be about creating an ideal speech situation. Different interests should take part in undisturbed communication in order to reach a consensus concerning goals for, and formation of, the planning process.

This paradigm shift in planning thought is often considered to be part of a greater historical discontinuity. The positivist, logical deductive method of the modernist planner stands against the qualitative, interpretative inquiry, often identified as a postpositivist approach, in the communicative perspective.

2.8.1.2 Paradigms of Urban Water Management

Increasing incidences of local water shortages and recent water quality disasters have led many to question the technical reliability and institutional capacity of urban water management. People are beginning to recognize that continuing to expand infrastructure and develop new water sources has become increasingly expensive and ultimately unsustainable, both economically and environmentally. There has also been a change of paradigm from supply side to demand side management. This is explained in detail in this book.

2.8.2 Land Use Planning

A good part of this section on land use planning is adapted from Chen et al. (2005) on urban areas. It is a good attempt to integrate land use planning into water resources management in the context of so-called "dynamic strategy planning."

2.8.2.1 Dynamic Strategy Planning for Sustainable Urban Land Use Management

The dynamic strategy planning for sustainable river basin land use management includes the following: (1) identification of a sustainable urban land use management system and components, (2) identification of the dynamic relationships among components, and (3) the DSR dynamic strategy planning procedure that stands for Driving forces (D), State of environmental and land use components (S), and Responses of land–water resources to allocation (R) for sustainable urban land use management.

2.8.2.1.1 Identification of a Sustainable Urban Land Use Management System and Its Components

The boundaries of an urban environment system are usually drawn politically rather than geographically. However, when it comes to water, even at a city scale, it can be conceptually divided into drainage zones according to geographical characteristics (Chen et al., 1997, 2000).

This section focuses on the direct impacts of land use on water resource quantity, and water quality, which are caused by human activities. Therefore, based on systems thinking, the principal components identified in this investigation include human activities, land resources, and water resources. Human activities and the attributes of land resources include land area and land use type, with the latter being divided into six categories: residential land use, industrial land use, paddy land use, dry-farming land use, forest land use, and other land use. Additionally, the attributes of water resources include the quantity and quality of water. Air resources are also an integral part of land use planning, but here we concentrate on water aspects of land use.

2.8.2.1.2 Identification of the Dynamic Relationships among the Components of Urban Land Use Management Systems

Various models were proposed for discussing sustainable development, including the DSR (Driving force, State, and Response) model, the PSR (Pressure, State, and Response) model (OECD, 1993), and the DPSIR (Driving force, Pressure, State, Impact, and Response) model (Denisov et al., 2000). Each model has its own characteristics and is suitable for different problem solving. Since there are DSR dynamic interactions among human activities, land, water, and air resources for top land use management, the concept of the DSR framework and the theory of system dynamics are applied to address the cause-and-effect relationships. The dynamic relationships of the components in the sustainable urban land use management system are identified as shown in Figure 2.3. Among the components of the urban land use management system, human activities are the initial forces influencing the sustainability of land resources, which are identified as the driving forces.

Land use needs water resources, and also pollutes the quality of river water and air. Therefore, situations of land area, land use types, water quantity, water quality, air quality, benefits, and populations are identified as states. In the states, land use causes resource consumption and pollution production. In addition, land use type, land area, water quality, water quantity, air quality, and population influence the characteristics of an urban system.

Decision makers may sometimes decide to respond based on the states to modify driving forces. Pollution abatement and resource allocation are identified as responses. In the responses, land resource allocation, water resource allocation, and water pollution abatement are implemented based on available quantity of land resources, available quantity of water resources, water quality standards, and air quality standards. Furthermore, land resource allocation, water resource allocation, and water

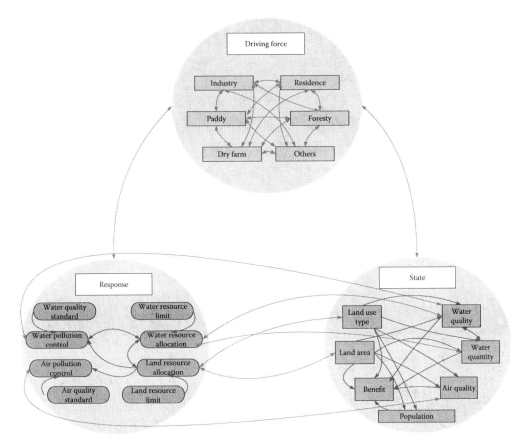

FIGURE 2.3 Dynamic relations among the components in the sustainable urban land use management system in the conceptual DSR framework. (Adapted from Chen, C. H. et al. 1997. *Proceedings of the National Science Council, Republic of China. Part A, Physical Science and Engineering* 21: 389–409; Chen, C. H. et al. 2000. *Water Science and Technology* 42: 389–396.)

pollution abatement influence one another. Land resource allocation also influences air pollution abatement, which could have an impact on water quality as well.

Furthermore, dynamic interactions exist among the driving forces, states, and responses in an urban land use management system. Between driving forces and states, human activities use land resources, water resources, and air resources and alter the states of land use type, land area, water quality, water quantity, air quality, and population. Between states and responses, if the states of land area and land use type are changed, it will lead managers to reallocate land resources.

2.8.2.1.3 DSR Dynamic Strategy Planning Procedure for Sustainable Urban Land Use Management

As shown in Figure 2.4, the procedure includes six phases and two DSR cycles. Through continuous and iterative modification of the data of the two DSR cycles, an optimal solution or a satisfactory alternative for sustainable urban land use management can be obtained based on the DSR dynamic strategy planning procedure.

1. *Identification of a river basin system, drainage zones, and reaches.* The boundaries of a river basin environment system are drawn geographically. The river consists of a mainstream and some branches.

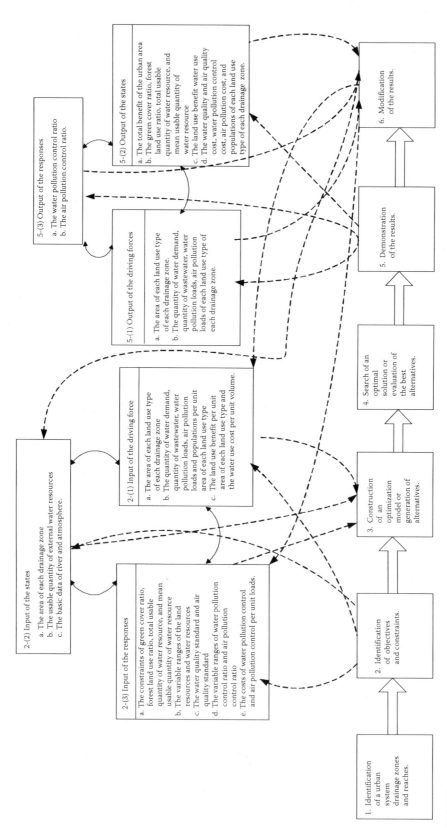

FIGURE 2.4 DSR dynamic strategy planning procedure for sustainable urban land use management. (Adapted from Chen, C. T. et al. 2005. *Science of the Total Environment* 346: 17–37. With permission.)

2. *Identification of objectives and constraints.* The objective function of sustainable land use management for an urban setting is to maximize the net benefit of urban land use within its water resources carrying capacity. The green cover ratio, forest land use ratio, total available quantity of water resources, and mean available quantity of water resources are used as constraints for controlling land and water resource uses. The water and air quality standards are used as constraints to control pollution loads to meet water and air quality requirements. The constraints used for ensuring urban development can be kept within its natural resources, including water-carrying capacity.

Since each human activity requires land resources and produces air and water pollution, each land use type can be assumed to be a related type of human activity. The loads, costs, and benefits of each land use type of each drainage zone are calculated by the loads, costs, and benefits per unit area of each land use type multiplying the land area of each land use type of each drainage zone. For the states, the area of each land use type of each drainage zone, available quantity of external water resources, basic river data, and basic atmosphere data should be inputted here.

3. *Construction of a system dynamics optimization model or generation of alternatives.* Based on the identification of objectives and constraints, a system dynamics optimization models objective function is to obtain maximum net benefit of an urban area.

4. *Searching for an optimal solution or evaluation of the best alternative.* The policy optimization function is combined with the system dynamics optimization model to search for an optimal solution in this investigation. Additionally, the alternatives generated in step 3 are evaluated by comparing the benefits of each alternative to find the best one.

5. *Demonstration of the results.* The results of the optimal solution or the best alternative are demonstrated in this step. For the driving forces of sustainable urban land use management, water demand quantity, wastewater quantity, and water pollution load of each land use type are obtained. For the states, allocations of land area and land use type are demonstrated, along with allocations of water resources, river water quality, populations, and benefits to the urban system. For the responses, water pollution abatement is obtained. The above data belong to each drainage zone and the whole urban area.

6. *Modification of the results.* If the optimal solution or the best alternative cannot be accepted by the decision makers, the phase feeds back to step 2 and proceeds through steps 3–6. The six steps and two DSR cycles are not finished until decision makers find appropriate responses to generate an acceptable alternative (Chen et al., 2005).

2.8.2.2 Urban and Regional Plans and the DSR

The DSR dynamic strategy planning procedure for sustainable urban land use management can be used for constructing urban and regional plans to obtain effective alternatives. The combined processes are shown in Figures 2.5 and 2.6. In the governmental generation process of an urban plan, as shown in Figure 2.5, planners gather related information about the selected area. Based on the development needs and the relevant information of the area, the land area and land use type are allocated to the planned region in advance. After gathering the opinions of relevant government sectors and stakeholders, the draft urban plan is proposed to the urban planning committee for review.

The DSR dynamic strategy planning procedure can be used in the analysis and review phases of the process to improve effectiveness and efficiency. During the analysis phase, planners can use this procedure to generate alternatives or evaluate existing alternatives to sketch a draft plan. In the review phase, according to the draft plan, members of the urban planning committee can propose different considerations and input relevant data in the DSR dynamic strategy planning procedure. Figure 2.6 shows a governmental generation process of a regional plan for an urban area and the applicant proposes amendments to the proposed land use in a land development plan.

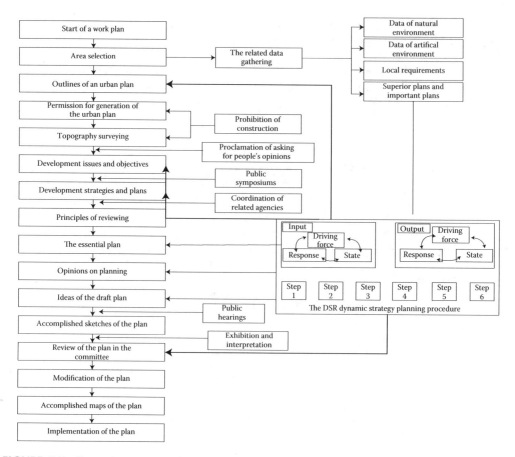

FIGURE 2.5 Generation process of an urban plan combined with the DSR dynamic strategy planning procedure. (Adapted from Chen, C. T. et al. 2005. *Science of the Total Environment* 346: 17–37. With permission.)

The related sectors of local government then review the plan in their authorized level of responsibility before issuing the land development permit. The DSR dynamic strategy planning procedure can also be used in the review phase. The different considerations from responsible authorities can be circularly integrated into the optimization model. When an accepted alternative is obtained, the land development permit is then issued. For top land use management, governments can obtain sustainable land use management strategies through the above procedure. The strategies can be used as a guideline for middle and lower land use management (Chen et al., 2005).

2.8.2.3 DSR Dynamic Decision Support System

The sustainable urban land use management DSR dynamic decision support system (called SRBLUM-DSRD-DSS) is developed based on the DSR dynamic strategy planning procedure to help decision makers generate an optimal land use management strategy or some alternatives to the strategy. Based on the DSR dynamic interactions among components of the system, the concept of object orientation can be used to establish the system dynamic optimization and simulation models of DSS. Each drainage zone can be identified as an object, and the contents of the components of the drainage zone are identified as object attributes, including all human activities, and attributes of land, water, and air resources.

In the conceptual structure, attributes are placed on the object to which they belong, and attributes of the same type are placed on the same plain. Horizontal arrows represent interactions among objects

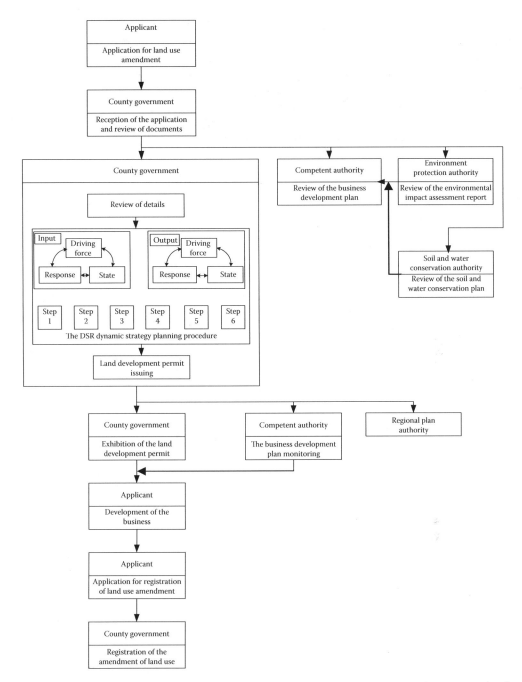

FIGURE 2.6 Generation process of a regional plan combined with the DSR dynamic strategy planning procedure. (From Chen, C. T. et al. 2005. *Science of the Total Environment* 346: 17–37. With permission.)

for an identical attribute. Meanwhile, vertical arrows represent interactions among attributes for an identical object. For example, the air quality of a drainage zone is influenced by the air pollution load of this drainage zone. The system dynamics optimization and simulation models are developed by using VENSIM or alternative models like STELLA based on the conceptual structure.

Through chain reaction effects, each component influences all other components in the structure by dynamic relationships. Since the attributes and processes are encapsulated in each object, their definitions and modifications will not be influenced by each other in the process of model development. Consequently, if the boundary of the urban land use management system is expanded, only some new objects need to be added to the DSS.

By using the conceptual object-oriented structure, the objectives, constraints, and variables of the DSS can be modified easily without rebuilding the system structure. Based on the concepts of cyclic modification in the DSR dynamic strategy planning procedure, the conceptual system structure of SRBLUM-DSRD-DSS developed here, as shown in Figure 2.7, includes (1) identification of objectives and constraints as well as data input, (2) search for an optimal solution or evaluation of the best alternative, and (3) demonstration of the results and data output.

The function of identifying objectives and constraints as well as data input corresponds to steps 1 and 2 of the DSR dynamic strategy planning procedure. The data of the driving forces, states, and responses are inputted by using an interactive interface. This interface has some classified tables for input, which are listed in the interface systemically. The function of searching for an optimal solution or evaluating the best alternative corresponds to steps 3 and 4 of the DSR dynamic strategy planning procedure (Chen et al., 2005).

2.9 URBAN WATER ASSESSMENT

In this section, the availability and demand of water resources assessment are described as follows. There are many insets that are adopted from Clausen (2004) in this section.

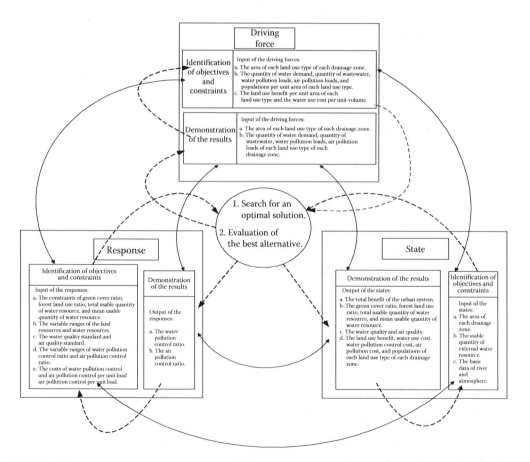

FIGURE 2.7 Conceptual system structure of a DSS for land use management. (From Chen, C. T. et al. 2005. *Science of the Total Environment* 346: 17–37. With permission.)

2.9.1 AVAILABILITY AND DEMAND

2.9.1.1 Importance of Water Assessments

The management of water resources requires an understanding of the nature and scope of the problem to be managed. How are all relevant water resources problems identified? How can we make sure that we acquire useful information that enables us to identify and assess existing and potential future water resources problems and solutions? Carrying out water resources assessments is a useful way of acquiring such information as a basis for management.

2.9.1.2 Need for a Water Knowledge Base

In many countries, available information about the water resources situation is scarce, fragmented, outdated, or otherwise unsuitable for management purposes. Without adequate access to scientific information concerning urban water within the hydrological cycle and associated ecosystems, it is not possible to evaluate the resource or to balance its availability and quality against demands. Hence, the development of a water knowledge base is a precondition for effective urban water management. It takes stock of the resources and demands and establishes the natural limits for management.

2.9.1.3 Objectives

The concept of urban water assessments is interpreted here to imply a holistic view of the water resources situation and its interaction with societal use in an urban area or region. The assessment should address the occurrence, in space and time, of both surface and groundwater quantities and associated qualities and give a tentative assessment of the water requirements for the assumed development. The objective of the assessment is not to solve the problems but to identify and list the problems and identify priority areas within which more detailed investigations may be carried out.

2.9.1.4 Demand as a Function of User Behavior and Preferences

It is important to stress that the water knowledge base must include data on variables that influence demand; only with such data can a flexible and realistic approach to assessing water demands be taken. If not considered in a context of water scarcity and competition, sectoral planners may be overly optimistic about possible development and associated water requirements. Effective demand management may influence demand figures significantly.

2.9.1.5 Importance of Monitoring and Gauging Systems

The assessment of water availability and quality, and their possible long-term changes through consumptive water use, climate, or land use change are highly dependent on reliable data from monitoring and gauging systems; this indicates the need for resources to be allocated for the investment, operation, and maintenance of this aspect of water infrastructure.

2.9.1.6 Environmental Impact Assessments

EIA plays a central role in acquiring information on the social and environmental implications—including water resources implications—of development programs and projects. The Integrated Urban Water Management (IUWM) approach implies that sectoral developments are evaluated for possible impacts on the water resource and that such evaluations are considered when designing as well as giving priority to development projects. EIAs are concerned not only with impacts on the natural environment but also with effects on the social environment. Hence, EIA touches the heart of the need for cross-sectoral integration involving project developers, water managers, decision makers, and the public, and provides a mechanism or tool to achieve this.

2.9.1.7 Risk Assessment Tools

Risks associated with IUWM come in different shapes—usually related to extreme climatic events, public health, and environmental damage (in addition to business-related risks). It is never possible to eliminate risk. Well-established techniques are available to undertake hazard (frequency and magnitude of events) and risk assessments. However, such assessments, which rely heavily on science, technology, and economics, neglect the question of what levels and types of risks are acceptable within civil society. This is a perceptual cultural issue that can only be addressed within a participatory approach to a broader issue of IUWM, in line with the overall issues of Integrated Water Resource Management (IWRM).

2.9.1.8 Risk Management

Risk management is about achieving an appropriate balance between the benefits of risk taking and the losses incurred, and about preparing the means by which residents' small commercial and industrial establishments and their properties can be safeguarded when adverse conditions arise.

2.9.1.9 Precautionary Principle

From an environmental point of view, the precautionary principle in risk management may be warranted in some instances. One key lesson, for example, is that actions to avoid potentially irreversible environmental damage should not be postponed on the ground that scientific research has not fully proved and quantified a causal link between cause and potential damage. The principle here is that a precautionary approach may reduce costs by preventing the damage rather than having to remedy the damage after the event.

2.9.2 Communication and Information Systems

2.9.2.1 Communication for Enhancement of Stakeholder Involvement

The principle of stakeholder participation in water management requires a serious increase of awareness among politicians, decision makers in the water-selected sectors, professionals, interest groups, and the public at large. Communication and information systems should address the question of opportunity cost and trade-offs between alternative water uses and projects and other social investments.

2.9.2.2 Information Needs for Stakeholder Involvement

In order to encourage stakeholder participation in water resources management and for the participatory process to be effective, the availability of timely and relevant information such as up-to-date registers and records of water uses and discharges, water rights, and the beneficiaries of such rights to all concerned is an essential precondition. In addition, the results of benchmarking and performance evaluations of service providers should be made publicly available as this contributes to the competitive and transparent provision of water services. Transparency of information can not only provide water service providers with incentives to improve their performance (e.g., benchmarking league tables) but can also allow civil society and governmental bodies to judge and push for performance improvements.

2.9.2.3 Stakeholder Communication Strategies

Concrete strategies for communication with all actors and stakeholders need to be devised through, for instance, public information sessions, expert panel hearings, citizen juries, and similar methods. However, the most appropriate method in each case needs to take account of social, political, cultural, and other factors.

Governance and Urban Water Planning

2.9.2.4 Openness and Transparency

Some countries have little experience of conducting water resources management in an open and transparent manner with full public access to information. A continuation of this approach will be counterproductive to assuring broad participation and private sector investment in water management.

2.9.2.5 Exchange of Information

Especially when dealing with different watercourses, openness and sharing of information are keys to the achievement of IUWM since all involved parties have "natural monopolies" in data collection and dissemination within their boundaries.

2.9.3 WATER ALLOCATION AND CONFLICT RESOLUTION

2.9.3.1 Issues in Allocation

To allocate water efficiently and effectively to competing users, market mechanisms (trading systems and/or full cost pricing through valuation) could be improved in conjunction with the formulation of appropriate regulatory systems. Conflict resolution mechanisms may be used to facilitate water sharing among competing users such as upstream and downstream stakeholders as well as demand allocation to different uses.

2.9.3.2 Allocation by Market-Based Instruments

Full cost pricing tools through valuation and enhanced water trading are needed to complement and correct the faulty market valuation processes because of the way in which the water market has been managed historically; not all water values (including social and environmental values) are or indeed can be reflected in market prices.

2.9.3.3 Using Valuation to Resolve Conflicts

The process of determining the value of water to various stakeholders could enhance their participation in decision making and contribute to resolving conflict. These tools would not only ensure that existing water supplies are allocated in a sustainable fashion to the highest-value uses but would also enable water managers to determine when the users are willing to pay the costs of investing in additional water-dependent services.

2.9.3.4 Resolution of Upstream–Downstream Conflicts

Conflicts among upstream and downstream users within a region tend to be pervasive and usually result in undue delays in the implementation of urban water development projects. It is important to note that resolving upstream–downstream conflicts requires acceptable estimates of water availability over time, taking into account return flows and the effects of catchments development on evaporation losses and runoff.

2.9.3.5 Conflict Management Techniques

A wide range of conflict management techniques, involving both consensus building or conflict prevention and conflict resolution, is available to assist stakeholders in their negotiations. In Chapter 10 of this book, an introduction about conflict resolution techniques is presented.

2.9.3.6 Valuation by Conflict Resolution Methods

All water services can be valued in an objective and quantitative manner, independent of the value systems of those involved. They could also link valuation directly to conflict resolution techniques.

In the absence of a market, values can be approximated through explicit valuation techniques that transform attributes into their monetary units, or they can be determined implicitly through conflict resolution methods.

2.9.3.7 Valuation Research on Environmental Benefits

There is a special need to develop further methodologies for valuing the benefits of ecological services provided by nature. The main problems appear to be in assigning economic values to nonmarket benefits, such as biodiversity and intrinsic value. In practical terms, as with many aspects of environmental planning, the first requirement is to broaden the scope of valuation exercises, through linking the expertise of economists to the analyses of hydrologists and ecologists.

2.9.4 REGULATORY INSTRUMENTS

A multitude of regulatory instruments is at the disposal of water authorities in setting up appropriate management structures and procedures. These fall into three main groups: direct controls, economic instruments, and encouraged self-regulation. In most situations, authorities will need to employ a mix of instruments to ensure effective and low-cost regulation.

2.9.4.1 Direct Controls

2.9.4.1.1 Executive Regulations

Executive regulations are needed for abstraction of water and discharge of wastewater and may order users—or certain categories of users—to obtain permits for the abstraction or discharge of water. The regulations would describe the procedures to be followed in applying for permits and the criteria for granting permits. As a general rule, it should be ensured that only executive regulations that are enforceable be implemented.

2.9.4.1.2 Water Rights Systems

While in certain urban governance around the world water is considered a national asset under public ownership, there are some countries that implicitly treat water as an unlimited resource, where it is in fact a "common resource" without clearly defined property rights. In other regions water rights are linked to land ownership, with inadequacies and conflicts occurring because of the nonstationary nature of water and interconnections within the hydrological cycle (who owns the water flowing in the river, and how can the necessary multiple use of water be accounted for?).

2.9.4.1.3 Standards and Guidelines

These standards have been widely applied to control the quantity of water withdrawn by users and the discharge of waste products into watercourses. The standards and guidelines also require specific technologies to be employed (technology standards) to reduce either water use or waste loads, and specify product standards, both for water provided for specific users and for goods that are potentially polluting (e.g., water efficiency standards).

2.9.4.1.4 Land Use Planning Controls

Some water authorities have long employed land use controls to protect their supply sources; for example, land uses may be regulated in upstream recharge areas and around reservoirs to prevent pollution, siltation, and changed runoff regimes. In the context of IWRM, the management of land use is as important as managing the water resource itself since it will affect flows, patterns of demand, and pollution loads.

2.9.4.2 Economic Instruments

2.9.4.2.1 Efficiency of Economic Tools

Economic tools may offer several advantages, such as providing incentives to change behavior, raising revenue to help finance necessary investments, establishing user priorities, and achieving management objectives at the least possible overall cost to society. Designing appropriate economic instruments requires simultaneous consideration of efficiency, environmental sustainability, equity, and other social concerns, as well as the complementary institutional and regulatory framework.

2.9.4.2.2 Water Prices, Tariffs, and Subsidies

According to the principle of managing water as an economic and social good, the recovery of full costs should be the goal for all water uses, unless compelling reasons indicate otherwise. At a minimum, full supply costs should be recovered in order to ensure the sustainability of investment and the viability of service providers.

2.9.4.2.3 Tariffs as Incentives

In the domestic sector, the scope for reducing water consumption may be relatively small because of the need to provide enough water to meet basic health and hygiene requirements. Nevertheless, reductions are possible and, overall, tariff or fee setting that sends the right price signals to water users is an important element of much-needed demand management. In irrigation, pricing may be used to encourage a shift from water-intensive crops to other crops.

2.9.4.2.4 Fee Structures

Water tariffs provide little incentives for the sustainable use of water if charged at a flat rate independent of the amount used. In such cases, setting the right fee structure may induce a more judicious use of the resource.

2.9.4.2.5 Fees for Wastewater Discharges

Fees should be set to reflect both the cost of environmental externalities and those associated with treating polluted wastewater or recipient waters. Fees can be related to both the quantity and quality of individual discharges and then adjusted carefully to create optimum incentives for polluters to introduce improved treatment technology, reuse water, and minimize the pollution of water resources.

2.9.4.2.6 Water Markets

Under the right circumstances, water markets can improve the efficiency of water resources allocation and help ensure that water is used for higher-value purposes. Taxes can be applied for both products involving high water consumption and products contributing to water pollution. For nonpoint pollution problems, especially those related to stormwater management, this option is a useful tool, since direct discharge control or treatment options are not always feasible. Hence, the reduction of pollution is achieved as a reduced product of waste response to higher-priced services and products. In recent years, the high costs of command and control regulation have encouraged the development of "self-regulatory" mechanisms, supported by appropriate procedures for performance monitoring. For example, professional organizations may produce best practice guidelines or governments may introduce "quality" hallmarking schemes; such schemes are now quite common in the environmental and product safety areas and may be a useful addition to the water sector toolbox.

2.9.5 TECHNOLOGICAL INSTRUMENTS

2.9.5.1 Technological Advances Toward Sustainability

In evaluating the range of available management tools, the role of and scope for technological advances should still be carefully considered as a factor that may help achieve sustainable water resources management. Traditional technologies such as rainwater harvesting can also play a key role.

2.9.5.2 Research and Development in Technology

Technological innovation and adaptation are key components of many efforts within the water sector. At the conceptual level, models and forecasting systems are being improved and they help to reduce uncertainties and risks in the use and management of urban water resources. Water-saving technologies allow more water for domestic use, improved and cost-effective methods for the treatment and reuse of wastewater in industries and domestic systems, aquifer recharge technologies, and human waste disposal systems.

2.9.5.3 Technology Assessment

What could be labeled as "auxiliary" technological achievements may also be usefully considered in water management. These are technologies that are developed for purposes other than water saving and water management, but may have considerable effects on the water sector.

2.9.5.4 Technological Choices

Technological choices must take account of specific conditions prevailing at the location of use. This means that the most advanced and modern technology is not necessarily the optimal choice in all cases. If the system cannot be sustained because of lack of spare parts, skilled manpower, or economic resources for operation, it is not the most appropriate solution. Moreover, high-cost technologies can prevent community and household involvement in water management.

2.10 CONCLUDING REMARKS

In this chapter, a new vision on urban water management, using stormwater and wastewater for different water usage in urban areas, is discussed. This new concept in urban planning and management, which is called WSUD, is a new paradigm in urban management, especially in arid and semiarid areas of the world. Since this new approach in urban planning and management needs communication between different decision makers in urban areas, in this chapter the institutional change in water governance is discussed. The change in paradigms and shifting from the Newtonian to holistic paradigm was introduced and its impact on shifting from supply management to demand management and using urban runoff and wastewater instead of discharging them to downstream water bodies was addressed. In the Newtonian paradigm, decision makers and planners look at technology as a way of salvation many shortcomings and lack of information. In this paradigm, the planners try to solve the problems by constructing new infrastructures, such as infrastructures for supplying demand in urban areas and surface water drainage from urban areas. The holistic view of planning requires developing tools and methods that are adaptive and sensitive to the nontechnical aspects of planning such as social, cultural, and environmental factors, and that can be used with the current state of local knowledge and water infrastructure.

The role of water in land use planning has been a challenge for planners and decision makers, and affects both quantity and quality of urban runoff and the other components of the urban water cycle. One of the main objectives of modern land use planning is the protection of ecologically valuable areas and land use that supports integrated water management. Special attention should be given to the effect of land use changes on the hydrological cycle and the protection of groundwater systems, especially discharge and recharge areas. Urbanization impacts on the environment are discussed in Chapter 8 of this book. Finally, an introduction on water resources assessment as a means of measuring, evaluating, monitoring, forecasting, and bringing consensus into IUWM is also presented.

REFERENCES

Agenda 21. 2002. World Summit on Sustainable Development, Johannesburg.

Bakker, K. 2004. *Good Governance in Restructuring Water Supply: A Handbook*. Federation of Canadian Municipalities, Canada.

Chen, C. H., Liaw, S. L., Wu, R. S., Chen, Q. L., and Chen, J. L. 1997. Development of a decision making theory and a decision support system for river basin water management. *Proceedings of the National Science Council, Republic of China. Part A, Physical Science and Engineering* 21: 389–409.

Chen, C. H., Wu, R. S., Liaw, S. L., Sue, W. R., and Chiou, I. J. 2000. A study of water–land environment carrying capacity for a river basin. *Water Science and Technology* 42: 389–396.

Chen, C. T., Liu, W., Liaw, S., and Yu, C. 2005. Development of a dynamic strategy planning theory and system for sustainable river basin land use management. *Science of the Total Environment* 346: 17–37.

Clausen, T. 2004. *Integrated Water Resources Management*. Global Water Partnership, South Africa.

Dahlenburg, J. 2003. *Water Sensitive Planning Guide*. Upper Parramatta River Catchment Trust, Sydney.

Denisov, N., Grenasberg, M., Hislop, L., Schipper, E. L., and Sbrensen, M. 2000. Cities environment reports on the internet: Understanding the CEROI template. Arendal, Norway 7, UNEP/GRID-Arendal.

Dublin Principles. 1992. International Conference on Water and the Environment, Dublin.

Federation of Canadian Municipalities. 2002. Policy on Municipal Infrastructure. *Adopted at FCM Annual Conference*. Available at http://www.fcm.ca/newfcm/Java/frame.htm.

Lloyd, S. D. 2001. Water Sensitive Design in the Australian Context. Synthesis of a conference held August 30–31, 2000, Melbourne, Australia. Cooperative Research Centre for Catchment Hydrology Technical Report 01/7.

McAlister, Tony. 2007. *National Guidelines for Evaluating Water Sensitive Urban Developments*. BMT WBM Pty Ltd., Australia.

McGranahan, G., Jacobi, P., Songsore, J., Surjadi, C., and Kjellén, M. 2001. *The Citizens at Risk: From Urban Sanitation to Sustainable Cities*. Earthscan, London.

O'Brien, J. J., McIntyre, E., Fortin, M., and Loudon, M. 2002. Governance and methods of service delivery for water and sewage systems. Commissioned Paper 17, The Walkerton Inquiry. Queen's Printer for Ontario, Toronto.

O'Connor, D. 2002. Report of the Walkerton Inquiry, Part I. The Events of May 2000 and elated Issues. Queen's Printer for Ontario, Toronto.

OECD. 1993. OECD core set of indicators for environmental performance reviews. Paris, France 7, OECD.

United Nations Conference on Environment and Development (UNCED). 1992. Report of the United Nations Conference on Environment and Development (Rio de Janeiro, June 3–14, 1992). Annex I: Rio Declaration on Environment and Development. UN Doc. A/CONF.151/26 (Vol. 1). Available at http://www.un.org/documents/ga/conf151/aconf15126-1annex1.htmi (accessed June 17, 2005).

United Nations Educational, Scientific and Cultural Organization World Water Assessment Program (UNESCO-WWAP). 2003. Water for People, Water for Life. The United Nations World Water Development Report. UNESCO and Berghahn Books, Barcelona.

Whelans and Halpern Glick Maunsell. 1994. Planning and Management Guidelines for Water Sensitive Urban (Residential) Design. Report prepared for the Department of Planning and Urban Development, the Water Authority of Western Australia and the Environmental Protection Authority.

World Resources Institute (WRI). 2003. Water Resources eAtlas. Watersheds of the World. AS19 Mekong. Available at http://multimedia.wri.org/watersheds_2003

3 Urban Water Hydrology

3.1 INTRODUCTION

Urbanization has adverse effects on the characteristics of local hydrology and stormwater quality. These effects have been realized in recent years and different approaches have been utilized to address and mitigate them. Urban stormwater quality concepts rely on empirical techniques and limited available data. Recently, considerable efforts have been made on urban stormwater quantity estimation and analysis. The emphasis of this chapter is on stormwater quantity analysis and related issues such as excess rainfall estimation, rainfall–runoff analysis, calculation of peak flow, and its occurrence time as well as hydrograph analysis. A detailed discussion on the effects of urbanization on stormwater quality and pollution is given in Chapter 7.

3.2 RAINFALL–RUNOFF ANALYSIS IN URBAN AREAS

Because of high imperviousness in urban areas, the mechanism of rainfall conversion to runoff is different from the other regions. For example, the accurate estimation of infiltrated rainfall into pervious surfaces in residential regions is very complicated and considerably affects the rainfall–runoff analysis.

The small variations in rainfall characteristics highly and quickly affect the response of urban areas, but these short-term variations do not affect the outflow of undeveloped regions. For precise estimation of a runoff hydrograph, the rainfall data should be available at time steps of 5 min or less. Also the peak discharge of urban areas is considerably higher because of faster movement of flow in these regions through channels and drainage systems. This means that for rainfall–runoff analysis in urban regions, one rain gauge is not sufficient. For explanation of the dynamics of moving storms, at least three rain gauges and a wind measurement are necessary (James and Scheckenberger, 1984).

3.2.1 DRAINAGE AREA CHARACTERISTICS

An urban drainage area is characterized by the area, shape, slope, soil type, land-use pattern, percent of imperviousness, roughness, and different natural or man-made storage systems. The main factors in the estimation of runoff volume in a residential region are area and imperviousness. Note that, for accurate runoff estimation, the hydraulically connected impervious areas should be considered. The areas that are connected directly to a drainage system and drain into it are called hydraulically connected impervious areas. A street surface with curbs and gutters, which collects runoff from the surface and drains into a storm sewer, is an example of a hydraulically connected impervious area. The areas where the runoff drains to pervious areas and does not directly enter the storm drainage system are called nonhydraulically connected areas. Rooftops or driveways are examples of this kind of areas.

Often the imperviousness is estimated based on the land use. Typical values are available in texts for the estimation of imperviousness, such as Table 3.1. The population density can also be used for imperviousness estimation in large urban areas, such as Equation 3.1 developed by Stankowski (1974):

$$I = 9.6 \text{PD}^{(0.479 - 0.017 \ln \text{PD})}, \tag{3.1}$$

where I is the percent imperviousness and PD is the population density (persons/km^2).

TABLE 3.1
Impervious Cover (%) for Various Land Uses

		Source				
Land Use	Density (Dwelling Units/km^2)	Northern Virginia (NVPDC, 1980)	Olympia (COPWD, 1995)	Puget Sound (Aqua Terra, 1994)	NRCS (USDA, 1986)	Rouge River (Kluiteneberg, 1994)
Low-density residential	<12.5	6	—	10	—	19
	12.5	—	—	10	12	
	25	12	—	10	20	
Medium-density residential	50	18	—	10	25	
	75	20	—	—	30	
	100	25	40	40	38	
High-density residential	125–175	35	40	40	—	38
Multifamily	Townhouse (>175)	35–50	—	60	65	—
Industrial	—	60–80	86	90	72	76
Commercial	—	90–95	86	90	85	56
Roadway	—	—	—	—	—	—

The use of population density to estimate impervious cover is attractive since it provides a rapid technique for generating a quantitative estimation of present and projected land surface cover. Graham et al. (1974) and Hicks and Woods (2000) have also developed other empirical relationships with different functional forms to relate population density to percentage of the impervious cover. These equations should be used with caution in different regions because they are developed based on the analysis of the available data in some regions with specific characteristics. The percentage of imperviousness should be calibrated during the rainfall–runoff modeling and analysis procedure.

The results of mentioned empirical relationships for the estimation of imperviousness are compared with the data gathered by Frederick and Graham in Figure 3.1 (Bird et al., 2001). Stankowski's relationship seriously underpredicts an impervious area at population densities greater than 400 persons/km^2, whereas Graham's clearly underpredicts the percentage of total impervious area (TIA) for population densities of under 500 persons/km^2. The Hicks and Woods (2000) relationship fits the overall response better than Graham or Stankowski.

A network of channels and ducts that form a sewer system are the main part of an urban drainage system that normally flows with gravity in open channels. The geometry and hydraulic characteristics of the drainage system are described by a set of parameters. For the design of a new drainage system, all of design parameters are determined considering the regional characteristics and the design needs. Evaluation of the life cycle and aging effect of system components and alterations made to an available drainage system is highly important for system development and improvement and needs a major site and future needs assessment.

3.2.2 Rainfall Losses

All of the possible losses during the rainfall-to-runoff conversion process are subtracted from the total rainfall to determine the excess rainfall. Rainfall loss includes depression storage (or "interception storage") on planted and other surfaces, infiltration, and evaporation. For a specific storm that happens in a short period, evaporation can be neglected, but in long-term analysis of the urban water budget the evaporation amount should be considered.

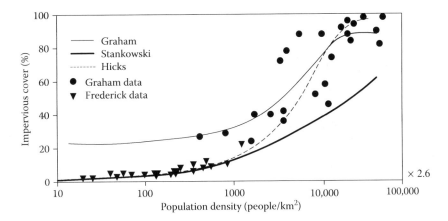

FIGURE 3.1 Comparison of the results on empirical relations for imperviousness estimation with the real data. (Adapted from Bird, S. L. et al., 2001. Estimating impervious cover from regionally available data. Poster at Southeastern Water Pollution Biologists Association Annual Meeting, Bowling Green, KY, October 30–November 1.)

Losses due to depression storage and infiltration cannot be separated over pervious areas. The amount of depression storage is highly variable in different areas; for example, it is about 16 mm for impervious surfaces and 63 mm for well-graded pervious surfaces (Tholin and Kiefer, 1960). In highly impervious areas, the depression storage is also related to the slope of the drainage area. Since it is very difficult to estimate depression storage, it is often determined during the rainfall–runoff model calibration process.

3.2.3 Time of Concentration

Two different definitions used for time of concentration (t_c) are as follows:

1. The travel time of a wave from the hydraulically most isolated point of the drainage area to the outlet point
2. The equilibrium time of the drainage area under a storm with steady excess rainfall

Note that the t_c is different from the travel time of a water increment through the drainage area.

After t_c, if the rainfall continues, as every portion of the drainage area contributes to the outlet flow, the drainage area is in equilibrium. The celerity of a wave is considerably more than the water increment so that the time of concentration (and drainage area equilibrium) occurs sooner than if it is considered based on overland flow. The wave speed in overland flow is usually estimated using the kinematic wave equation

$$c = kV = \gamma k y^{k-1}, \qquad (3.2)$$

where c is the wave speed (m/s), V is the average flow velocity (m/s), y is the water depth (m), and k, γ are the kinematic wave parameters that are estimated using the homogeneous flow momentum equation, and

$$q = yV = \gamma y^k, \qquad (3.3)$$

where q is the flow per unit width of the drainage area (m³/s/m).

Parameter k has been determined to be equal to 5/3 and 3 using Manning's equation, for turbulent flow and laminar flow, respectively. It can be demonstrated that the wave speed is about 1.7–3 times more than a water increment (parcel) speed. The values of parameter γ are also different for turbulent and laminar flows. The following equations are used to estimate the γ value in different flow conditions:

$$\gamma = \frac{\sqrt{S}}{n} \quad \text{Turbulent flow,} \tag{3.4a}$$

$$\gamma = \frac{gS}{3\nu} \quad \text{Laminar flow,} \tag{3.4b}$$

where S is the slope (as the slope is usually small, it is considered equal to angle in radians), n is Manning's roughness, g is the acceleration of gravity (m/s^2), and ν is the kinematic viscosity (m^2/s).

After the estimation of kinematic wave parameters, the time of concentration for overland flow is calculated as

$$t_c = \left(\frac{L}{\gamma i_e^{k-1}}\right)^{1/k}, \tag{3.5}$$

where L is the length of overland flow path (m) and i_e is the excess rainfall (m/s).

As Equation 3.5 shows, t_c and the excess rainfall are inversely related to each other; so, as the intensity of excess rainfall increases, the value of t_c decreases. The other available equations in the literature for the estimation of t_c, which are often based on water increments travel time, do not consider this important physical property of the drainage area (see McCuen et al., 1984; Ramsay and Huber, 1987, for more information).

EXAMPLE 3.1

Estimate the flow rate per unit width q (m^3/s/m) resulting from a rainfall event on an asphalt street surface for both laminar and turbulent flow conditions. Assume that the street slope and roughness are 0.5% and 0.011, respectively. The runoff depth is 0.5 cm, and temperature is 20°C.

Solution: For laminar flow, Equation 3.4b is used for the estimation of γ. ν is 1.003×10^{-6} m^2/s at a temperature of 20°C and γ is estimated as

$$\gamma = \frac{gS}{3\nu} = \frac{9.81 \times 0.005}{3 \times 1.003 \times 10^{-6}} = 1.63 \times 10^4\, \text{s}^{-1}\text{m}^{-1},$$

and as k is equal to 3, $q = 1.63 \times 10^4 \times (0.5 \times 10^{-2})^3 = 2.04 \times 10^{-3}$ m^3/s/m, for laminar flow. Equation 3.4a is used for the estimation of γ in turbulent flow conditions:

$$\gamma = \frac{S^{0.5}}{n} = \frac{0.005^{0.5}}{0.011} = 6.43\, \text{m}^{1/3}/\text{s}.$$

Then considering that k is equal to 5/3 for turbulent flow,

$$q = 6.43 \times (0.5 \times 10^{-2})^{5/3} = 9.4 \times 10^{-4}\, \text{m}^3/\text{s/m}.$$

Dynamic wave speed is used for the investigation of wave movement speed in open channels:

$$c = V \pm \sqrt{\frac{gA}{T}}, \qquad (3.6)$$

where A is the cross-sectional area in the channel and T is the surface width.

Downstream wave speeds that are specified by the positive sign in Equation 3.6 are apparently greater than the water velocity V. Downstream wave speeds for supercritical flow and upstream water speeds for subcritical flow are calculated by using the negative sign in Equation 3.6. If pressure flow occurs in closed channels, waves will move with very high speeds near the sound speed (Wylie and Streeter, 1978).

Various lag times are incorporated into the analysis of the urban hydrograph. Most of these parameters, such as time from the beginning of runoff to peak discharge, t_{peak}, can be determined by site measurements, but the time of concentration cannot be easily measured (Overton and Meadows, 1976) and is usually dealt with as a theoretical parameter. Lag times are commonly used in unit hydrograph (UH) analysis and some theoretical models.

In summary, after rainfall occurrence in any urban drainage area, waves are generated that move downstream through natural or man-made (sewer system) water paths. In urban drainage areas, the time of concentration that is equal to the time to equilibrium is less than the water travel time. For accurate estimation of peak flow and lag time to its occurrence as well as other characteristics of drainage area in response to rainfall, an exact estimate of t_c is necessary.

3.3 EXCESS RAINFALL CALCULATION

In this section, various methods used for the determination of the runoff quantity produced by a given storm event are investigated. Considerable parts of rainfall do not convert to runoff due to different lost sources. Plants intercept a part of rainfall. Some rainfall will remain in surface puddles. A part of rainfall infiltrates into the soil and a small amount of it evaporates before reaching the ground. The remaining rainfall after these losses will produce surface runoff.

The main sources of rainfall abstractions can be summarized as

- Interception
- Surface depression storage
- Evaporation
- Transpiration
- Infiltration

The evaporation and transpiration abstractions can be eliminated under design storm conditions, in an urban area. The relation between rainfall abstractions and overland flow (runoff) is presented in Figure 3.2. As mentioned before, the excess (effective) rainfall is equal to the total rainfall minus summation of all abstractions. The depth of excess rainfall produced per unit time is considered as the rate of excess (effective) rainfall. The excess rainfall rate is also equal to the rate of rainfall minus the rate of loss in the situation where the excess rainfall is the only source of storm runoff in the considered urban watershed. Thus the total volume of excess rainfall and the total volume of produced runoff are equal.

3.3.1 Interception Storage Estimation

The portion of the rainfall that is intercepted by flora before reaching the ground is named interception storage. The intercepted water is held by plant surfaces such as branches and leaves and, finally, evaporates without reaching the ground. Most of the interception occurs at the beginning of the storm, because the maximum water-holding capacity of leaves and branches is low and maximum interception is reached very soon.

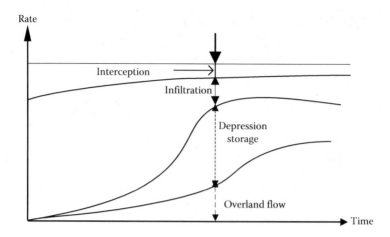

FIGURE 3.2 Relation between the interception, depression storage, infiltration, and overland flow.

The main factors in the determination of the amount of interception are the type and density of the plants as well as the amount of rainfall. Empirical methods are developed for the estimation of the losses due to interception. The following relationship is suggested by Horton (1919) for the estimation of interception storage:

$$\text{IS} = \alpha + \beta R_t^{\gamma}, \tag{3.7}$$

where IS is the interception storage depth (m), R_t is the total rainfall depth (m), and α, β, γ are the empirical constants.

Horton's equation is developed based on the data collected from summer storms, so Kibler et al. (1996) reanalyzed Horton's data and developed the following expression for the estimation of interception storage:

$$\text{IS} = k R_t^n, \tag{3.8}$$

where k and n are constant parameters that should be specified for each region. Values of α, β, γ, k, and n are presented in Table 3.2. It should be noted that Equations 3.7 and 3.8 are only applicable to the part of the urban area that is covered completely by plants.

Considering the interception loss during individual storm events is useless and estimation of losses due to interception is only significant in annual or long-term rainfall–runoff models (Viessman et al., 1989).

3.3.2 Estimation of Infiltration

The process of passing rainwater through the ground surface and filling the pores of the soil on both the surface (the availability of water for infiltration) and subsurface (the potential of the available water for infiltration) is called infiltration. In other words, infiltration is defined as the passage of water through the air–soil interface. As infiltration is the main source of rainfall loss, it is a very important process in urban stormwater management and, therefore, essentially all hydrologic methods explicitly account for infiltration. Infiltration rates are affected by factors such as time since the rainfall event began, soil porosity and permeability, antecedent soil moisture conditions, and the presence of vegetation. Urbanization usually decreases infiltration with a resulting increase in runoff volume and discharge.

TABLE 3.2
Values of Constant Parameters in Interception Equations for Different Types of Trees

Tree Type	Equation 3.7			Equation 3.8	
	α ($\times 10^{-4}$)	β ($\times 10^{-4}$)	γ	k	n
Apple	10.16	1.17	1.0	0.093	0.73
Ash	3.81	1.47	1.0	0.167	0.88
Beech	5.08	1.47	1.0	0.058	0.65
Chestnut	10.16	1.30	1.0	0.129	0.77
Elm	0.00	9.32	0.5	0.022	0.48
Hemlock	0.00	8.10	0.5	0.123	0.74
Maple	7.62	1.47	1.0	0.125	0.77
Oak	7.62	1.42	1.0	0.069	0.66
Pine	0.00	8.10	0.5	0.100	0.70
Willow	5.08	2.59	1.0	0.248	0.85

Source: Kibler, D. F. et al., 1996. *Urban Hydrology*, 2nd Edition. ASCE Manual of Practice, No. 28, New York. With permission.

The maximum possible rate of water infiltration is considered as infiltration capacity or the potential infiltration rate, which shows the capability of available water to infiltrate. For rainfall rates less than the infiltration capacity, the real rate of infiltration is equal to the rate of rainfall. If the rate of rainfall exceeds the infiltration capacity, the actual rate of infiltration will be equivalent to the infiltration capacity. In this situation, the rainwater that does not infiltrate to the soil fills the surface depressions and, finally, flows over the ground surface.

$$f = i \quad \text{if } f_f > i,$$

and

$$f = f_f \quad \text{if } f_f < i,$$

where f_f is the infiltration capacity, f is the actual rate of infiltration, and i is the rainfall rate.

In the situation where infiltration is the only or the dominant source of rainfall abstraction, the rate of excess rainfall, i_e, is calculated as

$$i_e = i - f. \tag{3.9}$$

The infiltration capacity estimation is a complicated process but some theoretical methods are developed for the determination of the infiltration capacity. The Richard equation is a sophisticated method for infiltration capacity estimation. This equation is nonlinear and its solving needs complex numerical algorithms. But in engineering practices much simpler methods are used for infiltration capacity estimation. The application of the Richard equation in different regions reveals the dependency between the infiltration capacity and the soil characteristics, especially the available humidity in the soil before and during the infiltration. Therefore, for realistic modeling of the soil infiltration capacity variation during the infiltration process, the initial moisture condition of the soil and the amount of infiltrated water into the soil from the start of rainfall should be considered.

3.3.2.1 Green and Ampt Model

The most physically based numerical infiltration model that is commonly used is the Green and Ampt (1974) model. All parameters incorporated into this model are determined from soil characteristics regarding their definition and physical basis. An illustration of a simplified model proposed by Green and Ampt is shown in Figure 3.3. It is assumed that the moisture of the soil underlying the pervious area is uniform and equals θ_i when rainfall starts. As rainwater penetrates into the soil, the soil saturation degree increases. The increase in soil moisture is greater near the ground surface and decreases as depth increases.

But the basic assumption behind the Green and Ampt equation is that water infiltrates into (relatively) dry soil as a sharp wetting front. The soil is saturated on top of the wetting front, and the initial degree of saturation is maintained below it. In other words, the volumetric water contents remain constant above and below the wetting front as it advances. By this assumption, two distinctive zones are formed below the ground surface during the infiltration process. A saturated zone is developed below the surface a little after the start of the rain. As more water infiltrates into the soil, the depth of this zone increases. The other zone with the initial moisture content is below the wetting front and it is assumed that this zone has unlimited depth (or, alternatively, the water table or the impervious bed rock is deep enough not to interfere with the infiltration process).

Darcy's law is applicable for the estimation of the infiltration capacity in the saturated zone adjacent to the soil surface:

$$f = -K_s \frac{dh}{dz} = -K_s \frac{(\psi_f + z_f) - (H + 0)}{z_f - 0} = -K_s \frac{\psi_f + z_f - H}{z_f}, \qquad (3.10)$$

where H is the the depth of ponding, K_s is the saturated hydraulic conductivity, f is the infiltration rate and is negative, ψ_f is the suction at the wetting front (negative pressure head), and z_f is the depth of the wetting front.

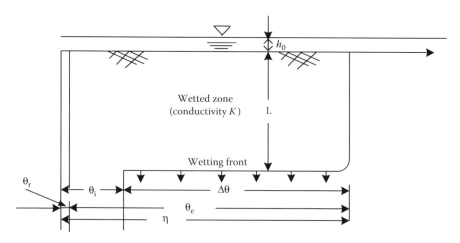

FIGURE 3.3 Subsurface moisture, the Green and Ampt model.

Note that the sign of f is negative because it is a vector pointing in the negative z direction. Furthermore, ψ_f and z_f are negative and the sign of H is positive. Because z_f is negative, F, which is the cumulative depth of infiltration, is negative and is calculated as

$$F = z_f(\theta_s - \theta_i) \quad \text{or} \quad z_f = \frac{F}{\theta_s - \theta_i}, \tag{3.11}$$

where θ_i is the initial moisture content and θ_s is the saturated moisture content.

Substituting Equation 3.11 into Darcy's equation gives the following equation:

$$f = \frac{dF}{dt} = -K\frac{\psi_f + [F/(\theta_s - \theta_i)] - H}{F/(\theta_s - \theta_i)} = -K\left(\psi_f + \frac{F}{\theta_s - \theta_i} - H\right)\frac{\theta_s - \theta_i}{F}. \tag{3.12}$$

By the assumption that H is neglectable in comparison with the other terms, Equation 3.12 is simplified to the Green and Ampt infiltration rate equation:

$$f = \frac{dF}{dt} = -K\left(\psi_f \frac{\theta_s - \theta_i}{F} + 1\right). \tag{3.13}$$

Rearranging Equation 3.13 gives the cumulative infiltration F as a function of infiltration rate f.

$$\frac{f}{K} + 1 = -\psi_f \frac{\theta_s - \theta_i}{F} \quad \Rightarrow \quad F = -\psi_f \frac{\theta_s - \theta_i}{1 + (f/K)}. \tag{3.14}$$

For F to be negative, $f < -K$.

Although the infiltration capacity of dry soil is usually high, this capacity decreases as the rain infiltration into the soil continues. The pending time t_p is defined as follows with the assumption that the rate of rainfall is constant during the pending time:

$$t_p = -\frac{\psi_f(\theta_e)(1 - \theta_i)}{(i^2/K) - i}, \tag{3.15}$$

where θ_e is the effective porosity.

In the application of the Green and Ampt method, some parameters should be determined according to the soil texture and land-use practices (Rawls et al., 1983). For estimating the Green and Ampt parameters, different tables such as Table 3.3 are available. The use of these tables is suggested only when there is no field data.

Although the Green and Ampt equations are used for homogeneous soils, these equations can be used for the evaluation of infiltration into layered soils by considering some modifications suggested by Rawls et al. (1993). Until the wetting front is in the first layer, there is no difference in the application of equations with homogeneous soil. When the wetting front enters the second layer, the suction head of the second layer is used in the equations and the hydraulic conductivity is estimated as $K = \sqrt{K_1 K_2}$. This procedure is followed for the subsequent layers.

TABLE 3.3
Average Soil Parameters for the Green and Ampt Model

Soil Type	Porosity (η)	Effective Porosity (θ_e)	Suction Head (ψ_f) (mm)	Hydraulic Conductivity (K) (mm/h)
Sand	0.437	0.417	49.5	117.8
Loamy sand	0.437	0.401	61.3	29.9
Sandy loam	0.453	0.412	110.1	10.9
Loam	0.463	0.434	88.9	3.4
Silt loam	0.501	0.486	166.8	6.5
Sandy clay loam	0.398	0.330	218.5	1.5
Clay loam	0.464	0.309	208.8	1.0
Silty clay loam	0.471	0.432	273.0	1.0
Sandy clay	0.430	0.321	239.0	0.6
Silty clay	0.479	0.423	292.2	0.5
Clay	0.475	0.385	316.3	0.3

EXAMPLE 3.2

For the following soil properties, determine the amount of water infiltrated into the soil. $K = 1.97$ cm/h, $\theta_i = 0.318, \theta_s = 0.518, \theta_e = 0.500, f = i = 7.88$ cm/h, $\psi_f = 9.37$ cm.

Solution: Using Equation 3.14,

$$F = -\psi_f \frac{\theta_s - \theta_i}{1 + (f/K)} = (-9.37)\frac{0.518 - 0.318}{1 + (-7.88/1.97)} = -0.625 \text{ cm}.$$

The ponding time is calculated using Equation 3.15 as

$$t_p = \frac{\psi_f(\theta_e)(1 - \theta_i)}{(i^2/K) - i} = -\frac{(-9.37)(0.500)(1 - 0.318)}{(7.88^2/1.97) - 7.88} = 0.54 \text{ h}.$$

3.3.2.2 Horton Method

An exponential decay function for the evaluation of infiltration during a storm event is suggested by Horton (1940) based on experimental data:

$$f_p = f_f + (f_0 - f_f)e^{-mt}, \tag{3.16}$$

where f_p is the rate of infiltration into the soil at time t, f_f is the final infiltration capacity, f_0 is the initial infiltration capacity, m is the decay constant, and t is the time from the beginning of rainfall.

Equation 3.16 is graphically represented in Figure 3.4, which explains the infiltration capacity variation with time regarding different values of m. Horton's equation is empirical and is commonly used in engineering practices despite its early years of development. Fitting Equation 3.16 to the measured infiltration data, three parameters f_0, f_f, and m are determined. Many hydrologists have a "feel" for the best values of these three parameters despite the lack of generally accepted listing of these parameters for different types of soils. An example of suggested values for Horton's equation parameters is given in Table 3.4. Examples 3.3 and 3.5 are adapted from Akan and Houghtalen (2003).

Urban Water Hydrology

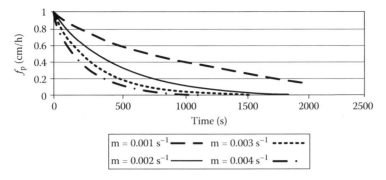

FIGURE 3.4 Horton infiltration capacity variations with the time.

TABLE 3.4
Assumed Horton Infiltration Parameters for Major Soil Types

Soil Type/Land Use (From Soil Mapping)	Initial Infiltration Rate (mm/h)	Final Infiltration Rate (mm/h)	Decay Rate (1/s)
Developed land—low or uncertain infiltration	6	3	0.0005
Forested land—low or uncertain infiltration	9	3	0.0001
Forested land—moderately well-drained soils or better	12.5	6.25	0.00005
Developed land—moderately well-drained soils or better	9	3	0.0005

EXAMPLE 3.3

The hydrograph and hyetograph of a 2-h storm event in an urban area are presented in column 3 of Table 3.5 and Figure 3.5, respectively. The Horton parameters for this area are estimated as $f_0 = 2.8$ cm/h, $f_f = 0.6$ cm/h, and $m = 1.1$/h. Determine the losses due to infiltration in the pervious areas and the excess rainfall, assuming that infiltration is the only source of rainfall loss.

Solution: The summary of calculations is tabulated in Table 3.5. The t_1 (column 1) and t_2 (column 2) represent the beginning and end of a time interval, respectively. f_p (column 5) is determined using Equation 3.16 at the midpoint of each time interval, t, tabulated in column 4. The actual infiltration rate, f, (column 6) is equal to the minimum of f_p and i. By subtracting the f from the i, the rate of effective rainfall i_e (column 7) is obtained. The infiltration capacity and excess rainfall are also illustrated in Figure 3.5. The total depth of infiltrated water is 2.28 cm, which is obtained by $\Delta t \Sigma f$ where Δt here is equal to 0.1. The total depth of excess rainfall is calculated in a similar way to be 0.77 cm.

TABLE 3.5
Summary of Calculations of Example 3.3

t_1 (h) (1)	t_2 (h) (2)	i (cm/h) (3)	t (h) (4)	f_p (cm/h) (5)	f (cm/h) (6)	i_e (cm/h) (7)
0	0.1	0.75	0.05	2.68	0.75	0.00
0.1	0.2	0.75	0.15	2.47	0.75	0.00
0.2	0.3	0.75	0.25	2.27	0.75	0.00
0.3	0.4	1.5	0.35	2.10	1.50	0.00
0.4	0.5	1.5	0.45	1.94	1.50	0.00

continued

TABLE 3.5 (continued)

t_1 (h) (1)	t_2 (h) (2)	i (cm/h) (3)	t (h) (4)	f_p (cm/h) (5)	f (cm/h) (6)	i_e (cm/h) (7)
0.5	0.6	2.25	0.55	1.80	1.80	0.45
0.6	0.7	2.25	0.65	1.68	1.68	0.57
0.7	0.8	2.25	0.75	1.56	1.56	0.69
0.8	0.9	2.75	0.85	1.46	1.46	1.29
0.9	1	2.75	0.95	1.37	1.37	1.38
1	1.1	2	1.05	1.29	1.29	0.71
1.1	1.2	2	1.15	1.22	1.22	0.78
1.2	1.3	2	1.25	1.16	1.16	0.84
1.3	1.4	1.5	1.35	1.10	1.10	0.40
1.4	1.5	1.5	1.45	1.05	1.05	0.45
1.5	1.6	1	1.55	1.00	1.00	0.00
1.6	1.7	1	1.65	0.96	0.96	0.04
1.7	1.8	1	1.75	0.92	0.92	0.08
1.8	1.9	0.5	1.85	0.89	0.50	0.00
1.9	2	0.5	1.95	0.86	0.50	0.00
		$\sum = 3.05$ cm			$\sum = 2.28$ cm	$\sum = 0.77$ cm

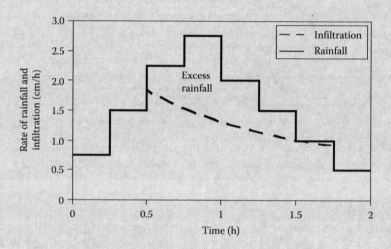

FIGURE 3.5 Hyetograph of rainfall in Example 3.3.

EXAMPLE 3.4

Given an initial infiltration capacity f_0 of 2.9 cm/h and a time constant m of 0.28 h^{-1} for a homogeneous soil, derive an infiltration capacity versus time curve if the ultimate infiltration capacity is 0.50 cm/h. For the first 8 h, estimate the total volume of water infiltrated in cm over the watershed.

Solution: Substituting the appropriate values into Equation 3.16 yields

$$f_p = 0.5 + (2.9 - 0.5)e^{-0.28t}.$$

For the times shown in Table 3.6, values of f are computed and entered into the table and, finally, the curve of Figure 3.6 is derived.

To find the volume of water infiltrated during the first 8 h, Equation 3.16 can be integrated over the range of 0–8:

$$V = \int \left[0.5 + (2.9 - 0.5)e^{-0.28t}\right] dt,$$

$$V = \left[0.5t + (2.4/(-0.28)e^{-0.28t}\right]_0^8 = 11.65 \text{ cm}.$$

TABLE 3.6
Calculations of Example 3.4

Time (h)	Infiltration (cm/h)	Time (h)	Infiltration (cm/h)
0	2.900	4	1.283
0.1	2.834	5	1.092
0.2	2.769	6	0.947
0.3	2.707	7	0.838
0.4	2.646	8	0.756
0.5	2.586	9	0.693
0.6	2.529	10	0.646
0.7	2.473	15	0.536
0.8	2.418	20	0.509
0.9	2.365	25	0.502
1	2.314	30	0.501
2	1.871	35	0.500
3	1.536	40	0.500

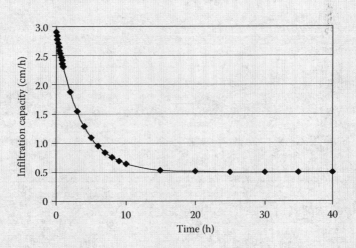

FIGURE 3.6 Infiltration curve for Example 3.4.

Example 3.5

Given the hyetograph of an 80-min rainfall event in Figure 3.7 and Table 3.7, calculate the hyetograph of excess rainfall and the rainfall–runoff conversion coefficient. Assume that the depression storage loss of this rainfall is 0.1 cm and the Horton infiltration parameters are determined as

$$f_0 = 0.51 \, \text{cm/h}, \quad f_f = 0.07 \, \text{cm/h}, \quad \text{and} \quad m = 0.8 \, \text{h}^{-1}.$$

Solution: Before the start of any infiltration losses, the depression storage is reached. The total volume of rainfall in the 20 and 30 min after the start of rainfall is 0.05 cm = [(0.1 + 0.2 cm/h) × 1/6 h] and 0.11 cm, respectively. So the depression storage is reached between 20 and 30 min rainfall increments. The exact time of depression storage is calculated as $t = 2 \times 1/6\,\text{h} + (0.05\,\text{cm}/0.35\,\text{cm/h}) = 2/6\,\text{h} + 1/7\,\text{h} = 28.57$ min.

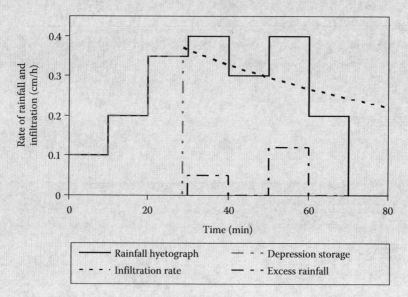

FIGURE 3.7 Rainfall and excess rainfall hyetographs of Example 3.5.

TABLE 3.7
Computation of Excess Rainfall

Time Interval (min)	Rainfall (cm/h)	Infiltration Capacity at Start of Time Interval (cm/h)	Average Infiltration Capacity (cm/h)	Excess Rainfall (cm/h)
0–10	0.1			
10–20	0.2			
20–28.57	0.35			
28.57–30	0.35	0.37	0.37	0[a]
30–40	0.4	0.36	0.35	0.05
40–50	0.3	0.33	0.31	0[a]
50–60	0.4	0.30	0.28	0.12
60–70	0.2	0.27	0.26	0[a]
70–80	0	0.24		

[a] Negative value set to zero.

Therefore infiltration starts at $t = 28.57$ min, as illustrated in Figure 3.7. The infiltration and excess calculations are summarized in Table 3.7 and presented in Figure 3.7. It should be noted that as negative excess rainfall is meaningful in the time increments when the rainfall rate is less than the infiltration capacity (e.g., 40–50 time increments), it is assumed that the rate of infiltration capacity decrease remains constant.

The rainfall volume is calculated using $\Delta t \Sigma i$ and is equal to 0.383 cm. The volume of excess rainfall is calculated in a similar way and is equal to 0.028 cm. Considering that the depression storage volume's equal to 0.10 cm, the actual infiltration volume is estimated as 0.255 cm. It should be noted that this volume and the area under the infiltration curve are not equal since the infiltration capacity has not been used all time of the rainfall event. The rainfall–runoff conversion coefficient is equal to the ratio of runoff (excess rainfall) to rainfall, which in this example is $0.028/0.383 = 0.07$.

3.3.2.3 Modified Horton Method

The major disadvantage of the Horton equation is that it does not consider the amount of water that has infiltrated in time t and it is assumed that the infiltration capacity is only dependent on time (Akan, 1992, 1993). In this way, the infiltration capacity is underestimated when the rate of rainfall becomes lesser than the infiltration capacity during the rainfall. Akan (1992) has mathematically modified the original Horton equation to overcome this problem. In the modified equation the infiltration capacity in time t is expressed as a function of the water volume that has infiltrated into the soil until time t as

$$f_p = f_0 - mF_e, \tag{3.17}$$

where F_e is the total depth of water infiltrated into the soil since the start of rainfall in excess of f_f. Based on the definition of F_e, it could be calculated as

$$F_e = \int_0^t (f_p - f_f)\, dt. \tag{3.18}$$

The integral of Equation 3.18 can be approximated for calculations over discrete time intervals Δt as

$$F_e = \sum (f_p - f_f) \Delta t. \tag{3.19}$$

A physical explanation of Equation 3.17 is that part of the infiltrating water filters deepen at a rate of f_f. Therefore, the accumulated water adjacent to the ground surface at a rate of $f_p - f_f$ affects the soil infiltration capacity.

When the rate of rainfall is less than the soil infiltration capacity during the infiltration process, the modified equation (Equation 3.17) becomes more reliable than the original equation. Using the modified equation, the ponding time formulation can be derived with the same physical meaning as the Green and Ampt ponding time. For a rainfall with a constant rate less than the initial infiltration capacity, the ponding time, t_p, is formulated as

$$t_p = \frac{f_0 - i}{K(i - f_f)}. \tag{3.20}$$

EXAMPLE 3.6

Determine the infiltration losses and excess rainfall from the given rainfall in Example 3.3 using the modified Horton equation.

Solution: The calculations are summarized in Table 3.8. The F_{e1} values corresponding to t_1 are tabulated in column 4, which for the first time interval $F_{e1} = 0$. The infiltration capacity in column 5 is calculated by using Equation 3.17, where F_{e1} is used instead of F_e so that the first entry in column 5 is the same as f_0 (the initial infiltration capacity). The smaller value of i and f_p is considered as f (column 6). The entry F_{e2} in column 8, which corresponds to the end of the time interval, is determined using $F_{e2} = F_{e1} + (f_p - f_f)\Delta t$. By subtracting the infiltration rate from rainfall rate, the rate of effective rainfall i_e is obtained. The calculations are repeated for all the time intervals in the same manner. It should be noted that the value of the F_{e2} in a time step is the F_{e1} for the next step. The total depth of filtered water is equal to $F = \Delta t \Sigma f = (0.1\,h)(27.31\,cm/h) = 2.73\,cm$, and the total depth of excess rainfall is equal to $(0.1\,h)(3.19\,cm/h) = 0.32\,cm$.

TABLE 3.8
Results of Example 3.6

t_1 (h) (1)	t_2 (h) (2)	i (cm/h) (3)	F_{e1} (cm) (4)	f_p (cm/h) (5)	f (cm/h) (6)	$(f - f_f)$ Δt (cm) (7)	F_{e2} (cm) (8)	i_e (cm/h) (9)
0	0.1	0.75	0	2.80	0.75	0.02	0.02	0.00
0.1	0.2	0.75	0.02	2.78	0.75	0.02	0.03	0.00
0.2	0.3	0.75	0.03	2.77	0.75	0.02	0.05	0.00
0.3	0.4	1.5	0.05	2.75	1.50	0.09	0.14	0.00
0.4	0.5	1.5	0.14	2.65	1.50	0.09	0.23	0.00
0.5	0.6	2.25	0.23	2.55	2.25	0.17	0.39	0.00
0.6	0.7	2.25	0.39	2.37	2.25	0.17	0.56	0.00
0.7	0.8	2.25	0.56	2.19	2.19	0.16	0.71	0.06
0.8	0.9	2.75	0.71	2.01	2.01	0.14	0.86	0.74
0.9	1	2.75	0.86	1.86	1.86	0.13	0.98	0.89
1	1.1	2	0.98	1.72	1.72	0.11	1.09	0.28
1.1	1.2	2	1.09	1.60	1.60	0.10	1.19	0.40
1.2	1.3	2	1.19	1.49	1.49	0.09	1.28	0.51
1.3	1.4	1.5	1.28	1.39	1.39	0.08	1.36	0.11
1.4	1.5	1.5	1.36	1.30	1.30	0.07	1.43	0.20
1.5	1.6	1	1.43	1.23	1.00	0.04	1.47	0.00
1.6	1.7	1	1.47	1.18	1.00	0.04	1.51	0.00
1.7	1.8	1	1.51	1.14	1.00	0.04	1.55	0.00
1.8	1.9	0.5	1.55	1.09	0.50	−0.01	1.54	0.00
1.9	2	0.5	1.54	1.10	0.50	−0.01	1.53	0.00
		$\Sigma = 3.05\,cm$			$\Sigma = 2.73\,cm$			$\Sigma = 0.32\,cm$

3.3.2.4 Holtan Method

The Holton method is formed based on the concept that the infiltration capacity is related to the available storage for holding water in the surface layer of the soil. Because of the increase of water infiltration into soil, the infiltration capacity decreases because of less storage availability. The Holtan method is applicable for agricultural lands, calculating the rainfall losses in wooded parts of urban areas and the areas covered by grass and plants. The main advantage of the model is the capability of recovery of soil infiltration capacity during periods of light or zero rainfall. The infiltration rate using the Holton method is expressed as (Holton and Lopez, 1971)

$$f_p = (GI)AS_\alpha^\beta + f_f, \qquad (3.21)$$

where f_p is the infiltration capacity in cm/h, GI is the "growth index" representing the relative maturity of the ground cover and can be determined using Figure 3.8, A is the cm/h/cm of available storage and is an empirical factor and can be estimated using Table 3.9, S_a is the soil storage capacity in cm of equivalent depth of pore space in the surface layer of the soil, f_f is the constant rate of percolation of water through the soil profile below the surface layer, and β is the empirical exponent, typically taken to be equal to 1.4.

The major advantage of the Holtan model over other infiltration equations is the use of cumulative infiltration instead of time as the independent variable. This offers several advantages in catchment-hydrology simulation.

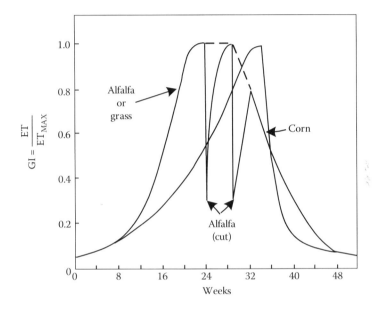

FIGURE 3.8 Estimation of GI index based on the plants growing situation.

TABLE 3.9
Estimation of Factor A Based on Land Use for the Holton Equation

Land Use or Cover	Betel Area Rating[a]	
	Poor Condition	Good Condition
Fallow[b]	0.10	0.30
Row crops	0.10	0.20
Small grains	0.20	0.30
Hay (legumes)	0.20	0.40
Hay (sod)	0.40	0.60
Pasture (bunchgrass)	0.20	0.40
Temporary pasture (sod)	0.40	0.60
Permanent pasture (sod)	0.80	1.00
Woods and forests	0.80	1.00

[a] Adjustment needed for (weeds) and (grazing).
[b] For fallow land only, "poor condition" means "after row crop" and "good condition" means "after sod."

The available storage S_a, as rainwater infiltrates into the soil during the rainfall-infiltration process, is reduced. The storage reduction over a discrete time interval Δt is calculated as

$$\Delta S_a = \Delta t (f_p - f_f). \qquad (3.22)$$

At the start of the calculations, the initial value of S_a should be determined as the product of the depth of the surface soil layer and initial moisture deficiency.

EXAMPLE 3.7

Find the losses of a 2-h rainfall with hourly intensity tabulated in column 3 of Table 3.10 due to infiltration using the Holton method. Assume that GI = 0.6, f_f = 0.5 cm/h, A = 0.4 cm/h/cm, and initially, S_a = 2.5 cm.

Solution: The calculations are summarized in Table 3.10. The initial value of S_a, for the first time step, is equal to 2.5 cm. By substituting S_{a1} for S_a in Equation 3.21, the infiltration capacity f_p (column 5) is calculated. The smaller value of i and f_p is set as the infiltration rate, f_f, in column 6. The reduction in available storage ΔS_a is estimated using Equation 3.22. The S_{a2} is estimated as $S_{a1} - \Delta S_a$. The same procedure is followed for all the time intervals. Finally, the total depth of water loss is equal to 2.12 cm and the total excess rainfall is estimated as 0.1 cm.

TABLE 3.10
Calculation Results of Example 3.7

t_1 (h) (1)	t_2 (h) (2)	i (cm/h) (3)	S_{a1} (cm) (4)	f_p (cm/h) (5)	f_f (cm/h) (6)	ΔS_a (cm) (7)	S_{a2} (cm) (8)	i_e (cm/h) (9)
0.0	0.1	0.5	2.50	1.37	0.50	0.00	2.50	0.00
0.1	0.2	0.5	2.50	1.37	0.50	0.00	2.50	0.00
0.2	0.3	1.0	2.50	1.37	1.00	0.05	2.55	0.00
0.3	0.4	1.0	2.55	1.39	1.00	0.05	2.60	0.00
0.4	0.5	1.2	2.60	1.41	1.20	0.07	2.67	0.00
0.5	0.6	1.2	2.67	1.45	1.20	0.07	2.74	0.00
0.6	0.7	1.6	2.74	1.48	1.48	0.10	2.84	0.12
0.7	0.8	1.6	2.84	1.53	1.53	0.10	2.94	0.07
0.8	0.9	2.0	2.94	1.59	1.59	0.11	3.05	0.41
0.9	1.0	2.0	3.05	1.64	1.64	0.11	3.16	0.36
1.0	1.1	1.5	3.16	1.70	1.50	0.10	3.26	0.00
1.1	1.2	1.5	3.26	1.76	1.50	0.10	3.36	0.00
1.2	1.3	1.3	3.36	1.81	1.30	0.08	3.44	0.00
1.3	1.4	1.3	3.44	1.86	1.30	0.08	3.52	0.00
1.4	1.5	1.0	3.52	1.90	1.00	0.05	3.57	0.00
1.5	1.6	1.0	3.57	1.93	1.00	0.05	3.62	0.00
1.6	1.7	0.8	3.62	1.96	0.75	0.03	3.65	0.00
1.7	1.8	0.8	3.65	1.97	0.75	0.03	3.67	0.00
1.8	1.9	0.3	3.67	1.98	0.25	−0.03	3.65	0.00
1.9	2.0	0.3	3.65	1.97	0.25	−0.03	3.62	0.00
		$\sum = 2.22$ cm			$\sum = 2.12$ cm			$\sum = 0.1$ cm

3.3.2.5 Simple Infiltration Models

Different models for the calculation of infiltration rate and losses from rainfall are available in the literature. The simplest infiltration model is referred to as the Φ-index method. In this method, as shown in Figure 3.9, the infiltration capacity that is equal to the index Φ has been considered constant. Consequently, initial rates are underestimated and final rates are overstated if an entire storm sequence with little antecedent moisture is considered. The best application is to large storms on wet soils or storms where infiltration rates may be assumed to be relatively uniform.

In this method, the total volume of the storm period loss is estimated and distributed uniformly across the storm pattern. This index is estimated using measured rainfall–runoff data. The Φ-index model is a crude approximation to represent the losses due to infiltration. To determine the Φ index for a given storm, the amount of observed runoff is determined from the hydrograph, and the difference between this quantity and the total gauged precipitation is then calculated. The volume of loss (including the effects of interception, depression storage, and infiltration) is uniformly distributed across the storm pattern, as shown in Figure 3.9.

Use of the Φ index for determining the amount of direct runoff from a given storm pattern is essentially the reverse of this procedure. Unfortunately, the Φ index determined from a single storm is not generally applicable to other storms, and unless it is correlated with basin parameters other than runoff, it is of little value. Where measured rainfall–runoff data are not available, the ultimate capacity of Horton's equation, f_f, might be considered as the Φ index.

Philip (1957) has suggested a two-parameter equation for the estimation of infiltration rate:

$$f_p = (0.5)(k)t^{-.05} + C, \tag{3.23}$$

where t is measured from the ponding time. Approximate expressions are proposed for the estimation of model parameters including k and C based on the soil characteristics (Youngs, 1968), but these parameters are usually determined during the calibration of the rainfall–runoff model. It should be noted that the Phillip equation is theoretical and is developed using an infinite series solution to Richards' equation for ideal conditions.

In the Kostiakov equation (Singh, 1992), which is similar to the Phillip equation, the infiltration rate f_p is related to time t:

$$f_p = (\mu)(\beta)t^{\beta-1}, \tag{3.24}$$

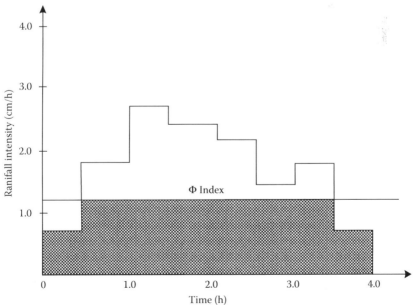

FIGURE 3.9 Representation of the Φ-index method.

where μ and β are calibration parameters where $\mu > 0$ and $0 < \beta < 1$. These parameters are determined based on rainfall and runoff observed data. This equation is often applied in irrigation studies.

3.3.3 Depression Storage

Depression storage is defined as the amount of total precipitation detained in and evaporated from depressions on both the impervious and pervious land surface. Depression storage is the water that does not run off or infiltrate. Surface type and slope, and the factors influencing evaporation, affect depression storage as well as soil moisture condition before the occurrence of rainfall (Figure 3.10). Because of its small magnitude, depression storage is not likely to be important in urban stormwater investigations.

The stored water in the depressions of impervious surfaces finally evaporates back to the atmosphere. After saturation of the soil surface and considerable reduction in infiltration rate, the small puddles and depressions in the pervious area are filled with rainwater. By percolating the water saturating the surface soil, deeper into the soil over time, some of the stored water infiltrates into the soil but most of it evaporates back to the atmosphere.

As modeling of depression storage is difficult and involves a small part of abstractions in comparison with the other abstractions, particularly infiltration, its accurate calculation is not necessary. In practice, various approximations of depression storage are used. For instance, the ASCE (1972) suggests depression storage losses of 1.6 mm and 6.25 cm for impervious and pervious areas, respectively. The typical values presented in Table 3.11 can be used for depression storage estimation.

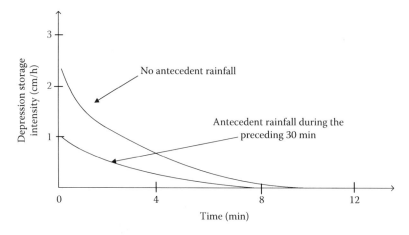

FIGURE 3.10 Depression storage intensity versus time for an impervious area.

TABLE 3.11
Empirical Estimates of Depression Storage (for Storms)

Soil Type	Depression Storage (mm)
Sand	5.1
Loam	3.8
Clay	2.5
Impervious areas	1.6
Pervious urban	6.25

Urban Water Hydrology

Linsley (1986) suggested the following equation for estimating the volume of depression storage as a function of the rainfall excess:

$$V_d = S_d(1 - e^{-K_d P_d}), \qquad (3.25)$$

where V_d is the volume of water in depression storage at any time, S_d is the depression storage capacity, K_d is a constant that for small P_e is equal to $1/S_d$, and P_d is the rainfall excess (rainfall reaching the ground minus infiltration).

3.3.4 Combined Loss Models

There are some models that estimate the total amount of rainfall loss due to interception, infiltration, and surface depression storage concurrently. Some of these models are discussed in this section.

3.3.4.1 Soil Conservation Service Method

The Soil Conservation Service (1972) developed a method for computing abstractions from storm rainfall and calculating excess rainfall (or runoff). All of the abstractions including interception, depression storage, evaporation, and infiltration are taken into account in the loss calculations of the SCS method.

3.3.4.1.1 Hydrological Soil Groups

Soil permeability highly affects the infiltration. For considering this fact in the estimation of infiltration, the SCS (1986) has categorized soils into four groups (A, B, C, and D) according to their minimum infiltration rate regarding soil textures. Soils in group A have the highest infiltration capacity and soils in group D have the lowest capacity.

- Group A includes sand, loamy sand, and sandy loam.
- Group B includes silt loam and loam.
- Group C includes sandy clay loam.
- Group D includes clay loam, silty clay loam, sandy clay, silty clay, and clay.

3.3.4.1.2 Runoff Curve Number

This is a basin parameter that varies from 0 to 100. The curve number (CN) associated with each group of soils (Table 3.12) is determined based on the cover (agricultural versus forest), management practice (tillage practice and mulching), hydrological condition (degree of grazing or percentage of area with good cover characteristics), the percentage of imperviousness, and the initial moisture condition of the soil. The CNs reported in Table 3.12 are for average moisture conditions (AMC II) before the beginning of a rainfall. For relatively wet (AMC III) or dry (AMC I) watershed soil moisture conditions, CNs should be modified due to available soil moisture conditions according to Table 3.13.

If several subareas with different CNs are included in the considered drainage area, a weighted average (based on area) or composite CN should be estimated for the entire area. In urban areas, the amount of CN depends on the percentage of impervious areas and the direct connection of the impervious surfaces to the drainage system. If the runoff from an impervious area flows directly into a drainage system or its runoff flows over a pervious area in concentrated form and then reaches the drainage system, it is referred to as a directly connected impervious area. But the runoff from an impervious area that spreads over a pervious area is called unconnected.

The CNs presented in Table 3.12 for urban land uses including commercial, industrial, and residential districts are composite values estimated based on the average percentage of imperviousness (second column) in these districts. It is supposed that all the impervious areas are directly connected

TABLE 3.12
Runoff CNs for Urban Land Uses

Cover Description		CNs for Hydrological Soil Group			
Cover Type and Hydrological Condition	Average Percent of Impervious Area	A	B	C	D
Fully Developed Urban Areas (Vegetation Established)					
Open space (lawns, parks, golf courses, cemeteries, etc.)					
Poor condition (grass cover <50%)		68	79	86	89
Fair condition (grass cover 50–75%)		49	69	79	84
Good condition (grass cover >75%)		39	61	74	80
Impervious Areas					
Paved parking lots, roofs, driveways, and so on (excluding right-of-way)		98	98	98	98
Streets and Roads					
Paved, curbs, and storm sewers (excluding right-of-way)		98	98	98	98
Paved, open ditches (including right-of-way)		83	89	92	93
Gravel (including right-of-way)		76	85	89	91
Dirt (including right-of-way)		72	82	87	89
Western Desert Urban Areas					
Natural desert landscaping (previous areas only)		63	77	85	88
Artificial desert landscaping (impervious weed barrier, desert shrub with 2.5–5 cm, sand or gravel mulch, and basin borders)		96	96	96	96
Urban Districts					
Commercial and business	85	89	92	94	95
Industrial	72	81	88	91	93
Residential Districts by Average Lot Size					
500 m^2 or less (town houses)	65	77	85	90	92
1000 m^2	38	61	75	83	87
1300 m^2	30	57	72	81	86
2000 m^2	25	54	70	80	85
4000 m^2	20	51	68	79	84
8000 m^2	12	45	65	77	82
Developing Urban Areas					
Newly graded areas (pervious areas only, no vegetation)		77	86	91	94

Source: Soil Conservation Service (SCS). 1986. *Urban Hydrology for Small Watersheds, Technical Release 55*, 2nd Edition. U.S. Department of Agriculture, Springfield, Virginia, June. NTIS PB87-101580 (microcomputer version 1.11. NTIS PB87-101598).

and their CN is equal to 98. The pervious surfaces are considered as open spaces in good hydrological condition.

For urban watersheds, if the percentage of imperviousness or the pervious area condition is different from what is assumed in Table 3.12, Figure 3.11 could be used to determine the CN. This table is applicable for urban watersheds where all of their impervious areas are directly connected or their

TABLE 3.13
Modification of CN for Different Soil Moisture Conditions

AMC (II) (Normal)	AMC (I) (Dry)	AMC (III) (Wet)
100	100	100
95	87	98
90	78	96
85	70	94
80	63	91
75	57	88
70	51	85
65	45	82
60	40	76
55	35	74
50	31	70
45	26	65
40	22	60
35	18	55
30	15	50
25	12	43
20	9	37
15	6	30
10	4	22
5	2	13

total impervious areas are more than 30%. For other urban watersheds, Figure 3.12 is used for the estimation of CN. For using Figure 3.12, in the right half of the figure, the percentage of total impervious areas and the ratio of total unconnected impervious area to total impervious areas are entered. Then it is moved left to the appropriate CN of the pervious area and is moved down to read the composite CN of the watershed.

Caution should be taken when applying Figure 3.12 because of the assumptions used in its development. In many instances, conveyance of flow from unconnected impervious areas may not

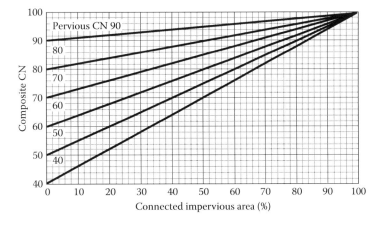

FIGURE 3.11 Composite CN with connected impervious area. (From Soil Conservation Service [SCS]. 1986. *Urban Hydrology for Small Watersheds, Technical Release 55*, 2nd Edition. U.S. Department of Agriculture, Springfield, Virginia, June. NTIS PB87-101580 [microcomputer version 1.11. NTIS PB87-101598].)

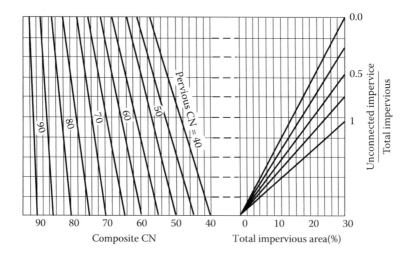

FIGURE 3.12 Composite CN with unconnected impervious area and total impervious area less than 30%. (From Soil Conservation Service [SCS]. 1986. *Urban Hydrology for Small Watersheds, Technical Release 55*, 2nd Edition. U.S. Department of Agriculture, Springfield, Virginia, June. NTIS PB87-101580 [microcomputer version 1.11. NTIS PB87-101598].)

exist or may be very direct. For example, portions of a rooftop may directly drain to a backyard that does not drain easily into the street gutter. However, the drainage path from the downspouts draining the front portion of the rooftop may be rather short, providing little opportunity for infiltration. Certainly, local knowledge of drainage design is needed to judge to what degree the unconnected impervious area acts as if it were hydraulically connected.

3.3.4.1.3 Runoff Equation

For the storm as a whole, the depth of excess precipitation or direct runoff P_d is always less than or equal to the depth or precipitation P; likewise, after runoff begins, the additional depth of water retained in the watershed, F_a, is less than or equal to some potential maximum retention S (Figure 3.13). There is some amount of rainfall I_a (initial abstraction before ponding) for which no runoff will occur, so the potential runoff is $P - I_a$. The hypothesis of the SCS method is the ratios of

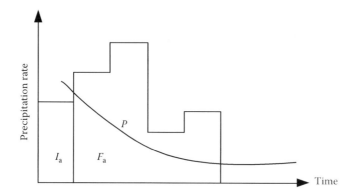

FIGURE 3.13 Variables in the SCS method of rainfall abstractions. (Adapted from Chow, V. T., Maidment, D. R., and Mays, L. W. 1988. *Applied Hydrology*. McGraw-Hill, New York.)

the two actual quantities to the two potential quantities are equal:

$$\frac{F_a}{S} = \frac{P_d}{P - I_a}. \tag{3.26}$$

From the continuity principle,

$$P = P_d + I_a + F_a. \tag{3.27}$$

Combining Equations 3.26 and 3.27 to solve for P_d gives the basic equation for computing the depth of excess rainfall or direct rainfall or direct runoff from a storm by the SCS method:

$$R = \frac{(P - I_a)^2}{(P - I_a) + S}. \tag{3.28}$$

The initial abstraction includes interception, depression storage, evaporation, and infiltration before forming runoff. Assuming that $I_a = 0.2S$ and substituting it into Equation 3.28, we obtain

$$R = \frac{(P - 0.2S)^2}{P + 0.8S}. \tag{3.29}$$

Because of the assumption made in developing Equation 3.29, it is only valid if $P > 0.2S$, otherwise $R = 0.1I_a$. S (mm) is estimated using Equation 3.30 which is an empirical relationship.

$$S = \frac{25{,}400 - 254\text{CN}}{\text{CN}}. \tag{3.30}$$

For given CN and P (mm) values, first S is calculated using Equation 3.30 and then runoff (mm) is determined using Equation 3.29. A graphical solution to Equation 3.29 is presented in Figure 3.14. If the total rainfall is given, the CN method can be used to determine the total excess rainfall. If the hyetograph of a rainfall is given, it can also be used to determine the rates of excess rainfall.

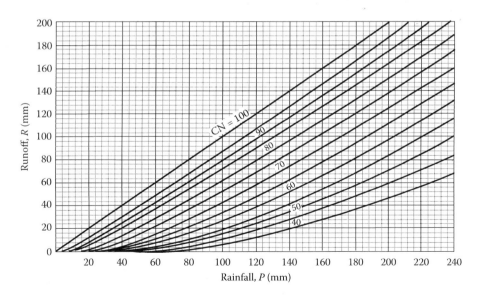

FIGURE 3.14 Graphical solution of Equation 3.29. (From Soil Conservation Service [SCS]. 1986. *Urban Hydrology for Small Watersheds, Technical Release 55*, 2nd Edition. U.S. Department of Agriculture, Springfield, Virginia, June. NTIS PB87-101580 [microcomputer version 1.11. NTIS PB87-101598].)

EXAMPLE 3.8

Determine the runoff volume of an urban watershed with an area of 1000 km^2 that includes 380 km^2 of open area with 65% grass cover, 200 km^2 of commercial district with 85% imperviousness, 150 km^2 of an industrial districts that is 72% impervious, and 270 km^2 residential region with an average lot size of 1000 m^2. The watershed is subjected to an 18-h rainfall with a total depth of 5.3 cm. Assume that the soil of the watershed belongs to group A based on its hydrological conditions.

Solution: From Table 3.12, CN = 49 for the open area, CN = 89 for the commercial district, CN = 81 for the industrial district, and CN = 61 for the residential region. The area-weighted averages of CNs for different districts are used as the CN representing the whole watershed:

$$CN = \frac{(380\,km^2)(49) + (200\,km^2)(89) + (150\,km^2)(81) + (270\,km^2)(61)}{1000\,km^2} = 65.0.$$

Knowing CN = 65.0 using Equations 3.30, S is estimated as 136.8 mm and R is estimated as 4.05 mm. The same value is obtained for the runoff by using Figure 3.14.

EXAMPLE 3.9

Determine the resulting runoff from the rainfall tabulated in column 3 of Table 3.14 over an urban residential region. The average size of lots in the considered area is 1500 m^2, and the soil of the district belongs to hydrological group D.

Solution: The CN of this urban watershed is estimated to be 86 from Table 3.12. Then S is determined using Equation 3.30 to be equal to 41.34 mm. The excess rainfall calculations of this example are summarized in Table 3.14. The considered time interval for calculation is 1 h as the rainfall rate is reported. The incremental rainfall depth accumulating over a time interval that is $\Delta P = i\Delta t$ is presented in column 4.

In columns 5 and 6, P_1 and P_2 ($P_1 + \Delta P$), which are the cumulative rainfall at t_1 and t_2, are presented. The R_1 and R_2 values given in columns 7 and 8 represent the cumulative runoff at time t_1 and t_2 and they are calculated using Equation 3.29 (or Figure 3.14) for the given CN and P. The rate of excess rainfall is equal to $\Delta R/\Delta t$, which is equal to ΔR as Δt is 1 h here. The total depth of runoff is equal to 0.55 cm.

TABLE 3.14
Summary of Calculations of Example 3.9

t_1 (h) (1)	t_2 (h) (2)	i (cm/h) (3)	ΔP (cm) (4)	P_1 (cm) (5)	P_2 (cm) (6)	R_1 (cm) (7)	R_2 (cm) (8)	ΔR (cm) (9)	i_e (cm/h) (10)
0	1	0.05	0.05	0	0.05	0.00	0.00	0.00	0.00
1	2	0.1	0.1	0.05	0.15	0.00	0.08	0.08	0.08
2	3	0.15	0.15	0.15	0.3	0.08	0.08	0.00	0.00
3	4	0.2	0.2	0.3	0.5	0.08	0.08	0.00	0.00
4	5	0.25	0.25	0.5	0.75	0.08	0.08	0.00	0.00
5	6	0.25	0.25	0.75	1	0.08	0.23	0.15	0.15
6	7	0.3	0.3	1	1.3	0.23	0.37	0.14	0.14
7	8	0.2	0.2	1.3	1.5	0.37	0.47	0.10	0.10
8	9	0.1	0.1	1.5	1.6	0.47	0.52	0.05	0.05
9	10	0.05	0.05	1.6	1.65	0.52	0.55	0.03	0.03

3.3.4.2 Other Combined Loss Models

An optional simple rainfall loss model is included in the U.S. Army Corps of Engineers products such as HEC-1 (1990) and HEC-HMS (1998). This rainfall loss model only considers a specified initial loss and a constant loss rate. Runoff forms after satisfaction of the initial loss and after that a constant rate is considered for rainfall loss. The constant loss rates could be determined based on the hydrological soil group (Table 3.15). This model has limitations similar to the Φ-index method.

3.4 RAINFALL MEASUREMENT

High-frequency information in the rainfall signal is transformed into high-frequency pulses in the runoff hydrograph since the highly impervious urban drainage area does not considerably damp such fluctuations, because a unique aspect of urban hydrology is the fast response of an urban drainage area to the rainfall input. Tipping bucket rain gauges are appropriate for representation of high frequencies, as opposed to weighing bucket gauges that are commonly used. Tipping bucket gauges could easily be set in a network. These gauges transmit the electrical pulses through a telephone or other communication devices or record them on a data logger at the site. Hyetographs in 5 min or less time intervals are necessary for modeling an urban watershed, although this may be rather relaxed for larger basins.

The outfall hydrograph is sensitive to the storm direction (James and Shtifter, 1981) except for very small drainage areas (such that the entire basin is covered completely by a storm). It is necessary to have more than one rain gauge to define the longitudinal movement (up or down the basin) in the simulation of the effects of a moving (or "kinematic") storm. The runoff hydrograph is more affected by longitudinal movement rather than movement of the storm across the flow direction. These effects on the urban flow rate are shown in Figure 3.15.

3.4.1 INTENSITY–DURATION–FREQUENCY CURVES: ADVANTAGES AND DISADVANTAGES

The total storm rainfall depth at a point, for a given rainfall duration and average return period, is a function of the local climate. Rainfall depths can be further processed and converted into rainfall intensities (intensity = depth/duration), which are then presented in intensity–duration–frequency (IDF) curves. Such curves are particularly useful in stormwater drainage design because many computational procedures require rainfall inputs in the form of average rainfall intensity. The three variables, frequency, intensity, and duration, are all related to each other. The data are normally presented as curves displaying two of the variables, such as intensity and duration, for a range of frequencies. These data are then used as the input in most stormwater design processes. IDF curves, which are the graphical symbols of the probability, summarize frequencies of rainfall depths or average intensities.

TABLE 3.15
Estimation of Constant Rate of Rainfall Loss Due to Hydrological Soil Group

Soil Group	Range of Loss Rate (cm/h)
A	0.75–1.13
B	0.38–0.75
C	0.125–0.375
D	0.0–0.125

Source: Adapted from Rawls, W. J. et al., 1996. Infiltration. *Hydrology Handbook*, Chapter 3, 2nd Edition. ASCE Manuals and Reports on Engineering Practice, No. 28, New York.

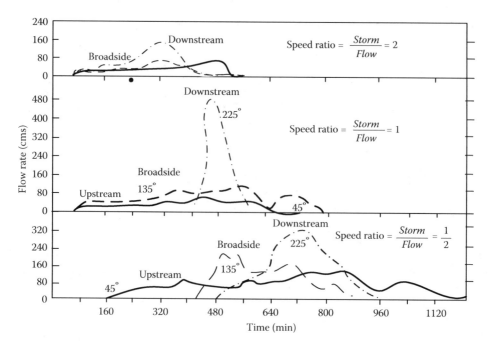

FIGURE 3.15 Effects of a storm moving in different directions on an urban catchment flow rate. (From Surkan, A. J., 1974. *Water Resources Research* 10(6): 1149–1160. Copyright 1974 American Geophysical Union. Reproduced by permission of American Geophysical Union.)

An important point that should be considered in the application of IDF curves is that the intensities obtained from this curves are averaged over the specified duration and do not show the actual precedent of rainfall. The developed contours for different return periods are the smoothed lines fitted to the results of different storms so that the IDF curves are completely hypothetical. Furthermore, the obtained duration is not the actual length of a storm; rather, it is only a 30-min period, say, within a longer storm of any duration, during which the average intensity was of the specified value. A detailed description of the method of developing IDF curves and the risks of inappropriate usages is given by McPherson (1978). An IDF curve for an urban area is presented in Figure 3.16.

The main disadvantage of the application of IDF curves in practice is that they do not assign a unique return period to a storm with specific depth or average intensity or vice versa. Many combinations of average intensity and return period can result in a given depth, and as the duration decreases, the return period for a given depth increases. Since the durations on IDF curves are not the actual representative of the storm durations, the curves can only be used to make an approximation of, say, the 30-year rainfall event on the basis of total depth. Instead of implication of IDF curves, a frequency analysis of storm depths on the time series of independent storm events could be performed, which is more time-consuming.

The IDF curves are used in conjunction with the rational method (which is introduced in the next section) to calculate the runoff from a particular watershed. Although the IDF data are commonly used in this case, it should be reiterated that

1. Since in the development of IDF curves the intensities are averaged over the indicated duration, these curves do not represent actual time histories of real storms.
2. A single curve is developed by using the gathered data from several different storms.
3. The duration is not representative of the total duration of an actual storm and most likely refers to a short period of a long storm.
4. The IDF curves cannot be used to estimate a storm event volume because the duration should be assumed at first.

Urban Water Hydrology

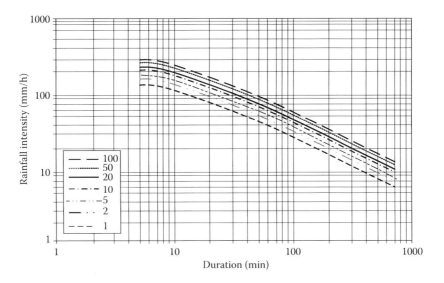

FIGURE 3.16 IDF curves for Kuala Lumpur. (Available at www2.water.gov.my/division/river/stormwater/Chapter_13.htm.)

3.5 ESTIMATION OF URBAN RUNOFF VOLUME

The most important parameters in an urban hydrological analysis are peak flow, runoff volume, or the complete runoff hydrograph and also the hydraulic grade line or flooding depths. Peaks and volumes are determined during the estimation of the resulted hydrographs.

3.5.1 Rational Method

The rational method is an empirical relation between rainfall intensity and peak flow, which is widely accepted by hydraulic engineers. It is used to predict the peak runoff from a storm event. Despite being one of the oldest methods, it is still commonly used especially in the design of storm sewers, because of its simplicity and popularity, although it contains some limitations that are not often treated. The peak runoff is calculated using the following formula:

$$Q_p = k \cdot C \cdot i \cdot A, \tag{3.31}$$

where Q_p is the peak flow (m³/s), C is the runoff coefficient, i is the rainfall intensity (mm/h), A is the drainage area (ha), and k is the conversion factor equal to 0.00278 for the conversion of ha mm/h to m³/s.

The rational method makes the following assumptions:

- Precipitation is uniform over the entire basin.
- Precipitation does not vary with time or space.
- Storm duration is equal to the time of concentration.
- A design storm of a specified frequency produces a design flood of the same frequency.
- The basin area increases roughly in proportion to increases in length.
- The time of concentration is relatively short and independent of storm intensity.
- The runoff coefficient does not vary with storm intensity or antecedent soil moisture.
- Runoff is dominated by overland flow.
- Basin storage effects are negligible.

It is important to note that all of these criteria are seldom met under natural conditions. In particular, the assumption of a constant, uniform rainfall intensity is the least accurate. So, in watersheds with an area larger than about 2.5 km², the rational method should not be applied to the whole of the watershed. In these cases, the watershed is divided into some subdrainage areas considering the influence of routing through drainage channels. The rational method results become more conservative (i.e., the peak flows are overestimated) as the area becomes larger, because actual rainfall is not homogeneous in space and time.

The runoff coefficient C, which is usually given as a function of land use, considers all drainage area losses (Table 3.16). When the watershed is composed of multiple land uses, an area-weighted runoff coefficient is usually used in Equation 3.31.

A better estimate of rainfall coefficient is obtained from site measurements. There is often a considerable variation in C values in the analysis of actual storm data. But since a rational method gives a coarse estimate of rainfall losses, it could be useful. One of the major problems in the

TABLE 3.16
Runoff Coefficients for the Rational Formula versus Hydrologic Soil Group (A, B, C, and D) and Slope Range

Land Use	A			B			C			D		
	0–2%	2–6%	6%+	0–2%	2–6%	6%+	0–2%	2–6%	6%+	0–2%	2–6%	6%+
Cultivated land	0.08[a]	0.13	0.16	0.11	0.15	0.21	0.14	0.19	0.26	0.18	0.23	0.31
	0.14[b]	0.18	0.22	0.16	0.21	0.28	0.2	0.25	0.34	0.24	0.29	0.41
Pasture	0.12	0.2	0.3	0.18	0.28	0.37	0.24	0.34	0.44	0.3	0.4	0.5
	0.15	0.25	0.37	0.23	0.34	0.45	0.3	0.42	0.52	0.37	0.5	0.62
Meadow	0.1	0.16	0.25	0.14	0.22	0.3	0.2	0.28	0.36	0.24	0.3	0.4
	0.14	0.22	0.3	0.2	0.28	0.37	0.26	0.35	0.44	0.3	0.4	0.5
Forest	0.05	0.08	0.11	0.08	0.11	0.14	0.1	0.13	0.16	0.12	0.16	0.2
	0.08	0.11	0.14	0.1	0.14	0.18	0.12	0.16	0.2	0.15	0.2	0.25
Residential lot size 500 m²	0.25	0.28	0.31	0.27	0.3	0.35	0.3	0.33	0.38	0.33	0.36	0.42
	0.33	0.37	0.4	0.35	0.39	0.44	0.38	0.42	0.49	0.41	0.45	0.54
Residential lot size 1000 m²	0.22	0.26	0.29	0.24	0.29	0.33	0.27	0.31	0.36	0.3	0.34	0.4
	0.3	0.34	0.37	0.33	0.37	0.42	0.36	0.4	0.47	0.38	0.42	0.52
Residential lot size 1300 m²	0.19	0.23	0.26	0.22	0.26	0.3	0.25	0.29	0.34	0.28	0.32	0.39
	0.28	0.32	0.35	0.3	0.35	0.39	0.33	0.38	0.45	0.36	0.4	0.5
Residential lot size 2000 m²	0.16	0.2	0.24	0.19	0.23	0.28	0.22	0.27	0.32	0.26	0.3	0.37
	0.25	0.29	0.32	0.28	0.32	0.36	0.31	0.35	0.42	0.34	0.38	0.48
Residential lot size 4000 m²	0.14	0.19	0.22	0.17	0.21	0.26	0.2	0.25	0.31	0.24	0.29	0.35
	0.22	0.26	0.29	0.24	0.28	0.34	0.28	0.32	0.4	0.31	0.35	0.46
Industrial	0.67	0.68	0.68	0.68	0.68	0.69	0.68	0.69	0.69	0.69	0.69	0.7
	0.85	0.85	0.86	0.85	0.86	0.86	0.86	0.86	0.87	0.86	0.86	0.88
Commercial	0.71	0.71	0.72	0.71	0.72	0.72	0.72	0.72	0.72	0.72	0.72	0.72
	0.88	0.88	0.89	0.89	0.89	0.89	0.89	0.89	0.9	0.89	0.89	0.9
Streets	0.7	0.71	0.72	0.71	0.72	0.74	0.72	0.73	0.76	0.73	0.75	0.78
	0.76	0.77	0.79	0.8	0.82	0.84	0.84	0.85	0.89	0.89	0.91	0.95
Open space	0.05	0.1	0.14	0.08	0.13	0.19	0.12	0.17	0.24	0.16	0.21	0.28
	0.11	0.16	0.2	0.14	0.19	0.26	0.18	0.23	0.32	0.22	0.27	0.39
Parking	0.85	0.86	0.87	0.85	0.86	0.87	0.85	0.86	0.87	0.85	0.86	0.87
	0.95	0.96	0.97	0.95	0.96	0.97	0.95	0.96	0.97	0.95	0.96	0.97

Source: McCuen, R. 2005. *Hydrologic Analysis and Design*, 3rd Edition. Pearson Prentice Hall, Upper Saddle River, NJ. With permission.

[a] For storm return periods less than 25 years.
[b] For storm return periods more than 25 years.

application of the rational method is considering a constant rate of loss during the rainfall, regardless of the total rainfall volume or initial conditions of the watershed. By an increase of imperviousness of the drainage area, this assumption becomes less important and the estimated runoff becomes closer to the observed value; in other words, Equation 3.31 gives almost exact values of runoff for a completely impervious area such as a rooftop.

The rainfall intensity (i) is determined using an IDF curve attributed to the studied watershed for a specified return period and duration equal to the time of concentration of the watershed. The time of concentration is equal to the watershed equilibrium time, which means that in this time the whole drainage area contributes to flow at the watershed end point. Thus, for times less than the watershed time of concentration, the total drainage area, A, should not be used in the estimation of watershed outflow. But after reaching the watershed equilibrium, a higher rainfall intensity, that occurred earlier, should be used. The appropriate estimation of t_c is the main issue in the application of the rational method. The time of concentration has an inverse relation with the rainfall intensity, so that the kinematic wave equation could be used for its estimation. Because of the relation between t_c and the unknown intensity of the rainfall, an iterative solution, which combines Equation 3.31 with the IDF curves, is utilized. This iterative procedure can be easily followed as it is described in Example 3.10.

The time of concentration is often approximated by constant overland flow inlet times, for more simplicity in the iterative process. These times vary between 5 and 30 min which are commonly used. For highly developed, impervious urban areas with closely spaced stormwater inlets, a value of 5 min is suitable for the time of concentration that increases up to 10 or 15 min for less developed urban areas of fairly lesser density. Values between 20 and 30 min are appropriate for residential regions with broadly spaced inlets. Longer values are appropriate for flatter slopes and larger areas.

The effect of channels in large drainage areas should also be considered in t_c estimation. For this purpose, the wave travel times in channels and conduits, t_r (based on the wave speed, using an estimate of the channel size, depth, and velocity), along the flow pathways should be added to overland flow inlet times to yield an overall time of concentration. Since sometimes the objective is the design of drainage channels, an iterative procedure is used to find the corresponding values of t_c and t_r.

EXAMPLE 3.10

Determine the 20-year peak flow at a stormwater inlet for a 100-ha urban watershed. Assume that the inlet time equals 30 min, the watershed soil belongs to group B, and the land-use pattern and slope are as given in Table 3.17. Use Figure 3.16 as the representative IDF curve of the watershed.

Solution: At first the runoff coefficients of different land uses are determined using Table 3.16, as shown below:

Open space: 0.08
Streets and driveways: 0.71

TABLE 3.17
Characteristics of the Watershed Considered in Example 3.10

Land Use	Area (ha)	Slope (%)
Open space	12	1.5
Streets and driveways	28	1
Residential lot size 1000 m²	47	2.1
Pasture	13	3

continued

Residential lot size 1000 m^2: 0.29
Pasture: 0.28.

The area-weighted runoff coefficient of the whole watershed is estimated as

$$\bar{C} = \frac{\sum A_i C_i}{\sum A_i} = \frac{12 \times 0.08 + 28 \times 0.71 + 47 \times 0.29 + 13 \times 0.28}{100} = 0.38.$$

From Figure 3.16, for a 30-min storm with a 20-year return period, the average intensity is estimated to be equal to 150 mm/h. The peak flow of this storm is calculated from Equation 3.31 as

$$Q_p = 0.00278 \times 0.38 \times 150 \times 100 = 15.85 \, \text{m}^3/\text{s}.$$

3.5.2 Coefficient and Regression Methods

Regression techniques are often used in combination with other methods such as synthetic UHs for the estimation of watershed runoff. For this purpose, some regression models are developed to relate the parameters of the hydrograph to physical parameters of the drainage area. The application of regression techniques for the estimation of peak flows is usually limited to large nonurban drainage areas. This technique has also been used by U.S. Geological Survey (USGS) for relating the frequency analysis results to drainage area characteristics. In the simplest application of these models, the measured values of desired independent variables such as rainfall and dependent variables such as runoff are related to each other through a simple regression. The usually fitted linear equation to the observed rainfall–runoff values has the following form:

$$R = C_m(P - S), \tag{3.32}$$

where R is the runoff depth, P is the rainfall depth, C_m is the slope of the fitted line (approximate runoff coefficient), and S is the depression storage (depth).

It should be noted that above-mentioned depths are forecasted, not flows. Depth and volume are often used interchangeably because they are equivalent (when multiplied by the drainage area). In this model, it is assumed that until the depression storage is filled by rainfall, the runoff depth will be zero. When S is negative, the equation can be rewritten to specify a positive interception on the runoff axis through adding a constant positive value to the right side of the equation. This situation may happen when there is a base flow that contributes to runoff even if the rainfall depth is zero. Slope C_m is approximately equivalent to the runoff coefficient, except that losses due to depression storage are not considered. Equation 3.32 becomes more accurate as the duration of analysis increases to month and year from individual storm events. It is suggested that this equation should not be used in situations with considerable carryover in the drainage area storage (Diskin, 1970).

3.5.3 USGS Urban Peak Discharge Formulae

A series of single-return-period regression equations has been developed by Sauer et al. (1983) using 269 urban watersheds data (Table 3.18). At least 15% of the considered watershed area is covered by residential, commercial, or industrial districts. The maximum change in the development of the gauged sites is 50% during the data recording period. The independent variables considered for the estimation of urban peak discharge for different return periods using regression equations are as follows:

A is the basin area (km^2).
SL is the channel slope (cm/m).

TABLE 3.18
Single–Return-Period Regression Equations for Estimation of Urban Peak Discharge

Regression Equations	log R^2	Standard Error of Regression Average Units	Standard Error of Regression Percent
$UQ2 = 0.47A^{0.41}SL^{0.17}(0.394RI2 + 3)^{2.04}(ST + 8)^{-0.65}$ $(13 - BDF)^{-0.32}IA^{0.15}RQ2^{0.47}$	0.93	0.1630	±38
$UQ5 = 0.71A^{0.35}SL^{0.16}(0.394RI2 + 3)^{1.86}(ST + 8)^{-0.59}$ $(13 - BDF)^{-0.31}IA^{0.11}RQ5^{0.54}$	0.93	0.1584	37
$UQ10 = 0.88A^{0.32}SL^{0.15}(0.394RI2 + 3)^{1.75}(ST + 8)^{-0.57}$ $(13 - BDF)^{-0.30}IA^{0.09}RQ10^{0.58}$	0.93	0.1618	38
$UQ25 = 0.90A^{0.31}SL^{0.15}(0.394RI2 + 3)^{1.76}(ST + 8)^{-0.55}$ $(13 - BDF)^{-0.29}IA^{0.07}RQ25^{0.60}$	0.93	0.1705	40
$UQ50 = 0.95A^{0.29}SL^{0.15}(0.394RI2 + 3)^{1.74}(ST + 8)^{-0.53}$ $(13 - BDF)^{-0.28}IA^{0.06}RQ50^{0.62}$	0.92	0.1774	42
$UQ100 = 0.92A^{0.29}SL^{0.15}(0.394RI2 + 3)^{1.76}(ST + 8)^{-0.52}$ $(13 - BDF)^{-0.28}IA^{0.06}RQ100^{0.63}$	0.92	0.1860	44
$UQ500 = 0.84A^{0.29}SL^{0.15}(0.394RI2 + 3)^{1.76}(ST + 8)^{-0.54}$ $(13 - BDF)^{-0.27}IA^{0.05}RQ500^{0.63}$	0.90	0.2071	49
Three-Parameter Equations			
$UQ2 = 45.59A^{0.21}(13 - BDF)^{-0.43}RQ2^{0.73}$	0.91	0.1797	±43
$UQ5 = 56.42A^{0.17}(13 - BDF)^{-0.39}RQ5^{0.78}$	0.92	0.1705	40
$UQ10 = 55.40A^{1.16}(13 - BDF)^{-0.36}RQ10^{0.79}$	0.92	0.1720	41
$UQ25 = 55.34A^{0.15}(13 - BDF)^{-0.34}RQ25^{0.80}$	0.92	0.1802	43
$UQ50 = 55.05A^{0.15}(13 - BDF)^{-0.32}RQ50^{0.81}$	0.91	0.1865	44
$UQ100 = 56.62A^{0.15}(13 - BDF)^{-0.32}RQ100^{0.82}$	0.91	0.1949	46
$UQ500 = 53.89A^{0.16}(13 - BDF)^{-0.30}RQ500^{0.82}$	0.89	0.2170	52

Source: From Sauer, V. B. et al., 1983. Flood characteristics of urban watersheds in the United States. USGS Water Supply Paper 2207, U.S. Government Printing Office, Washington, DC. With permission.

RI2 is the intensity (cm) of a 2-year, 2-h storm event.
ST is the basin storage (%) in lakes and reservoirs.
BDF is the basin development factor.
IA is the imperviousness (%).
RQi is the peak discharge before urbanization (m^3/s) for return period i, which is the same return period for the urban equation.

The dependent variable of these equations is UQi, which is the urban peak discharge (m^3/s) for return period i. The peak discharge before urbanization can be determined from available regional reports for the location where the urban peak discharge estimate is required. The most important index of urbanization is Basin Development Factor (BDF), which is highly significant in the application of regression equations. This index is easily determined from drainage maps and field examinations of the considered watershed and quantifies the efficiency of the drainage system. For the determination of BDF, the basin is subdivided into three parts. Then four aspects of the drainage system performance are evaluated in each part and a code is assigned to them as follows:

1. *Channel improvements*: A code of one (1) is assigned, if channel improvements such as straightening, enlarging, deepening, and clearing are prevalent for the main drainage channel and principal branches that drain directly into the main channel. If one or all of these

improvements are considered for the drainage system, a code of one (1) will be assigned. At least 50% of the main drainage channels and main branches should have improvements over natural conditions, in order for the improvements to be considered prevalent. If there are no channel improvements or they are not widespread, then a code of zero (0) is assigned to the drainage system.

2. *Channel linings*: A code of one (1) is assigned to the drainage system if more than 50% of main channels and branches have been lined with an impervious substance, such as concrete. Otherwise a code of zero (0) is assigned to the drainage system. The high percentage of channel linings also specifies the availability of channel improvements. Therefore, channel lining is an added issue that shows a more highly developed drainage system.

3. *Storm drains or storm sewers*: Enclosed drainage structures (usually pipes) named storm drains are regularly used as the secondary tributaries. Storm drains directly receive the drainage of drainage surfaces and empty into the main channel. The main channels can be open channels, or enclosed as box or pipe culverts. If more than 50% of the secondary tributaries are storm drains, a code of one (1) is assigned, otherwise a code of zero (0) is considered for the drainage system. Note that if more than 50% of the main channels and tributaries are enclosed, a code of one (1) would also be assigned to the channel improvements and channel lining issues.

4. *Curb and gutter streets*: For more than 50% of urbanization (including residential, commercial, and/or industrial regions) and availability of curbs and gutters in more than 50% of the streets and highways in the subarea, a code of one (1) is assigned. Often, collected surface water from curb and gutter streets empties discharges into storm drains.

The above procedures do not provide accurate measurements in the evaluation of a drainage system condition through assigning codes; therefore the condition of the field checking should be also considered for obtaining the best estimate of the drainage system. The BDF is equal to the summation of the assigned codes. As the basin is subdivided into three subareas and four drainage issues are evaluated with a code in each subarea, the maximum value of BDF would be 12 obtained for a completely developed drainage system. Conversely, a BDF of zero (0) would result for a drainage system with no development. It should be noted that a zero BDF does not necessarily mean that the basin is unaffected by urbanization. A BDF of zero (0) could be assigned to a partly urbanized basin that has some impervious areas, and have some improvements to secondary tributaries. The possible ranges of independent variables used in single-return-period urban equations (Table 3.18) are presented in Table 3.19.

TABLE 3.19
Minimum and Maximum Values of Independent Variables Used to Calibrate the Equations of Table 3.18

Variable	Minimum	Maximum	Units
A	0.52	256	km^2
SL	0.06	1.32	cm/m
R12	0.5	7	cm
ST	0	11	percent
BDF	0	12	—
IA	3.0	50	percent

Source: Sauer, R. T. et al., 1981. *Biochemistry*, 20: 35–91. With permission.

Urban Water Hydrology

EXAMPLE 3.11

An urban watershed with an area of 500 km² is subject to a development plan during a 20-year horizon. In the current situation, the percentage of imperviousness and watershed storage is about 15 and 3, respectively, and the BDF is estimated to be equal to 4. Evaluate the effect of the development plan on the peak discharge of this watershed if the percentage of imperviousness and watershed storage increases up to 28 and 5 after 20 years and the BDF becomes equal to 8. Assume that the average slope of the watershed is 6.5 cm/m and the intensity of a 2-year, 2-h rainfall is 6.5 cm. The peak discharges from different return periods before urbanization are given in Table 3.20.

Solution: Based on the given watershed characteristics and the previous urbanization peak discharges, the parameters of urban peak discharge equations are estimated for the current situation and also after the 20-year horizon. Then the peak discharges for different return periods are estimated using the equations of Table 3.18. The results are tabulated in Table 3.20. The urban peaks are considerably growing by an increase of urbanization especially for the smaller return periods.

TABLE 3.20
Estimated Peak Discharges for the Watershed of Example 3.11 in Different Urbanization Conditions

Return Period (yr)	Rural Peaks (m³/s)	Urban Peaks (m³/s)	
		Current Situation	After the 20-year Horizon
2	25.12	194.6	231.38
5	42.53	232.77	271.07
10	56.89	273.49	313.74
25	83.37	370.22	418.38
50	109.23	450.10	504.19
100	151.95	595.55	668.24
500	300.06	791.65	874.65

3.6 UNIT HYDROGRAPHS

A flood hydrograph for a basin can be simulated using a UH, defined as the direct runoff from a storm that produces one unit of rainfall excess. Using the UH method, the direct runoff hydrograph at the watershed outlet for given excess rainfall resulting from a particular storm event is calculated. As the spatial variations of physical characteristics of the watershed are not directly entered in the runoff calculations of UH, it is classified as a lumped method. The watershed characteristics through a mathematical procedure called a UH are combined. The UH method is commonly used in practice because of its simplicity and usefulness.

The principal concept underlying the application of a UH is that each basin has one UH that does not change (in terms of its shape) unless the basin characteristics change. Because the physical characteristics (such as drainage area, slope, etc.) of a basin typically remain unchanged, changes in the UH usually reflect changes in land-use patterns or urbanization. Given that the UH does not change in shape and represents stream flow response to 1 unit of runoff (excess rainfall) within a basin, flood hydrographs for actual storms can be simulated by multiplying the discharge ordinates from a UH by the excess rainfall computed from the observed rainfall record.

3.6.1 UH Development

A UH is a theoretical direct runoff hydrograph resulting from excess rainfall equal to unit depth with constant intensity for a specific watershed. A unit hydrograph is abbreviated as UH and a subscript is used to indicate the duration of the excess rainfall. For a gauged watershed, the UH is developed by analyzing the simultaneous records of rainfall and runoff. In ungauged watersheds the synthetic UH methods are used to develop a UH. In the development of a UH, different physical characteristics of the watershed are considered.

The derivation of a dimensionless hydrograph for all streams in the study area begins with the development of a station-average UH for each study site. The criteria for selecting a storm to develop a station-average UH for a site consisted of, to the extent possible, (1) concentrated storm rainfall that was fairly uniform throughout the basin during one period and (2) an observed hydrograph that had one peak (Figure 3.17). Storms in which the rainfall occurred over a long period punctuated by shorter periods of no rainfall and that resulted in complex, multipeak hydrographs should generally be avoided. For each observed hydrograph, the base flow should be removed by linear interpolation between the start of the rise and the end of the recession, where the rate of decreasing discharge generally becomes constant (indicating the end of the direct runoff segment in the observed hydrograph, Figure 3.17).

3.6.2 Espey 10-min UH

This empirical method is proposed by Espey et al. (1978) to obtain a synthetic 10-min hydrograph for urban watersheds. This model is developed based on the analysis of the runoff data gathered from 41 urban watersheds located in eight different states in the United States. As illustrated in Figure 3.18, nine parameters should be estimated for describing the Espey 10-min UH. The considered parameters are defined as follows:

Q_p is the peak discharge of the UH (m³/s).
T_p is the time of occurrence of the peak discharge (min).
T_B is the base time of the UH (min).
W_{50} is the width of the UH at $0.50Q_p$ (min).
W_{75} is the width of the UH at $0.75Q_p$ (min).
t_A is the time at which the discharge is $0.50Q_p$ on the rising limb (min).
t_B is the time at which the discharge is $0.75Q_p$ on the rising limb (min).

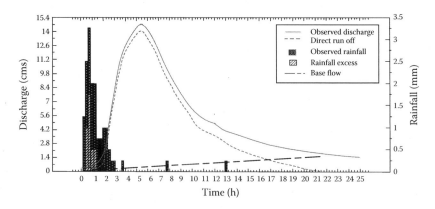

FIGURE 3.17 Observed and excess rainfall at rain gauge CRN01 and the resulting discharge and direct runoff at Mallard Creek below Stony Creek near Harrisburg, North Carolina, for the storm of December 12, 1996.

Urban Water Hydrology

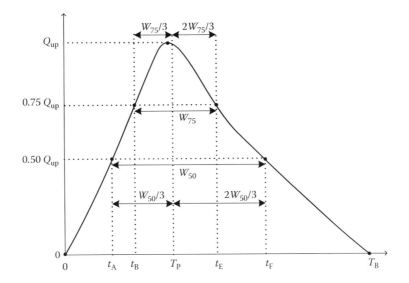

FIGURE 3.18 Elements of the Espey UH. (Adapted from Espey, W. H., Jr., Altman, D. G., and Graves, C. B., Jr. 1978. Monographs for ten-minute unit hydrographs for small urban watersheds. Technical Memorandum No. 32 [NTIS PB-282158], ASCE Urban Water Resources Research Program, ASCE, New York [also Addendum 3 in Urban Runoff Control Planning. EPA-600/9–78-035. EPA, Washington, DC], December.)

t_E is the time at which the discharge is $0.75Q_p$ on the falling limb (min).
t_F is the time at which the discharge is $0.50Q_p$ on the falling limb (min).

These parameters are estimated using the physical basin characteristics. The basin features useful in developing a UH are listed below:

L is the length of the main water path through the watershed from the upstream of the watershed to the watershed outlet (m).
S is the average slope of the main water path.
H is the difference between the main water path altitude at the outlet and $0.8L$ upstream (m).
A is the watershed drainage area (km^2).
I is the percentage of imperviousness in the watershed (%).
n is the Manning roughness coefficient of the main water path (Table 3.21).
ϕ is the dimensionless conveyance coefficient determined by using Figure 3.19.

The parameters of the Espey 10-min UH are calculated as

$$S = \frac{H}{0.8L}, \qquad (3.33)$$

$$T_p = \frac{4.1 L^{0.23} \phi^{1.57}}{S^{0.25} I^{0.18}}, \qquad (3.34)$$

$$Q_p = \frac{138.7 A^{0.96}}{T_p^{1.07}}, \qquad (3.35)$$

$$T_B = \frac{666.7 A}{Q_p^{0.95}}, \qquad (3.36)$$

TABLE 3.21
Manning Roughness Coefficient for Various Pipes and Open Channels

Conduit Material			Manning Coefficient
Closed conduits	Cast iron pipe		0.13
	Concrete pipe		0.13
	Corrugated metal pipe	Plain	0.024
		Paved invert	0.020
		Fully paved	0.015
	Plastic		0.013
	Vitrified clay		0.013
Open channels	Lined channels	Asphalt	0.015
		Concrete	0.015
		Rubble or rip rap	0.030
		Vegetal	0.040
	Excavated or dredged	Earth, straight and uniform	0.030
		Earth, winding, fairly uniform	0.040
		Unmaintained	0.100
Natural channels (minor streams)	Fairly regular section		0.050
	Irregular section with pools		0.100

$$W_{50} = \frac{105.1 A^{0.93}}{Q_p^{0.92}}, \tag{3.37}$$

$$W_{75} = \frac{45.1 A^{0.79}}{Q_p^{0.78}}, \tag{3.38}$$

$$t_A = T_p - \frac{W_{50}}{3}, \tag{3.39}$$

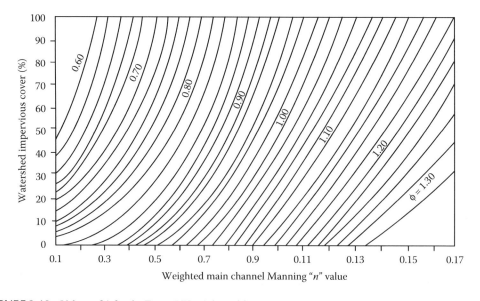

FIGURE 3.19 Values of ϕ for the Espey UH. (Adapted from Espey, W. H., Jr., Altman, D. G., and Graves, C. B., Jr. 1978. Monographs for ten-minute unit hydrographs for small urban watersheds. Technical Memorandum No. 32 [NTIS PB-282158], ASCE Urban Water Resources Research Program, ASCE, New York [also Addendum 3 in Urban Runoff Control Planning. EPA-600/9–78-035. EPA, Washington, DC], December.)

Urban Water Hydrology

$$t_B = T_p - \frac{W_{75}}{3}, \tag{3.40}$$

$$t_E = T_p + \frac{2W_{75}}{3}, \tag{3.41}$$

$$t_F = T_p + \frac{2W_{50}}{3}. \tag{3.42}$$

By the estimation of these parameters, seven points of the UH are determined and a graph constructed that is passing through these points. Finally, it should be checked that the area under the hydrograph is equal to 1 cm of direct runoff.

EXAMPLE 3.12

Calculate the Espey 10-min UH for an urban watershed with the following characteristics:

$$A = 1.13\,\text{km}^2, \quad H = 50\,\text{m}, \quad L = 3250\,\text{m}, \quad I = 51.2\%, \quad S = 0.019, \quad \text{and} \quad n = 0.014.$$

Solution: At first the conveyance coefficient (ϕ) is determined from Figure 3.19 using $n = 0.014$ and $I = 51.2\%$, equal to 0.6. Then Equations 3.33 through 3.42 are employed to obtain the seven points of the hydrograph as follows:

$$S = \frac{50}{(0.8)(3250)} = 0.019,$$

$$T_p = \frac{4.1(3250)^{0.23}(0.60)^{1.57}}{(0.019)^{0.25}(51.2)^{0.18}} = 15.66\,\text{min},$$

$$Q_p = \frac{138.7(1.13)^{0.96}}{(15.66)^{1.07}} = 8.2\,\text{m}^3/\text{s},$$

$$T_B = \frac{666.7(1.13)}{(8.2)^{0.95}} = 101.9\,\text{min},$$

$$W_{50} = \frac{105.1(1.13)^{0.93}}{(8.2)^{0.92}} = 16.96\,\text{min},$$

$$W_{75} = \frac{45.1(1.13)^{0.79}}{(8.2)^{0.78}} = 9.6\,\text{min},$$

$$t_A = 15.66 - \frac{16.96}{3} = 10\,\text{min},$$

$$t_B = 15.66 - \frac{9.6}{3} = 12.45\,\text{min},$$

$$t_E = 15.66 + 2\left(\frac{9.6}{3}\right) = 22.06\,\text{min},$$

$$t_F = 15.66 + 2\left(\frac{16.96}{3}\right) = 26.96\,\text{min}.$$

The discharge at t_A and t_F is $0.50Q_p = 4.1\,\text{m}^3/\text{s/cm}$ and at t_B and t_E, the discharge is $0.75Q_p = 6.16\,\text{m}^3/\text{s/cm}$. The developed 10-min UH of the watershed is depicted in Figure 3.20. Now it is checked that the area under the hydrograph is equal to 1 cm depth of the direct runoff. For this purpose,

continued

the discharges are read from the UH presented in Figure 3.20 at equal time increments and tabulated in Table 3.22.

The runoff volume is calculated as

$$\text{Runoff volume} = (41.18 \text{ m}^3/\text{s})(5 \text{ min})(60 \text{ s/min}) = 12,354 \text{ m}^3.$$

The runoff depth is obtained by dividing the runoff volume by the basin area (1,130,000 m²):

$$\text{Depth} = \frac{12,354 \text{ m}^3}{1,130,000 \text{ m}^2} = 0.0109 \text{ m} \simeq 1 \text{ cm}.$$

As the depth of the excess runoff is equal to 1, the hydrograph modification is not needed. The hydrograph ordinates should be adjusted if the estimated runoff depth turned out to be different from 1 cm.

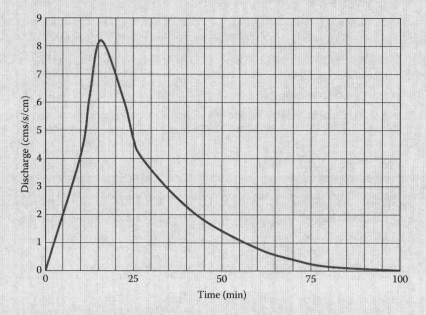

FIGURE 3.20 Developed Espey UH for the watershed of Example 3.12.

TABLE 3.22
Espey 10-min UH (Example 3.12)

Time (min)	Q (m³/s/cm)	Time (min)	Q (m³/s/cm)
0	0	55	1.1
5	2	60	0.8
10	4.10	65	0.5
15	8.1	70	0.4
20	7	75	0.2
25	4.7	80	0.15
30	3.7	85	0.08
35	2.9	90	0.04
40	2.2	95	0.01
45	1.8	100	0
50	1.4		

Urban Water Hydrology

3.6.3 SCS UH

The SCS curvilinear dimensionless UH procedure is one of the most well-known methods for deriving synthetic UHs in use today. The dimensionless UH used by the SCS is derived based on a large number of UHs from basins that varied in characteristics such as size and geographic location. The UHs are averaged and the final product is made dimensionless by considering the ratios of q/q_p (flow/peak flow) on the ordinate axis and t/t_p (time/time to peak) on the abscissa, where the units of q and q_p are flow/cm of runoff/unit area. This final, dimensionless UH, which is the result of averaging a large number of individual dimensionless UHs, has a time-to-peak located at approximately 20% of its time base and an inflection point at 1.7 times the time to peak. The dimensionless UH is illustrated in Figure 3.21. Figure 3.21 also illustrates the cumulative mass curve for the dimensionless UH. Table 3.23 provides the ratios for the dimensionless UH and the corresponding mass curve.

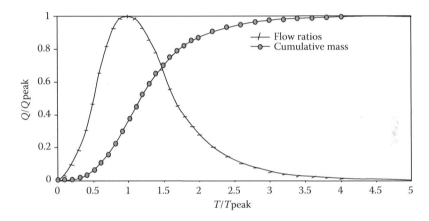

FIGURE 3.21   SCS dimensionless UH and mass curve (S hydrograph).

TABLE 3.23
Ratios for Dimensionless UH and Mass Curve

Time Ratios (t/t_p)	Discharge Ratios (q/q_p)	Mass Curve Ratios (Q_a/Q)	Time Ratios (t/t_p)	Discharge Ratios (q/q_p)	Mass Curve Ratios (Q_a/Q)
0.0	0.000	0.000	1.6	0.560	0.751
0.1	0.030	0.001	1.7	0.460	0.790
0.2	0.100	0.006	1.8	0.390	0.822
0.3	0.190	0.012	1.9	0.330	0.849
0.4	0.310	0.035	2.0	0.280	0.871
0.5	0.470	0.065	2.2	0.207	0.908
0.6	0.660	0.107	2.4	0.147	0.934
0.7	0.820	0.163	2.6	0.107	0.953
0.8	0.930	0.228	2.8	0.077	0.967
0.9	0.990	0.300	3.0	0.055	0.977
1.0	1.000	0.375	3.2	0.040	0.984
1.1	0.990	0.450	3.4	0.029	0.989
1.2	0.930	0.522	3.6	0.021	0.993
1.3	0.860	0.589	3.8	0.015	0.995
1.4	0.780	0.650	4.0	0.011	0.997
1.5	0.680	0.700	4.5	0.005	0.999
			5.0	0.000	1.000

The curvilinear UH may also be represented by an equivalent triangular UH. Figure 3.22 shows the equivalent triangular UH. Recall that the UH is the result of one unit of excess rainfall (of duration D) spread uniformly over the basin. Using the geometry of the triangle, the UH has 37.5% (or 3/8) of its volume on the rising side and the remaining 62.5% (or 5/8) of the volume on the recession side.

The following relationships are used in further developing the peak rate relationships. Note that the time base, T_b, of the triangular UH extends from 0 to 2.67 and that the time to peak, T_p, is at 1.0; thus the time base is 2.67 times the time to peak or

$$T_b = 2.67 T_p, \tag{3.43}$$

and that the recession limb time, T_r, is then 1.67 times the time to peak:

$$T_r = T_b - T_p = 1.67 T_p. \tag{3.44}$$

Using the geometric relationships of the triangular UH of Figure 3.22, the total volume under the hydrograph is calculated by (area under two triangles)

$$Q = \frac{q_p T_p}{2} + \frac{q_p T_r}{2} = \frac{q_p}{2}(T_p + T_r). \tag{3.45}$$

The volume, Q, is in cm (1 cm for a UH) and the time, T, is in hours. The peak rate, q_p, in cm/h, is

$$q_p = \frac{2Q}{T_p + T_r}. \tag{3.46}$$

To have the peak flow of the UH in terms of cms/cm, the drainage area, A (km^2), is added to the equation

$$q_p = \frac{2 \times 10^4 \times A \times Q}{3600 \times (T_p + T_r)}. \tag{3.47}$$

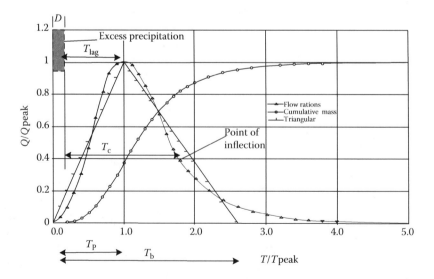

FIGURE 3.22 Dimensionless curvilinear UH and equivalent triangular hydrograph.

Note again that the recession limb time, T_r, is 1.67 times the time to peak, T_p. Substituting in these relationships from Equation 3.44, Equation 3.47 is rewritten as

$$q_p = \frac{2.08Q}{T_p}. \tag{3.48}$$

Because the above relationships were developed based on the volumetric constraints of the triangular UH, the equations and conversions are also valid for the curvilinear UH, which, proportionally, has the same volumes as the triangular representation. Note that the conversion constant (herein called the peaking factor) is the result of assuming that the recession limb is 1.67 times the rising limb (time to peak). This may not be applicable to all watersheds.

Steep terrain and urban areas may tend to produce higher early peaks and thus the values of the peaking factor may increase. Likewise, flat swampy regions tend to retain and store the water, causing a delayed, lower peak. In these circumstances values may tend toward 1.25 or lower (SCS, 1972; Wanielista et al., 1997). It would be very important to document any reasons for changing the constant from 2.08 because it will considerably change the shape of the UH.

The peak rate may also be expressed in terms of other timing parameters such as rainfall duration and basin lag time besides the time to peak. From Figure 3.22,

$$T_p = \frac{D}{2} + T_{\text{lag}}, \tag{3.49}$$

where D is the duration of the unit excess rainfall and T_{lag} is the basin lag time, which is defined as the time between the center of mass of excess rainfall and the time to peak of the UH.

The peak flow is now written as

$$q_p = \frac{2.08AQ}{(D/2) + T_{\text{lag}}}. \tag{3.50}$$

The SCS (1972) relates the basin lag time, T_{lag}, to the time of concentration, T_c, by

$$T_{\text{lag}} = 0.6T_c. \tag{3.51}$$

Combining this with other relationships, as illustrated in the triangular UH, the following relationships develop:

$$T_c + D = 1.7T_p \quad \text{and} \quad \frac{D}{2} + 0.6T_c = T_p. \tag{3.52}$$

From this, duration D can be expressed as

$$D = 0.133T_c. \tag{3.53}$$

Equations 3.43 through 3.53 provide the basis for the SCS dimensionless UH method. Equation 3.53 provides a desirable relationship between duration and the time of concentration, which should provide enough points to accurately represent the UH, particularly the rising limb.

Example 3.13

Develop a 30-min UH for a 3.43-km² urban basin.

Solution: First, using Equation 3.53, the time of concentration of the basin is determined as $0.5/0.133 = 3.76\,h$. Then the basin lag time is calculated as (Equation 3.51) $0.6 \times 3.76 = 2.26\,h$. Finally, the time-to-peak flow and peak flow are estimated as

$$T_p = \frac{D}{2} + T_{\text{lag}} = \frac{0.5}{2} + 2.26 = 2.51\,h,$$

$$q_p = \frac{2.08 A Q}{T_p} = \frac{2.08 \times 3.43 \times 1}{2.51} = 2.84\,m^3/s/cm.$$

Therefore, the peak discharge of the 30-min UH, which is $2.84\,m^3/s/cm$ of excess rainfall, occurs 2.51 h after the start of excess rainfall. For developing the 30-min UH, the values of t/t_p and q/q_p in Table 3.23 are multiplied by 2.51×60 and 2.84, respectively. The developed hydrograph is presented in Table 3.24.

TABLE 3.24
SCS UH of Example 3.13

t/T_p	q/q_p	t (min)	q (m³/s/cm)	t/T_p	q/q_p	t (min)	q (m³/s/cm)
0.00	0.000	0.00	0.00	1.70	0.460	256.02	1.31
0.10	0.300	15.06	0.85	1.80	0.390	271.08	1.11
0.20	0.100	30.12	0.28	1.90	0.330	286.14	0.94
0.30	0.190	45.18	0.54	2.00	0.280	301.20	0.80
0.40	0.310	60.24	0.88	2.20	0.207	331.32	0.59
0.50	0.470	75.30	1.33	2.40	0.147	361.44	0.42
0.60	0.660	90.36	1.87	2.60	0.107	391.56	0.30
0.70	0.820	105.42	2.33	2.80	0.077	421.68	0.22
0.80	0.930	120.48	2.64	3.00	0.055	451.80	0.16
0.90	0.990	135.54	2.81	3.20	0.400	481.92	1.14
1.00	1.000	150.60	2.84	3.40	0.029	512.04	0.08
1.10	0.990	165.66	2.81	3.60	0.021	542.16	0.06
1.20	0.930	180.72	2.64	3.80	0.015	572.28	0.04
1.30	0.860	195.78	2.44	4.00	0.011	602.40	0.03
1.40	0.780	210.84	2.22	4.50	0.005	677.70	0.01
1.50	0.680	225.90	1.93	5.00	0.000	753.00	0.00
1.60	0.560	240.96	1.59				

3.6.4 Gamma Function UH

The more general form of the SCS UH, described in the previous section, is the gamma function UH. Actually the SCS UH and the gamma UH are very similar in shape and virtually identical. The SCS UH is in tabular form, or in graphical form, and the gamma UH is in functional form and is much easier to use than the SCS method. In a more general form, Equation 3.48 is written as

$$q_p = \frac{P_r A Q}{T_p}, \tag{3.54}$$

TABLE 3.25
Relationship of Gamma Function UH Parameters

n	P_r (cms-h/km^2-cm)
1.5	0.67
2	1.02
2.5	1.28
3	1.50
3.5	1.69
4	1.86
4.5	2.02
5	2.17

where P_r is the peak rate factor that is dependent on the topographic characteristics of the watershed (Meadows, 1991). The following equation is used to determine the ordinates of the gamma function UH:

$$q_u = q_p \left[\frac{t}{T_p} \exp\left(1 - \frac{t}{T_p}\right) \right]^{n-1}, \qquad (3.55)$$

where n is a shape factor. The parameters n and P_r are related to each other as given in Table 3.25. Considering these relations in developing the UH using Equations 3.54 and 3.55 ensures that the developed hydrograph will result in a unit depth of excess rainfall. T_p and T_{lag} are estimated in a similar way as the SCS UH using Equations 3.49 and 3.51, respectively.

EXAMPLE 3.14

Develop a 25-min UH for a 1.8-km^2 watershed using the gamma function UH method. Assume that the time of concentration is 65 min and the peak rate factor P_r is 1.28 cms-h/km^2-cm.

Solution: For the determination of the time-of-peak discharge, first the watershed lag time is determined by using Equation 3.51 as

$$T_{lag} = 0.6 \left(\frac{65}{60} \right) = 0.65 \, h.$$

Then, with $D = 25 \, min = 0.42 \, h$, the time-to-peak discharge is calculated using Equation 3.49 as

$$T_p = \frac{0.42}{2} + 0.65 = 0.86 \, h = 52 \, min.$$

Then the peak discharge is calculated by using Equation 3.54 as

$$q_p = \frac{(1.28)(1.8)}{0.86} = 2.68 \, cms/cm.$$

According to Table 3.25, for $P_r = 1.28$ cms-h/km^2-cm, $n = 2.5$. Finally, Equation 3.55 is used to determine the ordinates of the UH. The calculation steps are tabulated in Table 3.26.

continued

TABLE 3.26
Calculation Steps of Example 3.14

t (min) (1)	t/T_p (2)	$1 - (t/T_p)$ (3)	$\exp(1 - (t/T_p))$ (4)	q_u (cms/cm) (5)
0	0.00	1.00	2.72	0.00
10	0.19	0.81	2.24	0.76
20	0.38	0.62	1.85	1.61
30	0.58	0.42	1.53	2.22
40	0.77	0.23	1.26	2.56
50	0.96	0.04	1.04	2.68
60	1.15	−0.15	0.86	2.64
70	1.35	−0.35	0.71	2.49
80	1.54	−0.54	0.58	2.28
90	1.73	−0.73	0.48	2.04
100	1.92	−0.92	0.40	1.79
110	2.12	−1.12	0.33	1.55
120	2.31	−1.31	0.27	1.32
130	2.50	−1.50	0.22	1.12
140	2.69	−1.69	0.18	0.94
150	2.88	−1.88	0.15	0.78
175	3.37	−2.37	0.09	0.48
200	3.85	−2.85	0.06	0.28
250	4.81	−3.81	0.02	0.09
375	7.21	−6.21	0.00	0.00

3.6.5 TIME–AREA UHS

This method is also called "the Clark method." The duration of a synthetic UH is generally dependent on the parameters used in the equations specific to the method. This differs from a derived UH, as UHs derived directly from gauged data have duration equal to the duration of excess precipitation from which they were derived. In this respect, Clark's (1943) UH is slightly different from other synthetic UH methods in that it has no duration; it is an instantaneous UH.

Clark's UH theory maintains the fundamental properties of a UH in that the sequence of runoff is the result of a unit depth of uniformly generated excess precipitation. However, the duration of the excess precipitation is considered to be infinitely small. Thus the Clark UH is generally referred to as the Clark "instantaneous unit hydrograph," or IUH. This excess precipitation is applied uniformly over a watershed that is broken into time–area increments.

Clark (1943) pointed out several advantages of his method, including the following:

1. The procedure is highly objective, using mathematically defined parameters based on observed hydrographs (except in the derivation of the time–area curve, which requires individual judgment). Thus, the procedure is repeatable by two individuals using the same data set.
2. The method does not require knowledge of spatial runoff distribution.
3. The ability to account for the shape of the drainage area and the capacity to produce large peak flows from concentrated runoff are included, subject to the accuracy of the developed time–area relationship.

4. The UH is the result of an instantaneous rainfall with the time of concentration defined as the time between the end of rainfall and a mathematically defined point on the hydrograph's falling limb. Therefore, personal judgments on effective rainfall duration and time of concentration, which inversely affects peak flow, are eliminated.

Also, the following theoretical and practical limitations of this method are noted:

1. Hydrographs are likely to underpredict on the recession side between the point of maximum recession rate and the point where subsurface flow begins to dominate.
2. The method may not be applicable to very large drainage areas. The application may cause too slow a rise and too rapid a recession, due to the use of the same storage factor for points that are both near and far from the outlet. While Clark does not provide a recommendation on the maximum size of the drainage basin, he does indicate that the basin may be subdivided to overcome this drawback.
3. Finally, Clark admits that while the method appears to account for the drainage basin shape and the capacity to produce high peak flows, it is possible that these influences are exaggerated by the method.

In this method, a *translation hydrograph* at the basin outlet is developed by translating (lagging) the excess hydrograph based on the travel time to the outlet. The translation hydrograph is routed through a single linear reservoir, and the resulting outflow represents the IUH for the basin. Figure 3.23 shows the components of the Clark method.

The translation hydrograph can be conveniently derived from a time–area relation. The area is the accumulated area from the basin outlet and the time is the travel time as defined by isochrones (contours of constant time of travel). Such a relationship can be expressed in dimensionless form with area as a percentage of the total basin area and time as a percentage of the time of concentration (t_c). The translation hydrograph can be obtained by determining from a time–area relation the portion of the basin that contributes runoff at the outlet during each time interval after the occurrence of the instantaneous burst of unit excess. The contributing area associated with a time interval (the unit depth and divided by the time interval) yields an average discharge. This is the ordinate of the translation hydrograph for that interval.

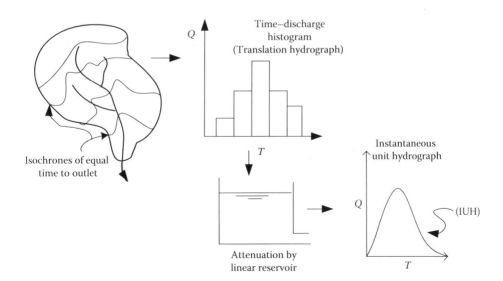

FIGURE 3.23 Components of the Clark model.

Isochrones for use in defining the translation hydrograph may be developed by estimating, for a number of points in the basin, overland flow and channel travel times to the basin outlet. A simpler approach is to assume a constant travel velocity and base the position of isochrones on the travel distance from the basin outlet, in which case the translation hydrograph reflects only the basin shape.

An even simpler approach is to use a translation hydrograph that is based on a standard basin shape, such as an ellipse. For many basins, storage effects represented by the linear reservoir cause substantial attenuation of the translation hydrograph such that the IUH is not very sensitive to the shape of the translation hydrograph. However, for a basin without a substantial amount of natural storage, such as a steep urban basin, the IUH will be much more sensitive to the shape of the translation hydrograph. For such a basin, the use of a standard shape may not be appropriate.

The routing of the translation hydrograph through a linear reservoir is based on simple storage routing by solving the continuity equation. An equation for the routing is

$$O(t) = C_a I + C_b O(t-1). \tag{3.56}$$

Coefficients C_a and C_b are defined by

$$C_a = \frac{\Delta t}{R + 0.5 \Delta t} \tag{3.57}$$

and

$$C_b = 1 - C_a, \tag{3.58}$$

where $O(t)$ is the ordinate of IUH at time t, I is the ordinate of the translation hydrograph for interval $t-1$ to t, R is the storage coefficient for the linear reservoir, and Δt is the time interval with which IUH is defined.

The two parameters for the Clark method are, t_c, the time of concentration (and time base for the translation hydrograph), and R, the storage coefficient for the linear reservoir. Values for these parameters, along with a time–area relation, enable the determination of a UH.

To calculate the direct runoff, the IUH can be converted to a UH of finite duration. Derivation of a UH of specified duration from the IUH is accomplished using techniques similar to those employed to change the duration of a UH. For example, if a 2-h UH is required, a satisfactory approximation may be obtained by first summing the ordinates of two instantaneous UHs, one of which is lagged 2 h. This sum represents the runoff from 2 cm of excess precipitation; to obtain the required UH, the ordinates must be divided by 2. This procedure is shown in Figure 3.24 and Table 3.27.

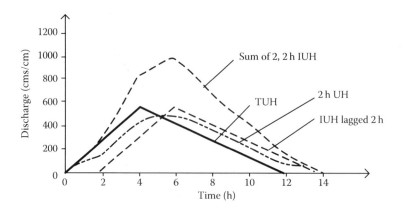

FIGURE 3.24 Conversion of IUH to UH with specific duration.

TABLE 3.27
Example of Developing 2-h UH from 2-h IUH

Time in Hours (1)	IUH Coordinate (cms/cm) (2)	IUH Lagged 2 h (3)	Sum of Columns 2 and 3 (4)	2 h UH (cms/cm) (5)
0	0	0	0	0
2	300	0	300	150
4	600	300	900	450
6	450	600	1050	525
8	300	450	750	375
10	150	300	450	225
12	0	150	150	75
14	0	0	0	0

EXAMPLE 3.15

Using the Clark method, develop the 15-min UH for an urban watershed with an area of 43 ha and a time of concentration of 45 min. The time–area curve for this watershed is given in columns 1 and 2 of Table 3.28. Assume that the storage coefficient is equal to 10 min.

TABLE 3.28
Summary of Time–Area UH Analysis in Example 3.15

t (min) (1)	A_c (ha) (2)	I (m^3/s) (3)	IUH (m^3/s) (4)	15-min Lagged IUH (m^3/s) (5)	Columns 5 and 6 (m^3/s) (6)	15-min UH (m^3/s) (7)
0	0	0.000	0.000	0.000	0	0
5	3.14	0.349	0.140	0.000	0.140	0.070
10	8.62	0.958	0.467	0.000	0.467	0.233
15	15.68	1.742	0.977	0.000	0.977	0.488
20	22.74	2.527	1.597	0.140	1.737	0.868
25	28.22	3.136	2.212	0.467	2.679	1.339
30	31.36	3.484	2.721	0.977	3.698	1.849
35	36.72	4.080	3.265	1.597	4.862	2.431
40	40.03	4.448	3.738	2.212	5.950	2.975
45	43	4.778	4.154	2.721	6.875	3.437
50		0.000	2.492	3.265	5.757	2.8785
55		0.000	1.495	3.738	5.233	2.616
60		0.000	0.897	4.154	5.051	2.077
65		0.000	0.538	2.492	3.030	1.515
70		0.000	0.323	1.495	1.818	0.909
75		0.000	0.194	0.897	1.091	0.545
80		0.000	0.116	0.538	0.654	0.327
85		0.000	0.070	0.323	0.393	0.196
90		0.000	0.042	0.194	0.236	0.118
95		0.000	0.025	0.116	0.141	0.070
100		0.000	0.015	0.070	0.085	0.042
105		0.000	0.009	0.042	0.051	0.025
110		0.000	0.000	0.025	0.025	0.012
115		0.000	0.000	0.015	0.015	0.007
120		0.000	0.000	0.000	0.000	0.000

continued

Solution: For $D = 15$ min, the rainfall intensity is $i_e = 1$ cm/15 min $= 0.0011$ cm/s $= 0.000011$ m/s. The translation hydrograph presented in column 3 is calculated by multiplying the rainfall intensity to the corresponding area in the outlet flow at time t. Then C_a and C_b are calculated as (using Equation 3.57)

$$C_a = \frac{5}{10 + 0.5 \times 5} = 0.4 \quad \text{and} \quad C_b = 1 - C_a = 1 - 0.4 = 0.6.$$

The ordinates of the IUH are calculated using Equation 3.56 (column 4). The outflow at time zero is equal to zero.

For the calculation of the 15-min UH, first the ordinates of two instantaneous UHs, one of which is lagged 15 min, are summed. The resulting hydrograph results in 2 cm excess rainfall; therefore the results are divided by 2 to develop the 15-min UH of the watershed (column 7).

Also, as a check, the area under the UH (that is the volume of excess rainfall) is calculated to correspond to 1 cm. The volume of runoff represented by the UH is equal to $\Delta t(\Sigma Q_u) =$ (5 min) (60 s/min) (25.03 m^3/s) $= 7508.25$ m^3. The basin area is (43 ha)(10,000 m^2/ha) $= 4{,}30{,}000$ m^2. Dividing the runoff volume by the basin area, the rainfall depth is obtained as $(7508.25)/(4{,}30{,}000) = 0.017$ m $= 1$ cm.

3.6.6 Application of the UH Method

The UH method is formed based on the assumption of a linear relation between the excess rainfall and the direct runoff rates. The following are the fundamental assumptions of the UH theory.

1. The duration of direct runoff is always the same for uniform-intensity storms of the same duration, regardless of the intensity. This means that the time base of the hydrograph does not change and that the intensity affects only the discharge.
2. The direct runoff volumes produced by two different excess rainfall distributions are in the same proportion as the excess rainfall volume. This means that the ordinates of the UH are directly proportional to the storm intensity. If storm A produces a given hydrograph and storm B is equal to storm A multiplied by a factor, then the hydrograph produced by storm B will be equal to the hydrograph produced by storm A multiplied by the same factor.
3. The time distribution of the direct runoff is independent of the concurrent runoff from antecedent storm events. This implies that direct runoff responses can be superimposed. If storm C is the result of adding storms A and B, the hydrograph produced by storm C will be equal to the sum of the hydrographs produced by storms A and B.
4. Hydrological systems are usually nonlinear due to factors such as storm origin and patterns and stream channel hydraulic properties. In other words, if the peak flow produced by a storm of certain intensity is known, the peak corresponding to another storm (of the same duration) with twice the intensity is not necessarily equal to twice the original peak.
5. Despite this nonlinear behavior, the UH concept is commonly used because, although it assumes linearity, it is a convenient tool to calculate hydrographs and it gives results within acceptable levels of accuracy.
6. An alternative to the UH theory is the kinematic wave theory and distributed hydrological models.

The main applications of UH can be summarized as follows:

1. Design storm hydrographs for selected recurrence interval storms (e.g., 50 yr) can be developed through convolution adding and lagging procedures.
2. Effects of land-use–land-cover changes, channel modifications, storage additions, and other variables can be evaluated to determine changes in the UH.
3. Effects of the spatial variation in precipitation can be evaluated.
4. Hydrographs of watersheds consisting of several sub-basins can be produced.

Urban Water Hydrology

EXAMPLE 3.16

Determine the ordinates of the direct runoff of the effective rainfall hyetograph shown in Figure 3.25. The ordinates of 10-min UH of the considered watershed are given in Table 3.29 (column 2).

Solution: At first, the depth of the excess rainfall produced during each of the 10-min periods is determined by multiplying the corresponding rainfall intensity by 1/6 h. Considering the UH assumptions, the direct runoff hydrograph is calculated as

$$DRH = 0.17UH + 0.25UH \text{ (lagged 10 min)} + 0.38UH$$
$$\text{(lagged 20 min)} + 0.17UH \text{ (lagged 30 min)} + 0.085UH$$
$$\text{(lagged 40 min)}.$$

The calculations are summarized in Table 3.29.

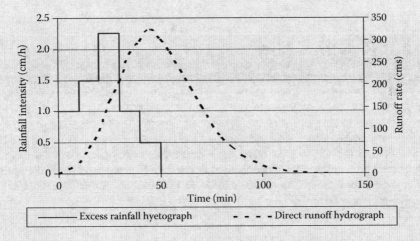

FIGURE 3.25 Characteristics of the rainfall event of Example 3.16.

TABLE 3.29
Application of the UH Method

t (min) (1)	UH (m³/s/cm) (2)	0.17UH (m³/s) (3)	0.25UH Lagged 10 min (m³/s) (4)	0.38UH Lagged 20 min (m³/s) (5)	0.17UH Lagged 30 min (m³/s) (6)	0.085UH Lagged 40 min (m³/s) (7)	Direct Runoff Hydrograph (m³/s) (8)
0	0	0	0	0	0	0	0
10	50	8.5	0	0	0	0	8.5
20	110	18.7	12.5	0	0	0	31.2
30	170	28.9	27.5	19	0	0	75.4
40	240	40.8	42.5	41.8	8.5	0	133.6
50	150	25.5	60	64.6	18.7	4.25	173.05
60	70	11.9	37.5	91.2	28.9	9.35	178.85
70	30	5.1	17.5	57	40.8	14.45	134.85
80	15	2.55	7.5	26.6	25.5	20.4	82.55

continued

TABLE 3.29 (continued)

t (min) (1)	UH (m³/s/cm) (2)	0.17UH (m³/s) (3)	0.25UH Lagged 10 min (m³/s) (4)	0.38UH Lagged 20 min (m³/s) (5)	0.17UH Lagged 30 min (m³/s) (6)	0.085UH Lagged 40 min (m³/s) (7)	Direct Runoff Hydrograph (m³/s) (8)
90	5	0.85	3.75	11.4	11.9	12.75	40.65
100	0	0	1.25	5.7	5.1	5.95	18
110	0	0	0	1.9	2.55	2.55	7
120	0	0	0	0	0.85	1.275	2.125
130	0	0	0	0	0	0.425	0.425
140	0	0	0	0	0	0	0

3.7 REVISITING A FLOOD RECORD

The annual flood records are not homogeneous because of the nonhomogeneity of the flood characteristics in an urban area due to land-use variability. Thus the assumption of homogeneity that is essential in a statistical flood frequency analysis is violated. Therefore prior to making a frequency analysis on flood records, the results of inaccurate estimates of flood for any return period and the effect of nonhomogeneity in the analysis results should be investigated.

There are no specific well-known procedures suggested for adjustment of flood records in the literature. Often multiparameter watershed models are used for adjustment of flood records but there is no single model or procedure widely accepted by the professional community and the results of different methods are not compared.

3.7.1 Urban Effects on Peak Discharge

As different methods are used for the estimation of peak discharge, the intensity of urban effects could be different. In the rational method, which is more popular in the estimation of urban peak discharge, urban development (increasing the impervious areas) affects the time of concentration and also the watershed runoff coefficient. Thus the rate of increase in the peak discharge for a specific return period cannot be measured as imperviousness is different.

Some regression-based equations are suggested for the estimation of peak discharge using the percentage of imperviousness. Sarma et al. (1969) provided a general equation:

$$q_p = 629.34 A^{0.723} (1 + U)^{1.516} P_E^{1.113} T_R^{-0.403}, \tag{3.59}$$

where A is the watershed area (km²), U is the fraction of imperviousness, P_E is the depth of excess rainfall (mm), T_R is the duration of excess rainfall (h), and q_p is the peak discharge (m³/s).

The effect of urbanization on the peak discharge could be evaluated through the relative sensitivity (S_R) of Equation 3.59, which is calculated as

$$S_R = \frac{\partial q_p}{\partial U} \cdot \frac{U}{q_p}. \tag{3.60}$$

For instance, the peak discharge will be increased by 1.52% in response to 1% change in U. Since peak discharge is independent of both the value of U and the return period, this method estimates the average influence of urbanization on peak discharge.

Urban Water Hydrology

Also, the following equation is suggested in the literature for evaluating the effect of urbanization on peak discharge:

$$f = 1 + 0.015U, \qquad (3.61)$$

where f is the relative increase in peak discharge for percentage of imperviousness of U. Thus a 1% increase in imperviousness will increase the peak discharge by 1.5%, which is similar to the result of Equation 3.60.

The SCS methods adjust flood records for urbanization (SCS, 1986) based on the percentages of imperviousness and the runoff CN as illustrated in Figure 3.26. In this method, the effect of return period on flood record adjustment is incorporated through the rainfall input.

The adjustment factors for imperviousness suggested by SCS are presented in Table 3.30. The effect of urbanization on the peak discharges would be the square of the adjustment factor assuming the same changes in each direction. The values of R_S that are also shown in Table 3.30 are approximate measures of the change in peak discharge due to the peak factors (f^2) (SCS, 1986). The urban development also changes the CN value that should be considered in the adjustment of

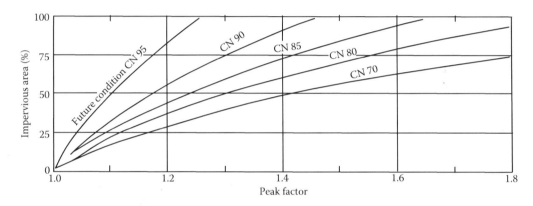

FIGURE 3.26 Suggested chart by SCS for adjustment of peak discharges based on CN and imperviousness.

TABLE 3.30
Peak Discharge Adjustment Factors for Urbanization

	SCS Chart Method				USGS Urban Equations			
CN	U	f	f^2	R_S	T	U	A	R_S
70	20	1.13	1.28	0.018	2 years	20	1.70	0.016
	25	1.17	1.37	0.019		25	1.78	0.016
	30	1.21	1.46	0.025		30	1.86	0.018
	35	1.26	1.59	0.026		35	1.95	0.020
	40	1.31	1.72	—		40	2.05	
80	20	1.10	1.21	0.013	100 years	20	1.23	0.010
	25	1.13	1.28	0.014		25	1.28	0.008
	30	1.16	1.35	0.019		30	1.32	0.008
	35	1.20	1.44	0.015		35	1.36	0.00
	40	1.23	1.51			40	1.41	

Source: From Soil Conservation Service (SCS). 1986. *Urban Hydrology for Small Watersheds, Technical Release* 55, 2nd Edition. U.S. Department of Agriculture, Springfield, Virginia, June. NTIS PB87-101580 (microcomputer version 1.11. NTIS PB87-101598.)

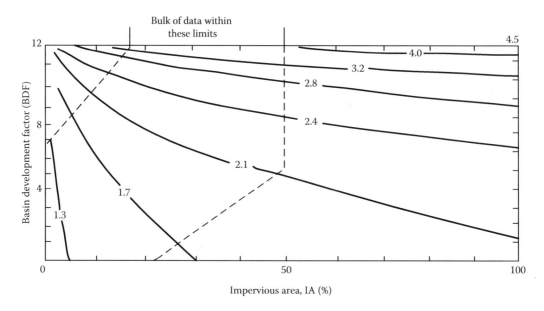

FIGURE 3.27 Ratio of the urban to rural 2-year peak discharge suggested by USGS.

peak discharge. However, the investigations on the SCS chart method recommend a change in peak discharge of 1.3–2.6% for a 1% change in urbanization.

USGS has proposed other equations for assessing the effects of urbanization on urban peak discharge. In this method, the adjustment is done based on the return period and the imperviousness (Table 3.30). Using Figures 3.27 and 3.28, the ratio of the urban-to-rural discharge can be estimated as a function of the percentage of imperviousness and BDF. For example, for a 2-year storm event, this ratio varies between 1 and 4.5, the value 4.5 corresponding to the complete development condition. For a 100-year event, the ratio has a maximum value of 2.7 for complete development.

In this way, a 1% change in urbanization will result in an average 1.75% change in peak discharge for the 2-year event and 0.9% for the 100-year event. The USGS equations suggest that the effects of urbanization are slightly lower for the less frequent storm events and slightly higher for the more frequent storm events in comparison with the other methods (Equations 3.60 and 3.61) that provide an effect of about 1.5%.

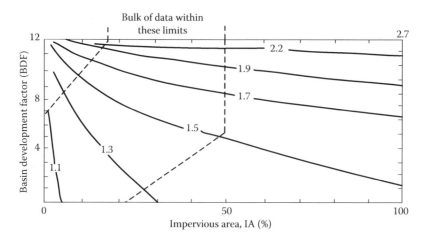

FIGURE 3.28 Ratios of the urban-to-rural 100-year peak discharge suggested by USGS.

3.7.2 FLOOD RECORD ADJUSTING

This method is developed based on the general trends of the data. The peak adjustment factor is determined using Figure 3.29 as a function of the exceedence probability for percentages of imperviousness less than 60%. Using this method, the greatest adjustment factors are obtained for more frequent events and the highest percentage of imperviousness.

The discharge must be adjusted from a partly urbanized watershed to a discharge for another urban condition. For this purpose, first the rural discharge is obtained by dividing the peak discharge in the existing condition to the peak adjustment factor. This way, the representative discharge of a nonurbanized condition is obtained. Then the resulting "rural" discharge is multiplied by the peak adjustment factor to obtain the peak discharge for the desired watershed condition. The exceedence probability is also necessary in using the proposed adjustment method of Figure 3.29. A plotting position formula could result in the best estimate of the probability for a flood record.

In a watershed that is experiencing a continuous variation in the level of urbanization, the following procedure could be utilized for adjusting a flood record (McCuen, 1998):

1. Determine the corresponding percentage of imperviousness for each recorded flood event and the percentage of imperviousness for which an adjusted flood record is needed.
2. Calculate the rank (i) and exceedence probability (p) for each event in the flood record using a plotting position formula or other methods.
3. Find the peak adjustment factor (f_1) from Figure 3.29, using the exceedence probability and the actual percentage of imperviousness. Then divide the measured peak from the actual level of imperviousness by f_1 to obtain the peak discharge in a nonurbanized condition.
4. Find the peak adjustment factor (f_2) from Figure 3.29, using the exceedence probability and the percentage of imperviousness for which a flood series is needed. Transform the nonurbanized peak to a discharge for the desired level of imperviousness.

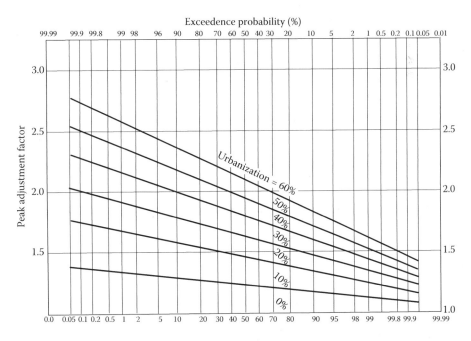

FIGURE 3.29 Peak adjustment factors for urbanizing watersheds. (From McCuen, R. H. 1998. *Hydrological Analysis and Design*. Prentice Hall, Englewood Cliffs, NJ. With permission.)

5. Compute the ultimate adjusted discharge (Q_a) using the following equation:

$$Q_a = \frac{f_2}{f_1} Q, \tag{3.62}$$

where Q is the measured discharge in the present condition.
6. Repeat steps 3, 4, and 5 for each recorded flood event and rank the adjusted series.
7. Repeat steps 2 through 6 until ranks of the measured (Q) and adjusted (Q_a) flood series are the same.

It should be noted that this procedure is not applicable to the logarithms of the flood peaks and when the adjusted series will be used to compute frequency curves such as log-normal or log-Pearson Type III.

Example 3.17

A 15-year record of annual maximum peak discharges for a watershed is given in Table 3.31. The percentage of impervious cover has increased from 13% to 27% from year 1 to year 15, as presented in Table 3.31. Adjust these flood records for the last year impervious cover condition. The mean and standard deviation of the logarithm of the record are 3.2480 and 0.4171, respectively. The mean skew of −0.5 is used for the entire watershed.

Solution: The flood records are adjusted for the desired condition of the impervious cover. For example, while in year 6 and year 10 the percentage of impervious cover was 18% and 22%, respectively, the values of flood records in these years are adjusted to a common percentage of 27%, which is the watershed state after year 14. As the return period for some of the previous events is modified from the measured record, the iterative process is required. These changes in the rank of the events change the exceedence probabilities. Since the adjustment factor is dependent on the exceedence probabilities, the adjustment factor also changes. The rank of the events did not change after the second adjustment; thus the procedure of adjustment is stopped. The adjusted series is presented in the last part of Table 3.31.

The mean and standard deviation of the logarithm of adjusted series are 3.2818 and 0.4139, respectively. The mean increased and the standard deviation decreased after adjustment at it was expected. The mean increased because the earlier events occurred when less impervious cover existed, which reduced the peak discharge. The standard deviation decreased because the measured data of peak discharge presented both natural variation and variation due to different levels of imperviousness. The adjusted flood frequency curve is generally higher than the curve for the measured series; the higher curve reflects the effect of the higher amount of imperviousness (27%). The adjusted flood frequency curve has also less sleep, which reflects the issue that the adjusted series is for a single level of imperviousness. The result of the analysis for the adjusted and unadjusted flood frequency curves are given in Table 3.32. The percentage increases in the 2-year, 5-year, 10-year, 15-year, 50-year, and 100-year flood magnitudes are also given in Table 3.32. In this table, p is the exceedence probability and K is the frequency factor of the Gumbel probability distribution that is fitted to the logarithm of peak flood records. The parameter (b) of the fitted probability distribution is −0.5. The corresponding values to considered exceedence probabilities are calculated as

$$\log_{10} Q = Y + KSy,$$

where Y and Sy are the mean and standard deviation of the considered data, respectively. The adjusted series represents the annual flood series for a constant urbanization condition (e.g., 27% imperviousness). As the adjusted series is not a measured series, its accuracy is dependent on the representativeness of Figure 3.29 for measuring the effects of urbanization in the considered watershed.

TABLE 3.31
Adjustment of Annual Flood Record for Urbanization

	Measured Series				Ordered Data		
Year	Urbanization (%)	Annual Peak (m³/s)	Rank	Exceedence Probability	Rank	Annual Peak (m³/s)	Year	Exceedence Probability
1991	13	1668	10	0.625	1	4978.34	2003	0.0625
1992	14	409.76	14	0.875	2	4897.79	1999	0.125
1993	16	260.82	15	0.9375	3	4306.64	1996	0.1875
1994	17	2552.97	8	0.5	4	3729.08	1995	0.25
1995	17	3729.08	4	0.25	5	3352.67	2004	0.3125
1996	18	4306.64	3	0.1875	6	2887.06	1998	0.375
1997	19	1611.02	11	0.6875	7	2671.96	2005	0.4375
1998	21	2887.06	6	0.375	8	2552.97	1994	0.5
1999	21	4897.79	2	0.125	9	2231.39	2000	0.5625
2000	22	2231.39	9	0.5625	10	1668	1991	0.625
2001	24	524.85	13	0.8125	11	1611.02	1997	0.6875
2002	25	603.19	12	0.75	12	603.19	2002	0.75
2003	26	4978.34	1	0.0625	13	524.85	2001	0.8125
2004	27	3352.67	5	0.3125	14	409.76	1992	0.875
2005	27	2671.96	7	0.4375	15	260.82	1993	0.9375

Iteration 1

			Correction Factor			Adjusted Series	
Year	Urbanization (%)	Measured Peak (m³/s)	Exist.	Ultimate	Peak	Rank	Exceedence Probability
1991	13	1668	1.287	1.549	2007.56	10	0.625
1992	14	409.76	1.258	1.457	474.58	14	0.875
1993	16	260.82	1.265	1.414	291.54	15	0.9375
1994	17	2552.97	1.4	1.585	2890.33	7	0.4375
1995	17	3729.08	1.452	1.66	4263.27	4	0.25
1996	18	4306.64	1.494	1.684	4854.34	3	0.1875
1997	19	1611.02	1.403	1.53	1756.85	11	0.6875
1998	21	2887.06	1.514	1.62	3089.19	6	0.375
1999	21	4897.79	1.593	1.713	5266.74	1	0.0625
2000	22	2231.39	1.486	1.567	2353.02	9	0.5625
2001	24	524.85	1.441	1.486	541.24	13	0.8125
2002	25	603.19	1.48	1.51	615.42	12	0.75
2003	26	4978.34	1.738	1.756	5029.9	2	0.125
2004	27	3352.67	1.639	1.639	3352.67	5	0.3125
2005	27	2671.96	1.603	1.603	2671.96	8	0.5

Iteration 2

1991	13	1668	1.287	1.549	2007.56	10	0.625
1992	14	409.76	1.258	1.457	474.58	14	0.875
1993	16	260.82	1.265	1.414	291.54	15	0.9375
1994	17	2552.97	1.412	1.603	2898.31	7	0.4375
1995	17	3729.08	1.452	1.66	4263.27	4	0.25
1996	18	4306.64	1.494	1.684	4854.34	3	0.1875
1997	19	1611.02	1.403	1.53	1756.85	11	0.6875
1998	21	2887.06	1.514	1.62	3089.19	6	0.375
1999	21	4897.79	1.629	1.756	5279.63	1	0.0625
2000	22	2231.39	1.486	1.567	2353.02	9	0.5625

continued

TABLE 3.31 (continued)

Year	Urbanization (%)	Measured Peak (m³/s)	Correction Factor		Adjusted Series		
			Exist.	Ultimate	Peak	Rank	Exceedence Probability
2001	24	524.85	1.441	1.486	541.24	13	0.8125
2002	25	603.19	1.48	1.51	615.42	12	0.75
2003	26	4978.34	1.696	1.713	5028.24	2	0.125
2004	27	3352.67	1.639	1.639	3352.67	5	0.3125
2005	27	2671.96	1.585	1.585	2671.96	8	0.5

TABLE 3.32
Computation of Flood Frequency Curves for the Watershed of Example 3.17 in Both Actual (Q) and Ultimate Development (Q_a) Conditions

p (1)	K (2)	$\log_{10} Q$ (3)	Q (m³/s) (4)	$\log_{10} Q_a$ (5)	Q_a (m³/s) (6)
0.99	−2.68572	2.127798	134	2.170088	148
0.90	−1.32309	2.696137	497	2.734117	542
0.70	−0.45812	3.056908	1140	3.092152	1236
0.50	0.08302	3.282612	1917	3.316144	2071
0.20	0.85653	3.605236	4029	3.636321	4328
0.10	1.21618	3.755242	5692	3.78519	6098
0.04	1.56740	3.901733	7975	3.930569	8523
0.02	1.77716	3.989222	9755	4.017395	10,409
0.01	1.95472	4.06328	11,569	4.090891	12,328

3.8 TEST OF THE SIGNIFICANCE OF THE URBAN EFFECT

A series of data used for frequency analysis should be sampled from a unique population. This assumption in hydrological terms means that during data gathering the watershed should not significantly change. But urban watersheds are continuously changing due to developmental practices. The variation of some factors such as land cover and soil moisture can highly affect the hydrological processes in the watershed. The variations of these factors bring about the variation in the annual maximum flood series which are close to the natural variations of storms. For solving this problem, these factors are usually considered as random parameters and, in this way the assumption of watershed homogeneity is not violated.

However, it should be noted that when major changes happen in land use, the watershed situation cannot be considered homogenous. The flood magnitudes or probabilities estimated from a flood frequency analysis are highly affected by these significant variations. Classical flood frequency analysis cannot be used for the determination of the accurate indicators of flooding for the changed watershed condition. For example, for a developing watershed, whose imperviousness increases from 5% to 45%, the duration and magnitude of the flood record are computed from a frequency analysis for 45% imperviousness. The results of this frequency analysis are not representative of this watershed, because many of the flood records that are used to derive the frequency curve correspond to a condition of much less urban development. Therefore, prior to making a flood frequency analysis, the homogeneity of data should be tested. The result of both hydrological and statistical assessments

Urban Water Hydrology

should suggest the necessity for adjustment. Among the available statistical methods for testing the nonhomogeneity of data, the most appropriate one should be selected due to watershed and data characteristics.

If the results of statistical analysis show the nonhomogeneity of data, then the adjustment of flood magnitudes is necessary before any frequency analysis. The results of the adjusted and unadjusted series could be significantly different. Two statistical tests that are commonly used to test the nonhomogeneity of flood records are introduced in the following sections. Both of these methods could be applied to records that have undergone a gradual change. Some other tests are also available for detecting a change in a flood record due to a rapid watershed change.

3.8.1 Spearman Test

If the annual measures of the gradual watershed change are available for each year of the flood record, the Spearman test could be applied. It is a bivariate test and needs both flood records and imperviousness data. For example, in a watershed with a 40-year flood record and the annual variations of percentage of impervious cover, the Spearman test could be used. The Spearman test includes the six general steps of a hypothesis test as follows (McCuen, 1998):

1. State the hypotheses. For testing the homogeneity of flood records in regard to urbanization, the null (H_0) and alternative (H_A) hypotheses are defined as follows:
   H_0: The urbanization does not affect the magnitude of the flood peaks.
   H_A: The magnitudes of the flood peaks are increased due to urbanization.
   The test is applied as a one-tailed test, as the statement of the alternative hypothesis indicates. The hypothesis statement also shows the expectation of a positive correlation.
2. Determine the test statistic. The Spearman test statistic is calculated using Equation 3.63. This statistic is named as the Spearman correlation coefficient (R_S) as shown below:

$$R_S = 1 - \frac{6 \sum_{i=1}^{n} d_i^2}{n^3 - n}, \quad (3.63)$$

   where n is the sample size and d_i is the ith difference in the ranks of the two series, which in this case are the series of annual flood peaks and the series of the average annual percentage of imperviousness. After ranking the measurements of each series from largest to smallest, the difference in ranks d_i is computed.
3. Specify the desired level of significance. Often the value of 5% is considered but in specific cases values of 1% and 10% can also be used. The level of significance is usually determined based on expert judgment rather than through a rational analysis of the situation.
4. Estimate the test statistic R_S for the available sample through the following steps:
   a. Find the rank of each event in each series, while keeping the two series in chronological order.
   b. Calculate the difference in the ranks of the two series for each record i.
   c. Using Equation 3.63, compute the value of the Spearman test statistic, R_S.
5. Determine the critical value of the test statistic from Table 3.33. The values in the table are for the upper tail. As the distribution of R_S is symmetric, the critical values for the lower tail could be obtained by multiplying the value obtained from Table 3.33 by -1.
6. Make a decision. For an upper one-tailed test, if the computed value in step 4 is larger than the critical value obtained in step 5, the null hypothesis is rejected. For a lower one-tailed test, the null hypothesis is rejected if the computed value is less than the critical value.

TABLE 3.33
Critical Values for the Spearman Correlation Coefficient for the Null Hypothesis $H_0 : |\rho_S|$ and Both the One-Tailed Alternative $H_A : |\rho_S| > 0$ and the Two-Tailed Alternative $H_A : \rho_S \neq 0$

	Level of Significance for a One-Tailed Test			
	0.050	0.025	0.010	0.005
Sample Size	Level of Significance for a Two-Tailed Test			
	0.100	0.050	0.020	0.010
5	0.900	1.000	1.000	1.000
6	0.829	0.886	0.943	1.000
7	0.714	0.786	0.893	0.929
8	0.643	0.738	0.833	0.881
9	0.600	0.700	0.783	0.833
10	0.564	0.648	0.745	0.794
11	0.536	0.618	0.709	0.818
12	0.497	0.591	0.703	0.780
13	0.475	0.566	0.673	0.745
14	0.457	0.545	0.646	0.716
15	0.441	0.525	0.623	0.689
16	0.425	0.507	0.601	0.666
17	0.412	0.490	0.582	0.645
18	0.399	0.476	0.564	0.625
19	0.388	0.462	0.549	0.608
20	0.377	0.450	0.534	0.591
21	0.368	0.438	0.521	0.576
22	0.359	0.428	0.508	0.562
23	0.351	0.418	0.496	0.549
24	0.343	0.409	0.485	0.537
25	0.336	0.400	0.475	0.526
26	0.329	0.392	0.465	0.515
27	0.323	0.385	0.456	0.505
28	0.317	0.377	0.448	0.496
29	0.311	0.370	0.440	0.487
30	0.305	0.364	0.432	0.478

For a two-tailed test, if the computed value lies in either tail, the null hypothesis is rejected, but the level of significance is twice that of the one-tailed test. If the null hypothesis is rejected, it could be concluded that the peak flood records are not independent of the urbanization and the watershed change highly affects the hydrologic situation of the watershed as well as the annual maximum flood series.

Example 3.18

Eleven annual maximum discharges and the corresponding imperviousness for a watershed are given in Table 3.34. Test the homogeneity of these flood records.

Solution: The homogeneity of data is tested using the Spearman test because of the gradual change in watershed imperviousness. Columns 4 and 5 of Table 3.34 present the ranks of the series, and the rank differences are calculated in column 6. Then the value of R_S is calculated as

$$R_S = 1 - \frac{6(288)}{11^3 - 11} = -0.310.$$

TABLE 3.34
Spearman Test for Examples 3.18

Year (1)	Annual Maximum Discharge Y_i (m³/s) (2)	Average Imperviousness X_i (%) (3)	Rank of Y_i r_{yi} (4)	Rank of X_i r_{xi} (5)	Difference $d_i = r_{yi} - r_{xi}$ (6)
1	123	40	11	5	6
2	453	35	5	10	−5
3	246	41	8	4	4
4	895	36	1	9	−8
5	487	37	4	8	−4
6	311	45	6	1	5
7	285	33	7	11	−4
8	206	39	9	6	3
9	803	42	2	3	−1
10	568	38	3	7	−4
11	189	44	10	2	8
					$\sum d_i^2 = 288$

Since the critical value of R_S is 0.536 for significance level of 0.05, from Table 3.33, which is more than the computed value of R_S, the null hypothesis is rejected. This means that urbanization has had a considerable effect on the annual maximum discharge series.

3.8.2 SPEARMAN–CONLEY TEST

If the annual data of watershed imperviousness change are not available, which is a common situation, the Spearman test cannot be used. In these cases, the Spearman–Conley test (Conley and McCuen, 1997) can be used to test for serial correlation. For applying the Spearman–Conley test, the following steps should be passed (McCuen, 1998):

1. State the hypotheses. For this test, the null and alternative hypotheses are defined as follows:

   H_0: There is no temporal relation between annual flood peaks.
   H_A: There is a significant correlation between consequence values of the annual flood series.
   The alternative hypothesis can be also expressed as a one-tailed test with an indication of positive correlation for a flood series supposed to be affected by watershed development. Since urban development increases the peak floods, a positive correlation coefficient is expected.

2. Determine the test statistic. The same statistic as the Spearman test (Equation 3.63) can be used for the Spearman–Conley test. However, the statistic is denoted as R_{sc}. In applying Equation 3.63, n indicates the number of pairs, which is 1 less than the number of annual maximum flood magnitude records. For the calculation of the R_{sc} value, a second series, X_t, is formed as $X_t = Y_{t-1}$. The value of R_{sc} is calculated with the differences between the ranks of the two series by using Equation 3.63.

3. Specify the desired level of significance. Again, this is usually determined by convention and a 5% value is commonly used.

4. Compute the test statistic R_{sc} for the available sample through the following steps:
   a. Create a second series of flood magnitudes (X_t) by offsetting the actual series (Y_{t-1}).
   b. Identify the rank of each event in each series, while keeping the two series in chronological order. The series are ranked in decreasing order so that a rank of 1 is used for the largest value in each series.
   c. Calculate the difference in ranks for each i.
   d. Compute the value of the Spearman–Conley test statistic R_{sc} using Equation 3.63.
5. Obtain the critical value of the test statistic. It should be noted that the distribution of R_{sc} is not symmetric and is different from that of R_S. The critical values for the upper and lower tails of the Spearman–Conley test statistic R_{sc} are given in Table 3.35. In this table, the number of pairs of values is used to compute R_{sc}.
6. Make a decision. For a one-tailed test, if the computed R_{sc} is greater in magnitude than the value of Table 3.35, the null hypothesis is rejected. It could be concluded that the annual maximum floods are serially correlated if the null hypothesis is rejected. This correlation reflects the impact of urbanization on the peak flood discharges.

TABLE 3.35
Critical Values for Univariate Analysis with the Spearman–Conley One-Tailed Serial Correlation Test

Sample Size	Lower-Tail Level of Significance (%)					Upper-Tail Level of Significance (%)				
	10	5	1	0.5	0.1	10	5	1	0.5	0.1
5	−0.8	−0.8	−1.0	−1.0	−1.0	0.4	0.6	1.0	1.0	1.0
6	−0.7	−0.8	−0.9	−1.0	−1.0	0.3	0.6	0.8	1.0	1.0
7	−0.657	−0.771	−0.886	−0.943	−0.943	0.371	0.486	0.714	0.771	0.943
8	−0.607	−0.714	−0.857	−0.893	−0.929	0.357	0.464	0.679	0.714	0.857
9	−0.572	−0.667	−0.810	−0.857	−0.905	0.333	0.452	0.643	0.691	0.809
10	−0.533	−0.633	−0.783	−0.817	−0.883	0.317	0.433	0.617	0.667	0.767
11	−0.499	−0.600	−0.746	−0.794	−0.867	0.297	0.406	0.588	0.648	0.745
12	−0.473	−0.564	−0.718	−0.764	−0.836	0.291	0.400	0.573	0.627	0.727
13	−0.448	−0.538	−0.692	−0.734	−0.815	0.287	0.385	0.552	0.601	0.706
14	−0.429	−0.516	−0.665	−0.714	−0.791	0.275	0.369	0.533	0.588	0.687
15	−0.410	−0.495	−0.644	−0.692	−0.771	0.270	0.358	0.521	0.574	0.671
16	−0.393	−0.479	−0.621	−0.671	−0.754	0.261	0.350	0.507	0.557	0.656
17	−0.379	−0.462	−0.603	−0.650	−0.735	0.256	0.341	0.491	0.544	0.641
18	−0.365	−0.446	−0.586	−0.633	−0.718	0.250	0.333	0.480	0.530	0.627
19	−0.354	−0.433	−0.569	−0.616	−0.703	0.245	0.325	0.470	0.518	0.613
20	−0.344	−0.421	−0.554	−0.600	−0.687	0.240	0.319	0.460	0.508	0.601
21	−0.334	−0.409	−0.542	−0.586	−0.672	0.235	0.313	0.451	0.499	0.590
22	−0.325	−0.399	−0.530	−0.573	−0.657	0.230	0.307	0.442	0.489	0.580
23	−0.316	−0.389	−0.518	−0.560	−0.643	0.226	0.301	0.434	0.479	0.570
24	−0.307	−0.379	−0.506	−0.549	−0.632	0.222	0.295	0.426	0.471	0.560
25	−0.301	−0.370	−0.496	−0.537	−0.621	0.218	0.290	0.419	0.463	0.550
26	−0.294	−0.362	−0.486	−0.526	−0.610	0.214	0.285	0.412	0.455	0.540
27	−0.288	−0.354	−0.476	−0.516	−0.600	0.211	0.280	0.405	0.448	0.531
28	−0.282	−0.347	−0.467	−0.507	−0.590	0.208	0.275	0.398	0.441	0.523
29	−0.276	−0.341	−0.458	−0.499	−0.580	0.205	0.271	0.392	0.434	0.515
30	−0.270	−0.335	−0.450	−0.491	−0.570	0.202	0.267	0.386	0.428	0.507

Urban Water Hydrology

EXAMPLE 3.19

Assuming the imperviousness data of Example 3.18 (column 3 of Table 3.34 is not available), test the homogeneity of the flood data.

Solution: The flood series from the second year are tabulated consequently in column 1 of Table 3.36. The offset values are given in column 2. As the record should be offset, one value is lost and thus for this test $n = 10$. The ranks of the two series are given in columns 3 and 4, and the differences in ranks are listed in column 5. The sum of the squares of the d_i values is equal to 86. The test statistic value is computed as

$$R_{sc} = 1 - \frac{6(200)}{10^3 - 10} = -0.212.$$

The critical value is 0.433 from Table 3.35, for a 5% level of significance and a one-tailed upper test. As the calculated value of R_{sc} is less than the critical value, the null hypothesis is rejected and it can be concluded that the flood data are temporally correlated. This serial correlation could be the result of urbanization.

TABLE 3.36
Summary of the Spearman–Conley Test for Example 3.19

Annual Maximum Discharge t Y_t (m³/s) (1)	X_t = Offset Y_t (m³/s) (2)	Rank of Y_i r_{yi} (3)	Rank of X_i r_{xi} (4)	Difference $d_i = r_{yi} - r_{xi}$ (5)
453	123	5	10	−5
246	453	8	5	3
895	246	1	8	−7
487	895	4	1	3
311	487	6	4	2
285	311	7	6	1
206	285	9	7	2
803	206	2	9	−7
568	803	3	2	1
189	568	10	3	7
				$\sum d_i^2 = 200$

3.9 HYDROLOGIC MODELING SYSTEM (HEC-HMS)

The hydrologic modeling system is a popular model employed for rainfall–runoff analysis in dendrite watershed systems. It is designed to be applicable in a wide range of geographic areas for solving the widest possible range of problems. This range includes large basin water supply and flood hydrology, and small urban or natural watershed runoff. Hydrographs produced by the program are used directly or in conjunction with other software for studies of water availability, urban drainage, flow forecasting, future urbanization impact, reservoir spillway design, flood damage reduction, floodplain regulation, and system operation.

The program features a completely integrated work environment including a database, data entry utilities, computation engine, and results reporting tools. A graphical user interface allows the user seamless movement between the different parts of the program. Program functionality and appearance are the same across all supported platforms.

3.9.1 Modeling Basin Components

The physical representation of watersheds or basins and rivers is configured in the basin model. Hydrological elements are connected in a dendritic network to simulate runoff processes. Available elements are sub-basin, reach, junction, reservoir, diversion, source, and sink (Figure 3.30). Computation proceeds from upstream elements in a downstream direction.

The loss rate methods include the following:

- The deficit constant
- Green–Ampt
- SCS CN
- Soil moisture accounting (SMA)
- Gridded SCS CN
- Initial constant
- Gridded SMA

The transform methods include the following:

- Clark
- Kinematic wave
- Modclark
- Synder
- SCS
- User-specified graph
- User-specified UH

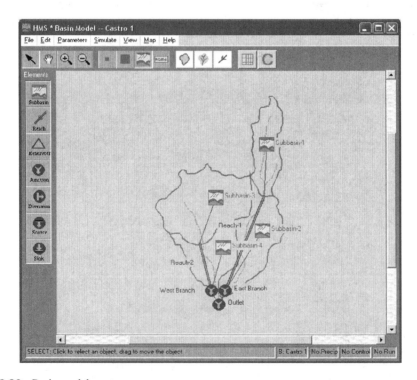

FIGURE 3.30 Basin model screen.

Urban Water Hydrology 117

3.9.2 ANALYSIS OF METEOROLOGICAL DATA

Meteorological data analysis is performed by the meteorological model and includes precipitation and evaporation. Six different historical and synthetic precipitation methods are included. One evaporation method is included at this time.

Different precipitation methods considered in the model are as follows:

- User hyetograph
- User gauge weighting
- Inverse-distance gauge weighting
- Girded precipitation, frequency storm
- Standard project storm–eastern U.S.
- SCS hypothetical storm (SMA)

3.9.3 RAINFALL–RUNOFF SIMULATION

The time span of a simulation is controlled by control specifications. Control specifications include a starting date and time, ending date and time, and computation time step (Figure 3.31).

3.9.4 PARAMETERS ESTIMATION

Most parameters for methods included in sub-basin and reach elements can be estimated automatically using the optimization manager. The observed discharge that must be available for at least one element upstream of the observed flow can be estimated. Different objective functions are available to estimate the goodness-of-fit between the computed results and the observed discharge. Two different search methods can be used to find the best fit between the computed results and the observed discharge. Constraints can be imposed to restrict the parameter space of the search method.

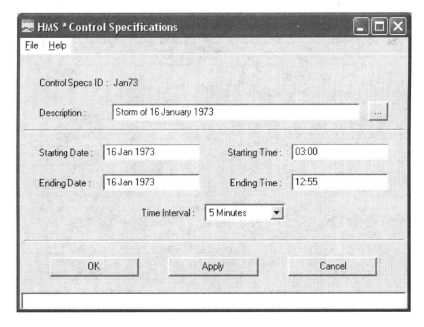

FIGURE 3.31 Control specification screen.

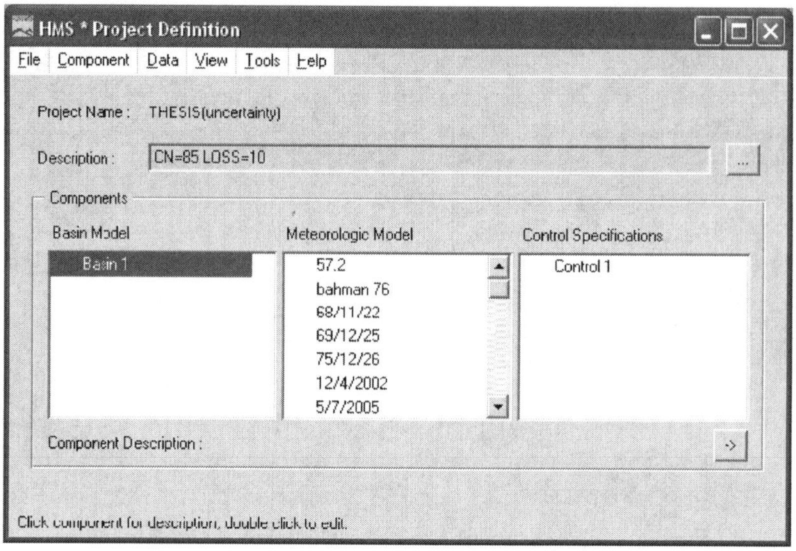

FIGURE 3.32 Project definition screen.

3.9.5 Starting the Program

3.9.5.1 Project Definition Screen

The project definition screen (Figure 3.32) is used to create and manage projects. It also provides access to components, analysis tools, and other data sets such as precipitation and discharge gauges. This part includes

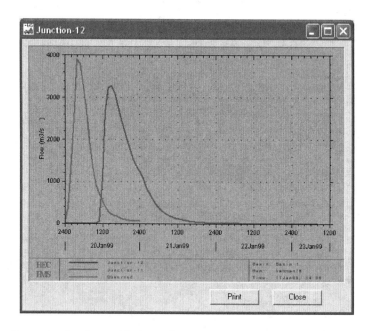

FIGURE 3.33 Comparing the observed hydrograph and outflow hydrograph.

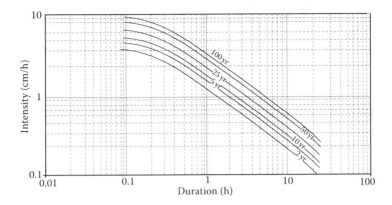

FIGURE 3.34 IFD curve for Problem 3.

- The File menu that contains items for creating and managing projects.
- The Components section of the Project Definition that contains three lists: the basin model, meteorologic models, and control specifications.
- The Data menu that contains items for accessing time series and other data managers.
- The View menu that contains items for accessing log files. The project log contains information messages generated since the project was opened.
- The Tools menu that contains items for working with simulation and optimization runs.

3.9.5.2 Input Data in HEC-HMS Model

Based on the different methods that are used in the basin model for the determination of loss rate, transform, base flow, and other items, or in the meteorologic method, diverse inputs are needed. As an illustration, CN, initial abstraction, and imperviousness are the inputs of the SCS CN method for determining the loss rate. Also, the storm depth is necessary in the SCS hypothetical storm model for the meteorologic model. Furthermore, the sub-basin area is needed.

3.9.5.3 Output Data in the HEC-HMS Model

The outflow discharge in the junctions and the output hydrograph from the last junction are the output data that are obtained from this model. Figure 3.33 compared the observed hydrograph and the outflow hydrograph in the last junction in a simulated basin.

PROBLEMS

1. Results of a practice for determining the Horton infiltration capacity in the exponential form are tabulated in Table 3.37. Determine the infiltration capacity exponential equation.

TABLE 3.37
Recorded Filtration Capacity Data of Problem 1

Time (h)	0.25	0.5	0.75	1.00	1.25	1.50	1.75	2.00
f_p (cm/h)	5.60	3.20	2.10	1.50	1.20	1.10	1.00	1.00

2. Given an initial infiltration capacity f_0 of 3.0 cm/h and a time constant m of 0.29 L/h, derive an infiltration capacity versus time curve if the ultimate infiltration capacity is

0.55 cm/h. For the first 10 h, estimate the total volume of water infiltrated over the watershed.
3. Estimate the time of concentration for shallow sheet flow on a 500 m section of asphalt roadway at a slope of 8% and manning roughness equal to 0.012. Assume a 25-year design frequency and the IDF curve of Figure 3.34. Assume the road soil belongs to group A.
4. The mass curve of a rainfall of 100 min duration is given in Table 3.38. (a) If the catchment has an initial loss of 0.6 cm and a Φ index of 0.6 cm/h, calculate the total surface runoff from the catchment. (b) If the direct runoff of catchment is 2 cm, determine the Φ for the basin.

TABLE 3.38
Mass Curve of Rainfall of Problem 4

Time from start of rainfall (min)	0	20	40	60	80	100
Cumulative rainfall (cm)	0	0.5	1.2	2.6	3.3	3.5

5. The data presented in Table 3.39 is the annual maximum series (Q_p) and the percentage of impervious area (I) for an urbanized watershed for the period from 1930 through 1977. Adjust the flood series to an eventual development of 50%. Estimate the effect on the estimated 2-, 10-, 25-, and 100-year floods.

TABLE 3.39
Data of Problem 5

Year	I (%)	Q_p (cms)	Year	I (%)	Q_p (cms)	Year	I (%)	Q_p (cms)
1930	21	1870	1946	35	1600	1962	45	2560
1931	21	1530	1947	37	3810	1963	45	2215
1932	22	1120	1948	39	2670	1964	45	2210
1933	22	1850	1949	41	758	1965	45	3730
1934	23	4890	1950	43	1630	1966	45	3520
1935	23	2280	1951	45	1620	1967	45	3550
1936	24	1700	1952	45	3811	1968	45	3480
1937	24	2470	1953	45	3140	1969	45	3980
1938	25	5010	1954	45	2410	1970	45	3430
1939	25	2480	1955	45	1890	1971	45	4040
1940	26	1280	1956	45	4550	1972	46	2000
1941	27	2080	1957	45	3090	1973	46	4450
1942	29	2320	1958	45	4830	1974	46	4330
1943	30	4480	1959	45	3170	1975	46	6000
1944	32	1860	1960	45	1710	1976	46	1820
1945	33	2220	1961	45	1480	1977	46	1770

6. Residents in a community at the discharge point of a 614-km^2 watershed believe that the recent increase in peak discharge rates is due to the deforestation by a logging company that has been occurring in recent years. Analyze the annual maximum discharges (Q_p) and an average forest coverage (FC) for the watershed data given in Table 3.40.
7. Analyze the data of problem 5 to evaluate whether or not the increase in urbanization has been accompanied by an increase in the annual maximum discharge. Apply the Spearman test with both a 1% and a 5% level of significance. Discuss the results.

TABLE 3.40
Data of Problem 6

Year	Q_p (cms)	FC (%)	Year	Q_p (cms)	FC (%)	Year	Q_p (cms)	FC (%)
1982	8000	53	1987	12,200	54	1992	5800	46
1983	8800	56	1988	5700	51	1993	14,300	44
1984	7400	57	1989	9400	50	1994	11,600	43
1985	6700	58	1990	14,200	49	1995	10,400	42
1986	11,100	55	1991	7600	47			

8. Consider two watersheds with different capacities for storage, such as sandy soil and clay with potential maximum retention of 7 and 5 cm, respectively and also initial abstraction before ponding is 1.5 cm. If a storm of 10 cm during 10 min occurs in both watersheds, determine the percentage of effective rainfall in each watershed.

9. A development project on a small upland watershed is shown in Figure 3.35. The developed portion of the area is 0.7 km² in which 21% is impervious area. The developed area is graded so that runoff is collected in grass-lined swales at the front of the lot and

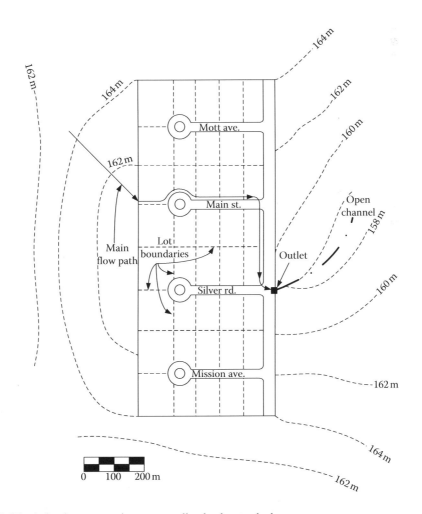

FIGURE 3.35 A development project on a small upland watershed.

drained to a paved swale that flows along the side of the main road. Flow from the paved swales passes through a pipe culvert to the upper end of a stream channel. The upper portion of the watershed with a maximum height of 164 m is a forest with B type soil (CN = 60) and has an area of 0.3 km². Estimate the peak runoff of this area if the design return period of the drainage system is 25 years. Use the IDF curve of Problem 3.

10. A rainfall hyetograph for a 70-min storm with a total depth of 10.3 cm is shown in Figure 3.36. The depth of direct runoff is 5 cm and the depth of water loses is 5.3 cm. (a) Estimate the Φ index for this basin. (b) Determine the excess rainfall and loss during the rainfall.

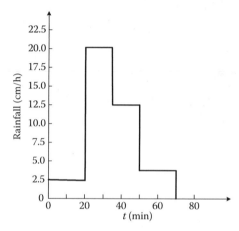

FIGURE 3.36 The hyetograph of rainfall of Problem 10.

11. Determine the infiltration losses and excess rainfall from the rainfall data given in Table 3.41 on a watershed using the modified Horton equation. The Horton infiltration capacity in the watershed is as follows: $f = 0.2 + 0.4e^{-0.6t}$.

TABLE 3.41
Hyetograph of Rainfall of Problem 11

Time (h)	i (cm/hr)	Time (h)	i (cm/hr)
0	1.2	2.00	2.0
0.25	1.2	2.25	1.7
0.50	1.2	2.50	1.7
0.75	1.8	2.75	1.7
1.00	1.8	3.00	0.9
1.25	1.8	3.25	0.9
1.50	2.0	3.50	0.4
1.75	2.0	3.75	0.4
		4.00	0

12. A watershed with an area of 230 km² is under further development during a 10-year horizon. After finishing the development project the percentage of imperviousness, watershed storage, and basin development factor will change from 25, 4 and 3 to 45, 6 and 9, respectively. Evaluate the effect of the urbanization on the peak discharge of this watershed. Assume that the average slope of watershed is 0.02 m/m. The IDF curve

Urban Water Hydrology

TABLE 3.42
Estimated Peak Discharges for Watershed of Problem 12 in Different Urbanization Conditions

Return Period (yr)	Rural Peaks (m³/s)
2	32.5
5	49.8
10	67.3
25	86.1
50	104.8
100	150.3
500	210.2

of Problem 3 could be applied in this watershed. The peak discharges from different return periods before urbanization are given in Table 3.42.

13. Calculate the Espey 10-min UH for an urban watershed with an area of 5 km², length of 1500 m, percentage of imperviousness of 45%, and a Manning's Roughness coefficient (n) of 0.012. The maximum difference between main water path altitude at the outlet and upstream (H) is about 60 m. If during urbanization the H and n decrease to 45 m and 0.01, how will the UH of this watershed be affected?

14. Develop the 1 h UH for an urban watershed with an area of 595 km² and a time of concentration of 10 h. The time of concentration and area for different regions of this watershed are given in Figure 3.37. Assume a storage coefficient of 45 min.

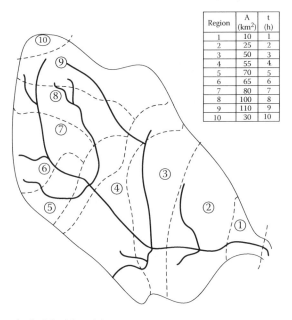

Region	A (km²)	t (h)
1	10	1
2	25	2
3	50	3
4	55	4
5	70	5
6	65	6
7	80	7
8	100	8
9	110	9
10	30	10

FIGURE 3.37 The watershed of Problem 14.

15. The ordinates of a 4-h unit hydrograph of a catchment are given in Table 3.43. Derive the flood hydrograph in the catchment for to the storm figures presented in Table 3.44.

TABLE 3.43
Ordinates of 4-h UH of Problem 15

Time (h)	0	2	4	6	8	10	12	14	16	18	20	22	24	26	28
Q (m^3/s)	0	25	50	85	125	160	185	160	110	60	36	25	16	8	0

TABLE 3.44
Hyetograph of Storm of Problem 15

Time from start of storm (h)	0	4	8	12
Accumulated rainfall (cm)	0	3.5	11.0	16.5

The storm loss rate (Φ index) for the catchment is estimated as 0.43 cm/h. The base flow at the beginning is 10 m^3/s and will increase by 1.5 m^3/s every 8 h until the end of the direct runoff hydrograph.

REFERENCES

Akan, A. O. 1992. Horton infiltration formula revisited. *Journal of Irrigation and Drainage Engineering, ASCE* 118: 828–830.

Akan, A. O. 1993. *Urban Stormwater Hydrology—A Guide to Engineering Calculations.* Technomic, Lancaster, PA (ISBN: 0-87762-966-6).

Akan, A. O. and Houghtalen, R. J. 2003. Urban hydrology. *Hydraulics and Stormwater Quality: Engineering Applications and Computer Modeling.* Wiley, New York (ISBN 0-471-43158-3).

American Society of Civil Engineering (ASCE). 1972. Design and construction of sanitary and storm sewers. Manual and Reports on Engineering Practice, No. 37, New York.

Aqua Terra Consultants. 1994. *Chambers Watershed HSPF Calibration,* prepared by D. C. Beyerlein and J. T. Brascher. Thurston County Storm and Surface Water Program, Thurston County, WA.

Bird, S. L., Exum, L. R., Alberty, S., Perkins, C., and Harrison, J. 2001. Estimating impervious cover from regionally available data. Poster at Southeastern Water Pollution Biologists Association Annual Meeting, Bowling Green, KY, October 30–November 1.

Chow, V. T., Maidment, D. R., and Mays, L. W. 1988. *Applied Hydrology.* McGraw-Hill, New York.

City of Olympia Public Works Department (COPWD). 1995. *Impervious Surface Reduction Study.* Olympia, WA.

Clark, C. O. 1943. Storage and the unit hydrograph. *Transactions of the American Society of Civil Engineers,* 110:1419–1446.

Conley, L. C. and McCuen, R. H. 1997. Modified critical values for Spearman's test of serial correlation. *Journal of Hydrological Engineering, ASCE* 2(3): 133–135.

Diskin, M. H. 1970. Definition and uses of the linear regression model. *Water Resources Research* 6(6): 1668–1673.

Espey, W. H., Jr., Altman, D. G., and Graves, C. B., Jr. 1978. Monographs for ten-minute unit hydrographs for small urban watersheds. Technical Memorandum No. 32 (NTIS PB-282158), ASCE Urban Water Resources Research Program, ASCE, New York (also Addendum 3 in Urban Runoff Control Planning. EPA-600/9–78-035. EPA, Washington, DC), December.

Graham, P. H., Costello, L. S., and Mallon, H. J. 1974. Estimation of imperviousness and specific curb length for forecasting stormwater quality and quantity. *Journal of Water Pollution Control Federation* 46(4): 717–725.

Green, W. H. and Ampt, G. A. 1974. Studies on soil physics. 1: The flow of air and water through soils. *Journal of Agriculture Science* 9: 1–24.

Hicks, R. W. B. and Woods, S. D., 2000. Pollutant load. Population growth and land use. *Progress: Water Environment Research Foundation* 11, 10.

Holtan, H. N. and Lopez, N. C., 1971. USADHL-70. Model of watershed hydrology. Technical Bulletin No. 14, 15 U.S. Department of Agriculture, Washington, DC.

Horton, R. E. 1919. Rainfall interception. *Monthly Weather Review*, 47: 603–623.

Horton, R. E. 1940. An approach toward physical interpretation of infiltration capacity. *Soil Science Society Proceedings* 5: 399–417.

James, W. and Scheckenberger, R. 1984. RAINPAK: A program package for analysis of storm dynamics in computing rainfall dynamics. *Proceedings of Stormwater and Water Quality Model Users Group Meeting*, Detroit, Michigan. EPA-600/9-85-003 (NTIS PB85-168003/AS), Environmental Protection Agency, Athens, GA, April.

James, W. and Shtifter, Z. 1981. Implications of storm dynamics on design storm inputs. *Proceedings of Stormwater and Water Quality Management Modeling and SWMM Users Group Meeting Niagara Falls Ontario*. McMaster University, Department of Civil Engineering, Hamilton, Ontario, pp. 55–78, September.

Kibler, D. F., Akan, A. O., Aron, G., Burke, C. B., Glidden, M. W., and McCuen, R. H. 1996. *Urban Hydrology*, 2nd Edition. ASCE Manual of Practice, No. 28, New York.

Kluiteneberg, E. 1994. *Determination of Impervious Area and Directly Connected Impervious Area*. Memo for the Wayne County Rouge Program Office, Detroit, MI.

Linsley, R. K. 1986. Flood estimates: How good are they? *Water Resources Research*, 22(9): 159–164.

Meadows, M. E. 1991. Extension of SCS TR-55 and development of single outlet detention pond performance charts for various unit hydrograph peak rate factors. Project Complementation Report, Vol. III, Submitted to U.S. Geological Survey, Reston, VA.

McCuen, R. H. 1998. *Hydrological Analysis and Design*. Prentice Hall, Englewood Cliffs, NJ.

McCuen, R. H., Rawls, W. J., and Wong, S. L. 1984. SCS urban peak flow method. *Journal of Hydrological Engineering, ASCE* 110(3): 290–299.

McCuen, R. H. 1998. *Hydrological Analysis and Design*. Prentice Hall, Englewood Cliffs, NJ.

McCuen, R. 2005. *Hydrologic Analysis and Design*, 3rd Edition. Pearson Prentice Hall, Upper Saddle River, NJ.

McPherson, C. B. 1978. *Mainstream and Critical Position*, University of Toronto Press, Toronto.

Northern Virginia Planning District Commission (NVPDC). 1980. *Guidebook for Screening Urban Nonpoint Pollution Management Strategies*. Northern Virginia Planning District Commission, Falls Church, VA.

Overton, E. D. and Meadows, M. E. 1976. *Stormwater Modeling*. Academic Press, New York.

Philip, J. R. 1957. An infiltration equation with physical significance. *Soil Science* 77(2): 153–157.

Ramsay, J. G. and Huber, M. I. 1987. *The Techniques of Modern Structural Geology, Volume 2: Folds and Fractures*, 700pp. Academic Press Inc., London.

Rawls, W. J., Brakensick, D. I., and Miller, N. 1993. Green–Ampt infiltration parameters from soils data. *Journal of Hydraulic Engineering, ASCE* 109: 62–70.

Rawls, W. J., Brakensiek, D. L., and Soni, B. 1983. Agricultural management effects on soil water processes. *Part L Soil Water Retention and Green and Ampt Infiltration Parameters Transactions of the ASAE* 26: 1747–1752.

Rawls, W. J., Goldman, D., Van Mullen, J. A., and Ward, T. J. 1996. Infiltration *Hydrology Handbook*, Chapter 3, 2nd Edition. ASCE Manuals and Reports on Engineering Practice, No. 28, New York.

Sarma, P. G. S., Delleur, J. W., and Rao, A. R. 1969. A program in urban hydrology. Part II. Technical Report No. 9, Waster Research Center, Purdue University.

Sauer, R. T., Pan, J., Hopper, P., Hehir, K., Brown, J., and Poteete, A. R. 1981. *Biochemistry*, 20: 35–91.

Sauer, V. B., Thomas, W. O., Jr., Stricker, V. A., and Wilson, K. B. 1983. Flood characteristics of urban watersheds in the United States. USGS Water Supply Paper 2207, U.S. Government Printing Office, Washington, DC.

Singh, V. J. 1992. *Elementary Hydrology*. Prentice Hall, Englewood Cliffs, NJ.

Soil Conservation Service (SCS). 1972. *National Engineering Handbook. Hydrology Section 4*, Chapters 4–10. United States Department of Agriculture, Soil Conservation Service, Washington, DC.

Soil Conservation Service (SCS). 1986. *Urban Hydrology for Small Watersheds, Technical Release 55*, 2nd Edition. U.S. Department of Agriculture, Springfield, Virginia, June. NTIS PB87-101580 (microcomputer version 1.11. NTIS PB87-101598).

Stankowski, S. J. 1974. Magnitude and frequency of floods in New Jersey with effects of urbanization. Special Report 38, USGS, Water Resources Division, Trenton, NJ.

Surkan, A. J. 1974. Simulation of storm velocity effects on flow from distributed channel networks. *Water Resources Research* 10(6): 1149–1160.

Tholin, A. L. and Keifer, C. J. 1960. Hydrology of urban runoff. *Transactions, ASCE*. 125: 1308–1379.

U.S. Army Corps of Engineers. 1990. *HEC-l Flood Hydrograph Package. User Manual*. Hydrologic Engineering Center, Davis, CA.

U.S. Army Corps of Engineers. 1998. *HEC-HMS Hydrological Modeling Systems User Manual*. Hydrologic Engineering Center, Davis, CA.

USDA. 1986. Natural Resources Conservation Service (NRCS). *Technical Release 55: Urban Hydrology for Small Watersheds*, 2nd Edition. United States Department of Agriculture, Washington, DC.

Viessman, W., Jr., Lewis, G. L., and Knapp, J. W. 1989. *Introduction to Hydrology*. Harper & Row, New York.

Wanielista, M., Kersten, R., and Eaglin, R. 1997. *Hydrology: Water Quantity and Quality Control*, 2nd Edition, Wiley and Sons, Inc., New York.

Wylie, E. B. and Streeter, V. L. 1978. *Fluid Transients*. McGraw-Hill, New York.

Youngs, E. G. 1968. An estimation of sorptivity for infiltration studies from moisture movement considerations. *Soil Science* 106(3): 157–163.

4 Urban Water Supply and Demand

4.1 INTRODUCTION

A major part of the sustainable approach to the urban water infrastructure process is the provision of continuous service. Historically, water supply, drainage, stormwater collection and disposal, sewage collection and treatment, and receiving water utilization are the main components of urban water systems that should be integrated into the provision of providing water services. The policies affecting the usage of receiving waters, including natural functions, in-stream uses, and withdrawals (e.g., for water supply), are the driving forces that control urban drainage and wastewater effluents.

There are some factors that cause the reduction of degraded urban water supply such as lower water consumption, conservation of natural drainage, water reuse and recycling, water contamination reduction, and preservation and/or enhancement of the receiving water ecosystem. This has led to the advocacy of the sustainable urban water system. Urban water systems should specifically meet the following basic objectives:

- Supply of safe and good-tasting drinking water to the inhabitants on a continuous basis
- Reclamation, reuse, and recycling of water and nutrients for use in farming, parks, or households in case of water scarcity
- Collection and treatment of wastewater in order to protect the inhabitants against diseases as well as the environment from harmful impacts
- Control, collection, transport, and quality enhancement of stormwater in order to protect the environment and urban areas from flooding and pollution

The first objective is addressed in this chapter. The second objective is discussed in Chapter 5. The third and fourth objectives are presented in Chapters 6 and 7. Most of the goals of sustainability have been reached or are within reach in North America and Europe, but are still far from being achieved in developing parts of the world. Poverty reduction and reduced child mortality through the millennium development goals (MDG) are the most important factors. The two specific goals related to water are

- Halving the proportion of people who are incapable of accessing or affording safe drinking water and sanitation
- Stopping the unsustainable exploitation of water resources by developing water management strategies at local, regional, and national levels, which promote both equitable access and adequate supplies

The populations of the urban areas are expected to increase dramatically, especially in Africa, Asia, and Latin America. The population of urban areas in Africa is expected to rise more than two times over the next 25 years. The urban population of the Caribbean and Latin America is expected to increase by almost 50% over the same period (WHO and UNICEF, 2000). Consequently, to meet the fast-growing necessities, the urban services will face great challenges over the coming decades. Water management with the provision of urban water supply services, including the basic requirements on urban water infrastructure, is the main concern of this chapter.

In the remaining part of this chapter, first an overview of water supply history and its variations in the recent years are given. Then different issues and challenges in the urban water supply are presented. Water demand projection which is an important factor in the urban water supply planning, is discussed and followed by a detail description on hydraulic analysis of water distribution networks. Both quantitative and qualitative aspects of water distribution networks are addressed in this chapter.

4.2 HISTORY OF WATER SUPPLY DEVELOPMENT

In formulating modern and unified drinking water systems, some valuable milestones are used. In about 3000 BC, drinking water was distributed through lead and bronze pipes in Greece. In 800 BC, the Romans built aqueduct systems that provided water for drinking, street washing, public baths, and latrines. At the beginning of the nineteenth century, the first public water supply systems in cities such as Philadelphia, USA and Paisley, Scotland were constructed. In the middle of the nineteenth century, water quality became a problem and hence filter systems were introduced in some cities. The spread of cholera necessitated the use of disinfection, and the first chlorination plants were installed around 1900 in Belgium and New Jersey, USA. During the twentieth century, advanced treatment of centrally supplied drinking water, including physical, biological, and chemical treatment, was presented by all large cities in North America and Europe. This was mainly done for surface waters; however, groundwater supplies in many countries have required and even now require minimal treatment or no treatment at all (e.g., in Slovenia and Denmark). An innovative treatment process called microfiltration of raw drinking water was introduced at the end of the twentieth century. Water treatment and water distribution were effectively governed by various laws and regulations, such as the U.S. Safe Drinking Water Act (SDWA) in USA (introduced in 1974, amended in 1986 and 1996) and the 1998 European Union Drinking Water Directive 98/83/EC. More extensive, globally developed guidelines on drinking water quality can be obtained from the World Health Organization (WHO, 2004).

4.3 WATER AVAILABILITY

A considerable uncertain environmental destruction appeared as a result of progress despite improvement in the quality of life. Since the beginning of nineteenth century, the population of the world has tripled, nonrenewable energy consumption has increased by a factor of 30, and the industrial production has multiplied by 50 times.

Currently, the world's urban population growth, which is a key issue in urban areas, is at four times the rate of the rural population. The urban population is projected to become more than 5 billion by the year 2025 and two-thirds of the world's population will be living in towns and cities (World Resources Institute, 1998). According to World Water Institute (1998), the rate of urban population growth is much higher in the developing countries. Developed regions include North America, Japan, Europe, and Australia and New Zealand. Developing regions include Africa, Asia (excluding Japan), South America and Central America, and Oceania (excluding Australia and New Zealand). The population projections in the future decades in the developed and developing countries show completely different behaviors. The results show about 2 billion population increase in the developing countries in the next 25 years. But the population remains constant in the developed countries in the same period. This indicates that the developed countries have reached their population growth limits. The rate of urbanization in different regions is also considerably different due to their rate of development. As an example, the average annual urban growth rate is about 4% per year in Asia, and Africa has the highest urban growth rate of about 5% per year (World Water Institute, 1998).

EXAMPLE 4.1

Estimate the population of a town in the year 2010 based on the past population data given in Table 4.1.

Solution: Since actual populations between the 10-year census intervals are not available, the data points by an assumed linear modification can be connected among each census. One can speculate on the reasons for some of the changes:

- The drop in population during the depression of the 1930s, most likely caused by people retreating significantly due to the drop in employment.
- The rapid augmentation in the war and postwar boom periods of the 1940s and 1950s.
- A slowdown in growth during the difficult time of the 1970s.

Are the 1980s being a replica of the 1930s? Does the drive to achieve more jobs through new industries succeed? Could the rate of growth in the coming years be as high as during the 1940–1960 rapid growth periods?

Answers to these questions will be available only with more information on the specific town and region. In fact no one, even with broad information, can make more than an informed, educated guess. The best we can do are the following projections:

High projection: Assume that the current growth rate in 1950 will continue for 5 years, followed by an increasing rate of about four-fifth of the maximum previously experienced growth rate to the year 2000:

$$P_{(1990-2000)} = \frac{66,200 - 59,400}{10} = 680 \text{ persons/year},$$

$$P_{(1950-1960)} = \frac{45,050 - 32,410}{10} = 1264 \text{ persons/year}.$$

Therefore the high population for 2010 is $66,200 + 5 \times 680 + 5 \times 4/5 \times 1264 = 74,656 = 74,700$ persons.

Medium projection: The current growth rate in 1950 will continue for the next 17 years. Therefore, the medium projection for 2010 is $66,200 + 10 \times 680 = 73,000$ persons.

Low projection: Assume that the current growth rate in 1950 will continue for 5 years, followed by a decreasing rate of about four-fifths of the maximum previously experienced drop rate to the year 2010. The drop rate of 1930–1940 is

$$\frac{22,480 - 26,210}{10} = -373.$$

Therefore the low population for 2010 is $66,200 + 5 \times 680 - 5 \times 4/5 \times 373 = 68,108 = 68,100$ in persons. Therefore, the estimated range of population growth is 4900 or about 7% of the present population.

TABLE 4.1
The Population Data of Example 4.1

Year	1900	1910	1920	1930	1940	1950	1960	1970	1980	1990	2000
Mid-year population	10,240	12,150	18,430	26,210	22,480	32,410	45,050	51,200	54,030	59,400	66,200

4.4 WATER DEVELOPMENT AND SHARE OF WATER USERS

An answer to the question—How development could be done in a way that is economically and ecologically sustainable?—is a vision of the future, and our planning schemes are environmentally responsible and sensitive to the major elements of our physical environment, namely air, water, and soil. Among these elements, water is of particular importance and should be considered in the plans for sustainable development of water resources development, water conservation, and waste and leakage prevention as follows:

1. Improving the efficiency of water systems
2. Improving the quality of water
3. Downstream environmental flow consideration and water withdrawal and usage within the limits of the system
4. Water pollution reduction considering the carrying capacity of the streams
5. Water discharge from groundwater considering the safe yield of the system

The total world freshwater supply is estimated to be 41,022 billion cubic meters (World Resources Institute, 1998). Only 8% of the total freshwater supply on earth has been used, with agriculture having the highest rate of 69% among the water uses. Eight percent of the total water consumption has been used in the domestic sector. Although the percentage of water consumption in the domestic sector is not high, the concentration of population in the urban areas has amplified the water shortages in this sector.

Water use in the world has increased by six times between the years 1900 and 1995, which is more than twice the rate of the population growth. As mentioned earlier, only 8% of all the available freshwater, that is, about 220 lpd on average per capita, remains for all other domestic uses. In developing countries, especially in mega cities, drinking water demand growth is faster than urban population growth. Despite the fact that the urban population uses only small amounts of the available water for consumption, delivery of sufficient water volumes is becoming a difficult logistic and economical problem. In spite of efforts made during the past several decades, about 1.2 billion people in underdeveloped and developing countries do not have access to safe drinking water supplies. The population of water-short areas will be approximately 65% of the world population by the year 2050 (Milburn, 1996). More recent studies have shown that the pace of population growth is slowing down but the total population is increasing where as the state of renewable water resources remains the same or even decreases due to quality issues (Knight, 1998).

Example 4.2

Consider an urban area with approximately 348,000 residents. Because of the water supply and distribution system, only 38% of the water supplied is used and the remaining is lost. To alleviate this problem, a rehabilitation and improvement program is considered.

(a) If the rehabilitation and improvement program reduced consumption losses by 1%, calculate the volume of water that would be saved in 1 year. The maximum water demand per capita is 350 m^3/year.
(b) Calculate the size of a city that could be served with the water saved, if the water demand is the same.

Urban Water Supply and Demand

> *Solution*:
>
> (a) Current consumptive loss = 62% of the water used. Water saved by a 1% reduction of consumptive losses is
>
> $$1\% \times 348{,}000 \text{ (person)} \times 350 \text{ (m}^3/\text{person.year)} \times 0.62 = 755{,}160 \text{ (m}^3/\text{year)}.$$
>
> (b) Population that could be supplied by the reduction of loss is
>
> $$\frac{755{,}160 \text{ (m}^3/\text{year)}}{350 \text{ (m}^3/\text{person·year)}} = 2158 \text{ person}.$$

4.5 MAN-MADE AND NATURAL RESOURCES FOR WATER SUPPLIES

There is a constant expansion in the gap between society's needs and the natural capacity of water supply. The supply that bridges this gap is an important element of the total artificial water cycle in the urban areas. Groundwater or surface resources such as lakes, reservoirs, and rivers are the origins of the necessary water in urban areas. It is called untreated or raw water and is usually transferred to a water treatment plant. The water distribution network starts after the treatment facilities. The degree of treatment will be determined by the raw water quality and the intended use of water. In the past decades, different water quality standards for municipal purposes were developed and used.

Quantity, quality, time variation, and price are the four major characteristics of water supply. In almost all urban areas, the time variation of available water resources does not follow demand variations. However, when the quantity and time variation of water resources conforms to the water use patterns in an urban area, storing or controlling water by man-made structures or tools is not needed. Certain facilities, therefore, should be implemented to store the excess water in the high-flow seasons for it to be consumed in the low-flow periods.

Planning for water development requires an assessment of the costs of initial investment and operation and maintenance costs that should be used in the economic studies. In the case of treating water quality for different water uses, the costs should also be incorporated into the urban water resources development studies. Generally, the following categories represent water supply methods:

- Large-scale and conventional methods of surface and groundwater resources development
- Nonconventional methods

Some of the conventional methods of water supply are the construction of large-scale facilities such as dams, water transfer structures, and well fields. Dams are the most important element of urban water resources systems in many large cities around the world.

4.6 SUPPLEMENTARY SOURCES OF WATER

Rainwater harvesting, bottled water, and wastewater reclamation and reuse are three important additional sources of water in urban areas. Rainwater harvesting, especially in places with relatively extreme rainfall and limited surface waters (e.g., small islands), is a supplementary or even primary water source at the household or small community level. In WSUD, the use of urban runoff for watering the open-space lawn and landscape or other uses in urban area parks is highly recommended. This method supports environmental sustainability by reducing water supply demands for those purposes and reducing urban runoff and its impacts. The most common collecting surfaces are roofs of buildings and natural and artificial ground collectors. Falkland (1991) suggested the following issues in the design of rainfall-harvesting systems.

In the quantity issues, insufficient storage tank volumes or collector areas often affect the rainwater collection systems. Leaching from tanks due to poor design, selection of materials, construction, or a combination of these factors is a major problem of the rainwater collection systems. In the quality issues, the rainwater quality in many parts of the world is good, but water quality problems may arise within the collection systems or air quality. Physical, chemical, and biological pollution of rainwater collection systems occur where inappropriate construction materials have been used or where maintenance of roofs and other catchment surfaces, gutters, pipes, and tanks is lacking.

In many parts of the world, rainwater tanks or other rainwater collection devices are used for centuries. Currently, this approach is used widely in Australia and India. Many farms in rural areas of Australia are not connected to water supply systems nor do they have adequate well water. These systems are encouraged by local water authorities in urban areas, and in some cases by offering rebates to customers who use rainwater for subpotable uses. The most possible reuse of rainwater in urban areas is for gardening, which accounts for 35–50% of domestic water use in many large cities of the world. A fairly simple system, with very low environmental risks, is used for reusing rainwater in the garden, and it is therefore encouraged by many water authorities (Karamouz et al., 2003). An example of a rooftop rainwater-harvesting system is presented in Figure 4.1. More details on application and design of rainwater harvesting systems is given in Chapter 5.

The collected rainwater can be used for toilet flushing (about 20% of domestic water use), the laundry, kitchen, bathroom, pools, and washing cars, so some potable water can be further saved. In some situations (e.g., in some rural areas), it may be possible to use rainwater for most domestic

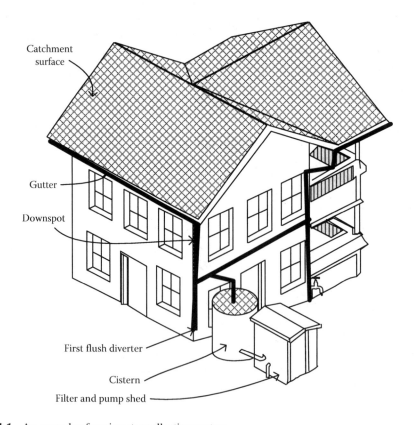

FIGURE 4.1 An example of a rainwater collection system.

uses, without relying on the public water supply. In all these cases, strict regulations for reclaimed water quality must be followed and safety systems employed, particularly in connection with drinking water, which should be protected by the so-called "multiple barrier system." In these cases, multiple barriers are used to control the microbiological pathogens and contaminants that may enter the water supply system, thereby ensuring clean, safe, and reliable drinking water.

4.7 WATER TREATMENT METHODS

Making raw water drinkable is generally done by water treatment. In parallel with reducing the high-quality sources of water, using lower quality water resources requires more treatment and enduring more water losses during the treatment.

An alternative system delivers raw source water, or just primarily treated water, directly to the consumer, who further treats this water in small, local treatment units near the point of consumption. The avoided deterioration of the water quality during transport and each consumer's treating the water to the level specifically needed for his/her requirements are the main advantages of this system. Obviously, for purposes such as drinking water, process water for industry, cooling water, irrigation water, or water for flushing toilets, the water quality is not the same. Small and locally used microfiltration units are quickly becoming competitive in price.

Adding pretreatment and some chlorine to the treated water ozonation and UV irradiation in order to maintain chlorine residuals during transport in the distribution network and prevent growth of bacteria originating from biofilms found in the pipe system is needed before discharging disinfected water into the distribution system. In many countries, a measure of safety is the residual chlorine concentration at the tap. In this chapter, through two simple examples, some of the sizing characteristics of water treatment are presented. The detailed discussion of water treatment is beyond the scope of this book and can be found in related textbooks such as Metcalf and Eddy (2003).

EXAMPLE 4.3

Assume a water treatment plant for a city of 100,000 people with maximum hourly and daily rates of water demand equal to 6.87×10^6 liters per hour (Lph) and 99.0×10^6 liters per day (Lpd), respectively. The detention time for coagulation/flocculation (A) is 25 min and the tank is 3.7 m deep. The detention time for the sedimentation tank (B) is 2 h and the tank is 5.0 m deep. The filtration rate (C) is 110 liters per minute (Lpm)/m^2. Select the appropriate dimensions for these three units. Three parallel sets of tanks are used to provide flexibility in operation and each set includes four filters. Consider that the widths for all the treatment units are equal to 18 m.

Solution: The required processing rate is the maximum daily water demand. Each tank will handle one-third of this flow or 33×10^6 Lpd (22,916 Lpm). Accordingly, the required capacity of the coagulation/flocculation tank is

$25 \text{ min} \times 22{,}916 \text{ Lpm} = 572.9 \times 10^3 \text{ L} = 572.9 \text{ m}^3$.

So, length A of the coagulation/flocculation tank is $572.9/(18 \times 3.7) = 8.6$ m.

The required capacity of the sedimentation tank is

$120 \text{ min} \times 22{,}916 \text{ Lpm} = 2749.9 \times 10^3 \text{ L} = 2749.9 \text{ m}^3$.

So, length B of the sedimentation tank is $2749.9/(18 \times 5) = 30.6$ m.

Each filter will handle one-twelfth of the total flow or 5729 Lpm. Thus the required area of each filter is $5729/110 = 52.1$ m^2.

And the length C of each filter box is $52.1/4.5 = 11.6$ m.

> **EXAMPLE 4.4**
>
> Calculate (a) the number of kilograms of chlorine is needed per day and (b) the capacity of the contact tank in the water treatment plant for the city of 100,000 people given in the previous example. The chlorine demand is 1 mg/L. It should be noted that at least 1.2 mg of chlorine must be added to every liter to overcome the chlorine demand of 1 mg/L and produce a free available chlorine concentration of 0.2 mg/L.
>
> *Solution*:
>
> (a) Since the treatment plant must be capable of operating at the maximum daily flow rate, the amount of chlorine needed can be calculated as
>
> $$\frac{\text{kg of chlorine}}{\text{day}} = \text{maximum daily flow rate} \times \frac{1.2 \text{ mg of chlorine}}{\text{liter}} \times \frac{\text{kg}}{1 \times 10^6 \text{ mg}}$$
>
> $$= 99.0 \times 10^6 \text{ Lpd} \times 1.2 \text{ mg/L} \times \frac{1}{10^6} \text{ kg/mg} = 118.8 \text{ kg/day}.$$
>
> (b) If we assume a minimum contact time of 40 min, then
>
> $$\text{Required capacity of the contact tank} = \text{flow rate} \times \text{contact time}$$
>
> $$= 99.0 \times 10^6 \text{ (Lpd)} \times (\text{day}/1440 \text{ min}) \times 40 \text{ (min)}$$
>
> $$= 2.75 \times 10^6 \text{ L} = 2750 \text{ m}^3.$$

4.8 WATER SUPPLY SYSTEM CHALLENGES

A number of key challenges required for the management of urban water systems, are common to all towns and cities. They include environmental, social, and economic dimensions but many of the underlying causes are interrelated and overlapping. One of the major challenges is reaching consensus among various stakeholders on the environmental, social, and economic goals and values of urban water systems. More extensive community input and greater understanding of water management options will improve the sustainability of current systems and make it faster.

Other major challenges suggested by Mays (1996) are as follows:

- Inadequate water flows from excessive and inefficient water use.
- Infection of surface waters and groundwater from uncontrolled or deficiently directed stormwater drainage and wastewater disposal.
- Consumers and ratepayers have increasing expectations about the provision and quality of water services but there is often a negative reaction to large rate increases or increased charges to fund the required infrastructure development.
- Lack of awareness and understanding of the value of urban water systems and the costs of improving water supplies, wastewater, and stormwater management.
- Poor recreational and bathing water quality, and poor information disclosure.
- Lack of investment and deferred maintenance, in part through incomplete pricing and inadequate financial contributions from new urban developments.
- Institutional and regulatory barriers to improved management.
- Potential risk of infrastructure failure.

4.9 WATER DEMAND

A problem for government authorities around the world is the provision of adequate water supply and hygiene to the rapidly growing urban population. Capacity expansion by finding new water resources

Urban Water Supply and Demand

TABLE 4.2
Definitions of Water Use Terms

Term	Definition
Consumptive use	The part of water withdrawn that is removed from the immediate water environment through natural or anthropogenic phenomenons such as evaporation, transpiration, incorporation into products or crops, consumption by humans or livestock
Conveyance loss	The amount of water that is lost during water transmission through a pipe, canal, conduit, or ditch by leakage or evaporation
Delivery and release	The quantity of water delivered to the consumption point and the amount released after use
In-stream use	Water that is used without withdrawn from any water resource for some purposes such as hydroelectric-power generation, navigation, water quality improvement, fish propagation, and recreation
Offstream use	Water withdrawn or derived from a water resource for different uses such as public water supply, industry, irrigation, livestock, thermoelectric-power generation
Public supply	Water withdrawn by public or private water suppliers and delivered to users
Return flow	Water release from usage point that reaches a water resource. Return flow could be used again in future
Reclaimed wastewater	The effluent of wastewater treatment-plant that has been diverted for beneficial use before it discharges to water bodies
Self-supplied water	Water withdrawn from a water resource by a user rather than being obtained from a public supply
Withdrawal	Water removed from the ground or delivered from a surface-water source for offstream use

or expanding existing sources is becoming more difficult and costly in many developed or developing parts of the world, and is often physically and economically infeasible. In some cities, the actual cost of water per cubic meter in second- and third-generation water supply projects has doubled compared to the first- and second-generation projects (Bhatia and Falkenmark, 1993).

Water demand is the scheduling of quantities that consumers expect to use per unit of time (for particular prices of water). Water use can be classified into two basic categories: consumptive and nonconsumptive uses; the first category removes water from the immediate water source, and the second one is related to water diversion or nondiversion from the water sources and immediate return of water to the source at the point of diversion in the same quantity as diverted and water quality standards (Table 4.2).

To estimate the total communal water use in a city, the study area should first be divided into homogeneous subareas. These homogeneous areas are usually selected based on pressure districts or land-use units and the water use rates can be assumed to be constant for different users within each subarea. Temporal (annual, seasonal, monthly, etc.) variation may also be considered in disaggregating the water uses.

4.9.1 Fluctuations in Water Use

The demands on a water system vary from year to year, season to season, day to day, and hour to hour. An example of short-term variation in residential water demand during summer and winter is shown in Figure 4.2. Note that a considerable increase in water consumption may result due to lawn watering during the early hours of summer evenings. Demand fluctuations are a part of the average daily demand. Records of water demand in similar areas can be analyzed statistically to yield ratios such as those given in Table 4.3.

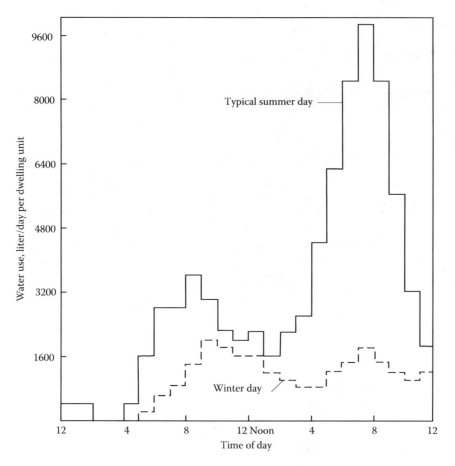

FIGURE 4.2 Residential water use fluctuations.

Providing the larger of the maximum hourly demand or the maximum daily demand plus the fire demand to any group of hydrants in the system is the basis for the designing of water distribution systems. The flow required to put out or at least contain a fire in an individual group of buildings can be estimated from an experimental formula (Insurance Services Office, 1974):

$$F = 224C\sqrt{A}, \quad (4.1)$$

where F is the required fire flow (L/min); C is a coefficient that takes into account the type of construction, existence of automatic sprinklers, and building exposures. Its value is 1.5 for wood

TABLE 4.3
Demand Variations

Average Daily Rate	Yearly 1.0 Summer 1.25 Winter 0.8
Maximum daily rate	1.5 (range 1.2–2.0)
Maximum hourly rate	2.5 (range 1.5–12)

frame construction, 1.0 for ordinary construction, 0.8 for noncombustible construction, and 0.6 for fire-resistive construction; and A is the total building area or floor space (m^2).

EXAMPLE 4.5

Calculate the water consumption (average daily rate, maximum daily rate, maximum hourly rate, and fire flow) of an industrial–commercial–residential city of 100,000 people. The total floor area of the largest office building complex downtown is 25,000 m^2 and average daily water consumptions are presented in Table 4.4. Assume that the coefficient C is 1.0 for this building, and determine the required capacity of the pipe distribution system.

Solution: From Table 4.4, the average daily consumption of water is 620 lpcd. Then

(a) average daily rate $= 620 \times 100{,}000 = 62.0 \times 10^6$ Lpd.
   From Table 4.3, assume that
(b) maximum daily rate $= 1.5 \times$ average daily rate $= 1.5 \times 62 \times 10^6 = 93.0 \times 10^6$ Lpd.
(c) maximum hourly rate $= 2.5 \times$ average hourly rate $= 2.5 \times 62 \times 10^6 = 155 \times 10^6$ Lpd $= 6.46 \times 10^6$ Lph.
(d) The required fire flow is determined using Equation 4.1:

$$F = 224C\sqrt{A} = 224C\sqrt{A} = 224(1.0)\sqrt{25{,}000} = 35.4 \times 10^3 \text{ Lpm}.$$

(e) The design flow for the pipe distribution network is the greater of either (1) the sum of the maximum daily demand and the fire flow or (2) the maximum hourly rate. We have
   (1) 93.0×10^6 (Lpd) $\times$ (1 day/1440 min) $+ 35.4 \times 10^3$ (Lpm) $= 64{,}583 + 35{,}400 = 99{,}983$ Lpm $= 6.0 \times 10^6$ Lph.
   (2) From part (c), maximum hourly rate $= 6.46 \times 10^6$ Lph. Therefore, the pipe capacity must be 6.46×10^6 Lph.

TABLE 4.4
Water Use in the City of Example 4.1

Use	Average Daily Consumption Per Person (lpcd)	Percentage of Total Use
Domestic	270	45
Commercial	110	15
Industrial	160	25
Other	80	15
Total	620	100

4.9.2 WATER QUANTITY STANDARDS IN URBAN AREAS

Two terminologies that are commonly used in the texts related to water quantity and quality management are criteria and standards. McKee (1960) has differentiated between standards and criteria as follows.

The term "standard" applies to any definite rule, principle, or measure established by authority. The fact that it has been established by authority makes a standard rather rigid, official, or quasi-legal; but this fact does not essentially mean that the standard is fair, equitable, or based on sound scientific knowledge, for it may have been established somewhat arbitrarily on the basis of inadequate technical data tempered by a caution factor of safety. Such arbitrary standards may be justified, where health is involved and where scientific data are sparse. A criterion designates a mean by which anything is tried, in forming a correct judgment concerning it. Unlike a standard, it carries no connotation of authority other than fairness and equity; nor does it imply an ideal condition. Where scientific data are being accumulated to serve as yardsticks of water quality, without considering legal authority, the term "criterion" is most applicable.

The water supply systems should satisfy some quantitative guidelines and standards, in urban areas. In urban areas where their water demand is supplied by surface water resources, the tributary watershed should be able to yield the estimated maximum daily demand 10 years into the future, and the storage capacity of an impounding reservoir should be equivalent to at least 30 days maximum daily demand 5 years into the future. Ideally, no mining of water is required for well supplies; that is, neither the static groundwater level nor the specific capacity of the wells ($m^3/s/m$ of drawdown) should decrease appreciably as the demand increases. Preferably these values should be constant over a period of 5 years except for minor variations that correct themselves within one week.

The domestic consumption of water per capita is the amount of water consumed per person for the purposes of ingestion, hygiene, cooking, washing of utensils, and other household purposes. The per capita water consumption can be measured (or estimated) through metered supply, local surveys, sample surveys, or total amount supplied to a community divided by the number of inhabitants. In order to achieve the sustainable development in supplying water demands, the per capita water use should be limited using some site-specific regulations or standards with or without limited increase in per capita water consumption.

Delivering required fire flows in different parts of the municipality with usage at the maximum daily rate should be considered in the designing of the layout of supply mains, arteries, and secondary distribution feeders. The water quantity standards should be given for the greatest effect that a break, joint division, or other kinds of main failure could have on the supply of water to a system. Some of these standards or regulations are as follows.

The recommended water pressure in a distribution system is about 450–520 kpa, which is considered to be adequate to compensate for local fluctuations in consumption. This level of pressure can provide for ordinary consumption in buildings up to 10 stories in height, as well as sufficient supply for automatic sprinkler systems for fire protection in buildings of four or five stories. The minimum pressure in the water distribution main should be 690 kpa for a residential service connection. The maximum allowable pressure is 1030 kpa (Walski et al., 2001).

In evaluating a system, pumps should be rated at their effective capacities when discharging at normal operating pressures. The pumping capacity in conjunction with storage should be adequate to maintain the maximum daily use rate plus maximum required fire flow with the single most important pump that is out of service.

Storage is frequently used to equalize pumping rates into the distribution system as well as provide water for fire fighting. In determining the fire flow from storage, it is necessary to calculate the rate of delivery during the specified period. Although the amount of storage may be more than adequate, the flow to a hydrant cannot surpass the carrying capacity of the mains, and the residual pressure at the point of use cannot be <140 kpa (Walski et al., 2001).

Fire hydrants are installed at spacings and in locations convenient for fire determined by the Fire Department. The linear distance between hydrants along streets in residential districts is normally 180 m with a maximum of 240 m and in high value districts is normally 90 m with a maximum of 150 m (Walski et al., 2001).

A gravity system delivering water without using a pump is desirable from a fire protection standpoint. Because of reliability, well-designed and properly protected pumping systems can be developed with a high degree of standards so that no distinction is made between the reliability of gravity and pumping systems. Two separate lines from different directions to all pumping stations and treatment facilities should provide electric power. The reader is referred to the water distribution textbooks such as Walski et al. (2001) for more detailed information about the quantitative standards in water distribution systems.

4.10 WATER DEMAND FORECASTING

Regression analysis is the most commonly used method for water use prediction. The basis for how the independent variables of the regression model should be selected is the available data about different factors affecting the water use and their relative importance in increasing or decreasing water uses. One of the most important factors in estimating water use in an urban area is the population of each subarea. A multiple regression method can also be used to incorporate more variables correlated with water use in municipal areas for estimating and forecasting the water demand in the future. Some of the variables are population, price, income, air temperatures, and precipitation (Baumann et al., 1998).

To forecast water demands, time series analysis has also been used. For this purpose, the time series of municipal water use and related variables are used to model the historical pattern of variations in water demand. Long memory components, seasonal and nonseasonal variations, trends, jumps, and outlier data should be cautiously identified and used in modeling water demand time series. More introductions to water demand forecasting and its different methods are presented in Chapter 5.

4.11 WATER STORAGE

Storage tanks are the main components of urban water distribution networks and are necessary to supply variable water demand, provide fire protection, and for emergency needs. Surface reservoirs, standpipes, and elevated tanks are three types of reservoirs used in the urban water distribution networks. Surface reservoirs are situated where they could provide adequate water pressure. They are usually covered to avoid contamination. Sufficient pressure will be provided by the natural elevation on a hill or through the use of pumps. Standpipes are basically tall cylindrical tanks. The upper portion of these tanks is used for storage to produce the necessary pressure head and the lower portion serves to support the structure. Standpipes over 15 m in height are uneconomical, and above this height elevated storage tanks are the preferred choice.

High-lift pumps at the treatment facilities are not normally designed to meet variation of residential water demand; the usual practice is to pump water into the distribution system at a fixed rate for a given period and allow a reservoir to either supply extra water if the demand exceeds this rate or receive water if the demand is less than the pumping rate. In this way, reservoirs accomplish their regulating function by hydrostatic pressure alone. In large cities, reservoirs could be located in the middle of several distribution areas. How the location of a reservoir affects its capability to balance operating pressures throughout a distribution system is shown in Figure 4.3. Note how high water use and the accompanying friction losses increase the slope of the pressure profile so that water starts to flow from the reservoir to the surrounding area.

The slope of the hydraulic grade line from the pump to the tank decreases by decreasing the demand and allowing water to enter the tank and recharge the storage. Recently, elevated tanks have become less popular, partially due to their increased cost and the availability of relatively inexpensive variable-speed pumps and controls, which make it possible to adjust pumping rates to varying demand.

The type and location of storage and storage size must be selected and determined based on the population (water demand) and the purpose of the storage. The sum of the three volumes for balancing, fire, and emergency is the storage capacity provided in a municipal water supply system. It will normally be about one day's average consumption. Volumes for the three purposes are calculated separately according to the required time period. Operating/equalizing storage is used to meet variable water demands while maintaining adequate pressure on the system. Where information on water demand is accessible, storage volume can be calculated or found graphically (e.g., from a mass diagram). When no information is available, operating storage is taken to be 15–25% of maximum daily consumption.

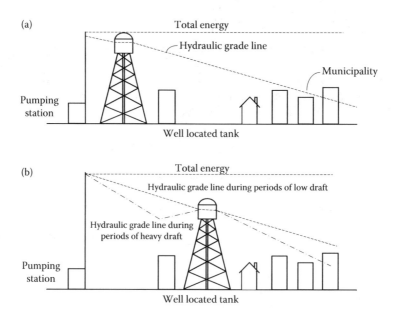

FIGURE 4.3 Effect of water storage reservoir location on pressure distribution.

Fire storage is calculated by taking the product of fire flow and fire duration. The values suggested by the National Fire Protection Association (NFPA) are given in Table 4.5. Fire-flow capacity may be elevated or lowered depending on the reliability of the water supply source. For instance, a municipality may increase its flow storage capacity if a water source such as a single well is used. Emergency storage of up to five times the maximum daily demand is suggested by the Insurance Advisory Organization to provide water during shutdowns for maintenance or repair to the system. This is rarely done in practice, and emergency storage is usually estimated to be one-quarter to one-third of the sum of the operating and fire capacity requirements (Gupta, 1989).

TABLE 4.5
Duration of Required Fire Flow

Required Fire Flow (L/s)	Duration (h)
160 or less	2
190	3
220	3
250	4
280	4
320	5
350	5
380	6
440	7
550	8
570	9
630 or more	10

Urban Water Supply and Demand

> **EXAMPLE 4.6**
>
> For the mixed industrial–commercial–residential city of 100,000 people used in Example 4.5, calculate the required storage capacity.
>
> *Solution*: From Example 4.5, the maximum daily consumption is 93.0×10^6 Lpd and the fire flow is 35,400 Lpm (590 L/s). The recommended flow duration is estimated, based on Table 4.5, to be equal to 9 h. Since there is no information available, the operating storage has been considered to be equal to 20% of the maximum daily water demand of the city.
>
> Operating storage = 93×10^6 L $\times$ 0.20 = 18.6×10^6 L.
> Fire-flow storage = 35,400 (Lpm) $\times$ 60 (min/h) $\times$ 9 h = 19.1×10^6 L.
> Emergency storage = 1/3 (operating storage + fire-flow storage)
>   = 1/3 (18.6×10^6 + 19.1×10^6)
>   = 12.6×10^6 L.
> Total storage required = $(18.6 + 19.1 + 12.6) \times 10^6$ L = 50,267 m^3.

4.12 WATER DISTRIBUTION SYSTEM

Providing the desired quantity of water to the appropriate place at a suitable time with acceptable quality is the basic function of water distribution systems. Distribution piping, storage tanks, and pumping stations are three major components of urban water distribution systems. These components can be further divided into subcomponents. For example, structural, electrical, piping, and pumping units are subcomponents of the pumping station component. The pumping unit can be divided into: a pump, a driver, controls, power broadcasting, and piping and valves. The definition of components depends on the level of detail of the required analysis and the available data. A hierarchy of building blocks is used to construct the urban water distribution system. The relationship between components and subcomponents is summarized in Figure 4.4.

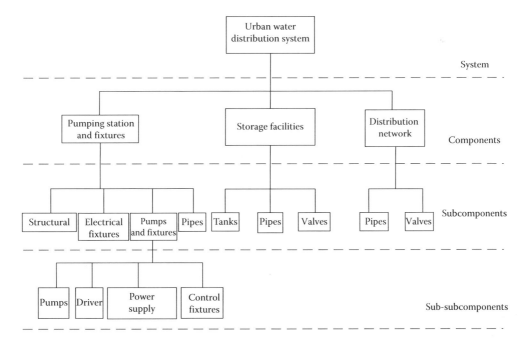

FIGURE 4.4 Different level of hierarchies in a water distribution system.

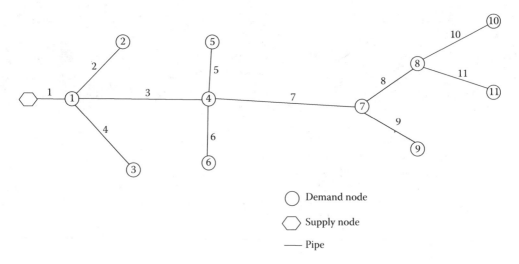

FIGURE 4.5 A typical branch distribution system.

Pipes, valves, pumps, drivers, power transmission units, controls, and storage tanks are seven sub-subcomponents that can be readily identified for analysis. The pumping unit subcomponent is composed of pipes, valves, a pump, a driver, power transmission, and control sub-subcomponents. The reliability of the urban water distribution systems is elevated by three subcomponents: pumping units, pipe links, and storage tanks. Distribution piping is either branched as demonstrated in Figure 4.5 or looped as presented in Figure 4.6, or is a combination of branched and looped.

4.12.1 System Components

The most widely used elements in the network are pipe sections and these contain fittings and other appurtenances, such as valves, storage facilities, and pumps. Pipes are the largest capital investment

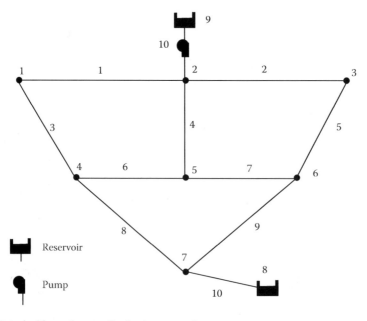

FIGURE 4.6 A typical looped water distribution network.

in a distribution system; they are manufactured in different sizes and are composed of different materials, such as asbestos cement, steel, cast or ductile iron, reinforced or prestressed concrete, polyethylene, polyvinyl chloride, and fiberglass. The AWWA (American Water Works Association) publishes standards for pipe construction, installation, and performance. It is called the C-series standards (continually updated).

Each end of a pipe is called a node. Junction nodes and fixed-grade nodes are two kinds of nodes in a water distribution network. Junction nodes, where the inflow or the outflow is known have lumped demand, which may vary with time. Nodes to which a reservoir is added are referred to as fixed-grade nodes. These nodes can take the form of tanks or large constant-pressure mains.

The flow or pressure in water distribution systems is regulated by control valves. If conditions exist for flow reversal, the valve will close and no flow will pass. The pressure-reducing or relief (pressure-regulating) valve (PRV) is the most common sort of control valve that is placed at pressure zone boundaries to dissipate pressure. There are many other types of valves, including isolation valves to shut down a segment of a distribution system; direction-control (check) valves to allow the flow of water in only one direction, such as swing check valves, rubber-flapper check valves, slanting check disk check valves, and double-door check valves; and air-release/vacuum-breaker valves to control flow in the main.

The PRV maintains a constant pressure on the downstream side of the valve for all flows with a pressure lower than the upstream head. In a water distribution network, when a high-pressure and a low-pressure water distribution systems are connected, the PRV permits flow from the high-pressure system if the pressure on the low side is not excessive. If the downstream pressure is greater than the PRV setting, then the pressure inside will close the valve. The PRV head loss depends on the downstream pressure and is independent of the flow in the pipe. Another type of check valve, a horizontal swing valve, operates under a similar principle. Pressure-sustaining valves operate similarly to PRVs monitoring pressure at the upstream side of the valve.

In water distribution systems, pumps are used to enhance the energy. The most common type of pump used in water distribution systems is the centrifugal pump. There are many different types of pumps (positive-displacement pumps, kinetic pumps, turbine pumps, horizontal centrifugal pumps, vertical pumps, and horizontal pumps).

In an urban water distribution system, to provide supply during outages of individual components, to equalize pump discharge near an efficient operating point in spite of varying demands, to provide water for fire fighting, and to dampen out hydraulic transients, water storage in the system is needed (Walski, 1996). Distribution is closely associated with the water tank. Tanks are usually used for water storage in a water distribution network and are made of steel. The water tank is used to supply water to meet the requirements during high system demands or during emergency conditions when pumps cannot adequately satisfy the pressure requirements at the demand nodes. The water storage tanks can be at ground level or at a certain elevation above the ground. If, at all times, a minimum quantity of water is kept in the tank, unexpectedly high demands cannot be met during critical conditions. The higher the pump discharge, the lower the pump head. Thus, during a period of peak demand, the amount of available head is low.

A wide array of metering devices are used for the measurement of water mains flow:

- Turbine meters have a measuring chamber that gets turned by the flow of water.
- Multijet meters have a multiblade rotor mounted on a vertical spindle within a cylindrical measuring chamber.
- Electromagnetic meters measure flow by means of the magnetic field generated around an insulated section of a pipe.
- Ultrasonic meters use sound-generating and sound-receiving sensors (transducers) attached to the sides of the pipe.
- Proportional meters utilize abstraction in the water line to divert a portion of water into a loop that holds a turbine or displacement meter. The diverted flow is proportional to the flow in the main line. This meter has a turbine meter in parallel with a multijet meter.

4.13 HYDRAULICS OF WATER DISTRIBUTION SYSTEMS

In a water distribution system, almost all of the principal components work under pressure. When the pipes are full and the water in the system is moving under gravitation force, it is also considered as pressure flow. The sewage force mains that receive discharge from a pumping station also carry flows under pressure. There is a range of minimum to maximum pressure in which these components operate. Due to frictional resistance by the pipe walls and fittings, the pressure is reduced as water flows through the system. This is measured in terms of the head/energy loss. The principles of hydraulics that are used in water distribution modeling are described in the next section.

4.13.1 Energy Equation of Pipe Flow

A pipeline segment is illustrated in Figure 4.7. The total energy at any point consists of potential or elevation head, pressure head, and velocity head. The hydraulic grade line demonstrates the elevation of pressure head along the pipe (i.e., it is a line connecting the points to which the water will rise in piezometric tubes installed at different sections of a pipeline). This concept is similar to the water surface in open-channel flow. The total head at different points of a pipe section is represented by the energy grade line. In a uniform pipe, the velocity head is constant. Thus the energy grade line is parallel to the hydraulic grade line.

Applying the energy equation between points 1 and 2 yields

$$Z_1 + \frac{P_1}{\gamma} + \alpha \frac{V_1^2}{2g} = Z_2 + \frac{P_2}{\gamma} + \alpha \frac{V_2^2}{2g} + h_f. \tag{4.2}$$

In Equation 4.2, h_f is the head loss along the pipeline due to friction. The energy gradient S_f is equal to h_f/L. Additional losses resulting from valves, fittings, bends, and so on are known as the minor losses, h_m, and have to be included when present. Then, in Equation 4.2, the term h_f will be replaced by the total head loss, h_{loss}, which is equal to the summation of h_f and h_m. Since the minor losses are localized, the energy grade line, represented by h_f/L, will have breaks wherever the minor losses occur. If a mechanical energy is added to the water by a pump or taken off by a turbine between the two points of interest, it should be added or subtracted from the left side of Equation 4.2. In a uniform

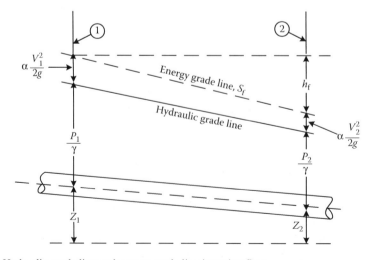

FIGURE 4.7 Hydraulic grade line and energy grade line in a pipe flow.

pipe, $V_1 = V_2$, and elevations Z_1 and Z_2 are generally known. To ascertain the pressure reduction; it is necessary to evaluate the head loss (and minor losses if present).

EXAMPLE 4.7

The diameter of a pipe varies between 200 and 100 mm from section A to section B, respectively. The pressure at section A is 8.15 m and there is a negative pressure equal to 2.5 m at section B. The velocity in section A is 1.8 m/s. If section B is 7 m higher than section A, calculate (a) the flow rate, (b) the velocity at section B, (c) the flow direction, and (d) the head loss in the system.

Solution:

(a) The pipe flow rate is calculated as

$$Q = A_A V_A = \frac{\pi}{4}(0.2)^2 1.8 = 0.057 \text{ m}^3/\text{s}.$$

(b) Since the flow rate in the pipe is constant, the flow velocity at section B is

$$V_B = \frac{Q}{A_B} = \frac{0.057}{(\pi/4)(0.1)^2} = 7.26 \text{ m/s}.$$

(c) For determining the flow direction, at first the total head at sections A and B should be calculated. For this purpose, it is assumed that the datum line corresponds with section A.
Total head at section A:

$$H_A = Z_A + \frac{P_A}{\gamma} + \frac{V_A^2}{2g} = 0 + 8.15 + \frac{1.8^2}{2 \times 9.81} = 8.32 \text{ m}.$$

Total head at section B:

$$H_B = Z_B + \frac{P_B}{\gamma} + \frac{V_B^2}{2g} = 7 + (-2.5) + \frac{7.26^2}{2 \times 9.81} = 7.19 \text{ m}.$$

Since the total head in section A is more than that in section B, the flow direction is from section A to section B.

(d) The system head loss is equal to the difference between the total head at sections A and B and is equal to

$$h_f = 8.32 - 7.19 = 1.13 \text{ m}.$$

4.13.2 Evaluation of Head Loss Due to Friction

4.13.2.1 Darcy–Weisbach Equation

The Darcy–Weisbach equation (1845) is the most general formula in the pipe flow application. It was obtained experimentally as

$$h_f = \frac{fL}{d} \frac{V^2}{2g}, \qquad (4.3)$$

where h_f is the head loss due to friction in the pipe (m), f is the friction factor, dimensionless, L is the length of the pipe (m), d is the internal diameter of the pipe (m), and V is the mean velocity of flow in the pipe (m/s).

The solution of Equation 4.3 requires the interim step of ascertaining an appropriate value of the friction factor, f, to be used in the equation.

4.13.2.1.1 Friction Factor for the Darcy–Weisbach Equation

The friction factor relation depends on the Reynolds number which is the state of flow. The pipe diameter is a characteristic dimension and the Reynolds number is given by

$$Re = \frac{Vd}{v}, \tag{4.4}$$

where V is the average velocity of flow (m/s), d is the internal diameter of the pipe (m), and v is the kinematic viscosity of the fluid (m^2/s).

The Reynolds number for laminar flow is given by the following expression as a function of the friction factor:

$$f = \frac{64}{Re} \quad \text{[For laminar flow } (Re < 2000)\text{]}. \tag{4.5}$$

Any friction factor relation cannot be applied in the critical region of Re between 2000 and 4000 the flow alternates between the laminar and turbulent flows ($Re > 4000$). In turbulent flow, the friction factor is a function of the Reynolds number and the relative roughness of the pipe surface. The roughness is characterized by a e/d parameter, where e is the average height of roughness along the pipe. The turbulent flow is further categorized into three zones as follows:

1. Flow in a smooth pipe, where the relative roughness e/d is very small.
2. Flow in a fully rough pipe.
3. Flow in a partially rough pipe, where both the relative roughness and viscosity are significant.

From the obtainable implicit relations for calculation of the friction factor for the Darcy–Weisbach equation, Moody (1944) has prepared a diagram of the friction factor versus the Reynolds number and the relative roughness. The Moody diagram is illustrated in Figure 4.8. For the application of the Moody diagram, the velocity of flow and the diameter of the pipe should be known so that the Reynolds number can be determined. When the velocity or diameter is unknown, the procedure has been discussed in the next section.

EXAMPLE 4.8

Determine the friction factor for water flowing at a rate of 0.028 cms in a cast iron pipe 50 mm in diameter at 80°F. The pipe roughness is $e = 2.4 \times 10^{-4}$ m.

Solution: Since the area of the pipes cross section is $A = (\pi/4)5^2 \times 10^{-4} = 0.002$ m^2, the velocity of flow is calculated as $V = Q/A = 0.028/0.002 = 14$ m/s.

At 80°F, the kinematic viscosity is 0.86×10^{-6} m^2/s and the Reynolds number is estimated as

$$Re = \frac{Vd}{v} = \frac{14 \times 50.0 \times 10^{-3}}{0.86 \times 10^{-6}} = 8.1 \times 10^5.$$

Since $Re > 4000$, the flow is turbulent. The equivalent roughness is $e = 2.4 \times 10^{-4}$ and the relative roughness is calculated as $e/d = 2.4 \times 10^{-4}/50 \times 10^{-3} = 0.005$. On the Moody diagram (Figure 4.8), the point of intersection of $Re = 8.1 \times 10^5$ and $e/d = 0.005$ is projected horizontally to the left to read $f = 0.03$.

Urban Water Supply and Demand

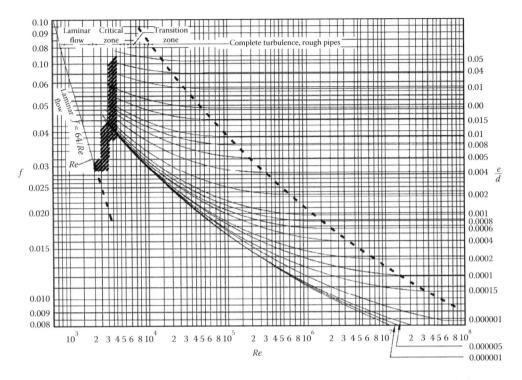

FIGURE 4.8 Moody diagram.

4.13.2.1.2 Application of the Darcy–Weisbach Equation
The Darcy–Weisbach Equation 4.3 has three applications:

1. It is used to compute the head loss, h_f, in a pipe of given size, d, that carries a known flow, V or Q.
2. It is used to determine the flow, V (or Q), through a pipe of given size, d, in which the head loss, h_f, is known.
3. It is used to establish the pipe size, d, to pass a given rate of flow, Q, within a known limit of head loss, h_f.

In the first case, the application of Equation 4.3 is direct. From known V and d values, Re can be computed and then f can be determined by the Moody diagram. Thus Equation 4.3 can be solved for h_f. In the second and third cases, Re, and f, cannot be determined. Since Re is unknown, many investigators have prepared particular diagrams between certain groups of variables in nondimensional form, instead of Re versus f, that enable direct determination of pipe size or flow. There is a trial-and-error procedure as follows:

1. A value of f is assumed near the rough turbulence if e/d is known.
2. Using the Darcy–Weisbach Equation 4.3, either V or d can be analyzed.
3. Re and revised f are determined.
4. Steps 2 and 3 are repeated until a common value of f is obtained.

> **EXAMPLE 4.9**
>
> Determine the head loss in a cast iron pipe that delivers water at a rate of 0.03 cms at 45°C between two points A and B. The pipe diameter and length are 150 mm and 300 m, respectively. If point B is 25 m higher than point A and both of the points have the same pressure, what is the pump head required to deliver water from point A to point B? The roughness of cast iron pipe is equal to 0.000244 m.
>
> *Solution*: Velocity of flow is $Q/A = 0.03/(\pi(0.15)^2/4) = 1.70$ m/s, and at 45°C, υ is 0.86×10^{-6} m²/s. Thus Re is calculated as
>
> $$Re = \frac{Vd}{\upsilon} = \frac{1.70(0.150)}{0.86 \times 10^{-6}} = 3.0 \times 10^5.$$
>
> Relative roughness is $e/d = 0.000244/0.15 = 0.0016$ and, from the Moody diagram, $f = 0.028$. Hence the head loss in the pipe is calculated as
>
> $$h_f = \frac{fL}{D}\frac{V^2}{2g} = (0.028)\left(\frac{300}{0.15}\right)\frac{1.7^2}{2(9.81)} = 8.25 \text{ m}.$$
>
> For determining the required pump head for water delivery, the energy equation is applied between points A and B:
>
> $$Z_A + \frac{P_A}{\gamma} + \frac{V_A^2}{2g} + h_p = Z_B + \frac{P_B}{\gamma} + \frac{V_B^2}{2g} + h_f.$$
>
> Since the pressure and velocity of both points are the same, the equation is summarized as $Z_A + h_p = Z_B + h_f$, and then
>
> $$0 + h_p = 25 + 8.25 \Rightarrow h_p = 33.25 \text{ m}.$$

4.13.2.2 Hazen–Williams Equation for the Friction Head Loss

The Hazen–Williams equation is another common formula for head loss in pipes. The Hazen–Williams formula is used in pipe designs, although it is accurate within a certain range of diameters and friction slopes. The Hazen–Williams equation in SI units is

$$V = 0.849 C R^{0.63} S^{0.54}, \tag{4.6}$$

where V is the mean velocity of flow (m/s), C is the Hazen–Williams coefficient of roughness, R is the hydraulic radius (m), and S is the slope of the energy gradient and is equal to h_f/L.

Jain et al. (1978) signified that an error of up to 39% can be involved in the evaluation of the velocity by the Hazen–Williams formula over a wide range of diameters and slopes. Two sources of error in the Hazen–Williams formula are the following:

1. The multiplying factor 0.849 should change for different R and S values for the same value of C.
2. The Hazen–Williams coefficient C is considered to be related to the pipe material only. Similar to the friction factor of Darcy–Weisbach, the Hazen–Williams coefficient also depends on pipe diameter, velocity, and viscosity.

Urban Water Supply and Demand

A nomogram based on Equation 4.6 is given in Figure 4.9 by Gupta (1989) to facilitate the solution. Equation 4.6 and the nomogram provide a direct solution to all types of pipe problems:

1. Computation of head loss
2. Assessment of flow
3. Determination of pipe size

The nomogram in Figure 4.9 is based on the coefficient $C = 100$. For pipes of a different coefficient, the adjustments are made as follows:

Adjusted discharge:
$$Q = Q_{100}\left(\frac{C}{100}\right). \tag{4.7}$$

Adjusted diameter:
$$d = d_{100}\left(\frac{100}{C}\right)^{0.38}. \tag{4.8}$$

Adjusted friction slope:
$$S = S_{100}\left(\frac{100}{C}\right)^{1.85}. \tag{4.9}$$

Here the subscript 100 refers to the value obtained from the nomogram.

EXAMPLE 4.10

Compute the head loss in Example 4.9 by using the Hazen–Williams formula.

Solution: For a new cast iron pipe, the Hazen–Williams coefficient is equal to 130. Then from Equation 4.6, $1.7 = 0.849(130)(0.15/4)^{0.63}S^{0.54}$ or $S = 0.020$.
Hence $h_f = SL = 0.020(300) = 6.0$ m.
The problem can also be solved using the nomogram (Figure 4.9) as follows:
At first a point is marked at 0.03 m^3/s on the discharge scale and another point is marked at 150 mm on the diameter scale. The straight line corresponding to these points meets the head loss for the 1000 scale at 35. Thus $S_{100} \times 10^3 = 35$ or $S_{100} = 0.035$.
This value of S should be adjusted for $C = 130$.
$S = S_{100}(100/C)^{1.85} = (0.035)(100/130)^{1.85} = 0.021$ and hence $h_f = SL = 0.021(300.0) = 6.3$ m.

4.13.2.3 Minor Head Loss

Local head losses occur at changes in pipe section, at bends, valves, and fittings in addition to the continuous head loss along the pipe length due to friction. These losses are important for less than 30-m-long pipes and may be ignored for long pipes. Minor losses are important when pipe lengths in water supply and wastewater plants are generally short. There are two ways to calculate these losses. In the equivalent length technique, a fictitious length of pipe is estimated that will cause the same pressure drop as any fitting or alteration in a pipe cross section. This length is added to the actual pipe length. In the second method, the loss is considered proportional to the kinetic energy head given by the following formula:

$$h_m = K\frac{V^2}{2g}, \tag{4.10}$$

where h_m is the minor head loss (m), K is the loss coefficient, and V is the mean velocity of flow (m/s). Some characteristic values of the minor head loss coefficient are given in Table 4.6.

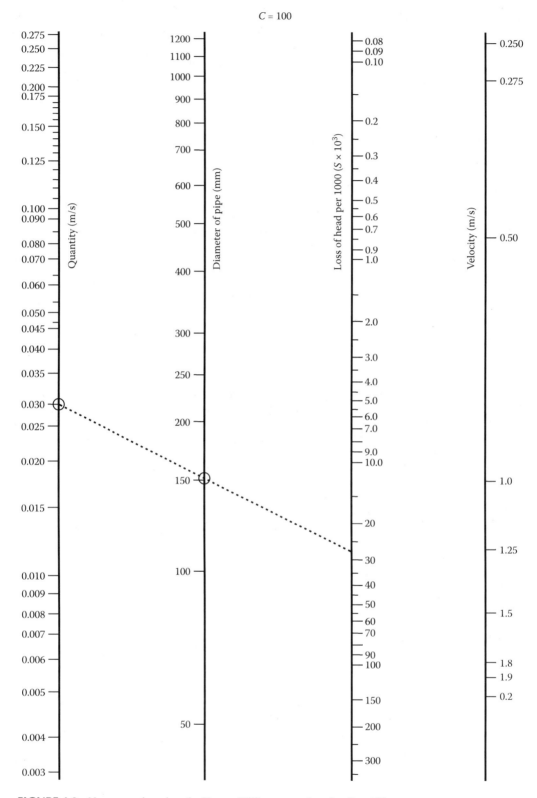

FIGURE 4.9 Nomogram based on the Hazen–Williams equation, for $C = 100$.

TABLE 4.6
Minor Head Loss Coefficients

Item	Loss Coefficient (K)
Entrance Loss from Tank to Pipe	
Flush connection	0.5
Projecting connection	1.0
Exit loss from pipe to tank	1.0
Sudden Contraction	
$d1/d2 = 0.5$	0.37
$d1/d2 = 0.25$	0.45
$d1/d2 = 0.10$	0.48
Sudden Enlargement	
$d1/d2 = 2$	0.54
$d1/d2 = 4$	0.82
$d1/d2 = 10$	0.90
Fittings	
90° bend-screwed	0.5–0.9
90° bend-flanged	0.2–0.3
Tee	1.5–1.8
Gate valve (open)	0.19
Check valve (open)	3.00
Glove valve (open)	10.00
Butterfly valve (open)	0.30

Source: Gupta, R. S. 1989. *Hydrology and Hydraulic Systems.* Prentice Hall, Englewood Cliffs, NJ.

4.14 PIPELINE ANALYSIS AND DESIGN

The analysis consists of determination of the head loss or the rates of flow through a pipeline of a given size. Selection of a pipe size that will carry a design discharge between two points with specified reservoir elevations or a known pressure difference forms the design situation. The problems can be solved by the Darcy–Weisbach equation. If minor losses are neglected, the Hazen–Williams equation leads to the direct solution of both the analysis and design problems.

EXAMPLE 4.11

Two reservoirs are connected by a 200-m-long cast iron pipeline, as shown in Figure 4.10. If the pipeline is proposed to convey a discharge of 0.2 m³/s, what is the required size of the pipeline ($\upsilon = 1.3$ mm²/s)?

Solution: The energy equation is applied between points 1 and 2:

$$Z_1 + \frac{P_1}{\gamma} + \frac{V_1^2}{2g} = Z_2 + \frac{P_2}{\gamma} + \frac{V_2^2}{2g} + h_{\text{loss}}$$

$$90 + 0 + 0 = 80 + 0 + 0 + h_{\text{loss}}$$

$$h_{\text{loss}} = 10 \text{ m}.$$

continued

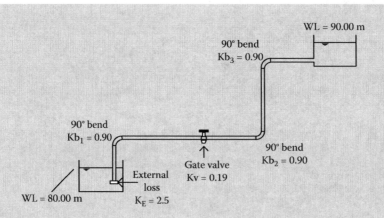

FIGURE 4.10 Pipe system connecting two reservoirs in Example 4.11.

Head loss is the summation of friction loss and minor losses ($h_{loss} = h_f + h_m$), which are calculated as follows:

Friction loss:

$$h_f = \frac{fL}{d}\frac{V^2}{2g} = \frac{fL}{d}\frac{Q^2}{[(\pi/4)d^2]^2 \times 2g} = \frac{fLQ^2}{12.1 d^5} = 0.66\frac{f}{d^5}.$$

Minor losses: The minor losses coefficients are listed in Table 4.7.

$$h_m = \sum \frac{KV^2}{2g} = \sum K \frac{Q^2}{[(\pi/4)d^2]^2 \times 2g} = 0.003\frac{\sum K}{d^4} = \frac{0.016}{d^4},$$

$$h_{loss} = h_f + h_m \Rightarrow 0.66\frac{f}{d^5} + \frac{0.016}{d^4} = 10,$$

$$10d^5 - 0.016d - 0.66f = 0.$$

The above equation should be solved by trial and error. In the first trial, $f = 0.03$ and it has been calculated that $d = 0.30$ m. Thus

$$V = \frac{4Q}{\pi d^2} = \frac{4(0.2)}{\pi(0.30)^2} = 2.83 \text{ m/s},$$

$$Re = \frac{Vd}{\upsilon} = \frac{2.83(0.30)}{1.3 \times 10^{-6}} = 6.5 \times 10^5,$$

$$\frac{e}{d} = \frac{2.44 \times 10^{-4}}{0.30} = 0.0008.$$

TABLE 4.7
Minor Losses Coefficient for Example 4.11

Item	K
Three 90° bends	2.7
Gate valve	0.19
External loss	2.5
Total	5.39

Urban Water Supply and Demand

$f = 0.022$ (from the Moody diagram) is substituted in equation $10d^5 - 0.016d - 0.66f = 0$. Solved by trial and error: $d = 0.29$ m. Thus

$$V = \frac{4Q}{\pi d^2} = \frac{4(0.2)}{\pi(0.29)^2} = 3.03 \text{ m/s},$$

$$Re = \frac{Vd}{\upsilon} = \frac{3.03\,(0.29)}{1.3 \times 10^{-6}} = 6.8 \times 10^5,$$

$$\frac{e}{d} = \frac{2.44 \times 10^{-4}}{0.29} = 0.0008.$$

$f = 0.022$ (from the Moody diagram). Since the amount of f has not changed, the required pipe diameter is 290 mm.

For lifting the water level and, at intermediate points for boosting the pressure, the pumps are commonly used in waterworks and wastewater systems. A situation of water supplied from a lower reservoir to an upper-level reservoir is indicated in Figure 4.11.

To analyze the system, the energy equation is applied between upstream and downstream ends of the pipe:

$$Z_1 + \frac{P_1}{\gamma} + \frac{V_1^2}{2g} + h_p = Z_2 + \frac{P_2}{\gamma} + \frac{V_2^2}{2g} + h_f + h_m. \tag{4.11}$$

Treating $V_1 = V_2$ yields

$$h_p = \left(Z_2 + \frac{P_2}{\gamma}\right) - \left(Z_1 + \frac{P_1}{\gamma}\right) + h_f + h_m, \tag{4.12}$$

$$h_p = \Delta Z + h_{\text{loss}}, \tag{4.13}$$

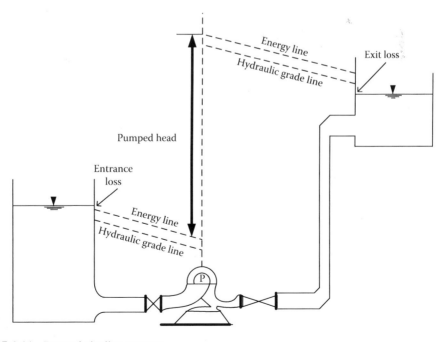

FIGURE 4.11 Pumped pipeline systems.

where h_p is the energy added by the pump (m), ΔZ is the difference in downstream and upstream piezometric heads or water levels, or total static head (m), h_f is the friction head loss and is equal to $(fL/D)(V^2/2g)$ (m), h_m is the minor head loss and $\sum KV^2/2g$ (m), and h_{loss} is the total of friction and minor head losses.

The energy head, h_p, and the brake horsepower of the pump are related as

$$\text{BHP} = \gamma \frac{Qh_p}{\eta}, \tag{4.14}$$

where BHP is the brake horsepower (kW), Q is the discharge through the pipe (m³/s), h_p is the pump head (m), and η is the overall pump efficiency.

EXAMPLE 4.12

A pumping system is employed to pump a flow rate of 0.15 m³/s. The suction pipe diameter is 200 mm and the discharge pipe diameter is 150 mm. The pressure at the beginning of the suction pipe A is −2 m and at the end of the discharge pipe B is 15 m. The discharge pipe is about 1.5 m above the suction pipe and the power of the pumping system engine is 35 kW. (a) Estimate the pumping system efficiency. (b) Determine the pressure at B if the power of the pumping system engine with the same efficiency is increased up to 40 kW.

Solution:

(a) For estimating the consumed pumping power, the energy equation between points A and B should be considered.

$$V_A = \frac{Q}{A_A} = \frac{0.15}{\pi/4(0.2)^2} = 4.77 \text{ m/s},$$

$$V_B = \frac{Q}{A_B} = \frac{0.15}{\pi/4(0.15)^2} = 8.49 \text{ m/s},$$

$$Z_A + \frac{P_A}{\gamma} + \frac{V_A^2}{2g} + h_p = Z_B + \frac{P_B}{\gamma} + \frac{V_B^2}{2g},$$

$$0 + (-2) + \frac{(4.77)^2}{2 \times 9.81} + h_p = 1.5 + 15 + \frac{(8.49)^2}{2 \times 9.81} \Rightarrow h_p = 21.01 \text{ m}.$$

Pumping system power, $P = \gamma Q h_p = 9.81 \times 0.15 \times 21.01 = 30.91$ kW.
Pumping system efficiency $= 30.91/35 = 0.88\%$.

(b) At first the effective pumping system power is calculated:

$$P = 0.88 \times 40 = 35.2 \text{ kW}.$$

Then the pumping head is calculated:

$$h_p = \frac{P}{\gamma Q} = \frac{35.2}{9.81 \times 0.15} = 23.92 \text{ m}.$$

Again the energy equation between points A and B is considered:

$$0 + (-2) + \frac{(4.77)^2}{2 \times 9.81} + 23.92 = 1.5 + P_B + \frac{(8.49)^2}{2 \times 9.81} \Rightarrow P_B = 17.91 \text{ m}.$$

4.14.1 PIPES IN SERIES

Several pipes of different sizes could be connected together to form pipes in series or a compound pipeline as illustrated in Figure 4.12. According to the continuity and the energy equations, the

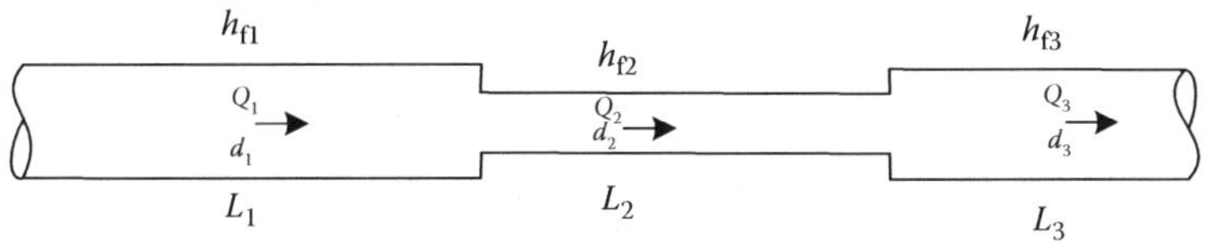

Urban Water Supply and Demand

FIGURE 4.12 Compound pipeline.

following relations apply to the pipes in series:

$$Q = Q_1 = Q_2 = Q_3 = \ldots \ldots \ldots \ldots \quad (4.15)$$

$$h_f = h_{f1} + h_{f2} + h_{f3} + \ldots \ldots \ldots \ldots \quad (4.16)$$

For analysis purpose, different pipes are replaced by a pipe of uniform diameter of equivalent length that will pass a discharge, Q, with the total head loss, h_f. This is known as the equivalent pipe. The procedure is shown by an example.

EXAMPLE 4.13

For a flow rate of 0.04 m³/s, determine the pressure and total heads at points A, B, C, and D for the series pipes shown in Figure 4.13 and Table 4.8. Assume a fully turbulent flow for all cases and the pressure head at point A to be 40 m. Determine the length of an equivalent pipe having a diameter of 1000 mm and f equal to 0.025.

Solution: The pressure and total heads are computed using the energy equation along the path beginning at point A. Given the pressure head and elevation, the total head at point A is

$$H_A = \frac{P_A}{\gamma} + Z_A = 40 + 20 = 60 \text{ m}.$$

Note that the velocity and the velocity head are

$$V_1 = \frac{Q}{A} = \frac{0.04}{\pi (0.3)^2/4} = 0.57 \text{ m/s}$$

and

$$\frac{V_1^2}{2g} = \frac{0.57^2}{2(9.81)} = 0.017 \text{ m}.$$

The velocity head is four orders of magnitude less than the static head, so it can be neglected. Neglecting the velocity head is a common assumption in pipe network analysis.

All energy loss in the system is due to friction. So, following the path of flow, the total heads at A, B, C, and D are

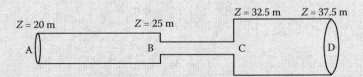

FIGURE 4.13 Compound pipe system of Example 4.13.

continued

$$H_B = H_A - h_f^{A-B} = 60 - f_1 \frac{L_1}{d_1} \frac{V_1^2}{2g} = 60 - 0.022 \left(\frac{2000}{0.3}\right) \frac{0.57^2}{2(9.81)}$$
$$= 60 - 2.4 = H_B = 57.6 \, \text{m}.$$

$$H_C = H_B - h_f^{B-C} = 57.6 - f_2 \frac{L_2}{d_2} \frac{V_2^2}{2g} = 57.6 - 0.025 \left(\frac{1000}{0.2}\right) \frac{\left(0.04/\pi(0.2)^2/4\right)^2}{2(9.81)}$$
$$= 57.6 - 10.3 \Rightarrow H_C = 47.3 \, \text{m}.$$

$$H_D = H_C - h_f^{C-D} = 47.3 - f_3 \frac{L_3}{d_3} \frac{V_3^2}{2g} = 47.3 - 0.021 \left(\frac{2000}{0.4}\right) \frac{\left(0.04/\pi(0.4)^2/4\right)^2}{2(9.81)}$$
$$= 47.3 - 0.5 \Rightarrow H_D = 46.8 \, \text{m}.$$

The pressure heads are

At point B:
$$H_B = \frac{P_B}{\gamma} + Z_B = 57.6 = \frac{P_B}{\gamma} + 25 \Rightarrow \frac{P_B}{\gamma} = 32.6 \, \text{m}.$$

At point C:
$$H_C = \frac{P_C}{\gamma} + Z_C = 47.3 = \frac{P_C}{\gamma} + 32.5 \Rightarrow \frac{P_C}{\gamma} = 14.8 \, \text{m}.$$

At point D:
$$H_D = \frac{P_D}{\gamma} + Z_D = 46.8 = \frac{P_D}{\gamma} + 37.5 \Rightarrow \frac{P_D}{\gamma} = 9.3 \, \text{m}.$$

Total head loss of the equivalent pipe is $h_{loss} = H_A - H_D = 60 - 46.8 = 13.2$.

$$h_{loss} = f_{eq} \frac{L_{eq}}{d_{eq}^5} \frac{Q_{eq}^2}{12.1} \Rightarrow 13.2 = 0.025 \left(\frac{L_{eq}}{0.3^5}\right) \frac{0.04^2}{12.1} \Rightarrow L_{eq} = 9703 \, \text{m}.$$

The calculations are summarized in Table 4.8.

TABLE 4.8
Computation for Pipes in Series

(1) Pipe	(2) Pipe Diameter (mm)	(3) Discharge (m³/s)	(4) Friction Slope	(5) Pipe Length (m)	(6) Head Loss (m)
1	300	0.04	0.022	2000	2.4
2	200	0.04	0.025	1000	10.3
3	400	0.04	0.021	2000	0.5
	300 (selected)	0.04	0.025 (selected)	9703 (determined)	13.2

EXAMPLE 4.14

For the series pipe system in Example 4.13, find the equivalent roughness coefficient and the total head at point D for a flow rate of 0.03 m³/s.

Solution: The equivalent pipe loss coefficient is equal to the sum of the pipe coefficient or

$$K_{eq}^s = \sum_{l=1}^{3} k_l = \sum_{l=1}^{3} \frac{8 f_l L_l}{g \pi^2 d_l^5}.$$

For this problem,

$$K_{eq}^s = K_1 + K_2 + K_3 = \frac{8f_1 L_1}{g\pi^2 d_1^5} + \frac{8f_2 L_2}{g\pi^2 d_2^5} + \frac{8f_3 L_3}{g\pi^2 d_3^5}$$

$$= \frac{8(0.022)(2000)}{9.81\pi^2 (0.3)^5} + \frac{8(0.025)(1000)}{9.81\pi^2 (0.2)^5} + \frac{8(0.021)(2000)}{9.81\pi^2 (0.4)^5} = 1496 + 6455 + 339 \Rightarrow K_{eq}^s = 8290.$$

Note that pipe 2 has the largest loss coefficient since it has the smallest diameter and highest flow velocity. As seen in the previous example, although it has the shortest length, most of the loss occurs in this section. The head loss between nodes A and D for $Q = 0.03 \text{ m}^3/\text{s}$ is then

$$h_f^{A-D} = K_{eq}^s Q^2 = K_1 Q^2 + K_2 Q^2 + K_3 Q^2 = 8290(0.03)^2 \Rightarrow h_f^{A-D} = 7.5 \text{ m}.$$

We can also confirm the result in the previous example by substituting $Q = 0.04 \text{ m}^3/\text{s}$ in which case

$$h_f^{A-D} = K_{eq}^s Q^2 = 8290(0.04)^2 = 13.26 \text{ m}$$

and

$$H_D = H_A - h_f^{A-D} = 60 - 13.26 \Rightarrow H_D = 46.74 \text{ m}.$$

That would be equivalent to the earlier result if the previous example was carried to two decimal places.

4.14.2 Pipes in Parallel

For the parallel or looping pipes of Figure 4.14, the continuity and energy equations provide the following relations:

$$Q = Q_1 + Q_2 + Q_3 + \cdots + Q_n. \tag{4.17}$$

$$h_f = h_{f1} = h_{f2} = h_{f3} = \cdots = h_{fn}. \tag{4.18}$$

A procedure similar to that used for pipes in series is also used in this case, as illustrated in the following example.

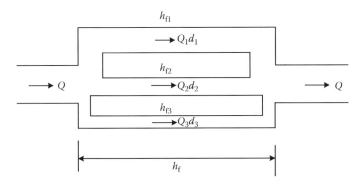

FIGURE 4.14 Parallel pipe system.

EXAMPLE 4.15

Given the data for the three parallel pipes in Figure 4.15, compute (1) the equivalent parallel pipe coefficient, (2) the head loss between nodes A and B, (3) the flow rates in each pipe, (4) the total head at node B, and (5) the diameter of an equivalent pipe with length 100 m and $C_{HW} = 100$.

Solution:

(1) The equivalent parallel pipe coefficient allows us to determine the head loss that can then be used to disaggregate the flow between pipes. The loss coefficient for the Hazen–Williams equation for pipe 1 is

$$K_1 = \frac{10.667 L_1}{C_1^{1.85} d_1^{4.87}} = \frac{10.667(91.5)}{120^{1.85}(0.356)^{4.87}} = 21.253.$$

Similarly, K_2 and K_3 equal 6.366 and 9.864, respectively.
The equivalent loss coefficient is

$$\left(\frac{1}{K_1}\right)^{1/n} + \left(\frac{1}{K_2}\right)^{1/n} + \left(\frac{1}{K_3}\right)^{1/n} = \left(\frac{1}{21.253}\right)^{0.54} + \left(\frac{1}{6.366}\right)^{0.54} + \left(\frac{1}{9.864}\right)^{0.54}$$

$$= 0.192 + 0.368 + 0.291 = \left(\frac{1}{K_{eq}^p}\right)^{0.54} \Rightarrow K_{eq}^p = 1.348.$$

(2) and (4) The head loss between nodes A and B is then

$$h_L = K_{eq}^p (Q_{total})^n = 1.348(0.283)^{1.85} = 0.131 \text{ m}.$$

(4) So the head at node B, H_B, is

$$H_A - H_B = h_L = 24.4 - H_B = 0.131 \text{ m} \Rightarrow H_B = 24.269 \text{ m}.$$

(3) The flow in each pipe can be computed from the individual pipe head loss equations since the head loss is known for each pipe ($h_L = 0.131$ m),

$$Q_1 = \left(\frac{1}{K_1}\right)^{1/n} h_L^{1/n} = \left(\frac{1}{21.253}\right)^{0.54} 0.131^{0.54} = 0.064 \text{ m}^3/\text{s}.$$

The flows in pipes 2 and 3 can be calculated by the same equation and are 0.123 and 0.097 m³/s, respectively. The sum of the three pipe flows equals 0.284 m³/s, which is approximately same as the inflow to node A. The results are tabulated in Table 4.9.

(5) The diameter of the equivalent pipe is estimated as

$$K_{eq} = \frac{10.667 L_{eq}}{C_{eq}^{1.85} d_{eq}^{4.87}} \Rightarrow 1.348 = \frac{10.667(100)}{100^{1.85}(d_{eq})^{4.87}} \Rightarrow d_{eq} = 0.685 \text{ m}.$$

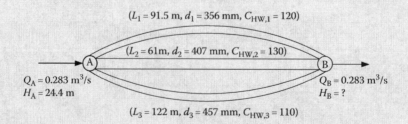

FIGURE 4.15 Pipes in parallel for Example 4.15.

TABLE 4.9
Computation of Parallel Pipe System

(1) Pipe	(2) Pipe Diameter (cm)	(3) Head Loss (m)	(4) Pipe Length (m)	(5) C_{HW}	(6) Discharge (m³/s)
1	356	0.131	91.5	120	0.064
2	407	0.131	61	130	0.123
3	457	0.131	122	110	0.097
	685 (computed)	0.131	100 (assumed)	100 (computed)	0.284

4.14.3 Pipe Networks

The analytical solution of pipe networks is quite complicated. In this system, the flow to an outlet comes from different sides. Three simple methods are the Hardy Cross method, the linear theory method, and the Newton–Raphson method. A most popular procedure of analysis is the Hardy Cross method. It involves a series of consecutive estimations and corrections to flows in individual pipes.

Using the Darcy–Weisbach equation for circular pipes,

$$h_f = \frac{fL^2}{d}\frac{V^2}{2g} = \frac{16\,fL}{\pi^2 d^5}\frac{Q^2}{2g}. \tag{4.19}$$

From the Hazen–Williams equation for circular pipes,

$$Q = 0.278\, C d^{2.63} \left(\frac{h_f}{L}\right)^{0.54} \tag{4.20}$$

or

$$h_f = \frac{10.7\,L}{C^{1.85} d^{4.87}} Q^{1.85}. \tag{4.21}$$

Both h_f equations can be expressed in the general form

$$h_f = KQ^n, \tag{4.22}$$

where K is given in Table 4.10 and n is 2.0 for the Darcy–Weisbach equation and 1.85 for the Hazen–Williams equation.

The sum of head losses around any closed loop is zero (energy conservation), that is,

$$\sum h_f = 0.$$

TABLE 4.10
Equivalent Resistance, K, for the Pipe

Formula	Units of Measurement	K
Hazen–Williams	Q, m³/s, L, m, d, m, h_f, m	$\dfrac{10.7\,L}{C^{1.85} d^{4.87}}$
Darcy–Weisbach	Q, m³/s, L, m, d, m, h_f, m	$\dfrac{fL}{12.1\, d^5}$

Consider that Q_a is an assumed pipe discharge that varies from pipe to pipe of a loop to satisfy the continuity of flow. If δ is the correction made in the flow of all pipes of a loop to satisfy the above equation, then by substituting $h_f = KQ^n$ in $\sum h_f = 0$,

$$\sum K(Q_a + \delta)^n = 0.$$

Expanding the above equation by the binomial theorem and retaining only the first two terms yields

$$\delta = -\frac{\sum KQ_a^n}{n \sum KQ_a^{n-1}} \qquad (4.23)$$

or

$$\delta = -\frac{h_f}{n \sum |h_f/Q_a|}. \qquad (4.24)$$

Equations 4.22 and 4.24 are used in the Hardy Cross procedure. The values of n and K are obtained based on the Darcy–Weisbach or Hazen–Williams equations from Table 4.10. The procedure is summarized as follows:

1. Divide the network into a number of closed loops. The computations are made for one loop at a time.
2. Compute K for each pipe using the appropriate expression from Table 4.10 (column 3 of Table 4.11).
3. Assume a discharge, Q_a, and its direction in each pipe of the loop (column 4 of Table 4.11). At each joint, the total inflow should be equal to total outflow. Consider the clockwise flow to be positive and the counterclockwise flow to be negative.
4. Compute h_f in column 5 for each pipe by Equation 4.22, retaining the sign of column 4. The algebraic sum of column 5 is h_f.
5. Compute h_f/Q for each pipe without considering the sign. The sum of column 6 is $\sum |h_f/Q|$.
6. Determine the correction, δ, by Equation 4.24. Apply the correction algebraically to the discharge of each member of the loop.
7. For common members among two loops, both δ corrections should be made, one for each loop.
8. For the adjusted Q, steps 4 through 7 are repeated until δ becomes very small for all loops.

TABLE 4.11
Iteration 1 of the Hardy Cross Procedure

(1) Loop	(2) Pipeline	(3) K	(4) Q_a (m³/s)	(5) $h_f = KQ_a^{1.85}$	(6) $1.852|h_f/Q|$	(7) Q Corrected $= Q_a + \delta$
1	1	5	1.8	14.850	15.279	+1.698
	2	8	−1.2	−11.213	17.305	−1.302
	3	8	−0.1	−0.112	2.083	−0.224
	Sum			3.525	34.667	—
2	3	8	0.1	0.112	2.083	−0.224
	4	6	1.1	7.158	12.052	+0.674
	5	5	−0.2	−0.254	2.350	−0.626
	Sum			7.016	16.485	—

Urban Water Supply and Demand

Example 4.16

Find the discharge in each pipe of the pipe network shown in Figure 4.16. The head loss in each pipe is calculated by using $h_f = KQ^{1.852}$, where the values of K for each pipe are given in Figure 4.16. The pressure head at point 1 is 100 m. Determine the pressure at different nodes.

Solution: Two first Hardy Cross iterations are summarized in Tables 4.11 and 4.12. In the first iteration, the flow in each pipe is assumed to be indicated in Figure 4.16.

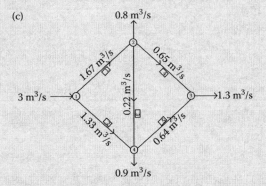

FIGURE 4.16 (a) Pipe network of Example 4.16; (b) assumed discharges; (c) discharge after the final iteration.

continued

TABLE 4.12
Iteration 2 of the Hardy Cross Procedure

(1) Loop	(2) Pipeline	(3) K	(4) Q_a (m³/s)	(5) $h_f = KQ_a^{1.85}$	(6) $1.852|h_f/Q|$	(7) Q Corrected $= Q_a + \delta$
1	1	5	1.698	13.330	14.538	1.677
	2	8	−1.302	−13.042	18.551	−1.323
	3	8	0.224	0.501	4.141	0.219
	Sum			0.789	37.230	—
2	3	8	−0.224	−0.501	4.141	−0.219
	4	6	0.674	2.890	7.940	0.658
	5	5	−0.626	−2.100	6.213	−0.642
	Sum			0.289	18.294	—

Iteration 1:
For loop 1,

$$\delta = (-3.525/34.667) = -0.102, \text{ adjusted } Q_1 = +1.8 + (-0.102) = 1.698.$$

For loop 2,

$$\delta = (-7.016/16.485) = -0.426, \text{ adjusted } Q_3 = 0.1 - (-0.102) + (-0.426) = -0.224.$$

Iteration 2:
For loop 1,

$$\delta = (-0.789/37.230) = -0.021, \text{ adjusted } Q_1 = +1.698 + (-0.021) = 1.677.$$

For loop 2,

$$\delta = (-0.289/18.294) = -0.016, \text{ adjusted } Q_3 = -0.224 - (-0.021) + (-0.016) = -0.219.$$

The Hardy Cross process converges to the final results after five iterations. The final results are tabulated in Table 4.13.

TABLE 4.13
Final Flows and Pressure Heads

Pipe	Flow (m³/s)	Head Loss (m)	Node	Pressure Head (m)
1	1.675	13.00	1	100 m (given)
2	1.326	13.49	2	$100 - h_1 = 87$
3	−0.222	0.49	3	$H_2 - h_4 = 87 - 2.73 = 84.27$
4	0.653	2.73	4	$100 - h_2 = 86.51$
5	0.647	2.23	—	—

4.15 WATER QUALITY MODELING IN A WATER DISTRIBUTION NETWORK

Prior to the passage of the U.S. SDWA of 1974, the focus of most water utilities was on treating water, even though it has long been recognized that water quality can deteriorate in the distribution system. However, after the SDWA was amended in 1986, a number of rules and regulations were amended which had a direct impact on water quality in distribution systems. To decrease what was considered an unreasonably high risk of waterborne illness, the U.S. EPA promulgated the Total Coliform Rule (TCR) and Surface Water Treatment Rule (SWTR) in 1989 (U.S. EPA, 1989a and b). More recently, there has been an increased focus on distribution systems and their importance in maintaining water quality in drinking water distribution systems. There has also been general agreement that the most vulnerable part of a water supply system is the distribution network.

This section discusses the general features associated with water quality modeling in water supply distribution networks. Water quality models and their potential for use in tracking and predicting water quality movement and changes in water quality are examined. Because a key aspect of assessing the performance of drinking water systems is the response of the network and the resulting network model to the pattern of water demands, new research for characterizing water systems is presented. If water quality models are to be used for predicting water quality in networks, they should be verified. Therefore the use of tracers for verifying water quality models including planning, conducting tracer studies in the field, and the analysis of results are discussed.

4.15.1 Water Quality Standards

Water quality is evaluated by the physical, chemical, and biological characteristics of water and its intended uses. Water to be used for public water supplies must be drinkable and without any pollution. The presence of any unknown substance (organic, inorganic, radiological, or biological) that tends to degrade the water quality or impairs the usefulness of the water is defined as pollution.

The basis for defining water quality requirements and standards in urban areas is the drinking water security. Pollution control and health authorities give more attention to the protection of surface and groundwater resources. The water pollution control agencies in different countries use classification of the surface water quality standards in urban areas as stream standard, effluent standards, or a combination of them. These standards are associated with particular numerical limits for designated beneficial uses of water resources. The discharge of a specific amount of conventional pollutants such as the emission of toxic pollutants from residential and industrial units is limited in these standards.

4.15.2 Water Quality Model Development

Modeling the movement of a contaminant or a chemical substance such as chlorine within the distribution systems, as it moves through the system from various points of entry (e.g., treatment plants) to water users, is based on three principles:

- Conservation of mass within differential lengths of the pipe
- Complete and instantaneous mixing of the water entering pipe junctions
- Appropriate kinetic expressions for the growth or decay of the substance as it flows through pipes and storage facilities

Advection (movement in the direction of flow) and dispersion (movement in the transverse direction due to concentration difference) are the two important mechanisms for transportation of a substance. The basic equation describing advection–dispersion transport is based on the principle of conservation of mass and Fick's law of diffusion. For a nonconservative substance, the principle of mass conservation within a differential section of a pipe, that is, the control volume, can be stated as

$$\begin{vmatrix} \text{Rate of change} \\ \text{of mass in} \\ \text{control volume} \end{vmatrix} = \begin{vmatrix} \text{Rate of change} \\ \text{of mass due} \\ \text{to advection} \end{vmatrix} + \begin{vmatrix} \text{Rate of change} \\ \text{of mass due} \\ \text{to dispersion} \end{vmatrix} + \begin{vmatrix} \text{Transformation} \\ \text{reaction} \\ \text{rate} \end{vmatrix}.$$

Considering the first-order reaction of the considered substance with other substances, conservation of mass is given by

$$\frac{\partial C(x,t)}{\partial t} = -V\frac{\partial C(x,t)}{\partial x} + E\frac{\partial^2 C(x,t)}{\partial x^2} - K_R[C(x,t)], \qquad (4.25)$$

where $C(x,t)$ is the substance concentration (g/m^3) at position x and time t, V is the flow velocity (m/s), E is the coefficient of longitudinal dispersion (m^2/s), and $K_R[C(x,t)]$ is a reaction rate expression in which the $(-)$ sign reflects the decrease in concentration due to decay rate.

Dispersion of common substances such as chlorine is negligible in water distribution systems and therefore Equation 4.25 could be reduced to

$$\frac{\partial C(x,t)}{\partial t} = -V\frac{\partial C(x,t)}{\partial x} - K_R[C(x,t)]. \qquad (4.26)$$

According to Equation 4.26, the rate at which the mass of a material changes within a small section of a pipe equals the difference in mass flow into and out of the section plus the rate of reaction within the section. It is assumed that the velocities in the links are known beforehand from the solution to a hydraulic model of the network. In order to solve Equation 4.26, it is important to know C at $x = 0$ for all times (a boundary condition) and the reaction rate expression, $K_R[C(x,t)]$. As chlorine is the most common and essential substance in water distribution networks, chlorine decay modeling is expressed in the following section.

Equation 4.27 represents the concentration of material leaving the junction and entering a pipe.

$$C_{ij} = \frac{\sum_k Q_{ki} C_{kj}}{\sum_k Q_{kj}}, \qquad (4.27)$$

where C_{ij} is the concentration at the start of the link connecting node i to j, in mg/L, C_{kj} is the concentration at the end of a link, in mg/L, and Q_{kj} is the flow from k to i.

Equation 4.27 implies that the concentration leaving a junction equals the total mass of a substance mass flowing into the junction divided by the total flow into the junction.

Storage tanks can be modeled as completely mixed, variable volume reactors in which the change in volume and concentration over time are

$$\frac{dV_s}{dt} = \sum_k Q_{ks} - \sum_i Q_{sj}, \qquad (4.28)$$

$$\frac{dV_s C_s}{dt} = \sum_k Q_{ks} C_{ks} - \sum_i Q_{sj} C_s + k_{ij}(C_s), \qquad (4.29)$$

where C_s is the concentration for the tank s, in mg/L, dt is the change in time, in s, Q_{ks} is the flow from node k to s, in m^3/s, Q_{sj} is the flow from node s to j, in m^3/s, dV_s is the change in volume of the tank at nodes, in m^3, V_s is the volume of the tank at nodes, in m^3, C_{ks} is the concentration of the contaminant at the end of links, in mg/m^3, and k_{ij} is the decay coefficient between nodes i and j, in s^{-1}.

There are currently several models available for modeling both the hydraulics and water quality in the drinking water distribution system. However, most of this discussion will be focused on a U.S. EPA developed hydraulic/contaminant propagation model called EPANET (Rossman et al., 1994), which is based on mass transfer concepts (transfer of a substance through another on a molecular scale). Another approach to water quality contaminant propagation to be discussed is the approach developed by Biswas et al. (1993). This model uses a steady-state transport equation that takes into account the simultaneous corrective transport of chlorine in the axial direction, diffusion in the radial

Urban Water Supply and Demand

direction, and consumption by first-order reaction in the bulk liquid phase. Islam (1995) developed a model called QUALNET, which predicts the temporal and spatial distribution of chlorine in a pipe network under slowly varying unsteady flow conditions. Boulos et al. (1995) proposed a technique called the event-driven method (EDM), which is based on the "next-event" scheduling approach, which can significantly reduce computing times.

4.15.3 Chlorine Decay

The chlorine decay in the pipe has two mechanisms. The first mechanism, which is considered as bulk decay, is the reaction of chlorine with other substances available in water. The second mechanism is the reaction of chlorine with substances of the pipe wall and is known as *wall decay*. The wall decay in distribution networks may be predominant where significant corrosion is present.

4.15.3.1 Bulk Decay

Bulk decay is mostly assumed to follow the first-order kinetics, which is formulated as

$$\frac{dC}{dt} = -k_b C \tag{4.30}$$

or

$$C_t = C_0 e^{-k_b t}, \tag{4.31}$$

where C_t is the concentration after time t, C_0 is the initial chlorine concentration, and k_b is the coefficient of bulk decay.

The bulk decay rate is measured by observing the chlorine concentration, at specified time intervals, from glass bottles that are previously filled with sample water. Then the coefficient of bulk decay is determined by the least-squares curve-fitting method. The bulk decay is a function of initial chlorine concentration as well as the temperature and total organic content (TOC) of water (Powell et al., 2000).

4.15.3.2 Wall Decay

Wall decay of chlorine is mostly due to reaction with a corrosion by-product. Rossman et al. (1994) assumed that the wall reaction rate is first order with respect to wall concentration and is formulated as

$$N = k_f (C - C_w) = k_w C_w, \tag{4.32}$$

where N is the flux of chlorine to the pipe wall (g/m^2/s), k_f is the mass transfer rate coefficient (m/s), C and C_w are the chlorine concentrations in the bulk and the wall, respectively, and k_w is the first-order rate coefficient for the wall reaction.

Equation 4.32 is rewritten to express C_w and N in terms of C:

$$C_w = \frac{k_f}{(k_w + k_f)} C, \tag{4.33}$$

$$N = \frac{k_w k_f}{(k_w + k_f)} C. \tag{4.34}$$

The mass transfer rate coefficient, k_f, is estimated as (Gupta, 1989):

$$k_f = Sh \left(\frac{D_m}{d} \right) \tag{4.35}$$

and

$$Sh = 0.023 Re^{0.83} Sc^{0.33} \quad \text{if } Re > 2300,$$

$$Sh = 3.65 + \frac{0.0668(d/L)ReSc}{1 + 0.04[(d/L)ReSc]^{2/3}} \quad \text{if } Re < 2300, \tag{4.36}$$

where Sh is the Sherwood number, Re is the Reynolds number, Sc is the Schmidt number ($= v/D_m$), D_m is the molecular diffusivity of chlorine in water, v is the water viscosity, L is the pipe length, and d is the pipe diameter.

4.15.3.3 Overall Decay Rate

A simple method to define the overall decay rate constant is to express it as a sum of bulk and wall decay rate constants. Thus,

$$k = k_b + k_w, \tag{4.37}$$

where k is the overall decay rate constant. However, Equation 4.37 does not consider the mass transfer rate of chlorine from the bulk to the pipe wall. The overall rate expression to account for bulk and wall reactions simultaneously within a section of a pipe is obtained as

$$\left(\frac{\pi}{4}d^2 L\right)\frac{\partial C}{\partial t} = -\left(\frac{\pi}{4}d^2 L\right)k_b C - N(\pi dL), \tag{4.38}$$

where L and d are the length and diameter of the pipe, respectively. Dividing Equation 4.38 by $(\pi/4)d^2 L$ and substituting the value of N from Equation 4.34 results in

$$\frac{\partial C}{\partial t} = -\left[k_b + \frac{k_w k_f}{(d/4)(k_w + k_f)}\right] C = \left[k_b + \frac{k_w k_f}{R(k_w + k_f)}\right] C, \tag{4.39}$$

where R is the hydraulic radius. Equation 4.39 describes time variation of chlorine along a single pipe. Thus, the overall decay rate constant is given by

$$k = \left[k_b + \frac{k_w k_f}{R(k_w + k_f)}\right]. \tag{4.40}$$

EXAMPLE 4.17

The discharge pipe of the pump illustrated in Figure 4.17 is 1500 m long and 300 mm in diameter. The pumping rate is constant and equal to 0.028 m³/s. Chlorine is completely mixed in the pump and the concentration of chlorine is kept at 1.5 mg/L in the pump. The overall decay rate of chlorine is 6.417×10^{-6} s^{-1} (Axworthy and Karney, 1996). (a) Determine the steady-state chlorine concentration at node 1 and time of achievement. (b) Determine the concentration at subnodes 150 m apart in different time steps before reaching the steady-state condition.

Solution:

(a) First the velocity of flow in the pipe is calculated as follows:

$$V = Q/(\pi d^2/4) = 4 \times 0.028/(\pi \times 0.3^2) = 0.396 \text{ m/s}.$$

Then the travel time from the pump to node 1 is calculated as $t = L/V = 1500/0.396 = 3788$ s. The chlorine concentration at node 1 would be zero from the start to 3788 s, and after that the chlorine concentration would be constant and equal to

$$C(1500, 3788) = C(0,0)e^{-6.417 \times 10^{-6} \times 3788} = 1.5 \times 0.976 = 1.464 \text{ mg/L}.$$

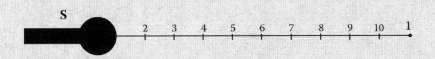

FIGURE 4.17 The schematic of the water distribution system of Example 4.17.

(b) Chlorine concentration is required at subnodes. Since the distance between subnodes is equal to 150 m, the travel time between two consequent subnodes is 378.8 s. Chlorine concentrations at subnodes are determined after every 378.8 s for a period of 3788 s as given in Table 4.14. Chlorine reaches node 2 after 378.8 s, where its concentration is 1.496 mg/L; it reaches node 3 after 757.6 s, where its concentration is 1.493 mg/L; and so on. As was seen earlier, chlorine reaches node 1 after 3788 s and its concentration reduces to 1.464 mg/L.

TABLE 4.14
Chlorine Concentration at Various Subnodes of Example 4.17 at Different Time Steps

Time	Chlorine Concentration (mg/L) at Subnode									
T (s)	S	2	3	4	5	6	7	8	9	10	1
0	0	0	0	0	0	0	0	0	0	0	0
378.8	1.0	1.496	0	0	0	0	0	0	0	0	0
757.6	1.0	1.496	1.493	0	0	0	0	0	0	0	0
1136.4	1.0	1.496	1.493	1.489	0	0	0	0	0	0	0
1515.2	1.0	1.496	1.493	1.489	1.485	0	0	0	0	0	0
1894	1.0	1.496	1.493	1.489	1.485	1.482	0	0	0	0	0
1.485	1.0	1.496	1.493	1.489	1.485	1.482	1.478	0	0	0	0
2651.6	1.0	1.496	1.493	1.489	1.485	1.482	1.478	1.475	0	0	0
3030.4	1.0	1.496	1.493	1.489	1.485	1.482	1.478	1.475	1.471	0	0
3409.2	1.0	1.496	1.493	1.489	1.485	1.482	1.478	1.475	1.471	1.468	0
3788	1.0	1.496	1.493	1.489	1.485	1.482	1.478	1.475	1.471	1.468	1.464

EXAMPLE 4.18

In the branched network of Figure 4.18, there is a source node and three demand nodes (nodes 1, 2, and 3). The pipe characteristics are given in Table 4.15. Assuming the fixed chlorine concentration at source node 1 is equal to 0.8 mg/L, obtain the steady-state concentration at different nodes. Assume that the overall decay rate constant is $6.417 \times 10^{-6} s^{-1}$ for all pipes.

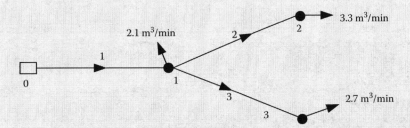

FIGURE 4.18 Branched network of Example 4.18.

continued

TABLE 4.15
Pipe Details for the Branched Network of Example 4.18

Pipe Number (1)	Length, L (m) (2)	Diameter, D (m) (3)	Discharge, Q (m³/min) (4)	Area, A (m²) (5)	Velocity, V (m/min) (6)	Travel Time, t (min) (7)
1	500	0.350	8.1	0.0962	84.20	5.938
2	350	0.200	3.3	0.0314	105.10	3.330
3	400	0.150	2.7	0.0177	152.54	2.622

Solution: The pipe area, flow velocity, and pipe travel time are calculated for all the pipes as tabulated in columns 5–7 of Table 4.15. The chlorine concentration at the upstream of pipe 1, C_{1u}, is the same as that for the source node, that is, 0.8 mg/L. Since the travel time for chlorine in pipe 1 is 5.938 min, the chlorine concentration would remain constant downstream of pipe 1, C_{1d}, after 5.938 min. Thus, chlorine concentration at the downstream end of pipe 1 is given by

$$C_{1d} = C_{1u} e^{-6.417 \times 10^{-6} \times 5.938 \times 60} = 0.8 \times 0.9977 = 0.7982 \text{ mg/L}.$$

Considering mixing at node 1, the chlorine concentration at node 1 (C_1) and upstream of pipes 2 (C_{2u}) and 3 (C_{3u}) can be obtained. Since there is only one supply pipe and no addition of chlorine at node 1, the chlorine concentration at the upstream ends of pipes 2 and 3 would be the same as that at the downstream end of pipe 1. Thus, $C_1 = C_{2u} = C_{3u} = 0.7982$ mg/L.

Now, travel times in pipes 2 and 3 are 3.330 and 2.622 min, respectively. Therefore, the steady-state chlorine concentration at the downstream ends of pipe 2 (C_{2d}) and pipe 3 (C_{3d}) observed after 3.330 and 2.622 min, respectively, would be

$$C_{2d} = C_{2u} e^{-6.417 \times 10^{-6} \times 3.330 \times 60} = 0.7982 \times 0.9987 = 0.7972 \text{ mg/L}$$

and

$$C_{3d} = C_{3u} e^{-6.417 \times 10^{-6} \times 2.622 \times 60} = 0.7982 \times 0.9990 = 0.7974 \text{ mg/L}.$$

Thus, the steady-state chlorine concentration at nodes 2 (C_2) and 3 (C_3) would be 0.7972 and 0.7974 mg/L, respectively. Total time to reach the steady-state chlorine concentration condition at nodes 2 and 3 would be 9.268 and 8.560 min, respectively.

4.15.4 STATISTIC MODEL OF SUBSTANCE CONCENTRATION

As is observed in the previous examples, substance concentration such as chlorine concentration remains constant after a certain time period, which is various at different points of the system. In the static or steady-state model, the initial variation of chlorine concentration is ignored and it is assumed that the substance concentrate is constant at all points. For determining the constant, initially, pipe discharges are determined through hydraulic analysis of the network and then the velocity and travel time in each pipe is computed. For determining the substance concentration at different nodes of a multi-source looped network having M sources, N demand nodes (labeled $M + 1, \ldots, M + N = J$), and X pipes, the following equations should be developed. The problem is formulated in the general form considering the concentration at all nodes to be unknown except a reference node with known concentration that is necessary for analysis.

To obtain the substance concentration at nodes, the concentrations both upstream and downstream of pipes is required. Thus, the total number of unknowns is $M + N + 2X$ and the same number of equations is required to obtain all unknowns. These equations are formulated as follows.

Urban Water Supply and Demand

Concentration upstream of a pipe. Substance concentration upstream of a pipe is the same as the concentration at the upstream node of that pipe. This gives

$$C_{xu} = C_j, \quad \text{for } x = 1,\ldots,X, \tag{4.41}$$

where C_{xu} is the chlorine concentration at the upstream end u of pipe x and j is the upstream node of pipe x. X number of equations are developed using Equation 4.41.

Chlorine decay rate equations for pipes. Chlorine concentrations at the upstream and downstream ends of a pipe are related through the chlorine decay rate equation. Thus,

$$C_{xd} = C_{xu} e^{-(k\Delta t)_x} = C_{xu} F_x \quad \text{for } x = 1,\ldots,X \tag{4.42}$$

where C_{xd} is the substance concentration at the downstream end d of pipe x and Δt is the travel time in the pipe. F_x is the decay factor for pipe x, which is equal to $e^{-(k\Delta t)_x}$. Equation 4.42 provides X number of equations.

Mass conservation equation at nodes. The mass balance equation at the source and demand nodes can be written as

$$\sum_{x \in I} C_{xd} Q_x - \sum_{x \in O} C_{xu} Q_x - C_j q_j = 0 \quad \text{for } j = 1,\ldots,M+N. \tag{4.43}$$

Equation 4.43 provides $M+N$ number of equations. The simultaneous solution of Equations 4.41 through 4.43 provides values of $M+N+2X$ unknowns. Booster chlorination stations are provided in WDNs to boost the chlorine concentration to the specified level or by the specified amount. Thus, at the booster point, either the final chlorine concentration is known or the amount of chlorine added is known in analysis, and accordingly, Equations 4.41 through 4.43 are modified and solved to obtain the unknown chlorine concentration.

4.15.5 Solution Models

There are several different numerical methods that can be used to solve contaminant propagation equations. Four commonly used techniques are the finite difference method (FDM), discrete volume method (DVM), time-driven method (TDM), and EDM.

The FDM approximates derivatives with finite-difference equivalents along a fixed grid of points in time and space. Islam (1995) used this technique to model chlorine decay in distribution systems. The DVM divides each pipe into a series of equally sized, completely mixed volume segments. At the end of each successive water quality time step, the concentration with each volume segment is first reacted and then transferred to the adjacent downstream segment. This approach was used in the models that were the basis for early U.S. EPA studies.

The TDM tracks the concentration and size of a nonoverlapping segment of water that fills each link of a network. As time progresses, the size of the most upstream segment in a link increases as water enters the link. An equal loss in the size of the most downstream segment occurs as water leaves the link. The size of these segments remains unchanged.

The EDM is similar to TDM, except that rather than updating an entire network at fixed time steps, individual link node conditions are updated only at those times when the leading segment in a link completely disappears through this downstream node.

As mentioned previously, the EPANET hydraulic model has been a key component in providing the basis for water quality modeling. There are many commercially available hydraulic models that incorporate water quality models as well. EPANET is a computer program based on the EPS approach to solving the hydraulic behavior of a network. In addition, it is designed to be a research tool for modeling the movement and fate of drinking water constituents within distribution systems. The model is available at the U.S. EPA website.

4.15.5.1 EPANET

EPANET for hydraulic simulation of water networks uses the Hazen–Williams formula, the Darcy–Weisbach formula, or the Chezy–Manning formula for calculating the head loss in pipes, pumps, valves, and minor loss. It is assumed that water usage rates, external water supply rates, and source concentrations at nodes remain constant over a fixed period of time, although these quantities can change from one period to another. The default period interval is one hour but can be set to any desired value. Various consumption or water usage patterns can be assigned to individual nodes or groups of nodes.

EPANET solves a series of equations for each link using the gradient algorithm. Gradient algorithms provide an interactive mechanism for approaching an optimal solution by calculating a series of slopes that lead to better and better solutions. Flow continuity is maintained at all nodes after the first iteration. The method easily handles pumps and valves.

For water quality simulation, the EPANET model uses the flows from the hydraulic simulation to track the propagation of contaminants through a distribution system. Water quality time steps are chosen to be as large as possible without causing the flow volume of any pipe to exceed its physical volume. Therefore the water quality time step dt_{wq} source cannot be larger than the shortest time of travel through any pipe in the network, that is,

$$dt_{wq} = \min \left\{ \frac{V_{ij}}{q_{ij}} \right\}, \quad \text{for all pipes } i,j, \tag{4.44}$$

where V_{ij} is the volume of pipe i, j and q_{ij} is the flow rate of pipe i, j. Pumps and valves are not part of this determination because transport through them is assumed to occur instantaneously. Based on this water quality time step, the number of volume segments in each pipe (n_{ij}) is

$$n_{ij} = \text{INT} \left\{ \frac{V_{ij}}{q_{ij} \, dt_{wq}} \right\}, \tag{4.45}$$

where INT(x) is the largest integer less than or equal to x. There is both a default limit 100 for pipe segments or the user can set dt_{wq} to be no smaller than a user-adjustable time tolerance. EPANET models both types of reactions using first-order kinetics. There are three coefficients used by EPANET to describe reactions within a pipe. These are the bulk rate constant k_b and the wall rate constant k_w, which must be determined empirically and supplied as input to the model. The mass transfer coefficient is calculated internally by EPANET.

EPANET can also model the changes in age of water and travel time to a node. The percentage of water reaching any node from any other node can also be calculated. Source tracing is a useful tool for computing the percentage of water from a given source at a given node in the network over time.

4.15.5.2 QUALNET

Islam (1995) developed a model called QUALNET, which predicts the spatial and temporal distributions of chlorine residuals in pipe networks under slowly varying unsteady flow conditions. Unlike other available models, which use steady-state or extended period simulation (EPS) of steady flow conditions, QUALNET uses a lumped-system approach to compute unsteady flow conditions and includes dispersion and decay of chlorine during travel in a pipe. The pipe network is first analyzed to determine the initial steady-state conditions. The slowly varying conditions are then computed by numerically integrating the governing equations by an implicit finite difference, a scheme subject to the appropriate boundary conditions. The one-dimensional dispersion equation is used to calculate the concentration of chlorine over time during travel in a pipe, assuming a first-order decay rate.

Numerical techniques are used to solve the dispersion, diffusion, and decay equation. Complete mixing is assumed at the pipe junctions. The model has been verified by comparing the results with those of EPANET for two typical networks. The results are in good agreement at the beginning of the simulation model for unsteady flow; however, chlorine concentrations at different nodes vary when the flow becomes unsteady and when reverse flows occur. The model may be used to analyze the propagation and decay of any other substance for which a first-order reaction rate is valid (Chaudhry, 1987). The water quality simulation process used in the previous models is based on a one-dimensional transport model in conjunction with the assumption that complete mixing of material occurs at the junction of pipes. These models consist of moving the substance concentrations forward in time at the mean flow velocity while undergoing a concentration change based on kinetic assumptions. The simulation proceeds by considering all the changes to the state of the system as the changes occur in chronological order. Based on this approach, the advective movement of substance defines the dynamic simulation model. Most water quality simulation models are interval oriented, which in some cases can lead to solutions that are either prohibitively expensive or contain excessive errors.

4.15.5.3 Event-Driven Method

Boulos et al. (1995) proposed a technique mentioned earlier called the EDM. This is extremely simple in concept and is based on a next-event scheduling approach. In this method, the simulation clock time is advanced to the time of the next event to take place in the system. The simulation scheduled is executed by carrying out all the changes to a system associated with an event, as events occur in chronological order. Since the only factors affecting the concentration at any node are the concentrations and flows at the pipes immediately upstream of the given node, the only information that must be available during the simulation is the different segment concentrations. The technique makes an efficient use of the water quality simulation process.

The advective transport process is dictated by the distribution system demand. The model follows a front tracking approach and explicitly determines the optimal pipe segmentation scheme with the smallest number of segments necessary to carry out the simulation process. To each pipe, pointers (concentration fronts), whose function is to delineate volumes of water with different concentrations, are dynamically assigned. Particles representing substance injections are processed in chronological order as they encounter the nodes. All concentration fronts are advanced within their respective pipes based on their velocities. As the injected constituent moves through the system, the position of the concentration fronts defines the spatial location behind which constituent concentrations exist at any given time. The concentration at each affected node is then given in the form of a time–concentration histogram.

The primary advantage of this model is that it allows for dynamic water quality modeling that is less sensitive to the structure of the network and to the length of the simulation process. In addition, numerical dispersion of the concentration front profile resolution is nearly eliminated. The method can be readily applied to all types of network configurations and dynamic hydraulic conditions and has been shown to exhibit excellent convergence characteristics.

4.15.6 Water Quality Monitoring

Monitoring in distribution systems provides the means for identifying variations in water quality spatially and temporally (Grayman et al., 2000). The resulting data can be used to track transformations that are taking place in water quality and can also be used to calibrate water quality models. Monitoring can be classified as routine or for special studies. Routine monitoring is usually conducted in order to satisfy regulatory compliance requirements. Special studies are usually conducted to provide information concerning water quality problems.

Routine monitoring in the United States is one of the requirements specified by the U.S. SDWA. Although from a historical viewpoint most attention has been focused on the performance of the treatment plant, a number of the SDWA maximum contaminant levels (MCL) must be met at the tap. Samples are collected by the use of continuous monitors or as grab samples. Continuous monitoring is generally performed by sensors or remote monitoring stations. Although continuous monitoring is frequently conducted at the treatment plant (turbidity and chlorine residuals), most distribution system monitoring is based on grab samples. Continuous monitoring is capital intensive and requires maintenance and calibration. Grab samples are labor intensive and provide data based on the time of collection.

Special studies might include the following:

- Measurement of disinfectant residuals in a distribution system
- Tracer studies to assist in calibrating water quality models

Some of the issues that should be considered in the preparation of a special study are as follows:

- Sampling locations
- Sampling frequency
- System operation
- Preparation of sampling sites
- Sampling collection procedures
- Analysis procedures
- Personnel organization and schedule
- Safety issues
- Data recording
- Equipment and supply needs
- Training requirement
- Contingency plans
- Communications
- Calibration and review of analytical instruments

4.15.7 Conducting a Tracer Study

A tracer study is a mechanism for determining the movement of water within a water supply system. In such a study, a conservative substance is injected into the water and the resulting concentration of the substance is measured over time as it moves through the system. Historically, tracer studies have been used to study the movement of water through water treatment plants and distribution systems. Although conceptually quite straightforward, a successful field tracer study requires careful planning and implementation in order to achieve useful results.

Tracer studies have been most commonly used as a means of determining the travel time through various components in a water treatment plant (Teefy and Singer, 1990; Teefy, 1996). Its most frequent usage is to ascertain that there is an adequate chlorine contact time in clearwells as required by U.S. EPA regulations. Generally, a conservative chemical such as fluoride, rhodamine WT, or calcium, sodium, or lithium chloride is injected into the influent of the clearwell as either a pulse or a step function. Subsequently, the concentration is monitored in the effluent and statistics such as the T10 value (time until 10% of the injected chemical reaches the outflow) are calculated.

Tracer studies have also been conducted in distribution system storage facilities to understand the mixing processes within the facility. Fluoride was injected into the influent of a reservoir in California and sampled both in the effluent and at an interior point to assess mixing (Grayman et al., 1996; Boulus et al., 1996). In a research study sponsored by AWWARF, Grayman et al. (2000) used fluoride and calcium chloride as tracers to study mixing in a variety of underground, ground level, and elevated tanks.

Within distribution systems, tracer studies provide a means to understand the movement of water throughout the distribution system. This can serve multiple purposes including (1) calculating travel time or water age, (2) calibrating a water distribution system hydraulic model, (3) defining zones served by a particular source and blending with water from other sources, and (4) determining the impacts of accidental or purposeful contamination of a distribution system. A variety of tracers has been used in distribution system studies including chemicals that are injected into the system (e.g., fluoride and calcium chloride), chemical feeds that are turned off (e.g., fluoride), and naturally occurring constituents that differ in different sources serving a system.

Grayman et al. (1988) studied blending in a multiple source system by measuring hardness, chloroform, and total trihalomethanes that differed significantly between the surface water source and the groundwater sources. Clark et al. (1993) turned off a fluoride feed to a water system and traced the resulting front of low-fluoride water as it moved through the system and subsequently restarted the fluoride feed and traced the front of fluoridated water. In an AWWARF-sponsored research project, Vasconcelos et al. (1996, 1997) conducted a series of tracer studies around the United States for use in developing models of chlorine decay. DiGiano and Carter (2001) turned off the fluoride feed at one treatment plant and changed the coagulant feed at another plant to simultaneously trace water from both plants and used the resulting feed measurements to calculate the mean constituent residence time in the system. Grayman (2001) discusses the use of tracers in calibrating hydraulic models. The Stage 2 DBPR (Department of Business and Professional Regulation) Initial Distribution System Evaluation Guidance Manual (U.S. EPA, 2003) recognized the use of tracers as a means of calibrating models and predicting residence time as a partial substitute for the required field monitoring.

A tracer study can be divided into three phases: planning, conducting the actual field work, and analysis of the results. In the planning phase, the logistics of the study are determined, including tracer selection and injection methods, system operation, personnel requirements, training and deployment, monitoring strategy, and assembling and testing of all equipment. An important part of the planning step is the use of a distribution system model to simulate the behavior of the tracer during the course of the actual tracer study so as to have a good understanding of what is expected to occur. The outcome of the planning phase should be a detailed work plan for conducting the tracer study. The actual field study should be preceded by a small-scale field test of the tracer injection and monitoring program. This step will identify potential problems prior to the actual full-scale field study. Following the field program, the data are analyzed, the methods and results are documented, and the tracer study is assessed in order to bring improvement in future studies. Specific procedures are described below.

Tracer Selection. A tracer should be selected to meet the specific needs of a project. Criteria that can influence the selection of a particular tracer include the following:

- Regulatory requirements
- Analytical methods for measuring tracer concentration
- Injection requirements
- Chemical composition of the finished water
- Cost
- Public perception

The most commonly used tracers in distribution systems are fluoride, calcium chloride, sodium chloride, and lithium chloride. Fluoride is a popular tracer for those utilities that routinely add fluoride as part of the treatment process. In this case, the fluoride feed can be shut off, and a front of low-fluoride water is traced. However, several states that regulate and require the use of fluoride will not permit the feed to be shut off, even for short periods. Availability of continuous online monitoring equipment for fluoride is limited so that measurements are generally performed manually in the field or laboratory. Fluoride can interact with coagulants that have been added during treatment and in some circumstances can interact with pipe walls leading to neoconservative behavior.

Calcium chloride has been used in many tracer studies throughout the United States. Generally, a food grade level of the substance is required. It can be monitored by measuring conductivity, or by measuring the calcium or chloride ion. The upper limit for concentration of calcium chloride is generally limited by the secondary MCL for chloride (250 mg/L).

Sodium chloride has many similar characteristics to calcium chloride in that it can be traced by monitoring for conductivity of the chloride or sodium ion. The allowable concentration for sodium chloride is limited by the secondary MCL for chloride and potential health impacts of elevated sodium levels.

Lithium chloride, a popular tracer in the United Kingdom, is less frequently used in the United States, in part because of the public perception of lithium as a medical pharmaceutical. Additionally, concentrations of lithium must be determined as a laboratory test.

In addition to the injection of these popular tracer chemicals, other methods have been used to induce a change in the water quality characteristics of the water that can then be traced through the distribution system. These include the following:

- Switching coagulants (i.e., switch from $FeCl_3$ to $(Al)_2(SO_4)_3$)
- Monitoring differences in source waters

Injection Procedures. The goal of a distribution system tracer study is to create a change in water quality in the distribution system that can be traced as it moves through the system. This is accomplished by injecting a tracer over a period of time (i.e., a pulse) and observing the concentrations at selected locations within the distribution system at some interval of time. Best results are usually obtained by creating a near-instantaneous change in concentration in the receiving pipe at the point of injection and maintaining this concentration at a relatively constant value for a period of time (typically several hours). A variation on this approach has been to inject a series of shorter pulses and monitor the resulting concentration in the distribution system. However, if the pulses are too short in duration or the distribution system is too large or complex, then the pulses may interact, resulting in confusion as to the movement of water in the system.

Injection may be accomplished by pumping the tracer into the pressurized system at a connection to a pipe or by feeding the tracer into a clearwell. For the pumping situation, the pump should be selected to overcome the system pressure and to inject at the required rate. It is desirable to have a flow meter both on the injection pump and on the receiving pipe so that the rate of injection and the resulting concentration in the receiving pipe are known. When feeding the tracer into a clearwell, the tracer should be well mixed in the clearwell, and the concentration of the tracer should be monitored in the clearwell effluent. In general, the volume of water in the clearwell should be minimized in order to generate an effluent tracer front that is as sharp as possible.

Monitoring Program. Tracer concentrations can be monitored in the distribution system by taking grab samples and analyzing them in the field or laboratory, or through the use of online monitors that analyze a sample at a designated frequency. When manual sampling is employed, a route for visiting monitoring locations is generally defined and the circuit is repeated at a predefined frequency. If online monitoring is employed, the instruments should be checked at frequent intervals during the tracer study in order to ensure that they are operating properly and we should take a grab sample that can later be compared with the automated measurement.

Since sampling is generally performed in time during tracer studies, monitoring locations should be selected to provide for accessibility at all times. Dedicated sampling taps and hydrants are frequently used as sampling sites. Alternatively, buildings such as fire stations or 24 h businesses can be used. When using hydrants, in order to facilitate grab samples, a faucet is frequently installed on one of the hydrant ports.

For both grab samples and online monitoring, precautions should be taken to assure that the sample is representative of water in the main serving the sampling site. The travel time from the main can be calculated based on the flow rate, the distance from the main to the sample site, and the diameter of the connecting pipe. For grab samples, either water should be allowed to run continuously at a reasonable rate during the study or the sampling tap should be flushed prior to each sample long enough to ensure water from the main. For online monitors, a constant flow rate should be maintained and the travel time from the main to the monitor is calculated and considered when analyzing the results.

Urban Water Supply and Demand

System Operation. The operation of the water system will have a significant effect on the movement of water in the distribution system during a tracer study. Therefore, information on the system operation should be collected both on a real-time basis and following the completion of the study. Real-time information can be used to ensure that the system is being operated in accordance with planning and to make modifications to monitoring in the case of significant variations in system operation. Additional system operation data can usually be obtained from a SCADA (Supervisory Control And Data Acquisition) system following the completion of the study and it can be used as part of the modeling process.

One category of data that is especially useful for both real-time assessment and poststudy analysis is flow measurements in key pipes. For example, if flow rates in the pipe receiving the tracer are much higher or lower than planned, then travel times will likely be faster or slower than anticipated. This may necessitate changes in monitoring schedules. Flow rates in the receiving pipe, and other inflows and outflows to the study area, are also very useful when modeling the distribution system.

4.16 CONCLUDING REMARKS

Water systems are complex and difficult to characterize. It has become conventional wisdom that water quality can change significantly as water moves through a water system. This awareness has led to the development of water quality/hydraulic models that can be used to understand the factors that affect these changes and to track and predict water quality changes in drinking water networks. Recent events have focused the water industries, attention on the issue of water system vulnerability and it is apparent that the most vulnerable portion of a water utility is the network itself. Therefore, interest has grown rapidly in the potential use of water quality hydraulic models for assessing water system security and for their potential for assisting in the protection of water systems from deliberate contamination by biological and chemical agents.

In order to use water quality/hydraulic models correctly, there are two features that must be understood. These are the characterization of system demands and the proper calibration of these models using tracer tests. A good understanding of both of these aspects could substantially improve the use of models for the system's protection.

PROBLEMS

1. Calculate the water consumption (average daily rate, maximum daily rate, maximum hourly rate, and flow rate) for a town of 10,000 people in semi-arid area with limited water supply resources. As it is an old urban area, the buildings do not have good resistance against fire. Almost more than 90% of the individual buildings in this city have an area less than 200 m^2 and the largest building complex is about 1000 m^2. The only industry in the town, a wool and textile mill with a production of 100 tons/month, has its own water supply. What will be the design rate of water distribution system in the study area? If there are data missing, make your own assumptions and give your reasons for them.
2. A small community with a population of 1000 has a trucked water supply system that provides water from a lake (3 km from a village). There are 200 houses, one hotel, one hospital, one school, one nursing station, and two general stores in the community and the largest area is about 2000 m^2. The total road system in the town is 2 km long. Each house is equipped with a water storage tank of 100 L capacity, with bigger tanks in the other establishments. The average water consumption is 40 Lpcd. It is quite common for winter storms to prevent the trucks from traveling to the lake for up to 3 days. Based on this information, determine (a) the size of the storage reservoir in the village and (b) the number of trucks required, if each truck has a 4000 L tank. The trucks also serve the purpose of providing fire protection. Make any assumption you feel are necessary to complete the assignment, giving reasons for each assumption.
3. Why are distribution pipes not sized according to maximum hourly demand plus fire flow instead of maximum daily demand plus fire flow?

4. Chlorination is the usual method for disinfecting water.
   a. Name the two parameters that control the extent of disinfection.
   b. Why is it necessary to guard against an overdose of chlorine?
   c. Assume that disinfection with chlorine follows a first-order reaction (the disinfection rate has a direct relation to its concentration). In a chlorinated water sample containing 1.0 mg/L chlorine, the initial concentration of viable bacteria is 100,000/(mL). At the end of a 5 min contact time, the number of viable bacteria has decreased to 10/(mL). What effect would a contact time of 10 min have on the bacteria count?
5. Water having a temperature of 15°C is flowing through a 150 mm ductile iron main at a rate of 19 L/s. Is the flow laminar, turbulent, or transitional?
6. Assuming that there are no head losses through the Venturi meter demonstrated in Figure 4.19, what is the pressure reading in the throat section of the Venturi? Assume that the discharge through the meter is 0.6 m³/s.

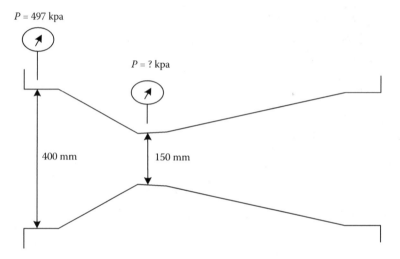

FIGURE 4.19 Venturi of Problem 6.

7. Does conservation of energy apply to the system represented in Figure 4.20? Based on the conservation of energy, the summation of head losses in a loop should be equal to

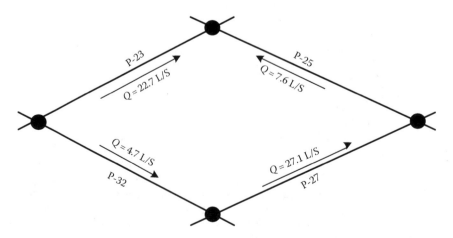

FIGURE 4.20 Pipeline of Problem 7.

TABLE 4.16
Characteristics of the Network of Problem 8

Pipe Label	Length (m)	Diameter (mm)	Hazen–Williams C-Factor
P-23	381.0	305	120
P-25	228.6	203	115
P-27	342.9	254	120
P-32	253.0	152	105

zero. Data describing the physical characteristics of each pipe are presented in Table 4.16 and neglect the minor losses in this loop.

8. Find the pump head needed to deliver water from the lower reservoir to the upper reservoir in Figure 4.21 at a rate of 0.3 m³/s. The suction pipe length, diameter, and roughness coefficient are 20 m, 200 mm, and 130, respectively. The discharge pipe length and diameter are 200 m and 300 mm and the Hazen–Williams roughness coefficient is 110. Minor head loss coefficients are given in Figure 4.21.

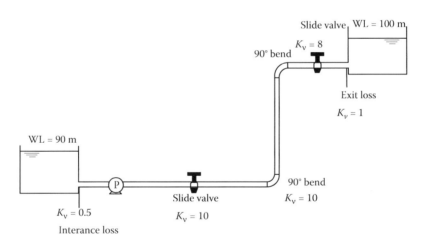

FIGURE 4.21 The reservoir system of Problem 8.

9. Manually find the discharge through each pipeline and the pressure at each junction node of the rural water system shown Figure 4.22. Physical data for this system are given in Table 4.17.
10. In the network depicted in Figure 4.23, determine the discharge in each pipe. Assume that $f = 0.015$.
11. A fire hydrant is supplied through three welded steel pipelines ($f = 0.012$) arranged in series (Table 4.18). The total drop in pressure due to friction in the pipeline is limited to 35 m. What is the discharge through the hydrant?
12. The characteristics of the pipe system illustrated in Figure 4.24 are presented in Table 4.19. (a) Determine the diameter of the equivalent pipe of this system with 600 m length and the Hazen–Williams roughness coefficient equal to 130. (b) Determine the Hazen–Williams roughness coefficient of an equivalent pipe of this system with the length and diameter equal to 600 m and 300 mm, respectively.

178   Urban Water Engineering and Management

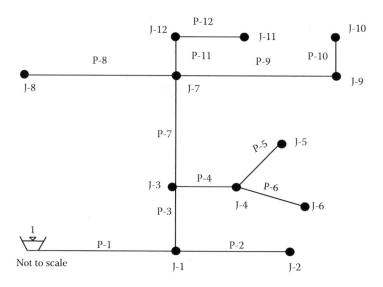

FIGURE 4.22 Pipeline of Problem 9.

TABLE 4.17
Physical Data of the Pipeline System of Problem 10

Pipe Label	Length (m)	Diameter (mm)	Hazen–Williams Roughness Coefficient (C)	Node Label	Elevation (m)	Demand (L/s)
P-1	152.4	254	120	R-1	320.0	N/A
P-2	365.8	152	120	J-1	262.1	2.5
P-3	1280.2	254	120	J-2	263.7	0.9
P-4	182.9	152	110	J-3	265.2	1.9
P-5	76.2	102	110	J-4	266.7	1.6
P-6	152.5	102	100	J-5	268.2	0.3
P-7	1585.0	203	120	J-6	269.7	0.8
P-8	1371.6	102	100	J-7	268.2	4.7
P-9	1676.4	76	90	J-8	259.1	1.6
P-10	914.4	152	75	J-9	262.1	0
P-11	173.7	152	120	J-10	262.1	1.1
P-12	167.6	102	80	J-11	259.1	0.9
				J-12	257.6	0.6

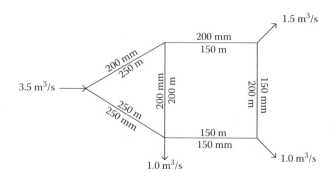

FIGURE 4.23 Layout of the network of Problem 10.

TABLE 4.18
Characteristics of Fire Hydrant System of Problem 12

Pipe	Diameter (mm)	Length (m)
1	150	1500
2	200	6000
3	300	48,000

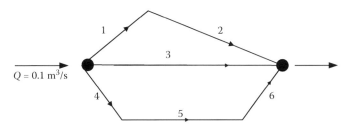

FIGURE 4.24 The layout of network of Problem 12.

TABLE 4.19
Characteristics of Pipe System of Problem 13

Pipe No.	Length (m)	Diameter (mm)	Hazen–Williams Roughness Coefficient
1	300	250	120
2	400	200	110
3	600	200	100
4	200	200	90
5	300	250	110
6	250	200	120

13. A cast iron pipe is employed to deliver a flow rate of $0.1 \text{ m}^3/\text{s}$, between two points that are 800 m apart. Determine the pipe size if the allowable head loss is 10 m and the Hazen–Williams roughness coefficient is 130.

REFERENCES

Axworthy, D. H. and Karney, B. W. 1996. Modeling low velocity/high dispersion flow in water distribution systems. *Journal of Water Resources Planning and Management, ASCE*, 122(3): 218–221.

Baumann, D. D., Boland, J. J., and Hanemann, W. M. 1998. *Urban Water Demand Management and Planning*. McGraw-Hill, New York.

Bhatia, R. and Falkenmark, M. 1993. *Water Resource Policies and the Urban Poor: Innovative Approaches and Policy Imperatives*. World Bank, Washington, DC.

Biswas, P., Lu, C., and Clark, R. M. 1993. A model for chlorine concentration decay in drinking water distribution pipes. *Water Research* 27(12): 1715–1724.

Boulos, P. F., Altman, T., Jarrige, P., and Collevati, F. 1995. Discrete simulation approach for network-water-quality models. *Journal Water Resources Planning and Management, ASCE* 121(1): 49–60.

Boulus, P. F., Grayman, W. M., Bowcock, R. W., Clapp, J. W., Rossman, L. A., Clark, R. M., Deininger, R. A., and Dhingra, A. K. 1996. Hydraulic mixing and free chlorine residual in reservoirs. *Journal of the American Water Works Association* 88(7): 48–59.

Chaudhry, M. H. 1987. *Applied Hydraulic Transients*, 2nd ed., Van Nostrand Reinhold, New York.

Clark, R. M., Grayman, W. M., Males, R. M., and Hess. A. F. 1993. Modelling contaminant propagation in drinking water distribution systems, *Journal of Environmental Engineering, ASCE*, 119(2): 349–364.

Falkland, A., Ed. 1991. *Hydrology and Water Resources of Small Islands: A Practice Guide*. United Nations Educational, Scientific, and Cultural Organization (UNESCO), Paris, France.

DiGiano, F. A. and Carter, G. T. 2001. Tracer studies to measure water residence time in a distribution system supplied by two water treatment plants. *Proceedings of the AWWA Annual Conference*.

Grayman, W. M., Clark, R. M., and Males, R. M. 1988. Modeling distribution system water quality: dynamic approach. *Journal Water Resources Planning and Management, ASCE* 114(3): 295–312.

Grayman, W. M., Deininger, R. A., Green, A., Boulus, P. F., Bowcock, R. W., and Godwin, C. C. 1996. Water quality and mixing models for tanks and reservoirs. *Journal of the American Water Works Association* 88(7): 60–73.

Grayman, W. M., Rossman, L. A., Arnold, C., Deininger, R. A., Smith, C., Smith, J. F., and Schnipke, R. 2000. *Water Quality Modeling of Distribution System Storage Facilities*. American Water Works Association Research Foundation, Denver, CO.

Grayman, W. M. 2001. *Use of Tracer Studies and Water Quality Models to Calibrate a Network Hydraulic Model. Curr Methods.* 1(1) Heastad Press, Waterbury, CT, pp. 38–42.

Gupta, R. S. 1989. *Hydrology and Hydraulic Systems*. Prentice Hall, Englewood Cliffs, NJ.

Islam, M. R. 1995. Modeling of chlorine concentration in unsteady flow in pipe networks, Ph.D. Thesis, Washington State University, WA.

Jain, A. K., Mohan, D. M., and Khanna P. 1978. Modified Hazed-Williams formula. *Journal of Environmental Engineering Diversion ASCE*, 104(1): 137–146.

Karamouz, M., Szidarovszky, F, and Zahraie, B. 2003. *Water Resources Systems Analysis*. Lewis Publisher, Boca Raton, Florida, 600pp.

Knight, P. 1998. *Environment-Population: Outlook Bleak on Water Resources*. World News Interpress Service, IPS. Washington, DC, USA. December 17.

Mays, L. W., Ed. 1996. *Water Resources Handbook*. McGraw-Hill, New York.

McKee, R. H. 1960. A new American paper pulp process, part I—pulping by the full hydrotropic process. *Paper Industry* 42(4): 255–257, 266.

Metcalf and Eddy 2003. *Wastewater Engineering, Treatment and Reuse*. 4th Edition, McGraw-Hill, New York.

Milburn, A. 1996. A global freshwater convention—the best means towards sustainable freshwater management. *Proceedings of the Stockholm Water Symposium*, 4–9 August, pp. 9–11.

Moody, L. F. 1944. Friction factors for pipe flow. *Transactions of the American Society of Mechanical Engineers* 66: 671–684.

Powell, J. C., Hallam, N. B., West, J. R., Forster, C. F., and Simms, J. 2000. Factors which control bulk chlorine decay. *Journal of Water Resources Planning and management, ASCE* 126(1): 13–20.

Rossman, L. A., Clark, R. M., and Grayman, W. M. 1994. Modeling chlorine residuals in drinking water distribution systems. *Journal of Environmental Engineering, ASCE* 120(4): 803–821.

Teefy, S. M. and Singer, P. C. 1990. Performance testing and analysis of tracer tests to determine compliance of a disinfection scheme with the SWTR. *Journal of the American Water Works Association* 82(12): 88–98.

Teefy, S. M. 1996. *Tracer Studies in Water Treatment Facilities: A Protocol and Case Studies*. American Water Works Association Research Foundation and American Water Works Association, Denver, CO.

U. S. EPA. 1989a. Exposure factors handbook. Tech. Rep. EPA/600/8-89/043, U.S. Environ. Protect. Agency, Washington, DC.

U. S. EPA. 1989b. Risk assessment guidance for superfund: Human health evaluation manual, Part A (Interim Final), Tech. Rep. EPA 9285.701A, U.S. Environ. Protect. Agency, Washington, DC.

U. S. EPA. 2003. LT1ESWTR disinfection profiling and benchmarking: Technical guidance manual. Office of Water. EPA 816-R-03-004.

Vasconcelos, J. J., Boulus, P. F., Grayman, W. M., Kiene, L., Wable, O., Biswas, P., Bhari, A., Rossman, L. A., Clark, R. M., and Goodrich, J. A. 1996. Characterization and modeling of chlorine decay in distribution systems. Report no. 90705, AWWA Research Foundation, 6666 West Quincy Avenue, Denver, CO.

Vasconcelos, J. J., Rossman, L. A., Grayman, W. M., Boulos, P. F., and Clark, R. M. 1997. Kinetics of chlorine decay. *Journal of the American Water Works Association* 89(7): 54–65.

Walski, T. M. 1996. Water distribution. In: L. W. Mays, Ed., *Water Resources Handbook*. McGraw-Hill, New York.

Walski, T. M., Chase, D. V., Savic, D. A., Grayman, W. M., Bechwith, S., and Koelle, E. 2001. *Advanced Water Distribution Modeling and Management.* Heastad Methods, Watertown, CT.

WHO and UNICEF, 2000. Global water supply and sanitation assessment, 2000 Report. WHO, Geneva, Switzerland.

WHO, 2004. *Guidelines for Drinking-water Quality*, 3rd Edition. WHO, Geneva, Switzerland (ISBN 924 154 6387).

World Resources Institute, 1998. *World Resources: A Guide to the Global Environment.* A joint program. The United Nations Development Program and the World Bank. Oxford University Press, New York.

5 Urban Water Demand Management

5.1 INTRODUCTION

Urban water demand management (UWDM) is a key strategy for achieving the MDGs for supplying potable water and sanitation for all the people in an urban area. In a national arena, IWRM strategies and water efficiency plans should be developed before implementing UWDM strategies. A balanced set of objectives of UWDM for the management and allocation of water resources are efficiency, equity, and sustainability. Successful implementation of UWDM as a component of IWRM plays a significant role in the reduction of poverty in society through more efficient and equitable use of the available water resources and by facilitating water service through municipal water supply agencies in a region.

UWDM covers a wide range of technical, economic, educational, capacity-building, and policy measures that are useful for municipal planners, water supply agencies, and consumers. There are different economic, social, and environmental reasons for UWDM in different countries in terms of meeting basic human needs and providing affordable access to minimum supplies of water for that purpose (GWP-TEC8, 2003). UWDM is normally referred to as one of the most useful tools in achieving IWRM (Goldblatt et al., 2000).

UWDM is particularly a major challenge in developing countries. Few important attempts are made by international agencies to allocate water to competing end users and sectors in an equitable fashion. This is particularly evident in municipalities, where there is a need to extend water supply coverage to districts that do not yet have service coverage while trying to increase the efficiency of the operation and maintenance of the water supply infrastructure in districts where service is provided.

Lack of access to safe drinking water supplies and sanitation services are recognized as a poverty indicator, but existing data are inadequate as far as equity issues are concerned. It is important to note that UWDM does not always promote water usage reduction. Important health advantages accrue from adequate access to clean safe water that could lead to reduction of waterborne diseases. In this chapter, first the concepts of water demand and demand management are introduced. Then the methodologies that could be applied to achieve the demand management goals are described.

5.2 BASIC DEFINITIONS OF WATER USE

As stated by Mays and Tung (1992), water demand (an analytical concept) is the scheduling of quantities that consumers use per unit of time for a particular price schedule. Water use can be classified into two basic categories: consumptive and nonconsumptive uses.

Consumptive use occurs when water is an end in itself. Domestic, agricultural, industrial, and mining water uses are consumptive. On the other hand, water in nonconsumptive use is a means to an end. Hydropower, transportation, and recreation are the main nonconsumptive water uses. They are also referred to as instream uses that serve the so-called environmental demand as well. In this type of water use, a use is made of a water body without withdrawing water from it except the water used on-site for maintaining swamps and wetlands for wildlife habitats and ditching and ponding.

In an overall water accounting system, the following water uses are identified by the USGS (Solley et al., 1993):

1. Water withdrawal for offstream purposes
2. Water deliveries at the point of use or quantities released after use
3. Consumptive use
4. Conveyance loss
5. Reclaimed wastewater
6. Return flow
7. Instream flow

Water use can also be classified as offstream use and instream use. Different water demands resulting from offstream uses can be classified as follows (Karamouz et al., 2003):

- Domestic or municipal water uses include residential (apartments and houses), commercial (stores and businesses), institutional (hospitals and schools), industrial, and other water uses such as fire fighting, swimming pools, lawning, and gardening.
- Industrial water uses include water needed in different industrial processes such as cooling water in steam–electric power generation units, refineries, chemical and steel manufacturing, textiles, food processing, pulp and paper mills, and also mining.
- Agricultural water uses include those for irrigation and for the drinking and care of animals.
- Miscellaneous water uses include fisheries, recreation, and mining.

These uses require withdrawal of water from surface or groundwater resources systems. Part of the water withdrawn may return to the system perhaps in a different location and time, with a different quality and period of time. The percentage of return flow is an important factor in evaluating the water use efficiency. This should also be considered in water resources management schemes. Table 5.1 shows a more detailed classification of water demands. Water use is also classified as municipal, agricultural, industrial, environmental, and infrastructure (public work).

Water demand changes from year to year, month to month, and also during the months in different years. There are many physical, economic, and social reasons for these variations. More pronounced climate change impacts have been observed in recent years in many parts of the world such as more intense floods, more precipitation, and even unusual droughts in some regions.

5.3 PARADIGM SHIFT IN URBAN WATER MANAGEMENT: TOWARD DEMAND MANAGEMENT

Many investigators have begun to question the technical reliability and institutional capacity of urban water management after increasing incidences of local water shortages and more frequent water quality problems. It has been recognized that continuing to expand the infrastructure and developing new water sources have become increasingly expensive and ultimately unsustainable economically and environmentally. There has also been a change of paradigm from the supply side to the demand side of management. A continuum of water management in different approaches to urban water management is described in Table 5.2.

5.3.1 Supply Management

For many years, the Newtonian paradigm of controlling the environment was the main framework for decision making for building a massive infrastructure in order to store, transfer, and distribute water. There are many dependable and therefore vulnerable subsystems that are expanded as well, consuming both water and other resources. These low-efficiency, expensive-to-operate, high-risk, low-resilience,

TABLE 5.1
Detailed Classifications of Water Uses

Water Demand	Purpose Classification	Use-Type Classification	Rate of Consumption
Drinking	Municipal	Withdrawal	Low
Domestic (cooking, washing, etc.)	Municipal	Withdrawal	Low
Fish and wildlife	Agricultural and Environmental	Withdrawal, Instream, Onsite	Moderate
Livestock	Agricultural, Municipal	Withdrawal	Moderate
Drainage	Agricultural	Withdrawal, Onsite	High
Irrigation	Agricultural, Municipal	Withdrawal	High
Wetland habitat	Agricultural, Environmental	Onsite	Moderate
Soil moisture conservation	Agricultural	Onsite	High
Utilization of estuaries	Agricultural	Instream, Onsite	—
Recreation and water sports	Municipal, Infrastructure	Instream	—
Esthetic enjoyment	Municipal, Infrastructure	Instream	—
Navigation	Infrastructure	Instream	—
Hydropower	Infrastructure	Instream	—
Mining	Industrial	Withdrawal	Moderate
Cooling	Industrial and Municipal	Withdrawal	High
Boiling	Industrial and Municipal	Withdrawal	High
Processing	Industrial and Municipal	Withdrawal	Moderate
	Industrial and Municipal	Withdrawal	High
Waste disposal	Industrial, Municipal, Agricultural	Instream	—

Source: Karamouz, M., Szidarovszky, F., and Zahraie, B. 2003. *Water Resources Systems Analysis.* Lewis Publishers, Boca Raton, FL.

and vulnerable systems that seem politically right have given social stature to municipalities. But the high entropy imposed on societies let decision makers realize that these systems are not sustainable. Supply-side management treats freshwater as a virtually limitless resource, resulting in a regime of water policy and practice, concerned primarily with securing sufficient quantities of water to meet the foreseeable demands. Underlying this approach is the assumption that current levels of water demand are insensitive to policy and behavioral changes (Renzetti, 2003; Shrubsole and Tate, 1994). This supply-side orientation rarely takes full account of environmental or economic impacts of municipal water services.

Some products of the supply-side approach are large, centralized engineering projects—dams, diversions, treatment, and distribution systems. Continuing to depend on expansion of these high-cost water systems puts an increasing, and often unnecessary, strain on the economic stability of municipal water utilities and the integrity of the local aquatic ecosystems (Shrubsole and Tate, 1994; Gleick, 2000).

The Holistic paradigm in urban water management follows the notion of sustainable water resources, looking at system dynamics (the limits to growth) and utilizing different elements of conservation to curb the increasing demand resulting from increasing population and expansion of urban and industrial sectors.

5.3.2 Demand Management

Demand-side management (DSM) is gaining recognition in a number of resource fields including transportation, energy, and, more recently, water. NRTEE (2003) explicitly recommends demand management as a key strategy for mitigating the downward trends in regional environmental quality.

TABLE 5.2
Continuum of Water Management

Characteristic	Supply-Side	Demand-Management	"Soft Path" for Water
Philosophy	Water resources are infinite and only limited by our capacity to access new sources or store larger volumes of water	Water resources are limited and need to be conserved and used efficiently	Water resources are limited and we need to fundamentally re-evaluate the way we develop, manage, and use water
Basic approach	Reactive-status quo	It is currently used as a short-term and temporary approach; however, when used in a comprehensive fashion it represents an incremental step toward a broader "soft path" approach	Long term with potential for fundamental change in resource use
Fundamental question	How can we meet the future projected needs for water?	How can we reduce current and future needs for water to conserve the resource, save money, and reduce environmental impacts?	How can we deliver the services currently provided by water in the most sustainable way?
Tools	Bigger, centralized, expensive engineering solutions (including dams, reservoirs, treatment plants, and distribution systems)	Any measure that increase the efficiency and/or timing of water use (including technologies, pricing, education, and policies)	Any measure that can deliver the services provided by the resource, and the resource itself, taking full costs into account
Planning process	Planners model future growth, extrapolate from current consumption, plan for an increase in capacity to meet anticipated future need, then find a new source of supply to meet that need	Planners model growth and account for a comprehensive water efficiency and conservation program to maximize the use of existing capacity	Planners model future growth, describe a desired future water-efficient state (or scenario), then 'backcast' to find a feasible and desirable way between the future and the present

Source: Gordon Foundation. 2004. Controlling Our Thirst: Managing Water Demands and Allocations in Canada. Available at: www.gordonfn.ca/resfiles/Controlling_Our_Thirst.doc.

It is relevant to note, however, that many national reports presented at the world water forum in Keyoto did not adequately address demand management in the urban water context.

Curran (2000) defines DSM, in the context of urban water systems, generally as "reducing the demand for a service or resource rather than automatically supplying more needed services or resources." DSM is commonly referred to as demand management; it involves any measure or group of measures that improve the efficiency and timing of water use. Some of the examples of demand-side approaches are pricing, education, water-efficient technologies, and regulatory regimes that promote reuse and recycling. Water demand management is "a measure that reduces average or peak withdrawals from surface or groundwater sources without increasing the extent in which wastewater is degraded" as defined by Brooks and Peters (1988). At the core of UWDM, it is recognized that developing new water resources may be too costly, whereas influencing consumers' demand is more cost effective. This is particularly true when environmental and economic costs of urban water services are taken into account. For a thorough discussion of UWDM and the relationships among demand tools, see Maas (2003).

For improving water management even in countries and regions with considerable water resources such as Canada, there are pressing issues. With high levels of urban water use, a growing number of

TABLE 5.3
Pressing Issues in Water Demand Management in Canada

High Urban Water Use	Supply Limitations	Capital Costs	Environment Impacts	Drinking Water Quality
Canadians are the 2nd highest urban water users in the world, using 2.5 times more water than the average European	A number of surface waters have reached or are close to, their limit for withdrawals	Water/wastewater infrastructure upgrades represent significant unmet capital costs—estimated at $23–$49 Canada-wide	Water withdrawls and wastewater returns are geographically concentrated, which amplifies their impact	Reduced water flow through the treatment process is less costly, allowing money to be used to meet higher drinking water standards and reducing the amount of water treated to this level
Total municipal water use increased by 6% during the 1990s	Groundwater extraction will be, or is already, depleting a number of aquifers	Stringent drinking water standard will further increase costs	Development projects (dams and diversion) destroy aquatic and land habitat, introduce nonnative species, and block fish migration	Less volume of wastewater increases the effectiveness of sewage treatment and decreases pollution in receiving waters that provide communities downstream with drinking water
The average Canadian uses 343 liters per capita per day (Lcd) at home	A large number of water sources are contaminated or at risk of contamination	Increasing peak water and/or wastewater treatment demands create additional capital costs	Ground and surface-water withdrawals can reduce surface water flows, altering marine habitat and damaging fish populations	Decreased demand reduces the need for additional water sources and protects groundwater sources from overpumping
Total residential water use increased by 21% during the 1990s	Increasing uncertainty about stream flows and lake levels due to global climate change			

Source: NRTEE. 2003. Environmental Quality in Canadian Cities: The Federal Role. Available at: www.nrtee-trnee.ca.

municipalities facing limitations of supply and/or infrastructure, increasing capital costs of infrastructure expansions, the environmental impacts of water withdrawals and wastewater discharges, and the consequent impacts on downstream drinking water quality are among such issues.

Many of the water management concerns in Canada outlined in Table 5.3 are becoming more common all over the world and will increase as municipal water use continues to increase. Although demand management is not the only solution to all urban water management problems, it can help mitigate many of these problems.

Adopting DSM in urban water management can help reduce, or at least curb, current urban water use and wastewater production. In the context of population growth and urbanization, this means increasing per-capita water use efficiency in order to stabilize or reduce total water use. DSM programs mitigate the pressures of excessive urban water use on municipal finances, infrastructure, and the aquatic ecosystems that these systems rely on, ultimately increasing water use efficiency.

5.4 THE "SOFT PATH" FOR WATER

"Soft paths" for water is an overarching, long-term approach to water planning that fundamentally re-evaluates the way freshwater resources are developed, managed, and used. DSM is the main issue in this concept (Gleick, 2002). The soft path searches for more sustainable and equitable strategies in water management. In this manner, an attempt has been made to increase the water productivity rather than finding additional supplies. In this way, the stakeholders participate in the decision making, and the environmental water demand is considered in developing water allocation schemes (Wolff and Gleick, 2002; Brooks, 2003).

The soft path is basically different from the conventional (or "hard path") water planning approach. In this approach, water is considered as the means to accomplish certain tasks such as household sanitation or agricultural production and it is not viewed as the end product. Brooks (2003) states that, "With some important exceptions, the demand used in common projections is not for water itself, but for services provided by water." In the soft path approach, the objective of the water planning and management strategies is changed from water supplying to satisfying demands for water-based services (Gleick, 2002).

The soft path employs some strategies and instruments to improve the existing water infrastructure and reduce or eliminate the need for supply-side developments. Demand-side measures such as efficient technologies, education, regulation, and the rational use of economic instruments should be applied to increase the productivity of current withdrawals in an equitable manner.

The main issue in soft path planning is recognition of the water needs that could be supplied with lower quantity and/or quality of water than is currently used. There is a high potential to increase water productivity. For example, the reclaimed wastewater can be used to flush toilets or dry sanitation systems can be used to completely eliminate water use.

Traditional methods commonly assume an ever-increasing demand. The future demand is commonly overestimated because these methods do not consider changes in technology, costs, prices, customer preferences, and market forces (Wolff and Gleick, 2002) that may affect the water demand variations in the future. The "backcasting" approach is often used in the soft path method for demand planning instead of forecasting future demand based on past trends. In backcasting, planning a preferred future is defined and then planners work backwards to find feasible paths to reach that future situation (Brooks, 2003).

Ecological water requirements are assessed and integrated into the backcasting process. For example, the volume of water required to meet basic human needs and those of aquatic ecosystems should be determined for the system (i.e., a watershed or aquifer). Then the remaining resource should be efficiently distributed and productivity is maximized for economic development and other social needs through the application of the soft path concept and a comprehensive demand management approach.

For creating an opportunity for action, a paradigm shift in water management, away from the entrenched supply-side focus toward a comprehensive DSM based on the soft path planning approach should be established. A schematic was proposed by Ashton and Haasbroek (2001) for the representation of the growth in water demand and management, presented in Figure 5.1. Two periods of demand I and demand II are defined to bring the normal growth of demand, as a result of population growth, to the level of sustainable limits of water supply.

By curbing the unaccounted-for water and using better fixtures and water conservation measures, end-use efficiency is increased in the demand I period. In the demand II period, the allocation efficiency, the reuse, and the recycle schemes are examined and adopted. Another observation on the holistic view of resource allocation is when Karamouz et al. (2001) examined the water transfer projects in the context of having an overstock of water resources in one region and high demand for water not only for consumptive use but also for environmental restoration of aquifers in another region.

Because of limitations in developing new water resources, water conservation is the best solution for meeting future water demands in recent years. This approach is accepted from economic and environmental aspects in different regions all over the world. Furthermore, the water demand management approach is an effective strategy that improves the efficiency and sustainability of water resources through economic, social, and environmental aspects of developments (Wegelin-Schuringa, 1999; Butler and Memon, 2005).

Urban Water Demand Management

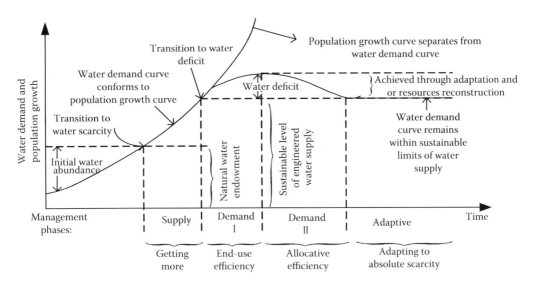

FIGURE 5.1 Schematic representation of different phases of growth of water demand and supply. (From Ashton, P. J. and Haasbroek, B. 2001. In A. R. Turton and R. Henwood Eds, *Hydropolitics in the Developing World: A Southern African Perspective*, 21pp. African Water Issues Research Unit [AWIRU] and International Water Management Institute [IWMI], Pretoria.)

Developing water saving technologies should be considered in developing water demand management strategies. These methods cover a wide range of solutions, such as dual-flush toilets, flow restrictors on showers and automatic flush controllers for public urinals, automatic timers on fixed garden sprinklers, moisture sensors in public gardens, and improved leakage control in domestic and municipal distribution systems. All these measures are practical, but regulations and incentives are needed for their implementation. The losses rate may vary from 10% to 60% of the supplied water and higher values are reported for the developing countries. To achieve the desired level of efficiency in water use, more attention should be given to decrease the amount of water losses and nonrevenue water (NRW), which generally includes

- Leakage from different components of water supply and distribution systems such as pipes, valves, and meters
- Leakage/losses from reservoirs and storage tanks (including evaporation and overflows)
- Water used in the treatment process (backwash, cooling, pumping, etc.) or for flushing pipes and reservoirs

The strategies used for demand management in rural and urban areas are completely different because of the scale and lifestyle effective factors. The costs of water and supplying services and technologies have a major influence on the level of water demand especially in the developing countries. But the distance from households to the standpipes and the number of persons served by a single tap or well are the main factors that control the water demand in rural areas.

5.5 URBAN WATER DEMAND MANAGEMENT: OBJECTIVES AND STRATEGIES

For developing effective UWDM plans, the key issue is to identify their objectives. Different objectives in different levels regarding the regional characteristics could be considered. However, in this section, the main 13 strategic objectives as stated by Chapman and the Water Services Interim Management Team (2001) that should be considered in developing UWDM strategies, are listed. More descriptions of the ways of achieving these goals are given in the next sections.

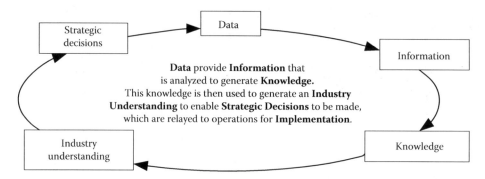

FIGURE 5.2 Schematic of bulk meter management system.

Strategic objective 1. Identify "champions" for water demand management among politicians and define their role of promoting UWDM within the community and within the municipalities. These champions will need to have the necessary passion and commitment to water conservation. Specific roles need to be defined for these champions, such as assisting during Water Week, appearing at various functions, lobbying other departments to conserve water, and many others.

Strategic objective 2. Raise the profile and priority of UWDM in the organization in order to achieve the defined objectives. As a key component of this strategy, giving the specific responsibilities to each manager within the water services cluster in accordance with this strategy has been proposed. The revised reporting lines will be addressed in the new structures that are developed as a part of the city-wide transformation.

Strategic objective 3. Prepare comprehensive business plans for each of the water divisions with respect to allocated water demand management initiatives. This will ascertain that each of the initiatives is well planned, executed, and reported on, thus ensuring a tightly managed progression of initiatives. The appropriate divisions within water services will be required to produce a business plan for each strategic objective consistent with this strategy.

Strategic objective 4. Implement a bulk meter management system (Figure 5.2) to ensure accuracy of measurement. The very cornerstone of water demand management is measurement because what is not measured cannot be managed effectively. Any decision-making process commences with data. With better data, better decisions will be made. With a bulk meter management system, the optimum accuracy of the large meters and the resultant income stream will be ensured. This will include a testing and replacement program.

A distinction is made between small domestic water meters and large, "bulk water," meters. A program to improve large (bulk) water meter accuracy will be developed, comprising the following:

- List the details of all existing meters by age or date of installation.
- Draft a program for replacement based on long-term asset management and best practice (sample testing), and so on.
- Thereafter, implement a meter replacement program to ensure optimized accuracy and cost benefits.

Strategic objective 5. Draft a municipal policy to ensure wise water use within the municipality. To achieve the full potential savings, a policy on water conservation will be developed to optimize water efficiency within the municipality to include buildings (rented and owned), amenities, and parks and forest departments.

Strategic objective 6. Develop appropriate water services by law to legislate the optimal use of water, incorporating the essential water demand management requirements to limit the inefficient and wasteful use of water.

Strategic objective 7. Minimize leakage and waste through the planning, design, construction, operation, and monitoring of suitably located district metering areas and pressure management systems. The leakage and pressure management strategies are described in the following sections.

Strategic objective 8. Promote the optimal use of water by consumers through education awareness programs and projects. The water restrictions awareness campaign needs to be analyzed regarding the effectiveness of the various methods in order to maximize future funding of such campaigns. Developing all programs on a regional level, using professional people wherever possible, is the objective. Offering "water audit" advice to property owners and managers (all user categories) will be considered to promote water savings as a service to our valued customers.

Strategic objective 9. Remove and prohibit all automatic flushing urinals (AFUs). Some wasteful practices and devices should be prohibited. In this regard, AFUs have been referred to as "the ultimate water wasters." About 70% of all the water passing through these water consumptions passes through while the building is unoccupied. The removal of these wasteful devices is one of the short-term actions that can be taken to reduce water demand. They are immediately effective and considered to be a "quick fix" action and require very short lead times to replace. The prohibition of AFUs will be achieved through the new water services by law.

Strategic objective 10. Minimize water losses in low-income housing through targeted plumbing repairs and education programs. A balanced approach will be sought for householders to take responsibility for their water consumption. A short-term measure is to limit the excessive losses occurring at the cost of the wider consumer body. The long-term solution should be that of universal metering and billing with appropriate credit control and customer management. The repair of private plumbing should ideally be the responsibility of the householder.

Each administration will organize a list of areas where such projects still need to be implemented, including manpower and budgetary requirements. An estimate of costs, water, and financial savings will be included. Some important aspects are

- Measures to ensure that degeneration does not occur.
- Occupants of repaired houses undertake to pay for services.
- Ongoing repairs become the responsibility of the homeowner.

Strategic objective 11. Continually optimize tariff structures to ensure universal metering and billing and a rate of payment that makes financial viability certain. The aim is to continuously reduce NRW (difference between the volume of water that is put into supply and the volume that is delivered to consumers) through improving meter accuracy, fixing leaks, and the elimination of book entry errors, of meter reading errors and estimates, of illegal and unmetered connections, and so on. These are ongoing functions of any water supply authority. A major aim of this strategy is to improve confidence in the NRW figures and prioritize tasks that need attention.

Strategic objective 12. Maximize the use of alternative sources of water. Some inside urban areas lend themselves to the exploitation of on-site alternative sources of water such as "gray water," rainwater, well-points, and boreholes utilization. Most administrations do supply suitably treated effluents for irrigating facilities such as golf courses, sports fields, and other recreational and landscaping uses.

Strategic objective 13. Make strong representations to draft a national performance standard for water fittings, appliances, and devices. The onus to select water efficient fittings should not only rest with the consumer, but the government should also control the use of fittings. Wasteful fittings are always cheaper and often have a similar esthetic appearance as water-efficient fittings. Once the national regulations are promulgated, they may be enforced through a number of measures. Investigating the retrofitting of households is required with a view to assisting the householder. Various rebate incentives are possible.

5.6 UWDM SCREENING

The primary screening of urban water demand is based on technical feasibility and institutional factors. Criteria for water demand management could include the following:

- Technical reliability—The existence of technology and results of its establishment elsewhere are considered.
- Environmental impact—All of the environmental impacts possible or difficult to mitigate are recognized.
- Institutional feasibility—It should be determined whether or not the agency has the water rights to perform the project by itself or the cooperation of other agencies is required and if so how this cooperation could be achieved?
- If any extensive institutional or administrative changes are required for the implementation of the demand management strategies, they should be identified and assessed before adopting UWDM strategies.

By considering the above factors, the alternatives that are inferior are eliminated. In this step, alternatives are eliminated as much as possible because the detailed evaluation phase can be expensive, such as preparing an environmental impact report on a new reservoir.

5.7 URBAN WATER DEMAND

Forecasting the water demand is a key factor in urban water resources planning as well as UWDM. "Urban water uses" include residential, commercial, institutional, industrial, and other water uses such as fire fighting, swimming pools, parks, and lawn sprinkling. The urban water use is usually measured as "per-capita water use," which is the volume of water used by each person per day. It is usually measured in liters per day or gallons per day per person or per metered service.

The withdrawal of water for municipal purposes is different in different regions. Average urban water use in the United States is about 2270 (L/day) per metered service. Water consumption of residential areas in eastern and southern states is about 790 (L/day), in central states it is about 1060 (L/day), and in western states it is about 1740 (L/day) per household service (Hammer and Hammer Jr., 1996). The water consumption per capita in mega cities of the Africa is less than 100 (L/day) but this amount for Auckland city located in the Oceania, Moscow in the Europe, New York in the United States, and Toronto in the Canada is more than 400 (L/day) (World Water Institute, 1998). In some regions, especially those with large residential lots and lawn, sprinkling has a significant influence on water demand. In such areas, about 50–75% of the total daily volume of water consumption may be attributed to irrigation.

Health standards in urban and rural areas are another determinant factor in municipal water demand. The quality of the drinking water has a direct effect on health, and the amount and quality of water used for bathing and cleaning also have a significant indirect impact on health conditions. Per-capita water use in most of the residential areas has been significantly increased in the past decades. Most new houses have more water fixtures, modern appliances, spacious lawns, and other conveniences that consume larger volumes of water.

Urban water demand is highly variable in time and space, and these variations should be considered in demand forecasting. The time horizon for water demand forecasting should be determined by regarding the useful time of water resources projects. In projects with a relatively long useful life, the demand forecasting time horizon should be extended to about 50 years. In medium-scale development plans, 15–25 years lead time could also be considered. It should be also considered that the prediction is for general planning purposes in a relatively large region rather than for construction design. These projections should be made for at least three levels, the extremes of which should encompass the conceivable minimum and maximum rates of change.

Urban Water Demand Management

The basic information needed for long-term water demand forecasting is the projection of population and economic activities. Other factors such as the technological changes in production processes, product outputs, wastewater treatment methods, social taste, and public policies with respect to water use and development should also be considered (Mays and Tung, 1992). Furthermore, the growing number of conflicts between water users should be incorporated in long-term water demand forecasting.

Therefore, the regional characteristics should be considered in the water demand forecasts. The land, water, and other natural resources characteristics are site specific and therefore the responses of industrial and agricultural sectors to development policies vary within and between different regions. The main differences between regions in response to development of water resources are listed as follows:

- Availability of water resources with desired quantity and quality
- Economic development level of the region
- Technological improvements in operation and utilization of water resources
- Costs of extraction
- Availability of skilled labor
- Costs of transportation
- Water resources markets
- Adaptive capacities of the environment

Water demand prediction with simple extrapolation of current trends may lead to serious errors. For instance, high water effluent standards and water prices may cause shifts in population. So in the long-term planning horizon, they may shift water demands from one region to another. As another example, encouraging the population to move to the regions where there are more job opportunities is caused by programs for supporting agricultural development (such as price-support programs). Therefore, the following data and information are needed for long-term forecasting of the regional water demand (Karamouz et al., 2003):

- Population projections based on the demographic studies and the studies done by the agencies responsible for multisectoral investment decisions
- Distribution of urban and rural populations among subregions
- Gross national product (GNP)
- Projected outputs of major manufacturing sectors for determining the regional distribution of activities based on the projected GNP
- Projected rates of unit water use based on technological advancements and relative share of instream and offstream uses
- Expected use of nonconventional water resources

It is not necessary to incorporate the water supply characteristics in the first estimation of the water demand projections. But when the costs of water are taken into account, a projection of demand must be modified by considering the supply curve of water for each planning region.

5.7.1 Urban Water Demand Estimation and Forecasting

To achieve a precise estimation of urban water use, the method of "disaggregate estimation of water use" could be used. In this method, the total delivery of water to urban areas is disaggregated into a number of classes of water use. Then, separate average rates of water use for each class are determined. In comparison with aggregate water use, disaggregate water uses within some homogeneous sectors are less variable and result in more precision in the estimation of water. For the application of this method, first the main components of water use should be identified. For this purpose, the following classification could be used (Karamouz et al., 2003):

1. Domestic
   - Washing and cooking
   - Toilet
   - Bath and shower
   - Laundry
   - House cleaning
   - Yard irrigation
   - Swimming pool
   - Car washing
   - Other personal uses (hobbies, etc.)
2. Public services
   - Public swimming pools
   - Governmental agencies and private firms
   - Educational services (such as schools, universities, dormitories, etc.)
   - Fire fighting
   - Urban irrigation of parks, golf courses, and so on
   - Health services (hospitals, etc.)
   - Public services (public baths, public toilets, etc.)
   - Cultural public services (libraries, museums, etc.)
   - Street cleaning and sewer washing
   - Entertainment and sport complexes (cinemas, clubs, etc.)
   - Food and beverage services (restaurants, etc.)
   - Accommodation services (hotels, etc.)
   - Barber shops and beauty parlors
3. Small industries (laundry, workshops, etc.)
4. Construction and public works
5. Water losses
   - Leakage from pipes, valves, meters, and so on
   - Evaporation in open reservoirs
   - Overflow of reservoirs
   - Disorderly condition of elements of water distribution networks such as cracked or leaky reservoirs, returning flow through nonreturning valves and pumps, and so on
   - Loss in production process (cooling, pumping, etc.)
6. Transportation
   - Taxies, buses, and other conveyances (stations, garages, etc.)
   - Ports and airports
   - Railways (stations and workshops)

Then the study area is divided into homogeneous subareas with an approximately constant rate of water use. These homogeneous areas are usually selected based on pressure districts or land-use units. Temporal (annual, seasonal, monthly, etc.) variation can also be considered in disaggregating the water uses.

5.7.1.1 Regression Analysis

The most widely used method in water use forecasting is regression analysis. The simple regression model can be formulated as follows:

$$Q_{t,i,j} = aX_{t,i,j} + b + \varepsilon_t, \tag{5.1}$$

Urban Water Demand Management

where $Q_{t,i,j}$ is the average rate of water use for disaggregated use i in subarea j in time t, $X_{t,i,j}$ is the independent variable for disaggregated use i in subarea j in time t, ε_t is the error term in time period t, and a and b are the regression coefficients.

Different factors affecting the water use and its relative importance in water use variations as well as the availability of the data should be considered in the determination of the independent variable, $X_{t,i,j}$. For instance, the population in each subarea is the most important factor in estimating water use in an urban area. In Equation 5.1, if the number of users in each subarea is considered as the independent variable, then the regression slope coefficient, a, would be the per-capita water use. The water use can also be estimated based on the number of connections to the water distribution system as follows:

$$Q_{t,i,j} = e \times NC_{i,j} \times W_{t,i,j}, \tag{5.2}$$

where $NC_{i,j}$ is the number of connections for disaggregated use i in subarea j, $W_{t,i,j}$ is the water use per connection for disaggregated use i in subarea j in time t, and e is the efficiency of the water distribution system, which is a function of leakage and other water losses in the system.

If more than one variable is likely to be correlated with water use in municipal areas, the multiple regression method is utilized for incorporating more variables in estimating and forecasting water use. The common variables in forecasting future water demand that are widely used in the literature are population, price, income, temperature, and precipitation. Depending on the independent variables employed, water demand forecasting models are classified into requirement models and demand models. The models that only consider the physical and ecological variables in demand forecasting are called requirement models, but demand models are based on economic reasoning and include only those variables that are associated significantly with water use and are expected to be causally related to water use (Mays and Tung, 1992). Linear and logarithmic water use models have been suggested by different investigators:

$$Q = a_1 x_1 + a_2 x_2 + \cdots + a_n x_n + \varepsilon, \tag{5.3}$$

$$Q = a_1 \ln x_1 + a_2 \ln x_2 + \cdots + a_n \ln x_n + \varepsilon. \tag{5.4}$$

Besides regression models, time series analysis has also been used to forecast future variations of water demands. For this purpose, the historical pattern of variations in water demand is recognized using the time series of municipal water use and related variables. In modeling the water demand time series, long memory components, seasonal and nonseasonal variations, jumps, and outliner data should be carefully identified and managed to achieve reliable results.

5.7.1.2 Parametric Models

These models are commonly used for demand forecasting at the household level; it is assumed that water demand is a function of price, weather, house and household characteristics, and any other distinguished (and observed) policy interventions taken during the study period (e.g., restrictions) (e.g., see Gaudin, 2006; Hewitt and Hanemann, 1995; Olmstead et al., 2003). The general form of these models for forecasting the total water demand by household i during time period t ($W_{i,t}$) can be formulated as

$$\ln(W_{i,t}) = \begin{cases} \beta_0 + \beta_1 \ln(ap_{i,t-1}) + \beta_2(\ln(ap_{i,t-1}) \times r_t) + \beta_3 r_t + \beta_4 br + \\ \beta_5 \ln(blpr_{i,t}) + \beta_6 outreb_{i,t} + \beta_7 inreb_{i,t} + \beta_8 wsr_{i,t} + \beta_9 Irr_t \\ \beta_{10} Hol_t + \beta_{11} amt_t + \beta_{12} prec_t + \phi_1 \ln(hin_t) + \phi_2 mage_t + \\ \phi_2 pph_t + \phi_3 hown_t + \phi_4 nho_t + \phi_5 oho + \phi_6 nbed_t + \varepsilon_{it} \end{cases}, \tag{5.5}$$

$$\varepsilon_{it} = \eta_i + \mu_{it},$$

TABLE 5.4
Definitions of the Variables Used in Equation 5.5

Variable	Definition	Units
Factor Under Utility Control		
ap	The average price of water	$/gallon
r	Indicator variable, equal to one if restrictions are in place at some point during the current bill period	0–1
blpr	Length of the current bill period	Days
outreb	Indicator variable, equal to one if the household participated in an outdoor rebate program	0–1
inreb	Indicator variable, equal to one if the household participated in an indoor rebate program	0–1
wsr	Indicator variable, equal to one if the household purchased a water Smart reader	0–1
Factor Outside of Utility Control		
Seasonal/weather related		
Irr	Indicator variable, equal to one if any portion of the bill period occurred during the irrigation season (May–Oct.)	0–1
Hol	Indicator variable, equal to one if special holidays occurred during some portion of the current bill period	0–1
amt	Average daily maximum temperature over the course of the current bill period	Centigrade
prec	Total precipitation over the course of the current bill period	mm
Economic–demographic (block-level)		
hin	Median household income	$
mage	Median age of homeowner	Years
pph	Median size of household	Persons
hown	Percentage of homes that are owner occupied	Percentage
nho	Percentage of newly build homes	Percentage
oho	Percentage of old homes	Percentage
nbed	Median number of bedrooms	Number of bedrooms

where ε represents the effects of unobserved factors on water demand. This term is composed of two parts. The first part, μ_{it}, refers to random unobserved influences, where the mean of μ_{it} is assumed to be zero. The second component, η_i, reflects differences between household characteristics that are unobserved from the analyst's perspective (e.g., lot size, irrigation technology, etc.). The other variables employed in Equation 5.5 are defined in Table 5.4.

This demand model includes two important terms:

1. A price–restrictions interaction term ($\ln(ap_{i,t-1}) \times r_t$), which incorporates the differences in price sensitivities when restrictions are in place. This approach has been suggested by Moncur (1987) and Michelson et al. (1999). However, it is usually omitted from analysis because of a lack of variation in the data set resulting in high colinearity between variables.
2. A block rate dummy variable (*br*) is considered. This term reflects the reasons other than the direct price that could result in difference in household consumption patterns under increasing block rate structures (Olmstead et al., 2003).

5.7.1.3 Data Analysis: Fixed Effects—Instrumental Variables

Estimating the demand for water by utilizing the regression technique is known as ordinary least squares (OLS). However, if the OLS is used for the estimation of Equation 5.5, it will produce biased results because some independent variables are correlated with the error term, ε_{it}. There are two main reasons for this problem.

1. The data set omits some factors relevant to determining household water demand. For instance, the data about the presence/absence of evaporative coolers (Hewitt and Hanemann, 1995) and the data about the lawn and the type of sprinkler systems employed in their maintenance are eliminated in this model. Since most of these "unobserved" effects (reflected in Equation 5.5 by η_i) are likely correlated with included variables, such as household income, the OLS will result in biased parameter estimates (Wooldridge, 2002).
2. Under block rate pricing, there is a complicated relationship between price and consumption. The price (either average or marginal) highly influences the consumption and also the level of consumption influences the price (both marginal and average). Because of this, $\ln(ap_{i,t-1})$ is correlated with ε_{it} through the unobserved individual effects, η_i, and possibly through, μ_{it-1}. This problem is well documented throughout the water demand literature (e.g., see Arbues et al., 2003; Pint, 1999).

There are very few studies that have tried to deal with both problems of correlation between independent variables and error term (e.g., see Arbues and Barberan, 2004; Pint, 1999). For solving this problem a fixed effects-instrumental variables (FE-IV) technique could be applied. The fixed effects component of this technique provides a solution to the unobserved effects problem utilizing the full advantage of the panel nature of the available data set. In these models, the deviations of water demand in each time period from the long-term household's average are estimated. In this approach, time invariant unobserved effects are "averaged-out" and therefore unbiased parameters are estimated for the remaining variables. The disadvantage of this method is that the parameters cannot be estimated for variables that are fixed in time but vary across households.

5.7.1.4 End-Use Analysis

"End-use analysis" defines the ways in which customers use water as great a level of disaggregation as possible. This can be achieved by use of customer surveys of water-using appliances (toilets, showers, and taps) and water-using practices (frequency of bath and shower use and frequency of clothes washing), through analysis of market research data for large appliances (e.g., clothes washers) or through industry sales statistics provided by manufacturers. In these models, the appliance stock and technical data with behavioral/usage data are combined. Demographic and land-use data as well as population data are also needed. Housing stock (dwelling type mix) and occupancy (number of persons per dwelling) also strongly influence demand. Modeled demand is correlated to historical demand data, for bulk water production, and metered customer data by sector. Depending on the nonresidential sectors' (commercial, industrial, and institutional sectors) proportion of total demand, they are included to varying degrees of disaggregation.

For a better understanding of the importance of end-use analysis in demand estimation, in an example, it is described how water demand is decreased due to toilets in residential dwellings in Sydney, Australia. The average flush volume of cisterns in Australia has been reduced significantly from about 11–13 L in 1980 to <4 L today, because of development of the dual-flush toilet (White, 1998, 1999). By replacing the older toilets and building new houses with this, less water will be used in toilets per person. In this way the per-capita demand in year 2001 has been decreased by about 20 L/person/day, resulting in a total of 24,000 ML water saving for Sydney in a year. By continuing the increase of toilets efficiency, in the year 2020 this will have bisected the per-capita demand for water in toilets. As the second stage of end-use modeling, a range of demand management measures are developed and assessed. By this method, potential levels of water conservation can be estimated.

5.7.1.5 Demand Models

A demand function relates the quantity of a commodity that a consumer is willing to purchase to the price, income, and other variables. If demand is considered as a function of price, demand curve is negatively sloped. This means that the lower the price the greater the water demand (Figure 5.3). A general form of demand models can be expressed as follows:

$$d = f(x_1, x_2, \ldots, x_k) + \varepsilon, \tag{5.6}$$

where d is the demand value, f is the function of variables $x_1, x_2, \ldots, x_k$, and ε is a random error (random variable) describing the joint effect of all the factors not explicitly considered by the variables.

Several explicit linear, semilogarithmic, and logarithmic models have been developed. For example, Agthe and Billings (1980) developed the following water demand function for Tucson, Arizona:

$$\ln(d) = -9.04 - 0.267 \ln(P) + 1.61 \ln(I) - 0.123 \ln(DIF) + 0.0897 \ln(W), \tag{5.7}$$

where d is the monthly water consumption of the average household in m³; P is the marginal price facing the average household in cents per m²; DIF is the difference between the actual water and sewer use bill minus what would have been paid if all water were sold at the marginal rate ($); I is the personal income per household ($/month); and W is the evapotranspiration minus rainfall (mm). The above equation implicitly relates demand to the hydrological index W. The positive coefficient of W shows that demand increases exponentially with W, which indirectly indicates increases in demand with the dryness of weather conditions. Equation 5.7 may be rearranged as

$$d = 0.000119 \, P^{-0.267} I^{1.61} (DIF)^{-0.123} W^{0.0897}, \tag{5.8}$$

or, in more general terms,

$$d = a' P^{b'} I^{c'} (DIF)^{d'} W^{e'}, \tag{5.9}$$

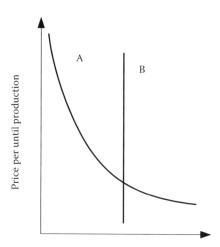

FIGURE 5.3 Demand functions. Curve A is a typical demand curve and curve B is the demand curve for a requirement for which the same quantity is demanded no matter what the price is.

where a', b', c', d', and e' are constants. The price elasticity (which will be described in the following sections) of demand for Equation 5.8 is -0.267. Therefore, changing the price, while keeping the other variables constant, results in different average demand values. Again, varying W, while keeping the other variables constant, gives a general relation of the average demand associated with the return period T.

5.8 URBAN WATER DEMAND MANAGEMENT FACTORS

The important factors in the development and implementation of demand management strategies in urban areas can be divided into two groups: factors under the utility control and factors that are out of control.

The main factors of domestic water demand management that are under utility control are listed below (Karamouz et al., 2003):

- Water loss reduction
  - Leakage detection and reduction
  - In-house retrofitting
  - Reduction of illegal connections
- Education and training
  - Public awareness
  - In-school education
  - Training and education of the staff in water-related agencies
- Economic incentives and water pricing
  - Water metering
  - Tariff structure
- Institutional measurements and effective legislation
  - Regulations for water demand management
  - Regulations on resale of water
- Water reuse

In addition to the various price and nonprice tools that utilities can employ to manage demand, there are a host of independent factors recognized to influence residential water requirements. A chief factor among these is weather. It is well documented because weather can impact short-term water demand decisions (particularly for landscape irrigation). Weather variables are typically controlled for regression-based studies which focus on price and nonprice tools (e.g., see Gutzler and Nims, 2005). But beyond the intuitive conclusion that hot-dry weather generates higher demands than cool-wet conditions, the exact nature of the weather/water demand relationship has a lot of uncertainty associated with variables such as humidity, precipitation, temperature, sunny hours, and wind.

Considering which climate variables could be used as predictand, is the main question; for example, what is more important: total precipitation over a month, the number of precipitation events, or the time between events? Questions of this nature are difficult to reply to for a variety of reasons, including issues of microclimate (i.e., weather conditions in one neighborhood may not match another). The presence of major outdoor water uses other than for irrigation (e.g., the use of evaporative coolers) and seperating the impact of weather from a broad spectrum of pricing and nonpricing demand management strategies are used during the hottest and driest seasons. The literature does not identify a preferred method for modeling weather variables. Also, research is frequently constrained by the fact that household-level consumption data are only available at a monthly scale, while weather variables alter daily.

5.8.1 Water Loss Reduction

A large percentage of the water is lost when transferred from treatment plants to consumers, in most water distribution systems. The amount of water that is lost or unaccounted-for is typically 20–30% and as high as 50% in some older systems. Leakage, metering errors, public usage such as fire fighting and pipe flushing, and unauthorized connections are among several causes of water loss in an urban water distribution network.

The major sources of water losses are leakage in distribution networks and home appliances as well as illegal connections. These water losses are called "unaccounted-for water," which is often used for measuring the efficiency of performance of a water supply system. Accounted-for water includes water consumption as recorded by customers' meters, water stored in service reservoirs, and authorized gratis use such as for flushing and sterilization of mains and routine cleaning of service reservoirs. The difference between the amount of water supplied from the water works and the total amount of accounted-for water is referred to as unaccounted-for water.

In many cities, unaccounted-for water or nonrevenue water (NRW) is estimated to be as high as 50% of water supplied. This might be because of

1. Inaccurately estimating the amount of water pumped or purchased due to not metering all water at the intake source or by using raw water or finished water meters that are inaccurate or improperly installed
2. Inaccurate customer metering
3. Bookkeeping errors
4. Nonmetered uses such as water used in the treatment process, city buildings, churches, watering a golf course, and so on
5. Water leaks

A certain level of unaccounted-for water cannot be avoided for technical and practical reasons but in some cities such as Singapore, it has been reduced to about 6% (United Nations, 1998).

5.8.1.1 NRW and Water Loss Indexes

The difference between the input amount of water of the distribution system and the amount of water billed to consumers is called "NRW." High levels of NRW reflect huge volumes of water being lost through leaks, not being invoiced to customers, or both. A high NRW level is normally a substitute for a badly operated water utility that lacks the governance, the autonomy, the accountability, and the technical and managerial skills necessary to provide reliable service to their population.

An annual water balance is normally used to assess NRW and its components. IWA (International Water Association) Task Forces produced an international "best practice" standard approach for water balance calculations (Table 5.5), with definitions of a wide range of terms involved (Hirner and Lambert, 2000; Alegre et al., 2000).

Definitions of the terms and principal components of Table 5.5 are

- System input volume: The annual input to a defined part of the water supply system.
- Authorized consumption: The annual volume of metered and/or nonmetered water taken by registered customers, the water supplier, and others implicitly or explicitly authorized to do so. It includes water exported, and leaks and overflows after the point of customer metering.
- NRW: The difference between system input volume and billed authorized consumption. NRW consists of unbilled authorized consumption and water losses.
- Water losses: The difference between system input volume and authorized consumption, consisting of apparent losses and real losses.
- Apparent losses: Consists of unauthorized consumption and metering inaccuracies.

TABLE 5.5
IWA "Best Practice" Standard Water Balance

Input volume	Authorized consumption	Billed	Billed metered consumption	Revenue water
			Billed nonmetered consumption	
		Unbilled	Unbilled metered consumption	Nonrevenue water
			Unbilled nonmetered consumption	
	Water losses	Apparent losses	Unauthorized connections	
			Metering inaccuracy	
			Raw water losses	
		Real losses	Distribution main losses	
			Burst, etc.	
			Background losses	

- Real losses: The annual volumes lost through all types of leaks, bursts, and overflows on mains, service reservoirs, and service connections, up to the point of customer metering. Corrosion, material defects, faulty installation, excessive water pressure, water hammer, ground movement due to drought or freezing and excessive dynamic loads and vibration from road traffic and construction are the main causes of leakage in water distribution networks. Leaks cause loss of economic and natural resources, and they put public health at risk. The basic economic loss is the cost of natural water, treatment, and transportation. Leakage leads to additional economic loss in the form of damage to the pipe network, such as erosion of pipe bedding and pipe breaks, and to the foundations of roads and buildings.

Real losses are where operational control is most relevant and can be classified as follows:

- Background losses from aggregation of a very large number of very small undetectable leaks (weeping joints, etc.)
- Typically low flow rates, over long periods, resulting in significant volumes
- Losses from observed leaks and bursts reported to the water supplier
- Typically high flow rates, short duration, and moderate volumes
- Losses from unreported bursts, found by active leakage control
- Medium flow rates, but duration and volume depend on active leakage control policy
- Overflows and leakage from reservoirs

All metered or assessed input data to the water balance are subject to errors and uncertainty. These errors accumulate in the calculated volumes of NRW and real losses. Quantification of these uncertainties is a practical approach to dealing with them. Uncertainty calculations have been standard practice for many years in hydrological measurements such as gauging of river flows, but until recently have not been used in water loss calculations. If it is considered necessary to improve the reliability of NRW or real losses estimates, the "entered value" component with the greatest variance should be the priority. Some of the main indices that are employed in leakage evaluation are as follows (Lambert, 2003):

5.8.1.1.1 Unavoidable Annual Real Losses
Real losses cannot be eliminated completely. The minimum achievable annual volume of real losses is known as unavoidable annual real losses (UARL). UARL can be assessed using a formula as follows (Lambert et al., 1999):

$$\text{UARL (L/day)} = (18 L_m + 0.8 N_c + 25 L_p) \times P, \tag{5.10}$$

where N_c is the number of service connections, L_m is the length of mains (km), L_p is the length of private pipes between the street property boundary and customer meters (km), and P is the average operating pressure (m). This formula is the most reliable predictor with real losses for systems with more than 5000 service connections, connection density (N_c/L_m) more than 20, and average pressure more than 25 m.

5.8.1.1.2 Infrastructure Leakage Index

UARL is used in the calculation of a new and important performance indicator, the infrastructure leakage index (ILI), which is the ratio of current annual real losses (CARL) to UARL.

$$\text{ILI} = \frac{\text{CARL}}{\text{UARL}}. \tag{5.11}$$

5.8.1.1.3 Economic Leakage Index

The extent of utilities investment in water supply and NRW controlling activities must be considered in an overall economic analysis to optimize monetary investment in resources and activities for NRW loss reduction, as compared to the cost of water saved arising from these programs (Figure 5.4). The minimum economically achievable ILI is called economic leakage index (ELI). Figure 5.5 shows the relationship between CARL and UARL.

Methods of leak detection do not always result in an immediate return on investment and unfortunately are complex and time consuming. The amount of ELI depends very much on local or national factors. This explains why policies and attitudes to leakage can legitimately differ from country to country and are due to factors such as

- Different ways of calculating leakage (e.g., night flow or unaccounted-for water)
- Different base levels (a function of differing operating environments)
- Different economic levels with a balance or imbalance in supply and demand and the expense of future resource developments
- Different levels of available water resources—in an area rich in available water resources, the environmental impact of abstraction will be low with a consequently lower justification for high levels of leakage control

There are two short-term and long-term levels of economic leakage control. In the short-term economic level, the cost of reducing leakage is equal to the cost of the deferred value of any capital

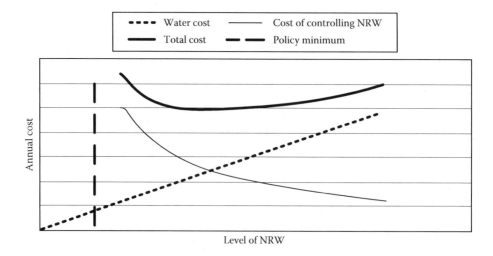

FIGURE 5.4 Trade-off between the water cost and NRW controlling cost and the optimum level of NRW.

Urban Water Demand Management

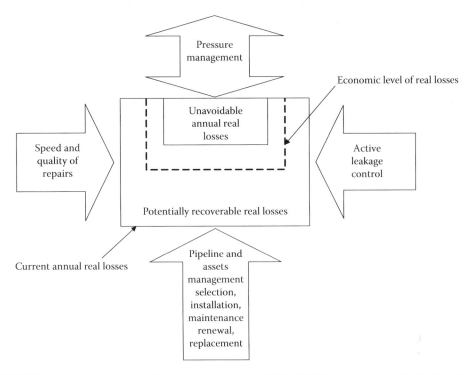

FIGURE 5.5 Four basic methods of managing real losses, CARL, UARL, and economic level of real losses.

investment for the extra volume plus its production cost. This assumes that repairs result in a permanent improvement in leakage in reality; repairs often result in increased pressures, which in turn promote further bursts.

It is also important to introduce the comprehensive leakage management policies and consequently reduce losses by better control of system pressures or mains replacement. The cost is offset by reduced costs of active leak control but may require a 20-year planning horizon.

5.8.1.2 Reduction of NRW

For reduction of NRW, the following issues can be considered:

- *Leakage detection*: A specific periodical program is needed for leakage detection. It can include visual inspection for leaks along transmission and distribution pipelines and also leak detection night tests for distribution mains. The leak detection tests are usually planned at nighttime because the pressure is usually higher in the system (because of less consumption) and therefore it is easier to detect the leaks (leakage is at the highest rate) and also it reduces the inconvenience to the customers. In order to locate the leaks, different mechanical, electronic, and computerized acoustic instruments such as stethoscopes, geophones, electronic leak detectors, and leak noise correlators can be used. The time frequency of testing is determined based on the percentage of unaccounted-for water and share of leakage, and also the size of water distribution systems.
- *Leakage control*: Utilizing high-quality pipes and fittings with more durable and corrosion-resistant materials considerably helps in controlling the amount of leakage. Minimizing the number of joints in the design of the water distribution network reduces the probable minor leaks. Teflon packing for repairing valve glands and seepout due to wear and tear and dezincification-resistant brass fittings can also reduce the minor leakage in the water

distribution network. Furthermore a sustainable physical loss control strategy must comprise four main elements (Figure 5.5) (Kingdom et al., 2006):

- *Active leakage control*: Monitoring network flows on a regular basis to identify the occurrence of new leaks earlier so that they can be detected and repaired as soon as possible.
- *Pipeline and asset management*: Managing network rehabilitation in an economical manner to reduce the need for corrective maintenance.
- *Speed and quality of repairs*: Repairing leaks in a timely and efficient manner (often requiring a thorough shake-up of working practices, organization, and stockkeeping of repair materials).
- *Pressure management*: Regulating network pressure through the judicious use of PRV (often an underestimated option for leakage reduction).

Using these four methods of leakage management, real losses can be reduced, but they cannot be reduced any further than the UARL. Also, the economic level of real losses is slightly greater than UARL.

- *Full and accurate metering policy*: The more accurate meters such as electromagnetic meters should be applied instead of conventional less precision meters such as venturi/dall tubes. An important issue that should be considered is developing a specific program for regular maintenance and replacement of meters. This can be done through bulk changing programs.
- *Proper accounting of water used*: A considerable amount of water in water distribution networks is used for the commissioning and filling of new mains, connections, and service reservoirs, for cleaning and flushing the distribution system during maintenance, and for fire fighting. Since these water consumptions are not usually recorded properly, they are considered in water losses.
- *Strict legislation on illegal drawoff*: The illegal drawoff should be detected and anyone who is responsible for it should be prosecuted.

5.8.2 Education and Training

An important issue in achieving the water demand management goals is public education and training. The results of investigations on the public education programs show that they are modestly advantageous, especially in the short term (Michelson et al., 1999; Syme et al., 2000). It is difficult to quantify the effectiveness of this variable. Therefore, it is a major challenge to

- Separate the effect of education programs from other pricing and nonprice programs.
- Distinguish between the different educational efforts and their effectiveness.
- Assess the long-term value of public education in promoting a conservation ethic.

A certain "critical mass" of educational programs is needed to generate significant benefits, but utilities soon reach a point of declining returns as extra efforts are then implemented (Michelson et al., 1999). Somewhat more attention has been presented to understand the effectiveness of technological modifications, especially indoor retrofitting of water-using devices such as toilets, showerheads, and washing machines. These kinds of studies are frequently based on engineering assumptions of anticipated reductions (Michelsen et al., 1999). The most common methods in public training and education that are applied in both developed and developing countries are as follows:

- Training programs in the mass media
- Billboards in streets and on public transport vehicles
- Sending training brochures with water bills

Urban Water Demand Management

Education of school children has been given priority in many countries. In these programs, basic information about the hydrologic cycle, costs of water supply and maintenance of water distribution networks, water pollution, and costs of wastewater treatment can be presented. Education and training can also increase the achievements of other water management strategies such as changing water tariffs and charging effluents.

Another important factor in water-related agencies is well-trained staff. Training employees to be technically competent and customer service oriented should be considered as an important objective in designing employee training programs.

5.8.3 Economic Incentives and Urban Water Pricing

A powerful tool used to reduce urban water use is pricing. In many countries, water and water-related services have been provided almost free, especially to domestic customers, or at a very low charge. The water charge in many urban areas do not even cover the cost of operation and maintenance of a water supply system and, therefore, those services have been highly subsidized.

Since the Dublin conference on water and the environment (ICWE, 1992), it has become generally accepted among water resources managers that water should be considered an economic good (the four Dublin principles, see Table 5.6). Two schools of thought could be distinguished in interpreting this issue (Van der Zaag and Savenije, 2000). The first school maintains that water should be priced at its economic value. The market will then ensure that the water is allocated to its best uses. The second school interprets "water as an economic good" to mean the process of integrated decision making on the allocation of scarce resources, which does not necessarily involve fiscal transactions.

Water pricing is not an instrument for water allocation, but rather an instrument to achieve financial sustainability. Only if the financial costs are recovered, an activity can remain sustainable. A good illustration of this premise is the "free water dilemma."

Free water dilemma: If water is for free, then the water provider does not receive sufficient payment for its services. Consequently, the provider is not able to maintain the system adequately and, hence, the quality of services will deteriorate. Eventually, the system collapses, and people have to drink unsafe water or pay excessive amounts of money to water vendors, while wealthy and influential people receive piped water directly into their houses, at subsidized rates. Thus the water-for-free policy often results in powerful and rich people getting water cheaply, while poor people buy water at excessive rates or drink unsafe water.

Hence water pricing is an important instrument to break the vicious circle of the "free water dilemma." But high attention should be paid to price determination and its impact on society. For this purpose, looking at both the costs and value of water is essential. Figure 5.6 represents the buildup of costs and values according to Rogers et al. (1997).

TABLE 5.6
Four Dublin Principals

1. Water is a finite, vulnerable, and essential resource that should be managed in an integrated manner
2. Water resources development and management should be based on a participatory approach, involving all relevant stakeholders
3. Women play a central role in the provision, management, and safeguarding of water
4. Water has an economic value and should be recognized as an economic good, taking into account affordability and equity criteria

Source: International Conference on Water and the Environment (ICWE). 1992. The Dublin statement on water and sustain-able development. ICWE, Dublin, Ireland.

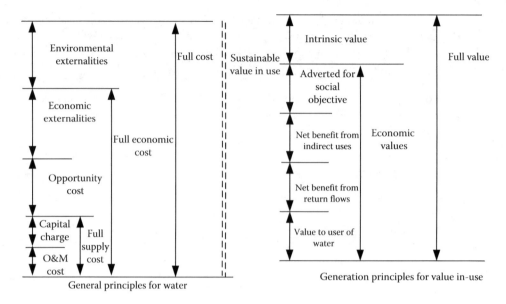

FIGURE 5.6 General principles for the cost and value of water. (From Rogers, R. W., Hayden, J. W., and Ferketish, B. J. 1997. *Organizational Change that Works*. DDI Press, Pittsburgh, PA.)

In the buildup of the costs, Rogers et al. (1997) distinguish the full supply cost (the financial costs related to the production of the water), the full economic cost, and the full cost. The distinction between the latter two is open to discussion. Some economists would say that the economic cost includes the full supply costs plus the opportunity cost (the cost of depriving the next best user of consuming the water). These economists consider all other impacts to be externalities. Of these, particularly the environmental externalities and the impacts on long-term sustainability are difficult to quantify in monetary terms. These economists are called "first school." Therefore Rogers et al. (1997) made a distinction between economic externalities and environmental externalities. In the broader definition of the "second school of thought," however, both types of externalities should be part of the economic decision problem.

A similar problem arises in the definition of the water assessment. The value to the user may be quantified by his/her willingness to pay, but there are additional benefits, such as benefits from return flows, multiplier effects from indirect uses, and in a broader sense the benefits of meeting societal objectives. The second aspect is often omitted by the "first school" economists, because it cannot be quantified in monetary terms; however, it is essential to the integrated decision-making process. The last part, the intrinsic value, consists of cultural, esthetic, and merit values of water, also very difficult to quantify in monetary terms. If we use the definition that economics is "about applying reason to choice," then the full cost and the full value as stated by Rogers et al. (1997) should be used for making allocation decisions.

It is obvious that a certain allocation of water is attractive when the full value is higher than the full cost. Determining these values and costs is precisely what is required in economic analysis. Once the decision has been made to allocate the water on economic grounds, then the next issue is to decide on the financing of the allocation. For the first school, this is no problem. The price should be the full economic cost, or the full cost, but that is not necessary. In principle, if society finds allocation a good idea, then society may decide to finance the allocation completely. This is common practice with security (police and defense), judiciary, and administration, and most countries subsidize education and health from government funds. But dissimilar to the above-mentioned sectors, the water segment in many countries is able to attain cost recovery. (In certain cities of Zimbabwe,

Urban Water Demand Management

the water account produces a surplus that these cities use to subsidize other sectors, such as basic health care.) The decision on how to allocate water resources on economic grounds comes first, and should be conceptually separated from the decision on how this allocation should be financed.

For water pricing, the following considerations are important:

- The institution responsible for the supply of the water should have sufficient autonomy to operate and maintain the system adequately and sustainably.
- Only when it has functional autonomy, including financial autonomy, can it perform its task in a sustainable fashion.
- There should be full cost recovery and preferably reservations for future investments.
- It is important to give due attention to equity considerations to prevent the weakest people from carrying too high a burden.
- The price should be "reasonable," allowing for full cost recovery, but in line with consumer's ability to pay off consumers.
- Those who can pay an economic price (industries and highly developed urban areas) should pay a high price and, by doing so, cross-subsidize the poorer strata of society.
- It is possible, in principle, to provide poor people with a minimum amount of water for free; it is, however, often considered more sustainable to ask for a nominal connection fee (within their ability to pay) or charge a subsidized "lifeline" rate, which gives them a claim on a proper service.

In applying this approach of a reasonable price, one automatically comes to increasing block tariffs, or a stepped tariff system as Kasrils (2001) calls it. By applying these increasing block tariffs, one can reach full cost recovery, institutional sustainability, and equity and, purely as a fringe benefit, send out a message to the large water consumers that water is precious and needs to be conserved. Only in this sense, as an afterthought, is water pricing a demand management tool.

5.8.3.1 Price Elasticity

In economic aspects, demand is considered to be a general concept that shows the willingness of users to purchase goods, services, or raw materials for the production of new goods. Commonly, it is considered that for a single consumer or group of consumers, there is an inverse relation between the water demanded and the price (cost) per unit. In other words, by increasing the water price, the quantity of demanded water decreases. It should be noted that the water requirement and demand are completely different. The water requirement is not affected by the price but the water demand is a function of different factors and price is one of them.

The responsiveness of consumers' purchases to price variations is considered as demand elasticity. But the most common concept in water demand analysis is price elasticity, which is the percentage change in water demand in response to 1% change in price. Young (1996) defines the price elasticity of water demand as a measure of the willingness of consumers to give up water use in the face of rising prices, or conversely, the tendency to use more as price falls.

The price elasticity of water demand, η_p, is formulated as follows (Mays, 2001):

$$\eta_p = \frac{\Delta d}{\bar{d}} \div \frac{\Delta p}{\bar{p}}, \qquad (5.12)$$

where $\bar{d}$ is the average water demand, $\bar{p}$ is the average price, Δd is the demand varation, and Δp is the price variation. If a continuous-demand function is considered, Equation 5.12 will be changed (Mays, 2001)

$$\eta_p = \frac{dd}{d} \div \frac{dp}{p}. \qquad (5.13)$$

TABLE 5.7
Summary of Some of the Price Elasticity Values

No.	Researchers	Research Area	Estimated Price Elasticity	Estimated Income Elasticity	Remarks
1	Howe and Linaweaver (1967)	Eastern U.S.	−0.860		
2	Howe and Linaweaver (1967)	Western U.S.	−0.52		
3	Wong (1972)	Chicago	−0.02	0.2	
4	Wong (1972)	Chicago suburb	−0.28	0.26	
5	Young (1973)	Tucson	−0.60 to −0.65		Exponential and linear models used
6	Gibbs (1978)	Metropolitan Miami	−0.51	0.51	Elasticity measured with the mean marginal price
7	Gibbs (1978)	Metropolitan Miami	−0.62	0.82	Elasticity measured with the average price
8	Agthe and Billings (1980)	Tucson	−0.27 to −0.71		Long-run model
9	Agthe and Billings (1980)	Tucson	−0.18 to −0.36		Short-run model
10	Howe (1982)		−0.06		
11	Howe (1982)	Eastern U.S.	−0.57		
12	Howe (1982)	Western U.S.	−0.43		
13	Hanke and de Mare (1982)	Malmo, Sweden	−0.15		
14	Jones and Morris (1984)	Metropolitan Denver	−0.40 to −0.44	0.40 to 0.55	Linear and log–log models
15	Moncur (1989)	Honolulu	−0.27		Short-run model
16	Moncur (1989)	Honolulu	−0.35		Long-run model
17	Jordan (1994)	Spalding County Georgia	−0.33		A price elasticity of −0.07 was also reported for nonrate structure, but increased price level

Mays (2001) has summarized different values of price elasticity of water demand reported in the literature as given in Table 5.7. Two different approaches have been employed for formulating the price elasticity of demand. One approach considers the average price and the second one uses marginal prices in price elasticity analysis. Agthe and Billings (1980) recommend the use of the marginal price because the application of average prices overestimates the results of elasticity.

The following schematic may depict the general relation of price and water demand based on the recommendation of the conclusions made by Jordan (1994):

$$\Uparrow \text{(price)} \Rightarrow \Downarrow \text{(water demand)} \, \& \, \Uparrow \text{(revenue)}$$

Moncur (1987) reported that an about 40% increase of the price resulted in a 10% decrease in the demand in Honolulu, Hawaii in the drought episodes of 1976–1978 and in 1984. This was achieved employing a price elasticity of −0.265.

Example 5.1

Interpret the meaning of the following: The price elasticity of water demand is equal to −0.8.

Solution: A price elasticity of −0.8 means that with a 1.0% increase in the price of water, the water demand will decrease by 0.8% if all other factors affecting the water demand remain constant. This will be valid only if the price–quantities data pairs are close to each other, or a smooth demand function fitted to the known data.

The development structure of utilities is probably the main factor that causes some difficulties in the study of price elasticity. For example, the reaction of customers who own homes or who pay water bills to the price changes is more severe than those who rent apartments or who do not pay water bills. Furthermore, the reaction of different groups of customers such as residential, commercial, and industrial to price changes is also different because of different demand patterns. Therefore, different price elasticity studies should be done for different groups of consumers.

The price elasticity of residential customers is more complicated than other customers because of a wide range of factors that are effective in residential water usage. For instance, many residents who rent housing are indifferent to demand regulations because they do not pay for water and, as another example, the patterns for indoor and outdoor water demand are significantly different and hence need different approaches to demand analysis. Furthermore, the climatic conditions of the study area and the time of the year (e.g., winter or summer) are two main factors that considerably affect the price elasticity.

Example 5.2

Consider the following demand function for urban water demand.

$$\ln Q = -1.5 - 0.2P + 0.4 \ln(I) - 0.2 \ln(F) + 0.4 \ln(H),$$

where Q is the annual water demand (m^3), P is the average water price ($ per m^3), I is the average yearly household income ($), F is the yearly precipitation (mm), and H is the average number of residents per meter. Determine the price elasticity of demand calculated at the mean price, $3.42 per m^3.

Solution: The demand function can be rewritten as

$$Q = e^{-1.50 - 0.2P} I^{0.4} F^{-0.2} H^{0.4},$$

and the derivative of Q with respect to price is

$$\frac{dQ}{dP} = e^{-1.5 - 0.2P} I^{0.4} F^{-0.2} H^{0.4}(-0.2) = -0.2Q.$$

Price elasticity of demand is calculated as

$$\varepsilon = \frac{P}{Q}\frac{dQ}{dP} = \frac{P}{Q}(-0.2Q) = -0.2(3.42) = -0.68.$$

The price elasticity of −0.68 indicates that if the price of water increases by 1.0%, the water demand will decrease about 0.68%.

5.8.3.2 Water Demand and Price Elasticity

The following typical equation is used to estimate the water demand based on price elasticity.

$$Q = cP^E, \qquad (5.14)$$

where Q is the quantity of demand for water, P is the price of water, c is a constant, and E is the elasticity of demand, which normally ranges between -1 and 0.

This equation is difficult to apply for the water sector as a whole, but for certain subsectors (urban water use, industrial water use, and irrigation) it may serve the purpose of analyzing the effects of tariff changes. The problem with the equation is that E is not a constant. It depends on the price, the water use, and it varies over time. So it is an equation with limited applicability. When we approach the more essential needs of the user (Figure 5.7), the primary uses of water have a special characteristic in that the elasticity becomes rigid (inelastic; E close to 0). People need water, whatever the price. For the most essential use of water (drinking), few alternative sources of water are available. For sectors such as industry and agriculture, demand for water is generally more elastic (E closer to -1) which is more in agreement with the general economic theory. This is because alternatives for water use exist in these sectors (e.g., introducing water saving production technologies and shifting to less water-demanding products/crops). For basic needs, however, demand is relatively inelastic or rigid. In urban water supply, elasticity is therefore generally close to 0, unless additional (nonfinancial) measures are taken. Poor consumers can often afford to use only small amounts of water (the basics), and any increase in tariffs will have little effect because they cannot do with less water. For large consumers (the ones that irrigate their gardens, own cars that need to be washed, etc.), the ability to pay is such that the need to save money on water is limited.

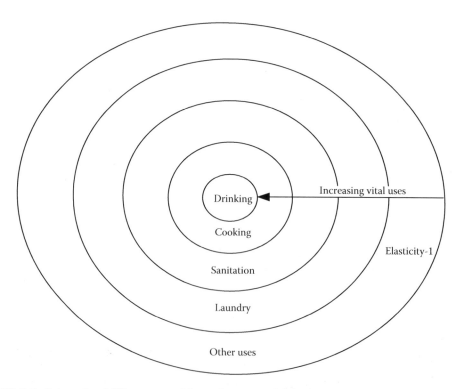

FIGURE 5.7 Schematic of different uses of domestic water and their elastic ties of demand.

In the latter case, awareness campaigns, regulation, policing, leak detection, renewal of appliances, and so on are often more effective than the price mechanism. The increasing block tariff system has been accepted by many societies as achieving the best compromise between efficiency and equity for domestic water supply, and poses an interesting paradox with neoclassical economics. It prices the highest value use (the most essential requirements such as drinking and cooking) lowest (first block at "lifeline" tariff), and the lowest value use (less essential uses such as washing a car) highest. The increasing block tariff system is a clear example of societies having to decide that neoclassical economics do not apply to the provision of domestic water services.

When the demand for water is inelastic, as is the case for urban water, the water provider may be tempted to raise tariffs, since this will always result in higher revenues, while water consumption drops only slightly. The provider may not be interested in curbing water demand through other means (e.g., through awareness campaigns or through subsidizing the retrofitting of houses with water-saving devices). It is therefore that water utilities should preferably remain publicly owned. If privatized they should operate within a stringent and effective regulatory environment.

5.8.4 Institutional Measurements and Effective Legislation

Fragmentation and overlapping of the responsibilities among the agencies involved in water resources management is an important obstacle to efficient water use in urban areas. In most countries, different organizations are responsible for the following tasks:

- Water resources development planning
- Water supply to the urban water distribution network
- Wastewater collection
- Water pollution control

The level of responsibilities of these agencies may also be different at national or more limited provincial, or city levels. However, the above tasks can be handled by different organizations but the decisions made in each of them can significantly affect other tasks and responsible agencies. Additionally, the urban sector is closely interrelated with other economy sectors. That is why any proposed program for urban water conservation will not be successful if it attempts to isolate the water supply and sanitation sector from other urban sectors. Therefore, to achieve economical use of water in urban areas, integrated management policies are required (United Nations, 1998). However, such policies can only be put into practice with the commitment and close cooperation of all the different organizations and agencies involved. For this purpose, conflict resolution issues should be incorporated in the development of planning and management policies.

By allowing the private sector to manage and handle the associated risk, the efficiency of implementing water resources development projects and utility management can be improved. But the governmental agencies should have the overall control through legislations and regulation. The injection of private sector efficiency and productivity also results in reduced public sector expenditure.

5.8.5 Water Reuse

By appropriately treating recycled wastewater and gray water, it can be used as a source of water to offset certain current or future potable water demands. For instance, golf courses can be irrigated with recycled water, removing them from the potable supply and freeing up water to support future growth. Incorporating dual water systems in new developments is considerably cheaper than retrofitting an already developed area.

Water reuse can be treated as follows:

- As an additional source of water to the supply system–quantitative value
- To reutilize a certain portion of water resources that is otherwise classified as wastewater–qualitative value

A summary of the allowed uses of recycled water corresponding with the degree of treatment is presented in Table 5.8 (California Code of Regulations, Title 22). Reused water has been utilized for agricultural and landscaping irrigation, industrial processes, and as cooling water, complying with environmental instream flow requirements, groundwater recharge, and direct consumptive use. Applications of reused water have been increasing as a consequence of severe droughts and water pollution control regulations (Lund and Israel, 1995).

TABLE 5.8
Recycled Water Criteria

Treatment Level	Allowed Uses of Recycled Water

Undisinfected secondary	*Surface Irrigation*:
	   Vineyards—no contact with the edible portion of crop
	   Orchards—no contact with the edible portion of crop
	   Pasture for animals—not producing milk for human consumption
	   Seed crops—not for human consumption
	   Ornamental nursery stock
	   Sod farms and Christmas trees
	   Fodder and fiber crops
	 Others:
	   Flushing sanitary sewers

Disinfected secondary	*Irrigation*:
	   Cemeteries
	   Freeway landscaping
	   Restricted access golf courses
	   Ornamental nursery stock
	   Sod farms
	   Pasture for livestock producing milk for human consumption
	   Nonedible vegetation where access is controlled—cannot be used for school yards, playgrounds, and parks
	   Impoundment
	   Landscape impoundments not utilizing decorative fountains
	 Cooling:
	   Industrial/commercial cooling that does not use cooling tower. Evaporative condensers or spraying
	 Others:
	   Industrial boilers
	   Nonstructural fire fighting, backfill
	   Soil compaction
	   Mixing concrete
	   Dust control
	   Cleaning roads and sidewalks
	   Industrial processes where it does not come in contact with workers

Disinfected secondary	*Irrigation*: Food crops—edible portion above the ground and not contacted by recycled water

continued

TABLE 5.8 (continued)

Treatment Level	Allowed Uses of Recycled Water

Disinfected tertiary	*Irrigation*:
	   Food corps
	   Parks and playgrounds
	   School yards
	   Residential landscaping
	   Unrestricted access golf courses
	 Impoundments:
	   Nonrestricted recreational
	 Cooling:
	   Cooling towers
	   Evaporative condensers
	   Spraying or mist cooling.
	 Others:
	   Flushing toilets/urinals
	   Industrial processes
	   Structural fire fighting
	   Decorative fountains
	   Commercial laundries
	   Commercial car washes—where public is excluded from the process
	   Consolidation of backfill around potable water pipelines
	   Artificial snowmaking for commercial outdoor use

Source: Unofficial California Code of Regulations (CCR) Title 22.

Water reuse is not an open option. It requires careful control of the treatment process and is more expensive than the use of freshwater, where it is readily accessible. In evaluating the cost of reuse as a water supply source, the costs of extra treatment, the redistribution system, and operation and maintenance should be considered. The treatment expenditure can make it impossible for small communities to reuse water, but large communities may be able to boost their water supply by 50% or more by reutilizing sewage. Water reuse can be classified into direct reuse and indirect reuse:

- Indirect reuse: Water is taken from a river, lake, or underground aquifer that contains sewage. The practice of discharging sewage to surface waters and its withdrawal for reuse allows the processes of natural purification to occur.
- Direct reuse: Planned and deliberate use of treated squander water for some beneficial purpose. Direct reuse of reclaimed waters is practiced for several applications without dilution in natural water resources.

Water can be reused for the following purposes:

Agricultural irrigation: The largest use of the reused water has been in agriculture. The original form of sewage treatment was sewage farming where sewage was disposed of on farmlands. Most sewage farms were replaced 70–90 years ago by biological treatment plants, which were discharged to the nearest watercourse (Dean and Lund, 1981). Reused water can also be used for urban irrigation. Irrigation with raw or partially treated sewage can conserve water and fertilize crops economically by capturing nutrients that would normally be wasted. This irrigation method is also an effective way to preclude contamination of nearby channels with the disease-causing organisms such as coliform and bacteria that the sewage contains. The most serious drawback of using sewage for irrigation is its role in transmitting infectious diseases to agricultural workers and the general public. The infections related to Ascaris and Trichuris worms are commonly associated with wastewater irrigation. These worms can be transmitted by eating uncooked vegetables that have been irrigated with such water. Wastewater irrigation has also been linked to transmission of enteric diseases such as cholera and typhoid, even in areas where these diseases are not endemic. The advantages of reclaiming wastewater and sewage for irrigation can be summarized as follows:

- Increasing crop yield
- Reducing pollution of freshwater resources by sewage discharge
- Decreasing the use of fertilizers
- Soil conservation
- Lessening the cost of sewerage and plumbing

Urban irrigation: Urban irrigation includes parks, golf courses, and landscape medians. Areas such as golf courses, cemeteries, and highway medians, where public access is restricted and where water is applied only during night hours without airborne drift or surface runoff into public areas, present limited exposure risk. Other urban uses include toilet flushing, fire protection, and construction.

Intentional reuse: Since the wastewater will soon become part of a potable water supply, the term "intentional" is used to demonstrate that at least part of the treatment has been given.

Groundwater recharge: One of the most generally accepted forms of water reuse is recharging underground aquifers with treated sewage. Water reuse by groundwater recharge is extensively believed to entail a lesser degree of risk than other means of reuse are entitled to, in which the recycling connection is more direct. The degree of treatment depends on the sort of application to the soil, soil formation and chemistry, depth to groundwater level, dilution available, and the residence time to the point of first extraction.

Recreation: Wastewater is sometimes used to fill lakes as a part of a recreational system. Biological treatment in a series of lagoons usually provides the required quality for recreational purposes.

Industrial use: The quantity of water used in many manufacturing processes and power generation is very large. The treated wastewater is a proper source of water for such industries, particularly for regions facing water scarcity. The water needed for cooling purposes, which does not have to be very pure, in some of the industries can also be supplied from treated wastewater.

5.8.6 Rainwater Harvesting

Through rainwater harvesting, water is taken out of the hydrological cycle for different uses. The usages of collected rainwater can be classified into two groups including agricultural use and human-related uses in which the latter is the main idea of this section. Since the water for human use should be convenient and clean, roofs are a good choice for a catchment surface because of their high elevation that protects them from contamination and damage. Tanks located close to homes provide high convenience for using the harvested rainwater. These systems are called rainwater roof catchment systems (RRCS), which are widely used throughout the world (Figure 5.8).

There are some advantages and disadvantages in using RRCS systems that should be considered before deciding on their development in a region. The main advantages of these systems are summarized by Schiller and Latham (2000) as follows:

- The high quality of collected water.
- The independency of the system, which make it suitable for scattered residential areas.
- The system could be developed by using local material and craftsmanship.
- Running the system does not need any energy.
- Ease of maintenance by the owner/user.

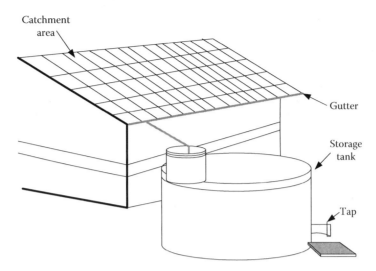

FIGURE 5.8 Typical RRCS.

- Convenience and accessibility of water.
- Saving time in water supply.

Some disadvantages or limitations are

- The high initial capital cost for system development.
- The volume of supplied water is limited depending on the rainfall amount and the building's roof area; therefore some other supplementary water sources should also be provided.
- In regions with long dry seasons, the required storage volume will be too high.
- The taste of rainwater, which is mineral-free, is not desired for most of the people.
- Mineral-free water may cause nutrition deficiencies in people who are already on mineral-deficient diets.

5.8.6.1 The Feasibility of an RRCS

As the first step in the planning for development of RRCS systems, the feasibility and efficiency of these systems from different technical, economic, and social angles, such as should be evaluated. The main point in the technical point of view is evaluating the supplied water through RRCS systems with the expected water usage and demand. The supplied water volume is highly dependent on the yearly rainfall amount and its temporal distribution. The water demand is related to the considered usages to be supplied by RRCS. For example, in a household, water is used for drinking, cleaning, cooking, and washing. The summation of these water usages in a rural area in developing countries for each person varies between 15 and 30 L/day.

For an accurate estimation of the water volume that could be supplied through an RRCS, reliable rainfall data should be provided. Then the annual water demand supposed to be supplied with RRCS systems is estimated based on the available water consumption data and is compared with the determined possible water supply based on rainfall amount and roof area. For example, for a building with a roof area of $30\,m^2$ and annual precipitation equal to $0.6\,m$, the possible water supply will be about $15\,m^3$ during a year. If the water supply is more than the water demand, then the RRCS construction is feasible from the technical point of view based on total maximum supply over a year. In the cases where water supply is less than water demand, depending on the amount of difference, some solutions such as increasing the rainfall catchment area or limiting the supposed demands from rainfall harvesting can also be considered. For example, rainwater could be used just for drinking and cooking that need high quality water, and the water required for cleaning and washing is supplied from other sources such as a well, a river, and so on.

The RRCS must be economically feasible to the household. Therefore, the costs of the proposed RRCS should be evaluated and compared with the costs of alternative water supply sources. The last step in evaluating the development of an RRCS system after demonstrating its feasibility from technical and economical aspects is social and community assessment. The success of an RRCS development project highly depends on this step in the strategies considered for proving people to accept this system as a source for their water demand supply (Schiller and Latham, 2000).

5.8.6.2 Design of an RRCS

The design stage of the project involves sizing the storage tank. There are a number of methods that can be used to determine tank volume. The more popular methods in storage tank sizing are introduced in the following paragraphs.

5.8.6.2.1 Dry Season Demand versus Supply

In this method, the storage tank is designed to supply the household water demand during the dry season of the year. Therefore dry season length is considered as a constraint in the design of a storage tank in an RRCS system. This method is appropriate for regions with definite dry periods during the year. Since the annual variations in annual rainfall are not considered in this method, it only gives a rough estimate of required storage tank volume and further considerations should be taken into account for determining the optimal storage tank volume.

5.8.6.2.2 Mass Curve Analysis

A more accurate method of sizing a tank is the mass curve technique. To achieve reliable results from this method, about 10 years of monthly rainfall data are needed. In the mass curve analysis method, the runoff production coefficient, which is called the runoff coefficient, should be estimated at first, which is defined as the amount of runoff entering the tank divided by total rainfall on the catchment roof. Through this definition, the water losses during water collection are also considered in runoff coefficient estimation. The runoff coefficient is estimated regarding the type of roof, the condition of gutters and piping, and the evaporation expected from the roof and tank. Approximate runoff coefficient values are given in Table 5.9. In the case where the exact value of the runoff coefficient could not be estimated, a low value such as 0.75 or 0.7 is considered, which is more conservative. For the calculation of storage tank size using the mass curve analysis method, the six steps listed below should be followed.

1. Preparing a 10-year series of monthly rainfall data starting from the beginning of a rainy season.

TABLE 5.9
Approximate Runoff Coefficient Values for RRCS Design

Type of Roof	Good Gutters	Poor Gutters
Metal	0.9	0.8
Other roofs	0.8	0.7

Urban Water Demand Management 217

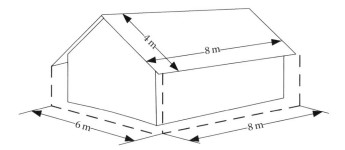

FIGURE 5.9 Roof area catchment.

2. Determining the monthly possible water supply through rainwater-harvesting systems which is equal to the rainfall multiplied by the runoff coefficient and the catchment area. In estimating the water supply, the effective roof area for collecting water should be considered which is not the roof area itself but is equal to the horizontal or ground area covered by the roof. For the example roof shown in Figure 5.9, the effective roof area for water collection is 48 m² (6 m × 8 m).
3. Developing the cumulative water supply series, since the start of calculations.
4. Estimating the monthly water demand proposed to be supplied using the rainfall collected water.
5. Developing the monthly stored water series, which is equal to the monthly water supply minus monthly water demand and could get both positive and negative values. Positive numbers indicate that the tank is being filled, whereas negative numbers mean that water is being taken out of the tank.
6. Developing the total stored water series as the cumulative sum of the monthly stored water.

The required tank volume in each year is determined by subtracting the least stored water amount during the dry season from the largest stored water volume during the wet season of the same year. The ultimate storage tank size is considered to be equal to the maximum amount of the required storage tanks in each year.

This procedure could also be followed graphically. The procedure is

a. The cumulative water supply and demand values are plotted on a supply graph (Figure 5.10). Demand must be less than supply.
b. Then the cumulative supply and demand curves are compared with each other. For this purpose, the demand curve is moved up and down placed on the top of the supply curve until it becomes tangent to one of the crests of the supply curve (Figure 5.10). Note that during this procedure the demand and supply axes must remain parallel to each other.
c. The largest vertical distance between the demand and supply curves in each year is considered as the required storage volume for that year of data. Similar to the previous procedure, this is repeated each year and the largest amount of storage volume is considered as the final storage tank size.

EXAMPLE 5.3

The 10-year rainfall data for a synoptic station in an urban area are given in Table 5.10. Considering a roof area of 50 m² and a runoff coefficient of 0.85, determine the required tank size for rain harvesting. Consider that the fixed monthly water demand is equal to 25 L/day.

continued

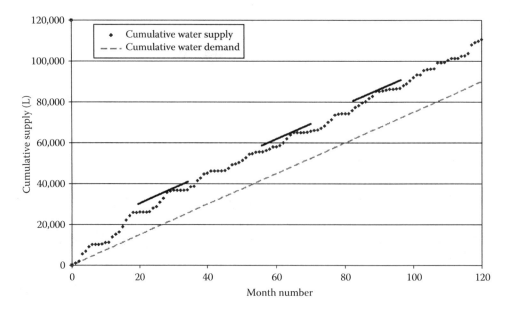

FIGURE 5.10 Cumulative supply graph.

Solution: At first, the average possible annual water supply from rainfall is calculated. Calculating that the average annual rainfall depth is equal to 249 mm, we have

$$\text{Annual water supply} = \text{area} \times \text{runoff coefficient} \times \text{average annual rainfall}$$

$$= 50\,\text{m}^2 \times 0.85 \times 0.249\,\text{m} = 10.6\,\text{m}^3.$$

Table 5.11 is developed to determine the storage tank size, as described before (note that only the last 5 years data are given in this table for summarizing the table). In this example, the largest storage difference occurs in the sixth year. Therefore the required tank volume is 3856 L, which is equal to the maximum difference between the wet- and dry-season storage volumes in 10 years of available data.

The results of the graphical method that are presented in Figure 5.10 are the same as those with the above procedure.

TABLE 5.10
Rainfall Data (mm) of Example 5.3

Year	Jan	Feb	Mar	Apr	May	Jun	Jul	Aug	Sep	Oct	Nov	Dec
1	23.8	20.6	88.5	30.1	53.9	22	0	0	5.4	14.4	5.8	61.1
2	31.6	23.9	61.5	76.9	54.6	32.1	0.4	6	0.6	2.7	3.7	39.4
3	18.1	50.7	45.7	66	17.1	8.3	0.8	0.5	0.2	1	38.6	6.8
4	63.8	33.2	42.8	10.1	24.4	4.1	0	0	1	29.8	37.8	13.2
5	14.7	26	28.4	43.7	3.2	13.6	5.1	0	17.6	15.8	22.6	3.8
6	13.7	33.1	44.3	42.7	26.3	0	2.2	0.1	3.1	11	10.8	5.6
7	20.5	25.3	43.4	26.7	55.6	14.1	2.3	0	0.2	35.6	38.2	19.3
8	29.1	16.7	34.3	26.2	56.7	1	2.3	13.6	7.4	3	3.8	7.6
9	28.6	18.2	39.1	33.7	33	0	48.7	10	4.2	8.2	64.6	2.8
10	4.6	22.6	20.7	0.9	0.1	25.2	5.1	24	102.7	26.8	15.4	19

TABLE 5.11
Mass Curve Analysis of Example 5.1

No.	Monthly Rainfall (mm)	Monthly Supply (L)	Cumulative Supply (L)	Monthly Demand (L)	Amount Stored (L)	Total Amount Stored (L)	Required Tank Volume (L)
1	13.7	582.25	58,701	750	−167.75	12,951	
2	33.1	1406.75	60,107.75	750	656.75	13,607.75	
3	44.3	1882.75	61,990.5	750	1132.75	14,740.5	
4	42.7	1814.75	63,805.25	750	1064.75	15,805.25	
5	26.3	1117.75	64,923	750	367.75	16,173	3856
6	0	0	64,923	750	−750	15,423	
7	2.2	93.5	65,016.5	750	−656.5	14,766.5	
8	0.1	4.25	65,020.75	750	−745.75	14,020.75	
9	3.1	131.75	65,152.5	750	−618.25	13,402.5	
10	11	467.5	65,620	750	−282.5	13,120	
11	10.8	459	66,079	750	−291	12,829	
12	5.6	238	66,317	750	−512	12,317	
13	20.5	871.25	67,188.25	750	121.25	12,438.25	
14	25.3	1075.25	68,263.5	750	325.25	12,763.5	
15	43.4	1844.5	70,108	750	1094.5	13,858	
16	26.7	1134.75	71,242.75	750	384.75	14,242.75	
17	55.6	2363	73,605.75	750	1613	15,855.75	2294.5
18	14.1	599.25	74,205	750	−150.75	15,705	
19	2.3	97.75	74,302.75	750	−652.25	15,052.75	
20	0	0	74,302.75	750	−750	14,302.75	
21	0.2	8.5	74,311.25	750	−741.5	13,561.25	
22	35.6	1513	75,824.25	750	763	14,324.25	
23	38.2	1623.5	77,447.75	750	873.5	15,197.75	
24	19.3	820.25	78,268	750	70.25	15,268	
25	29.1	1236.75	79,504.75	750	486.75	15,754.75	
26	16.7	709.75	80,214.5	750	−40.25	15,714.5	
27	34.3	1457.75	81,672.25	750	707.75	16,422.25	
28	26.2	1113.5	82,785.75	750	363.5	16,785.75	
29	56.7	2409.75	85,195.5	750	1659.75	18,445.5	3605.25
30	1	42.5	85,238	750	−707.5	17,738	
31	2.3	97.75	85,335.75	750	−652.25	17,085.75	
32	13.6	578	85,913.75	750	−172	16,913.75	
33	7.4	314.5	86,228.25	750	−435.5	16,478.25	
34	3	127.5	86,355.75	750	−622.5	15,855.75	
35	3.8	161.5	86,517.25	750	−588.5	15,267.25	
36	7.6	323	86,840.25	750	−427	14,840.25	
37	28.6	1215.5	88,055.75	750	465.5	15,305.75	
38	18.2	773.5	88,829.25	750	23.5	15,329.25	
39	39.1	1661.75	90,491	750	911.75	16,241	
40	33.7	1432.25	91,923.25	750	682.25	16,923.25	
41	33	1402.5	93,325.75	750	652.5	17,575.75	
42	0	0	93,325.75	750	−750	16,825.75	
43	48.7	2069.75	95,395.5	750	1319.75	18,145.5	1298
44	10	425	95,820.5	750	−325	17,820.5	
45	4.2	178.5	95,999	750	−571.5	17,249	
46	8.2	348.5	96,347.5	750	−401.5	16,847.5	
47	64.6	2745.5	99,093	750	1995.5	18,843	2515
48	2.8	119	99,212	750	−631	18,212	
49	4.6	195.5	99,407.5	750	−554.5	17,657.5	
50	22.6	960.5	100,368	750	210.5	17,868	

continued

TABLE 5.11 (continued)

No.	Monthly Rainfall (mm)	Monthly Supply (L)	Cumulative Supply (L)	Monthly Demand (L)	Amount Stored (L)	Total Amount. Stored (L)	Required Tank Volume (L)
51	20.7	879.75	101,247.8	750	129.75	17,997.75	
52	0.9	38.25	101,286	750	−711.75	17,286	
53	0.1	4.25	101,290.3	750	−745.75	16,540.25	
54	25.2	1071	102,361.3	750	321	16,861.25	
55	5.1	216.75	102,578	750	−533.25	16,328	
56	24	1020	103,598	750	270	16,598	
57	102.7	4364.75	107,962.8	750	3614.75	20,212.75	
58	26.8	1139	109,101.8	750	389	20,601.75	
59	15.4	654.5	109,756.3	750	−95.5	20,506.25	
60	19	807.5	110,563.8	750	57.5	20,563.75	

5.8.6.2.3 Mass Curve with Dimensionless Constant Analysis

This method is an extension of the mass curve analysis procedure. Continuing from (c) in the previous method:

d. Derive the curves for 70%, 50%, 30%, and 100% demand supply. This is done by repeating procedures (b) and (c) for each of the demands required. The resulting volumes are then expressed as a percentage of the supply.
e. The next step involves plotting these data. The storage required as a percentage of the supply is plotted along one axis and the demand as a percentage of the supply is plotted along the other axis (Figure 5.11).

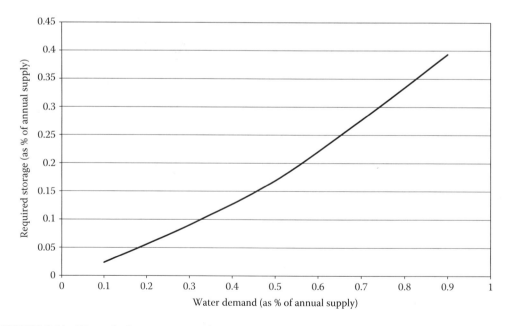

FIGURE 5.11 Dimensionless constant graph.

Urban Water Demand Management

Although this graph is derived from a particular roof, it will work for all demands, for any roof area and with any runoff coefficient. The graph can be taken into the field for rapid analysis of individual systems. The restriction on this graph is that the graph can be used only for the region where the rainfall data are applicable; hence it is a regional graph.

EXAMPLE 5.4

Derive the dimensionless demand graph of Example 5.3 and determine the required storage tank volume for a 40 m² roof with a runoff coefficient of 0.75 and a 6.5 m³ demand for a year.

Solution: The corresponding storage volumes for 90%, 70%, 50%, 30%, and 10% demand supplying 9.54 m³, 7.42 m³, 5.3 m³, 3.18 m³, and 1.06 m³ per year, respectively, when the annual supply is 10.6 m³, are calculated (Table 5.12). The results are plotted in Figure 5.11.

The average annual water supply of the considered roof in this example is equal to $40 \times 0.75 \times 0.249$ (the average depth of yearly rainfall) $= 7.47 \, \text{m}^3$.

TABLE 5.12
Results of Mass Curve Analysis with Different Demands

Demand % (as % of supply)	100	30	50	70	90
Storage required (m³)	1.06	3.18	5.3	7.42	9.54
Storage required as % of supply	2.3	9	16.9	27.7	39.4

The demand as a percentage of supply is $6.5/7.47 \times 100 = 93.7\%$; then from the percentage F1 demand-percentage storage curve (Figure 5.11), the storage required as a percentage of the supply is obtained as 37%. Therefore the tank volume required is $7.47 \, \text{m}^3 \times 37\% = 2.8 \, \text{m}^3$.

EXAMPLE 5.5

Consider a small village in a farming area with limited water available nearby. The rainfall data for this village are given in Table 5.13. The summation of supposed demands to be supplied by collected rainfall is 15 L/day per capita and 4 L of it is used for drinking and cooking. The considered region has 250 habitants and the average size of a household is 7. The percentages of pervious and impervious areas of the village are 20% and 70%, and 10% of the region is nonresidential. The average roof size in this region is 30 m². There is an alternative water resource with high potential of water supply about 1 km away from the village, but its water quality is low. The cost of materials used for the construction of an RRCS in this village is as follows:

Gutter materials cost ($/m)	1
Tank materials cost ($/m³ of tank capacity)	
Concrete	10
Reinforcing	5
Corrugated metal	30
Plastic	25
Miscellaneous	10

continued

TABLE 5.13
Rainfall (mm) for the Village of Example 5.5

Jan	Feb	Mar	Apr	May	Jun	Jul	Aug	Sep	Oct	Nov	Dec	Yearly Totals
3	113	30	46	0	22	0	0	0	1	32	99	346
15	18	7	32	2	0	0	2	9	27	78	37	227
364	28	76	11	4	0	0	0	0	14	21	58	576
91	39	12	19	10	0	0	14	23	58	46	24	336
173	128	14	35	5	0	1	0	60	71	121	102	710
284	73	22	70	8	0	0	0	0	0	20	164	641
53	174	71	31	76	0	0	0	13	69	59	57	603
75	136	125	0	0	0	0	8	28	11	67	189	639
293	73	24	30	10	0	0	0	8	126	53	27	644
131	23	60	0	1	0	0	7	0	93	72	29	416

A. Determine whether the installation of a rainwater collection system in this village is feasible.

B. Prepare a list of support services needed for user/owners including what must be done, who is responsible, means to supply service, and benefits.

Solution: There are no social conditions that would prevent from the building of the rainfall storage tanks. However, special loans for owners must be arranged and half of the costs of construction and the salary have to be paid by the national government because of the low income and poverty.

$$\text{Average rainfall supply} = 30 \times 513 \text{ (the average yearly rainfall)}/1000 = 15.4 \, m^3/\text{year}.$$

$$\text{Annual demand (all uses)} = 7 \times 15 \times 365/1000 = 38.3 \, m^3.$$

Therefore the average water supply is much higher than annual demand. The most important uses are drinking and cooking so that the results are recalculated using these values only.

$$\text{Annual demand (drinking and cooking only)} = 7 \times 4 \times 365/1000 = 10.2 \, m^3.$$

Demand is much less than supply.

Design of a tank using dry season collection versus the demand method:

$$\text{Demand for 6 months} = 10.2 \times 6/12 = 5.1 \, m^3.$$

$$\text{Mean monthly rainfall} = 513/12 = 42.8 \, mm.$$

Dry season is from April to September (6 months); the average rainfall during the dry season is: $27 + 12 + 2 + 0 + 3 + 14 = 58$ mm.

$$\text{Average amount supplied} = 58/1000 \times 30 = 1.7 \, m^3.$$

Therefore the tank size is $5.1 - 1.7 = 3.4 \, m^3$.

It should be noted that the "yearly totals" column in Table 5.13 indicates the wide variation in annual rainfall amounts. A conservative design would be arrived at by using data from the driest season on record. This would result in a tank size of $5.1 \, m^3 - (8/1000 \times 30) = 4.9 \, m^3$ (a 44% increase using data from the tenth year). We will use a $5 \, m^3$ volume for cost calculations.

Cost of tanks:

$$\text{Cost per } m^3 = \$10 \text{ (cement)} + \$5 \text{ (reinforcing)} = \$15.$$

$$\text{Cost for 5} \, m^3 = 5 \times \$15 = \$75.$$

Length of gutters (assuming square building with side $= x$)

$x^2 = 30, x = 5.5$ m.
Length of gutters $= 4 \times 5.5 = 22$ m.
Cost of gutters $= 22 \times \$1 = \$22$.
Miscellaneous: \$10
Total cost $= \$75 + 22 + 10 = \$107.00$.
Cost to the owner/user $= \$107/2 = \$53.50$.

This is a substantial amount of money in many areas—repayment in installments may be required.

Other benefits:

- Water obtained is of better quality. Health of the family is improved; medical costs are reduced.
- Carrying of heavy loads is reduced.
- Water carrier freed for productive work—farming, and so on.

The installation of an RRCS is feasible in this village through providing

a. Capital to the household in the form of loans.
b. Storage water only for drinking and cooking purposes.

The list of support services needed for user/owner RRCS is as presented in Table 5.14.

TABLE 5.14
List of Support Services Needed for User/Owner RRCS

Support Service	Who is Responsible	Means	Benefits
Design of RRCS	Central agency	Hire and train technician or engineer	– Good quality systems – Most appropriate materials – Lower cost
Training in construction	Central agency	Local instruction/demonstration projects	– Lower cost – User/owner does most of the construction work
Construction	Designer, user/owners, technicians	Designer provides additional information Technician handles specialized jobs (placing concrete, plumbing, etc.) User/owner provides labor and assistance, depending on skill level	– Skilled personnel ensures good quality system – User/owner has personal input – Cost reduced
User education	Central agency and others	Instruction of all family members on the importance of clean water, need to use water carefully, need for maintenance, hygienic handling of water, what to do in droughts	– System used properly – Continuous water supply – Need for maintenance understood – Water and health connection reinforced
Funding arrangements (Note: user/owner should pay some of the cost)	Central agency	Low-interest loans Grants from government Grants from international agencies Credit union	– User/owners have pride of ownership – Personal investment made – Better operation of the system is an individual responsibility
Check of maintenance requirements	Local officials, central agency	Check all aspects of system maintenance. Encourage user/owners to do maintenance. Eliminate problems that prevent maintenance	– State of systems maintained – Defects in construction design identified – Maintenance improved – The system is more reliable – Water quality maintained

5.9 DEMOGRAPHIC CONSIDERATIONS

Data limitations are a common impediment to assessing the impact of demographic characteristics on residential water demand. Researchers rarely have data sets that allow them to match household-level consumption data with demographic data about the people and house associated with a residential water account. Nonetheless, research to date is sufficient to suggest that household water demand is influenced by heterogeneity associated with differences in wealth (income), family size and age distribution, and household preferences toward water use and conservation (Cavanaugh et al., 2002; Hanke and de Mare, 1982; Jones and Morris, 1984; Lyman, 1992; Renwick and Green, 2000; Syme et al., 2000). Similarly, housing characteristics useful in explaining residential water demand can include the type of dwelling (e.g., single family home versus apartment), age of house, size of house/lot, and the water-using technologies featured (Billings and Day, 1989; Cavanaugh et al., 2002; Lyman, 1992; Mayer et al., 1999; Renwick and Green, 2000). Considering these influences is difficult not only due to the aforementioned lack of the relevant household/account level data, but also given that many features of a home (e.g., size) are likely to be correlated with household features, particularly income.

5.10 EVALUATING WATER CONSERVATION

For the evaluation of the effectiveness of outdoor watering restrictions, the results of voluntary and mandatory programs are compared. The reported results in the literature show the significant savings (sometimes up to 30% or more) from mandatory restrictions. But the result of voluntary restrictions is highly variable and, most of the time, the resulted water demand reduction is much less than what is obtained through implementation of mandatory programs (see Renwick and Green, 2000).

Combining the restriction programs with other price and nonprice efforts is a major challenge in assessing their impacts. Few studies have considered both of these policies in their analyses (e.g., Renwick and Green, 2000; Michelson et al., 1999). Furthermore, the studies that include both of them have usually omitted two important factors. First, the sensitivity to restrictions is heavily dependent on the distribution of users (Goemans, 2006). As an example, in communities with dominant large water users, reductions in response to restrictions are more than in municipals with a relatively small number of these types of consumers. Second, when restrictions are imposed, the response of households to changes in price is different (Howe and Goemans, 2002).

Least cost planning (LCP) uses an economic evaluation of options, as the aim is to minimize the total social cost of meeting service needs. As true economic analysis is difficult, evaluation in LCP often uses what is termed as a total resource cost (TRC) test to compare direct costs of conservation by demand management programs with the cost of supply. This test includes all costs and benefits to both the utility and its consumers in the analysis. Decreases in consumer bills spent on the conserved resource are not included in the TRC test as the utility sees these savings as a cost of foregone revenue. Equity issues between groups are not addressed by the TRC test, and impartiality between future and current generations is only addressed concerning the discount rate applied to future costs. Urban water conservation has the potential to reduce both the economic cost and the ecological impacts of providing urban water services. There are economic cost savings by eliminating or reducing the size of capacity augmentations as well as the operating and maintenance costs of treating and distributing potable water, and collecting and treating sewage. Some ecological impacts of urban water systems may be included in the evaluation of water conservation if monetary values can be agreed upon.

The comparison of supply conservation is often constructed in terms of unit cost, with conserved water being equated to an equivalent increase in supply. This recognizes that supply for a new development can be obtained by increases in capacity or by raising the efficiency of existing and future water users. The preferred cost measure is levelized unit cost, which is the present value of all costs of a measure or option over the present value of all water supplied (or conserved). A number of authors have described calculating unit costs using the present value of a physical water flow. Then, a water supply scheme marginal capacity cost using the term average incremental cost (AIC) has been computed. This AIC uses the present value of a physical water flow in the calculation of a unit cost, but only includes future capital costs of supply and corresponding capacity increases. Levelized unit cost (L) is similar but slightly conceptually broader as it can account for all capital and operating expenditure by water service providers or their customers in providing for increased flows or for reduced demand.

$$L = \frac{\text{present value of costs}}{\text{present value of water saved or supplied}}. \tag{5.15}$$

Demand management measures may affect various parameters other than average costs in an urban water system (Maddaus, 1999). On the water supply side, the parameters include (1) peak day demand, (2) peak hourly demand, and (3) amount of potable supply consumed per capita. Conservation measures might also avoid energy use for hot water, quantities of detergents required, and stormwater infrastructure.

Many externalities cannot be effectively valued and included in present value calculations. For those that can, "market" values are often inconsistent and/or insufficient. External costs that can be valued in surrogate markets include the cost to other water uses for water taken from a particular catchment, CO_2 releases, and the costs resulting from waterborne illnesses. Other concerns include impacts such as phosphate and micronutrient loss from agricultural land effluent volumes released to the ocean and some pollution releases to the environment. These externalities need to be included in Integrated Resource Planning (IRP) as parameters beyond least cost. It is also important to remember that evaluation is not a snapshot activity but should be a dynamic learning process. In evaluation, the assumptions should always be questioned, and the knowledge gained from previous analyses should be included in subsequent iterations.

5.11 FUTURE DEMAND SUPPLY

Many options are available for the water planner to meet future demands. For instance, water demands can be reduced through water conservation, certain potable uses may be replaced with recycled (reclaimed) wastewater, new wells may be drilled, new reservoirs constructed, or new water purchased from regional water suppliers (wholesalers). Evaluating all the options can be a costly and time-consuming task. Therefore, a manageable number of alternatives or combinations of resources should be developed that meet certain objectives and criteria and will generally meet future demands in some fashion. This section describes how to compile the options discussed in earlier chapters into resource combinations.

5.11.1 Demand-Side Management

Long-term water conservation and wastewater recycling are available resources to reduce demands in normal years. During droughts, short-term water conservation programs can help balance available water supply and demand.

Long-term conservation. Water conservation reduces demand for water through enhancements in efficiency and diminishing water waste. Water consumption can be reduced 10–20% over 10–20 years by a carefully planned and implemented long-term water conservation program. Review of previous conservation efforts, and identification of new measures with a screening for feasibility (including water savings estimates and cost-effectiveness) will lead to a list of viable conservation measures. These measures should then be consolidated into three to four programs that can be combined with supply options to meet future demands. Each program could consist of several, up to 10 or 15, measures. Suggested groupings are listed below:

- Current codes and regulations—These programs, including the U.S. Energy Policy Act and landscaping regulations, should be accounted for separately since the decision has

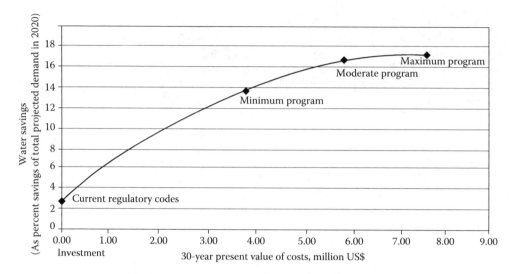

FIGURE 5.12 Example of investment at different conservation program levels.

already been made to implement these programs. These changes are not yet reflected in historical water use patterns, because they are likely to be recent. Ordinarily, they will not be accounted for in the base water demand. These programs can be used to produce what can be called a "net demand" projection. Net demand means the base demand less the effects of implemented conservation programs.

- Minimum program—Contains water conservation measures currently being planned for the next 10 years.
- Moderate program—Contains cost-effective water conservation programs that save more water than the minimum program.
- Maximum program—Consists of an aggressive conservation program that goes beyond the moderate program and is made of measures that save most water.

Each of these programs will have a time stream of costs and water savings that can be integrated with various supply options. An example of benefit of a small utility in water savings from investments in water conservation at varying program levels is shown in Figure 5.12. The optimal plan selected by the utility for implementation, based on the moderate program, gives ∼16% water savings in 2020 to stay under their fixed supply threshold of 11.48 million liters per day (MLD).

Short-term conservation. The utility's drought history should be examined to gauge the effectiveness of prior short-term programs. Different levels of voluntary to mandatory programs should be formulated to deal with possible water supply shortages. Shortages can also be treated as a resource as managed scarcities are a way to meet demand. A typical array of shortage management programs is depicted in Table 5.15. These programs have been used to achieve 5–35% cutbacks for a number of months to years depending on drought severity, location, and system usage patterns.

5.11.2 Supply-Side Resources

Resource alternatives for water supply range from groundwater to surface water to conjunctive use and commonly include

- Increased surface water storage
- Increased groundwater extraction
- Conjunctive use by combining surface water and groundwater resources
- Supplemental supply options in dry years

Urban Water Demand Management

TABLE 5.15
Model Drought Demand Management Plan

Water Storage	Action	Expected Range of Use Reduction (%)
Moderate	Voluntary/public education	5–10
Server	Restricted uses	10–20
Critical	Allocation/rationing	20–35

Source: Adapted from Maddaus, W. O. and Macy, P. P. 1989. *Benefit and Cost Analysis of Water Conservation Programs.* Brown and Caldwell Consulting Engineers, Pleasant Hill, CA.

- Water transfers and exchanges
- Additional treatment capacity

5.11.3 Overview of Evaluating Resource Combinations

The overall IRP evaluation process must be rigorous and fair and address all reasonable issues raised by the public, technical advisory committee members, and other stakeholders. Within IRP, the resource combinations are water supply and demand reduction scenarios or alternatives that improve water supply reliability at some cost and may have environmental and other impacts. The evaluation process helps decision makers understand the trade-offs between meeting future demands and improving supply reliability and associated costs. If the planned facilities are dependent on the growth of water demand, then reduction in future water use can affect the timing of construction of these facilities. Figure 5.13 illustrates how water conservation could affect the timing of capital facilities and assist in delaying infrastructure investments. In this case, a facility needed in 2020 could be delayed for about 7 years. In the example presented, demand reduction would reduce peak day demands by about 20%. The resultant money savings to the utility are the difference in the present value of the costs associated with building the facility in 2027 instead of 2020.

5.12 CONFLICT ISSUES IN WATER DEMAND MANAGEMENT

Water resources systems are mostly multiobjective and there are many objectives that are usually in conflict. Supplying demands is one of the main objectives in most of the water resources systems. The common conflict issue in water resources systems planning and operation happens when the

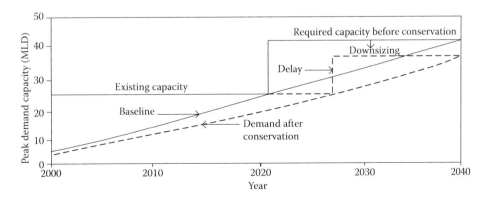

FIGURE 5.13 Example of delaying and/or downsizing a capital facility.

system should supply water to different demand points for different purposes and there are some constraints on water quality, flood control, and so on. In many mega cities of the world, water demands are raised to such a level that resources are not capable of supplying water requirements. Reduction of water demands by different water management schemes should be considered (Karamouz et al., 2003).

Besides the differences in the objectives of water users, there are also important issues from an institutional point of view. For instance, the Department of Agriculture is interested in supplying irrigation demands and water needed for livestock. Industries are interested in supplying water required for manufacturing processes, cooling systems, and so on. In municipalities, responsible agencies are more concerned with increasing the reliability in supplying urban and rural water demands. Therefore, the alternatives for water demand management should be selected and applied by different decision makers. These decision makers usually come from different agencies and so they might have different levels of authority to apply their preferred policies. The following example illustrates how the demand management alternatives can be selected using the Nash theorem, discussed in Chapter 10, where there are conflicting issues between different organizations.

Example 5.6

Assume that city A is located near river X and municipal water demands of this city are supplied from this river. Industrial complex B is also located near this river. Water required for manufacturing procedures and cooling systems of the industries located in this complex is also supplied by a water diversion system implemented downstream of the diversion point for city A. The remaining flow in the river supplies irrigation demands of agricultural lands located downstream of the industrial complex B. In the current circumstance, the river is capable of supplying total water demands in high-flow seasons. For simplicity, average monthly stream flow in the river and municipal, agricultural, and industrial water demands are considered in this example as displayed in Table 5.16.

According to this table, about 33% of water demands cannot be supplied by the river, which is the only source of water supply in this system. The following options are proposed for demand management by responsible agencies:

- The Department of Water Supply has proposed the following options for the reduction of municipal water demands:
  - Public education.
  - Reduction of unaccounted-for water by leakage control.
  - Urban irrigation by extracted water from groundwater aquifers.

TABLE 5.16
Average Monthly Streamflow in River X and Water Demands (Example 5.6)

	Volume (MCM)
Streamflow in river X	40
Municipal water demand	20
Industrial water demand	15
Agriculture water demand	25

TABLE 5.17
Estimated Costs and Expected Percentage of Reduction in Water Demands of Each Sector

Option for Demand Management	Percentage of Reduction in Water Demand of Each Sector	Estimated Cost (Monetary Unit)
Public education in the city and vicinities	1.5	0.8
Reduction of unaccounted-for water by leakage control	15	20.5
Urban irrigation by extracted water from a groundwater aquifer	4	3.5
Implementation of a system for recycling water in the cooling systems in the complex	45	12
Modification of the manufacturing process	12	18
Implement of the drip and pressure irrigation system (for increasing the irrigation efficiency)	38	9.5
Education and training programs for farmers	8	0.6

- Industries have proposed the following options for the reduction of industrial water demand:
  - Implementation of a system for recycling the water of the cooling systems in the complex.
  - Modification of the manufacturing process.
- The Department of Agriculture has proposed the following options for the reduction of agricultural water demands:
  - Implementation of the drip and pressure irrigation system (for increasing the irrigation efficiency).
  - Education and training programs for farmers.

Table 5.17 presents the estimated costs and expected percentage of reduction in water demands of each sector. Each of the above-mentioned agencies has proposed the favorite range of investment for demand management based on the costs of each of these options and the benefits associated with supplying water demands with higher reliability. The consequences of these proposals are summarized in Figures 5.14 through 5.16. As can be observed in these figures, the most favorable range of investment by each organization is revealed by a utility equal to one. Other investment levels are ranked by lower values of utility. The highest possible level of investment is assigned by each organization for zero utility. Find the optimal set of demand management options to be able to supply the whole demands of the system by the river.

Solution: The nonsymmetric Nash theorem has been used (more details of this method are given in Chapter 10) and the optimization problem for finding the optimal set of demand management options is formulated as

$$\text{Maximize} \prod_{i=1}^{3} (f_i - d_i)^{w_i},$$

where w_i is the relative authority of each agency, f_i is the utility function, and d_i is the disagreement point.

continued

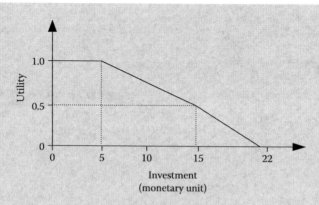

FIGURE 5.14 Utility function of the Department of Water Supply for investing money in municipal water demand management.

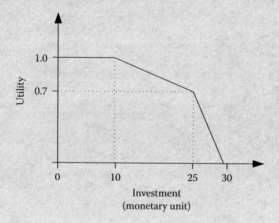

FIGURE 5.15 Utility function of industries for investing money in industrial water demand management.

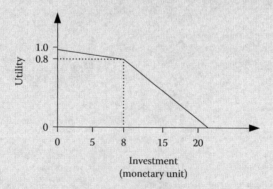

FIGURE 5.16 Utility function of the Department of Agriculture for investing money in agricultural water demand management.

Urban Water Demand Management

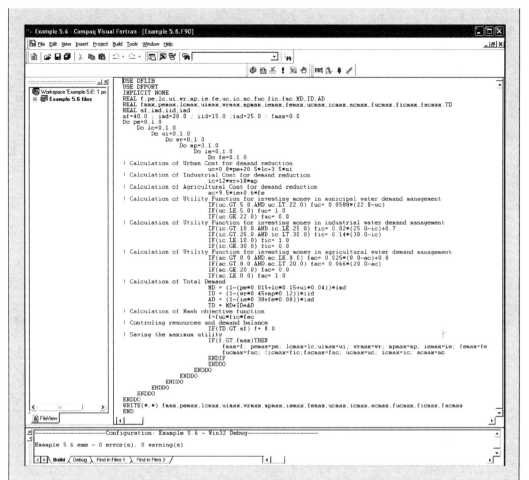

FIGURE 5.17 Source code of a Fortran program developed for solving Example 5.6.

In this example, the relative authority of each agency is considered to be equal to one, so the problem is simplified to a symmetric problem. The disagreement point is also set to zero.

$$\text{Subject to } d_i \leq f_i \leq f_i^* \quad \forall i, \quad \sum_{i=1}^{3} \sum_{j=1}^{NO_i} R_{i,j} \geq 20,$$

where $R_{i,j}$ is the volume of expected water demand reduction for option j ($j = 1, \ldots, NO_i$), proposed for sector i, f_i^* is the ideal point of agency i, and NO_i is the total number of water demand management options proposed for sector i. The above equation displays that the total water demand reduction of the system should be greater than or equal to 20 MCM, which is the mean monthly deficit in the current situation. In order to find the optimal solution of the Nash product, in this example, a Fortran code, which is shown in Figure 5.17, is generated for searching all feasible space. After running the program, the optimal solution of the problem is obtained as shown in Table 5.18. As can be seen in this table, total reduction in water demands is 21.25 MCM, which is higher than the 20 MCM deficit in the current situation. The total estimated cost for municipal, industrial, and agricultural demand reduction is 20.5, 12, and 10.1, respectively.

continued

TABLE 5.18
Optimal Set of Water Demand Management Options—Example 5.6

Sector	Water Demand Management Option	Reduction in Water Demand (MCM)	Utility	Total Water Demand Reduction (MCM)
Municipal	Reduction of unaccounted-for water by leakage control	3	0.09	3
Industrial	Implementation of a system for recycling water in the cooling systems in the complex	6.75	0.96	6.75
Agriculture	Implementation of the drip and pressure irrigation system (for increasing the irrigation efficiency)	9.5	0.66	11.5
	Education and training program for farmers	2		
	Total reduction in water demand			21.25

5.13 APPLICATION OF UWDM FOR SUSTAINABLE WATER DEVELOPMENT

UWDM strategies could be employed as an important factor in achieving sustainable water development goals in urban areas. For achieving the UWDM objectives after considering all the factors that affect it, an appropriate action plan should be provided. The UWDM action plans should be developed in provincial and municipal levels.

5.13.1 PROVINCIAL ACTION PLAN

The first step in developing a provincial action plan to achieve UWDM goal is reviewing the water sector subsidies and linking infrastructure funding to demand management. Financial policies should reorient public funds away from infrastructure expansion and toward demand management programs. The provincial government can promote full cost accounting for water utilities and ensure effective comprehensive conservation programs by reducing supply-side subsidies and promoting demand management programs directly.

The second step is capacity building to implement demand management. Permanent budgetary allocations should be provided for demand management staff, training, planning, and implementation. By dedicating financial resources to demand management, the importance of these strategies is recognized and it is acknowledged that many DSM measures such as education, promoting best practices, integrating new technologies, and program evaluation should be continued.

Then, as the third step, the integrated planning is employed for all watersheds and groundwater aquifers affected by urbanization. In this step, comprehensive long-term solutions should be developed to decrease or eliminate the undesired effects of urbanization on water resources which are discussed in detail in Chapter 7. The building and plumbing codes are modified to require water-efficient fixtures in all new constructions and renovations. The benefits of these amendments are not immediate; however, they ensure that the residential sector incorporates a high-efficiency infrastructure over time, dispersing costs and providing inducements for further conservation-based innovation.

In step 4, it is ensured that suitable "future proofing" (such as metering and research and development of reuse and recycling opportunities) is undertaken. Future proofing and investing in research and development of water demand management strategies will promote widespread and cost-effective implementation of appropriate technologies as conditions and circumstances change and as new technologies appear.

A water use efficiency task force is developed in step 5 to advise the government on existing and emerging water management issues and to organize water conservation efforts. This committee should include representations from all levels of government and different departments (such as environment, energy, municipal affairs, and public works), professional and industry associations, citizen groups, and NGOs. This task force could promote the institutional capacity to assess and implement provincial policies.

The last step is improving data standardization, collection, analysis, and availability. Appropriate systems and strategies should be developed for data gathering and analysis. The collected data should help toward better understanding of the nature of end use, the health of source ecosystems, minimum flow requirements for ecosystems, and water and cost savings associated with conservation measures. Improving databases through teamwork with related federal departments and providing general water withdrawal statistics is a vital step in the understanding of water use and the potential for water conservation (Brandes and Reynolds, 2004).

5.13.2 Municipal Action Plan

The main issue in developing the municipal action plan is creating participatory decision-making and planning processes. These processes could happen in different forms. It is necessary to ensure that an appropriate expression of community values is achieved and all of the stakeholders have participated in finding innovative, long-term, and sustainable solutions. Then the local water supply availability should be evaluated. In determining the water resources availability, probable impacts of climate change and long-term, integrated planning for water management should be considered. Knowing the short-term (e.g., 10 years) and long-term (e.g., 100 years) water withdrawal limits and the ecological water requirements is necessary to water supply planning. DSM should be considered as fundamental to water resources planning, and all opportunities to provide a cost-effective reduction in demand must be investigated prior to expansion of supply-side infrastructure. Finally, permanent budgetary allocations are developed for demand management staff, training, planning, and implementation (Brandes and Reynolds, 2004).

5.14 CONCLUDING REMARKS

The huge increase of world population, particularly those living in cities, and also limitation of water resources with acceptable quantity and quality, especially near the urban areas, have forced a great paradigm shift in urban water management from supply management to demand management. Providing a strategic view is unavoidable in achieving demand management goals. The main factors in developing these strategies such as public participation, training, economic issues, and providing necessary infrastructures are discussed in this chapter.

In the demand management approach, the system is seeking to recognize the potentials of water demand decreasing. Therefore, at first a realistic value of water demand should be provided. Different methods in demand estimation are discussed in this chapter. The main points in the selection of the demand estimation method are characteristics of the society and the climate of the region as well as available data related to system characteristics. Inaccurate estimation of water demand will affect the total urban water demand strategies. After determining the actual water demand, the main sources of extra water consumptions are identified and appropriate strategies are employed to reduce them as much as possible.

Studies have demonstrated that a considerable part of consumed demand (in some cases, more than 30%) is related to the urban water distribution system; therefore, evaluation and monitoring of distribution system performance in a quantified approach that is discussed in this chapter will help decision makers and managers to allocate needed financial, labor, and equipments to avoid leakages, especially in emergency situations. Furthermore, providing the alternative water resources such as treated wastewater and rain harvesting will help to decrease the probable water shortages, increase the residents' water security, water allocation conflicts, and system reliability and resiliency (which are discussed in detail in Chapter 11).

Taking the regional characteristics of urban area into account is necessary for the success of UWDM strategies. Through employing demand management approaches, the pressure on the available urban water resources decreases and the urban water system goes toward more sustainability.

PROBLEMS

1. Consider a small city with a river as a major alternate source of water supply. People on the edge of the city are not being served by a central water system. The rainfall data for this city are given in Table 5.19. Background data for the city are as follows.

Typical demand values (L/day/person)	
Drinking and cooking	4.7
Hygiene	7.9
Washing	5.8
Population in the design area (persons)	20,000
Average household size	5
Average roof area (m^2)	40
Roof material (% of each kind in the region)	
Pervious (thatched, etc.)	5
Impervious (metal, clay, etc.)	80
Nonresidential %	15
Other water sources	
Distance (km)	2.0
Quantity available (L/day/household)	900
Quality (contamination)	High
Minerals	High
Gutter materials cost ($/m)	0.8
Tank materials cost ($/m^3 of tank capacity)	
Concrete	12
Reinforcing	4.5
Corrugated metal	35
Plastic	28
Miscellaneous	7

TABLE 5.19
Rainfall (mm) for the City of Problem 1

Jan	Feb	Mar	Apr	May	Jun	Jul	Aug	Sep	Oct	Nov	Dec	Yearly Totals
21.4	68.2	27.5	42.8	71.0	1.9	0.0	0.0	13.0	0.4	25.4	21.9	293.5
37.0	47.6	57.0	31.8	41.1	12.2	0.0	3.0	33.0	14.1	20.0	49.2	346.0
40.8	93.0	42.8	25.9	6.8	5.0	0.0	0.0	5.3	16.9	42.6	25.6	304.7
16.8	18.9	39.0	68.5	84.8	10.1	66.4	1.2	15.3	24.5	17.1	19.8	382.4
41.1	46.3	28.0	3.5	27.9	16.4	13.0	0.1	12.3	26.2	26.9	29.7	271.4
24.8	16.6	18.8	17.7	14.6	0.2	0.0	3.0	4.2	0.7	15.1	27.3	143.0
16.8	20.3	43.1	48.0	36.0	1.7	2.0	0.0	2.6	24.5	47.9	30.5	273.4
26.8	25.2	18.0	16.2	3.9	3.0	0.0	7.3	0.0	14.2	26.8	24.9	166.3
16.6	15.1	21.4	49.2	27.9	34.4	10.7	26.0	7.9	11.1	35.4	55.2	310.9
5.5	22.3	97.3	20.8	32.9	0.0	40.2	2.7	2.2	27.6	18.2	13.8	283.5

A. Determine whether the installation of a rainwater collection system in this city is feasible.
B. Prepare a list of support services needed for users/owners including what must be done, who is responsible, means to supply service, and benefits.

There are no social conditions that would prevent from the building of the rainfall storage tanks. However, special loans for owners must be arranged because they are not served by the central water system and the people who live in edge of city have low income.

2. Estimate the storage tank size of the city described in Problem 1, using the curve mass method.
3. Historical data for water use, population, price, and precipitation in a city for a 10-year time horizon are given in Table 5.20. Formulate a multiple regression model for estimating the water use data and comment on selecting the independent variables.

TABLE 5.20
Historical Data for Water Use, Population, Price, and Precipitation for a 10-Year Time Horizon in the City of Problem 3

Year	Population	Price (Monetary Unit/m^3)	Annual Precipitation (mm)	Water Use (m^3/year)
2000	21,603	0.25	1015	2,540,513
2001	22,004	0.27	810	2,754,901
2002	23,017	0.29	838	2,722,911
2003	23,701	0.30	965	2,832,270
2004	24,430	0.31	838	2,904,727
2005	25,186	0.32	787	3,004,690
2006	26,001	0.33	720	3,309,927
2007	26,825	0.35	695	3,382,633
2008	27,781	0.31	820	3,375,392
2009	28,576	0.34	800	3,406,259

4. The monthly data of water price and consumption of a city is given in Table 5.21. Determine the price elasticity. Discuss the effectiveness of applying pricing strategies for decreasing water consumption.

TABLE 5.21
Price and Water Consumption Data of Problem 4

Consumption (MCM)	260	320	185	200	280	310	290	250	210
Price ($/cm)	13	10	20	17	12	9	12	13	18

5. Consider a city with about 1 million habitants. The daily water consumption per capital is about 180 L, the average temperature is 28°C, and the weather humidity is <35% in most days. The capital daily income is about $60. There are limited surface water resources near this city. There is a big aquifer below the city but its mineral concentration is high and in some places it is contaminated by septic tanks. Discuss water demand strategies that could be effective in this city.
6. The linear demand model derived by Hanke and de Marg (1982) for Malmo, Sweden, is

$$Q = 64.7 + 0.00017(Inc) + 4.76(Ad) + 3.92(Ch) - 0.406(R)$$
$$+ 29.03(Age) - 6.42(P),$$

where Q is the quantity of metered water used per house per semiannual period (m^3). *Inc* is the real gross income per house per annum (in Swedish crowns; actual values reported per annum and interpolated values used for mid-year periods). *Ad* is the number of adults per house per semiannual period. *Ch* is the number of children per house per semiannual period. *R* is the rainfall per semiannual period (mm). *Age* is the age of the house, with the exception that it is a dummy variable with a value of 1 for houses built in 1968 and 1969, and a value of 0 for houses built between 1936 and 1946. *P* is the real price in Swedish crowns per m^3 of water per semiannual period (includes all water and sewer commodity charges that are a function of water use). Using the average values of $P = 1.5$ and $Q = 81.4$ for the Malmo data, determine the elasticity of demand and interpret the result.

7. Determine the elasticity of demand for the water demand model in problem 6 using $P = 1.5$ and $Q = 75.2$; $P = 3.0$ and $Q = 75.2$; $P = 2.25$ and $Q = 45$; and $P = 2.25$ and $Q = 100$.

8. Consider a city with current population of 1 million (year 2009). We are planning to develop the water supply facilities of this city during the next 20 years. The water consumption per capita for this city is presently 320 L/day. The share of different consumption components relative to water consumption per capita is given in Table 5.22. Available water supply resources of this city include two surface reservoirs with an average water allocation capacity 100 and 40 MCM/yr. The population growth rate of this city has been determined as 2.3% per year. How would the state of water resources availability be in the year 2029 regarding to water consumption?

TABLE 5.22
The Share of Different Consumption Components Relative to Water Use Per Capita for the City of Problem 8

Share of Use (%)	Water Use Per Capita	Users
56	180	Residential
9	30	Public
8	25	Industrial–commercial
9	30	Public green space
17	55	Losses
100	320	Sum

9. In order to fill the water supply and demand gap of the city of Problem 8 for the year 2029, the following demand management strategies are recommended:
    a. Water withdrawn from a groundwater source up to 20 MCM/year.
    b. Reduction of network losses from 17% of total water supply to 15% by implementing a pressure management program, improving main breaks management and leakage management.
    c. Reduction of landscaping water demand up to one third of the present figure by utilizing plants resistant to water shortage and irrigation with recycled water.
    d. Reduction of water demand in residential, public, industrial and commercial sectors through enforcing legislation and public education up to 10%.

    Discuss how the application of these strategies will be useful in mitigating water shortages of this city in 2029.

10. Consider a small city with a population of 100,000. The water consumption per capita in this city is about 250 L/day. The available water supply system of this city includes a surface reservoir with an annual water supply capacity of 15 MCM. The population growth rate has been determined equal to 2% and the migration rate is about 0.5%. In the current rate of water consumption per capita will increase 1% each year. There is also

an aquifer near the city with a water supply capacity of 5 MCM/yr. The treatment plant of this city has limitation on the TDS intake of 300 mg/L. The TDS of surface water and groundwater are 250 and 400 mg/L, respectively. The cost of water supply from surface water is about 0.1$/L, where the cost of water withdrawn from groundwater is three times the surface water. The water supply manager of the city has planned some demand management studies with estimated reduction in water consumption and cost as given in Table 5.23. Develop an optimization model for water supply development and demand management schemes in this city for the next 50 years. The objective of the optimization model is to minimize the water deficit and water supply cost. The rate of return on investments is considered equal to 5%. Make any other assumption needed to solve this problem.

TABLE 5.23
Considered Water Demand Management Strategies in the City of Problem 10

Demand Management Strategy	Maximum Reduction in Consumption	Cost ($) per Each Liter Reduction in Consumption
Training programs	2%	0.02
Leakage reduction	8%	0.5
Using of water consumption reduction strategies	1.5%	0.6
Use of recycled water for landscaping/Parks irrigation	6%	0.2

11. A survey on a water distribution network with 25% loss shows that 18% of the losses are due to physical losses of the network and the remaining is not a real loss. 50% of physical losses are apparent and reported by people to the water supply agency. The survey result shows that 25% of nonapparent losses could be avoided by low-cost rapid surveys on the system. Determine the efficiency of this water supply system through estimation of different leakage indices.

REFERENCES

Agthe, D. E. and Billings, R. B. 1980. Dynamic models of residential water demand. *Water Resource Research* 3: 476–480.
Alegre, H., Hirner, W., Baptista, J. M., and Parena, R. 2000. *Performance Indicators for Water Supply Services*. IWA Manual of Best Practice, IWA Publishing, London.
Arbues, F., Garcia-Valinas, M. A., and Martinez-Espineira. R. 2002. Estimation of residential water demand: A state of the art review. *Journal of Socio-Economics* 32(1): 81–102.
Arbues, Fernando and Ramon Barberan. 2004. Price impact of urban residential water demand: A dynamic panel data approach. *Water Resources Research* 40: 24–30.
Ashton, P. J. and Haasbroek, B. 2001. Water demand management and social adaptive capacity: A South African case study. In A. R. Turton and R. Henwood, Eds, *Hydropolitics in the Developing World: A Southern African Perspective*, 21pp., African Water Issues Research Unit (AWIRU) and International Water Management Institute (IWMI), Pretoria.
Billings, R. B. and Agthe, D. E. 1980. Price elasticities for water: a case of increasing block rates. *Land Economics* 56(1): 73–84.
Billings, R. B. and Day, W. M. 1989. Demand management factors in residential water use: The Southern Arizona experience. *Journal American Water Works Association* 81(3): 58–64.

Brandes, O. M. and Reynolds, E. 2004. *Developing Water Sustainability Through Urban Water Demand Management: A Provincial Perspective*. The POLIS Project on Ecological Governance, University of Victoria.

Brooks, D. B. and Peters. 1988. *Water: The Potential for Demand Management in Canada*. Science Council of Canada, Ottawa.

Brooks, D. B. 2003. Water scarcity: An alternative view and its implications for policy and capacity building. *Natural Resources Forum* 27: 99–107.

Butler, D. and Memon, F. Eds. 2005. *Water Demand Management*. IWA Press, London, U.K.

Cavanaugh, B. P. and Polen, S. M. 2002. Add decision analysis to your COTS selection process. *The Journal of Defense Software Engineering*, available at http://www.stsc.hill.af.mil/crosstalk/2002/04/phillips.html.

Chapman, C. and the Water Services Interim Management Team. 2001. *Water Demand Management Implementation Strategy*. City of Cape Town, South Africa.

Curran, D. and Leung, M. 2000. *Smart Growth: A Primer*, Vancouver: Smart Growth, British Columbia, Canada.

Dean, R. B. and Lund. E. 1981. *Water Reuse: Problems and Solution*. Academic Press, New York.

Gaudin, S. 2006. Effect of price information on residential water demand. *Applied Economics* 38: 383–393.

Gibbs, K. 1978. Price variables in residential water demand models. *Water Resource Research* 14(1): 15–18.

Gleick, P. H. 1998. The World's Water 1998–1999: The Biennial Report on Freshwater Resources. Island Press, Washington, DC.

Gleick, P. 2000. The changing water paradigm—A look at the twenty-first century water resources development. *International Water Resources Association, Water International* 25(1): 127–138.

Goemans, H. B. Bounded communities: Territoriality, territorial attachment, and conflict. In M. Kahler and B. F. Walter, Eds, *Territoriality and Conflict in Era of Globalization*, pp. 25–61. Cambridge University Press, Cambridge, U.K.

Goldblatt, M., Ndamba, J., and van der Merwe, B. 2000. *Water Demand Management: Towards Developing Effective Strategies for Southern Africa*, Gland, Switzerland, IUCN.

Gordon Foundation. 2004. Controlling our thirst: managing water demands and allocations in Canada. Available at: www.gordonfn.ca/resfiles/Controlling_Our_Thirst.doc.

Gutzler, D. S. and Nims, J. S. 2005. Interannual variability of water demand and summer climate in Albuquerque, New Mexico. *Journal of Applied Meteorology* 44: 1777–1787.

Hammer, M. J. and Hammer, M. J., Jr. 1996. *Water and Wastewater Technology*, Third Edition. Prentice Hall, Englewood Cliffs, NJ.

Hanke, S. H. and de Marg, L. 1982. Residential water demand: A pooled time series, cross section study of Mamlo. *Water Resources Bulletin* 18(4): 621–625.

Hewitt, J. A. and Hanemann. W. M. 1995. A discrete/continuous choice approach to residential water demand under block rate pricing. *Land Economics* 71(2): 173–92.

Hirner, W. and Lambert, A. 2000. Losses from water supply systems: Standard terminology and recommended performance measures. Available at: www.iwahq.org.uk/bluepages.

Howe, C. W. and Linaweaver, F. P. 1967. The impact of price on residential water demand and its relation to system design and price structure. *Water Resource Research* 3(1): 13–32.

Howe, C. W. 1982. The impact of price on residential water demand: some new insights. *Water Resource Research* 18(4): 713–716.

Howe, C. W. and White, S. 1999. Integrated resource planning for water and wastewater: Sydney case studies. *Water International* 24(4): 356–362.

Howe, C. W. and Geomans, C. 2002. Effectiveness of water rate increases following watering restrictions. *Journal of the American Water Works Association* 94(10): 28–32.

International Conference on Water and the Environment (ICWE). 1992. *The Dublin Statement on Water and Sustainable Development*. ICWE. Dublin, Ireland.

Jones, C. V. and Morris, J. R. 1984. Instrumental price estimates and residential water demand. *Water Resource Research* 20(2): 197–202.

Jordan, I. L. 1994. The effectiveness of pricing as a stand-alone water conservation program. *Journal of the American Water Resources Association* 30(5): 871–877.

Karamouz, M., Zahraie, B., Shahsavari, M., Torabi, S., and Araghi-Nejad, Sh. 2001. An integrated approach to water resources development of Tehran region in Iran. *Journal of American Water Resources Association* 37(5): 1301–1311.

Karamouz, M., Szidarovszky, F., and Zahraie, B. 2003. *Water Resources Systems Analysis*. Lewis Publishers, Boca Raton, FL.

Kasrils, R. 2001. Minister of water affairs and forestry, Debate on the President's State of the Nation Address, Parliament, Cape Town.

Kingdom, B., Liemberger, R., and Marin, P. 2006. The challenge of reducing non-revenue water (NRW) in developing countries. How the private sector can help: A look at performance-based service contracting, water supply and sanitation sector board discussion paper series, Paper No. 8.

Lambert, A., Brown, T. G., Takizawa, M., and Weimer, D. 1999. A review of performance indicators for real losses from water supply systems. *AQUA* 48(6): 227–237.

Lambert, A. 2003. Assessing non-revenue water and its components: A practical approach. The IWA water loss task force, Water 21, Article No. 2.

Lund, J. R. and Israel, M. 1995. Optimization of transfers in urban water supply planning, *Journal of Water Resources Planning and Management*, 121(1): 41–48.

Lyman, R. A. 1992. Peak and off-peak residential water demand. *Water Resources Research* 28(9): 2159–2167.

Maas, T. 2003. *What the Experts Think: Understanding Urban Water Demand Management in Canada*. POLIS Project on Ecological Governance. University of Victoria.

Maddaus, W. O. and Macy, P. P. 1989. *Benefit and Cost Analysis of Water Conservation Programs*. Brown and Caldwell Consulting Engineers. Pleasant Hill, CA.

Maddaus, W. O. 1999. Realizing the benefits of water conservation. *Water Resources Update* 1114: 8–17.

Mayer, P. W., DeOreo, W. B., Opitz, E. M., Kiefer, J. C., Davis, W. Y., Dziegielewski, B., and Nelson. J. O. 1999. *Residential End Uses of Water*. American Water Works Research Foundation, Denver, CO.

Mays, L. W. and Tung, Y. K. 1992. *Hydrosystems Engineering and Management*, McGraw-Hill, New York.

Mays, L. W. 1996. *Water Resources Handbook*. McGraw-Hill, New York.

Mays, L. W. 2001. *Stormwater Collection Systems Design Handbook*. McGraw-Hill, New York.

Mays, L. W. 2001. *Water Resources Engineering*. Wiley, New York.

Michelsen, A. M., McGuckin, J. T., and Stumpf, D. 1999. Nonprice water conservation programs as a demand management tool, *Journal of the American Water Resources Association*, 35(3): 593–602.

Moncur, J. E. T. 1987. Urban water pricing and drought management. *Water Resources Research* 23(3): 393–398.

Moncur, J. E. T. 1989. Drought episodes management: The role of price. *Water Resources Bulletin. The American Water Resources Association* 25(3): 499–505.

NRTEE. 2003. Environmental quality in Canadian cities: The federal role. Available at: www.nrtee-trnee.ca.

Olmstead, S. W., Hanemann, M., and Stavins, R. N. 2003. Does price structure matter? Household water demand under increasing-block and uniform prices. NBER Research Paper, March, Cambridge, MA.

Pint, E. M., 1999. Household responses to increased water rates during the California drought. *Land Economics* 75(2): 246–266.

Renwick, M. B. and Green, R. D. 2000. Do residential water demand side management policies measure up? An analysis of eight California water agencies, *Journal of Environmental Economics and management* 40: 37–55.

Renzetti, S. 2003. *Incorporating Demand-Side Information into Water Utility Operations and Planning*. Department of Economics, Brock University, St. Catharines.

Rogers, R. W., Hayden, J. W., and Ferketish, B. J. 1997. *Organizational Change that Works*. DDI Press, Pittsburgh, PA.

Schiller, E. J. and Latham, B. G. 2000. *Rainwater Roof Catchment Systems. Information and Training for Low-Cost Water Supply and Sanitation*. World Bank, Washington, DC.

Shrubsole, D. and Tate, D. 1994. *Every Drop Counts*. Cambridge Water Resource Association, Cambridge, UK.

Solley, W. B., Pierce, R. R., and Perlman, H. A. 1993. Estimated use of water in the United States, U.S. Geological Survey Circular 1081, U.S. Government Printing Office, Washington, DC.

Syme, G. J., Nancarrow, B. E., and Seligman, C. 2000. The evaluation of information campaigns to promote voluntary household water conservation. *Evaluation Review* 24(6): 539–578.

Young, R. A. 1996. *Measuring Economic Benefits for Water Investments and Policies*. The World Bank, Washington, DC.

Unofficial California Code of Regulations (CCR). Title 22. Available at: http://www.dtsc.ca.gov/LawsRegsPolicies/Title22/.

van der Zaag, P. and Savenije, H. G. 2000. Towards improved management of shared river basins: Lessons from the Maseru Conference. *Water Policy* 2: 47–63.

Wegelin-Schuringa, M. 1999. Water demand management and the urban poor. Special Paper, IRC International Water and Sanitation Centre, Delft, the Netherlands.

World Resources Institute. 1998. *World Resources: A Guide to the Global Environment. A Joint Publication by the World Resources Institute, the United Nations Environment Program and The World Bank.* Oxford University Press, London, U.K.

Wong, S. T. 1972. A model on municipal water demand: A case study of Northern Illinois. *Land Economics*, 48(1): 34–44.

Wooldridge, J. M. 2002. *Econometric Analysis of Cross Section and Panel Data.* The MIT Press, Cambridge.

6 Urban Water Drainage System

6.1 INTRODUCTION

Urban water drainage systems are one of the key infrastructures in urban areas and play an important role in people's health and safety. The appropriate operation of these systems is related to their design and operation. The urban poor are affected the most by problems related to poor drainage in cities of developing countries. Low-value marginal land that is prone to flooding, or steep-sided hillsides, is often inhabited by poor communities that are unattractive for improvement. Although the consequences of flooding can be destructive, generally, the disadvantages of the substandard drainage system are balanced by the benefits of living near sources of employment and urban services (Parkinson, 2002).

An accurate estimate of runoff is very important in urban drainage system planning and management. The estimation of the peak rate of runoff, runoff volume, and the temporal and spatial distribution of flow is the basis for planning, design, and construction of a drainage system. Urban drainage is usually designed based on the concept of draining water from urban surfaces as quickly as possible through pipes and channels, but the peak flow and the cost of the drainage system is increased by this approach (Tucci, 1991). There are also certain peak increases at minor drainage levels, which highly impact the condition of the major drainage that should be controlled. Incorrect hydrology calculations result in undersized or oversized design of structures. However, it should be noted that the results of all urban runoff analysis only provide an approximation of the actual runoff. In this chapter, the sound design and management of urban drainage systems are introduced. This chapter is particularly useful for developing urban areas. The presented design procedures are based on fundamental hydrological and hydraulic design concepts.

6.2 URBAN PLANNING AND STORMWATER DRAINAGE

Urban stormwater drainage systems are one of the fundamental components of the urban infrastructure. Appropriate design of these systems minimizes flood damages and disruptions in urban areas during storm events and also protects the urban water resources environment. The primary objective of drainage systems is to collect excess stormwater from street gutters, convey the excess stormwater through storm sewers and along the street, and discharge it into a detention basin or the nearest receiving water body (FHWA, 1996). Water-quality Best Management Practice (BMP) should be exercised throughout the collection, treatment, and disposal processes. There are also some other objectives considered in the planning and management of drainage systems. These objectives could be listed as follows:

- Preserving the operation of the natural drainage system
- Providing safe passage of all means of transportation during storms
- Safeguarding public and managing floods during storms
- Protecting the urban stream and local rivers environment
- Design and operation of these systems, at least capital and maintenance costs

Pollution and a wide range of problems connected to waterborne diseases in poorly drained areas are due to urban runoff mixed with sewage from overflowing sewers. Focally contaminated wet soils provide ideal circumstances for the spread of intestinal worm infections, and flooded septic tanks and leach pits that provide breeding sites for mosquitoes. Infiltration of polluted water into low-pressure water-distribution systems, contaminating drinking water supplies, could cause outbreaks of diarrhea and other gastrointestinal diseases (Parkinson, 2002).

Poor solid waste management causes many problems associated with the operation of stormwater drainage systems. Municipal agencies that are responsible for solid waste management usually do not have adequate resources and equipment for drain cleaning. The different urban authorities responsible for operating and maintaining the diverse components of the drainage network often have poor communication and coordination.

The urban drainage problems are amplified by residential housing and commercial developments that increase the impervious areas, which intensifies urban runoff. There is often little control over new developments that can restrict the drainage of runoff by downstream flow constrictions such as encroachment into floodplain. Urban drainage engineers do not consider the existence of waterways and wetlands and are usually insensitive to the use of drainage through natural systems (Parkinson, 2002).

A more modern, sustainable, and cost-effective approach that utilizes natural channels and flood basins for drainage and flood control could be developed based on the result of evaluation of conventional approaches to drainage system design and considering the advantages and disadvantages of traditional approaches. This approach requires improved catchments planning and greater coordination among relevant institutions responsible for drainage, irrigation, and land use in the urban and suburban areas.

Design, operation, and maintenance of urban drainage systems are a major challenge for urban authorities and engineers. The effectiveness of stormwater management systems and the efficiency of urban management are directly linked. There are many issues that should be emphasized even though improvements in technology have provided effective tools for planning and for operationally sustainable and more cost-effective urban drainage systems. Parkinson (2002) summarizes these issues as follows:

1. Coordination among responsible urban authorities and agencies
2. Collaboration among private and public organizations as well as NGOs (nongovernmental organizations) to promote effective partnership with civil society and the private sector
3. Capability building for improved planning, design, and operation of urban drainage systems

6.2.1 Best Management Practices

Different practices could be considered for collected stormwater management, depending on the drainage area and stormwater characteristics. Some commonly employed BMPs are briefly described in the following sections (see McAlister, 2007 for more details).

6.2.1.1 Sediment Basins

Sediment basins are used for flow control and also water-quality treatment purposes. They are usually employed as an inlet pond to a constructed wetland or bioretention basin. Sediment basins are more effective in removing coarse sediments. The rate of sediment removal is usually between 70% and 90%; however, it is dependent on the basin area and the design discharge (Figure 6.1).

The exceeding runoff from the design flow is directed to a bypass channel through a secondary spillway. This prevents the resuspension of sediments previously trapped in the basin. Sediment basins should be designed with sufficient sediment storage capacity to ensure acceptable frequencies of desalting.

FIGURE 6.1 Typical sediment basin (http://www.ia.nrcs.usda.gov/features/images/Gallery/FiltrexxSoxxBasin.jpg).

6.2.1.2 Bioretention Swales

Bioretention swales provide both flow conveyance and storage in the swale and water-quality treatment through the bioretention area in the base of the swale (Figure 6.2). A high rate of water-quality treatment can be provided through bioretention areas for small to modest flow rates. In cases where the cross section of the swale is large relative to the flow rate, a limited flow detention capacity can be provided. The recommended longitudinal slope for bioretention swales is between 1% and 4%. In these slopes, the flow capacity is maintained without creating high velocities, potential erosion

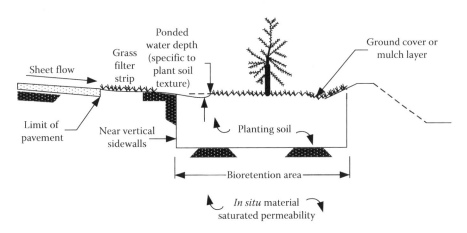

FIGURE 6.2 Structure of a typical bioretention swale (http://www.fhwa.dot.gov/environment/ultraurb/fig6b.gif).

of the bioretention or swale surface, and safety hazards. Pollutant removal is achieved through sedimentation and filtration through the filtration media and biological processes. The pollutant removal efficiency in a bioretention swale is dependent on the filter media, landscape planting species, and the hydraulic detention time of the system.

6.2.1.3 Bioretention Basins

The objectives of bioretention basins are flow control and water-quality treatment. Bioretention basins can be used on lots where there are several buildings and the lot is under single ownership, and larger basins are frequently used as part of urban development plans. The pollutant removal efficiency is dependent on the employed filtration media and the relative magnitude of the extended detention component of the basin. The maximum annual pollutant removal efficiency at the same time as flow control can be achieved if the extended detention component of the basin is used for small-to-medium runoff events. Pollutant removal is achieved through sedimentation and water filtration through the filter media and biological processes.

6.2.1.4 Sand Filters

The operation of sand filters is similar to bioretention systems except that they do not usually support vegetation owing to the filtration media being free draining (Figure 6.3). Sand filters are usually employed in confined spaces without sustainable vegetation such as below the ground. Sand filters typically include three separate chambers for sedimentation, sand filtration, and overflow. The medium-to-coarse sediments and pollutants are removed in the sedimentation chamber and the sand filter chamber then removes much of the medium-to-coarse sediment as well as some of the finer particulate and dissolved pollutants. To maintain removal efficiency, the sand filter should be regularly maintained to prevent crust formation.

6.2.1.5 Swales and Buffer Strips

Flow conveyance function is done along the swale, and then, water-quality treatment is provided through sedimentation and contact of flowing water with swale vegetation. The greatest effect of water-quality treatment functionality is on small to modest flow rates. A limited flow detention capacity can also be used when the cross section of the swale is large relative to the flow rate. The pollutant removal efficiency is dependent on the longitudinal slope, the vegetation height, and the area and length of the swale. The recommended longitudinal slope is the same as bioretention swales. Swales cannot effectively treat fine pollutants, but can provide pretreatment for downstream measures.

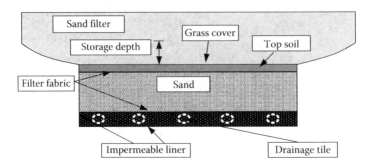

FIGURE 6.3 Schematic structure of a typical sand filter (http://www.civil.ryerson.ca/Stormwater/menu_1/bin/pic_20.gif).

6.2.1.6 Constructed Wetlands

Constructed wetland systems are shallow, extensively vegetated water bodies (Figure 6.4). Extended detention, fine filtration, and biological pollutant uptake processes are employed in these systems for pollutants removal. Wetlands are composed of three major parts including an inlet zone (sedimentation basin), a macrophyte zone, and a high-flow bypass channel. The depth of the macrophyte zone varies between 0.25 and 0.5 m and the theoretical detention time is between 48 and 72 h. Wetlands store runoff during the rainfall and slowly release it after finishing the event and in this way provide a flow control function. When flows exceed the design flow of a wetland for protecting wetland vegetation and also avoiding resuspension of trapped pollutants, excess water is directed around the macrophyte zone via a bypass channel.

6.2.1.7 Ponds and Lakes

Ponds and lakes are usually formed by a simple dam wall with a weir outlet structure or are created by excavating below the natural surface level. Ponds and lakes collect water for stormwater reuse schemes and also act as part of a flood detention system. They also participate in pollutants removal through sedimentation, absorption of nutrients, and ultraviolet disinfection. Ponds cannot be considered as "stand-alone" stormwater treatment measures and require pretreatment via constructed wetlands or other measures. The historical runoff and/or predevelopment flows for a range of flood events are considered in the design of pond outlets. The minimum average turnover period between 20 and 30 days should be considered in the design of a pond or lake to prevent water-quality problems.

6.2.1.8 Infiltration Systems

Stormwater infiltration systems infiltrate stormwater into surrounding soils so that their performance efficiency is dependent on local soil characteristics (Figure 6.5). The best application of infiltration systems is in sandy-loam soils with deep groundwater. Stormwater infiltration systems effectively decrease the volume and magnitude of peak discharges from impervious areas. A vital component in employing infiltration systems is pretreatment that avoids clogging, which deteriorates the infiltration effectiveness over time.

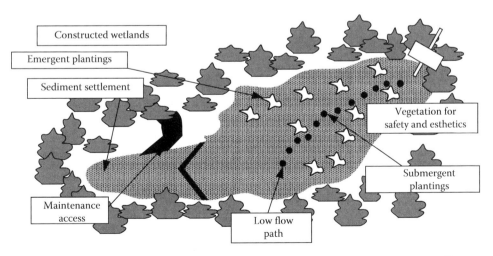

FIGURE 6.4 Different components of a typical constructed wetland (http://www.civil.ryerson.ca/Stormwater/menu_1/bin/pic_21.gif).

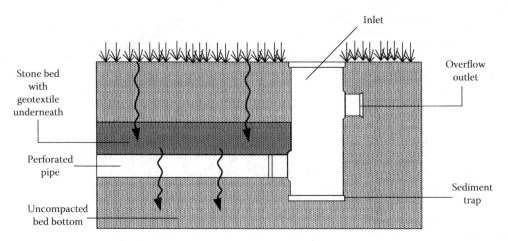

FIGURE 6.5 Structure of a constructed infiltration bed for stormwater drainage (http://www.csc.temple.edu/t-vssi/images/survey_BMP/diagrams/infiltration_bed_large.jpg).

6.2.1.9 Aquifer Storage and Recovery

In an aquifer storage and recovery (ASR) system, water recharge to underground aquifers is increased either by pumping or by gravity feed. The stored water can be pumped during dry periods for subsequent reuse, thereby providing a low-cost alternative to large surface water storages. Usually, treated stormwater or recycled water is used for this purpose to protect groundwater from deterioration of quality or aquifer properties. For this reason, ASR systems typically incorporate a constructed wetland, detention pond, dam, or tank, part or all of which acts to remove pollutants and provide a temporary storage role. The level of pretreatment of stormwater is dependent on the quality of the groundwater and its current use. The viability of an ASR scheme is dependent on local hydrology, the underlying geology of an area, and the presence and nature of aquifers. If the salinity of an aquifer is greater than the injection water, then this may influence the viability of recovering water from the aquifer. A typical ASR system using the collected rainfall from the rooftop is illustrated in Figure 6.6.

6.2.1.10 Porous Pavement

Porous paving surfaces allow stormwater to be filtered by a coarse sub-base and may allow infiltration to the underlying soil (Figure 6.7). Porous paving can be utilized to promote a variety of water management objectives, including

- Reduction of peak stormwater discharges from paved areas
- Increasing the groundwater recharge
- Improving stormwater quality
- Reduction of the area of land allocated for stormwater management

6.3 OPEN-CHANNEL FLOW IN URBAN WATERSHEDS

The detailed procedure of estimation of rainfall excess (or runoff) resulted from a rainfall event, discussed in Chapter 3. The processes for carrying the produced excess rainfall to the watershed outlet and/or desired point in the watershed, which often occurs in the form of open-channel flow, is discussed in this section.

The flow in most of the components of the urban drainage system is in the form of open-channel flow. The main characteristic of this type of flow is a free surface at atmospheric pressure. At first, the excess rainfall flows as "overland flow" with very small depth over the roofs, lawns, driveways, and

Urban Water Drainage System 247

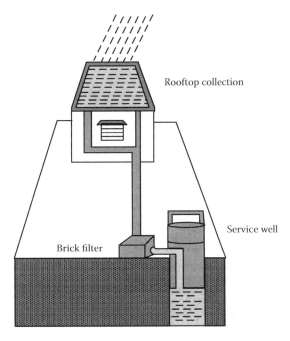

FIGURE 6.6 Schematic of an aquifer storage and recovery system using collected rainfall from the roof.

street pavements. Then the rainfall excess is drained into a channel such as a street gutter, a ditch, a drainage channel, or a storm sewer, which are usually partly full. The excess rainfall finally reaches the watershed outlet in the form of a channel flow. The driving force in all types of open-channel flows is the gravitational force.

6.3.1 OPEN-CHANNEL FLOW

The open-channel flow is classified owing to the criterion considered in the evaluation of flow characteristics. If the flow characteristics at a specific point do not change with time, the flow is steady; otherwise it is considered as an unsteady flow. When the considered criterion is space, if the

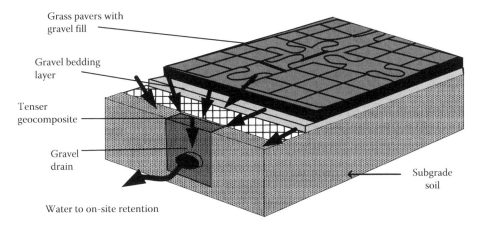

FIGURE 6.7 Structure of a typical porous pavement (Available at http://www.tensar.co.uk/images/graphic4.gif).

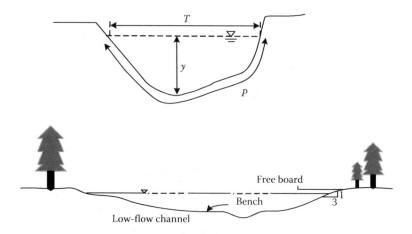

FIGURE 6.8 Elements of a channel section.

flow characteristics do not change along the channel, the flow is considered uniform; otherwise it is called a nonuniform flow. A nonuniform flow is also classified into gradually varied and rapidly varied flows depending on whether the variations along the channel are gradual or rapid. The hydraulic and geometric elements of the open-channel flow are depicted in the channel cross section of Figure 6.8.

The total energy head in each flow section is equal to the total energy per unit weight of the flowing water. The total energy head is measured vertically from a fixed horizontal datum, as discussed in Chapter 4 (Figure 6.9). By assuming a hydrostatic distribution for the flow pressure, the total energy head of the flow is formulated as

$$H = z + y + \frac{V^2}{2g}, \tag{6.1}$$

where H is the total energy head, z is the elevation head, y is the flow depth, g is the gravitational acceleration, and $V^2/2g$ is the velocity head.

The elevation head z, which corresponds to the potential energy, is measured as the vertical distance between the channel bottom and the assumed datum. Flow depth y is called the pressure energy, and the kinetic energy of the flow is quantified by the velocity head equal to $V^2/2g$. The energy grade

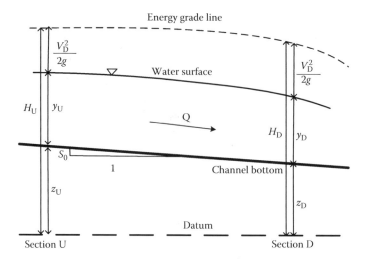

FIGURE 6.9 Definition of energy head and energy grade line.

Urban Water Drainage System

line shows the variations of total energy head along the channel. The hydraulic head is equal to the summation of the elevation head and the pressure energy head, and in an open channel the hydraulic head corresponds to the water surface.

6.3.1.1 Open-Channel Flow Classification

Various forces such as viscous, inertial, and gravitational forces affect the flow condition in open channels. The Reynolds number, Re, is defined as the ratio of inertial forces to viscous forces as explained in Chapter 4. The Re is calculated for open channels as

$$Re = \frac{VR}{\nu}, \tag{6.2}$$

where ν is the kinematic viscosity of water, V is the flow velocity, and R is the hydraulic radius.

Same as the pipes, for $Re < 500$, the flow is considered in the laminar state, and when $Re > 4000$, the flow is turbulent. If $500 < Re < 4000$, then a transitional flow occurs. Another dimensionless parameter that is used in determination of the flow state is the Froude number, Fr, which is equal to the ratio of the inertial to gravitational forces and is calculated as

$$Fr = \frac{V}{\sqrt{gy}}, \tag{6.3}$$

where y is the flow depth.

The flow is at the critical, subcritical, and supercritical states when Fr is equal to 1.0, less than 1.0, and more than 1.0, respectively. The flow in an open channel is significantly affected by the flow state as just described.

6.3.1.2 Hydraulic Analysis of Open-Channel Flow

The St. Venant continuity and momentum equations that are used for the description of unsteady, gradually varied flow in open channels are formulated as

$$\frac{\partial Q}{\partial x} + \frac{\partial A}{\partial t} = q_{\text{LAT}} \tag{6.4}$$

and

$$\frac{\partial Q}{\partial t} + \frac{\partial (Q^2/A)}{\partial x} + gA\frac{\partial y}{\partial x} = gA(S - S_\text{f}), \tag{6.5}$$

where Q is the flow discharge, A is the flow area, x is the displacement in the flow direction, t is the time, q_{LAT} is the lateral inflow, S is the channel bottom slope, and S_f is the friction slope.

The friction slope that equals the slope of the energy grade line is calculated using the Darcy–Weisbach formula as

$$S_\text{f} = \frac{fV^2}{8gR}, \tag{6.6}$$

where f is the friction factor and R is hydraulic radius. In practice, for the following equation determination of the friction factor for laminar flow, is employed:

$$f = \frac{C}{Re}, \tag{6.7}$$

where C is the resistance factor for the laminar flow. A similar formulation for the estimation of the friction factor for the transitional and turbulent flow regimes are available, but in practice, the Manning formula is commonly used for evaluation of the friction slope in these situations as

$$S_f = \left(\frac{nQ}{A}\right)^2 \frac{1}{R^{4/3}}, \quad (6.8)$$

where n is Manning's coefficient.

The values of Manning's coefficient are dependent on the channel material and also on flow depth as presented in Table 6.1. In turbulent flow, the dependency of Manning's coefficient on flow depth decreases, and could be neglected. In this situation, the roughness factor is constant for a given channel material.

The St. Venant equations (Equations 6.4 and 6.5) are valid for unsteady, gradually varied flow conditions. In steady flows, as the flow characteristics remain constant with time, all of the partial differential terms with respect to time t in Equations 6.4 and 6.5 could be dropped. Also, by assuming that $q_{LAT} = 0$, Equation 6.4 could be simplified to

$$\frac{\partial Q}{\partial x} = 0. \quad (6.9)$$

Equation 6.9 means that the discharge is constant under the steady-state flow conditions without lateral inflow. Similarly, Equation 6.5 is simplified to

$$\frac{d(Q^2/A)}{dx} + gA\frac{dy}{dx} = gA(S - S_f). \quad (6.10)$$

The term Q/A is substituted with V, and finite difference discretization is performed on Equation 6.10 to obtain Equation 6.11, which is commonly used in practice for the evaluation of steady-state flow conditions:

$$\left(y_1 + \frac{V_1^2}{2g}\right) - \left(y_2 + \frac{V_2^2}{2g}\right) = \Delta x(S_f - S), \quad (6.11)$$

where subscripts 1 and 2 correspond to the upstream and downstream sections, respectively, and Δx is the distance between the two considered sections. In Equation 6.11, S_f is the average friction slope between the upstream and downstream sections. Considering that the channel bottom slope, S, is equal to $(z_1 - z_2)/\Delta x$ and also head loss, h_f, is obtained by $(\Delta x)(S_f)$, Equation 6.11 is rearranged as

$$h_f = H_1 - H_2. \quad (6.12)$$

TABLE 6.1
Manning's Coefficient, n, for Different Depths of Flow over Different Materials

Channel Material	Depth Range		
	0–0.15 m	0.15–0.60 m	>0.6 m
Concrete	0.015	0.013	0.013
Grouted riprap	0.010	0.030	0.028
Soil cement	0.025	0.022	0.0211
Asphalt	0.019	0.016	0.016
Rare soil	0.023	0.020	0.020
Rock cut	0.045	0.035	0.025

If the steady flow is also uniform (the flow characteristics are constant along the channel), Equation 6.10 can be simplified further by dropping all the differential terms with respect to x. What finally remains is $S_f = S$. This state of flow (steady, uniform) is called normal flow. In this condition, the energy grade line, the water surface, and the channel bottom are parallel to each other, and the Manning formula (Equation 6.13) can be used for the description of the water surface profile.

$$Q = \frac{1}{n} A R^{2/3} S^{1/2}. \tag{6.13}$$

However, it should be noted that ideally normal flow can occur only in long, prismatic channels without any flow control structures. But in practice, for the design of stormwater channels, the flow is considered normal with the constant design discharge equal to the peak discharge.

For the description of the relationship between the flow area, A, and the discharge, Q, at a given channel section, rating curves are developed. The field measured data should be used for the development of these curves, but in the cases where the field data are not available or accessible, Equation 6.13 can be used for the estimation of a rating curve. Considering that n and S are constant for a given channel, and R and A are related to each other, the channel rating curve can be formulated as

$$Q = eA^m, \tag{6.14}$$

where e and m are constant, which are determined analytically. It should be noted that e and m are dependent on the channel flow and their average values should be used for flow analysis.

6.3.2 Overland Flow

The overland flow occurs on impervious and pervious surfaces in urban watersheds such as roofs, driveways, and parking lots as well as lawns. The driving force in this kind of flow is gravitational, and flow over these surfaces occurs in the form of a sheet with very low depth.

Overland flow can be modeled as an open-channel flow with a very shallow depth so that the same equation as for the open-channel flow can be applied. The flow depth in overland flow is very small, and this results in a low Re; therefore the overland flow can be classified as laminar when only Re is considered. But some other factors should be considered in the determination of the flow state. For instance, different surface coverage such as rocks, grass, and litter cause some disturbances in overland flow, which cannot be considered as laminar flow despite the low Re value (Akan, 1993). In this way, it is more logical to consider overland flow as turbulent flow. Different equations are employed for the description of the overland flow in laminar or turbulent flow conditions.

If the overland flow is turbulent, the Manning equation can be employed. Manning's coefficients used in overland flow analysis are different from those for channel flow because of the small depth and the turbulent nature of the flow. All the factors affecting the flow resistance in overland flow are aggregated into the Manning's coefficient that is referred to as the effective Manning's roughness factor. In Table 6.2, the effective n values for different types of land surfaces are given (Woolhiser, 1975; Engman, 1983).

6.3.2.1 Overland Flow on Impervious Surfaces

For the analysis of the overland flow on impervious surfaces, the basin soil is considered to be dry at the start of rainfall. At $t = 0$, excess rainfall and infiltration are produced. It is assumed that the rainfall would continue with a constant rate so that the discharge at the basin outlet would increase until basin equilibrium time (time of concentration) is reached, and then it would remain constant. The time of concentration for the overland flow is estimated as

$$t_c = \frac{L^{1/m}}{(\alpha i_0^{m-1})^{1/m}}, \tag{6.15}$$

TABLE 6.2
Effective Manning n Values for Overland Flow

Surface Type	n
Concrete	0.01–0.013
Asphalt	0.01–0.015
Bare sand	0.01–0.016
Graveled surface	0.012–0.3
Bare clay-loam (eroded)	0.012–0.033
Fallow, no residue	0.008–0.012
Conventional tillage, no residue	0.06–0.12
Conventional tillage, with residue	0.16–0.22
Chisel plow, no residue	0.06–0.12
Chisel plow, with residue	0.10–0.16
Fall disking, with residue	0.30–0.50
No till, no residue	0.04–0.10
No till (20–40% residue cover)	0.07–0.17
No till (60–100% residue cover)	0.17–0.47
Sparse vegetation	0.053–0.13
Short grass prairie	0.10–0.20

where t_c is the time of concentration, L is the length of the overland flow on the watershed, and $i_0 = i - f$ is the constant rate of rainfall excess. α and m are constants which are dependent on the flow condition that is either laminar or turbulent. For laminar flow, $m = 3.0$ and

$$\alpha = \frac{8gS}{C\nu}, \qquad (6.16)$$

where C is the laminar flow resistance factor, ν is the kinematic viscosity of the water, and S is the basin slope.

For turbulent flow, $m = 5/3$ and

$$\alpha = \frac{S^{1/2}}{n}. \qquad (6.17)$$

As the actual rainfall occurs in a finite duration, t_d, the shape and the peak discharge of the runoff hydrograph are dependent on the relation between t_d and t_c.

1. If $t_c < t_d$ (Figure 6.10a), the rising limb of the runoff hydrograph before reaching the equilibrium time is calculated as

$$q_L = \alpha(i_0 t)^m, \qquad (6.18)$$

where q_L is the discharge produced per unit width at $x = L$, the basin outlet. The discharge will remain constant between the equilibrium time and the end of rainfall and is equal to

$$q_L = i_0 L. \qquad (6.19)$$

By the end of rainfall at t_d, the discharge gradually reduces. The flow variations with time are calculated as

$$t = t_d + \frac{L/(\alpha y_L^{m-1}) - (y_L/i_0)}{m}, \qquad (6.20)$$

Urban Water Drainage System

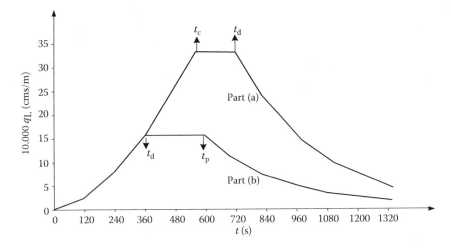

FIGURE 6.10 Overland flow hydrographs.

where y_L is the flow depth at the downstream of the basin and the corresponding discharge is calculated as

$$y_L = \left(\frac{q_L}{\alpha}\right)^{1/m}. \tag{6.21}$$

2. When the time of concentration is greater than the rainfall duration (Figure 6.10b), the basin does not reach the equilibrium condition. The rising limb is calculated using Equation 6.18 before the rainfall ends. After that, the runoff hydrograph becomes flat with a constant discharge until t_p. The discharge is calculated as

$$q_L = \alpha(i_0 t_d)^m, \tag{6.22}$$

where

$$t_p = t_d - \left(\frac{t_d}{m}\right) + \frac{L}{m\alpha(i_0 t_d)^{m-1}}. \tag{6.23}$$

The runoff gradually decreases during the period when $t > t_p$, and Equations 6.20 and 6.21 can again be applied to compute the falling limb of the runoff hydrograph. It should be noted that in this case, the flat portion of the runoff hydrograph between t_d and t_p does not mean reaching the equilibrium condition. In this period of time, the depths and discharges are decreasing at the upstream sections, while a constant flow rate is observed at the downstream of the watershed. A detailed description of this phenomenon can be provided using a characteristic method, which is beyond the scope of this text.

EXAMPLE 6.1

Consider a parking lot with length 50 m and slope 0.005. The effective Manning's coefficient of the parking lot is 0.020. Determine the runoff hydrograph for a constant rainfall excess rate equal to 15 mm/h at the downstream end of the parking lot. Solve the example for two rainfall durations of 15 and 5 min.

continued

Solution: Considering that the flow is turbulent, α and t_e are calculated from Equations 6.17 and 6.15, respectively, using the excess rainfall of 15 mm/h = 4.167×10^{-6} m/s, $m = 5/3 = 1.667$, $1/m = 0.6$, and $(m-1)/m = 0.4$,

$$\alpha = \frac{(0.005)^{0.5}}{0.020} = 3.54 \text{ m}^{1/3}/\text{s},$$

$$t_c = \frac{50^{0.6}}{(3.54)^{0.6}(4.167 \times 10^{-6})^{0.4}} = 695 \text{ s} = 11.59 \text{ min}.$$

1. For $t_d = 15$ min $= 900$ s, $t_c < t_d$. Thus, between time zero and 695 s, Equation 6.18 is used to calculate the rising limb of the hydrograph. The resulting q_L values are given in column 2 of Table 6.3, part (a). The discharge at equilibrium time (695 s) is equal to 20.83×10^{-5} m^3/s/m and remains constant until the rain ceases at $t = t_d = 900$ s. The q_L values in column 3 of Table 6.3 are arbitrarily chosen from the peak flow to zero to calculate the falling limb. Then, columns 4 and 5 are calculated using Equations 6.21 and 6.20.
2. For $t_d = 5$ min $= 300$ s, $t_c > t_d$. The rainfall ceases before reaching the basin equilibrium. Therefore, the peak discharge occurs at $t = 300$ s.

Equation 6.18 is used for calculating the rising limb for $0 < t < 300$ s. As shown in column 2 of part (b) of Table 6.3, the maximum discharge is 5.14×10^{-5} m^3/s/m. Then the t_p is estimated using Equation 6.23 as

$$t_p = 300 - \frac{300}{1.667} + \frac{50}{(1.667)(3.54)(4.167 \times 10^{-6})^{0.667}(300)^{0.667}} = 850 \text{ s}.$$

The discharge will remain constant at 5.14×10^{-5} m^3/s/m between 300 and 850 s. After 850 s, flow rates decrease. Equations 6.20 and 6.21 are used to calculate the falling limb, and the results are tabulated in part (b) of Table 6.3. The q_L values in column 3 are used to calculate the falling limb. Then, column 4 is calculated using Equation 6.21, and column 5 is calculated using Equation 6.20.

TABLE 6.3
Kinematic Wave Model Example

	Rising Limb	Falling Limb		
(1)	(2)	(3)	(4)	(5)
t (s)	q_L (10^{-5} m^3/s/m)	q_L (10^{-5} m^3/s/m)	y_L (10^{-3} m)	t (s)
(a)				
0	0.00	20.83	2.90	695
150	1.62	20.83	2.90	900
300	5.14	15	2.38	1033
450	10.01	10	1.86	1088
600	16.30	5	1.23	1461
695	20.83	0.5	0.31	2710
(b)				
0	0.00	5.14	1.25	850
100	0.82	3	0.91	1071
200	2.61	2	0.71	1263
300	5.14	1	0.47	1634
		0.5	0.31	2106

Urban Water Drainage System

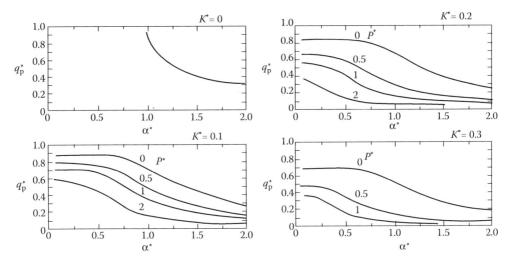

FIGURE 6.11 Overland flow from pervious surfaces. (From Akan, A. O. 1985. *Journal of Hydraulic Engineering, ASCE* 111: 1057–1067. With permission.)

6.3.2.2 Overland Flow on Pervious Surfaces

The analytical solutions governed for impervious surfaces can also be used for pervious surfaces when there is a constant rate of rainfall excess. But for variable excess rainfall rates, an analytical solution cannot be governed. Akan (1985) has developed a graphical solution for the determination of the peak overland flow discharge over pervious surfaces (Figure 6.11). Before using this chart, the following dimensionless parameters should be calculated:

$$q_p^* = \frac{q_p}{iL}, \tag{6.24}$$

$$\alpha^* = \frac{L^{0.6}}{i^{0.4} t_d \alpha^{0.6}}, \tag{6.25}$$

$$K^* = \frac{K}{i}, \tag{6.26}$$

$$P^* = \frac{P_f \phi (1 - S_i)}{i t_d}, \tag{6.27}$$

where q_p is the peak discharge per unit width at the basin outlet, i is the constant rate of rainfall, L is the length of the overland flow path, K is the hydraulic conductivity of soil, P_f is the soil suction head, ϕ is the effective soil porosity, S_i is the initial degree of saturation, t_d is the duration of rainfall, and

$$\alpha = \frac{1}{n} S^{0.5}. \tag{6.28}$$

It should be noted that Figure 6.11 is only applicable for basins with uniform characteristics and when the only source of rainfall loss is infiltration.

EXAMPLE 6.2

Determine the peak discharge resulting from a 1 h rainfall with a constant intensity equal to 1.5×10^{-5} m/s that happens on a pervious rectangular basin with the following characteristics: $L = 120$ m,

continued

$S = 0.002$, and $n = 0.02$. The infiltration parameters are $\varphi = 0.414$, $K = 1.5 \times 10^{-6}$ m/s, and $P_f = 0.25$ m. At the beginning of the rainfall, the basin soil is 0.65% saturated.

Solution: For utilizing Figure 6.11, variables α, α^*, K^*, and P^* are calculated using Equations 6.28 and 6.25 through 6.27, respectively.

$$\alpha = \frac{(0.002)^{0.5}}{0.02} = 2.24 \text{ m}^{1/3}/\text{s}.$$

$$\alpha^* = \frac{(120)^{0.6}}{(1.5 \times 10^{-5})^{0.4}(3600)(2.23)^{0.6}} = 0.26,$$

$$K^* = \frac{1.5 \times 10^{-6}}{1.5 \times 10^{-5}} = 0.1,$$

$$P^* = \frac{(0.25)(0.414)(1.0 - 0.65)}{(1.5 \times 10^{-5})(3600)} = 0.67.$$

Then $q_p^* = 0.74$ using Figure 6.11. Finally, q_p is calculated as

$$q_p = (0.74)(1.5 \times 10^{-5})(120) = 1.33 \times 10^{-3} \text{ m}^3/\text{s/m}.$$

6.3.3 Channel Flow

Channel flow occurs in different elements of an urban watershed including gutters, ditches, and storm sewers. The flow in these structures is unsteady and nonuniform during and after a rainstorm. The St. Venant equations (Equations 6.4 and 6.5) could be employed for simulation of this flow condition. The models that use these equations are referred to as "dynamic models," as they could explain the dynamic nature of the system. There is no analytical solution for the St. Venant equation and complicated computer programs are developed for solving them. But as this procedure takes a lot of time and is costly, in practice less complex methods are employed.

The simplified methods are categorized into two groups including hydraulic and hydrological models. In hydraulic models, the St. Venant equations are employed but hydrological models use an assumed relationship instead of the momentum and continuity equations. The main disadvantage of simpler models is that some phenomena that affect the flow such as downstream backwater are not considered. Unsteady open-channel flow calculations are referred to as flood routing or channel routing. Two types of flood routing have been described in this section.

6.3.3.1 Muskingum Method

In this method, the outflow hydrograph at the downstream of a channel reach is calculated for a given inflow hydrograph at the upstream. The uniform short channels are modeled as a single reach, but the long channels with various characteristics in length are often divided into several reaches and each is modeled separately. When more than one reach is considered, the calculations start from the most upstream reach and then proceed in the direction of the downstream. The outflow from the upstream reach is used as the inflow for the next downstream reach.

The continuity equation for a channel reach is written as

$$\frac{dS}{dt} = I - Q, \tag{6.29}$$

where S is the storage water volume in the channel reach, I is the upstream inflow rate, and Q is the downstream outflow rate.

As there are two unknown variables including S and Q in Equation 6.29, a second relationship should be developed to solve the flood routing problem. In the Muskingum method, a linear

Urban Water Drainage System

relationship is assumed between S, I, and Q as

$$S = K[XI + (1 - X)Q], \tag{6.30}$$

where K is the travel time constant and X is the weighing factor that varies between 0 and 1.0.

Parameters K and X are constant for a channel and do not have physical meaning. These parameters are determined during the model calibration. Equation 6.29 can be rewritten in the finite difference form over a discrete time increment as

$$\frac{S_2 - S_1}{\Delta t} = \frac{I_1 + I_2}{2} - \frac{Q_1 + Q_2}{2}, \tag{6.31}$$

where $\Delta t = t_2 - t_1$, in which the variables with subscripts 1 and 2 are related to the beginning and the end of the time increment. Equation 6.30 can be written for times t_1 and t_2 to provide a relationship between S and Q in these times. By substituting the obtained relations into Equation 6.31, the following equation is obtained:

$$Q_2 = C_0 I_2 + C_1 I_1 + C_2 Q_1, \tag{6.32}$$

where

$$C_0 = \frac{(\Delta t/K) - 2X}{2(1 - X) + (\Delta t/K)}, \tag{6.33}$$

$$C_1 = \frac{(\Delta t/K) + 2X}{2(1 - X) + (\Delta t/K)}, \tag{6.34}$$

$$C_2 = \frac{2(1 - X) - (\Delta t/K)}{2(1 - X) + (\Delta t/K)}. \tag{6.35}$$

Note that $C_0 + C_1 + C_2 = 1.0$.

Q_2 is the only unknown variable in routing Equation 6.32. I_2 and I_1 are obtained from the given inflow hydrograph, and Q_1 is obtained from the initial condition of the channel reach or from the previous time step computation.

EXAMPLE 6.3

The inflow hydrograph for a channel reach with an initial steady flow equal to $45 \, \text{m}^3/\text{s}$ is given in Figure 6.12. Calculate the outflow hydrograph in 1-h time increments. The Muskingum parameters are determined as $K = 3 \, \text{h}$ and $X = 0.15$.

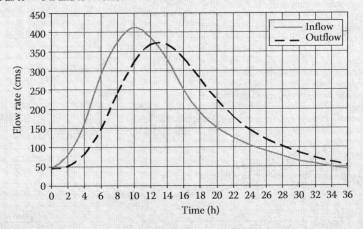

FIGURE 6.12 Inflow and outflow hydrographs of Example 6.3.

continued

Solution: Using Equations 6.33 through 6.35, the weighting factors are estimated as being $C_0 = 0.016$, $C_1 = 0.311$, and $C_2 = 0.673$. The routing calculations are given in Table 6.4. The I_1 and I_2 values for each time are calculated using the inflow hydrograph presented in Figure 6.12. The beginning and end time of each 1-h time increment are tabulated in columns 2 and 3 of Table 6.4. Because the initial flow in the channel is equal to 45 m^3/s, Q_1 of the first time period has been set as 45 m^3/s. The routed outflow hydrograph is also presented in Figure 6.12. Three main points can be observed in Figure 6.12 by a comparison of the inflow and outflow hydrographs:

1. The peak outflow rate is less than the peak inflow rate.
2. The peak outflow occurs later than the peak inflow.
3. The volumes of inflow and outflow hydrographs are equal.

In any unsteady channel-flow situation, these observations are true. The only exceptions are channel flows with very high Froude numbers or channels with lateral flows.

TABLE 6.4
Summary of Example 6.3 Calculations

(1) Time Step	(2) t_1 (h)	(3) t_2 (h)	(4) I_1 (cm)	(5) I_2 (cm)	(6) Q_1 (cm)	(7) Q_2 (cm)
1	0	1	45	62	45	45
2	1	2	62	80	45	51
3	2	3	80	120	51	61
4	3	4	120	160	61	81
5	4	5	160	225	81	108
6	5	6	225	290	108	147
7	6	7	290	333	147	195
8	7	8	333	375	195	241
9	8	9	375	394	241	285
10	9	10	394	412	285	321
11	10	11	412	399	321	351
12	11	12	399	385	351	367
13	12	13	385	358	367	372
14	13	14	358	330	372	367
15	14	15	330	290	367	354
16	15	16	290	250	354	332
17	16	17	250	220	332	305
18	17	18	220	190	305	277
19	18	19	190	170	277	248
20	19	20	170	150	248	222
21	20	21	150	138	222	198
22	21	22	138	125	198	178
23	22	23	125	115	178	160
24	23	24	115	105	160	145
25	24	25	105	98	145	132
26	25	26	98	90	132	121
27	26	27	90	84	121	111
28	27	28	84	77	111	102
29	28	29	77	71	102	94
30	29	30	71	65	94	86
31	30	31	65	61	86	79
32	31	32	61	57	79	73
33	32	33	57	53	73	68
34	33	34	53	49	68	63
35	34	35	49	47	63	58
36	35	36	47	45	58	54

6.3.3.2 Muskingum–Cunge Method

The major disadvantage of the Muskingum method is that parameters K and X do not have a physical basis and are difficult to estimate. Cunge (1969) has suggested a new method called the Muskingum–Cunge method for solving this problem. In the Muskingum–Cunge method, K and X are defined based on the physical characteristics of the channel as

$$K = \frac{L}{mV_0}, \qquad (6.36)$$

$$X = 0.5\left(1 - \frac{Q_0/T_0}{S_0 m V_0 L}\right), \qquad (6.37)$$

where Q_0 is a reference discharge, T_0 is the top width corresponding to the reference discharge, V_0 is the cross-sectional average velocity corresponding to the reference discharge, S_0 is the slope of the channel bottom, L is the length of the channel reach, and m is the exponent of the flow area A in Equation 6.14.

Different values can be considered for reference discharge such as dry-weather discharge in the channel, the peak of the inflow hydrograph, or the average inflow rate. After the determination of K and X from Equations 6.36 and 6.37, the weighting coefficients C_0, C_1, and C_2 are calculated from Equations 6.33 through 6.35, and then Equation 6.32 is used to calculate the outflow hydrograph. In the Muskingum–Cunge method, X can also take negative values. It should be noted that the Muskingum–Cunge and Muskingum methods are basically different; the Muskingum method is a hydrological routing method but the Muskingum–Cunge method is a hydraulic method that uses an approximation of the St. Venant equation.

For obtaining more accurate results, it is suggested that Δt should be selected as less than one-fifth of the time between the beginning and the peak of the inflow hydrograph (Ponce and Theruer, 1982), and also the length of the channel reach for Muskingum–Cunge computations should satisfy the following inequality:

$$L \leq 0.5\left(mV_0\Delta t + \frac{Q_0/T_0}{mV_0 S_0}\right). \qquad (6.38)$$

The selected reference discharge for calculating K and X parameters highly affects the results of the Muskingum–Cunge method. Ponce and Yevjevich (1978) have suggested Equation 6.39 for updating the reference discharge at every time step to eliminate the effect of the reference discharge on the model parameters.

$$Q_0 = \frac{I_1 + I_2 + Q_1}{3}, \qquad (6.39)$$

and T_0, V_0, X, K, C_0, C_1, and C_2 are recalculated using the updated reference discharge at each time step.

6.4 COMPONENTS OF URBAN STORMWATER DRAINAGE SYSTEMS

Accumulated stormwater on a street surface interrupts traffic and may cause harmful accidents due to the decrease of skid resistance and visibility reduction because of water splash (Johnson and Chang, 1984). Therefore, the reduction of stormwater from the street surfaces is a major issue in effective urban stormwater management.

Urban stormwater drainage systems within the district comprise three primary components: (1) street gutters and roadside swales, (2) stormwater inlets, and (3) storm sewers (and appurtenances such as manholes, junctions, etc.). Street gutters and roadside swales collect the runoff from the street (and adjacent areas) and convey the runoff to a stormwater inlet while maintaining the street's level-of-service.

Inlets collect stormwater from streets and other land surfaces, provide transfer of the flow into storm sewers and often provide maintenance access to the storm sewer system. Storm sewers convey stormwater in excess of a street's or a swale's capacity and discharge it into a stormwater management facility or a nearby receiving water body.

All of these components should be designed properly to achieve the stormwater drainage system's objectives. In this section, the basic and important concepts related to the design of different curbs, gutters, and inlets of urban drainage systems' components are discussed.

6.4.1 General Design Considerations

Keeping the water depth and spread on the street below a permissible value for a design-storm return period is the key issue in the design of street drainage systems. Rainfall events vary greatly in magnitude and frequency of occurrence. Major storms produce large flow rates but rarely occur. Minor storms produce smaller flow rates but occur more frequently. For economic reasons, stormwater collection and conveyance systems are not normally designed to pass the peak discharge during major storm events but a compromised storm, the capital investment economically and meet, which justifies the local standards as well.

Stormwater collection and conveyance systems are designed to pass the peak discharge of a storm event with minimal disruption to street traffic. To accomplish this, the spread of water on the street is limited to some maximum, mandated value during the storm event. As the volume of the traffic in the streets increases, storm with larger return periods should be considered. Typical design return periods and permissible spread for different kinds of roads are given in Table 6.5.

Inlets must be strategically placed to pick up the excess gutter or swale flow once the limiting spread of water is reached. The inlets direct the water into storm sewers, which are typically sized to pass the peak flow rate from the minor storm without any surcharge. The magnitude of the minor storm is established by local ordinances or criteria, and the 2-, 5-, or 10-year storms are most commonly used.

Sometimes, storms will occur that surpass the magnitude of the design-storm event. When this happens, the spread of water on the street exceeds the allowable spread and the capacity of the storm sewers. Street flooding occurs and traffic is disrupted. However, proper design requires that public safety be maintained and the flooding be managed to minimize flood damage. Thus, local ordinances also often establish the return period for the major storm event, generally the 100-year storm. For this event, the street becomes an open channel and must be analyzed to determine whether or not the consequences of the flood are acceptable with respect to flood damage and public safety.

Besides traffic flow, streets serve another important function in the urban stormwater collection and conveyance system. The street gutter or the adjacent swale collects excess stormwater

TABLE 6.5
Minimum Design Frequency and Spread for Different Types of Roads

Road Classification	Design Frequency	Design Spread
Arterial	10 years	Shoulder
	50 years (sag vertical curve)	Shoulder + 90 cm
Collector streets	10 years	Shoulder
	50 years (sag vertical curve)	$\frac{1}{2}$ Driving lane
Residential street	10 years	$\frac{1}{2}$ Driving lane

Note: Check 100 years storm for all road classifications. The water level should remain less than 45 cm.

from the street and adjacent areas and conveys it to a stormwater inlet. According to FHWA (1984), proper street drainage is essential to

- Maintain the street's level-of-service.
- Reduce skid potential.
- Maintain good visibility for drivers (by reducing splash and spray).
- Minimize inconvenience/danger to pedestrians during storm events.

Streets geometry is another important factor in the design of drainage systems. The longitudinal slope of curbed streets should not be <0.3% and should be >0.5%. An appropriate cross slope is necessary to provide efficient drainage without affecting the driving comfort and safety. In most situations, a cross slope of 2% is preferred. Higher cross slopes could be used for multilane streets, but it should not be >4%.

An additional design component of importance in street drainage is gutter (channel) shape. Most urban streets contain curb and gutter sections. Various types exist, which include spill shapes, catch shapes, curb heads, and roll gutters. The shape is chosen for functional, cost, or esthetic reasons and does not dramatically affect the hydraulic capacity. Swales are common along some urban and semiurban streets, and roadside ditches are common along rural streets. Their shapes are important in determining hydraulic capacity.

6.4.2 Flow in Gutters

Street slope can be divided into two components: longitudinal slope and cross slope. The longitudinal slope of the gutter essentially follows the street slope. The hydraulic capacity of a gutter increases as the longitudinal slope increases. The allowable flow capacity of the gutter on steep slopes is limited in providing for public safety. The cross (transverse) slope represents the slope from the street crown to the gutter section. A compromise is struck between large cross slopes that facilitate pavement drainage and small cross slopes for driver safety and comfort. Usually, a minimum cross slope of 1% for pavement drainage is recommended. For increasing the gutter capacity, composite sections are often preferred because the gutter cross slopes are steeper than street cross slopes.

For evaluating the drainage capacity of the street gutters, four major steps should be followed:

1. Calculating the theoretical gutter flow capacity to convey the minor storm based on the allowable spread.
2. Repeat step 1 based on the allowable depth.
3. Calculating the allowable street gutter flow capacity by multiplying the theoretical capacity (calculated in step 2) by a reduction factor used for safety considerations. The minimum of the capacities calculated in step 1 and this step is the allowable street gutter capacity.
4. Calculating the theoretical major storm conveyance capacity based on the road inundation criteria. Reducing the major storm capacity by a reduction factor to determine the allowable storm conveyance capacity.

The flow should be routed through gutters to determine the flow depth and spread of water on the shoulder, parking lane, or pavement section under design flow conditions. The design discharge is calculated using the rational method. For more simplicity in calculations, the flow in a gutter is assumed to be steady and uniform at the peak design discharge. This approach usually results in more conservative results.

6.4.2.1 Simple Triangular Gutters

Since gutter flow is assumed to be uniform for design purposes, Manning's equation is appropriate with a slight modification to account for the effects of a small hydraulic radius. The adopted Manning

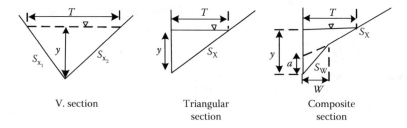

FIGURE 6.13 Various gutter sections.

formula for simple triangular gutter sections, presented in Figure 6.13, is

$$Q = \frac{0.56}{n} S_X^{5/3} S^{1/2} T^{8/3} \qquad (6.40)$$

or

$$T = \left(\frac{Qn}{0.56 S_X^{5/3} S^{1/2}} \right)^{3/8}, \qquad (6.41)$$

where Q is the gutter flow, T is the top width, n is Manning's roughness, S_X is the cross slope, and S is the longitudinal bottom slope.

The water depth is calculated using the water top width, T, as

$$y = S_X T. \qquad (6.42)$$

Note that the flow depth must be less than the curb height during the minor storm. For calculating the flow depth y in the gutter, Equation 6.40 is rewritten as

$$Q = \frac{0.56 y^{8/3} S^{1/2}}{n S_X}. \qquad (6.43)$$

The flow area A is calculated using the gutter geometry as

$$A = \frac{1}{2} S_X T^2. \qquad (6.44)$$

The gutter velocity at peak capacity may be found from the continuity equation ($V = Q/A$).

EXAMPLE 6.4

The design discharge for a triangular gutter with $S = 0.015$, $S_X = 0.025$, and $n = 0.015$ is estimated as $0.073 \, \text{m}^3/\text{s}$. Determine the flow depth and spread for this gutter.

Solution: Equation 6.41 is applied to calculate the top width as

$$T = \left[\frac{(0.073)(0.015)}{0.56(0.025)^{5/3}(0.015)^{1/2}} \right]^{3/8} = 0.44 \, \text{m}.$$

Then the water depth from Equation 6.42 is obtained as

$$y = (0.025)(0.44) = 0.011 \, \text{m}.$$

6.4.2.2 Gutters with Composite Cross Slopes

Gutters with composite cross slopes are often used to increase the gutter capacity. In composite gutter sections, such as those illustrated in the right-hand side of Figure 6.13, the gutter flow is divided into two fractions including discharge in the depressed section (Q_d) and discharge in the nondepressed section (Q_f). It can be shown that nondepressed discharge and gutter flow are related to each other as follows:

$$Q = \frac{Q_f}{1 - E_0}, \qquad (6.45)$$

in which

$$E_0 = \frac{1}{1 + \{(S_W/S_X)/[1 + (S_W/S_X)/(T/W - 1)]^{8/3} - 1\}} \qquad (6.46)$$

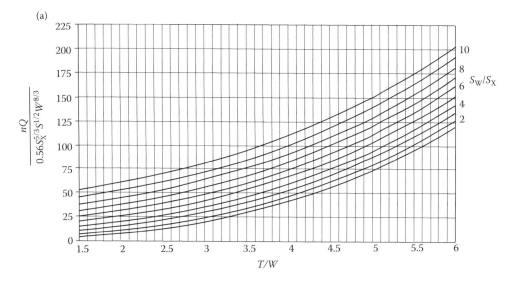

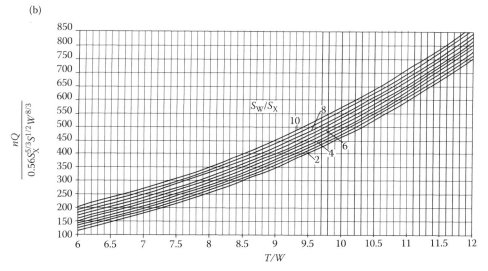

FIGURE 6.14 (a) Spread calculations for $T/W < 6.0$. (b) Spread calculations for $T/W > 6.0$. (From Akan, A. O., 2000. *Journal of Transportation Engineering, ASCE* 126(5): 448–450. With permission.)

and

$$S_W = S_X + \frac{a}{W}. \qquad (6.47)$$

The applied geometric variables in the above equations are illustrated in Figure 6.13. Equation 6.41 can be solved using the diagrams presented in Figure 6.14 (Akan, 2000). Also y, the flow depth at the gutter, and A, its flow area, can be calculated from the gutter geometry as

$$y = a + TS_X \qquad (6.48)$$

and

$$A = \frac{1}{2}S_X T^2 + \frac{1}{2}aW. \qquad (6.49)$$

EXAMPLE 6.5

Determine the flow in a composite gutter with the following geometry when $T = 1.75$ m, $W = 0.4$ m, $S = 0.01$, $S_X = 0.022$, and $a = 0.07$ m. Assume that the Manning roughness factor is 0.015.

Solution: The cross slope of the depressed part of gutter S_W is calculated using Equation 6.47 as

$$S_W = 0.022 + \frac{0.07}{0.40} = 0.197.$$

According to Figure 6.13, $T_f = T - W = 1.75 - 0.4 = 1.35$ m. For calculating the nondepressed part of gutter discharge, Equation 6.40 is rewritten for the triangular part of the composite gutter having T_f top width as

$$Q_f = \frac{0.56 T_f^{8/3} S_X^{5/3} S^{1/2}}{n} = \frac{(0.56)(1.35)^{8/3}(0.022)^{5/3}(0.01)^{1/2}}{(0.015)} = 0.014 \, \text{m}^3/\text{s}.$$

Using Equation 6.46, with $S_W/S_X = 0.197/0.022 = 8.95$, $T/W = 1.75/0.4 = 4.4$, and $(T/W - 1) = 4.4 - 1.0 = 3.4$,

$$E_0 = \frac{1}{1 + \{(8.95)/[1 + (8.95/3.4)]^{8/3} - 1\}} = 0.771.$$

The total gutter flow is calculated using Equation 6.45 as

$$Q = \frac{0.014}{1 - 0.771} = 0.061 \, \text{m}^3/\text{s}.$$

Figure 6.14 can be also used to solve this problem. With $T/W = 4.4$ and $S_W/S_X = 8.95$, the figure yields

$$\frac{nQ}{0.56 S_X^{5/3} S^{1/2} W^{8/3}} = 115$$

and

$$Q = \frac{115(0.56)(0.022)^{5/3}(0.01)^{1/2}(0.40)^{8/3}}{(0.015)} = 0.064 \, \text{m}^3/\text{s}.$$

The results of the two methods are similar to each other.

6.4.2.3 Gutter Hydraulic Capacity

Stormwater flowing along the streets exert momentum forces on cars, pavements, and pedestrians. To limit the hazardous nature of heavy street flows, it is necessary to set limits on flow velocities and depths. As a result, the allowable gutter hydraulic capacity is determined as the smaller value of

$$Q_A = Q_T$$

or

$$Q_A = RQ_F,$$

where Q_A is the allowable street hydraulic capacity, Q_T is the street hydraulic capacity limited by the maximum water spread, Q_F is the gutter capacity when the flow depth equals allowable depth, and R is the reduction factor.

For example, Guo (2000) has defined two sets of reduction factors, one for the minor storm events, and the other for the major events. The reduction factor is equal to one for a street slope <1.5%, and then decreases in a nonlinear behavior as the street slope increases.

It is important for street drainage designs that the allowable street hydraulic capacity be used instead of the calculated gutter full capacity. Thus, wherever the accumulated stormwater flow rate on the street is close to the allowable capacity, a street inlet shall be installed.

6.4.2.4 V-Shaped Sections

For safety and maintenance reasons, the depths and side slopes of V-shaped sections should be as shallow as possible. Street-side sections collect the initial runoff and transport it to the nearest inlet or major drainage way. For effective performance, the velocity, depth, and cross-slope geometries of these sections should be limited. The usually considered limitations include a maximum 2-year flow velocity equal to 0.9 m/s, a maximum flow depth equal to 0.3 m, and a maximum side slope of each side equal to 5H:1V. In practice, flatter side slopes are usually used.

Equation 6.40 can be used to calculate the flow rate in a V-section with an adjusted slope that is calculated as

$$S_x = \frac{S_{x1}S_{x2}}{S_{x1} + S_{x2}}, \tag{6.50}$$

where S_x is the adjusted side slope, S_{x1} is the right-side slope, and S_{x2} is the left-side slope.

6.4.3 Pavement Drainage Inlets

Stormwater inlets are a vital component of the urban drainage system. Inlets collect street surface runoff from the street, drain into an underground drainage system, and can provide maintenance access to the storm sewer system. They can be made of cast iron, steel, concrete, and/or precast concrete and are installed on the edge of the street adjacent to the street gutter or at the end of a swale.

Roadway geometrical features often dictate the location of pavement drainage inlets. In general, inlets are placed at all low points (sumps or sags) in the gutter grade, median breaks, intersections, and crosswalks. In other words, the drainage inlets are spaced so that the spread under the design (minor) storm conditions will not exceed the allowable flow spread (Akan and Houghtalen, 2003). Inlets used for the drainage of streets are divided into four major classes (Figure 6.15):

- Curb opening inlets.
- Grate inlets.

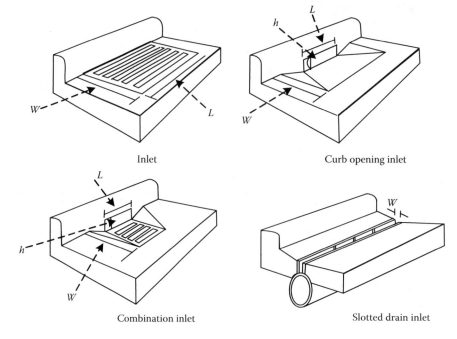

FIGURE 6.15 Four different types of inlets. (From Johnson, F. L. and Chang, F. M. 1984. *Drainage of Highway Pavements*. Hydraulic Engineering Circular, No. 12, Federal Highway Administration, McLean, VA.)

- Slotted drains, in which the performance is the same as curb opening inlets, can be considered as weirs lateral inflow.
- Combination inlets usually consist of some combination of three types of inlets. When curb is involved in a combination, its opening may extend upstream of the grate, and when the grate is involved, it is usually placed at the downstream end.

Each of these inlets can be used with or without a gutter depression. Inlets placed on continuous grades rarely intercept all of the gutter flow during the minor (design) storm. The efficiency of an inlet performance is calculated as

$$E = \frac{Q_i}{Q}, \qquad (6.51)$$

where E is the efficiency of inlet performance, Q is the total gutter flow rate, and Q_i is the intercepted flow rate.

The nonintercepted flow by an inlet, which is called bypass or carryover, is estimated as

$$Q_b = Q - Q_i, \qquad (6.52)$$

where Q_b is the carryover (bypass) flow rate.

The ability of an inlet to intercept flow (i.e., hydraulic capacity) on a continuous grade generally increases with increasing gutter flow, but the capture efficiency decreases. In other words, even though more stormwater is captured, a smaller percentage of the gutter flow is captured. In general, the inlet capacity depends on the inlet type and geometry, the flow rate (depth and spread of water), the cross (transverse) slope, and also the longitudinal slope.

6.4.3.1 Inlet Locations

Generally, highways drainage inlets are designed to collect the portion of stormwater collected on the pavement. The drained runoff from large areas toward the highways is intercepted by building roadside channels and inlets before reaching the pavement. However, in urban areas, runoffs from both pavement and off-pavement areas are often intercepted by the same stormwater inlets.

The inlet location may be dictated either on the basis of physical demands, hydraulic requirements, or both. In all instances, the inlet location is coordinated with physical characteristics of the roadway geometry, utility conflicts, and feasibility of underground pipe layout.

Logical locations for inlets include sag configurations, near street intersections, at gore islands, crosswalks, entrance and exit ramp gores, and superelevation transitions. Inlets with locations not established by physical requirements should be located on the basis of hydraulic demand. In other words, the distance between drainage inlets should be determined in such a way that the spread does not exceed the allowable value under the design-storm condition.

As mentioned before, the design discharge is calculated using the rational method. The main issue in using the rational method is an accurate estimation of the time of concentration. Because of the small drainage areas related to each inlet, the time of concentration is usually <5 min. But at most sites, rainfall data for durations <5 min are not recorded. Therefore a time of concentration of 5 min is considered in the design of municipal drainage systems. This assumption leads to equal spacing for similar inlets. The carryover from the upstream inlet should be added to the stormwater runoff generated over the street section between the two adjacent inlets. The resulting runoff volume should be considered as the design discharge of the downstream inlet. For more simplicity, the bypass flow from the upstream inlet is added to the peak flow rate obtained for the street section between two adjacent inlets. It should be noted that this approximated situation is completely different from the actual situation since the actual process is unsteady, and the peak street runoff and the bypass flow from upstream do not necessarily happen at the same time.

EXAMPLE 6.6

Curb-opening inlets are placed along a 10-m-wide street. The length of the inlets is 0.8 m and the inlets are placed on a continuous grade and the runoff coefficient is 0.85. The inlets drain stormwater to a triangular gutter with $S_X = 0.025$, $S = 0.015$, and $n = 0.015$. Determine the maximum allowable distance between inlets considering that the rainfall intensity is equal to 10 mm/h and the allowable spread is 1.8 m.

Solution: First the location of the most upstream inlet is specified. This inlet collects the runoff from a surface of $A_i = W_s L_d$, where W_s is the street width and L_d is the distance from the roadway crest to the inlet. As it is the first inlet, there is no carryover discharge from the upstream. The collected surface water at this inlet is calculated using the rational method as

$$Q = (0.85)(0.000003)(10)L_d = 0.00003 L_d.$$

If $S_X = 0.025$, $S = 0.015$, $n = 0.015$, and $T = 1.8$, then

$$0.00003 \times L_d = \frac{(0.56)(1.8)^{8/3}(0.025)^{5/3}(0.015)^{1/2}}{(0.015)} = 0.047 \, \text{m}^3/\text{s}.$$

Finally, $L_d = 0.047/(0.00003) = 1567$ m, which is the distance of the initial inlet from the crest. This inlet drains a flow of $Q = 0.047 \, \text{m}^3/\text{s}$, which corresponds to a spread of 1.8 m. Since the flow rate of next inlets is more than $0.047 \, \text{m}^3/\text{s}$, because of bypass flow from upstream inlets, their distance is less than 1567 m. Therefore the maximum distance between inlets is equal to 1567 m.

6.4.4 Surface Sewer Systems

The collected stormwater through inlets is conveyed by buried pipes to a point where stormwater is discharged to the receiving water body. In addition to pipes, different relevant structures such as inlets, manholes, junction chambers, transition structures, flow splitters, and siphons are included in a sewer system. An introduction on appurtenant structures is given here. For detailed descriptions of these components, see Brown et al. (1996).

For easy and convenient access to sewer systems for maintenance and repair, manholes (or access chambers) are used. Furthermore, manholes provide the appropriate air circulation in the sewer system. Manholes also serve as flow junctions and provide flow transitions for changes in pipe size, slope, and alignment. Often precast or cast-in-place concrete are used for manhole construction. A variety of configurations are used for manhole construction (Figure 6.16).

The diameter of the chamber is determined by the number and diameter of the sewer pipes coming into the manhole and the working space required. For deep manholes, the chamber should be large enough to provide benching or a landing adequate for two persons to stand upon. When a manhole is sited on a curve, or where additional pipes enter at the sides, a larger size may be required. However, the depth and diameter of usual manholes varies between 1.5 and 4.0 m, and 1.2 and 1.5 m, respectively. Manholes are recommended

- At intervals of up to 90 m, or 200 m, for man entry pipe runs
- Whenever there is a significant change of direction in a sewer
- Where another sewer is connected with the main pipe of a sewer
- Where there is a change of size or gradient of the pipeline

In the cases where the altitude of the incoming pipe is significantly higher than the outgoing pipe, drop manholes are utilized (Figure 6.17). For very large joining storm sewers that cannot

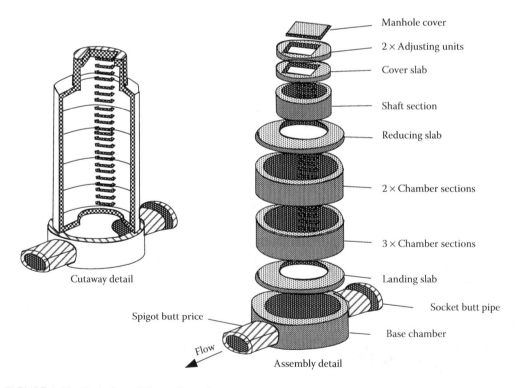

FIGURE 6.16 Typical manhole configuration.

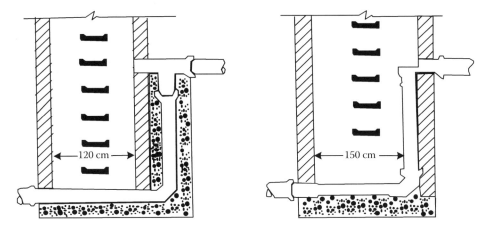

FIGURE 6.17 Schematic of drop manholes.

be accommodated by standard manholes, a junction chamber is utilized, which is made of the same material as manholes. To streamline the joining flows, channels are typically built into these chambers. A riser structure is placed in some junction chamber utilizations. The riser is extended to the street surface for more accessibility and interception of the surface flow.

Transition sections are used in the cases where the transition of pipe sizes is made without a manhole. As a significant amount of energy loss occurs where the pipe size suddenly changes, in these sections, gradual expansion from one size to another is provided.

Flow splitters are appurtenant junction structures that divide and derive an incoming flow into two or more downstream storm sewers. Other appurtenants called flow deflectors are used for minimizing the energy losses in flow splitters. Deposition of suspended material in stormwater makes it difficult and costly to maintain the flow splitters. The flow is carried under an obstruction, like a stream, through inverted siphons (or depressed sewers). Commonly, at least two-barrel siphons are constructed with each other. In systems with high potential for back-flood, because of high tides or high stages in the receiving system, flap gates are placed near storm sewer outlets. These gates could be made of different materials that are available and economical in the construction site.

6.4.5 Open Drainage Channel Design

The collected stormwater is transmitted open channels to accepting sources that can be an outfall through channel, a detention or retention basin, or a storm drainage inlet. Surface drainage systems include a wide range of structures such as roadside conduits, branches, and stormwater channels. Whether using a natural or constructed channel, hydraulic analyses must be performed to evaluate flow characteristics including flow regime, water surface elevations, velocities, depths, and hydraulic transitions for multiple flow conditions. Open-channel flow analysis is also necessary for underground conduits to evaluate hydraulics for less-than-full conditions.

The surface drainage channels are highly erodible because they are usually unlined or have flexible liners such as grass. The maximum allowed velocity method or the tractive force method is commonly used in the design of channel liners. The main assumption in using the maximum allowed velocity method is that until the average flow velocity of the channel is less than the maximum allowed velocity, erosion will not occur. In using the tractive force, the channel section is designed so that the resulting tractive force (shear force) that could move the particles on the channel bed does not exceed the values resisting the particle motion.

Trapezoid cross sections are commonly used in man-made open channels (Figure 6.18). Some geometrical parameters should be estimated in the design of an open channel. The main parameters include the longitudinal bottom slope, the bottom width b, side slopes that should be <0.5, and the

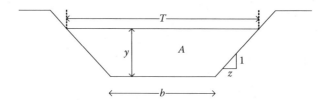

FIGURE 6.18 Schematic of a trapezoidal open channel section.

section depth y. The longitudinal bottom slope is usually governed by the site topography and the characteristics of *in situ* soil dictate the section sides' slope. The bottom width and the depth of the section are calculated after determination of the longitudinal slope and the side slopes in order to convey the resulting runoff under design situation without exceeding the allowed shear forces on the channel bed.

Similar to gutters, the open channels are designed for steady flow equal to the peak discharge, although in the actual situation the surface drainage channel flow is not steady or uniform. A freeboard should be added to the obtained flow depth under design condition to consider the deviations from the usual flow condition. The amount of appropriate freeboard is site dependent, but a 15-cm freeboard is usually used in the design of roadside channels (Young and Stein, 1999).

Another important characteristic of open-channel flow is the state of the flow, often referred to as the flow regime. The flow regime is determined by the balance of the effects of viscosity and gravity relative to the inertia of the flow. The Froude number (Fr) is considered in the design of channels for definition of flow regime. In surface drainage channels, it is preferred that flow be subcritical so that the Froude number should be <1.0 under design conditions. In Froude number close to one (critical flow), the flow condition may fluctuate between subcritical and supercritical states. In this situation, instabilities occur because of considerable changes in the discharge value.

6.4.5.1 Design of Unlined Channels

At first, the cross-sectional characteristics of a trapezoidal channel are formulated using the parameters illustrated in Figure 6.18 as follows:

$$A = (b + zy)y, \tag{6.53}$$

$$T = b + 2zy, \tag{6.54}$$

$$D = \frac{(b + zy)y}{b + 2zy}, \tag{6.55}$$

$$R = \frac{(b + zy)y}{b + 2y\sqrt{1 + z^2}}, \tag{6.56}$$

where D is hydraulic depth and R is hydraulic radius. These expressions are substituted into the Manning formulation and rearranged to obtain

$$\frac{nQ}{S^{1/2}y^{8/3}} = \frac{(b/y + z)^{5/3}}{(b/y + 2\sqrt{1 + z^2})^{2/3}}. \tag{6.57}$$

When applying Manning's equation, the choice of the roughness coefficient, n, is the most subjective parameter. The appropriate Manning roughness factor n is determined regarding the channel material. Typical values of the Manning roughness factor for open channels are given in Table 6.6. Obstructions in the channel will cause an increase in depth above normal depth and must be taken into account. Sediment and debris in channels increase the roughness coefficients and should be accounted for.

Urban Water Drainage System

TABLE 6.6
Manning's Roughness Coefficients (n) for Artificial Channels

Category	Lining Type	Depth Ranges		
		0–15 cm	15–60 cm	>60 cm
Rigid	Concrete	0.015	0.013	0.013
	Grouted riparp	0.040	0.030	0.028
	Stone masonry	0.042	0.032	0.030
	Soil cement	0.025	0.022	0.020
	Asphalt	0.018	0.016	0.016
	Bare soil			
Unlined	Bare soil	0.023	0.020	0.020
	Rock cut	0.045	0.035	0.025
Temporary	Woven paper net	0.016	0.015	0.015
	Fiber glass roving	0.028	0.022	0.019
	Straw with net	0.065	0.033	0.025
	Curled wood mat	0.066	0.035	0.028
	Synthetic mat	0.036	0.025	0.021
Gravel riprap	2.5 cm D^{50}	0.044	0.033	0.030
	5 cm D^{50}	0.66	0.041	0.034
Rock PEI PARO	15 cm D^{50}	0.104	0.069	0.035
	30 cm D^{50}		0.078	0.040

Source: USDOT (U.S. Department of Transportation). 1986. *Programmatic Environmental Assessment of Commercial Space Launch Vehicle Programs.* Washington, DC, February.

As there are two unknown variables in Equation 6.57, a trial-and-error procedure should be employed to solve it. A graphical solution to Equation 6.57, which is suggested by Akan (2001), is presented in Figure 6.19. For using this figure, first the value of b/y should be approximated using

$$\frac{b}{y} = 1.186 \left[\frac{nQ}{k_n S_0^{1/2} y^{8/3}} - \frac{z^{5/3}}{(2\sqrt{1+z^2})^{2/3}} \right]^{0.955}. \quad (6.58)$$

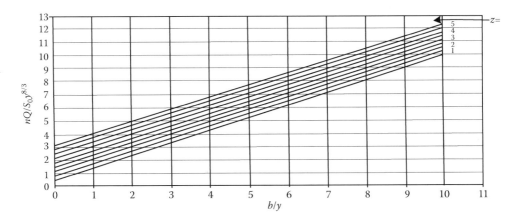

FIGURE 6.19 Graphical solution for the determination of open-channel geometry. (From Akan, A. O. 2001. *Canadian Journal of Civil Engineering*, 28(5): 865–867. With permission.)

6.4.5.2 Grass-Lined Channel Design

The tractive force approach is usually used in the design of grass-lined channels and the procedure is similar to unlined channels. The resulting shear stresses on the channel bed of the designed channel section should not exceed the maximum allowed value. The hydraulic characteristics of a grass lining are dependent on the type, maturity, and condition of the grass, and these factors should be incorporated in the design procedure.

The degree of retardance of grass lining and the hydraulic radius govern the Manning roughness coefficient of the channel according to the following equation (Chen and Cotton, 1988):

$$n = \frac{(3.28R)^{1/6}}{C_n + 19.97 \log[(3.28R)^{1.4} S^{0.4}]}, \qquad (6.59)$$

where C_n is the dimensionless retardance factor. For the determination of this factor, first the retardance class of the vegetal cover (A–E) is determined using Table 6.7 and then the value of C_n and permissible shear stress are estimated using Table 6.8.

The design procedure of grass-lined channels and that of unlined channels in cohesive soils are the same because of tractive forces on the channel bottom so that the maximum allowed flow depth (y_{allow}) can be calculated as

$$y_{\text{allow}} = \frac{C_s \tau}{\gamma S}, \qquad (6.60)$$

where C_s is the corrosion factor for the channel curvature, which is 1.0, 0.90, 0.75, and 0.6 for straight, slightly sinuous, moderately sinuous, and very sinuous channels, respectively.

As is obvious from Table 6.7, the maturity of the grass is important in the value of retardance factor. For incorporating this issue in the design of channels lined with grass, two stages should be considered (Chow, 1959; French, 1985). In the first stage, the channel is designed based on its stability in the situation with the minimum retardance factor of grass. Then the channel design is modified for flow situation with the maximum retardant factor. As the Manning's coefficient value is related to the hydraulic radius of the channel, an iterative procedure is again necessary for the design of a grass-lined channel using mathematical expressions (French, 1985). For avoiding this iterative procedure, dimensionless design charts presented in Figures 6.20 and 6.21 are developed by Akan and Hager (2001) for the estimation of channel bottom width and flow depth, respectively. The following variables should be estimated for using these charts:

$$Q_B = \frac{1.22 Q}{S^{1/2} y^{5/2} z}, \qquad (6.61)$$

$$\beta = C_n + 19.97 \log(5.28 S^{0.4} y^{1.4}), \qquad (6.62)$$

$$Q_Y = \frac{1.22 Q z^{3/2}}{S^{1/2} b^{5/2}}, \qquad (6.63)$$

$$\alpha = C_n + 19.97 \log\left(\frac{5.28 b^{1.4} S^{0.4}}{z^{1.4}}\right), \qquad (6.64)$$

$$B = \frac{b}{zy}, \qquad (6.65)$$

and

$$Y = \frac{zy}{b}. \qquad (6.66)$$

TABLE 6.7
Classification of Vegetal Covers as to Degrees of Retardance

Retardance	Cover	Condition
A	Weeping love grass	Excellent stand tall (average 75 cm)
	Yellow bluestem ischaemum	Excellent stand tall (average 90 cm)
B	Kudzu	Very dense growth uncut
	Bermuda grass	Good stand tall (average 30 cm)
	Native grass mixture: little bluestem, bluestem, blue gamma, other short and long stem Midwest love grasses	Good stand (unmoved)
	Weeping love grass	Good stand (tall, average 60 cm)
	Lespedeza sericea	Good stand not woody (tall, average 47.5 cm)
	Alfalfa	Good stand uncut (average 27.5 cm)
	Weeping love grass	Good stand, unmoved (average 32.5 cm)
	Kudzu	Dense growth, uncut
	Blue gamma	Good stand uncut (average 32.5 cm)
C	Crab grass	Fair stand, uncut (25–120 cm)
	Bermuda grass	Good stand, moved (average 15 cm)
	Common lespedezea	Good stand, uncut (average 27.5 cm)
	Grass–legume mixture, summer (orchard grass redtop, Italian ryegrass, and common lespedezea)	Good stand, uncut (15–20 cm)
	Centipede grass	Very dense cover (average 15 cm)
	Kentucky blue grass	Good stand, headed (15–30 cm)
D	Bermuda grass	Good stand (cut to 6 cm)
	Common lespedezea	Excellent stand uncut (average 12 cm)
	Buffalo grass	Good stand, uncut (7.5–15 cm)
	Grass–legume mixture: fall, spring (orchard grass redtop, Italian ryegrass, and common lespedezea)	Good stand, uncut (10–12.5 cm)
	Lespedeza sericea	After cutting to 5 cm (very good before cutting)
E	Bermuda grass	Good stand cut to 4 cm
	Bermuda grass	Burned stubble

Source: USDOT (U.S. Department of Transportation). 1986. *Programmatic Environmental Assessment of Commercial Space Launch Vehicle Programs.* Washington, DC, February.

TABLE 6.8
Retardance Factors and Permissible Shear Stresses of Grass-Lined Channels

Retardance Class	C_n	Permissible τ (N/m^2)
A	15.8	177.2
B	23.0	100.5
C	30.2	47.9
D	34.6	28.7
E	37.7	16.7

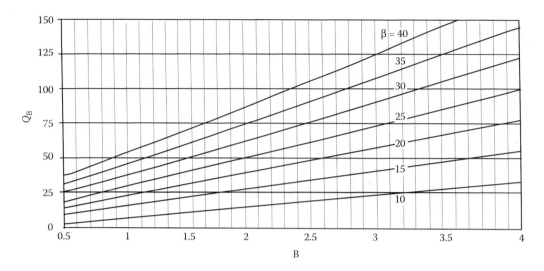

FIGURE 6.20 Design chart for bottom width calculation. (From Akan, A. O. and Hager, W. 2001. *Journal of Hydraulic Engineering, ASCE* 127: 236–238. With permission.)

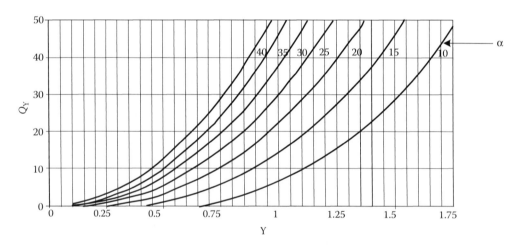

FIGURE 6.21 Analysis chart for flow depth calculation. (From Akan, A. O. and Hager, W. 2001. *Journal of Hydraulic Engineering, ASCE* 127: 236–238. With permission.)

EXAMPLE 6.7

Determine the depth and width of a straight ($C_s = 1.0$) trapezoidal channel lined with Bermuda grass which is proposed to carry $Q = 0.2\,\text{m}^3/\text{s}$. The slope of the channel bottom is $S = 0.005$ and its side slopes are the same and equal to $z = 2.0$.

Solution: According to Table 6.7, the lowest retardance class for Bermuda grass is E, which corresponds to $C_n = 37.7$ and $\tau = 16.7\,\text{N/m}^2$ using Table 6.8. The maximum allowable flow depth is obtained as

$$Y_{\text{allow}} = \frac{(1.0)(16.7)}{(9800)(0.005)} = 0.34\,\text{m}.$$

Considering $y = 0.34$ m, and employing Equation 6.61, the channel carrying capacity is calculated as

$$Q_B = \frac{1.22 \times 0.2}{(0.005)^{1/2}(0.34)^{5/2}(2)} = 25.60 \text{ cms}$$

and

$$\beta = 37.7 + 19.97 \log[5.28(0.005)^{0.4}(0.34)^{1.4}] = 20.65.$$

B is estimated to be equal to 1.2, by using Figure 6.20 and the above parameters. Using Equation 6.65, the channel width is determined as $b = (1.2)(2)(0.34) = 0.82 \approx 0.85$ m. It should be noted that the obtained value for channel bed width is the minimum allowable value and any greater values can also be used. For Bermuda grass, the highest retardance class is B that yields $C_n = 23$. The channel design should be justified to accommodate the flow with increased depth under the highest retardance conditions. In this stage, Equations 6.63 and 6.64 are employed:

$$Q_Y = \frac{1.22(0.2)(2)^{3/2}}{(0.005)^{1/2}(0.85)^{5/2}} = 14.65,$$

$$\alpha = 23 + 19.97 \log\left[\frac{5.28(0.85)^{1.4}(0.005)^{0.4}}{2^{1.4}}\right] = 8.66.$$

Y is estimated as 1.27 using Figure 6.21. Using Equation 6.66, under high retardance conditions, the flow depth is $y = (1.27)(0.85)/(2) = 0.54$ m. In other words, the channel will still accommodate the higher flow, if the grass is not mowed to the height specified in retardance class D. For completing the design, a freeboard is added to the channel depth. For example, with a freeboard of 0.06 m, the depth of the channel section is estimated to be equal to $0.54 + 0.06 = 0.60$ m. Usually, before finalizing the design, the Froude number is checked to be far enough from 1.0. In this example, the Froude numbers are lower than 1 in both stages of the design.

6.5 CULVERTS

A culvert is a short, closed (covered) conduit that conveys stormwater runoff under an embankment, usually a roadway. The primary purpose of a culvert is to convey surface water, but if properly designed it may also be used to restrict flow and reduce downstream peak flows. In addition to the hydraulic function, a culvert must also support the embankment and/or roadway, and protect traffic and adjacent property owners from flood hazards to the extent practicable. In detention reservoirs, culverts are also used as outlet conduits. Culverts have different shapes such as circular, rectangular (box), or elliptical as well as arch and pipe arch (Figure 6.22).

Different materials such as concrete, corrugated aluminum, and corrugated steel are used for the construction of culverts. Reinforced concrete pipes (RCP) are recommended for use (1) under a roadway, (2) when pipe slopes are <1%, or (3) for all flowing streams. RCP must be used for culverts designed for a 100-year storm, if the culverts lie in public lands or easements. RCP and fully coated corrugated metal pipes can be used in all other cases. High-density polyethylene (HDPE) pipes may also be used where permitted by the Director. For reduction of corrosion and flow resistance of the culverts, they are sometimes lined with another material, such as asphalt.

An important factor in the hydraulic performance of culverts is the inlet configuration. Different types of prefabricated and constructed-in-place inlet facilities are commonly used. The culvert inlet is composed of projecting culvert barrels, concrete headwalls, end sections, and culvert ends mitered to conform to the fill slope. Headwalls may be used for a variety of reasons, including increasing the efficiency of the inlet, providing embankment stability, providing embankment protection against erosion, providing protection from buoyancy, and shortening the length of the required structure.

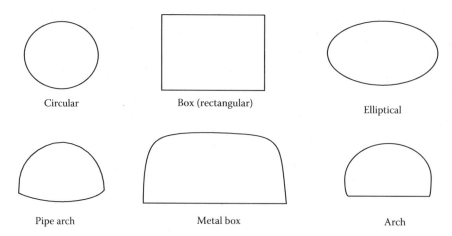

FIGURE 6.22 Different shapes of culverts. (From Normann, J. M., Houghtalen, R. J., and Johnston, W. J. 1985. *Hydraulic Design of Highway Culverts*. Hydraulic Design Series, No. 5, Federal Highway Administration, McLean, VA. With permission.)

Headwalls are required for all metal culverts and where buoyancy protection is necessary. If high headwater depths are to be encountered, or the approach velocity in the channel will cause scour, a short channel apron shall be provided at the toe of the headwall. Wing walls must be used where the side slopes of the channel adjacent to the entrance are unstable or where the culvert is skewed to the normal channel flow. Where inlet conditions control the amount of flow that can pass through the culvert, improved inlets can greatly increase the hydraulic performance of the culvert. Various standard inlet types are depicted in Figure 6.23.

The characteristics of the flow in a culvert depend on upstream and downstream conditions, inlet geometry, and barrel characteristics. It should also be considered that the flow condition for a specific culvert may vary over time. The inlet (upstream) or the outlet (downstream) condition controls the flow variations in a culvert. If the conveyance capacity of a culvert barrel is higher than the entrance flow

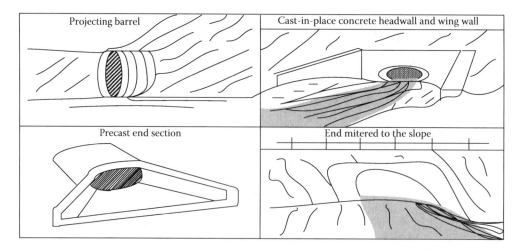

FIGURE 6.23 Standard inlet types. (From Normann, J. M., Houghtalen, R. J., and Johnston, W. J. 1985. *Hydraulic Design of Highway Culverts*. Hydraulic Design Series, No. 5, Federal Highway Administration, McLean, VA. With permission.)

Urban Water Drainage System

of the inlet, the inlet controls the flow condition. Otherwise, the flow condition depends on the outlet situation. Different flow situations are discussed in the following subsections.

6.5.1 Inlet Control Flow

Inlet control occurs when the culvert barrel is capable of conveying more flow than the inlet will accept. This typically happens when a culvert is operating on a steep slope. The control section of a culvert is located just inside the entrance. Critical depth occurs at or near this location, and the flow regime immediately downstream is supercritical.

As presented in Figure 6.24a and 6.24c, under supercritical conditions, a culvert is partially full. In the situation where the downstream of the culvert is submerged, a hydraulic jump occurs and thereafter the culvert flows full, as shown in Figure 6.24b. If the inlet is unsubmerged, the inlet performance is similar to that of a weir. Otherwise, the inlet acts as an orifice.

If the inequality of Equation 6.67 is satisfied, then the inlet is considered as unsubmerged:

$$\frac{Q}{AD^{0.5}g^{0.5}} \leq 0.62, \tag{6.67}$$

where Q is the discharge, A is the cross-sectional area of the culvert, D is the interior height of the culvert, and g is the gravitational acceleration.

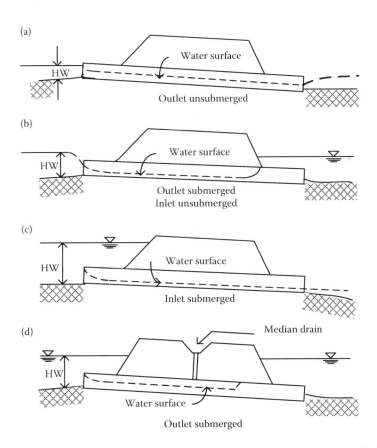

FIGURE 6.24 Types of inlet control flow in culverts. (From Normann, J. M., Houghtalen, R. J., and Johnston, W. J. 1985. *Hydraulic Design of Highway Culverts*. Hydraulic Design Series, No. 5, Federal Highway Administration, McLean, VA. With permission.)

Equation 6.68 is employed for hydraulic analysis of the flow in culverts in the situation where inlets are not submerged:

$$\frac{HW}{D} = \frac{y_c}{D} + \frac{V_c^2}{2gD} + K_1 \left(\frac{Q}{AD^{0.5}g^{0.5}} \right)^{M_1} + k_s S, \qquad (6.68)$$

where HW is the headwater depth above the upstream invert of the culvert, y_c is the critical depth, V_c is the velocity at critical depth, $k_s = 0.7$ for mitered inlets and -0.5 for nonmitered inlets, S is the culvert bottom slope, and K_1 and M_1 are empirical constants.

The critical depth y_c can be approximated for circular culverts. The corresponding area A_c and velocity V_c of the culvert are calculated from the geometry of a circular culvert as

$$A_c = \frac{(\theta - \sin\theta)D^2}{8}, \qquad (6.69)$$

$$V_c = \frac{8Q}{(\theta - \sin\theta)D^2}, \qquad (6.70)$$

where θ (in radians) is

$$\theta = 2\pi - 2\arccos\left(\frac{2y_c}{D} - 1\right). \qquad (6.71)$$

The critical depth for rectangular culverts is estimated as

$$y_c = \left(\frac{Q^2}{gb^2}\right)^{1/3}, \qquad (6.72)$$

where b is the width of the box culvert. Considering that $V_c^2/2g = 0.5 y_c$, Equation 6.68 for box culverts can be rewritten as

$$\frac{HW}{D} = \frac{3}{2D}\left(\frac{Q^2}{gb^2}\right)^{1/3} + K_1 \left(\frac{Q}{AD^{0.5}g^{0.5}} \right)^{M_1} + k_s S. \qquad (6.73)$$

Another equation that can be used for unsubmerged inlets is

$$\frac{HW}{D} = K_2 \left(\frac{Q}{AD^{0.5}g^{0.5}} \right)^{M_2}, \qquad (6.74)$$

where K_2 and M_2 are empirical constants. Although the application of Equation 6.74 is simpler than Equation 6.68, since Equation 6.68 has a stronger theatrical base, it is often preferred. However, both of the equations are used in practice depending on the available data for the determination of the empirical coefficients for the inlet conditions.

The following inequality should be satisfied if the inlet is submerged:

$$\frac{Q}{AD^{0.5}g^{0.5}} \geq 0.70. \qquad (6.75)$$

Equation 6.76 is used for flow calculations in submerged inlets:

$$\frac{HW}{D} = c\left(\frac{Q}{AD^{0.5}g^{0.5}}\right)^2 + Y + k_s S, \qquad (6.76)$$

Urban Water Drainage System

where S is the slope, $k_s = 0.7$ for mitered inlets and -0.5 otherwise, and c and Y are empirical constants.

The inlet condition will change from the unsubmerged to the submerged if

$$0.62 < (Q/AD^{0.5}g^{0.5}) < 0.7. \tag{6.77}$$

In the transition situation, the results of the submerged and unsubmerged inlet equations should be interpolated. The constant values used in the above equations including $M_1, K_1, M_2, K_2, c,$ and Y are determined regarding to inlet configuration as given in Table 6.9.

EXAMPLE 6.8

Determine the headwater depth for a culvert that conveys a flow of 0.9 cms under inlet control conditions. The culvert is circular and has a square edge inlet with a headwall and a diameter of 1.2 m and a slope of 0.025, and the inlet is not mitered.

Solution: At first it should be determined whether the inlet is submerged or not:

$$\frac{Q}{AD^{0.5}g^{0.5}} = \frac{0.9}{[\pi(1.2)^2/4](1.2)^{0.5}(9.81)^{0.5}} = 0.23 < 0.62.$$

Therefore, the inlet is unsubmerged and $K_1 = 0.3155$ and $M_1 = 2.0$ regarding to inlet configuration. For the determination of headwater depth, the following variables are calculated:

$$y_c = \frac{(1.01)(0.9)^{0.5}}{9.81^{0.25} 1.2^{0.25}} = 0.52 \text{ m},$$

$$\theta = 2(3.14) - 2 \arccos\left[\frac{2(0.52)}{1.2} - 1\right] = 2.87 \text{ rad} = 164°,$$

and

$$V_c = \frac{8(0.9)}{(2.87 - \sin 2.87)(1.2)^2} = 1.9 \text{ m/s}.$$

Then Equation 6.68 is used to calculate headwater depth with $k_s = -0.5$ for a nonmitered inlet,

$$\frac{HW}{D} = \frac{0.52}{1.2} + \frac{1.9^2}{2(9.81)(1.2)} + 0.3155(0.23)^{2.0} - (0.5)(0.025) = 0.591,$$

and finally, $HW = 0.591(1.2) = 0.71$ m.

6.5.2 OUTLET CONTROL FLOW

Outlet control flow occurs when the culvert barrel is not capable of conveying as much flow as the inlet opening will accept. The control section for outlet control flow in a culvert is located at the barrel exit or further downstream. Either subcritical or pressure flow exists in the culvert barrel under these conditions. In the outlet control situation, if the culvert is partially full, the culvert flow will be subcritical. Conditions A, D, and E in Figure 6.24 show the typical outlet control flow conditions.

TABLE 6.9
Culvert Inlet Control Flow Coefficients

Shape and Material	Inlet Edge Description	K_1	M_1	K_2	M_2	c	Y
Circular concrete	Square edge with headwall	0.3155	2.0			1.2816	0.67
	Groove end with headwall	0.2512	2.0			0.9402	0.74
	Groove end projecting	0.1449	2.0			1.0207	0.69
Circular corrugated metal	Headwall	0.2512	2.0			1.2204	0.69
	Mitered to slope	0.2113	1.33			1.4909	0.75
	Projecting	0.4596	1.50			1.7807	0.82
Circular	Beveled ring, 45° bevels	0.1381	2.50			0.9660	0.69
	Beveled ring, 33° bevels	0.1381	2.50			0.7825	0.64
Rectangle box	30° to 75° wingwall flares	0.1475	1.00			1.1173	0.57
	90° and 15° wingwall flares	0.2243	0.75			1.2880	0.67
	0° wingwall flares	0.2243	0.75			1.3621	0.82
Corrugated metal box	90° Headwall	0.2673	2.00			1.2204	0.69
	Thick wall projection	0.3025	1.75			1.3492	0.64
	Thin wall projection	0.4596	1.50			1.5971	0.57
Horizontal ellipse concrete	Square edge with headwall	0.3220	2.0			1.2816	0.67
	Groove end with headwall	0.1381	2.5			0.9402	0.74
	Groove end projecting	0.1449	2.0			1.0207	0.69
Vertical ellipse concrete	Square edge with headwall	0.3220	2.0			1.2816	0.67
	Groove end with headwall	0.1381	2.5			0.9402	0.74
	Groove end projecting	0.3060	2.0			1.0207	0.69
Rectangular box	45° wingwall flare d = .043 D			1.623		0.9950	0.80
	18° to 33.7° wingwall flare d = .083 D			1.547		0.8018	0.83
	90° headwall with 1.9 cm chamfers			1.639		1.2075	0.79
	90° headwall with 45° bevels			1.576		1.0111	0.82
	90° headwall with 33.7° bevels			1.547		0.8114	0.865
	1.9 cm chamfers; 45° skewed headwall			1.662		1.2944	0.73
	1.9 cm chamfers; 30° skewed headwall			1.697		1.3685	0.705
	1.9 cm chamfers; 15° skewed headwall			1.735		1.4506	0.73
	45° bevels: 10°–45° skewed headwall			1.585		1.0525	0.75
Rectangular box with 1.9 cm chamfers	45° nonoffset wingwall flares			1.582		1.0916	0.803
	18.4° nonoffset wingwall flares			1.569		1.1624	0.806
	18.4° nonoffset wingwall flares with 30° skewed barrel			1.576		1.2429	0.71
Rectangular box with top bevels	45° wingwall flares-offset			1.582		0.9724	0.835
	33.7° wingwall flares-offset			1.576		0.8144	0.881
	18.4° wingwall flares-offset			1.569		0.7309	0.887
Circular	Smooth tapered inlet throat			1.699		0.6311	0.89
	Rough tapered inlet throat			1.652		0.9306	0.90
Rectangular	Tapered inlet throat			1.512		0.5764	0.97
Rectangular concrete	Side tapered-less favorable edges			1.783		1.5005	0.85
	Side tapered-more favorable edges			1.783		1.2172	0.87
	Slope tapered-less favorable edges			1.592		1.5005	0.65
	Slope tapered-more favorable edges			1.592		1.2172	0.71

Source: Akan, A. O. and Houghtalen, R. J. 2003. *Urban Hydrology, Hydraulics and Stormwater Quality: Engineering Applications and Computer Modeling.* Wiley, Belgium, (ISBN 0-471-43158-3).

6.5.2.1 Full-Flow Conditions

By the assumption that the velocity heads at the upstream and downstream of the culvert are equal, the energy equation for a full-flow culvert can be written as

$$\text{HW} = \text{TW} - SL + \left(1 + k_c + \frac{2gn^2 L}{R^{4/3}}\right) \frac{Q^2}{2gA^2}, \quad (6.78)$$

where TW is the tailwater depth measured from the downstream invert of the culvert. S is the culvert slope, L is the culvert length, g is the gravitational acceleration, n is the Manning roughness factor, R is the hydraulic radius, A is the cross-sectional area, and k_c is the empirical constant which is determined regarding to type of structure and design of entrance as given in Table 6.10.

It should be noted that friction losses as well as entrance and exit losses are considered in developing Equation 6.78.

TABLE 6.10
Entrance Loss Coefficient

Type of Structure and Design of Entrance	K_c
Pipe, concrete	
Projecting from fill socket and (groove-end)	0.2
Projecting from fill, square cut end	0.5
Headwall or headwall and wingwalls	
Socket end of pipe (groove-end)	0.2
Square-edge	0.5
Rounded (radius = $\frac{1}{2}D$)	0.2
Mitered to conform to fill slope	0.7
End-section conforming to fill slope	0.5
Beveled edges, 33.7° or 45° bevels	0.2
Side- or slope-tapered inlet	0.2
Pipe or pipe-arch, corrugated metal	
Projecting from fill (no headwall)	0.9
Headwall or headwall and wingwalls square-edge	0.5
Mitered to conform to fill slope, paved or unpaved slope	0.7
End-section conforming to fill slope	0.5
Beveled edges, 33.7° or 45° bevels	0.2
Side- or slope-tapered inlet	0.2
Box, reinforced concrete	
Headwall parallel to embankment (no wingwalls)	
Square-edged on 3 edges	0.5
Rounded on three edges to radius of 1/12 barrel dimension, or beveled edges on 3 sides	0.2
Wingwalls at 30° to 75° to barrel	
Square-edged at crown	0.4
Crown edge rounded to radius of 1/12 barrel dimension or beveled top edge	0.2
Wingwalls at 10° to 25° to barrel	
Square-edged at crown	0.5
Wingwall parallel (extension of sides)	
Square-edged at crown	0.7
Side- or slope-tapered inlet	0.2

Source: Akan, A. O. and Houghtalen, R. J. 2003. *Urban Hydrology, Hydraulics and Stormwater Quality: Engineering Applications and Computer Modeling.* Wiley, Belgium, (ISBN 0-471-43158-3).

6.5.2.2 Partly Full-Flow Conditions

By using gradually varied flow calculations, an accurate relationship can be developed between culvert discharge and headwater elevation for partially full outlet control culvert flow conditions. In these calculations, the downstream depth has been considered equal to the higher value of tailwater depth TW and the critical depth y_c. If calculated water surface profile intersects the top of the culvert barrel, full-flow equations are employed for upstream of the intersection point. Equation 6.79 is used for the calculation of head loss h_f in the full-flow part of the culvert:

$$h_f = \left(1 + k_c + \frac{2gn^2 L_f}{R^{4/3}}\right) \frac{Q^2}{2gA^2}, \quad (6.79)$$

where L_f is the length of the culvert with full flow.

As gradually varied flow calculations are too time-consuming and need considerable computational effort, an approximate method is suggested for partially full outlet control culvert flow. The headwater elevation in this method is estimated as

$$\text{HW} = H_D - SL + \left(1 + k_c + \frac{2gn^2 L}{R^{4/3}}\right) \frac{Q^2}{2gA^2}, \quad (6.80)$$

in which R and A are calculated for full-flow culvert. Also, H_D is equal to the higher value of the tailwater depth TW or $(y_c + D)/2$.

As shown in Figure 6.25d, the more accurate results of Equation 6.80 are obtained when the least part of the culvert is full. It should be noted that the results of the approximate method are less reliable when the entire length of the culvert is partially full. In this situation, only when HW $>$ 0.75D, the approximate method can be applied. For headwater elevations $<$ 0.75D, gradually varied flow calculations are necessary to obtain acceptable results.

EXAMPLE 6.9

A rectangular box culvert has the following characteristics: $D = 0.8$ m, $b = 0.75$ m, $L = 15$ m, $n = 0.012$, and $S = 0.15$. Determine the headwater depth considering tailwater and the discharge equal to 0.45 m and 0.95 m^3/s, respectively. Consider k_c is equal to 0.7.

Solution: As the first step, the critical depth is calculated using Equation 6.72 as

$$y_c = \left[\frac{0.95^2}{(9.81)(0.75)^2}\right]^{1/3} = 0.547 \text{ m}.$$

Then $(y_c + D)/2 = (0.547 + 0.8)/2 = 0.674$ m, which is higher than TW $= 0.45$ m. Thus, the approximate method can be used in this example. Headwater is estimated as

$$\text{HW} = 0.674 - (0.015)(15) + \left[1 + 0.7 + \frac{(2)(9.81)(0.012)^2(15)}{(0.19)^{4/3}}\right] \frac{0.95^2}{2(9.81)(0.6)^2} = 0.716.$$

The calculated result is acceptable because 0.716 m $>$ (0.75)(0.8) m.

6.5.3 SIZING OF CULVERTS

Even though the flow condition (inlet or outlet control) and the nature of the flow (full or partly full) determine which equation should be used for description of the water surface profile in a culvert,

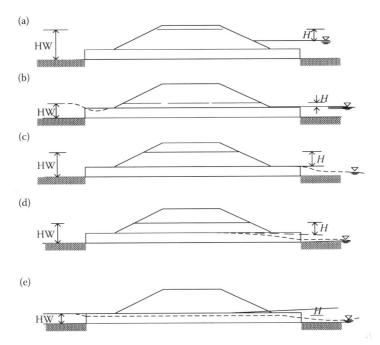

FIGURE 6.25 Types of outlet control flow in culverts. (From Normann, J. M., Houghtalen, R. J., and Johnston, W. J. 1985. *Hydraulic Design of Highway Culverts.* Hydraulic Design Series, No. 5, Federal Highway Administration, McLean, VA. With permission.)

usually the determination of flow type in a culvert is not that simple. So, designing (sizing) a culvert is a very complicated and difficult procedure.

Some hydraulic issues should be considered in the culvert design procedure for determining the flow condition. In mild slope culverts, the flow condition is often controlled by the outlet condition. In this situation, if TW > D, full flow will occur in the culvert; otherwise the culvert flow is partly full. In steep culverts, the inlet condition controls the culvert flow. Only when TW > D in the culverts with a steep slope, full flow will occur.

Drainage culverts are designed to convey a design discharge. Often a culvert is sized to avoid exceeding the headwater elevation from allowable tailwater elevation in the design discharge condition. The "minimum performance" approach is commonly used for sizing drainage system culverts because in this method the identification of the flow in the culvert is not needed. In the minimum performance approach, the culvert situation under both inlet and outlet control conditions is checked and the more conservative results are used for culvert design. As this method acts conservatively, it sometimes leads to a culvert size larger than what is needed.

6.5.4 Protection Downstream of Culverts

The riprap can be used for erosion protection downstream of the conduit and culvert outlets that are in line with major drainage channels. Inadequate protection at conduit and culvert outlets has long been a major problem. Scour resulting from highly turbulent, rapidly decelerating flow is a common problem at conduit outlets. The riprap protection design is suggested for conduit and culvert outlet Froude numbers up to 2.5 where the channel and conduit slopes are parallel with the channel gradient and the conduit outlet invert is flushed with the riprap channel protection.

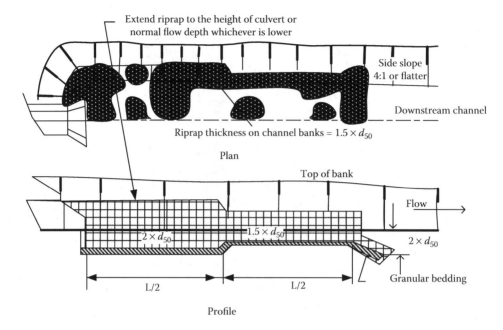

FIGURE 6.26 Culvert and pipe outlet erosion protection.

Figure 6.26 shows typical riprap protection of culverts and major drainage way conduit outlets. The additional thickness of the riprap just downstream from the outlet is to assure protection from flow conditions that might precipitate rock movement in this region. The rock size and length of riprap protection is determined based on the flow and culvert characteristics.

6.6 DESIGN FLOW OF SURFACE DRAINAGE CHANNELS

6.6.1 Probabilistic Description of Rainfall

Deterministic models do not result in accurate rainfall predictions because rainfall events are highly uncertain and their characteristics vary in time and space. For incorporating these uncertainties in hydrological studies, rainfall events are treated as random processes, and probabilistic methods are employed for the determination of their occurrence and magnitudes.

6.6.1.1 Return Period and Hydrological Risk

The probability of equaling or exceeding a rainfall event with a specified duration and depth in any one year is called the exceedence probability, p. The return period is defined as the inverse of the return period T_r as

$$p = \frac{1}{T_r}. \tag{6.81}$$

For example, if the 50-year, 18-h rainfall is 10 cm, then the probability that a depth of 10 cm or higher will be produced over an 18-h rainfall event in any given year is 2%. It should be noted that rainfall is considered a completely random process. This means that if a 50-year event is exceeded this year, the probability of its exceedence in the next year is still 2%.

Often, the design capacity of a stormwater structure is determined based on the rainfall events, and because of the random nature of the rainfall, the design capacity might be exceeded with a low probability at any given time. It can be concluded that there is always a hydrological risk the design should correspond to. The probability that, during the service life of the structure, the design event will be exceeded one time or more is considered a hydrological risk, HR, which is calculated as

$$\text{HR} = 1 - \left(1 - \frac{1}{T_r}\right)^n, \tag{6.82}$$

where T_r is the return period of the design event, and n is the life span of the considered structure.

EXAMPLE 6.10

Determine the hydrological risk of a roadway culvert with 25 years expected service life designed to carry a 50-year storm.

Solution: Using Equation 6.82,

$$\text{HR} = 1 - \left(1 - \frac{1}{50}\right)^{25} = 0.40 = 40\%.$$

EXAMPLE 6.11

Determine the design return period for the culvert of the previous example, if the acceptable hydrological risk is 0.15 (or 15%).

Solution: Equation 6.82 is employed again for determination of the return period of the design rainfall event as

$$0.15 = 1 - \left(1 - \frac{1}{T_r}\right)^{25}.$$

T_r has been calculated as 154 years, whereas in practice, a return period of 150 years is considered for the determination of culvert size.

Less return periods may be considered because of economic reasons that lead to much higher hydrological risks. In other words, flooding from a stormwater structure is not necessarily the result of poor structure design.

6.6.1.2 Frequency Analysis

Frequency analysis derives meaningful information from the available historical data. For instance, a frequency analysis should be carried out on the rainfall records at a specific rain gauge to determine the corresponding return period of a rainfall with known depth and duration.

For frequency analysis, at first, annual maximum series of rainfall depths for the specified duration are developed. It should be noted that "duration" is not necessarily equal to the full duration of historic storms. As an example, a historical storm may last for 50 min, but it can be used in the frequency analysis of 40-min storms. If the rainfall in a 40-min portion of this rainfall is greater than all of the 40-min rainfall events, the rainfall during this 40-min portion is considered as representative of the 40-min rainfalls in that year.

In the next step, the best theoretical probability distribution that fits the rainfall annual maximum series is specified. The best-fitting probability distribution is determined based on the goodness of fit, which is quantified using statistical tests such as the chi-square test. However, experience shows that the "Gumbel distribution" (the extreme value type I distribution), most of the time, fits the maximum rainfall data well, so that in practice this distribution is often used for frequency analysis of rainfall

TABLE 6.11
Frequency Factor K for the Gumbel Probability Distribution

N	T_r (years)				
	5	10	25	50	1M
15	0.967	1.703	2.632	3.321	4.005
20	0.919	1.625	2.517	3.179	3.836
25	0.888	1.575	2.444	3.088	3.729
30	0.866	1.541	2.393	3.026	3.653
35	0.851	1.516	2.354	2.979	3.598
40	0.838	1.495	2.326	2.943	3.554
45	0.829	1.478	2.303	2.913	3.520
50	0.820	1.466	2.293	2.889	3.491
75	0.792	1.423	2.220	2.912	3.400
100	0.779	1.401	2.187	2.770	3.349
∞	0.719	1.305	2.044	2.592	3.137

Source: Haan, C. T. 1977. *Statistical Methods in Hydrology*. The Iowa State University Press, Ames, IA.

data. The cumulative probability function of the Gumbel distribution is

$$p = 1 - e^{-e^{-b}}, \tag{6.83}$$

where b is the distribution parameter.

The maximum rainfall depths at a specific return period, T, for a given rainfall duration, t_d, could be expressed as

$$R_T = \overline{R} + Ks, \tag{6.84}$$

where R_T is the rainfall depth for a specified return period (T_r), $\overline{R}$ is the mean of annual maximum rainfall depths, s is the standard deviation of annual maximum rainfall depths, and K is the frequency factor.

The frequency factor, K, is determined based on the fitted probability distribution characteristics, the assigned period, and the length of the employed annual maximum rainfall series. The K values of the Gumbel probability distribution for different combinations of return periods and the length of time series are shown in Table 6.11.

EXAMPLE 6.12

Determine the 15-min storm depths and the average intensities with return periods of 5, 10, 25, 50, and 100 years for a 25-year annual maximum series of 15-min storm depths given in the first and third columns of Table 6.12. Suppose that the extreme value type I distribution (the Gumbel distribution) fits the annual maximum series.

Solution: The mean and the standard deviation of the annual maximum depths are calculated as

$$\overline{P} = \frac{\sum P_i}{n} = \frac{40.752}{25} = 1.630 \text{ cm}$$

and

$$s = \sqrt{\frac{\sum (P_i - \overline{P})^2}{n}} = 0.690 \text{ cm}.$$

Urban Water Drainage System

For the return periods of $T_r = 5, 10, 25, 50$, and 100 years, the frequency factors are determined as 0.888, 1.575, 2.444, 3.088, and 3.729, respectively, from Table 6.11. Then, Equation 6.84 is used to determine the 15-min, 5-year depth as

$$P_T = 1.63 + (0.888)(0.690) = 2.243\,\text{cm}.$$

Likewise, for $T_r = 10$ years,

$$P_T = 1.63 + (1.575)(0.690) = 2.717\,\text{cm}.$$

The rainfall depths corresponding to return periods of 25, 50, and 100 years are calculated in the same way and are equal to 3.32, 3.76, and 4.20 cm, respectively.

The rainfall depth is divided by the duration to find an average intensity. In this example, the duration is 15 min $= 0.25\,\text{h}$ so that the calculated P_T is divided by 0.25 h. The average intensities of 8.97, 10.87, 13.28, 15.04, and 16.80 cm/h are obtained for the return periods of 5, 10, 25, 50, and 100 years, respectively.

TABLE 6.12
Mean and Standard Deviation Example

P_j (cm)	$P_i - \bar{P}$	P_j (cm)	$P_i - \bar{P}$
2.742	1.112	1.548	−0.082
2.671	1.041	1.264	−0.366
2.657	1.027	1.201	−0.429
2.601	0.971	1.196	−0.434
2.526	0.896	1.105	−0.526
2.394	0.763	1.014	−0.616
2.156	0.526	1.009	−0.621
2.116	0.486	0.860	−0.770
1.990	0.360	0.849	−0.781
1.985	0.355	0.779	−0.851
1.636	0.006	0.708	−0.922
1.610	−0.020	0.585	−1.045
1.550	−0.081		

6.6.2 Design Rainfall

The design runoff event that a stormwater structure must carry out should be specified as the first step in sewer system development. It is assumed that the drainage system performs in its full capacity under design runoff so that the system may fail if the actual runoff is more than the designed value. Note that the system failure may correspond to physical damages to the system or a hydraulic failure in which the system does not perform as it is designed. An example of hydraulic failure is that water backs up and overtops the embankment if the storm runoff exceeds the design discharge of the associated culvert.

Often the runoff data are not available because stream gauges are too expensive and difficult to be used in urban areas. Therefore, instead of runoff data, rainfall data are used for the design of drainage systems. Thus for the estimation of the design runoff the rainfall data should be converted to runoff data through an appropriate rainfall–runoff model.

6.6.2.1 Extracting Design Rainfall from Historical Data

The approaches used in practice for extracting a design runoff event from historical rainfall data are divided into two groups including continuous simulation methods and single-event methods. These

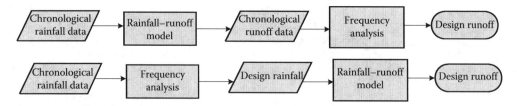

FIGURE 6.27 Design runoff formulation methods.

approaches have fundamental differences but frequency analysis of hydrological data and application of a rainfall–runoff model are included in both of them. General procedures of these methods are shown in Figure 6.27. The application of rainfall–runoff models is discussed in Chapter 3, and in the preceding sections a discussion on frequency analysis is given.

In the application of the continuous simulation approach, the input of the rainfall–runoff model is a chronological series of rainfall and the model output is a chronological series of runoff. A frequency analysis is performed on the resulting runoff series to determine the return period corresponding to different magnitudes of runoff events. The design runoff is selected based on the results of this frequency analysis. A subsurface flow factor should be incorporated into the rainfall–runoff model to consider the effects of water movement in the soil even during dry periods (zero rainfall). After the initial startup of the model, it would determine the initial basin soil moisture condition before each storm event. This issue plays an important role in the determination of design runoff by conversion of rainfall data because the resulting runoff from a rainfall event is significantly affected by the initial soil moisture condition.

In detention basin projects, the continuous simulation approach is preferred because the sequence of rainfall events and the intervened time are important. For example, the filled capacity of a detention basin can be considered at the time of the next storm occurrence in this approach. But for benefiting from the advantages of this approach, a sophisticated rainfall–runoff model should be employed. These kinds of rainfall–runoff models need considerable data and their application is not easy. Furthermore, as there are large amounts of data, the simulation is often too time-consuming, especially when the data of several decades, which include hundreds of storms, are analyzed. The difficulty in the application of this approach increases by decreasing the computational time increments to achieve more accurate results.

The IDF curves or relationships are considered in the single-event approach to select a design rainfall. The selected design rainfall is entered into a rainfall–runoff model to estimate the design runoff. In this approach, only a single rainfall–runoff process for a single storm event is simulated; therefore it is much simpler than the continuous simulation approach. Also a simple rainfall–runoff model can be employed and it is not necessary to take into account the soil moisture condition. But the initial condition of the basin preceding the design rainfall occurrence should be specified.

Continuous rainfall records obtained synthetically by using selected single rainfall events are also used in some recently developed hybrid methods. As the continuous simulation approach is too time-consuming and costly, and needs extensive watershed data to yield accurate results, the single-event design-storm method is more common in practice because of its simplicity and less data needed. A single-event design storm is identified by the corresponding return period, storm duration, and depth (or average intensity) as well as its spatial and temporal distribution. The selection procedure of these characteristics is described in the following subsections.

6.6.3 Design Return Period

Structural or hydraulic failures of urban stormwater structures may cause a lot of troubles and difficulties for the public, flood damages of different magnitudes as well as safety concerns. But as limited financial resources are available for sewer system construction, sonic risk of failure associated with the considered design return period is allowed in the design of infrastructures related to stormwater collection. The hydrological risk because of considering larger design return periods decreases, but this results in larger sewer structures, which are not economical.

There are a wide range of issues such as the importance of the structure, the cost, the level of protection provided by it, and the consequence of its failure that should be considered in selecting a design return period. The ideal design return period could be determined through a cost–benefit analysis. But in practice, for the establishment of appropriate return periods (standards) for various systems, past experiences of failures and the resulting costs in the project location are considered. The design return periods of culverts are highly dependent on the traffic volume of the street and usually typical values of 5–10, 10–25, and 25–50 years are used for streets carrying low, intermediate, and high traffic volumes, respectively. By increasing the carrying capacity of the sewer structures, the design return period increases as design return periods of 2–5, 2–25, and 10–100 years are used for street gutters, storm sewers, and detention basins, respectively, and 50–100 year return periods are used for the design of major highway bridges. The design return periods of specific projects are determined based on the site characteristics and the local drainage regulations.

6.6.4 Design-Storm Duration and Depth

The design-storm duration is dependent on the project type. The design-storm duration should result in the largest peak discharge for a specified return period because the storm sewers are designed for conveying the peak flow. In a similar way, the duration that causes the largest detention volume is considered in the design of detention ponds. By considering different values and evaluating their effects on the peak discharge and/or the detention volume, the design-storm duration is specified.

IDF (Intensity–Duration–Frequency) curves show the relationships between the average intensity (or depth), duration, and the return period as mentioned in Chapter 3. The average intensity is obtained from the local IDF curves, after selecting the design return period and duration.

Alternatively, the rainfall intensity for the area within the district can be approximated by the following equation:

$$I = \frac{28.5 P_1}{(10 + t_c)^{0.786}}, \tag{6.85}$$

where I is the rainfall intensity (mm per hour), P_1 is the 1-h point rainfall depth (mm), and t_c is the time of concentration (minutes).

6.6.5 Spatial and Temporal Distribution of Design Rainfall

The spatial distribution of rainfall over a watershed is not uniform, which should be considered in rainfall–runoff analysis, especially in large watersheds. For incorporating the effect of watershed size on rainfall distribution, a reduction factor is applied to the design rainfall. The rainfall intensity also varies during a single storm and these variations affect the produced runoff rate. So, for a complete description of a design storm, the temporal variations of its intensity should also be considered. Different methods such as SCS rainfall patterns are used for the specification of temporal variations in the design rainfall.

6.7 STORMWATER STORAGE FACILITIES

Detention and retention basins are used for stormwater runoff quantity control to mitigate the effects of urbanization on runoff flood peaks. As is the case with major drainage systems, the development of multipurpose, attractive detention facilities is recommended. These structures are safe, maintainable, and viewed as community assets rather than liabilities.

The storage facilities are divided into two groups. On-site runoff storage facilities are planned on an individual-site basis. Larger facilities that have been identified and sized as a part of some overall regional plan are categorized as "regional" facilities. In addition, the regional definition can also be applied to storage facilities that address moderately sized watersheds to encompass multiple land development projects.

On-site storage facilities are usually designed to control runoff from a specific land development site and are not located or designed with the idea of reducing downstream flood peaks along a major drainage system. The total volume of runoff detained in the individual on-site facility is quite small, and the detention time is relatively short. Therefore, unless design (i.e., sizing and flow release) criteria and implementation are applied uniformly throughout the urbanizing or redeveloping watershed, their effectiveness diminishes rapidly along the downstream reaches of waterways. The application of consistent design and implementation criteria is of paramount importance, if large numbers of on-site detention facilities are to be effective in controlling peak flow rates along major drainage systems (Glidden, 1981; Urbonas and Glidden, 1983).

The principal advantage of on-site facilities is that developers can be required to build them as a condition of site approval. Major disadvantages include the need for a larger total land area for multiple smaller on-site facilities as compared to larger regional facilities serving the same tributary catchment area. If the individual on-site facilities are not properly maintained, they can become a nuisance to the community and a basis for many complaints to municipal officials. It is also difficult to ensure adequate maintenance and long-term performance at the design levels. Prommesberger (1984) inspected approximately 100 on-site facilities built, or required by municipalities to be built, as a part of land developments over about a 10-year period. He concluded that a lack of adequate maintenance contributed to a loss of continued function of the facilities. He also concluded that a lack of local institutional structures contributed to the facilities no longer being in existence after their initial construction.

Some facilities are designed during watershed planning process. These facilities which are developed in a staged regional plan, are called regional facilities. These are often planned and located as part of the District's master planning process. They are typically much larger than on-site facilities. The main disadvantage of the regional facilities is the lack of an institutional structure to fund them. Another disadvantage of regional facilities is that they can leave substantial portions of the stream network susceptible to increased flood peaks, and plans must be developed to take this condition into account. In addition, to promote water-quality benefits, some form of on-site stormwater management is necessary upstream of the regional facilities. Examples include minimized directly connected impervious areas (MDCIA) that promote flow across vegetated surfaces utilizing "slow-flow" grassed swales and grading lots to contain depressions.

More economical and hydrologically reliable results can be achieved through stormwater management planning for an entire watershed that incorporates the use of regional facilities. Regional facilities also potentially offer greater opportunities for achieving multiobjective goals such as recreation, wildlife habitat, enhanced property value, open space, and others.

There are several types of stormwater storage facilities, where they are classified as on-site or regional, namely

1. *Detention*: Detention facilities provide temporary storage of stormwater that is released through an outlet that controls flows to preset levels. Detention facilities typically flatten and spread the inflow hydrograph, lowering the peak to the desired flow rate. Often these facilities also incorporate features designed to meet water-quality goals.
2. *Retention*: Retention facilities store stormwater runoff without a positive outlet, or with an outlet that releases water at very slow rates over a prolonged period. These differ in nature and design from "retention ponds" that are used for water-quality purposes.
3. *Conveyance (channel) storage*: Conveyance, or channel routing, is an often-neglected form of storage because it is dynamic and requires channel storage routing analysis. Slow-flow

and shallow conveyance channels and broad floodplains can retard the buildup of flood peaks and alter the time response of the tributaries in a watershed.
4. *Infiltration facilities*: Infiltration facilities resemble retention facilities in most respects. They retain stormwater runoff for a prolonged period of time to encourage infiltration into the groundwater. These facilities are difficult to design and implement because so many variables come into play.
5. *Other storage facilities*: Storage can occur at many locations in urban areas such as in random depressions and upstream of railroad and highway embankments. Special considerations typically apply to the use and reliance upon such conditions.

Detention and retention facilities can be further subdivided into

1. *In-line storage*: A facility that is located in-line with the drainage channel/system and captures and routes the entire flood hydrograph. A major disadvantage with in-line storage is that it must be large enough to handle the total flood volume of the entire tributary catchment, including off-site runoff, if any.
2. *Off-line storage*: A facility that is located off-line from the drainage way and depends on the diversion of some portion of flood flows out of the waterway into the storage facility. These facilities can be smaller and potentially store water less frequently than in-line facilities.

Irrespective of which type of storage facility is utilized, the designer is encouraged to create an attractive, multipurpose facility that is readily maintainable and safe for the public, under different weather condition (dry or wet). Designers are also encouraged to consult with other specialists such as urban planners, landscape architects, and biologists during planning and design.

6.7.1 Sizing of Storage Volumes

The reservoir routing procedure is usually employed for the sizing of detention storage volumes. This method is more complex and time-consuming than the use of empirical equations or the rational method. Its use requires the designer to develop an inflow hydrograph to the facility. This is an iterative procedure that is described as follows (Guo, 1999):

1. *Site selection*: The detention facility's location should be based on criteria developed for the specific project. Regional storage facilities are normally placed where they provide the greatest overall benefit. The location, geometry, and nature of these facilities are determined regarding multiuse objectives such as the use of the detention facility as a park or for open space, preserving or providing wetlands and/or wildlife habitat, or other uses and community needs.
2. *Hydrological determination*: The storage basin inflow hydrograph and the allowable peak discharge from the basin for the design-storm events should be determined. The allowable peak discharge is limited by the local criteria or by the requirements spelled out in the available master plans.
3. *Initial sizing*: Based on the inflow hydrograph and basin characteristics, the initial size of the detention storage volume is estimated.
4. *Initial configuration of the facility*: The initial configuration of the facility should be based on site constraints and other goals for its use. Then a stage–storage–discharge relationship is developed for the facility.
5. *Outlets design*: The initial design of the outlets entails balancing the initial geometry of the facility against the allowable release rates and available volumes for each stage of the hydrological control. This step requires the sizing of outlet elements such as a perforated plate for controlling the releases, orifices, weirs, the outlet pipe, spillways, and so on.

6. *Preliminary design*: The results of steps 3, 4, and 5 provide a preliminary design of the overall detention storage facility. In the preliminary design step, which is an iterative procedure, the size and shape of the basin and the outlet works design are checked using a reservoir routing procedure and then modified based on the design goals. The modified design is again checked and further modifications are made if needed. At the end of this step, the storage volume and nature and sizes of the outlet works are finalized.

7. *Final design*: The final design phase of the storage facility is completed after the hydraulic design has been finalized. This phase includes structural design of the outlet structure, embankment design, site grading, a vegetation plan, accounting for public safety, spillway sizing, assessment of dam safety issues, and so on.

6.8 RISK ISSUES IN URBAN DRAINAGE

There are three typical urban drainage risk issues suggested by Hauger et al. (2005), including (1) flooding of urban drainage systems, (2) dissolved oxygen (DO) depletion in streams, and (3) discharge of chemicals to receiving waters. Table 6.13 shows these key issues.

6.8.1 FLOODING OF URBAN DRAINAGE SYSTEMS

The primary concern when sewers are flooded in urban areas is economic losses including damage to buildings, damaged infrastructure, delayed traffic, and so on. Injuries and infections are of secondary importance in comparison with damage and other economic losses of flood in urban areas. The main source of risk is stormwater, which is the consequence of high-intensity precipitation. The total impact cannot be evaluated easily, because there are associated uncertainties. It is also difficult to get insurance information about the extent of damage.

Traditional indicators of risk are directed toward the hazard. For many years, IDF curves were used as indicators because of the extended flexibility in utilizing them (Marsalek et al., 1993). These days, with computational power and sophisticated simulation models, more elaborate indicators have been introduced such as for design-storm hyetographs and historical rainfall series (DWPC, 2005). Water level is another widely used indicator obtained from the simulation models. The advantage of water level is that it is highly correlated with damages. While water stays below street level, the damage is limited but when the water level rises, the economic loss increases rapidly. Water level is also used as the accepted service level of the system, and a return period for exceeding a given water level is defined.

Water level and rain intensity are both indicators of the hazards. There are certain advantages with respect to risk reduction when indicators of the vulnerability are used. Design, building material, and travel access are all indicators of the vulnerability and are therefore more suitable for damage reduction. Detention basins and ponds are interventions that establish a barrier to keep a part of the storm or wastewater away from the receiving water.

In many European cities, sewer systems are built 50 years ago or more and it is time to rehabilitate and upgrade them (Hauger et al., 2005). Climate change also adds to the need to upgrade the systems that have significant economical consequences. Evaluation of climate change effects on precipitation reveals different changes of behavior, which are different from region to region (see also Ashley et al., 2004; Grum et al., 2005).

Generally speaking, the pipe diameters have to be enlarged and more storage is needed, which involves major capital investment. This follows what is called "hazard strategy." But the urban planners are more geared toward what is called "vulnerability strategy." It manages the increased volumes of stormwater by determining the areas with less flooding damages and then allowing these areas to be flooded in a controlled manner. This strategy is similar to a traditional strategy used in managing fluvial flooding of floodplains.

TABLE 6.13
Three Typical Risks in Urban Drainage with Respect to Indicators and Intervention Options

	Pluvial Flooding of Urban Drainage Systems	Sewage Discharge into a Stream (DO Depletion in a Stream)	Discharge of Chemicals to Receiving Water
Primary concern/consequence	Economic loss (building interface structure, productive time)	Loss or degradation of habitat	Degradation of environmental quality Loss of recreational value
Risk source	Precipitation, high-intensity rain events	Organic matter in unwanted stormwater and wastewater	Chemicals that are dispersed after use
Indicator	IDF curves Return period for exceeding a specified water (service) level	Number of overflows Overflow volume Overflow duration	Inherent properties of chemicals Predicted environmental concentrations (PEC)
Risk object	Building, interface structure	Aquatic life in a stream, a stream as a recreational area	Receiving water, aquatic ecosystem
Indicator	Building design Building materials Location of the main interface structure routes	Minimum oxygen concentration DO demand curves as a function of duration and return period	Predicted number of effect concentrations (PNEC) Fauna index
Analysis	Simulated water level for a rain event with a defined return period Model simulation with historical series	Long rain series used to calculate minimum oxygen concentration as a function of return period and duration	Calculate PEC and comparison with PNEC
Definition of acceptance	Return period for exceeding a given water level must be higher than a defined value	Demand for daily and 1-year minimum oxygen concentration interpolated to a curve showing oxygen concentration as a function of return period	PEC (multiplied with an appropriate safety factor) must be lower than PNEC Fauna index must live up to a target classification
Risk reduction options for hazards	Expected for surveillance and forecasting, improved early warning of hazards that cannot be reduced	Decentralized cleaning before discharge, keeping the first and most polluted part of water (first flush) away from streams	Reduce usage of chemicals, substitute hazardous chemicals with harmless ones
Vulnerability	Adjusted construction methods Use of robust materials Design (no cellar, no toilets in cellar) and space management point out the areas that are allowed to be flooded (urban flood plans)	Retrofit the stream to increase robustness Increase reaeration of the stream	Improve dilution rate Contaminated sediment management (cleanup)
Barrier	Prevent sewer from flooding (increase capacity and storage; source control; keep stormwater out of the sewer system; infiltration reuse)	Keep wastewater and stormwater away from the stream (basins, infiltration, and increased capacity of the pipe system)	Treatment before discharge; source control; keeping water commissioning hazardous chemicals away from recipient

6.8.2 DO Depletion in Streams Due to Discharge of Combined Sewage

Often overflow structures are provided in combined sewer systems to minimize flooding inside the urban area. This solution causes other problems when a mixture of stormwater and wastewater is discharged directly without any treatment into receiving waters. The risk source is the organic matter and other compounds present in the storm or wastewater that cause water-quality changes in the risk object (the ecosystem). Changes can be decreased in the complexity of the food web and substitution of sensitive species with more tolerant ones, and a reduction of the recreational function of the area.

Hazard indicators such as the overflow frequency, overflow volume, and overflow duration are widely used (Marsalek et al., 1993), but the concentration of easily degradable organic matter in overflowing water is rarely used. Classification of water quality using fauna indices or the DO concentration is used as indicators for vulnerability. DO concentration has been used in the definition of the requirements on water quality in Denmark for many years (DWPC, 1985). This principle is similar to the Predicted No Effect Concept (PNEC) that is widely used in chemical risk assessment. With respect to risk reduction, the situation is opposite to the previous example. The risk object is a natural system, which as per tradition shall remain as undisturbed as possible. Unless the risk object is heavily modified by urban activity, the risk reduction potential in changing the physical conditions of the risk object is very limited. The risk reduction potential lies in reducing the risk source, and the possibilities are either avoiding that the combined sewage reaches the stream or reducing the pollution level in the combined sewage. Retrofitting a water body toward the original state results in much better conditions including increased possibilities for breeding in the stream and that the stream can receive a larger volume of stormwater considering the consequences related to water quality. Part of the retrofitting can be increased by natural reaeration, which would increase the potential for degradation of organic substances in the stream (Hauger et al., 2005).

6.8.3 Discharge of Chemicals

The discharge of chemicals into receiving waters is a major risk to human and aquatic ecosystems. They end up in the receiving waters via wastewater and surface runoff (Chocat and Desbordes, 2004). Indicators for the chemicals' hazard include toxicity, carcinogenity, and bioaccumulation as Predicted Environmental Concentrations (PEC).

Vulnerability is used as PNEC. Also the indicators PEC and PNEC define the acceptance level, which is a requirement that PEC must be lower than PNEC. A safety factor is used in estimating these factors. The major options include efficient treatment before discharge, reduction of usage of hazardous chemicals or less harmful substitutes, and controlling the fate of chemicals in receiving waters such as increasing the dilution rate or contaminated sediments the cleanup (Hauger et al., 2005).

6.9 URBAN FLOODS

Through the development of new urban areas, infiltration reductions and water retention characteristics of the natural land should be considered. The increasing quantity as well as the short duration of drainage produces high flood peaks that exceed the conveyed secure capacities of urban streams. The total economic and social damages in the urban areas tend to make significant impacts on the national economies as flood frequencies increase. Recent events in North America have been devastating in urban areas such as New Orleans.

With uncontrolled urbanization in the floodplains, a sequence of small flood events can be managed, but with higher flood levels, damage increases and the municipalities' administrations have to invest in population relief. Structural solutions have higher costs, but better chances of being implemented when damages are greater than their development or due to intangible social aspects. Nonstructural measures have lower costs, but have less chances of being implemented because they are not politically attractive in developing regions (Tucci, 1991).

6.9.1 Urban Flood Control Principles

Certain principles of urban drainage and flood control are proposed by Tucci (1991) as follows:

- Promoting an urban drainage master plan.
- Involvement of the whole basin in evaluating flood control.
- Any city developments should be within urban drainage control plans.
- Priority should be given to source control and keeping the flood control measures away from downstream reaches.
- The impact caused by urban surface washoff related to urban drainage water quality should be reduced.
- Improvement of nonstructural measures for floodplain control such as real-time flood forecasting, flood zoning, and flood insurance.
- Public participation in urban drainage management.
- Consideration of full recovery investments.

In developing countries, a set principle for urban drainage practices is not established due to fast and unpredictable developments (Dunne, 1986). Furthermore, the local regulations are neglected in the urbanization of preurban areas such as unregulated developments and invasion of public areas.

In many developing urban areas, the above principles are not incorporated because of the following:

- Lack of sufficient funds.
- Lack of appropriate garbage collection and disposal that decreases the water quality and the capacity of the urban drainage network due to filling.
- There is no preventive program for risk area occupation.
- Lack of adequate knowledge on how to deal with floods.
- Lack of institutional organization in urban drainage at a municipal level.

6.10 SANITATION

According to Ujang and Henze (2006), only 5% of global wastewater, mainly in developed countries, is properly treated. As a result, the majority of the world's population is still struggling with waterborne diseases resulting from the lack of treatment facilities or substandard design and operation. Therefore, the quality of water resources has been rapidly degraded, particularly in poor developing countries. At the beginning of the twenty-first century, about 1.1 billion people lived without access to clean water, 2.4 billion without appropriate sanitation, and 4 billion without sound wastewater disposal.

In recent years, the "conventional sanitation" approach has received widespread criticism and the so-called sustainable sanitation has been proposed. Sustainable sanitation is intended to achieve the MDG by 2015 by providing water supply and adequate sanitation for developing countries. Sustainable sanitation is flexible and can be applied in any community—poor or rich, urban or rural, water-rich or water-poor countries—and requires lower investment costs compared to conventional sanitation approaches. The implementation of sustainable sanitation is much easier in developing countries where water infrastructures are still at the developing stages. Ujang and Henze (2006) stated that in some developing countries, there are no available public facilities, and so it is an excellent opportunity to start a new infrastructure with a new framework.

6.11 WASTEWATER MANAGEMENT

It has been estimated that it will cost $16 billion a year until 2015 to meet the regional MDG targets for water supply and sanitation to cover all the unserved population. This is a significant number, but not an impossible figure. The Millennium Development Task Force defines basic sustainable sanitation for securing access to safe, hygienic, and convenient facilities and services for sludge disposal. At the same time it ensures a clean and healthy living environment both at home and in the neighborhood. This will allow low-cost options that can be adopted while governments build their full sewerage.

The specific technologies that can be tailor-made to a specific urban and rural setting vary from place to place. As stated by Lohani (2005), in dispersed, low-income rural areas, the appropriate technology may be a simple pit latrine, whereas in a congested urban area with a reliable water service, it may be a low-cost sewerage system.

There are some advantages to centralized schemes but decentralized wastewater treatment schemes are more sustainable. The latter is more suitable for achieving sustainability when compared with the centralized technology. It makes very small and community-scale technologies are more easily achievable. Its application in remote and rural areas needs an improved technology with integration with other technology sectors (energy and food production) to improve overall sustainability.

There is a high potential in the use of on-site, small, and community-scale technology in developing countries for achieving sustainability outcomes. It is likely to succeed if the technology is modified to be of lower cost and adaptable to local climates based on the same principles and engineering (Ho, 2005). The technology and technology choice are the key issue, even though technology by itself does not provide the full answer. Even though the classical wastewater management concept has been applied for many years in populated areas of developing countries, there are some disadvantages in utilizing small-scale systems (Ho, 2004).

Decentralized wastewater management systems, with the wastewater treated close to where it is generated, are being considered by various institutions including the World Bank as an alternative to the traditional centralized system. However, the degree of technological sophistication that should be applied is in question. The development and application of high-tech on-site treatment plants designed and fabricated by modern industrial techniques can be a solution. When mass-produced, the costs of manufacturing such package plants can be kept at a relatively low level. The plant should produce an effluent that is hygienically safe and that can subsequently be utilized for toilet flushing, washing clothes, cleaning floors, or watering lawns (Wilderer, 2004).

6.11.1 Technologies for Developing Countries

Many cities in developing countries are facing surface water and groundwater pollution problems. This deterioration of water resources needs to be controlled through effective and feasible concepts of urban water management. The Dublin Principles, Agenda21, Vision21, and the MDG provide the basis for the development of innovative, holistic, and sustainable approaches. Highly efficient technologies are available; the inclusion of these into a well-thought-out and systematic approach is critical for the sustainable management of nutrient flows and other pollutants into and out of cities. Based on cleaner production principles, there are three steps:

- Minimizing wastewater generation by drastically reducing water consumption and waste generation.
- The treatment and optimal reuse of nutrients and water at the smallest possible level, such as at the on-plot and community levels. Treatment technologies recommended make the best use of side products via reuse.
- Enhancing the self-purification capacity of receiving water bodies (lakes, rivers, etc.), through intervention.

The success of this so-called three-step strategic approach requires systematic implementation by providing specific solutions to given situations. This, in turn, requires appropriate planning and legal and institutional responses.

A sanitation system that provides Ecological Sanitation (EcoSan) is a cycle—a sustainable, closed-loop system, which closes the gap between sanitation and agriculture. The EcoSan approach is resource-minded and represents a holistic concept toward ecologically and economically sound sanitation. The underlying aim is to close (local) nutrient and water cycles with as less expenditure on material and energy as possible to contribute to a sustainable development. EcoSan is a systemic approach and an attitude; single technologies are only means to an end and may range from near-natural wastewater treatment techniques to compost toilets, simple household installations to complex, mainly decentralized systems. They are picked from the whole range of available conventional, modern, and traditional technical options, combining them to EcoSan systems (Langergraber and Muellegger, 2005).

Whether the classical system of urban water supply and sanitation is appropriate to satisfy the needs of the developing world and whether this system meets the general criteria of sustainability is questionable. Recovery and reintroduction of valuable substances, including water, into the urban cycle of materials are impossible because of mixing and dilution effects inherent in the system. Decentralized water and wastewater management should be seriously taken into account as an alternative.

Source separation of specific fractions of domestic and industrial wastewater, separate treatment of these fractions, and recovery of water and raw materials including fertilizer and energy are the main characteristics of modern high-tech on-site treatment/reuse systems. Mass production of the key components of the system could reduce the costs of the treatment units to a reasonable level.

On-site units can be installed independently of the development stage of the urban sewer system. In conjunction with building new housing complexes, a stepwise improvement of the hygienic sanitation in urban and preurban areas could be achieved. Remote control of the satellite systems using modern telecommunication methods would allow reliable operation and comfort for the users. Intensive research is required, however, to develop this system and bring it to a standard allowing efficient application worldwide (Wilderer, 2004).

Often the most appropriate technical solution to wastewater collection is simplified sewerage small-diameter sewers laid in-block at fairly flat gradients. In many situations, in developing countries, more appropriate technologies will have to be used. These include waste stabilization ponds, up-flow anaerobic sludge blanket reactors, wastewater storage, and treatment reservoirs (or some combination of these). High-quality effluents can be produced that are especially suitable for crop irrigation and fish culture. In the twenty-first century, "from sewage to soya" will have to become more common as the world's population increases (Mara, 2001).

Developing countries occupy regions where the climate is warm most of the time. Even in subtropical areas, low temperatures do not persist for long periods. This is the main factor that makes the use of anaerobic technology applicable and less expensive, even for the treatment of low-strength industrial wastewaters and domestic sewage. Emphasis should be given to domestic sewage treatment and to the use of compact systems in which sequential batch reactors (SBR) or dissolved-air flotation (DAF) systems are applied for the posttreatment of anaerobic reactor effluents.

Experiments on bench- and pilot-plants have indicated that these systems can achieve high performance in removing organic matter and nutrients during the treatment of domestic sewage at ambient temperatures (Foresti, 2001). China is facing serious water pollution problems. Only 15% of the municipal wastewater is appropriately treated each year. A lack of funds is one major obstacle facing the construction of wastewater treatment plants. Conventional treatment technology such as the activated sludge process has both relatively high capital and operational costs. An alternative to this dilemma is to develop and apply the technologies with lower costs or better performance. Four technologies, including natural purification systems, highly efficient anaerobic processes, advanced biofilm reactors, and membrane bioreactors are promising (Qian, 2000).

The up-flow anaerobic sludge blanket (UASB) or the anaerobic baffled reactor (ABR) as an anaerobic pretreatment system, and the reed bed or the stabilization pond with supporting media as a posttreatment system, may also be feasible. Results obtained in pilot- and full-scale treatment plants clearly reveal that the anaerobic treatment is indeed a very attractive option for municipal wastewater pretreatment at temperatures exceeding 20°C in tropical and subtropical regions. The UASB system has been commonly employed as an anaerobic pretreatment system.

The ABR provides another potential for the anaerobic pretreatment. The effluents from the anaerobic treatment system should be posttreated to meet discharge standards. Because of the advantages of the reed bed system when it is employed for tertiary treatment, this system can be considered as a posttreatment system. Another cost-effective system, the stabilization pond packed with attached-growth media, is also a potential posttreatment system (Yu et al., 1997).

6.11.2 Wetlands as a Solution for Developing Countries

In developing countries, the effective use of natural and artificial wetlands for water purification is particularly valuable and exploitable for the protection of water quality in catchments, rivers, and lakes. Constructed wetlands are potentially good, low-cost, appropriate technological treatment systems for domestic wastewater in rural areas. They can be integrated into agricultural and fish production systems where the products are useable and/or recycled for optimal efficiency.

However, currently, constructed wetlands are rarely installed. The reasons for this are discussed, drawing attention to the limitations of aid programs from donor countries and the need for in-house research, training and development. A cost-effective environmentally friendly municipal wastewater treatment system using constructed wetlands was developed. The wetland was monitored for 12 months to understand the removal efficiencies of various pollutants present in the municipal wastewater during different seasons in a year. Mathematical models to predict the behavior of the system were also developed. The results demonstrate that the system performs well and can be adopted in upcoming cities in developing countries where sufficient land is cheaply available (Jayakumar and Dandigi, 2003).

A low-tech variation of the system using wetlands for wastewater treatment for developing countries consisting of conversion of existing drains into several kilometers of gravel and redlined channels resulted in the conclusion that it can treat wastewater to high enough standards to enable its safe reuse in agriculture. The gravel is found to filter the water while the reeds act like a constructed wetland to absorb and break down organic pollutants into elemental compounds.

6.11.3 Example: Caribbean Wastewater Treatment

Although wastewater treatment in Caribbean countries is not well documented, the available information suggests that the development of the wastewater treatment around the world is located in different levels. In Havana Cuba, for example, there are no centralized wastewater treatment plants. The method used is to dispose of the wastewater of communities primarily through septic tanks. In the Dominican Republic also, there is poor wastewater disposal. The rural areas use latrines to dispose of the wastewater. In poor areas, the waste is directly deposited in rivers, thereby contributing to disease. In the urban areas of the country, the wastewater goes to septic tanks, which filter the solid wastes and let the water go to the ground.

The house owner is then responsible for emptying the collected waste, which can be a costly and difficult task. While there is an effort to plan for a centralized wastewater treatment plant, currently the treatment by individual house owners and businesses is sporadic. There are, however, some 300 oxidation ditch treatment facilities on the island, and the City of Havana, which receives its water from the Almandares Watershed, has efforts under way to manage and get public participation for the improvement of the watershed. In Haiti, no sanitary sewers exist. There are some pit latrines used, but these tend to be in the crowded urban areas or among the rich (Diaz and Barkdoll, 2006).

There are some septic tanks as well for waste collection. Some institutions have their own septic tanks for their office buildings, but maintenance and disposal are somewhat lacking. In Jamaica, despite its image as a paradise, wastewater treatment and disposal is lacking as well. Even in the few existing treatment plants, operation is not diligent, thereby causing effluent standards to remain unmeet. Again, poverty is a prime cause of the substandard treatment. A high tourism load, while bringing in some income, also puts a burden on the treatment infrastructure (Diaz and Barkdoll, 2006).

6.11.4 EXAMPLE: THE LODZ COMBINED SEWERAGE SYSTEM

6.11.4.1 Development and Implementation of a Sewerage System

The idea of sewerage implementation had to wait till around 1920 when the Lodz in Poland authorities decided to use an old plan from 1909 called "Lindley's plan." So, with a lot of enthusiasm, adequate construction plans were made and work started in 1925.

Until 1939, thanks to the excellent coordination maintained by Eng. Stefan Skrzywan, the main trunk and several side sewers were finished. Also, regulation work on the main segments of urban rivers was performed. The high quality of the construction and materials used allowed for further proper maintenance of sewers and their equipment for many years. The proposed treatment plant was not finished and only the mechanical stage was constructed before 1939 (Zawilski, 2005).

After World War II, the development of the city sewer system was continued, however, without a clear concept. First of all, the combined system was no longer applied and therefore some fragments of Lindley's system were not finished. It is not clear what reason prevailed but enforcement of the combined system might have stemmed from difficult economic conditions and tendencies to build sanitary sewers first. A common practice became the connection of sanitary sewers to the older combined sewerage. The combined system was not promoted apart from some provisional regulations of combined sewer overflow (CSO) weir levels.

Recently, a more modern wastewater treatment plant has been included in the combined system. The capacity of the total system is still sufficient; however, on some areas, urban floods are to be observed more and more frequently. It occurs due to an increasing number of areas of sealed surfaces and due to some bottleneck segments overloaded by improperly organized connections of storm sewers. Since 1999, a modern monitoring of CSOs has been available, and thus information about the spill frequency has been collected. As it turned out, most of the monitored CSOs are activated too frequently (up to 30–40 spill events per year), causing serious pollution of urban rivers and especially of the main receiver city area. Therefore, a plan for the combined system renovation was elaborated in 2003 (Zawilski, 2004).

6.11.4.2 Upgrading the Old Sewerage System

The decision taken in the past, almost a century ago appeared to be essential for the future modernization of the sewerage systems, which have to meet new European water law regulations. The main problems that appeared during the preparation of the modernization plan are summarized in Table 6.14.

Some technical and nontechnical problems often make implementation of modernization plans difficult or impossible. For Lodz, introduction of storage tanks close to each CSO had been established. In Figures 6.28 and 6.29, an example of a CSO upgrading is presented. A storage facility and a separator of screening have been proposed. The rest of the free land was utilized for both facilities (Zawilski, 2005).

There are no such favorable circumstances in the case of other investment proposals. In general, optimal solutions from the theoretical viewpoint turn out to be impossible to be applied in local conditions. Therefore, quasi-optimal solutions have to be implemented instead. Further optimization should be continued following an assessment of the enlarged storage capacity and using a possible introduction of RTC (Zawilski, 2005).

TABLE 6.14
Problems Detected in the Process of Upgrading the Old Lodz Sewerage System

Idea for Sewerage Upgrading	Problems	Remarks
Storage tanks for reducing CSOs	No space inside the urban infrastructure	Lack of free space or complicated property relationships; collisions with the existing pipelines and planned roads
Storage tanks for reducing CSOs	Deep sewerage causes deep storage tanks and emptying pumping stations	An alternative solution with shallow tanks requires high pumping capacity
Evaluating and improving the CSO hydraulics	Nontypical construction of CSO chambers	Some CSOs have very long weirs and can exchange water with nearby riverbeds
Detention of stormwater inside sewers	Unfavorable level conditions for the idea to be applied	RTC is necessary to be implemented
Separation of screenings from overflowing sewage	No space on overflow sewers; difficult construction conditions in river valleys	An alternative solution with facilities inside CSO chambers is even more difficult for implementation due to the shape of overflow weirs
Disconnection of runoff from the combined system	No space for new storm sewers in streets; unfavorable level conditions for outlets into urban rivers	Also existing house connections were built into the combined system
Retention of stormwater through infiltration into the ground	Variable and not fully recognized local geological conditions	No distinct favorable effect in dense builtup areas

Source: Zawilski, M. 2005. *Proceedings of 10th International Conference on Urban Drainage*, Copenhagen, Denmark, pp. 21–26.

6.12 StormNET: STORMWATER AND WASTEWATER MODELING

StormNET is one of the most advanced, powerful, and comprehensive stormwater and wastewater modeling package available for analyzing and designing urban drainage systems, stormwater sewers,

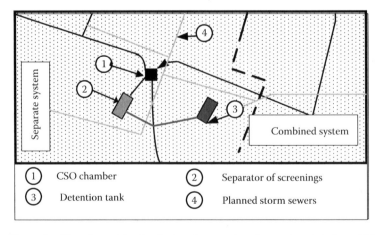

FIGURE 6.28 Example of inserting a detention tank and a separator of screenings into the existing infrastructure of an urban area in Lodz. (From Zawilski, M. 2005. *Proceedings of 10th International Conference on Urban Drainage*, Copenhagen, Denmark, pp. 21–26.)

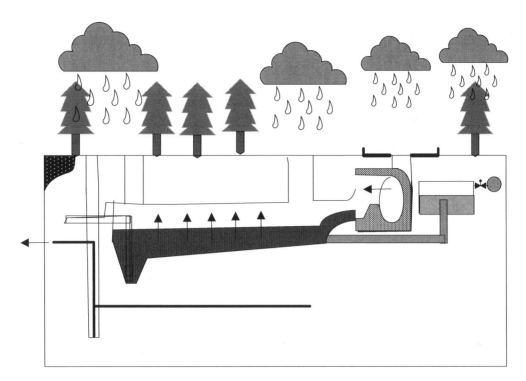

FIGURE 6.29 Scheme of a designed detention tank cooperating with the existing combined sewer. (From Zawilski, M. 2005. *Proceedings of 10th International Conference on Urban Drainage*, Copenhagen, Denmark, pp. 21–26.)

and sanitary sewers. StormNET is the only model that combines complex hydrology, hydraulics, and water quality in a completely graphical, easy-to-use interface.

Graphical symbols are used to represent network elements such as manholes, pipes, pumps, weirs, ditches, channels, and detention ponds. The graphical representation of the model can be the output at any drawing scale. Pipes can be curvilinear and lengths automatically computed. Scanned TIFF aerial orthophoto images and maps; Mr. SID high-resolution orthophotos; ArcGIS; AutoCAD; and MicroStation files of streets, parcels, and buildings can be imported and displayed as a background image.

StormNET provides a variety of modeling elements to select from

- Watershed sub-basins
- Inlets and catchments
- Detention ponds
- Complex outlet structures
- Standpipes, weirs, and orifices
- Stormwater and wastewater sewers
- Pumps and lift stations
- Manholes and junctions
- Rivers, streams, and ditches
- Culverts and bridges

StormNET is a link-node-based model that performs hydrology, hydraulic, and water-quality analysis of stormwater and wastewater drainage systems, including sewage treatment plants and water-quality control devices. A link represents a hydraulic element (i.e., a pipe, channel, pump, standpipe, culvert, or weir) that transports flow and constituents. There are numerous different link element types supported by StormNET. A node can represent the junction of two or more links, the location of a flow or pollutant input into the system, or as a storage element such as a detention pond, settling pond, or lake.

6.12.1 Hydrology Modeling Capabilities

StormNET accounts for various hydrological processes that produce runoff from urban areas, including

- Time-varying rainfall
- Evaporation of standing surface water
- Snow accumulation and melting
- Rainfall interception from depression storage
- Infiltration of rainfall into unsaturated soil layers
- Percolation of infiltrated water into groundwater layers
- Interflow between groundwater and the drainage system
- Nonlinear reservoir routing of overland flow

Spatial variability in all of these processes is achieved by dividing a study area into a collection of smaller, homogeneous sub-basin areas, each containing its own fraction of pervious and impervious subareas. Overland flow can be routed between subareas, between sub-basins, or between entry points of a drainage system.

6.12.2 Hydraulic Modeling Capabilities

StormNET also contains a flexible set of hydraulic modeling capabilities used to route runoff and external inflows through the drainage system network of pipes, channels, storage/treatment units, and diversion structures. These include the ability to

- Handle networks of unlimited size.
- Use a wide variety of standard closed and open conduit shapes as well as natural channels.
- Model special elements such as storage/treatment units, flow dividers, pumps, weirs, and orifices.
- Apply external flows and water-quality inputs from surface runoff, groundwater interflow, rainfall-dependent infiltration/inflow, dry-weather sanitary flow, and user-defined inflows.
- Utilize either kinematic wave or full hydrodynamic wave flow routing methods.
- Model various flow regimes, such as backwater, surcharging, reverse flow, and surface ponding.
- Apply user-defined dynamic control rules to simulate the operation of pumps, orifice openings, and weir crest levels.

6.12.3 Detention Pond Modeling Capabilities

StormNET enables accurate routing in complex detention pond situations. In some situations, downstream conditions can cause backwater effects that influence the performance of a detention pond outlet structure. For example, an upstream pond may discharge to another downstream pond that is similar in elevation or influenced by downstream flooding. Such situations can result in a decrease in outlet discharges or flow reversal back into the pond and can be difficult to model properly. Most approaches attempt to simplify the problem using overly conservative assumptions about the downstream water surface conditions that result in oversized detention facilities and increased costs. Still other methods simply ignore the downstream effects, thereby causing overtopping of the resulting undersized ponds. However, StormNET's interconnected pond routing allows you to easily model these complex situations with confidence.

6.12.4 Water-Quality Modeling Capabilities

StormNET can also compute the runoff and routing of pollutants. The following processes can be modeled for any number of user-defined water-quality constituents:

- Dry-weather pollutant buildup over different land uses
- Pollutant washoff from specific land uses during storm events
- Direct contribution of rainfall deposition
- Reduction in dry-weather buildup due to street cleaning
- Reduction in washoff load due to BMPs
- Entry of dry-weather sanitary flows and user-specified external inflows at any point in the drainage system
- Routing of water-quality constituents through the drainage system
- Reduction in constituent concentration through treatment in storage units or by natural processes in pipes and channels

6.12.5 Typical Applications of StormNET

StormNET's analysis engine has been used in thousands of sewer and stormwater studies throughout the world. Typical applications include

- Design and sizing of drainage system components for flood control
- Design and sizing of detention facilities for flood control and water-quality protection
- Flood plain mapping of natural channel systems
- Designing control strategies for minimizing CSO
- Evaluating the impact of inflow and infiltration on sanitary sewer overflows (SSOs)
- Generating nonpoint source pollutant loadings for waste load allocation studies
- Evaluating the effectiveness of BMPs in reducing wet-weather pollutant loadings

6.12.6 Program Output

The output summary report of StormNET contains the following sections:

1. Listing of any error conditions encountered during the simulation
2. Summary listing of input data, if requested
3. Summary of data read from each rainfall file used during the simulation
4. Description of each control rule action taken during the analysis, if requested
5. System-wide reporting of mass continuity errors for
   - Runoff quantity and quality
   - Groundwater flow
   - Conveyance system flow and water quality
6. Sub-basin runoff summary table that lists the following for each sub-basin:
   - Total precipitation
   - Total run-on from other sub-basins
   - Total evaporation
   - Total infiltration
   - Total runoff
   - Runoff coefficient (ratio of total runoff to total precipitation)

7. Node depth summary table that lists the following for each node:
   - Average water depth
   - Maximum water depth
   - Maximum hydraulic head (HGL) elevation
   - Elapsed time when the maximum water depth occurred
   - Total volume of flooding
   - Total minutes flooded
8. Conduit flow summary table that lists the following for each conduit:
   - Maximum flow rate
   - Elapsed time when the maximum flow rate occurred
   - Maximum flow velocity
   - Ratio of conduit equivalent length used in the simulation to the conduit's defined original length
   - Ratio of the maximum flow to the conduit's design flow (i.e., full normal flow)
   - Conduit design flow (i.e., full normal flow)
   - Total minutes that the conduit was surcharged (either flowed full or had flow greater than the full normal flow)
9. Flow classification summary table (for dynamic wave routing only) that lists the following for each conduit:
   - Fraction of time that the conduit was in each of the following flow categories:
     – Dry on both ends
     – Dry on upstream end
     – Dry on downstream end
     – Subcritical flow
     – Supercritical flow
     – Critical flow at upstream end
     – Critical flow at downstream end
   - Average Froude number of the flow
   - Average change in flow over all time steps
10. Element IDs of the five nodes with the highest individual flow continuity errors
11. Element IDs of the five nodes or conduits that most often determined the size of the time step used for the flow routing
12. Range of routing time steps taken and the percentage of these that were considered steady state

PROBLEMS

1. A sewer has to be laid in a place where the ground has a slope of 0.002. If the present and ultimate peak sewage discharges are 40 and 165 L/s, respectively, design the sewer section.
2. The layout of a sanitary sewer system is as shown in Figure 6.30. Data on area, length, and elevations are given in Table 6.15. The present population density, 100 persons/ha, is expected to increase to 250 persons/ha by conversation of the dwellings to apartments. The peak rate of sewage flow is 1600 Lpd per person. Design the sewer system.
3. Storm sewers need to be installed in a new development. Four inlets are proposed with pipes running from inlet A to B, then to C, and finally to D. Data associated with each pipe and inlet are listed below. The size all of the pipes, assuming that rainfall intensity can be described by the equation $i = 150/(t_d + 20)$. Use the following standard pipe sizes: 38.1 cm (minimum), 45.72 cm, 53.34 cm, 60.96 cm, 76.2 cm, 91.44 cm, and so on.

   Inlet A: drainage area = 12,000 m², $t_0 = 10$ min, $C = 0.30$
   Inlet B: drainage area = 16,000 m², $t_0 = 12$ min, $C = 0.40$

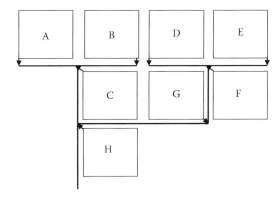

FIGURE 6.30 Layout of the sanitary sewer system of Problem 2.

TABLE 6.15
Characteristics of the Drainage Area of the Sanitary System of Problem 2

			Elevation (m)	
Block	Area (m^2)	Length (m)	Upstream	Downstream
A	8000	118.87	30.94	29.62
B	10,000	106.68	30.68	29.62
C	6000	100.58	29.62	28.43
D	5200	70.10	30.08	29.73
E	4800	89.92	30.63	29.73
F	8400	91.44	29.73	28.74
G	22,800	198.12	28.74	28.43
H	14,000	167.64	28.43	26.34

Inlet A: drainage area = 8000 m^2, t_0 = 10 min, C = 0.50
Inlet A: drainage area = 8000 m^2, t_0 = 10 min, C = 0.30
Pipe AB: length = 100 m, slope = 0.01, n = 0.013
Pipe BC: length = 135 m, slope = 0.02, n = 0.013
Pipe CD: length = 100 m, slope = 0.07, n = 0.013
Pipe DE: length = 165 m, slope = 0.01, n = 0.013

In your computations, take into consideration the various flow paths that lead to the determination of the time of the concentration. For instance, in determining the time of concentration for pipe BC, one flow path is from local inflow to inlet B and the other flow path is the local inflow time to inlet A plus the flow time through pipe AB. The entire drainage area only contributes to flow for the longest time of concentration.

4. Does a storm sewer pass its greatest flow rate when it is just flowing full? Explain. Determine the spacing to the first and second inlets that will drain a section of highway pavement if the runoff coefficient is 0.9 and the design rainfall is 15 cm/h. The pavement width is 10 m (S_X = 0.02, S_L = 0.02, and n = 0.016), the efficiency of inlet is 0.35, and the allowable spread is 2 m.
5. Using the Muskingum method, route the inflow hydrograph given in Table 6.16, assuming (a) K = 4 h and X = 0.12 and (b) K = 4 h and X = 0.0. Plot the inflow and outflow hydrographs for each case, assuming the initial outflow equals the inflow.
6. A storm event occurred in a watershed that produced a rainfall pattern of 5 cm/h for the first 10 min, 10 cm/h in the second 10 min, and 5 cm/h in the next 10 min. The watershed

TABLE 6.16
Inflow Hydrograph of Problem 6

Time (h)	0	2	4	6	8	10	12	14	16	18
Inflow (L/s)	0	135	675	1350	945	567	351	203	68	0

is divided into three sub-basins (Figure 6.31) with the 10-min UHs as specified in the Table 6.17. Sub-basins A and B has a loss rate of 2.5 cm/h for the first 10 min and 1.0 cm/h afterward. Sub-basin C has a loss rate of 1 cm/h for the first 10 min and 0.0 cm/h thereafter. Determine the storm hydrograph at point 2. Assume a lag time of 20 min.

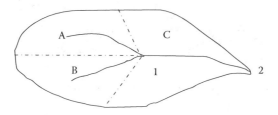

FIGURE 6.31 Schematic of watershed of Problem 7.

TABLE 6.17
Inflow Hydrographs of Problem 7

Time (min)		0	10	20	30	40	50	60	70	80	90	100
Inflow (L/s)	A	0	40	70	110	75	50	30	20	10	5	0
	B	0	30	65	100	75	50	35	20	7	3	0
	C	0	50	80	105	90	75	50	35	20	10	0

7. Determine the flow in a composite gutter with $W = 0.5$ m, $S = 0.015$, $S_X = 0.03$, and $a = 0.08$ m. Assume that T is equal to 2 m and the Manning coefficient is 0.01.
8. Consider a impervious area with the length of 40 m, slope of 0.03. The effective Manning coefficient is 0.03. Determine the runoff hydrograph for a constant rainfall excess rate of 20 mm/h with 10 min duration at the exit point of the area.
9. Determine the peak discharge resulting from a 2-h rainfall with a constant intensity equal to 2.5×10^{-5} mm/s that happens on an impervious rectangular basin with the following characteristics: $L = 200$ m, $S = 0.005$, and $n = 0.04$. The infiltration parameters are $\Phi = 0.5$, $K = 3 \times 10^{-6}$ m/s, and $P_f = 0.25$ m. At the beginning of the rainfall, the basin soil is 0.50% saturated.
10. Determine the flow depth and spread of a triangular gutter with $S = 0.02$, $S_X = 0.03$, and $n = 0.02$ where the peak flow is estimated as 0.1 m³/s.
11. Determine the hydrological risk of a roadway culvert to be flooded with 50 years expected service life designed to carry a 100-year storm.
12. Curb-opening inlets are placed along a 20-m wide street. The length of the inlets is 1 m that are placed on a continuous grade and the runoff coefficient is 0.9. The inlets drain the stormwater into a triangular gutter with $S_X = 0.03$, $S = 0.02$, and $n = 0.03$. Determine the maximum allowable distance between inlets considering that the rainfall intensity is equal to 15 mm/h and the allowable spread is 3 m.

13. Determine the depth and width of a straight trapezoidal channel ($Cs = 1.0$) lined with weeping love grass which is expected to carry $Q = 0.75$ m^3/s. The slope of the channel bottom is $S = 0.02$ and its side slopes are the same and equal to $z = 1.5$.
14. Determine the headwater depth for a culvert that conveys a flow of 0.5 m^3/s under inlet control conditions. The culvert is circular and has a square edge inlet which is mitered with a headwall and a diameter of 2 m and a slope of 0.01.

REFERENCES

Akan, A. O. 1985. Similarity solution of overland flow on pervious surface. *Journal of Hydraulic engineering, ASCE* 111:1057–1067.

Akan, A. O. 1993. *Urban Stormwater Hydrology—A Guide to Engineering Calculations*. Technomic, Lancaster, PA (ISBN: 0-97762-966-6).

Akan, A. O. 2000. Spread calculation in composite gutter sections. *Journal of Transportation Engineering, ASCE* 126(5): 448–450.

Akan, A. O. 2001. Tracive force channel design aid. *Candian Journal of Civil Engineering* 28(5): 865–867.

Akan, A. O. and Hager, W. 2001. Design aid for grass-lined channels. *Journal of Hydraulic Engineering, ASCE* 127: 236–238.

Akan, A. O. and Houghtalen, R. J. 2003. *Urban Hydrology, Hydraulics and Stormwater Quality: Engineering Applications and Computer Modeling*. Wiley, Belgium (ISBN 0-471-43158-3).

Ashley, R. M., Blamforth, D. J., Saul, A. J., and Blanksby, J. D. 2004. Flooding in the future—predicting climate change, risks and responses in urban areas. *Proceeding of 6th ICUD Modeling*, Dresden, pp. 105–113.

Brown, S. A., Stein, S. M., and Warner, J. C. 1996. *Urban Drainage Design Manual*. Hydraulic Engineering Circular 22, Federal Highway Administration, Washington, DC.

Chen, Y. H. and Cotton, B. A. 1988. Design of Roadside Channels with Flexible Linings. Hydraulic Engineering Circular No. 15 (HEC-15), FHWA, Publication No. FHWA-IP-87-7, USDOT/FHWA, McClean, VA.

Chocat, B. and Desbordes, M. 2004. *Proceedings of Novatech 2004: Sustainable Techniques and Strategies in Urban Water Management*, Lyon, France.

Chow, V. T. 1959. *Open Channel Hydraulics*. McGraw-Hill, New York.

Cung, J. A. 1969. On the subject of a flood propagation computation method (Muskingum method). *Journal of Hydraulic Research* 7: 205–230.

Danish Water Pollution Committee (DWPC). 2005. Danish Water Pollution Committee Publication No. 27. Functional practice for sewer systems during rain (in Danish), Final draft, Dansk Ingeniør Forening.

Diaz, J. and Barkdoll, B. 2006. Comparison of wastewater treatment in developed and developing countries. *Proceedings of World Environmental and Water Resources Congress*, Ohama, NE.

Dunne, T. 1986. Urban hydrology in the Tropics: Problems solutions, data collection and analysis. In: *Urban Climatology and its Application with Special Regards to Tropical Areas. Proceedings of the Mexico Tech Conf. Nov. 1984 World Climate Programme*, WMO.

Engman, E. T. 1983. Roughness coefficients for routing surface runoff. *Proceedings of the Conference of Frontiers in Hydraulic Engineering*, ASCE, pp. 560–565.

Federal Highway Administration (FHWA). 1984. *Drainage of Highway Pavements*. Hydraulic Engineering Circular 12, Federal Highway Administration, McLean, VA.

Federal Highway Administration (FHWA). 1996. *Urban Drainage Design Manual*. Hydraulic Engineering Circular 22, Federal Highway Administration, Washington, DC.

Foresti, E. 2001. Perspectives on anaerobic treatment in developing countries. *Water Science and Technology* 44(8): 141–148.

French, R. H. 1985. *Open Channel Hydraulics*. McGraw-Hill, New York.

Glidden, M. W. 1981. *The Effects of Stormwater Detention Policies on Peak Flows in Major Drainage Ways*. Master of Science thesis, Department of Civil Engineering, University of Colorado.

Grum, M., Jørgensen, A. T., Johansen, R. M., and Linde, J. J. 2005. The effects of climate change in urban drainage: An evaluation based on regional climate model simulations. *10th ICUD*, Copenhagen, Denmark.

Guo, J. C. Y. 1999. *Storm Water System Design*. University of Colorado at Denver, Denver, CO.

Guo, J. C. Y. 2000. Street storm water conveyance capacity. *Journal of Irrigation and Drainage Engineering* 136(2): 119–124.

Haan, C. T. 1977. *Statistical Methods in Hydrology*. The Iowa State University Press, Ames, IA.

Hauger, M. B., Mouchel, J. M., and Mikkelsen, P. S. 2005. Indicators of hazard, vulnerability and risk in urban drainage. *Proceedings of 10th International Conference on Urban Drainage*, Copenhagen, Denmark, pp. 21–26.

Ho, G. 2004. Small water and wastewater systems: Pathways to sustainable development? *Water Science and Technology* 48(11–12): 7–14.

Ho, G. 2005. Technology for sustainability: The role of onsite, small and community scale technology. *Water Science and Technology* 51(10): 15–20.

Jayakumar, K. V. and Dandigi, M. N. 2003. A cost effective environmentally friendly treatment of municipal wastewater using constructed wetlands for developing countries. *World Water and Environmental Resources Congress*, pp. 3521–3531.

Johnson, F. L. and Chang, F. M. 1984. *Drainage of Highway Pavements*. Hydraulic Engineering Circular, No. 12, Federal Highway Administration, McLean, VA.

Langergraber, G., and Muellegger, E. 2005. Ecological sanitation—a way to solve global sanitation problems? *Environment International* 31(3): 433–444.

Lohani, B. N. 2005. Advancing sanitation and wastewater management agenda in Asia and Pacific Region. Sanitation and Wastewater Management—The Way Forward Workshop, Manila, Philippines, September 19–20.

Mara, D. 2001. Appropriate wastewater collection, treatment and reuse in developing countries. *Proceedings of the Institution of Civil Engineers: Municipal Engineer* 145(4): 299–303.

Marsalek, J., Barnwell, T. O., Geiger, W., Grottker, M., Huber, W. C., Saul, A. J., Schilling, W., and Torno, H. C. 1993. Urban drainage systems: Design and operation. *Water Science and Technology* 27(12): 31–70.

McAlister, T. 2007. *National Guidelines for Evaluating Water Sensitive Urban Developments*. BMT WBM Pty Ltd, Queensland, Australia.

Normann, J. M., Houghtalen, R. J., and Johnston, W. J. 1985. *Hydraulic Design of Highway Culverts*. Hydraulic Design Series, No. 5, Federal Highway Administration, McLean, VA.

Parkinson, J. 2002. Stormwater management and urban drainage in developing countries. Available at http://www.sanicon.net/titles/topicintro.php3?topicId=5.

Ponce, V. M. and Yevjevich, V. 1978. Muskingum–Cunge method with variable parameters. *Journal of the Hydraulic Division, ASCE* 104: 1663–1667.

Ponce, V. M. and Theruer, F. D. 1982. Accuracy criteria in diffusion routing. *Journal of the Hydraulic Division, ASCE* 108: 747–757.

Prommesberger, B. 1984. Implementation of stormwater detention policies in the Denver Metropolitan Area. *Flood Hazard News* 14(1): 10–11.

Qian, Y. 2000. Appropriate technologies for municipal wastewater treatment in China. *Journal of Environmental Science and Health, Part A: Toxic/Hazardous Substances and Environmental Engineering* 35(10): 1749–1760.

Tucci, C. E. M. 1991. Flood control and urban drainage management. Available at http://www.cig.ensmp.fr/~iahs/ maastricht/s1/TUCCI.htm.

Ujang, Z. and Henze, M. 2006. *Municipal Wastewater Management in Developing Countries—Principles and Engineering*, p. 352. Published by IWA, London, U.K.

Urbonas, B. and Glidden, M. W. 1983. Potential effectiveness of detention policies. *Flood Hazard News* 13(1): 9–11.

USDOT (U.S. Department of Transportation). 1986. *Programmatic Environmental Assessment of Commercial Space Launch Vehicle Programs*. Washington, DC, February.

Wilderer, P. A. 2004. Applying sustainable water management concepts in rural and urban areas: Some thoughts about reasons, means and needs. *Water Science and Technology* 49(7): 8–16.

Woolhiser, D. A. 1975. Simulation of unsteady overland flow. In: K. Mahmood and V. Yevjevich, Eds, *Unsteady Flow in Open Channels*, Vol. II, pp. 485–508. Water resources publications, Fort Collins, CO.

Yu, H., Tay, J.-H., and Wilson and Francis. 1997. Sustainable municipal wastewater treatment process for tropical and subtropical regions in developing countries. *Water Science and Technology* 35(9): 191–198.

Young, G. K. and Stein, S. M. 1999. Hydraulic design of drainage for highways. In: L. W. Mays, Ed., *Hydraulic Design Handbook*. McGraw-Hill, New York.

Zawilski, M. 2004. Modernization problems of the sewerage infrastructure in a large polish city. In: J. Marsalek et al., Eds, *Enhancing Urban Environment by Environmental Upgrading and Restoration*. NATO Science Series, IV. *Earth and Environmental Sciences*, Springer, the Netherlands, 43: 151–162.

Zawilski, M. 2005. Trends in the combined sewerage development in the past and their importance for the present infrastructure. *Proceedings of 10th International Conference on Urban Drainage*, Copenhagen, Denmark, pp. 21–26.

7 Environmental Impacts of Urbanization

7.1 INTRODUCTION

Urbanization affects different aspects of the environment, especially the quantity and quality of water, soil, and air resources. Urbanization can be characterized as an increase in human population density, which leads to an increase in per capita consumption of natural resources and extensive modification of the natural landscape. This will create a built-up environment that is not usually sustainable over the long term and often continues to expand into natural areas (McDonnell and Pickett, 1990). The landscape alterations accompanying urbanization tend to be more long-lasting impacts than other factors. Generally, in urbanizing watersheds, water pollution and stormwater runoff are related to human habitation and the resultant increase in human land uses. The main focus of this chapter is on the effects of urbanization on water quality and the impact of urbanization on ecosystems. Some sections of this chapter are about the impacts of urban areas on the atmosphere and groundwater. Air pollution in urban areas is one of the main impacts of urbanization and a major source of acid rain in urban areas. Another main pollution source of surface water and groundwater in urban areas is stormwater. Stormwater causes considerable pollution of receiving waters. The stormwater in urban areas washes off all pollutants on the land surface during and after precipitation. These pollutants move through the drainage system and discharge to the receiving waters. There are many concerns about the faith of the pollution loads of stormwater as a major component of the urban water cycle. Therefore, a selection of this chapter is devoted to urban stormwater quality modeling.

Another water body affected by urbanization is groundwater. The quantity and quality of water in the aquifers under urban areas are affected by urbanization. Aquifers and groundwater quantity are affected by urbanization through reduced recharge of freshwater due to pavements or other impervious surfaces that prevent natural infiltration. The aquifer recharge may be affected by the increasing water leaks from leaking water mains and leaking wastewater from absorption wells. The water quality of the aquifers in urban areas is also affected by urbanization activities and the groundwater is polluted by infiltrating effluents, leakage from underground tanks, or accidental spills.

Table 7.1 shows the impacts of urbanization including chemical effects such as degraded water quality; physical effects such as altered hydrology, degraded habitat, and modified geomorphology; and biological effects.

7.2 URBANIZATION EFFECTS ON THE HYDROLOGIC CYCLE

This section discusses the changes in runoff and stream flow as a common event in urbanizing watersheds that can often cause changes in basin hydrology. Hydrological change also influences the whole range of environmental features that affect the aquatic biota-flow regime, aquatic habitat structure, water quality, biotic interactions, and food sources (Karr, 1991). Therefore runoff and the stream-flow regime are important due to their negative effects on aquatic life.

Increasing the impervious surfaces in the urbanized landscape is one of the most important impacts of watershed development. Impervious surfaces are defined as any portion of the built-up environment that affects and changes the natural hydrological regime. Impervious surfaces tend to inhibit or prevent

TABLE 7.1
Impacts of Urbanization on the Environment

Environmental Concern	Potential Impact	Cause/Source
Increase in flood peak or bank full stream flows	Downstream damages and degradation of aquatic habitat and/or loss of sensitive species	Increased stormwater runoff volume due to an increase in basin imperviousness
Increase in runoff-driven flooding frequency and duration	Degradation of aquatic habitat and/or loss of sensitive species	Increased stormwater runoff volume due to an increase in basin imperviousness
Increase in wetland water level fluctuations	Degradation of aquatic habitat and/or loss of sensitive species	Increased stormwater runoff due to an increase in basin imperviousness
Decrease in dry season base flows	Reduced aquatic habitat and less water for human consumption, irrigation, or recreational use	Water withdrawals and/or less natural infiltration due to an increase in basin imperviousness
Stream bank erosion and stream channel enlargement	Degradation of aquatic habitat and increased fine sediment production	Increase in stormwater runoff-driven stream flow due to an increase in basin imperviousness
Stream channel modification due to hydrological changes and human alteration	Degradation of aquatic habitat and increased fine sediment production	Increase in stormwater runoff-driven stream flow and/or channel alterations such as levees and dikes
Streambed scour and incision	Degradation of aquatic habitat and loss of benthic organisms due to washout	Increase in stormwater runoff-driven stream flow due to an increase in basin imperviousness
Excessive turbidity	Degradation of aquatic habitat and/or loss of sensitive species due to physiological and/or behavioral interference	Increase in stormwater runoff-driven stream flow and subsequent stream bank erosion due to an increase in basin imperviousness
Fine sediment deposition	Degradation of aquatic habitat and loss of benthic organisms due to fine sediment smothering	Increase in stormwater runoff-driven stream flow and subsequent stream bank erosion due to an increase in basin imperviousness
Sediment contamination	Degradation of aquatic habitat and/or loss of sensitive benthic species	Stormwater runoff pollutants
Loss of riparian integrity	Degradation of riparian habitat quality and quantity, as well as riparian corridor fragmentation	Human development encroachment and stream road crossings
Proliferation of exotic and invasive species	Displacement of natural species and degradation of aquatic habitat	Encroachment of urban development
Elevated water temperature	Lethal and nonlethal stress to aquatic organisms—reduced DO levels	Loss of riparian forest shade and direct runoff of high-temperature stormwater from impervious surfaces
Low DO levels	Lethal and nonlethal stress to aquatic organisms	Stormwater runoff containing fertilizers and wastewater treatment system effluent

continued

TABLE 7.1 (continued)

Environmental Concern	Potential Impact	Cause/Source
Lake and estuary nutrient eutrophication	Degradation of aquatic habitat and low DO levels	Stormwater runoff containing fertilizers and wastewater treatment system effluent
Bacterial pollution	Human health (contact recreation and drinking water) concerns, increases in diseases to aquatic organisms, and degradation of shellfish harvest beds	Stormwater runoff containing livestock manure, pet waste, and wastewater treatment system effluent
Toxic chemical water pollution	Human health (contact recreation and drinking water) concerns as well as bioaccumulation and toxicity to aquatic organisms	Stormwater runoff containing toxic metals, pesticides, herbicides, and industrial chemical contaminants

Source: Shaver, E. et al. 2007. *Fundamentals of Urban Runoff Management: Technical and Institutional Issues.* North American Lake Management Society in cooperation with the U.S. Environmental Protection Agency.

infiltration and groundwater recharge. Impervious areas also tend to have less evapotranspiration than natural areas.

The runoff coefficient reflects the fraction of rainfall volume that is converted to runoff. The runoff coefficient tends to closely track the percentage of impervious surface area in a given watershed, except at low levels of development where vegetation cover, soil conditions, and slope factors also influence the partitioning of rainfall. Impervious surfaces are hydrologically active, meaning they generate surface runoff instead of absorbing precipitation (Novotny and Chester, 1981).

The total fraction of a watershed that is covered by impervious surface areas is typically referred to as the percentage of total impervious area (%TIA). The %TIA of a watershed is a landscape-level indicator that integrates several concurrent interactions influencing the hydrological regime as well as the water quality (US-EPA, 1997). Another impervious term commonly used in urban watershed studies, especially in modeling, is effective impervious area (%EIA). The %EIA is that portion of the impervious surface that is directly connected (via open channels or stormwater piping) to the natural drainage network (Alley and Veenhuis, 1983). Another useful indicator of landscape-scale changes in a watershed condition is the fraction of the basin that is covered by natural vegetation. In many areas, forest cover is the key parameter, but in other regions, meadow or grass could be the key natural vegetation. In any case, native vegetation tends to be adapted to local climate conditions and soil characteristics, making it the land cover that best supports the natural hydrological regime. In general, urbanization tends to reduce natural vegetation land cover, while increasing the impervious surface area associated with the variety of land uses present in the built environment. In most regions, the fraction of the watershed covered by natural vegetation is inversely correlated with imperviousness.

Typical hydrographs for a stream before and after urbanization are illustrated and discussed in Chapter 1. The hydrograph emphasizes the higher peak flow rate of urbanized basins compared to natural landscape conditions. In addition, there is typically less "lag time" between rainfall and runoff when more impervious surfaces exist. The net effect of these urban watershed changes is that a higher proportion of rainfall is translated into runoff, which occurs more rapidly, and the resulting flood flows are therefore higher and much more "flashy" than natural catchments (Hollis, 1975). The effects of urbanization on the stream shape and the floodplain are illustrated in Figure 7.1. Increased peak discharge raises the floodplain level, flooding areas that were previously not at risk. An important measure of the degree of urbanization in a watershed is the level of impervious surfaces. As the level of imperviousness increases in a watershed, more rainfall is converted to runoff. Figure 7.2 illustrates this transformation.

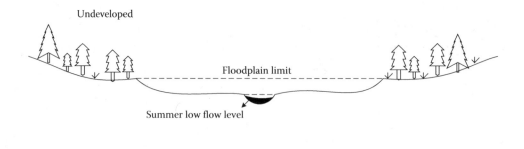

FIGURE 7.1 Effects of urbanization on stream slope and floodplain. (Adapted from Schueler, T. R. 1987. *Controlling Urban Runoff: A Practical Manual/or Planning and Designing Urban RMP's*. Metropolitan Information Center, Metropolitan Washington Council of Governments, Washington, DC.)

As discussed above, the hydrological impacts of watershed urbanization include the following:

- Greater runoff volume from impervious surfaces
- Higher flood recurrence frequency
- Less lag time between rainfall, runoff, and stream flow response

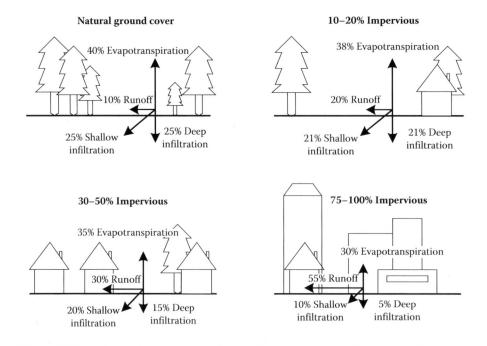

FIGURE 7.2 Effects of imperviousness on runoff and infiltration. (From Arnold, C. L. and Gibbons, C. J. 1996. *Journal of the American Planning Association* 62(2): 243–258. © American Planning Association, Chicago, IL. With permission.)

- Higher peak stream flow for a given size storm event
- More bank full or higher stream flows—flashier flows
- Longer duration of high stream flows during storm events
- More rapid recession from peak flows
- Lower wet and dry season base flow levels
- Less groundwater recharge
- Greater wetland water-level fluctuation

All of these characteristics represent alterations in the natural hydrological regime to which aquatic biota have adapted over the long term. These are significant hydrological changes that can negatively impact aquatic biota directly or indirectly. Direct impacts include washout of organisms from their preferred habitat and the physiological stress of swimming in higher velocity flows. Indirect impacts are centered on the degradation of instream habitat that occurs as a result of the higher urban stream flows.

7.3 URBANIZATION IMPACTS ON WATER QUALITY

Waterways and receiving waters near urban and suburban areas are often adversely affected by urban stormwater runoff. The degree and type of impact varies from location to location, but it is often significant relative to other sources of pollution and environmental degradation. Urban stormwater runoff affects water quality, water quantity, habitat and biological resources, public health, and the esthetic appearance of urban waterways. As stated in the National Water Quality Inventory 1996 Report to Congress (US-EPA, 1997), urban runoff is the leading source of pollutants causing water-quality impairment related to human activities in ocean shoreline waters, and the second leading cause in estuaries across the United States. Urban runoff was also a significant source of impairment in rivers and lakes. The percentage of total impairment attributed to urban runoff is substantial. This impairment constitutes approximately 13,000 km^2 (5000 square miles) of estuaries, 0.56 million ha (1.4 million acres) of lakes, and 48,000 km (30,000 miles) of rivers.

US-EPA (2005) has classified the adverse impacts of urban runoff on receiving waters into three categories as follows:

- Short-term changes in water quality during and after storm events including temporary increases in the concentration of one or more pollutants, toxics, or bacteria levels
- Long-term water-quality impacts caused by the cumulative effects associated with repeated stormwater discharges from a number of sources
- Physical impacts due to erosion, scour, and deposition associated with increased frequency and volume of runoff that alters aquatic habitat

According to Horner et al. (1997) (the Terrene Institute's Fundamentals of Urban Runoff Management), pollutants associated with urban runoff, which are potentially harmful to receiving waters, are listed in Table 7.2. These pollutants degrade the water quality in receiving waters near urban areas, and often contribute to the impairment of use and exceedences of criteria included in water-quality standards. The quantity of these pollutants per unit area delivered to receiving waters tends to increase with the degree of development in urban areas.

Esthetic impacts in the form of debris and litter floating in urban waterways and concentrated on stream banks and beaches are quite visible to the general public. Stormwater is the major source of floatables that include paper and plastic bags and packaging materials, bottles, cans, and wood. The presence of floatable and other debris in receiving waters during and following storm events reduces the visual attractiveness of the waters and detracts from their recreational value. Nuisance algal conditions including surface scum and odor problems can also be attributed to urban stormwater

TABLE 7.2
Urban Runoff Pollutant Sources and Constituents

Pollutant Source	Solids	Nutrients	Pathogens	Oxygen Demand	Metals	Oils	Organics
Soil erosion	✓	✓		✓	✓		
Fertilizers		✓					
Human waste	✓	✓	✓	✓			
Animal waste	✓	✓	✓	✓			
Vehicle fluids	✓		✓	✓	✓	✓	
Internal combustion							✓
Vehicle wear	✓			✓	✓		
Household chemicals	✓	✓		✓	✓	✓	✓
Industrial processes	✓	✓		✓	✓	✓	✓
Paints and preservatives					✓	✓	
Pesticides				✓	✓	✓	

Source: Shaver, E., et al. 2007. *Fundamentals of Urban Runoff Management: Technical and Institutional Issues.* North American Lake Management Society in cooperation with the U.S. Environmental Protection Agency.

in many instances. Based on available information and data, the following general statements can be made about urban stormwater impacts.

- Impacts on water quality in terms of water chemistry, especially in rivers, tend to be transient.
- Impacts on habitat and aquatic life are generally more profound, and are easier to see and quantify compared to water chemistry.
- Impacts from several sources including municipal discharges to urban waterways diffused from agricultural runoff and rural areas that affect urban waterways.
- Impacts are often interrelated and cumulative. For example, both degraded water quality and increased water quantity join to impact habitat and biological resources.

7.3.1 Water Pollution Sources

Stormwater runoff from urbanized areas carries a wide variety of pollutants from diverse and diffuse sources. Land-use activities in the drainage area where runoff is collected and atmospheric deposition from areas outside the watershed of the receiving water body are the main sources of runoff contamination. The impact of stormwater runoff pollutants on the receiving water quality depends on a number of factors, including pollutant concentrations, the mixture of pollutants present in the runoff, and the total load of pollutants delivered to the water body.

As mentioned above, urban land use (industrial, residential, and commercial), human activities (industrial operations, residential lawn care, and vehicle maintenance), and characteristics of the basin drainage determine the pollutants and level of pollutants found in stormwater runoff (Pitt et al., 1995). Atmospheric deposition of pollutants is typically divided into wet fall and dry fall components. These inputs can come from local sources, such as an automobile exhaust, or from distant sources such as coal or oil power plant emissions. Regional industrial and agricultural activities can also contribute to atmospheric deposition as dry-fall. Precipitation also carries pollutants from the atmosphere to earth as wet-fall. Depending on the season and location, atmospheric deposition can be a significant source of pollution in the urban environment. Another important source of urban runoff pollution is wastewater discharge that depends on the interaction between surface water collection systems and waste water collection systems. These systems can be separate or combined. Separate storm sewer systems convey only stormwater runoff that is discharged directly into receiving waters with or without treatment. Stormwater can also bypass the stormwater infrastructure and flow into receiving waters as diffuse runoff from parking lots, roads, and landscaped areas. In cases where a separate storm sewer system is present, sanitary sewer flows are conveyed to the municipal wastewater treatment plant (WWTP). In a combined sewer system, stormwater runoff may be combined with sanitary sewer flows for conveyance. During low flow periods, flows from combined sewers are treated by the WWTP prior to discharge into receiving waters. During large rainfall events, however, the volume of water conveyed in combined sewers often exceeds the storage and treatment capacity of the wastewater treatment system. As a result, discharges of untreated stormwater and sanitary wastewater directly into receiving streams can occur in these systems. This type of discharge is known as combined sewer overflow (CSO). Urban streets are typically significant source areas for most contaminants in all land-use categories. Parking lots and roads are generally the most critical source areas in industrial and commercial areas. Along with roads, lawns, landscaped areas, and driveways can be significant sources of pollution in residential areas. In addition, roofs can contribute significant quantities of pollutants in all land-use types (Bannerman et al., 1993). Due to the diffuse nature of many stormwater discharges, it is difficult to quantify the range of pollutant loadings to receiving waters that are attributable to stormwater discharges.

Water pollutants are typically quantified by concentrations and loadings. Concentration is the mass of pollutant per unit volume of water sample, usually expressed as mg/L or μg/L. Loading is the mass of pollutants delivered to a water body over a period of time and is usually given on an annual basis as kg/year. When ascribed to a particular land use, loading is sometimes termed yield or simply export per unit area of the land use (kg/ha-year). It represents the cumulative burden over the extended period and hence the potential chronic effects on receptor organisms. Tables 7.2 and 7.3 show the pollutants commonly found in stormwater, their sources, and their constituents.

Excessive nutrient levels in urban runoff can stimulate algal growth in receiving waters and cause nuisance algal blooms when stimulated by sunlight and high temperatures. The decomposition that follows these algal blooms, along with any organic matter (OM) carried by runoff, can lead to depletion of DO levels in the receiving water and bottom sediments. This can result in a degradation of habitat conditions (low DO), offensive odors, loss of contact recreation usage, or even fish kills in extremely low DO situations.

Nitrate is the form of nitrogen found in urban runoff that is of most concern. The nitrate anion (NO_3) is not usually adsorbed by soil and therefore moves with infiltrating water. Nitrates are presented in fertilizers, human wastewater, and animal wastes. Nitrate contamination of groundwater can be a serious problem, resulting in contamination of drinking water supplies (CWP, 2003). High nitrate levels in drinking water can cause human health problems.

Phosphates (PO_4) are the key form of phosphorus found in stormwater runoff. Phosphates in runoff exist as soluble reactive phosphorus (SRP) or orthophosphates, polyphosphates, and as organically bound phosphate. The poly form of phosphates is the one that is found in some detergents. Orthophosphates are found in sewage and in natural sources. Organically bound phosphates are also found in nature, but can also result from the breakdown of phosphorus-based organic pesticides. Very high concentrations of phosphates can be toxic.

Chlorides are salt compounds found in runoff that result primarily from road deicer applications during winter months. Sodium chloride (NaCl) is the most common example. Although chlorides in urban runoff come primarily from road deicing materials, they can also be found in agricultural runoff and wastewater. Small amounts of chlorides are essential for life, but high chloride levels can cause human illness and can be toxic to plants or animals.

Metals are among the most common stormwater pollutant components. These pollutants are also referred to as trace metals (e.g., zinc, copper, lead, chromium, etc.). Many trace metals can often be found at potentially harmful concentrations in urban stormwater runoff (CWP, 2003). Metals are typically associated with industrial activities, landfill leachate, vehicle maintenance, roads, and parking areas (Davies, 1986; Field and Pitt, 1990; Pitt et al., 1995).

TABLE 7.3
Pollutants Commonly Found in Stormwater

Pollutant Category	Specific Measures
Solids	Total suspended solids (TSS)
	Turbidity (NTU)
Oxygen-demanding material	Biochemical oxygen demand (BOD)
	Chemical oxygen demand (COD)
	Organic matter (OM)
	Total organic carbon (TOC)
Phosphorus (P)	Total phosphorus (TP)
	Soluble reactive phosphorus (SRP)
	Biologically available phosphorus (BAP)
Nitrogen (N)	Total nitrogen (TN)
	Total kjeldahl nitrogen (TKN)
	Nitrate + nitrite-nitrogen ($NO_3 + NO_2$-N)
	Ammonia-nitrogen (NH_3-N)
Metals	Copper (Cu), lead (Pb), zinc (Zn), cadmium (Cd), arsenic (As), nickel (Ni), chromium (Cr), mercury (Hg), selenium (Se), silver (Ag)
Pathogens	Fecal coliform bacteria (FC)
	Enterococcus bacteria (EC)
	Total coliform bacteria (TC)
	Viruses
Petroleum hydrocarbons	Oil and grease (OG)
	Total petroleum hydrocarbons (TPH)
Synthetic organics	Polynuclear aromatic hydrocarbons (PAH)
	Pesticides and herbicides
	Polychlorobiphenols (PCB)

Source: United States Environmental Protection Agency (US-EPA). 1999. Preliminary data. Summary of urban storm water best management practices. EPA-821-R-99-012.

Hydrocarbons are normally attached to sediment particles or OM carried in urban runoff. The increase in vehicular traffic associated with urbanization is frequently linked to air pollution, but there is also a negative relationship between the level of automobile use in a watershed and the quality of water and aquatic sediments. This has been shown for polycyclic aromatic hydrocarbon (PAH) compounds (Hoffman et al., 1982; Van Metre et al., 2000; Stein et al., 2006). In most urban areas, in stormwater runoff, hydrocarbon concentrations are generally less than 5 mg/L, but concentrations can increase up to 10 mg/L in urban areas that include highways, commercial zones, or industrial areas (CWP, 2003). Hydrocarbon "hot spots" in the urban environment include gas stations, high-use parking lots, and high-traffic streets (Stein et al., 2006). A study in Michigan showed that commercial parking lots contribute over 60% of the total hydrocarbon load in an urban watershed (Steuer et al., 1997). Lopes and Dionne (1998) found that highways are the largest contributor of hydrocarbon runoff pollution.

Microbial pollution includes bacteria, protozoa, and viruses that are common in the natural environment, as well as those that come from human sources (Field and Pitt, 1990; Young and Thackston, 1999; Mallin et al., 2000). Many microbes are naturally occurring and beneficial, but others can cause diseases in aquatic biota and illness or even death in humans. Some types of microbes can be pathogenic, while others may indicate a potential risk of water contamination, which can limit swimming, boating, shellfish harvest, or fish consumption in receiving waters. Microbial pollution is almost always found in stormwater runoff, often at very high levels, but concentrations are typically highly variable (Pitt et al., 2004). Sources of bacterial pollution in the urban environment include failing septic systems, WWTP discharges, CSO events, livestock manure runoff, and pet waste, as well as natural sources such as wildlife. Pesticides, herbicides, and other organic pollutants are often found in stormwater flowing from residential and agricultural areas.

7.3.2 Water Quality Modeling of Urban Stormwater

The watershed assessment process provides the framework for evaluating watershed conditions and quantifying watershed characteristics (US-EPA, 2005). The objectives of the watershed assessment effort, pollution source information, and the water-quality data available largely determine what will be the most appropriate method for quantifying pollutant loading. In general, the approach chosen should be the simplest approach that meets the objectives of the watershed management program. Pollutant-loading estimates are generally developed using a model or models. Models can be useful tools for watershed and receiving-water assessments because they facilitate the analysis of complex systems and provide a method for estimating pollutant loading for a large array of land-use scenarios. Models are only as good as the data used for calibration and verification. There will always be some uncertainties present in all models and these uncertainties should be quantified and understood prior to using the selected model. Many models utilize literature-based values for water-quality concentrations to estimate pollutant loads (US-EPA, 2005). Models have also become a standard part of most total maximum daily load (TMDL) programs (US-EPA, 1997). There are several recognized approaches used for estimating pollutant loadings for a drainage area or watershed basin as discussed in the next sections.

7.3.2.1 Unit-Area Loading

The unit-area loading method is a simple method for the estimation of pollutant loading for a specific land use based on empirically published yield values. As mentioned earlier in this chapter, loading is the mass of pollutants delivered to a water body over a period of time and is usually given on an annual basis as kg/year. When ascribed to a particular land use, loading is sometimes termed yield or simply export per unit area of the land use (kg/ha-year). Table 7.4 presents typical loadings for a number of pollutants and land uses. Although this table presents no ranges or statistics on the possible dispersion of these numbers when measurements are made, the variation is usually substantial from place to place in the same land use and from year to year at the same place. This method is least likely to give accurate results because of the general lack of fit between the catchments of interest and the data collection location(s). To apply this method, consult a reference like Table 7.4, select the area loading rate for each land use, multiply by the areas in each use, and sum the total loading for the pollutant of interest. This method can be improved by producing some measure of uncertainty or error in the estimates. To do so, it is necessary to establish ranges of area loadings from the published literature or actual sampling, estimate maximum and minimum and mean or median values of each pollutant, and then evaluate to determine whether uncertainty or error could change the conclusions. Table 7.4 presents loading rate ranges based on unpublished data collected in the Pacific Northwest (PNW). The PNW regional data provide values for total phosphorus and total nitrogen for most land uses and all pollutants in road runoff, except fecal coliform. Accordingly, the regional data have narrower ranges than the remainder. Data such as that shown in Table 7.4 should be used with caution, because the concentrations of most pollutants vary considerably depending on regional characteristics in land use and climate. The use of published yield or unit-area loading values from specific sources, rather than for land-use categories, is also feasible. For example, a study in Maryland (Davis et al., 2001) examined the loading rates of metals (zinc, lead, copper, and cadmium) from several common sources in the urban environment. These included building siding and rooftops as well as automobile brakes, tires, and oil leakage. Loading estimates (mean, median, maximum, and minimum) were developed for each of these sources for all four metals (Davis et al., 2001). These types of data can be very useful for a variety of management scenarios.

TABLE 7.4
Pollutant Loading (kg/ha-year) Ranges for Various Land Uses

Land-Use Category		TSS	TP	TN	Pb	In	Cu
Road	Minimum	281	0.59	1.3	0.49	0.18	0.03
	Maximum	723	1.5	3.5	1.1	0.45	0.09
	Median	502	1.1	2.4	0.78	0.31	0.06
Commercial	Minimum	242	0.69	1.6	1.6	1.7	1.1
	Maximum	1369	0.91	8.8	4.7	4.9	3.2
	Median	805	0.8	5.2	3.1	3.3	2.1
Single family	Minimum	60	0.46	3.3	0.03	0.07	0.09
Low density	Maximum	340	0.64	4.7	0.09	0.2	0.27
Residential	Median	200	0.55	4	0.06	0.13	0.18
Single family	Minimum	97	0.54	4	0.05	0.11	0.15
High density	Maximum	547	0.76	5.6	0.15	0.33	0.45
Residential	Median	322	0.65	5.8	0.1	0.22	0.3
Multifamily	Minimum	133	0.59	4.7	0.35	0.17	0.17
Residential	Maximum	755	0.81	6.6	1.05	0.51	0.34
	Median	444	0.7	5.6	0.7	0.34	0.51
Forest	Minimum	26	0.1	1.1	0.01	0.01	0.02
	Maximum	146	0.13	2.8	0.03	0.03	0.03
	Median	86	0.11	2	0.02	0.02	0.03
Grass	Minimum	80	0.01	1.2	0.03	0.02	0.02
	Maximum	588	0.25	7.1	0.1	0.17	0.04
	Median	346	0.13	4.2	0.07	0.1	0.03
Pasture	Minimum	103	0.01	1.2	0.004	0.02	0.02
	Maximum	583	0.25	7.1	0.015	0.17	0.04
	Median	343	0.13	4.2	0.01	0.1	0.03

Source: Horner, R. R. et al. 1997. In: L. A. Roesner, Ed., *Effects of Watershed Development and Management on Aquatic Ecosystems.* ASCE. With permission.

7.3.2.2 Simple Empirical Method

The "Simple Method" was first developed by Schueler (1987) and further refined by the Center for Watershed Protection (CWP, 2003). This method requires data on the watershed drainage area, impervious surface area, stormwater runoff pollutant concentrations, and annual precipitation. With the Simple Method, land use can be divided into specific types, such as residential, commercial, industrial, and roadways. Using these data, the annual pollutant loads for each type of land use can be calculated. Alternatively, generalized pollutant values for land uses such as new suburban areas, older urban areas, central business districts, and highways can be utilized. Stormwater pollutant concentrations can be estimated from local or regional data or from national data sources. As has been discussed, stormwater pollutant concentrations tend to be highly variable for a number of reasons. Because of this variability, it is difficult to establish different concentrations for each land use. The original Simple Method uses Nationwide Urban Runoff Program (NURP) data for the representative pollutant concentrations. Utilizing a more recent and regionally specific database would, in general, be more accurate for this purpose. If no regional or local data exist, the Simple Method can be utilized using median urban runoff value data (US-EPA, 1983).

The Simple Method estimates pollutant loads for chemical constituents as a product of annual runoff volume and pollutant concentration, as (CWP, 2003)

$$L = 1.04 \times R \times c \times A, \tag{7.1}$$

where L is the annual load (kg), R is the annual runoff (mm), c is the pollutant concentration (mg/L), and A is the area (ha).

The mean pollutant concentrations c should be obtained from local data. In the absence of such data, the c values given in Table 7.5 can be adopted. However, these values should be used carefully since most of the data are from the metropolitan Washington, DC, area. The Nationwide Urban Runoff program (NURP) values should be employed for cities other than Washington, DC, and Baltimore.

The Simple Method formula for estimation of bacteria pollutant loads is (CWP, 2003)

$$L = 1.0104 \times R \times c \times A, \qquad (7.2)$$

where L is the annual load (Billion Colonies), R is the annual runoff (mm) obtained from Equation 7.3, c is the bacteria concentration (MPN/100 mL), and A is the area (ha).

The Simple Method calculates annual runoff as a product of annual runoff volume and a runoff coefficient (R_V). Runoff volume is calculated as (CWP, 2003)

$$R = P \times P_j \times R_V, \qquad (7.3)$$

where R is the annual runoff (mm), P is the annual rainfall (mm), P_j is the fraction of annual rainfall events that produce runoff (usually 0.9), and R_V is the runoff coefficient.

In the Simple Method, the runoff coefficient is calculated based on impervious cover in the subwatershed. This relationship is based on empirical data. Although there is some variability in the data, watershed imperviousness does appear to be a reasonable predictor of R_V. The following equation represents the best-fit line for the data set ($N = 47$, $R^2 = 0.71$) based on data collected by

TABLE 7.5
Flow-Weighted Mean Pollutant Concentration c Values in mg/L for the Simple Method

Pollutant	New Suburban NURP Sites (Washington, DC)	Older Urban Areas (Baltimore)	Central Business District (Washington, DC)	National NURP Study Average	Hardwood Forest (Northern) (Virginia)	National Urban Highway Runoff
Phosphorus						
Total	0.28	1.08	—[a]	0.48	0.15	—
Ortho	0.12	0.28	1.01	—	002	—
Soluble	0.18	—	—	0.18	004	0.59
Organic	0.10	0.82	—	0.13	0.11	—
Nitrogen						
Total	2.00	13.8	2.17	3.31	0.78	—
Nitrate	0.48	8.9	0.84	0.98	0.17	—
Ammonia	0.28	1.1	—	—	0.07	—
Organic	1.25	—	—	—	0.54	—
TKN	1.51	7.2	1.49	2.35	0.81	2.72
COD	35.8	183.0	—	90.8	>40.0	124.0
BOD (5-day)	5.1	—	38.0	11.9	—	—
Metals						
Zinc	0.037	0.397	0.250	0.178	—	0.380
Lead	0.018	0.389	0.370	0.180	—	0.550
Copper	—	0.105	—	0.047	—	—

Source: Schueler, T. R. 1987. *Controlling Urban Runoff: A Practical Manual or Planning and Designing Urban RMP's.* Metropolitan Information Center, Metropolitan Washington Council of Governments, Washington, DC.

[a] Nonexisting.

Schueler (1987). The volumetric runoff coefficient R_V depends on many factors, but the watershed imperviousness is the most important. An empirical relationship between the two volumetric runoff coefficients and watershed imperviousness is expressed as

$$R_V = 0.05 + 0.009 A_I, \tag{7.4}$$

where A_I is the impervious fraction (%).

Equation 7.1 can be used for a time period of any length. When annual pollutant export is sought, P represents the annual rainfall in mm. If 90% of the rainfall events produce runoff in a year, then $P_j = 0.90$. When pollutant export due to a single event is sought, P becomes the rainfall depth for the event, and $P_R = 1.0$. The model is appropriate for areas smaller than 280 ha.

Another method for the estimation of the amount of annual runoff is suggested by Heaney et al. (1977):

$$R = \left[0.15 + 0.75\left(\frac{A_I}{100}\right)\right] P - 19.36 D_S^{0.5957}, \tag{7.5}$$

where R is the annual runoff (mm), A_I is the imperviousness (%), P is the annual precipitation (mm), and D_S is the depression storage (mm).

The depression storage D_S is determined from

$$D_S = 6.35 - 4.76\left(\frac{A_I}{100}\right). \tag{7.6}$$

EXAMPLE 7.1

Assume that a 1-ha woodland area with an estimated imperviousness of about 6% is being developed into a town-house community with an estimated imperviousness of about 50%. The annual rainfall is about 1 m. Assume that 80% of the rain events produce runoff under both the predevelopment and postdevelopment conditions. Determine the increase in annual total nitrogen loads due to this development.

Solution: For the predevelopment condition,

$$R_V = 0.05 + 0.009(6) = 0.104,$$

$$R = 1000 \times 0.8 \times 0.104 = 83.2 \text{ mm}.$$

Also, from Table 7.5, $c = 0.78$ mg/L for woodland. Then,

$$L = (1.04)(83.2)(0.78)(1) = 67.5 \text{ kg/year}.$$

For the postdevelopment condition,

$$R_V = 0.05 + 0.009(50) = 0.5,$$

$$R = 1000 \times 0.8 \times 0.5 = 400 \text{ mm},$$

and from Table 7.5, $c = 2$ mg/L for suburban areas. Thus,

$$L = (1.04)(400)(2)(1) = 832 \text{ kg/year}.$$

Therefore, as a result of this development, the annual nitrogen load will increase from 67.5 to 832 kg.

Limitations of the Simple Method: The Simple Method should provide reasonable estimates of changes in pollutant export resulting from urban development activities. However, several caveats should be kept in mind when applying this method. The Simple Method is most appropriate for assessing and comparing the relative storm flow pollutant load changes of different land use and stormwater management scenarios. It provides estimates of storm pollutant export that are probably close to the "true" or unknown value for a development site, catchments, or subwatershed. However, it is very important not to overemphasize the precision of the results obtained. The Simple Method provides a general planning estimate of likely storm pollutant export from areas at the scale of a development site, catchment, or subwatershed. More sophisticated modeling may be needed to analyze larger and more complex watersheds. Chandler (1993, 1994) utilized the Simple Model in four case-study comparisons with more complex models, including the EPA Stormwater Management Model (SWMM) and the Hydrologic Simulation Program FORTRAN (HSPF) model. Chandler (1993, 1994) concluded that there was no compelling reason for using complex models when estimating annual pollutant loading under most situations.

In addition, the Simple Method only estimates pollutant loads generated during storm events. It does not consider pollutants associated with base flow volume. Typically, base flow is negligible or nonexistent at the scale of a single development site, and can be safely neglected. However, catchments and subwatersheds do generate base flow volume. Pollutant loads in base flow are generally low and can seldom be distinguished from natural background levels. Consequently, base flow pollutant loads normally constitute only a small fraction of the total pollutant load delivered from an urban area. Nevertheless, it is important to remember that the load estimates refer only to storm event-derived loads and should not be confused with the total pollutant load from an area. This is particularly important when the development density of an area is low.

7.3.2.3 US-EPA Method

Heaney et al. (1977) developed a set of relationships for the EPA, for estimation of the mean annual pollutant loading of urban stormwater runoff, for separate storm sewer system areas, and also unsewered urban areas,

$$L = 0.0441 \times A \times K_1 \times R \times F(p) \times K_s, \tag{7.7}$$

where L is the annual pollution load (kg/year), A is the area (ha), K_1 is the pollutant-loading factor, R is the rainfall (mm/year), $F(p)$ is the function related to population density, and K_s is the factor related to intervals of street sweeping.

The loading factor K_1 depends on the land use and the pollutant being considered. Values of K_1 are tabulated in Table 7.6 for different pollutants. The population density is a function of land use and evaluated differently for different land uses (Akan, 1993):

$$F(p) = \begin{cases} 1.0 & \text{commercial and industrial areas} \\ 0.142 + 0.00P^{0.54} & \text{residential areas} \\ 0.142 & \text{others} \end{cases}, \tag{7.8}$$

where P is the population density (persons/m^2).

The factor related to intervals of street sweeping depends on the sweeping interval S_n as

$$K_s = \begin{cases} 1.0 & S_n > 20 \text{ days} \\ \dfrac{S_n}{20} & S_n < 20 \text{ days} \end{cases}. \tag{7.9}$$

For the combined sewer system, the annual pollutant loading for runoff is obtained using the following equation:

$$L = 0.0441 \times A \times K_2 \times R \times F(p) \times K_s, \tag{7.10}$$

TABLE 7.6
Loading Factor K_1 for Storm Sewer Areas

Land Use	BOD_5	SS	VS	PO_4	NO_3
Residential	0.799	16.3	9.4	0.0336	0.131
Commercial	3.200	22.2	14.0	0.0757	0.296
Industrial	1.210	29.1	14.3	0.0705	0.277
Others	0.113	2.7	2.6	0.0099	0.060

Source: Heaney, J. F. et al. 1977. Cost Assessment and Impacts, Report ERA-800/2-77-084. Environmental Protection Agency, Washington, DC.

TABLE 7.7
Loading Factor K_2 for Combined Sewer Areas

Land Use	BOD_5	SS	VS	PO_4	NO_3
Residential	3.290	67.2	36.9	0.1390	0.540
Commercial	13.20	91.6	57.9	0.3120	1.220
Industrial	5.000	120.0	59.2	0.2910	1.140
Others	0.467	11.1	10.6	0.0411	0.250

Source: Heaney, J. F. et al. 1977. Cost Assessment and Impacts, Report ERA-800/2-77-084. Environmental Protection Agency, Washington, DC.

where L is the annual pollution load (kg/year), A is the area (ha), K_2 is the loading factor (tabulated in Table 7.7), R is the rainfall (mm/year), $F(p)$ is the function related to population density, and K_s is the factor related to intervals of street sweeping.

Example 7.2

Consider a 30-ha area with commercial land use and separate storm sewer and sanitary sewer systems. The average annual precipitation is about 1 m. Assume that about 40% of the area is covered by impervious surfaces and the streets are swept monthly. Estimate the mean annual pollution loading and concentration of nitrate.

Solution: The nitrate loading factor from Table 7.6 for commercial areas and separated storm sewer and sanitary sewer is 0.296.

For monthly sweepings $K_s = 1$ (Equation 7.9).
For commercial areas $F(p) = 1$ (Equation 7.8).
From Equation 7.6,
$$D_S = 6.35 - 4.76 \times 0.4 = 4.45.$$

From Equation 7.5,
$$R = (0.15 + 0.75 \times 0.4) \times 1000 - 19.36(4.45)^{0.5957} = 402.89 \text{ mm}.$$

$$\text{Runoff volume} = 3{,}00{,}000 \times 0.40289 = 1{,}20{,}867 \text{ m}^3.$$

From Equation 7.7,

$$L = 0.0441 \times 30 \times 0.296 \times 402.89 \times 1 \times 1 = 157.77 \text{ kg/year}.$$

The concentration of nitrate is obtained by dividing the pollution load by runoff volume:

$$C = \frac{157.77}{1{,}20{,}867} = 0.0013 \text{ kg/m}^3 = 1.30 \text{ mg/L}.$$

7.3.2.4 Computer-Based Models

There are a wide variety of computer models available today that can be used for surface water and stormwater quality assessments. Many of these models are available in the public domain and have been developed and tested by resource agencies. Regionally or locally specific versions of many of these models are also common. In comparison with the approaches outlined previously, computer-based models provide a more complex approach to estimating pollutant loading and also often offer a means of evaluating various management alternatives (US-EPA, 2005). Detailed coverage of these models is beyond the scope of this chapter. The US-EPA Handbook for Developing Watershed Plans (US-EPA, 2005) contains a comprehensive discussion of computer-based models in Chapter 8 of that publication.

Examples of comprehensive computerized models include the SWMM; Better Assessment Science Integrating Point and Non-Point Sources (BASINS); the HSPF; Source Loading and Management Model (SLAMM); Storage, Treatment, and Overflow Runoff Model (STORM); and Spatially Referenced Regression on Watershed Attributes (SPARROW). These are only a few of the computer-based pollutant-loading estimation models available. In general, computer-based models contain hydrological and water-quality components and have statistical or mathematical algorithms that represent the mechanisms generating and transporting runoff and pollutants. The hydrological components of both SWMM and HSPF stem from the Stanford Watershed Model, first introduced almost 25 years ago, and produce continuous hydrograph simulations. In addition to these relatively complex computer-based models, there are numerous "spreadsheet" level models that have been developed by local and regional water-quality practitioners. In almost all cases, computer-based models need to be calibrated and validated using locally appropriate water-quality data (US-EPA, 2005), which, depending on the watershed under study, can be a time-consuming and relatively costly effort. Most computer-based models structure the water-quality components on a mass balance framework that represents the rate of change in pollutant mass as the difference between pollutant additions and losses. Soil erosion is usually calculated according to the Universal Soil Loss Equation (USLE). Losses are represented by a first-order washoff function (i.e., loss rate is considered to be a function of pollutant mass present); other losses are modeled in mathematically similar ways. For example, both OM decomposition and bacterial die-off are considered first-order reactions. Some models, like SWMM, have both a receiving water and a runoff component. These models treat some of the transformation processes that can occur in water [e.g., DO depletion according to the Streeter–Phelps equation or fecal coliform (FC) die-off using the Mancini equation]. However, no model can fully represent all of these numerous and complex processes. The BASINS model is a physical process-based analytical model developed by the US-EPA and typically used for watershed-based hydrological and water-quality assessments. For example, BASINS were used to model the East Fork of the Little Miami River (Tong and Chen, 2002). The HSPF model can be used as a component of the BASINS model (Bergman et al., 2002) or as a stand-alone model (Im et al., 2003). The SPARROW model is a statistical-regression, watershed-based model developed by the USGS (Smith et al., 1997) and is used primarily for water-quality modeling (Alexander et al., 2004). Many computer-based models utilize regression equations to describe pollutant characteristics (Driver and Tasker, 1990).

There are also a number of so-called buildup and washoff models that simulate pollutant buildup on impervious surfaces and use rainfall data to estimate washoff loading. The main limitation of these models is that model-controlling factors can greatly vary with surface characteristics, so calibration with actual field measurements is needed. These models can work well with calibration and can model intrastorm variations in runoff water quality, which is a key advantage. These models are often used for ranking or prioritizing, but not for predicting actual runoff water quality. SLAMM was developed to evaluate the effects of urban development characteristics and runoff control measures on pollutant discharges. This model examines runoff from individual drainage basins with particular land-use and control practices (Burton and Pitt, 2002).

Most models require substantial local data to set variable parameters in the calibration and verification phases. They also require considerable technical skill and commitment from personnel. Therefore, only those prepared to commit the resources to database development and expertise should embark on using these models. Most models used today also utilize the geographical information system (GIS) for data input and presentation of results. In many situations, the use of computer-based models may not be merited, but in other cases, it may be helpful in determining the magnitude of the water-quality problem or aid in finding a solution. Computer models can also extend data collected and enhance findings. In addition, they can be quite useful in running a variety of scenarios to help frame the water-quality problem. Examples of this include worst-case, full build-out scenarios or potential BMPs scenarios to estimate the effectiveness of a range of treatment options. In any case, model selection should be linked to the project objectives and must be compatible with the available data. In almost all cases, using the simplest model that will meet the project objectives is likely the best course to take. In all cases, models should be calibrated and verified with independent, local, or regionally specific data. A good example of a watershed-scale, computer-based model dealing with multiple water-quality parameters and their impact on receiving waters is the Sinclair–Dyes Inlet TMDL Project in the Puget Sound, Washington (Johnston et al., 2003). This model has a watershed component (HSPF) that is linked to a receiving-water component (CH3D). It includes dynamic loading from the contributing watershed and hydrodynamic mixing in the receiving waters. This model can be reviewed at www.ecy.wa.gov/programs/wq/tmdl/sinclair-dyes_inlets/index.html.

7.4 PHYSICAL AND ECOLOGICAL IMPACTS OF URBANIZATION

One of the important physical impacts of urbanization is erosion of natural streams. Stream bank erosion is a natural phenomenon and a source of both sediments and nutrients. However, urbanization can greatly accelerate the process of stream bank erosion. As the amount of impervious area increases, a greater volume of stormwater is discharged directly into receiving waters, often at a much higher velocity. The increased volume and velocity of the runoff can overwhelm the natural carrying capacity of the stream network. In addition, streams in urbanized areas can experience an increase in bankfull flows. Because bankfull flows are highly erosive, substantial alterations in stream channel morphology can be observed. Sediments from eroding banks (and upland construction) are deposited in areas where the water slows, causing buildup, destruction of benthic habitat, and a decreased stream capacity for flood waters. This ultimately results in a greater potential for further erosion.

As discussed above, the geomorphological impacts of watershed urbanization include the following:

- Stream channel enlargement and instability
- Stream bank erosion and fine sediment production
- Stream channel incision or down-cutting
- Stream bed scour and fine sediment deposition
- Riparian buffer (lateral) encroachment
- Riparian corridor (longitudinal) fragmentation
- Canalization and floodplain encroachment

Environmental Impacts of Urbanization

- Increased sediment yields, especially during construction
- Simplification of the natural drainage network, including loss of headwater channels and wetlands and lower drainage density
- Modification of natural instream pool–riffle structure
- Fish and amphibian migration barriers such as culverts and dams

Degradation of aquatic habitat is one of the most significant ecological impacts of the changes that accompany watershed urbanization. The complex physical effects from elevated urban stream flows and stream channel alterations can damage or destroy stream and wetland habitats. Biological degradation is generally manifested more rapidly than physical degradation. Aquatic biota tend to respond immediately to widely fluctuating water temperatures, water quality, reduced OM inputs or other food sources, more frequently elevated stream flows, greater wetland water-level fluctuations, or higher sediment loads.

The relationship between stormwater discharge and the biological integrity of urban streams is shown in Figure 7.3 (Masterson and Bannerman, 1994). As it can be seen, habitat is impacted by changes in both water quality and quantity, and the volume and quality of sediment on the habitat should be considered.

Schueler (1995) proposed a direct relationship between watershed imperviousness and stream health, and found that stream health impacts tend to begin in watersheds with only 10–20% imperviousness (the 10% threshold). As shown in Figure 7.4, sensitive streams can exist relatively unaffected by urban stormwater with good levels of stream quality where impervious cover is less than 10%, although some sensitive streams have been observed to experience water-quality impacts at as low as 5% imperviousness. Impacted streams are threatened and show physical habitat changes (erosion and channel widening) and decreasing water quality where impervious cover is in the range of 10–25%.

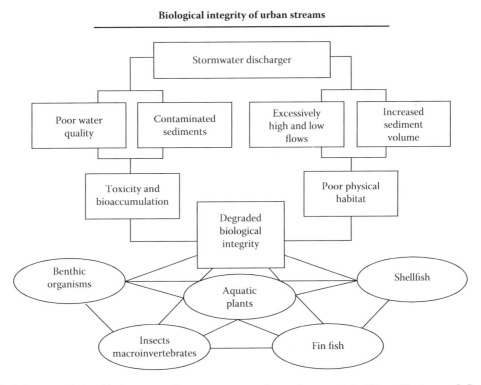

FIGURE 7.3 Relationship between urban stormwater and aquatic ecosystem. (From Masterson, J. P. and Bannerman, R. T. 1994. *National Symposium on Water Quality*, November.)

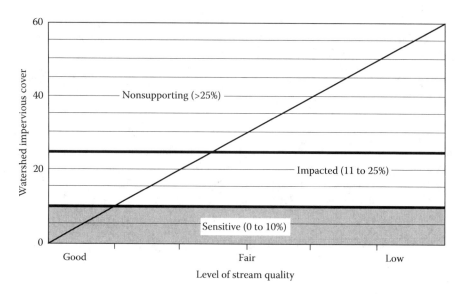

FIGURE 7.4 Relationship between impervious cover and stream quality. (Adapted from United States Environmental Protection Agency [US-EPA]. 1999. Preliminary data. Summary of urban storm water best management practices. EPA-821-R-99-012.)

Streams in watersheds where the impervious cover exceeds 25% are typically degraded, have a low level of stream quality, and do not support a rich aquatic community.

The physical impacts to streams due to urbanization and changes in watershed hydrology also cause many habitat changes. As shown in the comparison of healthy and eroding stream banks in Figure 7.5, loss of depth, sediment deposition, loss of shoreline vegetation, and higher temperatures

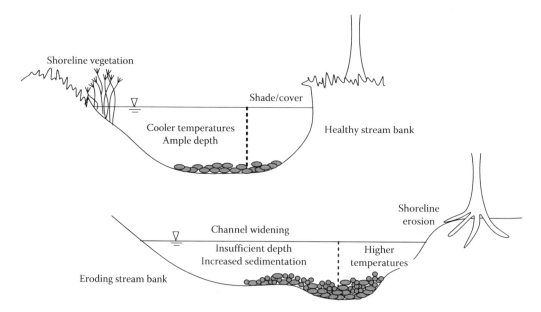

FIGURE 7.5 Comparison of a healthy stream bank and an eroding bank. (Adapted from Corish, K. A. 1995. Metropolitan Washington Council of Governments. Prepared for U.S. Environmental Protection Agency, Office of Wetlands, Oceans, and Watersheds, Grant no. X-818188-01-6.)

are combined to impact habitat. Schueler (1987) states that sediment pollution in the form of increased suspended solids can cause the following harmful impacts on aquatic life:

- Increased turbidity
- Decreased light penetration
- Reduced prey capture for sight-feeding predators
- Clogging of gills/filters of fish and aquatic invertebrates
- Reduced spawning and juvenile fish survival

Sediment is also a carrier of metals and other pollutants, and a source of bioaccumulating pollutants for bottom-feeding organisms. The rate of bioaccumulation is widely variable based on site-specific conditions including species, concentration, pH, temperature, and other factors.

7.5 URBANIZATION IMPACTS ON RIVERS AND LAKES

Urban areas affect the water quality in rivers through discharges of wastewaters, and thereby modify temperature, suspended solids, OM, turbidity, and faucal pollution indicators in rivers.

The self-purification processes happening in rivers and regulating soluble compounds in river water consist of flow turbulence, evaporation, and absorption to sediments, primary production, and OM oxidation (Chapman et al., 1998). Turbulence contributes to volatilization of chemicals and higher DO levels. Evaporation may increase pH, electrical conductivity, and precipitation of soluble materials. Consequences of sediment absorption are reduced concentrations of nutrients-dissolved organic carbon (DOC), soluble metals, and organic micropollutants. Finally, OM oxidation in the water column or anoxic sediment reduces pH, increases dissolved nutrients, reduces DO and DOC, and potentially increases soluble metal concentrations by adsorption (Chapman, 1992).

Current stormwater monitoring and impact assessment programs indicate that the most likely cause for degradation of biological integrity in receiving waters is a combination of physical habitat degradation, changes in the hydrological regime, food web disruptions, and long-term exposure to anthropogenic contaminants (Pitt and Bozeman, 1982). However, chronic or acute exposure to potentially toxic contaminants may be especially problematic for benthic organisms such as macroinvertebrates and for organisms that have a benthic life stage (e.g., salmonids during their embryonic development stage). Acute toxicity of aquatic biota due to exposure to stormwater runoff in receiving waters is rare.

Lakes and reservoirs are typified by long hydraulic residence times and relatively low capacity for decontamination. Lakes are often used for water supply, recreation, and receiving waters for urban effluents near urban areas. Thus, a delicate balance among the various water uses must be achieved. Lakes are typically thermally stratified, with the highest density water located at the bottom of the lake (hypolimnion) and lower density water found in the upper layer (epilimnion).

The epilimnion is completely combined with wind and waves, and is warmer than the hypolimnion. There is a clear physical division between both layers, identified as a thermocline, of which thickness typically varies from 10 to 20 m. In shallow lakes, stratification does not occur because of wind-induced mixing (Chapman et al., 1998). During the year, as temperature changes, many lakes undergo alternating periods of vertical mixing and stratification. Stratification can also be caused by dissolved solids (particularly chloride), which impede vertical mixing. One of the main characteristics of lakes and reservoirs is a significant water loss through evaporation. The main water-quality problems encountered in lakes and reservoirs are summarized in Table 7.8.

Acidification is one of the most important problems observed in lakes in temperate climates. It results from acidic depositions caused by air pollution and occurs in lakes with low alkalinity, hardness, conductivity, and dissolved solids. These conditions are normally found in lakes located

TABLE 7.8
Main Urban Water Quality Problems in Lakes and Reservoirs

Process	Cause	Effects on Water Quality
Acidification	Atmospheric deposition of acidity	Decrease in pH, increased concentrations of heavy metals, loss of biota
Increased salinity	Water balance modifications, soil leaching, industrial and municipal discharges	Increased TDS, increased treatment cost if dissolved solids exceed 1500–2000 mg/L
Eutrophication	Excess of nutrients	Increased algae and plant production; consumption of the hypolimnion oxygen; release of Fe, Mn, and metals from the hypolimnion: loss of diversity in the higher trophic levels; favorable conditions for reproduction of mosquitoes and spread of shistosomiasis
Pathogen contamination	Sewage discharges	Spread of disease including infection by bacteria, virus, protozoa and helminthes
Increased toxicity	Industrial and municipal discharges	Increased concentrations of trace organic and metal toxicants; introduction of endocrine disrupters; toxic effects through bioaccumulation and biomagnification; and fish tumors and loss of biota

in zones with noncarbonated soils (crystalline rocks or quartz sandstones) in temperate climates. Eutrophication is the process through which excess nutrients are responsible for increase in overall algal biomass, especially during "bloom" periods. This is due to increased loading of the nutrient that had previously been in short supply relative to their potential. This limiting nutrient is usually either phosphorus or nitrogen, but most often, and most consistently, it is phosphorus in freshwater lakes. Eutrophication degrades lake ecosystems in several ways. The filamentous algae are poorer food than diatoms to herbivores because of their structure and, sometimes, bad taste and toxicity (Walker, 1987).

7.6 IMPACTS OF URBANIZATION ON GROUNDWATER

Urbanization can have a major impact on groundwater recharge. As shown earlier in Figure 7.2, both shallow and deep infiltration decrease as watersheds undergo development and urbanization. Groundwater recharge is reduced along with a lowering of the water table. This change in watershed hydrology alters the base-flow contribution to stream flow, and it is most pronounced during dry periods. Ferguson (1990) points out that "base flows are of critical environmental and economic concern for several reasons. Base flows must be capable of absorbing pollution from sewage treatment plants and nonpoint sources, supporting aquatic life dependent on stream flow, and replenishing water-supply reservoirs for municipal use in the seasons when [water] levels tend to be lowest and water demands highest." Base flows on Long Island, New York, were substantially impacted by the construction of stormwater conveyance systems during the period of rapid development between the 1940s and 1970s. As shown in Table 7.9, a steady decline in the average percentage of base flow was observed for streams in urbanized sewered areas relative to streams in unsewered or rural areas (US-EPA, 1997).

Two types of urban effluent discharges, nonintentional (accidental) and intentional wastes, are reaching the groundwater. Nonintentional discharges are much more significant than expected. Infiltration of liquids from storage tanks and lagoons, sewage reuse in agriculture or landscape irrigation, landfill leaching, and infiltration of polluted water from channels and rivers are typical examples

TABLE 7.9
Average Percentage Base Flow of Selected Streams on Long Island by Area in New York

Years	Urbanized Sewered Area (% Flow from Base Flow)		Urbanized Unsewered Area (% Flow from Base Flow)		Rural Unsewered Area (% Flow from Base Flow)	
	Stream 1	Stream 2	Stream 1	Stream 2	Stream 1	Stream 2
1948–1953	—	86	84	94	96	95
1953–1964	63	69	89	89	95	97
1964–1970	17	22	83	84	96	97

Source: United States Environmental Protection Agency (US-EPA). 1997. Compendium of tools for watershed assessment and TMDL development. EPA-841-B-97-006.

of such discharges. Intentional discharges are primarily from dumping solid waste, contaminated liquids, chemicals and hazardous wastes on the ground and/or into the rivers and other water bodies.

7.6.1 Unintentional Discharges into Groundwater Aquifers

Foster et al. (2000) stated that

1. A high amount of urban human activities may potentially contaminate aquifers through nonintentional recharge, but only a few are responsible for the most severe problems.
2. The severity of the pollution caused is not directly proportional to the source size, because extended discharges from small operations (e.g., machine shops) can cause severe impacts.
3. The larger and more complex industries pose lower risks of accidental or intentional spills because of certain mandatory monitoring and quality control programs.
4. The concentration of pollutants in groundwater depends on the pollutant dispersion and persistence, more than on the type of pollution source.

The soil vadose zone retains a large part of the pollutants depending on the type of chemical compounds (concentration, volatility, and density). Furthermore, it depends on the geological properties of the soil material (physical heterogeneity and hydraulic characteristics), reactions involved (absorption, ion exchange, precipitation, oxidation, reduction, and biological transformation), and the hydrological conditions (retention time, flow pattern, and evaporation). In the following sections, aquifer pollution is discussed.

Any controlled or noncontrolled landfill can be a source of aquifer contamination. Old landfills or garbage dumps, which were built without adequate guidelines and without provisions for hazardous waste separation, are where most of such contamination is derived from. Landfills built from 1950 to 1970 are not equipped with geomembranes and are sources of leakages. In developing countries even the use of geomembranes are limited. In humid districts such problems are worse, where rain generates larger amounts of leakages and carrier plumes. In some arid climates, leakages may be smaller in volume, but may have higher concentrations of contaminants. The potential impacts of sanitary landfills on aquifers can be estimated from the type of solid wastes deposited and the amounts of rainfall (Nicholson et al., 1983).

Particularly in developing countries, tanks and lagoons are used for a dual purpose: to treat, to retain for evaporation or store liquid wastes (municipal, industrial, and mining drainage), or to use for flood control. Such storage facilities are generally less than 5 m deep and their hydraulic residence time varies from 1 to 100 days. Approximately all of these structures experience some leakage; even if the site is initially waterproof, the magnitude depends on the type and quality of construction and the maintenance program. The total volume of leakage was estimated at 10–20 mm/d by various authors (Miller and Scalf, 1978; Geake and Foster, 1986). Other sources of groundwater contaminants are due to confined hazardous wastes. Various trace of metals and organics may leak from these sites to the aquifers (NRC, 1998).

Given the population growth and the lack of sewers in certain urban areas in developing countries, wastewater may leak from septic tanks, latrines, sewers, wells, or directly discharge into soils (Lewis et al., 1988). Such leaks contribute to aquifer recharge and pollution.

Pollutants associated with sewage leakage include biodegradable OM, nitrogen compounds, phosphorus, microorganisms (including those causing typhoid, tuberculosis, cholera, and hepatitis), suspended solids, and trace organic compounds. Among these pollutants, nitrates are the most movable and persistent, which is the reason for their frequent presence in polluted aquifers.

Underground tanks are used for storing various liquids, but most often gasoline. Leaks usually develop due to corrosion and poor connections, and there is a close association between the frequency and size of the leaks and the age of the tank (Kostecki and Calabrese, 1989; Cheremisinoff, 1992). Generally, tanks that are more than 20 years old are very likely to leak, particularly if not properly maintained. With better design, construction, operation, and maintenance, leakage problems can be significantly reduced. By using cathodic protection or double-wall steel or plastic-reinforced fiberglass tanks, leaks can be controlled.

Many cases of groundwater contamination are caused by leaking gas station tanks. For example, in the United States, they account for at least one out of 30 leaks (Bedient et al., 1994). This pollution problem is exacerbated by the fact that gasoline tanks are widely distributed in cities, reflecting the local fuel demand rather than the need for environmentally suitable locations. Even though these leaks are usually small, they occur over long periods of time and produce pollution plumes of a great extent. Leaks should be detected by standard procedures and the tanks should be sealed to avoid such pollution. When pumping stations are coupled with service stations, significant amounts of organic solvents may also be spilled on soils, causing additional risks.

The conveyance of contaminated water in open channels and rivers is a frequent source of aquifer pollution and recharge. The significance of such an impact, which can be determined only by monitoring each case, depends on the rate and quality of the seepage flow. Industrial activities can seriously pollute the subsoil and groundwater depending on the type, volume, and the way of handling liquid and solid wastes. In industries using more than 100 kg of toxins per day (hydrocarbons, organic synthetic solvents, heavy metals, etc.), such risks are particularly significant. The pollutants involved are related to the type of industry as discussed by Foster et al. (2000).

In many cities and surrounding urban areas, small industries and service operations (mechanical factories, dry cleaning services, etc.) handling toxic substances such as chlorinated solvents, aromatic hydrocarbons, pesticides, and so on are potential polluters. Since they frequently store or discharge such wastes into soils instead of recycling or disposal at appropriate confined sites, they are considered as a major threat. Nevertheless, it is difficult to control small industries and services dealing with chemicals or detergents, because they often move, displace, or operate intermittently. It may be difficult to enforce such controls without sound regulations. To a small extent, the interment sites of humans and animals represent sources of microbiological contamination of aquifers. Watertight caskets must be used to avoid this problem, but they may not be affordable in under developed countries.

7.6.2 Intentional Discharges into Groundwater Aquifers

In urban areas with intensive land-use changes and impervious surfaces, aquifers may be used as water-storage facilities, also called "aquifer storage recovery systems, ASR." This is possible because any aquifer is essentially a water reservoir, from which water can be withdrawn when required. Depending on its intended use, stored water can be of varying quality (Table 7.10). In 1988, the first system of this nature began operating in the United States.

TABLE 7.10
Objectives of ASR

Quantity Objectives	Quality Objectives
Temporary water storage during various seasons of the year	Control of nutrient leaching
Long-term storage	Enhancing water well production
Storage for emergencies or as strategic reserves	Retarding water supply system expansion
Daily storage	Storing reclaimed water
Reduction of disinfection by-products	Soil treatment
Restoration of the phreatic levels	Refining water quality
Pressure and flow maintenance in the distribution network	Stabilizing aggressive waters
Improvement of the quality	Hydraulic control of pollution plumes
Prevention of saline intrusion	Water temperature maintenance to support fisheries
Water supply for agriculture	Reducing environmental effects caused by spills
	Compensating for salinity leaching from soils

Source: Oron, G. 2001. *Management of Effluent Reclamation via Soil-Aquifer-Treatment Procedures.* WHO expert consultation on Health Risks in Aquifer Recharge by Recycled Water, Budapest, November 8–9.

A well-defined methodology has been established and addresses design, construction, and operation for designing ASRs. As stated by Bouwer (1989), water-quality improvements during water injection depend on the combination of soil filtration properties and the method of water recharge. By infiltration through soils or direct injection, aquifers can be recharged. The latter requires better water quality than what is applied through soil infiltration to avoid well blockage and entry of pollutants into the subsoil and aquifer. The injected water must match the drinking water standards, or at least has the same quality as that of the aquifer (Crook et al., 1995).

Some critical considerations of ASR applications stated by Oron (2001) in conjunction with reclaimed water include the following:

- The risk of introducing recalcitrant pollutants into aquifers and soils, if there is no pretreatment of nonresidential discharges to the drainage system
- The leaching of dangerous pollutants originating in households
- The dilemma of sterilizing the effluent prior to recharge knowing that it could be fatal to the beneficial soil microorganisms

7.7 OVERALL STATE OF IMPACTS

The impacts of urbanization on various water resources and components of the environment in urban areas have been discussed in this chapter. The urbanization impacts are focused on water and other elements interacting with urban water, and the impacts of urban areas on the atmosphere, wetlands, surface waters, soils, and groundwater. As discussed, stormwater is one of the main pollution sources of urban areas. Furthermore, stormwater causes much pollution into the receiving surface and groundwater. BMPs are imposed and regulated in many municipalities such as City of San Diego in California.

The change in watershed hydrology due to urban development also causes channel widening and scouring, and sediment deposit in the urban streams. Visible impacts include eroded and exposed stream banks, fallen trees, sedimentation, and turbid conditions. The increased frequency of flooding in urban areas also poses a threat to public safety and property. Aquatic and riparian habitats in urban streams are also affected by both water quality and quantity impacts.

Higher levels of pollutants, increased flow velocities and erosion, alteration of riparian corridors, and sedimentation associated with stormwater runoff negatively impact the integrity of aquatic ecosystems. These impacts include the degradation and loss of aquatic habitat, and reduction in the numbers and diversity of fish and macro-invertebrates. Public health impacts are for the most part related to bacteria and disease-causing organisms carried by urban stormwater runoff into waters used for water supplies, fish, and recreation.

While water-quality impacts are often unobserved by the general public, other stormwater impacts are more visible. Stream channel erosion and channel bank scour provide direct evidence of water quantity impacts caused by urban stormwater. Urban runoff increases directly with imperviousness and the degree of watershed development. As urban areas grow, urban streams are forced to accommodate larger volumes of stormwater runoff that recur on a more frequent basis. This leads to stream channel instability.

Water supplies can potentially be contaminated by urban runoff, posing a public health threat. Bathers and others coming in contact with contaminated water at beaches and recreational sites can also be subject to health hazard. Beach closures caused by urban runoff have a negative impact on the quality of life, and can impede economic development as well. Similarly, the bacterial contamination of shellfish beds poses a public health threat, and shellfish bed closures negatively impact the fish industry and local economies.

Urban groundwater is affected by urbanization regarding both quantity and quality. The urbanization may also affect the aquifer characteristics, because of concentration and weight of infrastructures constructed in urban areas. Aquifers and groundwater tables are affected by urbanization through either reduced recharge (increased runoff) or increased recharge (leaking water mains, leaking sewers, stormwater infiltration, and certain absorption wells), and pollution by infiltrating effluents or accidental spills, depending on local circumstances as well as leakage from underground tanks of petroleum products.

A concluding remark could be that urbanization has a major impact on environmental quality of urban life and the surrounding ecosystem. It is important to differentiate between causes of a degraded environment by urbanization and the effects of BMP in an urban environment that has been enhanced through remediation activities.

PROBLEMS

1. Assume that an 8000-km^2 residential area is 50% impervious and has a population density of 100 people/km^2. The area is serviced by a new combined sewer system. The street sweeping interval is approximately 10 days. The annual precipitation is 68 mm. Determine the total annual loading (kg/year) and the average concentration (mg/L) of BOD$_5$, SS, VS, PO$_4$, and N in the stormwater runoff from this area.
2. Using the same problem statement as mentioned for the estimation of nitrate load (Example 7.2), estimate the total annual loading (kg/year) and the average concentration (mg/L) of BOD$_5$, SS, VS, and PO$_4$ in the stormwater runoff from this area.
3. Using the unit-area loading model for computing the pollution loads of the previous problem, compare the results with those found in the above problem and comment on any differences.

4. Discuss the parameters affecting runoff pollution and determine the multipliers (such as 1.04 in Equation 7.1, 1.0104 in Equation 7.2, and 0.0441 in Equations 7.7 and 7.10) of the formulations proposed in this chapter. If we use inches instead of millimeter and acres instead of hectares in Equations 7.1, 7.2, 7.7, and 7.10, determine the modified multipliers that should be used.

REFERENCES

Akan, A. O. 1993. *Urban Storm Water Hydrology—A Guide to Engineering Calculations*. Technomic, Lancaster, PA.

Alexander, R. B., Smith, R. A., and Schwarz, G. E. 2004. Estimates of diffuse phosphorus sources in surface waters of the United States using a spatially referenced watershed model. *Water Science and Technology* 49(3): 1–10.

Alley, W. A. and Veenhuis, J. E. 1983. Effective impervious area in urban runoff modeling. *Journal of Hydrological Engineering (ASCE)* 109(2): 313–319.

Arnold, C. L. and Gibbons, C.J. 1996. Impervious surface coverage: The emergence of a key environmental indicator. *Journal of the American Planning Association* 62(2): 243–258.

Bannerman, R., Owens, D. W., Dodds, R. B., and Hornewer, N.J. 1993. Sources of pollutants in Wisconsin stormwater. *Water Science and Technology* 28: 241–259.

Bedient, P. H., Rifai, H. S., and Newell, C.J. 1994. *Ground Water Contamination: Transport and Remediation*. Prentice Hall, Englewood Cliffs, NJ.

Bergman, M., Green, W., and Donnangelo, L. 2002. Calibration of storm loads in the South Prong Watershed, Florida, using BASINS-HSPF. *Journal of the American Water Resources Association* 38(3): 1423–1436.

Burton, G. A. and Pitt, R. E. 2002. *Stormwater Effects Handbook*. Center for Watershed Protection, Lewis Publishers, CRC Press, Boca Raton, FL.

Chapman, D., Ed. 1992. *Water Quality Assessment*. WHO/UNESCO/UNEP. Chapman & Hall, London.

Chapman, P. M., Wang, F., Janssen, C., Persoone, G., and Allen, H. E. 1998. Ecotoxicology of metals in aquatic sediments: Binding and release, bioavailability, risk assessment, and remediation. *Canadian Journal of Fisheries and Aquatic Sciences* 55: 2221–2243.

Cheremisinoff, P. 1992. *A Guide to Underground Storage Tanks: Evaluation, Site Assessment and Remediation*. Prentice Hall, Englewood Cliffs, NJ.

Corish, K. A. 1995. Clearing and grading strategies for urban watersheds. Metropolitan Washington Council of Governments. Prepared for U.S. Environmental Protection Agency, Office of Wetlands, Oceans, and Watersheds, Grant #X-818188-01-6.

CWP. 2003. Impacts of impervious cover on aquatic systems. Watershed Protection Research Monograph No. 1.

Davies, P. H. 1986. Toxicology and chemistry of metals in urban runoff. In: J. Urbonas and B. Roesner, Eds, *Urban Runoff Quality—Impacts and Quality Enhancement Technology*. ASCE, New York.

Driver, N. E. and Tasker, G. D. 1990. Techniques for estimation of storm-runoff loads, volumes and selected constituent concentrations in urban watersheds in the United States. Water Supply Paper 2363, U.S. Geol. Surv., Reston, VA.

Field, R. and Pitt, R. 1990. Urban storm-induced discharge impacts: US Environmental Protection Agency Research Program Review. *Water Science Technology* 22: 10–11.

Foster, G. D., Roberts, E. C., Gruessner, B., and Velinsky, D. J. 2000. Hydrogeochemistry and transport of organic contaminants in an urban watershed of Chesapeake Bay. *Applied Geochemistry* 15: 901–916.

Geake, A. K. and Foster, S. D. 1986. Groundwater recharge controls and pollution pathways in the alluvial aquifer of metropolitan Lima, Peru; results of detailed research at La Molina Alta and at San Juan de Miraflores wastewater reuse complex. British Geological Survey, Wallingford, U.K.

Heaney, J. F., Huber, W. C., Sheikh, H., Medina, M. A., Doyle, J. R., Pelt, W. A. and Darling, J. E. 1977. Nationwide evaluation of combined sewer overflows and urban stormwater discharges, Vol. II. Cost Assessment and Impacts, Report ERA-800/2-77-084. Environmental Protection Agency, Washington, DC.

Hoffman, E. J., Latimer, J. S., Mills, G. L., and Quinn, J. G. 1982. Petroleum hydrocarbons in urban runoff from a commercial land-use area. *Journal of the Water Pollution Control Federation* 54: 1517–1525.

Hollis, G. E. 1975. The effect of urbanization on floods of different recurrence interval. *Water Resources Research* 66: 84–88.

Horner, R. R., Booth, D. B., Azous, A., and May, C. W. 1997. Watershed determinants of ecosystem functioning. In: L. A. Roesner, Ed., *Effects of Watershed Development and Management on Aquatic Ecosystems*. ASCE, Washington, DC.

Im, S., Brannan, K., and Mostaghimi, S. 2003. Simulating hydrologic and water-quality impacts in an urbanizing watershed. *Journal of the American Water Resources Association* 42(2): 1465–1479.

Johnston, R. K., Wang, P. F., Halkola, H., Richter, K. E., Whitney, V. S., May, C. W., Skahill, B. E., Choi, W. H., Roberts, M., and Ambrose, R. 2003. An integrated watershed-receiving water model for Sinclair and Dyes Inlets, Puget Sound, Washington. *Proceedings of the Estuarine Research Federation (ERF) Conference*, Seattle WA.

Karr, J. R. 1991. Biological Integrity: A long neglected aspect of water resources management. *Ecological Applications* 1(1): 66–84.

Kostecki, P. T. and Calabrese, E. 1989. *Petroleum Contaminated Soils: Remediation Techniques, Environmental Fate and Risk Assessment*, Vol. 1. Lewis Publishers, Chelsea, MI.

Mallin, M., Williams, K., Esham, E., and Lowe, R. 2000. Effect of human development on bacteriological water quality in coastal watersheds. *Ecological Applications* 10(4): 1047–1056.

Marsalek, J., Karamouz, M., Jiménez-Cisneros, B. E., Malmquist, P. A., Goldenfum, J., and Chocat, B. 2006. *Urban Water Cycle Processes and Interactions*. UNESCO, Paris.

Masterson, J. P. and Bannerman, R. T. 1994. Impacts of stormwater on urban streams in Milwaukee County, Wisconsin. *National Symposium on Water Quality*, November.

McDonnell, M. J. and Pickett, S.T. 1990. Ecosystem structure and function along urban–rural gradients: An unexploited opportunity for ecology. *Ecology* 71(4): 1232–1237.

Miller, D. and Scalf, M. 1974. New priorities for groundwater quality protection. *Ground Water* 12: 335–347.

Nicholson, R. V., Cherry, J. A., and Reardon, E. J. 1983. Migration of contaminants in groundwater at a landfill: A case study. *Journal of Hydrology* 63: 131–176.

Novotny, V. and Chester, G. 1981. *Handbook of Non-Point Pollution: Sources and Management*. Van Nostrand Reinhold, New York.

Oron, G. 2001. *Management of Effluent Reclamation via Soil-Aquifer-Treatment Procedures*. WHO expert consultation on Health Risks in Aquifer Recharge by Recycled Water, Budapest, November 8–9.

Pitt, R. and Bozeman, M. 1982. Sources of urban runoff pollution and its effects on an urban creek. EPA-600/52-82-090, U.S. Environmental Protection Agency, Cincinnati, OH.

Pitt, R., Field, R., Lalor, M., and Brown, M. 1995. Urban stormwater toxic pollutants: Assessment, sources, and treatability. *Water Environment Research* 67(3): 260–275.

Pitt, R., Maestre, A., and Morquecho, R. 2004. The National Stormwater Quality Database (NSQD) Version 1.1 Report. University of Alabama, Department of Civil and Environmental Engineering, Tuscaloosa, AL.

Schueler, T. R. 1987. *Controlling Urban Runoff: A Practical Manual/or Planning and Designing Urban RMP's*. Metropolitan Information Center, Metropolitan Washington Council of Governments, Washington, DC.

Schueler, T. R. 1995. The architecture of urban stream buffers. *Watershed Protection Techniques* 1(4): 155–163.

Shaver, E., Horner, R., Skupien, J., May, C., and Ridley, J. 2007. *Fundamentals of Urban Runoff Management: Technical and Institutional Issues*. North American Lake Management Society in cooperation with the U.S. Environmental Protection Agency.

Smith, R. A., Schwarz, G. E., and Alexander, R. B. 1997. Regional interpretation of water-quality monitoring data. *Water Resources Research* 33(12): 2781–2798.

Stein, E. D., Tiefenthaler, L. L., and Schiff, K. 2006. Watershed-based sources of polycyclic aromatic hydrocarbons in urban stormwater. *Environmental Toxicology and Chemistry* 25(2): 373–385.

Tong, S. and Chen, W. 2002. Modeling the relationship between land use and surface water quality. *Journal of Environmental Management* 66: 377–393.

Van Metre, P. C., Mahler, B. J., and Furlong, E. T. 2000. Urban sprawl leaves its PAH signature. *Environmental Science and Technology* 34(19): 4064–4070.

United States Environmental Protection Agency (US-EPA). 1983. Results of nationwide urban runoff program. EPA-PB/84-185552.

United States Environmental Protection Agency (US-EPA). 1997. Compendium of tools for watershed assessment and TMDL development. EPA-841-B-97-006.

United States Environmental Protection Agency (US-EPA). 1999. Preliminary data summary of urban storm water best management practices. EPA-821-R-99-012.

United States Environmental Protection Agency (US-EPA). 2005. Handbook for developing watershed plans to restore and protect our waters. US-EPA Technical Publication, EPA-841-B-05-005.

Van Metre, P. C., Mahler, B. J., and Furlong, E. T. 2000. Urban sprawl leaves its PAH signature. *Environmental Science and Technology* 34(19): 4064–4070.

Walker, W. W. 1987. Phosphorus removal by urban runoff detention basins. *Lake and Reservoir Management* 3: 314–328.

Young, K. and Thackston, E. 1999. Housing density and bacterial loading in urban streams. *Journal of Environmental Engineering* 26(2): 1177–1180.

8 Tools and Techniques

8.1 INTRODUCTION

Engineering and management aspects of urban water require utilizing technical tools and techniques that are classified here as simulation, optimization, and economic techniques. Engineering applications primarily use analysis and design models. The overlay between engineering and management tools in urban water is simulation models, which can be used to test the performance of urban water systems. Another simulation technique that has received a great deal of attention is system dynamics utilizing object-oriented programming, which are discussed in Chapter 10, to simulate in a physical–mathematical fashion the behavior of an urban system under different design and operation alternatives. There are certain inserts in this chapter from the Karamouz et al. (2003) book that are adopted for urban planning and management. In the remaining of this chapter, stochastic simulation methods as well as neural networks are introduced. Then the conventional and evolutionary optimization techniques are introduced. It is followed by an introduction to fuzzy modeling. Finally, the basic economic models for urban water systems are described.

8.2 SIMULATION TECHNIQUES

Mathematical formulation of urban water systems is the basis of the simulation techniques. These techniques are useful methods for analyzing the urban water systems due to their mathematical simplicity and versatility. Simulation models are used for the evaluation of system performance under a given set of inputs and operating conditions. They do not identify optimal decisions compared to optimization models but they can merely be used for simulating a large number of scenarios. By using the simulation models, a very detailed and realistic representation of the complex physical, economic, and social characteristics of a water resources system can be considered in the form of mathematical formulations. The concepts inherent in the simulation approach are easier to comprehend and communicate than other modeling concepts.

Furthermore, the simulation methods can solve urban water planning models with highly nonlinear relationships and constraints that cannot be handled by constrained optimization procedures. The simulation models may deal with steady-state or transient conditions. For example, to study the operation of an urban water system over a relatively long period of time during which no major alterations in the system occur, we need to perform a steady-state analysis. The study of an, we reject to perform urban water system, such as culverts and diversion systems that are designed based on flood occurrence, lies in the area of transient analysis. The simulation models are categorized into two groups, namely, deterministic and stochastic. If the system is subject to random input events, or generates them internally, the model is called stochastic. The model is deterministic if no random components are involved. In this chapter, stochastic simulation is discussed first, and in deterministic simulation, artificial neural network (ANN) and fuzzy set theory applications are presented. Other deterministic applications are described throughout the book when we made the system representations in terms of equations, simple models, and heuristics.

8.2.1 Stochastic Simulation

To assess the effects of random elements on the sequence of events, stochastic simulation models can be used. The basis for stochastic simulation of urban water systems is generating independent random numbers with a uniform distribution. The generated random number cannot be used directly, because most of the random elements in nature have different distributions. So a transformation method is necessary in order to obtain random values from other than uniform distributions.

Assume a discrete random variable with values $x_1, x_2, \ldots$ and probabilities of occurrence as $p_1, p_2, \ldots$. Divide the probability interval of $[0, 1]$ into subintervals $[0, p_1), [p_1, p_1 + p_2), [p_1 + p_2, p_1 + p_2 + p_3)$, and so on, and let $I_1, I_2, \ldots$ denote these intervals. If U is a uniform variable in $[0, 1]$, then $P(U \in I_k) = p_k$. If x_k is the random variable that we wish to generate, its values are the x_k numbers, $k = 1, 2, \ldots$, and

$$P(X = x_k) = P(U \in I_k) = (p_1 + p_2 + \cdots + p_k) - (p_1 + p_2 + \cdots + p_{k-1}) = p_k.$$

For example, assume that X is a Bernoulli variable with probability p. That is, the value of X is either 1 or 0. The value of 1 can be considered as decision about doing something or success in maintaining service in supplying the demand, whereas the value of 0 can be considered as doing nothing or failing to provide adequate service, say, in supplying the demand:

$$P(X = x_k) = p \quad \text{and} \quad P(X = 0) = q = (1 - p).$$

In that case $I_1 = [0, p), I_2 = [p, 1)$. So, from a uniform random number, X is 1 if $U < p$ and X is zero otherwise.

The evolution of the Bernoulli variable to normal distribution, which is widely used in urban water studies, is discussed by Karamouz et al. (2003). In the following inserts, this process is explained briefly.

The binomial variable can be generated by considering n independent Bernoulli variables with the same parameter p as $X_1, X_2, \ldots, X_n$; then $X = X_1 + X_2 + \cdots + X_n$ is a binomial variable with parameters n and p. Then, the above-mentioned simple procedure is used n times, and the resulting random numbers are added. Then

$$P_k = p(x = k) = \binom{n}{k} p^k q^{n-k}.$$

If a binomial variable $n \to \infty, p \to 0$ so that np is kept at constant λ level, the distribution is called Poisson. Equation 8.1 shows that the probabilities of Poisson variables have only one parameter λ and take on values 0, 1, 2, and so on. Equation 8.2 proposes the recursive relation that proceeds from the general method. So, exponential random variables can be effectively used here as well:

$$p_k = P(X = k) = \left(\frac{\lambda^k}{k!}\right) e^{-\lambda}, \tag{8.1}$$

$$p_k = p_{k-1} \frac{\lambda}{k}. \tag{8.2}$$

Assume a number of independent exponential random variables with the same λ parameters as $X_1, X_2, \ldots$, so $E(X_k) = 1/\lambda$ for all k, and add the X_k values until

$$\sum_{i=1}^{k-1} X_i < 1 \leq \sum_{i=1}^{k} X_i. \tag{8.3}$$

Then $X = k$ gives a random value of the Poisson variable with the same parameter value λ.

Continuous random variables are generated by using their cumulative distribution functions. Let X be a random variable with distribution function $F(x)$ and let U be uniform in $[0, 1]$. Let u be a random value of U; then the solution of the equation $F(x) = u$ for x gives a random value of X. In order to see this, consider the distribution function of X at any real value t:

$$P(X \leq t) = P(F(X) \leq F(t)) = P(U \leq F(t)) = F(t),$$

since U is uniform in $[0, 1]$. Here we assume that F is strictly increasing, and therefore equation $F(x) = u$ has a unique solution for all $0 < u < 1$. The events $U = 0$ and $U = 1$ occur with zero probability.

As an example, if X is an exponential with parameter λ, then $F(x) = 1 - e^{-\lambda x}$; so we have to solve the equation

$$1 - e^{-\lambda x} = u,$$

which implies that

$$x = -\frac{1}{\lambda} \ln(1 - u). \tag{8.4}$$

A random variable is called *uniform* in a finite interval $[a, b]$ if its distribution function is

$$F(x) = \begin{cases} 0 & \text{if } x \leq a, \\ \frac{x-a}{b-a} & \text{if } a < x \leq b, \\ 1 & \text{if } x > b. \end{cases} \tag{8.5}$$

By differentiation,

$$f(x) = \begin{cases} \frac{1}{b-a} & \text{if } a < x < b, \\ 0 & \text{otherwise.} \end{cases} \tag{8.6}$$

Assume that $\alpha, \beta \in [a, b]$; then

$$P(\alpha \leq X \leq \beta) = F(\beta) - F(\alpha) = \frac{\beta - a}{b - a} - \frac{\alpha - a}{b - a} = \frac{\beta - \alpha}{b - a}, \tag{8.7}$$

showing that this probability depends only on the length $\beta - \alpha$ of the interval $[\alpha, \beta]$ and is independent of the location of this interval. That is why this variable is called *uniform*.

A continuous random variable is called *exponential* if

$$F(x) = \begin{cases} 0 & \text{if } x \leq 0, \\ 1 - e^{-\lambda x} & \text{if } x > 0, \end{cases} \tag{8.8}$$

where $\lambda > 0$ is a given parameter. It can be easily seen that

$$f(x) = \begin{cases} \lambda e^{-\lambda x} & \text{if } x > 0, \\ 0 & \text{otherwise.} \end{cases} \tag{8.9}$$

The exponential distribution has the forgetfulness property, which means the following: Let $0 < x < x + y$; then

$$P(X > x + y | X > y) = P(X > x), \tag{8.10}$$

showing that if X represents the lifetime of any entity, then at any age y, surviving an additional x time periods does not depend on its age. Because of the forgetfulness property, exponential variables are seldom used in lifetime modeling. For such purposes, the gamma or the Weibull distribution is used, which will be introduced next.

Consider two parameters, $\alpha, \lambda > 0$. The gamma variable with these parameters is defined by the following density function:

$$f(x) = \begin{cases} \dfrac{\lambda e^{-\lambda x}(\lambda x)^{\alpha-1}}{\Gamma(\alpha)} & \text{if } x > 0, \\ 0 & \text{otherwise,} \end{cases} \qquad (8.11)$$

where

$$\Gamma(\alpha) = \int_0^\infty x^{\alpha-1} e^{-x}\, dx$$

is the gamma function. Let $\alpha = n$ be an integer, and let $X_1, X_2, \ldots, X_n$ be independent exponential variables with the same parameter λ. Then $X_1 + X_2 + \cdots + X_n$ follows the gamma distribution with parameters $\alpha = n$ and λ.

The distribution function of gamma variables cannot be given in a closed form simple expression except in very special cases. They are tabulated; so function tables are available. Alternatively, numerical integration can be used.

Consider two parameters $\lambda, \beta > 0$. The Weibull distribution with these parameters is defined by the distribution function

$$F(x) = \begin{cases} 1 - e^{-\lambda x^\beta} & \text{if } x \geq 0, \\ 0 & \text{otherwise.} \end{cases} \qquad (8.12)$$

By differentiation,

$$f(x) = \begin{cases} \lambda \beta x^{\beta-1} e^{-\lambda x^\beta} & \text{if } x > 0, \\ 0 & \text{otherwise.} \end{cases} \qquad (8.13)$$

It is easy to see that if Y is an exponential with parameter λ, then $X = Y^{1/\beta}$ is a Weibull variable with parameters λ and β.

Random values for a Weibull variable can be obtained simply using exponential variables and the definition of the Weibull distribution. It is well identified that if Y is an exponential with parameter α, then $X = Y^{1/\beta}$ with $\beta > 0$ is a Weibull variable with parameters α and β. Then a random Weibull value can be obtained as follows:

$$x = \left[-\frac{1}{\lambda} \ln(1-u) \right]^{1/\beta}, \qquad (8.14)$$

where u is a uniform value in $[0, 1]$.

It is well known that if X_k ($k = 1, 2, \ldots, n$) are independent exponential variables with the same λ parameter, then $X_1 + X_2 + \cdots + X_n$ is a gamma variable with parameters $\alpha = n$ and $\beta = 1/\lambda$. Hence, we have to generate independent identical variable values $u_1, u_2, \ldots, u_n$; then

$$X = \sum_{k=1}^n x_k = -\frac{1}{\lambda} \sum_{k=1}^n \ln(1-u_k) = -\frac{1}{\lambda} \ln[(1-u_1),(1-u_2),\ldots,(1-u_n)] \qquad (8.15)$$

follows the given gamma distribution.

Tools and Techniques

In statistical methods, we often use distributions arising from the normal variable. The *standard normal* variable is given by the density function

$$\varphi(x) = \frac{1}{\sqrt{2\pi}} e^{-x^2/2} \quad (-\infty < x < \infty). \tag{8.16}$$

The distribution function cannot be given by an analytic form. It is tabulated and is usually denoted by $\Phi(x)$. A general *normal* variable is obtained as follows. Let μ and $\sigma > 0$ be two given parameters, and Z a standard normal variable. Then $X = \sigma Z + \mu$ follows a normal distribution with parameters μ and σ. It is easy to see that, in general,

$$F(x) = \Phi\left(\frac{x-\mu}{\sigma}\right) \tag{8.17}$$

and

$$f(x) = \frac{1}{\sigma\sqrt{2\pi}} e^{-((x-\mu)^2)/2\sigma^2}. \tag{8.18}$$

In the case of normal distribution, the general method cannot be used easily, since the standard normal distribution function Φ is not given in analytic form, it is only tabulated. Therefore the solution of the equation

$$\Phi\left(\frac{x-\mu}{\sigma}\right) = u$$

requires the use of a function table, which can be included in the general software. An easier approach is offered by the central limit theorem, which implies that if $u_1, u_2, \ldots, u_n$ are independent uniform numbers in $[0, 1]$, then for large values of n,

$$Z = \frac{\sum_{k=1}^n u_k - (n/2)}{\sqrt{n/12}} \tag{8.19}$$

is a standard normal value, so if μ and σ^2 are given, then

$$X = \sigma Z + \mu \tag{8.20}$$

follows the normal distribution with mean μ and variance σ^2.

Consider now any process that depends on random elements. Let $X_1, X_2, \ldots, X_n$ denote the random variables of the process. Assume that the outcome is a function of these variables, $Y = f(X_1, \ldots, X_n)$, where function f might be a given function, or represents a computational procedure such as given by neural networks. For any random values $x_1^{(i)}, \ldots, x_n^{(i)}$ of $X_1, X_2, \ldots, X_n$, the corresponding values of Y can be obtained as $y^{(i)} = f(x_1^{(i)}, \ldots, x_n^{(i)})$. Let N be the number of simulations, so $i = 1, 2, \ldots, N$. The values $y^{(1)}, y^{(2)}, \ldots, y^{(N)}$ can be considered as a random sample of Y; so standard statistical methods can be used to examine the characteristics of the random outcome. In most cases, we are interested in $E(Y)$, which is estimated by the sample mean

$$E(Y) \approx \frac{1}{N} \sum_{i=1}^N y^{(i)} = \bar{y}. \tag{8.21}$$

So the sample mean is accepted as the expected value. The accuracy of this estimation can be characterized by its variance:

$$\mathrm{Var}(\bar{y}) = \frac{1}{N} \mathrm{Var}(f(X_1, \ldots, X_n)), \tag{8.22}$$

which converges to zero as $N \to \infty$. If N is sufficiently large, then $\bar{y}$ can be considered as a normal variable as the result of the central limit theorem. Therefore we have for any $\varepsilon > 0$,

$$P(|\bar{y} - E(Y)| < \varepsilon) = P(-\varepsilon < \bar{y} - E(Y) < \varepsilon) = P\left(-\frac{\varepsilon\sqrt{N}}{\sigma} < \frac{\bar{y} - E(Y)}{\sigma/\sqrt{N}} < \frac{\varepsilon\sqrt{N}}{\sigma}\right),$$

where $\sigma^2 = \text{Var}(f(X_1, \ldots, X_n))$. Let Φ denote again the standard normal distribution function; then this probability can be further simplified as

$$\Phi\left(\frac{\varepsilon\sqrt{N}}{\sigma}\right) - \Phi\left(-\frac{\varepsilon\sqrt{N}}{\sigma}\right) = 2\Phi\left(\frac{\varepsilon\sqrt{N}}{\sigma}\right) - 1. \quad (8.23)$$

This is the probability that $\bar{y}$ approximates the true expectation $E(Y)$ within the error bound ε. Note that as $N \to \infty$, this probability value converges to $2\Phi(\infty) - 1 = 1$.

A comprehensive summary of random number generators and simulation methods is presented by Rubinstein (1981). In Table 8.1, the expectations and variances of the most popular distribution types are given.

8.2.2 STOCHASTIC PROCESSES

Assume that a certain event occurs at random time point $0 \leq t_1 < t_2 < \cdots$. These events constitute a *stochastic process*. For instance, the times when earthquakes, floods, rainfalls, and so on occur

TABLE 8.1
Expectations and Variances of the Most Popular Distribution Types

Distribution	$E(x)$	$\text{Var}(x)$
Bernoulli	p	pq
Binomial	Np	npq
Geometric	$\dfrac{1}{p}$	$\dfrac{q}{p^2}$
Negative binomial	$\dfrac{r}{p}$	$\dfrac{rq}{p^2}$
Hypergeometric	$n\dfrac{s}{N}$	$n\dfrac{s}{N}\left(1 - \dfrac{s}{N}\right)\left[1 - \dfrac{n-1}{N-1}\right]$
Poisson	λ	λ
Uniform	$\dfrac{a+b}{2}$	$\dfrac{(b-a)^2}{12}$
Exponential	$\dfrac{1}{\lambda}$	$\dfrac{1}{\lambda^2}$
Gamma	$\dfrac{\alpha}{\lambda}$	$\dfrac{\alpha}{\lambda^2}$
Weibull	$\lambda^{-1/\beta}\Gamma\left(1 + \dfrac{1}{\beta}\right)$	$\lambda^{-2/\beta}\Gamma\left(1 + \dfrac{2}{\beta}\right) - \mu^2$
Normal	μ	σ^2
Chi-square	n	$2n$
T	0 (if $n > 1$)	$\dfrac{n}{n-2}$ (if $n > 2$)
F	$\dfrac{m}{m-2}$ (if $m > 2$)	$\dfrac{m^2(2m + 2n - 4)}{m(m-2)^2(m-4)}$ (if $m > 4$)

Source: Karamouz, M., Szidarovszky, F., and Zahraie, B. 2003. *Water Resources Systems Analysis*. Lewis Publishers, CRC Press Company, Boca Raton, FL.

define stochastic processes. The process can be completely defined mathematically if we know the distribution functions of $t_1, t_2 - t_1, t_3 - t_2, \ldots$. A very important characteristic of the stochastic process is the number of events $N(t)$ that occur in the time interval $[0, t]$. The *Poisson process*, which is defined as follows, is the most frequently used stochastic process as expressed by Karamouz et al. (2003):

i. The number of events is zero at $t = 0$. This shows that the process starts at $t = 0$.
ii. The number of events occurring in mutually exclusive time intervals is independent and this is called "the independent increment assumption." For example, for time intervals of $0 \leq t_1 < t_2 < t_3 < t_4$, this condition states that the number of events in interval $[t_1, t_2]$ [which is $N(t_2) - N(t_1)$] is independent of the number of events in interval $[t_3, t_4]$, which is $N(t_4) - N(t_3)$.
iii. The distribution of $N(t + s) - N(t)$ depends on only s and is independent of t. This means that the distribution of the number of events occurring in any given time interval depends only on the length of the interval and not on its location. This condition is known as the stationary increment assumption.
iv. In a small interval of length Δt, the probability that one event occurs is a multiplier of Δt.
v. In a small interval of length Δt, the probability that at least two events take place is approximately zero.

Conditions (iv) and (v) can be expressed in the mathematical format as follows:

$$\lim_{\Delta t \to 0} \frac{P(N(\Delta t) = 1)}{\Delta t} = \lambda, \tag{8.24}$$

where $\lambda > 0$ is a constant:

$$\lim_{\Delta t \to 0} \frac{P(N(\Delta t) \geq 2)}{\Delta t} = 0. \tag{8.25}$$

Equation 8.26 denotes a Poisson distribution with parameter λt for $N(t)$ based on the above-mentioned assumption:

$$P(N(t) = k) = \frac{(\lambda t)^k}{k!} e^{-\lambda t}. \tag{8.26}$$

The distribution functions of different time intervals ($X_1 = t_1, X_2 = t_2 - t_1, X_3 = t_3 - t_2, \ldots$), which can be denoted by $F_1, F_2, F_3, \ldots$, are shown in Equations 8.27 and 8.28.

$$F_1(t) = P(t_1 < t) = 1 - P(t_1 \geq t) = 1 - P(N(t) = 0) = 1 - e^{-\lambda t}. \tag{8.27}$$

From Equation 8.27, therefore, X_1 is exponentially distributed with expectation $1/\lambda$.

For all $k \geq 2$ and $s, t \geq 0$, the distribution function of X_k is the same as that of X_1. All variables $X_1, X_2, X_3, \ldots$ are independent of each other and they are exponential with the same parameter λ as illustrated in Equation 8.28.

$$P(X_k > t \mid X_{k-1} = s) = P(0 \text{ event occurs in } (s, s+t) \mid X_{k-1} = s) = P(0 \text{ event occurs in } (s, s+t))$$

$$= P(0 \text{ event in } (0, t)) = P(N(t) = 0) = e^{-\lambda t}. \tag{8.28}$$

The distribution functions of the above three distributions cannot be presented in simple equations; therefore they are tabulated, and so their particular values can be looked up from these tables. For more details and tables, see Milton and Arnold (1995) or Ross (1987).

8.2.3 Artificial Neural Networks

ANNs provide a convenient means of either (1) simulating a system for which there is a large data set, but no known mathematical model exists, or (2) simplifying an excessively complex model. ANNs are global nonlinear function approximates, and are powerful and easy to use (StatSoft, 2002). These models are used in different urban water resource management fields such as simulating urban water demand based on the hydroclimatic and socioeconomic parameters and urban runoff based on the rainfall and basin characteristics.

The architectures of ANNs are adapted from biological neural networks, which can recognize patterns and learn from their interaction with the environment. It is believed that the powerful functionality of a biological neural system is attributed to the parallel distributed processing nature of a network of cells, known as neurons. An ANN emulates this structure by distributing the computation to small and simple processing units, called artificial neurons, or nodes. In ANNs, neurons receive a number of inputs from either the original data or other neurons. Each connection to a given neuron has a particular strength or weight that can have positive or negative values. Neurons with similar characteristics are arranged into a layer. A layer can be viewed as a group of neurons, which are connected to other layers and the external environment, which have no interconnections. There are basically three types of layers. The first layer connecting the input variables is called the input layer. The last layer connecting the output variables is called the output layer. Layers between the input and output layers are called hidden layers; there can be more than one hidden layer.

A simple neural network is shown in Figure 8.1, where we have the input variables in the nodes of the input layer and the output variables in the nodes of the output layer. So two input and output variables are assumed. In order to give a large level of flexibility in model formulation, the hidden layers in the middle; its nodes correspond to the hidden or artificial variables are introduced.

Let z_j denote the artificial variables corresponding to the jth neuron of the hidden layer. In the case of linear neural networks, this variable is calculated as follows:

$$z_j = \sum_{i=1}^{m} w_{ij} x_i, \tag{8.29}$$

where m is the number of input variables. The coefficients w_{ij} are constants, which are determined by the training procedure to be explained later. The relations between the output and hidden variables are also linear:

$$y_k = \sum_{j=1}^{l} \bar{w}_{jk} z_j, \tag{8.30}$$

where l is the number of hidden variables. The coefficients $\bar{w}_{jk}$ are also estimated by the training process. In most applications, more than one hidden layers are introduced, and similar relations to

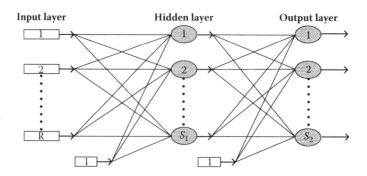

FIGURE 8.1 Simple neural network.

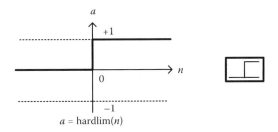

FIGURE 8.2 Hard-limit transfer function.

Equations 8.17 and 8.18 are assumed between the variables of the consecutive hidden layers. In the case of nonlinear neural network relations (transfer functions), Equations 8.17 and 8.18 have certain nonlinear elements; the parameters of the nonlinear transfer functions are also determined by the training process.

There are many transfer functions used in developing neural networks but here the three most popular ones are introduced. The hard-limit transfer function shown in Figure 8.2 limits the output of the neuron to either 0, if the net input argument $n < 0$, or 1, if $n \geq 0$. This function is used to create neurons that make classification decisions (Demuth and Beale, 2002).

The linear transfer function is shown in Figure 8.3. Neurons of this type are used as a linear approximator and are also employed in the output layer. The sigmoid transfer function depicted in Figure 8.4 takes the input, which may have any value between plus and minus infinity, and restricts the output to the range 0–1. This transfer function is commonly used in back propagation networks, in part because it is differentiable (Demuth and Beale, 2002).

ANNs can be trained using back propagation algorithm (BPA) during supervised learning. The BPA, also called the generalized delta rule, provides a way to calculate the gradient of the error function efficiently using the chain rule of differentiation. In this algorithm, network weights are moved along the negative of the gradient of the performance function. The term "back propagation" refers to the manner in which the gradient is computed for nonlinear multilayer networks. There are

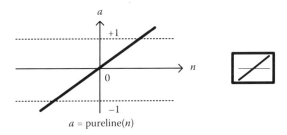

FIGURE 8.3 Linear transfer function.

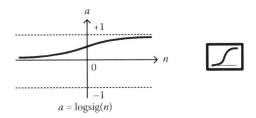

FIGURE 8.4 Log-sigmoid transfer function.

several performance functions that may be selected to evaluate network performance, such as root mean squared error (rmse), mean absolute error (mae), and sum squared error (sse). Assume that sse performance function is described by the following equation:

$$E = \sum_{k=1}^{S_2} (a_k^2 - T_k)^2, \qquad (8.31)$$

where a_k^2 is the output of the kth neuron of the output layer and T_k is its corresponding target. S_2 is the number of neurons in the output layer. The weights in the training process are updated through each iteration (which is usually called an epoch) in the steepest descent direction as follows:

$$\Delta w_{kj} = -\alpha \frac{\partial E}{\partial w_{kj}}, \qquad (8.32)$$

where α is the learning rate, whose value usually varies between 0 and 1.

The rmse and mae performance functions are formulated in Equations 8.33 and 8.34, respectively:

$$\text{rmse} = \sqrt{\frac{\sum_{k=1}^{S_2}(a_k^2 - T_k)^2}{S_2}}, \qquad (8.33)$$

$$\text{mae} = \frac{\sum_{k=1}^{S_2} |a_k^2 - T_k|}{S_2}, \qquad (8.34)$$

where the applied variables are the same as Equation 8.31.

8.2.3.1 The Multilayer Perceptron Network (Static Network)

The multilayer perceptron (MLP), also called the feed forward network, implements mappings from the input pattern space to the output space. Once the network weights are fixed, the states of the units are totally determined by inputs independent of initial and past states of the units. Thus, MLP is a static network in which information is transmitted through the connections between its neurons only in the forward direction and the network has no feedback, which covers its initial and past states. MLP can be trained with the standard BPA. Figure 8.1 shows a typical three-layer MLP, which, mathematically, can be expressed in the following formulations:

$$a_j^1(t) = F\left(\sum_{i=1}^{R} w_{j,i}^1 p_i(t) + b_j^1\right), \quad 1 \le j \le S_1, \qquad (8.35)$$

$$a_k^2(t) = G\left(\sum_{j=1}^{S_1} w_{k,j}^2 a_j^1(t) + b_k^2\right), \quad 1 \le k \le S_2, \qquad (8.36)$$

where t denotes a discrete time, R is the number of input signals, and S_1 and S_2 are the numbers of the hidden and output neurons, respectively. w^1 and w^2 are the weight matrices of the hidden and output layers, b^1 and b^2 are the bias vectors of the hidden and output layers, p is the input matrix, and a^1 and a^2 are the output vectors of the hidden and output layers. F and G are the activation functions of the hidden and output layers, respectively (Karamouz et al., 2007).

EXAMPLE 8.1

The studies in a city have shown that the water consumption is a function of three climatic variables including precipitation, weather humidity, and average temperature. The monthly data of water consumption, monthly rainfall, monthly average humidity, and temperature of this city are given in Tables 8.2 through 8.5. The studies have also demonstrated that the relation of water consumption with these parameters is complicated and cannot be formulated in regular form. Develop an MLP model to simulate the monthly water consumption of this city as a function of monthly precipitation, humidity, and temperature.

Solution: The input data of the MLP model should be completely random. For this purpose, the trend of city water consumption is estimated as a line with the following equation:

$$y = 0.1331x + 58.542.$$

TABLE 8.2
Data of Monthly Water Consumption (MCM) of Example 8.1

Year	1992	1993	1994	1995	1996	1997	1998	1999	2000	2001	2002	2003
Jan	46.3	50.9	52.6	53.8	57.3	63.0	63.8	64.8	66.9	68.0	66.4	66.9
Feb	48.2	52.5	53.3	54.7	58.5	64.1	65.2	65.8	67.8	68.6	68.6	67.2
Mar	50.1	54.0	54.8	56.4	59.4	66.2	67.4	66.8	69.4	70.2	69.2	69.1
Apr	50.3	55.9	56.9	59.2	61.3	65.6	68.6	68.1	71.0	69.1	66.8	69.9
May	55.2	61.3	63.3	65.0	68.7	72.5	75.8	75.8	76.7	73.6	72.0	73.1
June	63.1	68.8	70.7	71.9	77.9	78.1	84.1	81.5	80.5	76.3	80.2	79.8
July	69.1	71.8	75.4	78.5	82.2	82.0	88.4	83.7	84.1	77.5	83.6	86.0
Aug	67.6	71.6	74.4	78.3	82.3	80.6	85.5	83.9	81.6	76.4	82.7	84.3
Sep	61.7	66.9	66.7	71.8	78.3	74.9	80.0	80.4	76.9	72.2	78.8	79.1
Oct	56.4	57.9	57.8	61.8	70.0	69.5	71.8	72.9	69.5	66.2	73.5	73.1
Nov	53.4	53.7	54.7	58.2	65.8	65.0	67.9	68.7	67.5	64.2	69.3	70.4
Dec	51.2	52.6	54.0	57.2	63.9	63.2	66.3	67.2	67.7	65.0	67.2	68.7

TABLE 8.3
Data of Monthly Precipitation (mm) of Example 8.1

Year	1992	1993	1994	1995	1996	1997	1998	1999	2000	2001	2002	2003
Jan	36	36.1	31.1	7.3	57.2	22.7	34.9	52.2	48.6	35	29.2	36.2
Feb	24.5	40.1	0.3	30.1	27.9	0	8	64.5	4	16.2	22.6	6.1
Mar	39.5	12.7	43.2	22	25.5	85.7	27.1	31.7	3	7.4	4.9	13.8
Apr	1.5	70	3.6	16.5	52.3	9.9	1.2	1.8	0.6	0	8	8.8
May	6	2.6	9.8	0.3	9.6	5	3	0.4	0	0	0	0
June	0	0	0	0	0	0.4	0	0	0	0	0	0
July	0	0	0	0	0	0	0	0	0	0	0	0
Aug	0	0	0	0	0	0	0	0	0	0	0	0
Sep	0	0	0	0	0	0	0	0	0	0	0	0
Oct	0	11.8	23.6	0	1.5	8.5	0	0.1	1.9	2.4	0	0
Nov	17	4.1	89.3	0.4	1.5	56.2	0.2	11.2	13.8	6.9	0.3	3
Dec	29.3	0.2	26.1	70.4	1.2	47.8	14.5	66.6	83.6	123.4	49.6	47.3

continued

TABLE 8.4
Data of Monthly Average Temperature of Example 8.1

Year	1992	1993	1994	1995	1996	1997	1998	1999	2000	2001	2002	2003
Jan	10.1	10.7	14.8	13.2	13.3	13.5	11.7	14.1	12.3	12.2	12.1	13.5
Feb	13.3	13.8	15	15.5	16.3	12.7	14.9	15.9	14.3	15.2	15.1	16.2
Mar	16.3	18.7	19.7	19.6	19.5	16.6	18.6	19	19.2	21.3	21	19.3
Apr	24.9	24.2	26.9	24.6	24.5	23.8	26	26.9	29	27.4	25	26.4
May	31.2	29.8	31.2	31.5	32.9	31.7	32.4	33.3	33.4	32.7	32.7	32.6
June	36.4	34.5	34.9	35.6	35.5	36.3	37.4	37.3	36.2	35.8	35.6	36.7
July	37	36.5	36	36.5	37.9	37.1	37.5	37.7	38.8	37.3	38	37.3
Aug	37.3	36	35.5	36.6	37.3	35.5	38.5	38.1	38.4	37.6	37.1	36.9
Sep	32.6	32.2	32.9	31.9	33.1	33	34.1	33.8	32.6	33.3	33.5	33.2
Oct	25.3	27.1	27.5	26	26	27.9	26.9	28.7	26.6	27.6	28.6	28.4
Nov	19	18.5	20.8	18.9	19.7	19.9	21.5	19.4	18	18.5	19.1	19.1
Dec	12.9	15	11.3	12.4	16.8	13.7	16.7	13.4	13.8	16.2	14.2	14.4

TABLE 8.5
Data of Monthly Average Humidity (%) of Example 8.1

Year	1992	1993	1994	1995	1996	1997	1998	1999	2000	2001	2002	2003
Jan	64	77	68	74	77	64	73	71	69	77	69	67
Feb	55	65	50	66	68	39	61	69	50	59	62	51
Mar	52	58	50	55	60	55	57	55	34	48	45	45
Apr	37	54	42	45	51	49	42	35	30	30	43	40
May	30	40	30	32	38	33	31	25	23	20	23	26
June	23	26	22	24	28	26	27	21	19	16	22	23
July	24	28	27	25	29	21	28	26	25	21	20	21
Aug	28	30	26	24	29	23	33	30	27	35	20	33
Sep	35	32	35	32	27	25	33	32	31	32	28	25
Oct	39	43	47	39	43	47	38	38	37	43	42	46
Nov	62	52	67	47	51	66	52	53	58	43	50	55
Dec	75	61	71	68	64	77	54	73	82	78	65	74

The values estimated by the above equation are subtracted from the observed data. The remaining values are random and are used as input of the MLP model (Figure 8.5). The other data are also checked to determine whether there is any trend. Then the randomized consumption data and climatic data are standardized by subtracting the average of the data and then dividing by the standard division of data. This will help us to get better results from the MLP model, because the transfer functions are more sensitive to variations between 0 and 1.

For simulating monthly water consumption, a three-layer neural network with six neurons in the hidden layer and log-sig and linear transit functions in the hidden and output layers is employed as shown in Figure 8.6.

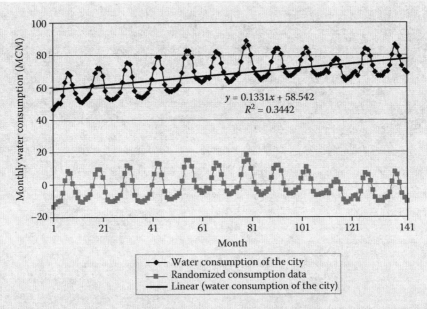

FIGURE 8.5 Randomization of monthly water consumption values of the city.

FIGURE 8.6 Structure of the developed MLP model for simulation of monthly water consumption in the city of Example 8.1.

One hundred series of data are used for model calibration (ClimaticCal and ConsumCal) and the remaining data are considered for model validation (ClimaticVal and ConsumVal). The model is trained in 3500 steps (epochs) and its goal is to achieve an error of less than 0.00001.

The simulated and recorded monthly consumption data are compared in Figure 8.7. As shown in this figure, about 30% of the data have been simulated with less than 5% error, and 60% of the data have been simulated with less than 10% error.

continued

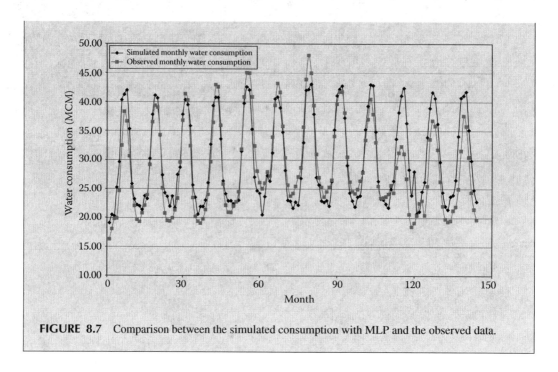

FIGURE 8.7 Comparison between the simulated consumption with MLP and the observed data.

8.2.3.2 Temporal Neural Networks

Static networks only process input patterns that are spatial in nature, that is, input patterns that can be arranged along one or more spatial axes such as a vector or an array. In many tasks, the input pattern comprises one or more temporal signals, as in speech recognition, time series prediction, and signal filtering (Bose and Liang, 1996). Time delay operators, recurrent connections, and the hybrid method are different approaches that are used to design temporal neural networks. Different types of temporal neural networks are introduced in the following subsection based on Karamouz et al. (2007).

8.2.3.2.1 The Tapped Delay Line

The tapped delay line (TDL) is based on a combination of time delay operators in sequential order. The architecture of the TDL is shown in Figure 8.8 that corresponds to a buffer containing the N most recent inputs generated by a delay unit operator D. Given an input variable $p(t)$, D operating on $p(t)$ yields its past values $p(t-1), p(t-2), \ldots, p(t-N)$, where N is the TDL memory length. Thus, the output of the TDL is an $(N+1)$-dimensional vector, made up of the input signal at the current time and the previous input signals. TDLs help networks to process dynamically through the flexibility in considering a sequential input.

8.2.3.2.2 The Adaptive Linear Filter

The adaptive linear (ADALINE) network is similar to the perceptron, but its activation function is linear. A schematic representation of this network is shown in Figure 8.8. The linear activation function allows ADALINE networks to take on any output value, whereas similar to the perceptron it can only solve linearly separable problems. Nevertheless, the ADALINE network is a widely used ANN found in practical applications, especially in adaptive filtering. In ADALINE the well-known LMS (least mean squares) learning rule is used, which is much more powerful than the perceptron learning rule. The LMS learning rule minimizes the mean squared error and thus moves the decision

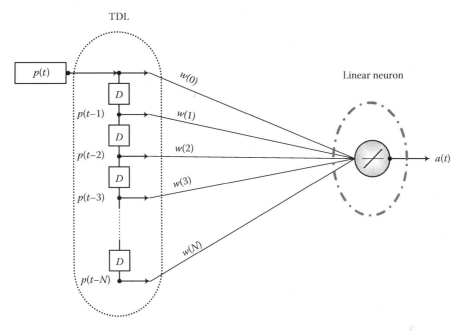

FIGURE 8.8 Architecture of the adaptive ADALINE.

boundaries as far as it can from the training patterns. The adaptive filter is produced by combining a TDL and an ADALINE network. This filter is adjusted at each time step and finds weights and biases that minimize the network's sum-squared error for recent input and target vectors. Therefore, the ADALINE output $a(t)$ is obtained by using the equation

$$a(t) = \sum_{k=0}^{N} w(k)p(t-k) + b, \qquad (8.37)$$

where $w(k)$ is the kth member of the weight vector, b is the bias of the linear neuron, and t denotes a discrete time.

8.2.3.2.3 The Recurrent Neural Network
For involving dynamic behavior, dependence of the initial and past states and serial processing, cyclic connections are described by directed loops in the network graph. Once cyclic connections are included, a neural network, often called a recurrent neural network (RNN), becomes a nonlinear dynamic system (Bose and Liang, 1996).

The recurrent scheme differs from the MLP, in the way in which its outputs recur either from the output layer or from the hidden layer back to its input layer. There are three general models of RNN, depending on the architecture of the feedback connections: the Jordan RNN (Jordan, 1986), which has feedback connections from the output layer to the input layer; the Elman RNN (Elman, 1990), which has feedback connections from the hidden layer to the input layer; and the locally RNN (Fransconi et al., 1992), which uses only local feedback (Figure 8.9). The network employs its output signals from the hidden layer to train the networks. The input layer is divided into two parts: the true input units and the context unit, which are a TDL memory of S_1 neurons from the hidden layer. The

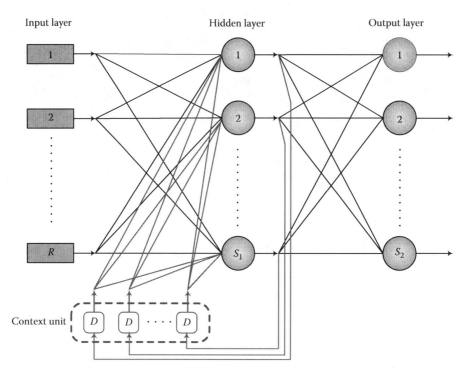

FIGURE 8.9 An Elman RNN.

Elman RNN is formulated by the following equations:

$$a_j^1(t) = F\left(\sum_{i=1}^{R} w_{j,i}^1 p_i(t) + \sum_{c=1}^{S_1} w_{j,c}^C a_c^1(t-1) + b_j^1\right), \quad 1 \leq j \leq S_1, \tag{8.38}$$

$$a_k^2(t) = G\left(\sum_{j=1}^{S_1} w_{k,j}^2 a_j^1(t) + b_k^2\right), \quad 1 \leq k \leq S_2, \tag{8.39}$$

where t denotes a discrete time, R is the number of input signals, and S_1 and S_2 are the numbers of hidden neurons and output neurons, respectively. w^1 and w^C are the weight matrices of the hidden layer for real inputs and the context unit, w^2 is the output layer weight matrix, b^1 and b^2 are the bias vectors of the hidden and output layers, p is the input matrix, and a^1 and a^2 are the hidden and output layers' output vectors. F and G are the hidden and output layers' activation functions, respectively. It is noteworthy that the modifiable connections are all feed forward, and the weights from the context units can be trained in exactly the same manner as the weights from the input units to the hidden units by the conventional back propagation method (Fauset, 1994).

8.2.3.2.4 The Input-Delayed Neural Network

To predict a temporal pattern, the input-delayed neural network (IDNN) can be used as a capable alternative. The IDNN consists of a static network (in this case an MLP) and some TDL is attached to its input layer (Coulibaly et al., 2001). In other words, the IDNN is a time delay neural network (TDNN), and only the input layer of it owns the memory. Figure 8.10 shows a generic the IDNN in which each input variable is delayed several time steps and fully connected to the hidden layer. The

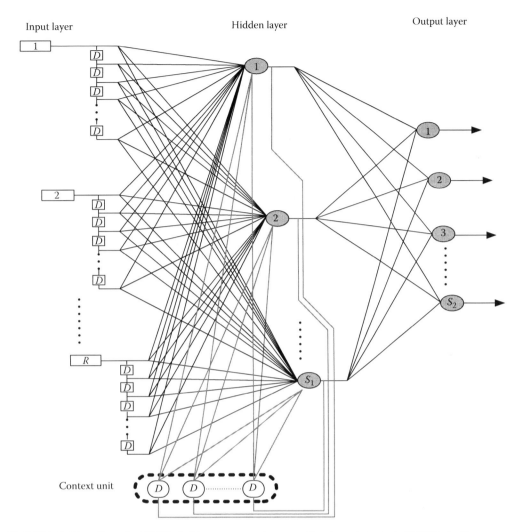

FIGURE 8.10 Generic IDNN (without recurrent connections) and typical IDRNN (with recurrent connections).

output of the IDNN is given by

$$a_j^1(t) = F\left(\sum_{d=0}^{D}\sum_{i=1}^{R} w_{j,i,d}^1 p_{i,d+1}(t) + b_j^1\right), \quad 1 \leq j \leq S_1, \tag{8.40}$$

$$a_k^2(t) = G\left(\sum_{j=1}^{S_1} w_{k,j}^2 a_j^1(t) + b_k^2\right), \quad 1 \leq k \leq S_2, \tag{8.41}$$

where t denotes a discrete time, D is the time delay memory order, R is the number of input signals, and S_1 and S_2 are the numbers of hidden neurons and output neurons, respectively. w^1 and w^2 are the weight matrices of hidden and output layers, b^1 and b^2 are the bias vectors of the hidden and output layers, p is the input matrix, and a^1 and a^2 are the hidden and output layers' output vectors. F and G are the hidden and output layers' activation functions, respectively.

An IDNN is a dynamic network and its static memory makes it capable of processing the temporal relationship of input sequences. An experimental study presented by Clouse et al. (1997) has even shown that the simple IDNN outperforms the general TDNN in situations where the input sequence is not short.

EXAMPLE 8.2

For the city of Example 8.1, simulate the water consumption using the IDNN model.

Solution: The same procedure as in the previous example is followed for preparing input data of the IDNN model. The developed structure of the IDNN model is shown in Figure 8.11. TDL memory of the developed model is 2 and there are 12 neurons in its hidden layer. The transit functions of the hidden and output layers are tan-sig and linear, respectively. The model is trained in 2000 steps (epochs) and the model goal is to achieve less than 0.00001 error.

FIGURE 8.11 Structure of the developed IDNN model for the simulation of monthly water consumption in the city of Example 8.2.

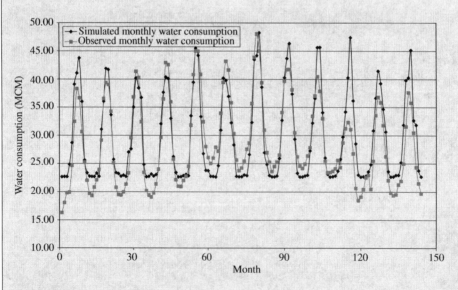

FIGURE 8.12 Comparison between the simulated consumption with IDNN and the observed data.

> The simulated and recorded monthly consumption data are compared in Figure 8.12. About 26% of the data have been simulated with less than 5% error, and 62% of the data have been simulated with less than 10% error.

8.2.3.2.5 The Time Delay Neural Network

In many tasks, the input pattern comprises one or more temporal signals, as in speech recognition, time series prediction, and signal filtering (Bose and Liang, 1996). A simple, commonly used method converts the temporal signal at the input into a spatial pattern by using a TDL, which provides the signal values at different time instants. This network is called the IDNN, which was introduced earlier. A further generalization is obtained by also replacing the internal connection weights in the network by the TDL (Waibel, 1989; Atiya and Parlos, 1992; Wan, 1993). The resulting network is sometimes called the TDNN or the spatiotemporal network. A schematic representation of the network is shown in Figure 8.13. To model this network, three-dimensional weight matrices should

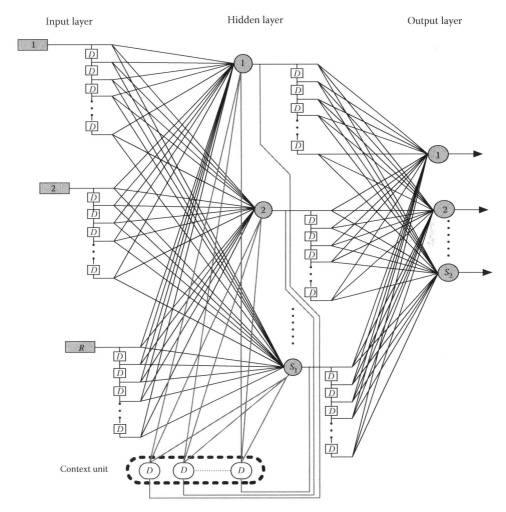

FIGURE 8.13 Generic TDNN (without recurrent connections) and typical TDRNN (with recurrent connections).

be considered for each layer, and the third dimensions are provided to represent temporal patterns. In the TDNN, the input vectors of each layer turn to two-dimensional matrices, and the second dimension represents discrete time. During the system modeling process, after each time step, the columns of the input matrix are shifted once and the instant input signals are substituted for the first column of the input matrix. Simultaneously, the instant outputs of each layer are substituted for the first column of the input matrix of the next layer. A three-layer TDNN can be described by the following equations:

$$a_{j,0}^1(t) = F\left(\sum_{d_1=0}^{D_1}\sum_{i=1}^{R} w_{j,i,d_1}^1 p_{i,d_1+1}(t) + b_j^1\right), \quad 1 \leq j \leq S_1, \tag{8.42}$$

$$a_k^2(t) = G\left(\sum_{d_2=0}^{D_2}\sum_{j=1}^{S_1} w_{k,j,d_2}^2 a_{j,d_2}^1(t) + b_k^2\right), \quad 1 \leq k \leq S_2, \tag{8.43}$$

where t denotes discrete time, D_1 and D_2 are the time delay memory order of the input and hidden layers, R is the number of input signals, and S_1 and S_2 are the numbers of hidden neurons and output neurons, respectively. w^1 and w^2 are the weight matrices of hidden and output layers, b^1 and b^2 are the bias vectors of hidden and output layers, p is the input matrix, and a^1 and a^2 are the hidden and output layers' output vectors. F and G are the hidden and output layers' activation functions, respectively. The TDNN output layer can be trained with the standard BPA. But only the instant error signal is back propagated through the current time route of each layer to train their pervious layers.

8.2.3.2.6 The Time Delay Recurrent Neural Network
In this study, a TDNN and the Elman recurrent network are combined together to examine how it performs in rainfall forecasting. A major feature of this architecture is that the nonlinear hidden layer receives the contents of both the input time delays and the context unit. Consequently, the time delay recurrent neural network (TDRNN) has both static and adaptive memory, which makes it suitable for complex sequential input learning (Coulibaly et al., 2001). Coulibaly et al. (2001) used this approach to forecast multivariate reservoir inflow and compared it with Elman RNN and IDNN networks results. The TDRNN used by Coulibaly et al. (2001) just uses time delay operators in its input layer, which also refers to the input-delayed recurrent neural network (IDRNN). Figures 8.10 and 8.13 show IDRNN and TDRNN. TDRNN can be described by the following equations:

$$a_{j,0}^1(t) = F\left(\sum_{d_1=0}^{D_1}\sum_{i=1}^{R} w_{j,i,d_1}^1 p_{i,d_1+1}(t) + \sum_{c=1}^{S_1} w_{j,c}^C a_c^1(t-1) + b_j^1\right), \quad 1 \leq j \leq S_1, \tag{8.44}$$

$$a_k^2(t) = G\left(\sum_{d_2=0}^{D_2}\sum_{j=1}^{S_1} w_{k,j,d_2}^2 a_{j,d_2}^1(t) + b_k^2\right), \quad 1 \leq k \leq S_2, \tag{8.45}$$

where t denotes discrete time, D_1 and D_2 are the time delay memory order of input and hidden layers, R is the number of input signals, and S_1 and S_2 are the numbers of hidden neurons and output neurons, respectively. w^1 and w^c are the weight matrices of the hidden layer for real inputs and the context unit, w^2 is the output layer's weight matrix, b^1 and b^2 are the bias vectors of the hidden and output layers, p is the input matrix, and a^1 and a^2 are the hidden and output layers' output vectors. F and G are the hidden and output layers' activation functions, respectively. The TDRNN output layer can be trained with the standard BPA. But only the instant error signal is back propagated through the current time route of each layer to train their previous layers.

8.2.4 MONTE CARLO TECHNIQUE

The common global optimization methods are completely based on random sample generation methods. For the generation of evaluation points in a more systematic way, adaptive random sampling (ARS) techniques have been developed. The main idea of ARS is that the new generated point around the actual one should improve the objective function; otherwise it is rejected (Rubinstein, 1981). Price (1965) proposed an advanced ARS strategy, which is the controlled random search technique. Price (1965) introduced the concept of an evolving population of feasible points. This concept is the basis of most modern global optimization methods. At each step, a simplex is formed from a sample of the population, which is reformed by reflecting one of its vertices through its centroid.

Monte Carlo simulation techniques are a group of ARS methods. A specific category of Monte Carlo techniques is the multistart strategy, which consists of running several independent trials of a local search algorithm. In the ideal case, these methods aim at starting the local search once in every region of attraction of local optima that may be identified via clustering analysis (Solomatine, 1999). A simple step-by-step procedure for using Monte Carlo simulation to calculate a desired quantity u_k in the e_kth quintile from variable a, which has m quintiles, is proposed by Fujiwara et al. (1988) as follows:

1. Select the random independent variables to produce a.
2. Generate random numbers and calculate m independent values for a.
3. Label the calculated values of a as $Y_1, Y_2, \ldots, Y_m$ and sort them in ascending order $Y_1 \leq Y_2 \leq \cdots \leq Y_m$.
4. Determine Y_i where i is the first $i > m \cdot e_k$. Y_i is the e_kth quintile of a_k.
5. Let $Z_1 = Y_i$; replicate the above procedure to determine $Z_2, \ldots, Z_N$.
6. Calculate the mean and variance of sample Z_i as follows:

$$\overline{Z}_N = \frac{\sum_{j=1}^{N} Z_j}{N}, \tag{8.46}$$

$$S_N^2 = \frac{\sum_{j=1}^{N} (Z_j - \overline{Z}_N)^2}{N-1}. \tag{8.47}$$

7. Calculate the relative precision (RP) using the equation

$$\text{RP} = \frac{\delta(N, \alpha)}{\overline{Z}_N}, \tag{8.48}$$

where $\delta(N, \alpha)$ is calculated as follows:

$$\delta(N, \alpha) = t_{N-1, 1-(\alpha/2)} \left(\frac{S_N^2}{N} \right), \tag{8.49}$$

where $t_{N-1, 1-(\alpha/2)}$ is the $[1 - (\alpha/2)]$ quintile of t distribution with $N - 1$ degrees of freedom.

8. The Monte Carlo simulation terminates when the RP becomes less than a given value β ($0 < \beta < 1$), and the desired quintile u_k is $\overline{Z}_N$; otherwise the N is increased until the desired precision is reached.

8.3 OPTIMIZATION TECHNIQUES

The variables used in the urban water planning and management problem are classified into two groups: *decision variables* and *state variables*. The decision variables define the best combination of actions, which is considered as the optimal management strategies. The state variables, which are usually dependent variables, take the given characteristics and guide the algorithm toward feasible solutions in deriving the management strategies.

The state variables are updated by using the simulation component and the optimal values for all the decision variables determined through the optimization component. Different forms of the objective function can be considered for an optimization problem. Some of the common objective functions in urban water planning and management are: costs, including the capital costs associated with operation costs over the time horizon of a project; maximum water allocation; and minimum water deficit. However, the exact form of the objective function is determined based on the nature of an individual problem.

A set of constraints considering the technical, economic, legal, or political limitations of the project should be considered in evaluating the optimal solution. There may also be some constraints in the form of either equalities or inequalities on decision and state variables.

The optimization problem can be solved through manual trial-and-error adjustment or using a formal optimization technique. Although the trial-and-error method is simple, it is tedious and also does not guarantee reaching the optimal solution. On the contrary, an optimization technique can be used to identify the optimal solution, which is feasible in terms of satisfying all the constraints.

The commonly used optimization techniques are

1. Linear programming (LP) (e.g., Lefkoff and Gorelick, 1987)
2. Nonlinear programming (NLP) (e.g., Wagner and Gorelick, 1987; Ahlfeld et al., 1988)
3. Mixed integer linear programming (MILP) (e.g., Willis, 1976, 1979)
4. Mixed integer nonlinear programming (MINLP) (e.g., McKinney and Lin, 1995)
5. Differential dynamic programming (DIFFDP) (e.g., Chang et al., 1992; Culver and Shoemaker, 1992; Sun and Zheng, 1999)

These methods are considered as "gradient" methods, because the gradients of the objective function regarding the variables to be optimized are repetitively calculated in the process of finding the optimal solution. Besides the computational efficiency of these techniques, they have some significant limitations. First, when the objective function is highly complex and nonlinear with multiple optimal points, a gradient method may be trapped in one of the local optima and cannot reach the globally optimal solution. Second, sometimes, numerical difficulties in gradient methods result in instability and convergence problems.

More recently, evolutionary optimization methods such as simulated annealing (SA), genetic algorithms (GA), and Tabu search (TS), which are based on heuristic search techniques, have been introduced. These methods imitate natural systems, such as biological evolution in the case of GA, to identify the optimal solution, instead of using the gradients of the objective function. The evolutionary methods generally require intensive computational efforts; however, they are increasingly being used nowadays because they are able to identify the global optimum, are efficient in handling discrete decision variables, and can be easily applied in different simulation models.

8.3.1 Linear Method

LP is the simplest single criterion optimization method. LP is applicable for continuous variables when both the objective function and constraints are linear. Because of computational efficiency, LP has been implemented in a wide range of practical optimization examples.

Consider an optimization problem with n decision variables, $x_1, x_2, \ldots, x_n$, and linear objective functions and constraints. The objective function is formulated as follows:

$$f(x) = c_1 x_1 + c_2 x_2 + \cdots + c_n x_n, \tag{8.50}$$

where $c_1, c_2, \ldots, c_n$ are real numbers.

It is assumed that all decision variables are non-negative. If any of the decision variables have only negative values, it is reformed to a new nonnegative variable as follows:

$$x_i^- = -x_i. \tag{8.51}$$

If a decision variable can be both positive and negative, then two new nonnegative variables are defined as follows:

$$x_i^+ = \begin{cases} x_i & \text{if } x_i \geq 0, \\ 0 & \text{otherwise,} \end{cases} \tag{8.52}$$

and

$$x_i^- = \begin{cases} -x_i & \text{if } x_i < 0, \\ 0 & \text{otherwise,} \end{cases} \tag{8.53}$$

and

$$x_i = x_i^+ - x_i^-. \tag{8.54}$$

All of the constraints should be rewritten as $\leq$-type conditions. Therefore, constraints with $\geq$ inequalities are multiplied by (-1) and equality constraints are considered as two inequality constraints as follows:

$$a_{i1}x_1 + a_{i2}x_2 + \cdots + a_{in}x_n = b_i \rightarrow \begin{cases} a_{i1}x_1 + a_{i2}x_2 + \cdots + a_{in}x_n \leq b_i \\ a_{i1}x_1 + a_{i2}x_2 + \cdots + a_{in}x_n \geq b_i, \end{cases} \tag{8.55}$$

where the second relation is again multiplied by (-1) to be as $\leq$-type conditions. The LP problem with the above characteristics can be rewritten into its primal form as follows:

$$\begin{aligned} \text{Maximize} \quad & c_1x_1 + c_2x_2 + \cdots + c_nx_n, \\ \text{Subject to} \quad & x_1, x_2, \ldots, x_n \geq 0, \\ & a_{11}x_1 + a_{12}x_2 + \cdots + a_{1n}x_n \leq b_1, \\ & a_{21}x_1 + a_{22}x_2 + \cdots + a_{2n}x_n \leq b_2, \\ & \qquad \vdots \\ & a_{m1}x_1 + a_{m2}x_2 + \cdots + a_{mn}x_n \leq b_m. \end{aligned} \tag{8.56}$$

In the above formulation, the maximization is only considered, because minimization problems can be reformed to maximization problems by multiplying the objective function by (-1). In the very simple case of the two decision variables problem, Equation 8.56 can be solved by the graphical approach.

Example 8.3

There are three methods for wastewater treatment in an urban area. The methods remove 1.2, 2.5, and 4 g/m^3 amounts of the pollutants, respectively. The third technology is the best, but because of some limitations in allocating enough area, it cannot be applied to more than 40% of the wastewater being treated. The costs of applying the treatment methods are 6, 4, and 3 dollars/m^3. If the discharge of wastewater that should be treated in a day is about 1200 m^3, and at least 2 g/m^3 pollutant has to be removed, determine the optimal combination of proposed methodologies in order to treat the wastewater of the urban areas.

Solution: A simple LP formulation can be given to model this optimization problem. Let x_1, x_2 be the amount of wastewater (in m^3) for which technologies 1 and 2 are used. Then $1200 - x_1 - x_2$ m^3 is the amount where technology variant 3 is applied. Clearly,

$$x_1, x_2 \geq 0,$$

$$x_1 + x_2 \leq 1200.$$

The condition that the third technology variant cannot be used in more than 50% of the treated waste water requires that

$$1200 - x_1 - x_2 \leq 480,$$

that is,

$$x_1 + x_2 \geq 720.$$

The total removed pollutant amount in g is

$$1.2x_1 + 2.5x_2 + 4(1200 - x_1 - x_2),$$

which has to be at least 2 g/m^3 of the total 1200 m^3 being treated, so

$$1.2x_1 + 2.5x_2 + 4(1200 - x_1 - x_2) \geq 2000.$$

This inequality can be rewritten as

$$2.8x_1 + 1.5x_2 \leq 2800.$$

The total cost is given as

$$6x_1 + 4x_2 + 3(1200 - x_1 - x_2) = 3x_1 + x_2 + 3600.$$

Since 3600 is a constant term, it is sufficient to minimize $3x_1 + x_2$ or maximize $-3x_1 - x_2$. Constraint (10–12) is a $\geq$-type relation, which is multiplied by (-1) in order to reduce it to a $<$-type constraint. After this modification is done, we have the following primal-form problem:

$$\begin{aligned}
\text{Maximize} \quad & f = -3x_1 - x_2, \\
\text{Subject to} \quad & x_1, x_2 \geq 0, \\
& x_1 + x_2 \leq 1200, \\
& -x_1 - x_2 \leq -480, \\
& 2.8x_1 + 1.5x_2 \leq 2800.
\end{aligned}$$

Each constraint can be represented in the two-dimensional space as a half-plain, the boundary of which is determined by the straight line representing the constraint as an equality. For instance, the equality $x_1 + x_2 = 1200$ is a straight line with the x_1-intercept 1200 and also with the x_2-intercept 1200. Since

Tools and Techniques

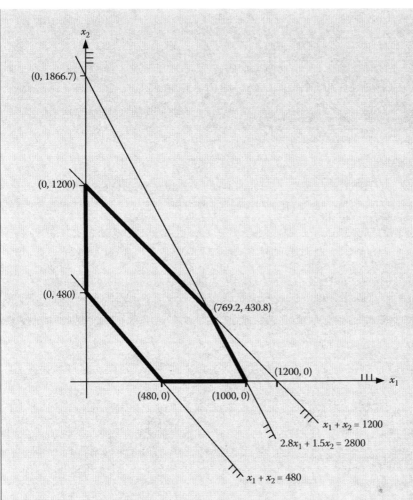

FIGURE 8.14 Feasible decision space of Example 8.3.

the origin, the (0, 0) point, satisfies the inequality, the half-plain that contains the origin represents the points that satisfy this inequality constraint. The intercept of the half-plains of all constraints gives the feasible decision space, as it is demonstrated in Figure 8.14. Note that this set has five vertices: (480, 0), (1000, 0), (769.2, 430.8), (0, 1200), and (0, 480) and the set is the convex hull, the smallest convex set containing the vertices.

From the theory of LP we know that if the optimal solution is unique, then it is a vertex, and in the case of multiple optimal solutions, there is always a vertex among the optimal solutions. Therefore it is sufficient to find the best vertex, which gives the highest objective function value. Simple calculation displays that

$$f(480, 0) = -1440,$$
$$f(1000, 0) = -3000,$$

continued

$$f(769.2, 430.8) = -2738.4,$$
$$f(0, 1200) = -1200,$$
$$f(0, 480) = -480.$$

Since -480 is the largest value, the optimal decision is: $x_1 = 0$, $x_2 = 480$, and $1200 - x_1 - x_2 = 720$. The minimal cost is therefore $3x_1 + x_2 + 3600 = \$4080$.

8.3.1.1 Simplex Method

In the case of more than two decision variables, so that the graphical representation of the problem is impossible, the *simplex method* is employed. This method searches the feasible decision space in a systematic way to find the optimal solution. The main idea of the simplex method can be summarized as follows: Any vertex of the feasible decision space corresponds to a basis of the column-space of the coefficient matrix of the linear constraints. The variables associated with the columns of the basis are called "basic variables." It can be revealed that moving from one vertex to any of its neighbors is equivalent to exchanging one of the basic variables with a nonbasic variable. In the simplex method, in each step the objective function increases, which can be ensured by the following conditions. Let a_{ij} be the (i,j)-element of the coefficient matrix, b_i the ith right-hand side number, and c_j the coefficient of x_j in the objective function. Then x_i is replaced by x_j if $c_j > 0$, $a_{ij} > 0$, and

$$\frac{b_i}{a_{ij}} = \min_l \left\{ \frac{b_l}{a_{lj}} \right\}. \tag{8.57}$$

Artificial variables called surplus are introduced to all $\geq$- and $=$-type constraints to find an initial basic solutions. For $\leq$-type constraints the slack variables play the same role. In the first phase of the simplex method, all artificial variables are removed from the basis by successive exchanges. In the second phase of the simplex method, the initial feasible basic solution is successively improved until reaching the optimum. Lingo is one of the professional software packages that implement the simplex method.

EXAMPLE 8.4

Based on the population growth analysis, the wastewater of a city is estimated to increase over time as depicted in Figure 8.15. Four phases for treatment plant expansion capacity are studied for treatment of

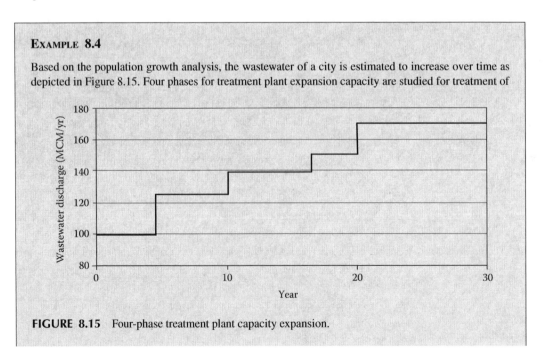

FIGURE 8.15 Four-phase treatment plant capacity expansion.

TABLE 8.6
Initial Investment and Capacity of Different Phases

Treatment Plant Expansion Projects	Initial Investment	Capacity
A	3	15
B	2	20
C	2.5	10
D	5	25

the future wastewater of this city in a 20 year planning time horizon. The capacity and initial investment of these projects are represented in Table 8.6. The interest rate is considered to be 5%. Find the optimal sequence of implementing projects.

Solution: The present value of the initial investment of the projects is presented in Table 8.7. In order to find the best timing for the construction of different phases, minimization of the present value of the cost is considered as the objective function of an LP model as follows:

$$\text{Minimize } C = 2.32 X_{11} + 1.8 X_{12} + 1.32 X_{13} + 1.08 X_{14}$$
$$+ 1.55 X_{21} + 1.2 X_{22} + 0.88 X_{23} + 0.72 X_{24}$$
$$+ 1.93 X_{31} + 1.5 X_{32} + 1.1 X_{33} + 0.9 X_{34}$$
$$+ 3.87 X_{41} + 2.99 X_{42} + 2.2 X_{43} + 1.79 X_{44},$$

Subject to

$$15 X_{11} + 20 X_{21} + 10 X_{31} + 25 X_{41} \geq 30,$$
$$15 X_{12} + 20 X_{22} + 10 X_{32} + 25 X_{42} \geq 20,$$
$$15 X_{13} + 20 X_{23} + 10 X_{33} + 25 X_{43} \geq 20,$$
$$15 X_{14} + 20 X_{24} + 10 X_{34} + 25 X_{44} \geq 10,$$
$$X_{11} + X_{12} + X_{13} + X_{14} \leq 1,$$
$$X_{21} + X_{22} + X_{23} + X_{24} \leq 1,$$
$$X_{31} + X_{32} + X_{33} + X_{34} \leq 1,$$
$$X_{41} + X_{42} + X_{43} + X_{44} \leq 1,$$

$$X_{ij} = \begin{cases} 1 & \text{if the project } i \text{ is constructed in the } j\text{th time period,} \\ 0 & \text{otherwise.} \end{cases}$$

In this problem, the constraints are defined in order to treat the wastewater in different years. They also ensure that each project can be constructed only once. The optimal timing of the implementation of the projects is found to be as projects D at year 5, A at year 10, C at year 16, and B at year 20. The present value of the costs of construction of the projects is also estimated to be 7.49 units.

TABLE 8.7
Present Value of Initial Investment for Different Phases

Project	$t = 5$ Years	$t = 10$ Years	$t = 16$ Years	$t = 20$ Years
A	2.32	1.8	1.32	1.08
B	1.55	1.2	0.88	0.72
C	1.93	1.5	1.1	0.9
D	3.87	2.99	2.2	1.79

8.3.2 Nonlinear Methods

When the objective function and/or one or more constraints are nonlinear, then the problem becomes NLP. Because of the nonlinear nature of practical optimization problems, NLP has been widely applied in solving practical problems. In the problems with discrete decision variables, MILP or MINLP must be used. In the case of only two variables, they can be solved by the graphical approach, but it should be considered that in some cases the feasible decision space does not have vertices or the optimal solution is not a vertex. In such cases, the curves of the objective function with different values have to be compared to find the optimal solution.

Example 8.5

For illustrative purposes, consider the previous example; however, assume that instead of minimizing the linear function $3x_1 + x_2$ a nonlinear function

$$f(\underline{x}) = x_1^2 + x_2^2$$

is minimized.

Solution: In order to find the feasible solution that minimizes this objective function, we first illustrate the solutions, which give certain special objective function values. The points with zero objective function values satisfy the equation

$$f(\underline{x}) = x_1^2 + x_2^2 = 0$$

when the only solution is $x_1 = x_2 = 0$, that is, the origin. The points with the objective function value 10,000 satisfy the equation

$$f(\underline{x}) = x_1^2 + x_2^2 = 10{,}000.$$

This is a circle with the origin being its center and having the radius 100. This circle has no feasible point, so 10,000 is a very small value for the objective function. By increasing the value of $f(\underline{x})$, we always get a similar circle with the same center, but with increasing radius. By increasing the radius of the circle we see that the smallest feasible radius occurs when the circle and the line $x_1 + x_2 = 480$ just touch each other. The tangent point is (220, 220), which is therefore the optimal solution of the problem. In more complicated and higher dimensional cases, computer software has to be used. There are several professional packages available to perform this task, so they are usually used in solving practical problems.

Linearization is the most popular nonlinear optimization method. In this method, all nonlinear constraints or objective functions are linearized. The resulting LP problem can be solved by the simplex method. Let $g(x_1, \ldots, x_n)$ be the left-hand side of a constraint or the objective function and $(x_1^0, \ldots, x_n^0)$ be a feasible solution. g is linearly approximated by the linear Taylor's polynomial as follows:

$$g(x_1, \ldots, x_n) \approx g(x_1^0, \ldots, x_n^0) + \sum_{r=1}^{n} \frac{\partial g}{\partial x_r}(x_1^0, \ldots, x_n^0)(x_r - x_r^0). \quad (8.58)$$

The optimal solution of the resulting LP problem is $(x_1^1, \ldots, x_n^1)$; then the original problem is linearized again around this solution and the new LP problem is solved. This iterative procedure is followed until the optimal solution in two iterations does not change. It is known from the theory of NLP that the solutions of the successive LP problems converge to the solution of the original nonlinear optimization problem. Further information about NLP can be found at Hadley (1964).

8.3.3 Dynamic Programming

In water resources planning, the dynamism of the system is a major concern that has to be taken into account in modeling and optimization. Dynamic programming (DP) is the most popular tool in optimizing dynamic systems. Consider the following optimization problem for supplying water for an urban area from a multireservoir system:

$$\min Z = \sum_{t=1}^{T} \text{Loss}\left(\sum_{s=1}^{X} R_{s,t}\right), \qquad (8.59)$$

where T is the time horizon, X is the total number of reservoirs in the system, and Loss is the cost of operation based on the ratio of supplied water ($R_{s,t}$) from reservoir s in month t to the monthly urban water demand. The continuity or a mass balance of the contents of the reservoir considering regulated and unregulated release (seepage) from the beginning of the month to the next is also included in the model:

$$S_{s,t+1} - S_{s,t} + R_{s,t} = I_{s,t}, \qquad (8.60)$$

$$R_{s,t} = R^R_{s,t} + R^U_{s,t}, \qquad (8.61)$$

where $R_{s,t}$ is the release from reservoir s in month t and I_t is the inflow volume to the reservoir in month t. $R^R_{s,t}$ and $R^U_{s,t}$ are regulated and unregulated releases from reservoirs, respectively. Additional constraints on maximum and minimum allowable release and storage during any season for all sites can be stated as

$$R^{\max}_{s,t} \geq R_{s,t} \geq R^{\min}_{s,t} \quad t = 1, \ldots, T \quad s = 1, \ldots, X, \qquad (8.62)$$

$$S^{\max}_{s,t} \geq S_{s,t} \geq S^{\min}_{s,t} \quad t = 1, \ldots, T \quad s = 1, \ldots, X, \qquad (8.63)$$

$$\left|S_{s,t+1} - S_{s,t}\right| \leq SC_s \quad t = 1, \ldots, T \quad s = 1, \ldots, X, \qquad (8.64)$$

where SC_s is the maximum allowable change in the storage of reservoir s within each month considering dam stability and safety conditions. Because of many decision variables and constraints, finding the direct solution is computationally difficult. Through the DP procedure, this unique optimization problem is changed to solve many smaller-sized problems in the form of recursive function.

$$f_{t+1}(S_{1,t+1}, \ldots, S_{X,t+1}) = \min \left(\text{Loss}\left(\sum_{s=1}^{X} R_{st}\right) + f_t\left(S_{1,t}, \ldots, S_{X,t}\right)\right),$$

$$S_{1,t} \in \Omega_{1,t} \qquad (8.65)$$

$$\ldots$$

$$\ldots$$

$$S_{X,t} \in \Omega_{X,t}.$$

The initial conditions are

$$f_1(S_{1,1}, \ldots, S_{X,1}) = 0, \quad S_{1,1} \in \Omega_{1,1}, \ldots, S_{X,1} \in \Omega_{X,1}, \qquad (8.66)$$

where $f_t(S_{1,t}, \ldots, S_{X,t})$ is the total minimum loss of operation from the beginning of month 1 to the beginning of month t, when the storage volume at the beginning of month t is $S_{1,t}$ at site 1, $S_{R,t}$ at site 2, ..., $S_{X,t}$ at site X. $\Omega_{s,t}$ is the set of discrete storage volumes that will be considered for the beginning of month t at site s.

The DP sequences are time or space dependent. So the "stages" of the problem are time if for instance the operation of the water distribution system in a daily time scale is considered or space if for example the design of a storm or sewer collection system is considered.

In most applications the optimization Equation 8.65 cannot be solved analytically and recursive functions are determined only numerically. In such cases, discrete dynamic programming (DDP) is used, where the state and decision spaces are discretized. The main problem in using this method is known as the "curse of dimensionality." Because of this, the number of states that should be considered in evaluation of different possible situations drastically increases with the number of discretizations. Therefore DDP is used only up to four or five decision and state variables.

Different successive approximation algorithms such as differential dynamic programming (DIFF DP), discrete differential dynamic programming (DDDP), and state incremental dynamic programming (IDP) are developed to overcome this problem. In applying any of these methods, the user should provide an initial estimation of the optimal policy. This estimation is improved in each step until convergence occurs. However, as is usually the case in nonconvex optimization, the limit might only be a local optimum, or the approximating sequence is divergent.

EXAMPLE 8.6

Assume a reservoir that supplies water for an urban area. The monthly water demand of this urban area (D_t) is about 10 million cubic meters. The total capacity of the reservoir is 30 million cubic meters. Let S_t (reservoir storage at the beginning of month t) take the discrete values 0, 10, 20, and 30. The cost of operation (Loss) can be estimated as a function of the difference between release (R_t) and water demand as follows:

$$\text{Loss}_t = \begin{cases} 0, & R_t \geq D_t, \\ (D_t - R_t)^2, & R_t > D_t. \end{cases}$$

(a) Formulate a forward moving deterministic DP model for finding the optimal release in the next three months.
(b) Formulate a backward moving deterministic DP model for finding the optimal release in the next three months.
(c) Solve the DP model developed in part (a), assuming that the inflows to the reservoir in the next three months ($t = 1, 2, 3$) are forecasted to be 10, 50, and 20, respectively. The reservoir storage in the current month is 20 million cubic meters.

Solution:
(a) The forward deterministic DP model is formulated as follows:

$$f_{t+1}(S_{t+1}) = \min[\text{Loss}(R_t) + f_t(S_t)], \quad t = 1, 2, 3$$

$$S_t \in \Omega_t,$$

$$S_t \leq S\max,$$

$$f_1(S_1) = 0,$$

$$\text{Loss}(R_t) = \begin{cases} 0, & R_t = 10, \\ (10 - R_t)^2, & R_t \neq 10, \end{cases}$$

$$R_t = S_t + I_t - S_{t+1},$$

$$S_{\max} = 30\,\text{MCM},$$

where Ω_t is the set of discrete storage volumes (states) that will be considered at the beginning of month t at site s. S_{t+1} is the storage at the beginning of month $t + 1$.

(b) The backward deterministic DP model is formulated as follows:

$$f_t(S_t) = \min [\text{Loss}(R_t) + f_{t+1}(S_{t+1})], \quad t = 1, 2, 3.$$

The constraints are as mentioned above.

(c) Loss is calculated using the forward formulation for three months. The results are shown in Table 8.8.

TABLE 8.8
DP Model Solution

Month	Inflow	S_t	S_{t+1}	R_t	Loss_t	R_t^*
1	10	20	0	30	400	10
			10	20	100	
			20	10	0	
			30	0	100	
2	50	20	0	70	3600	40
			10	60	2500	
			20	50	1600	
			30	40	900	
3	20	30	0	50	1600	20
			10	40	900	
			20	30	400	
			30	20	100	

Note: R_t^* is the optimal policy for each month.

8.3.4 Evolutionary Algorithms

The family of Evolutionary Algorithms (EAs), which are developed based on the mechanism of natural evolution, has brought considerable modifications to random searching. In EAs, the searching procedure is implemented in stages called "generations." In each one of these techniques, a population of randomly generated points is developed by applying the selection, recombination, and mutation operators. Some of the most popular EA techniques are briefly introduced in the following sections.

8.3.4.1 Genetic Algorithms

GAs are stochastic search methods that imitate the process of natural selection and mechanism of population genetics. They were first introduced by Holland (1975) and later further developed and popularized by Goldberg (1989). GAs are used in a number of different application areas ranging from function optimization to solving large combinatorial optimization problems. The general algorithm of GAs is described in Figure 8.16. As is illustrated in Figure 8.16, the GA begins with the creation of a set of solutions referred to as a population of individuals. Each individual in a population consists of a set of parameter values that completely describes a solution. A solution is encoded in a string called a chromosome, which consists of genes that can take a number of values. Initially, the collection of solutions (population) is generated randomly and, at each iteration (also called generation), a new generation of solutions is formed by applying genetic operators including selection, crossover and mutation, analogous to the ones from natural evolution. Each solution is evaluated using an objective function (called a fitness function), and this process is repeated until some form of convergence in fitness is achieved. The objective of the optimization process is to minimize or maximize the fitness.

```
BEGIN /* genetic algorithm*/
    Generate initial population;
    Compute fitness of each individual;
    WHILE NOT finished DO LOOP
        BEGIN
            Select individuals from old generations for mating;
            Create offspring by applying recombination and/or
mutation
            to the selected individuals ;
                Compute fitness of the new individuals;
            Kill old individuals to make room for new chromosomes
and         insert offspring in the new generation;
            IF Population has converged THEN finishes := TRUE ;
        END
END
```

FIGURE 8.16 General algorithm of GA.

The first step in developing a new generation is selection. Through the selection process, a population of chromosomes is selected via some stochastic selection process. The aim of this process is to propagate the better or fitter chromosomes. The simplest procedure to select chromosomes is known as tournament selection. New chromosomes are created in a process known as crossover. Crossover may be performed using different methods including single-point, double-point, and uniform methods. The crossover is performed on each selected pair with a specified probability, referred to as crossover probability, which is typically between 0.5 and 0.7. After a pair is selected for crossover, a random number between 0 and 1 is generated and compared with the crossover probability. The crossover will be performed only if the produced random number is less than the crossover probability. In the crossover process, the genetic material of the selected chromosomes is exchanged.

The last step in the production of a new generation is mutation. Through the mutation process, some diversity is added to the GA population to prevent from converging to a local minimum. The probability of selection of any gene for mutation is controlled by the mutation probability. Carroll (1996) suggests the following formulation for estimating the mutation probability, p_{mutate}.

$$p_{\text{mutate}} = \frac{1}{n_{\text{popsiz}}}, \qquad (8.67)$$

where n_{popsiz} is the size of the population. Similar to the crossover, a random number between 0 and 1 is generated. Only when this number is smaller than the mutation probability, mutation is performed.

The GA raises again a couple of important features. First it is a stochastic algorithm; randomness has an essential role in the GA. Both selection and reproduction need random procedures. A second very important point is that the GA always considers a population of solutions. Keeping in memory more than a single solution at each iteration offers a lot of advantages. The algorithm can recombine different solutions to get better answers and so it can use the benefits of assortment. The robustness of the algorithm should also be mentioned as something essential for the algorithm success. Robustness refers to the ability to perform consistently well on a broad range of problem types. There is no particular requirement on the problem before using GAs, so it can be applied to resolve any problem. All those features make the GA a really powerful optimization tool.

EXAMPLE 8.7

Find the maximum point of the function $f(x) = 2x - x^2$ in the [0, 2] interval using the GA.

Solution: First the decision space is discretized and each is encoded. For illustration purposes, we have four-bit representations. The population is provided in Table 8.9. The initial population is selected next. We have chosen six chromosomes; they are listed in Table 8.10, where their actual values and the corresponding objective function values are also presented.

The best objective function value is obtained at 1001, but there are two chromosomes with the second best objective value: 1010, 0110. Randomly selecting the first, we have the chromosome pair 1001 and 1010. A simple crossover procedure is performed by interchanging the two middle bits to obtain 1000 and 1011. The new population is presented in Table 8.11.

The best objective function value is obtained at 1000, and there is a chromosome with the second best objective value: 1001. We have the chromosome pair 1000 and 1001. By interchanging their middle bits we get 1001 and 1000, which are similar to those of their parents. The resulting new population is illustrated in Table 8.12.

Selecting the pair 1000 and 1001, we cannot use the same crossover procedure as before, since they have the same middle bits. Since the two chromosomes differ in only one bit, interchanging any substring of them will result in the same chromosomes. Therefore, mutation is used by reversing its second bit to have 1100. The resulting modified population is depicted in Table 8.13.

TABLE 8.9
Population of Chromosomes

Binary Code	Value	Binary Code	Value
0000	0	0110	0.75
0001	0.125	1010	1.25
0010	0.25	1100	1.5
0100	0.5	0111	0.875
1000	1	1011	1.375
0011	0.375	1101	1.625
0101	0.625	1110	1.75
1001	1.125	1111	1.875

TABLE 8.10
Initial Population

Chromosomes	Value	Objective Function
1010	1.25	0.9375
1100	1.5	0.75
1001	1.125	0.984375
1111	1.875	0.234375
0110	0.75	0.9375
0011	0.375	0.609375

continued

TABLE 8.11
First Modified Population

Chromosomes	Value	Objective Function
1000	1	1
1011	1.375	0.859375
1111	1.875	0.234375
1110	1.75	0.4375
0010	0.25	0.4375
1001	1.125	0.984375

TABLE 8.12
Second Modified Population

Chromosomes	Value	Objective Function
1001	1.125	0.984375
1000	1	1
1010	1.25	0.9375
1011	1.375	0.859375
1110	1.75	0.4375
0010	0.25	0.4375

TABLE 8.13
Third Modified Population

Chromosomes	Value	Objective Function
1001	1.125	0.984375
1000	1	1
1100	1.5	0.75
1001	1.125	0.984375
1000	1	1
1010	1.25	0.9375

It is interesting to examine how the best objective function is evolving from population to population. Starting from the initial table with the value 0.98, it increased to 1, and then it remained the same. Since this value is the global optimum, it will not increase in further steps. In practical applications we stop the procedure if no, or very small, improvement occurs after a certain (user-specified) number of iterations. More details of GA and possible applications can be found in Goldberg (1989).

8.3.4.2 Simulation Annealing

SA is developed based on the analogy between the physical annealing process of solids and optimization problems. In SA, one can simulate the behavior of a system of particles in thermal equilibrium using a stochastic relaxation technique developed by Metropolis et al. (1953). This stochastic relaxation technique helps the SA procedure escape from the local minimum. SA starts at a feasible solution, q_0 (a real-valued vector representing all decision variables), and objective function

$J_0 = J(q_0)$. A new solution q_1 is randomly selected from the neighbors of the initial solution and the objective function $J_1 = J(q_1)$ is evaluated. If the new solution has a smaller objective function value $J_1 < J_0$ (in minimization problems), the new solution is definitely better than the old one and therefore it is accepted and the search moves to q_1 and continues from there. On the other hand, if the new solution is not better than the current one, $J_1 < J_0$, the new solution may be accepted, depending on the acceptance probability defined as

$$\Pr(\text{accept}) = \exp\left\{\frac{-(J_1 - J_0)}{T}\right\}, \tag{8.68}$$

where T is a positive number that will be discussed later.

For a given value of T, the acceptance probability is high when the difference between the objective function values is small. To decide whether or not to move to the new solution q_1 or stay with the old solution q_0, SA generates a random number U between 0 and 1. If $U < \Pr(\text{accept})$, SA moves to q_1 and resumes the search. Otherwise, SA stays with q_0 and selects another neighboring solution. At this step, one iteration has been completed.

The control parameter T plays an important role in reaching the optimal solution through SA because the acceptance probability is strongly influenced by the choice of T. When the T is high the acceptance probability calculated is close to 1, and the new solution will very likely be accepted, even if its objective function value is considerably worse than that of the old solution. On the other hand, if T is low, approaching zero, then the acceptance probability is also low, approaching zero. This effectively precludes selection of the new solution even if its objective function value is only slightly worse than that of the old solution. Therefore, no new solution for which $J_1 > J_0$ is likely to be selected. Rather, only "downhill" points ($J_1 < J_0$) are accepted; therefore, a low value of T leads to a descending search.

Successful performance of the SA method requires that T be assigned a high initial value, and that it be reduced gradually throughout the sequence of calculations by the reduction factor λ. The number of iterations under each constant T should be sufficiently large. But higher initial T and larger number of iterations mean longer simulation time and may severely limit computational efficiency. SA stops either when it has completed all the iterations or when the termination criterion (a specific value of objective function) is reached.

As a rule of thumb, to balance the computational efficiency and quality of the SA method, the initial T should be chosen in such a way that the acceptance ratio is somewhere between 70% and 80%. It is recommended that a number of simulation runs should be executed first to estimate the acceptance ratio. Second, the number of iterations under a constant T is about 10 times the size of the decision variables. In the literature, the amount of 0.9 is suggested to be a reasonable number for λ as the reduction factor (Egles, 1990).

SA has many advantages including easy implementation, not requiring much computer memory and coding and providing guarantee for identification of an optimal solution if an appropriate schedule is selected. The last point becomes much more important in the cases where the solution space is large and the objective function has several local minima or changes dramatically with small changes in the parameter values. Examples of the SA application in the hydrological literature include Dougherty and Marryott (1991) and Wang and Zheng (1998).

8.3.5 MULTICRITERIA OPTIMIZATION

In multicriteria optimization problems, more than one objective function is considered. Different multicriteria decision making (MCDM) processes are developed for solving these kinds of problems in continuous and discrete situations. Consider an optimization problem with I criteria including $f_1,\ldots,f_I$, and set X illustrates which decisions can be made. In the presence of multiple criteria, we should represent the set of all possible outcomes, H, as follows:

$$H = \{(f_1(x),\ldots,f_I(x)) \mid x \in X\}. \tag{8.69}$$

This set is called the criteria space, which shows what can be achieved by selecting different alternatives. Any point of $\underline{f} \in H$ is called dominated, if there is another point $\underline{f}^1 \in H$ such that $\underline{f}^1 \geq \underline{f}$ and if there is strict inequality in at least one component. Similarly, a point $\underline{f} \in H$ is called nondominated, if there is not another point $\underline{f}^1 \in H$. In MCDM, nondominated points will serve as the solutions of the problem. In the case of single-objective optimization problems, optimal solutions are selected. Optimal solutions are always better than any nonoptimal solution, and in the case of several optimal solutions, they are equivalent to each other in the sense that they give the same objective function value. In the case of nondominated solutions, these properties are no longer valid.

If in a single-objective optimization problem there are multiple optimal solutions, it does not matter which one is selected, since they give the same objective function value. In multicriteria problems, the different nondominated solutions give different objective function values, so it is very important to decide which one of them is chosen, since they lead to different outcomes. To single out a special point from the set of nondominated solutions, additional preference information is needed from the decision makers to decide on trade-offs between the objectives.

Weighting is the most popular MCDM method, which reforms the multicriteria problem to a single-criterion problem as follows:

$$\begin{cases} \text{Maximize} & F(x) = [f_1(x), f_2(x), \ldots, f_I(x)] \\ \text{Subject to} & x \in X \end{cases} \Rightarrow \begin{cases} \text{Maximize} & F(x) = \sum_{k=1}^{I} W_k f_k(x) \\ \text{Subject to} & x \in X, \end{cases} \tag{8.70}$$

where W_k is the weight of the kth objective function that is a nonzero positive real number. Since there are both minimization and maximization criteria in a multicriteria optimization model; in the weighting method, all the criteria are reformed to maximization by multiplying the minimizing criteria by -1. The weight of different criteria is determined based on the decision maker's judgment and criteria priorities. Some methods for MCDM are TOPSIS (technique for order preference by similarity to ideal solution) and AHP (analytical hierarchy process).

EXAMPLE 8.8

Assume that there are four feasible water treatment options for the treatment of urban wastewater. These technologies are evaluated by their costs and convenience of usage. The cost data are given in $1000 units and the convenience of usage is measured in a subjective scale between 0 and 100, where 100 is the best possible measure. The data are given in Table 8.14. Discuss the optimal solution of this problem.

Solution: The cost data are multiplied by (-1) in order to maximize both criteria. The criteria space has four isolated points in this case as displayed in Figure 8.17. It is clear that none of the points dominates any other point. Since lower cost is always accompanied by lower convenience, increasing any objective should result in decreasing the other one. Therefore any one of the four technologies can be accepted as the solution.

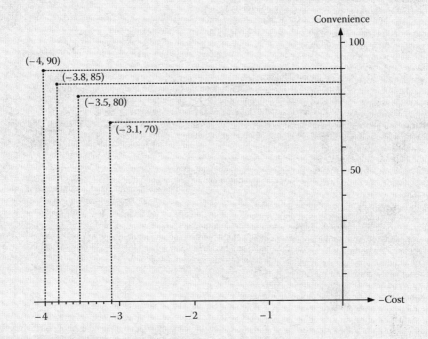

TABLE 8.14
Data for Discrete Problem

Option	− Cost	Convenience
1	−4	90
2	−3.8	85
3	−3.5	80
4	−3.1	70

FIGURE 8.17 Criteria space in the discrete case.

EXAMPLE 8.9

Assume that two pollutants are removed by a combination of three alternative technologies. They remove 3, 2, and 1 g/m^3, respectively, of the first kind of pollutant and 2, 1, and 3 g/m^3 of the second kind. If 1000 m^3 of each pollutant has to be treated in a day, determine the optimal combination of methodologies.

Solution: The total removed amounts of the two pollutants are

$$3x_1 + 2x_2 + 1(1000 - x_1 - x_2) = 2x_1 + x_2 + 1000$$

and

$$2x_1 + x_2 + 3(1000 - x_1 - x_2) = -x_1 - 2x_2 + 3000.$$

continued

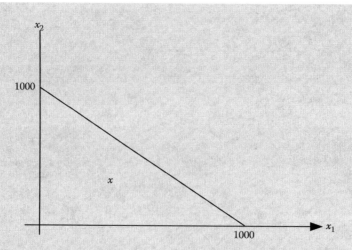

FIGURE 8.18 Feasible decision space for Example 8.9.

If no condition is given on the usage of the third technology, there is no constraint on the minimum amount to be removed, and no cost is considered, then this problem can be mathematically formulated as

$$\text{Maximize} \quad f_1(\underline{x}) = 2x_1 + x_2, \quad f_2(\underline{x}) = -x_1 - 2x_2,$$
$$\text{Subject to} \quad x_1, x_2 \geq 0,$$
$$x_1 + x_2 \leq 1000.$$

The feasible decision space is exhibited in Figure 8.18.

The criteria space can be determined as follows:

$$f_1 = 2x_1 + x_2 \quad \text{and} \quad f_2 = -x_1 - 2x_2.$$

Therefore,

$$x_1 = \frac{2f_1 + f_2}{3} \quad \text{and} \quad x_2 = \frac{-f_1 - 2f_2}{3}.$$

The constraints $x_1, x_2 \geq 0$ can be rewritten as

$$\frac{2f_1 + f_2}{3} \geq 0 \quad \text{and} \quad \frac{-f_1 - 2f_2}{3} \geq 0,$$

that is,

$$2f_1 + f_2 \geq 0 \quad \text{and} \quad f_1 + 2f_2 \leq 0.$$

The third constraint $x_1 + x_2 \leq 1000$ now has the form

$$\frac{2f_1 + f_2}{3} + \frac{-f_1 - 2f_2}{3} \leq 1000,$$

that is, $f_1 - f_2 \leq 3000$.

Hence, set H is characterized by the following inequalities:

$$2f_1 + f_2 \geq 0,$$
$$f_1 + 2f_2 \leq 0,$$
$$f_1 - f_2 \leq 3000,$$

and is illustrated in Figure 8.19. Note that H is the triangle with vertices $(0, 0)$, $(1000, -2000)$, and $(2000, -1000)$. The point $(0, 0)$ is nondominated, but it is not better than the point $(1000, -2000)$, which is dominated, since $1000 > 0$ gives a higher value in the first objective function. The points on the linear

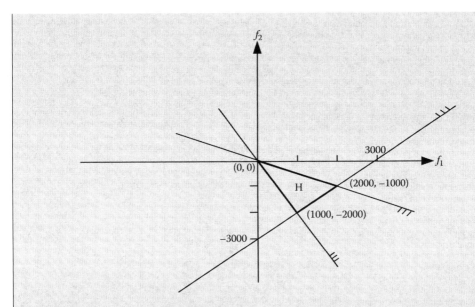

FIGURE 8.19 Criteria space for Example 8.9.

segment connecting the vertices (0, 0) and (2000, −1000) cannot be dominated by any point of H, since none of the objectives can be improved without worsening the other. This, however, does not hold for the other points of H, where both objectives can be improved simultaneously. Therefore all points between (0, 0) and (2000, −1000) can be accepted as the best solutions.

We can say that (0, 0) is better than (2000, −1000), if a 1000 unit loss in the second objective is not compensated by a 2000 unit gain in the first objective. These kinds of decisions cannot be made based on only the concepts being discussed above. Depending on the additional preference and/or trade-off information obtained from the decision makers, different solution concepts and methods are available.

8.4 FUZZY SETS AND PARAMETER IMPRECISION

Two approaches of randomization or fuzzification are used for evaluating the parameter uncertainty in water resources and hydrological modeling. In the previous sections, the randomization concepts and probabilistic methods were discussed. In this section, the fundamentals of fuzzy sets and fuzzy decisions will be introduced.

Let X be a set of certain objects. A *fuzzy set* A in X is a set of ordered pairs as follows:

$$A = \{[x, \mu_A(x)]\}, \quad x \in X, \tag{8.71}$$

where $\mu_A: X \mapsto [0, 1]$ is called the membership function and $\mu_A(x)$ is the grade of membership of x in A. In classical set theory, $\mu_A(x)$ equals 1 or 0, because x either belongs to A or does not.

The basic concepts of fuzzy sets are as follows:

i. A fuzzy set A is *empty* if $\mu_A(x) = 0$ for all $x \in X$.
ii. A fuzzy set A is called *normal* if

$$\sup_{x} \mu_A(x) = 1. \tag{8.72}$$

For normalizing a nonempty fuzzy set, the normalized membership function is computed as follows:

$$\mu_A(x) = \frac{\mu_A(x)}{\sup_x \mu_A(x)}. \quad (8.73)$$

iii. The *support* of a fuzzy set A is defined as

$$S(A) = \{x \mid x \in X, \mu_A(x) > 0\}. \quad (8.74)$$

iv. The fuzzy sets A and B are equal if, for all $x \in X$,

$$\mu_A(x) = \mu_B(x). \quad (8.75)$$

v. A fuzzy set A is a *subset* of B ($A \subseteq B$) if, for all $x \in X$,

$$\mu_A(x) \leq \mu_B(x). \quad (8.76)$$

vi. A' is the *complement* of A if, for all $x \in X$,

$$\mu_{A'}(x) = 1 - \mu_A(x). \quad (8.77)$$

vii. The *intersection* of fuzzy sets A and B is defined as

$$\mu_{A \cap B}(x) = \min\{\mu_A(x); \mu_B(x)\} \quad \text{(for all } x \in X\text{)}. \quad (8.78)$$

Note that (v) and (vii) imply that $A \subseteq B$ if and only if $A \cap B = A$.

viii. The *union* of fuzzy sets A and B is given as

$$\mu_{A \cup B}(x) = \max\{\mu_A(x); \mu_B(x)\} \quad \text{(for all } x \in X\text{)}. \quad (8.79)$$

ix. The *algebraic product* of fuzzy sets A and B is denoted by AB and is defined by the relation

$$\mu_{AB}(x) = \mu_A(x)\mu_B(x) \quad \text{(for all } x \in X\text{)}. \quad (8.80)$$

x. The *algebraic sum* of fuzzy sets A and B is denoted by $A + B$ and has the membership function

$$\mu_{A+B}(x) = \mu_A(x) + \mu_B(x) - \mu_A(x)\mu_B(x) \quad \text{(for all } x \in X\text{)}. \quad (8.81)$$

xi. A fuzzy set A is *convex*, if for all $x, y \in X$ and $\lambda \in [0, 1]$,

$$\mu_A[\lambda x + (1 - \lambda)y] \geq \min\{\mu_A(x); \mu_A(y)\}. \quad (8.82)$$

If A and B are convex, then it can be demonstrated that $A \cap B$ is also convex.

xii. A fuzzy set A is *concave* if A' is convex. In the case where A and B are concave, the $A \cup B$ will also be concave.

Tools and Techniques

xiii. Consider $f: X \mapsto Y$, a mapping from set X to set Y, and A a fuzzy set in X. The fuzzy set B *induced* by mapping f is defined in Y with the following membership function:

$$\mu_B(y) = \sup_{x \in f^{-1}(y)} \mu_A(x), \qquad (8.83)$$

where $f^{-1}(y) = \{x \mid x \in X, f(x) = y\}$.

EXAMPLE 8.10

Fuzzy sets A and B are defined in $X = [-\infty, \infty]$ by using the membership functions given below (Figure 8.20):

$$\mu_A(x) = \begin{cases} x & \text{if } 0 \le x \le 1, \\ -x+2 & \text{if } 1 < x \le 2, \\ 0 & \text{otherwise.} \end{cases} \quad \mu_B(x) = \begin{cases} x-1 & \text{if } 1 \le x \le 2, \\ 1 & \text{if } 2 \le x \le 3, \\ -x+4 & \text{if } 3 \le x \le 4, \\ 0 & \text{otherwise.} \end{cases}$$

Determine the membership functions of $A \cap B$, $A \cup B$, AB, and $A + B$ fuzzy sets.

Solution: Consider Equation 8.78, $\mu_{A \cap B}(x) = \min\{\mu_A(x); \mu_B(x)\}$; therefore $A \cap B$ is

$$\mu_{A \cap B}(x) = \begin{cases} x-1 & \text{if } 1 \le x \le 1.5, \\ -x+2 & \text{if } 1.5 < x \le 2, \\ 0 & \text{otherwise.} \end{cases}$$

From Equation 8.79 we have $\mu_{A \cup B}(x) = \max\{\mu_A(x); \mu_B(x)\}$; thus $A \cup B$ is as follows:

$$\mu_{A \cup B}(x) = \begin{cases} x & \text{if } 0 \le x \le 1, \\ -x+2 & \text{if } 1 < x \le 1.5, \\ x-1 & \text{if } 1.5 < x \le 2, \\ 1 & \text{if } 2 < x \le 3, \\ -x+4 & \text{if } 3 < x \le 4, \\ 0 & \text{otherwise.} \end{cases}$$

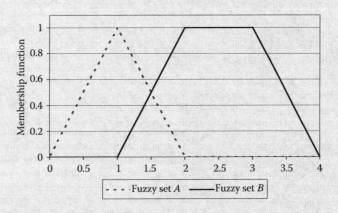

FIGURE 8.20 Membership functions of A and B.

continued

Using Equation 8.80, the product of A and B is determined as follows:

$$\mu_{AB}(x) = \begin{cases} (x-1)(-x+2) & \text{if } 1 \leq x \leq 2, \\ 0 & \text{otherwise.} \end{cases}$$

Finally, the summation of fuzzy sets is determined as follows:

$$\mu_{A+B}(x) = \begin{cases} x & \text{if } 0 \leq x \leq 1, \\ 1 - (x-1)(-x+2) & \text{if } 1 < x \leq 2, \\ 1 & \text{if } 2 < x \leq 3, \\ -x + 4 & \text{if } 3 < x \leq 4, \\ 0 & \text{otherwise.} \end{cases}$$

The membership functions of $A \cap B$, $A \cup B$, AB, and $A + B$ fuzzy sets are illustrated in Figure 8.21.

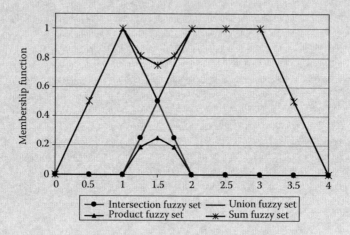

FIGURE 8.21 Operations of fuzzy sets $A \cap B$, $A \cup B$, AB, and $A + B$.

Example 8.11

Consider fuzzy set A defined in $X = [-2, 2]$ as follows:

$$\mu_A(x) = \begin{cases} \dfrac{2+x}{2} & \text{if } -2 \leq x \leq 0, \\ \dfrac{2-x}{2} & \text{if } 0 < x \leq 2, \\ 0 & \text{otherwise.} \end{cases}$$

Determine the membership function of the fuzzy set induced by the function $f(x) = x^2$.

Solution: Since $f([-2, 2]) = [0, 4]$, $\mu_B(y) = 0$ if $y \notin [0, 4]$. If $y \in [0, 4]$, then $f^{-1}(y) = \{\sqrt{y}, -\sqrt{y}\}$, and since $\mu_A(\sqrt{y}) = (2 - \sqrt{y})/2 = \mu_A(-\sqrt{y})$,

$$\mu_B(y) = \sup_{x \in f^{-1}(y)} \mu_A(x) = \frac{2 - \sqrt{y}}{2}.$$

Both $\mu_A(x)$ and $\mu_B(y)$ are illustrated in Figure 8.22.

Uncertain model parameters can be considered as fuzzy sets, and the result on any series of operations leads to a fuzzy set. Uncertainty in constraints and in the objective functions can also be expressed by fuzzy sets. See Karamouz et al. (2003) for more details.

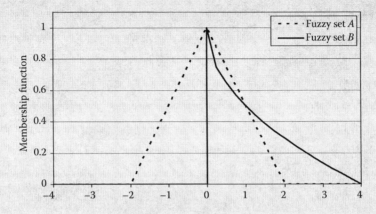

FIGURE 8.22 Illustration of $\mu_A(x)$ and $\mu_B(y)$.

8.5 URBAN WATER SYSTEMS ECONOMICS

Economics includes analytical tools that are used to determine the allocations of scarce resources by balancing the competing objectives. Different important and complex decisions should be made in the process of planning, design, construction, operation, and maintenance of urban water resources systems. Economic considerations besides the technological and environmental concerns play an important role in the decision making about urban water resources planning and management.

The economic analysis of urban water infrastructure development projects has two sides. They create value and they also encounter costs. The preferences of individuals for goods and services are considered in the value side of the analysis. The value of a good or service is related to the willingness to pay. The costs of economic activities are classified into fixed and variable costs. The range of operation or activity level do not affect the fixed costs such as general management and administrative salaries and taxes on facilities, but variable costs are determined based on the quantity of output or other measures of activity level.

EXAMPLE 8.12

There are two choices, A and B, for the trenchless replacement of mains in an urban water distribution system. $20,000 per week should be paid to the traffic control authorities for traffic jams during the project. The other costs of the two trenchless replacement methods are summarized in Table 8.15. Compare the two methods in terms of their fixed, variable, and total costs when 20 km of mains should be replaced. For the selected method, how many meters of the mains should be replaced before starting to make a profit, if there is a payment of $30 per meter of mains replacement?

Solution: As is depicted in Table 8.16, the plant setup is a fixed cost, whereas the payments for traffic jams and rent are variable costs.

continued

TABLE 8.15
Costs of Choices A and B for Trenchless Replacement of Example 8.10

Cost	Method A	Method B
Time taken to complete the project	10 week	6 week
Weekly rental of the site	$5000	$1000
Cost to set up and remove equipment	$150,000	$250,000
Payment for traffic jam	$20,000/week	$20,000/week

TABLE 8.16
Analysis of Costs of the Two Choices for Mains Replacement in Example 8.10

Cost	Fixed	Variable	Choice A	Choice B
Rent		✓	$50,000	$6000
Plant setup	✓		$150,000	$250,000
Traffic jam payment		✓	10($20,000) = 200,000	6($20,000) = 120,000
Summation	$400,000	$376,000		

As can be seen from Table 8.16, although choice B has higher fixed costs, it has less total cost than choice A. The project will make profit at the point where total revenue equals total cost of the kilometers of mains replaced. For choice B, the variable cost per meter of replacement is calculated as follows:

$$\frac{\$6000 + \$120,000}{20,000} = \$6.3.$$

Total cost = Total revenue when the project starts to make a profit. Let X be the meters of the mains that should be replaced before starting to make a profit; we will therefore have:

$$\$250,000 + \$6.3X = \$30X \text{ and } X \text{ is determined as } 10,549 \text{ m}.$$

Therefore, if choice B is selected, the project will begin to make a profit after replacing 10,549 m of the mains.

Costs are also classified as private costs and social costs. The private costs are directly experienced by the decision makers but all costs of an action are considered in the social costs without considering who is affected by them (Field, 1997). The difference between social and private costs is called external costs or usually environmental costs. Since the firms or agencies do not normally consider these costs in decision making, they are called "external." However, it should be noted that these costs may be real costs for some members of society. Open access resources such as reservoirs in urban areas are the main source of external or environmental costs.

For example, consider three cities around a lake that use the water of the lake and discharge the sewage back into the lake. Each city should pay about $30,000 per year for water treatment. Suppose that a new tourism complex wants to start operating near the lake. Because of the untreated sewages discharged to the lake by this tourism complex, each of these cities should spend $10,000 more per year for water treatment. In this case, the external cost is $30,000 ($10,000 × 3).

Similarly, external benefits are defined. An external benefit is a benefit that is experienced by somebody outside the decision about consuming resources or using services. For example, consider that a private power company constructed a dam for producing hydropower. The benefits experienced by people downstream, such as mitigating both floods and low flows, are classified as external benefits.

Many values of an action cannot be measured in commensurable monetary units as is desired by economists. For example, the effects on human beings physically (through loss of health or life), emotionally (through loss of national prestige or personal integrity), and psychologically (through environmental changes) cannot be measured by monetary units. These sorts of values are called intangible or irreducible (James and Lee, 1971).

In most of the engineering activities, the capital investment is committed for a long period; therefore the effects of time should be considered in economic analysis. This is because of the different values of the same amount of money spent or received at different times. The future value of the present amount of money will be larger than the existing amount due to opportunities that are available to invest the money in various enterprises to produce a return over a period. The interest rate is the rate at which the value of money increases from the present to the future. On the contrary, the discount rate is the amount by which the value of money is discounted from the future to the present.

A summary of the relations between general cash flow elements using discrete compounding interest factors is presented in Table 8.17 using the following notation:

i: Interest rate
N: Number of periods (years)
P: Present value of money
F: Future value of money
A: End of the cash flow period in a uniform series that continues for a specific number of periods
G: End-of-period uniform gradient cash flows

More details of time-dependent interest rates and interest formulae for continuous compounding are given in DeGarmo et al. (1997) and Au and Au (1983).

TABLE 8.17
Relations between General Cash Flow Elements Using Discrete Compounding Interest Factors

To Find	Given	Factor by Which to Multiply "Given"	Factor Name	Factor Function Symbol
F	P	$(1+i)^N$	Single payment compound amount	$\left(\dfrac{F}{P}, i\%, N\right)$
P	F	$\dfrac{1}{(1+i)^N}$	Single payment present worth	$\left(\dfrac{P}{F}, i\%, N\right)$
F	A	$\dfrac{(1+i)^N - 1}{i}$	Uniform series compound amount	$\left(\dfrac{F}{A}, i\%, N\right)$
P	A	$\dfrac{(1+i)^N - 1}{i(1+i)^N}$	Uniform series present worth	$\left(\dfrac{P}{A}, i\%, N\right)$
A	F	$\dfrac{i}{(1+i)^N - 1}$	Sinking fund	$\left(\dfrac{A}{F}, i\%, N\right)$
A	P	$\dfrac{i(1+i)^N}{(1+i)^N - 1}$	Capital recovery	$\left(\dfrac{A}{P}, i\%, N\right)$
P	G	$\dfrac{(1+i)^N - 1 - Ni}{i^2(1+i)^N}$	Discount gradient	$\left(\dfrac{P}{G}, i\%, N\right)$
A	G	$\dfrac{(1+i)^N - 1 - Ni}{i(1+i)^N - i}$	Uniform series gradient	$\left(\dfrac{A}{G}, i\%, N\right)$

Source: DeGarmo, E. P. et al. 1997. *Engineering Economy*. Prentice Hall, Upper Saddle River, NJ.

Example 8.13

A wastewater collection network project in a city produces benefits, as expressed in Figure 8.23: the $100,000 profit in year 1 is increased in 10 years on a uniform gradient to $1,000,000. Then it reaches $1,375,000 in year 25 with a uniform gradient of $25,000 per year and it remains constant at $1,375,000 each year until year 40, the end of the project life span. Assume an interest rate of 4%. What is the present worth of this project?

Solution: First, the present value of the project benefits in years 1–10 are calculated as follows:

$$100{,}000\left(\frac{P}{G},4\%,10\right) + 100{,}000\left(\frac{P}{A},4\%,10\right) = 100{,}000 \times 41.9919 = \$4{,}199{,}190.$$

The same procedure is repeated in years 11–25:

$$1{,}025{,}000\left(\frac{P}{A},4\%,15\right)\left(\frac{P}{F},4\%,10\right) + 25{,}000\left(\frac{P}{G},4\%,15\right)\left(\frac{P}{F},4\%,10\right)$$

$$= 1{,}025{,}000 \times 11.11839 \times 0.67556 + 25{,}000 \times 69.735 \times 0.67556 = 8{,}876{,}672.$$

The present value of benefits in years 26–40:

$$1{,}375{,}000\left(\frac{P}{A},4\%,15\right)\left(\frac{P}{F},4\%,25\right) = 1{,}375{,}000 \times 11.11838 \times 0.37512 = 5{,}734{,}749.$$

The total present worth is equal to $18,810,611, the summation of the above three values.

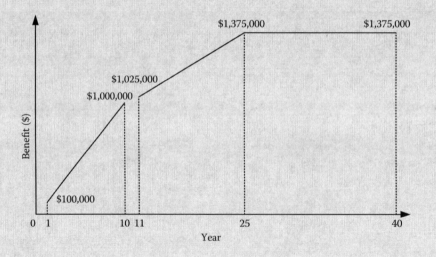

FIGURE 8.23 Cash flow diagram for Example 8.13.

8.5.1 Economic Analysis of Multiple Alternatives

Through economic analysis it is determined whether a capital investment and the cost associated with the project within the project lifetime can be recovered by revenues or not. It should also be determined how attractive the savings are regarding the involved risks and the potential alternative uses. The five most common methods including the Present Worth, Future Worth, Annual Worth, Internal Rate of

Return, and External Rate of Return methods are briefly explained in this section. The life span of different projects should be incorporated into economic analysis through the repeatability assumption.

In the Present Worth method, the alternative with the maximum present worth (PW) of the discounted sum of benefits minus costs over the project life span, namely Equation 8.84, is selected.

$$\text{PW} = \sum_{t=1}^{N} (B_t - C_t) \left(\frac{P}{F}, i\%, t\right), \tag{8.84}$$

where C_t is the cost of the alternative at year t, B_t is the profit of the alternative at year t, N is the study period, and i is the interest rate.

In the case of cost alternatives, the one with the least negative equivalent worth is selected. Cost alternatives are all negative cash flows except for a possible positive cash flow element from disposal of assets at the end of the project's useful life (DeGarmo et al., 1997).

In the Future Worth method, all benefits and costs of alternatives are converted into their future worth and then the alternative with the greatest future worth or the least negative future worth is selected. Two choices are available for selecting the study period including the repeatability assumption and the coterminated assumption. Based on the latter assumption, for the alternatives with a life span shorter than the study period, the estimated annual cost of the activities is used during the remaining years. For the alternatives with a life span longer than the study period, a re-estimated market value is normally used as a terminal cash flow at the end of the project's coterminated life as stated by DeGarmo et al. (1997).

In the Annual Worth (AW) method, the costs and benefits of all alternatives are considered in the annual uniform form and, similar to the other methods, the alternative with the greatest annual worth or the least negative worth is selected. Because in this method the worth of alternatives is evaluated annually, the projects with different economic lives are compared without any change in the study period.

The internal rate of return (IRR) method is commonly used in economic analysis of different alternatives. IRR is defined as the discount rate that when considered, the net present value or the net future value of the cash flow profile will be equal to zero. The PW and AW methods are usually used for finding the IRR. In this method, the minimum attractive rate of return (MARR) is defined and projects with an IRR less than the considered threshold are rejected.

The main assumption of this method is that the recovered funds, which are not consumed each time, are reinvested in IRR. In such cases where it is not possible to reinvest the money in the IRR rather than the MARR, the external rate of return (ERR) method should be used. The interest rate, e, external to a project, at which the net cash flows produced by the project over its life span can be reinvested, is considered in this method.

Example 8.14

Two different types of pumps (A and B) can be used in a pumping station of an urban water distribution system. The costs and benefits of each of the pumps are displayed in Table 8.18. Compare the two pumps economically using the MARR = 10% and considering that the time period of the study is equal to 12 years.

Solution: Since the life spans of the pumps are less than the study period, the repeatability assumption is used. At first the PW of the benefits and costs of the pumps are calculated.

PW of costs of A = $20,000 + ($20,000 − $500)(P/F, 10%, 5) − $500(P/F, %%, 10) = $31,916
PW of benefits of A = $10,000(P/A, 10%, 10) = $61,440

continued

TABLE 8.18
Data Used in Example 8.12

Economic Characteristics	Pump A	Pump B
Initial investment	$20,000	$70,000
Annual benefits	$10,000	$15,000
Salvage value	$500	$1000
Useful life	5	10

PW of costs of B = $70,000 − $1000(P/F, 10%, 10) = $69,614
PW of benefits of B = $15,000(P/A, 10%, 10) = $92,170

For selecting the economically best alternative, the graphical method is employed in this example. The location of projects on a graph showing the present value of benefits versus the present value of costs is determined as points A and B and then the slope of the line connecting the points is estimated and compared with a 45° slope line where the points on it have equal present values of costs and benefits (Figure 8.24). When the slope of the line connecting two projects is less (more) than 45°, the project with less (more) initial cost would be selected. As shown in Figure 8.24, pump A, which has less initial cost, should be selected.

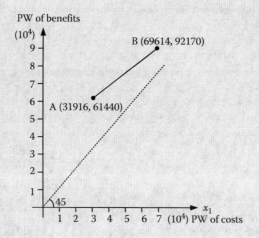

FIGURE 8.24 Comparing the two pumps in Example 8.14 by the rate of return method.

8.5.2 Economic Evaluation of Projects Using Benefit–Cost Ratio Method

The benefit–cost ratio method, which considers the ratio of benefits to costs, has been widely used in the economic analysis of water resources projects. The ratio is estimated based on the equivalent worth of discounted benefits and costs in the form of annual worth, present worth (PW), and/or future worth, to consider the time value of money. Equations 8.85 and 8.86 are basic formulations of the benefit–cost ratio method.

$$\frac{B}{C} = \frac{B}{I + C}, \tag{8.85}$$

$$\frac{B}{C} = \frac{B - C}{I}, \tag{8.86}$$

where B is the net equivalent benefits, C is the net equivalent annual cost including operation and maintenance costs, and I is the initial investment.

If the benefit–cost ratio of a project is less than or equal to 1, it is not considered. Through the following equation the associated salvage value of the investment is considered in the benefit–cost method:

$$\frac{B}{C} = \frac{B-C}{I-S}, \qquad (8.87)$$

where S is the salvage value of investment.

Example 8.15

Assume that the annual costs of operation and maintenance of pumps A and B are equal to $4000 and $6000, respectively. Compare the two pumps from an economic aspect, using the benefit–cost ratio method.

Solution: For considering the salvage value of the pumps, Equation 8.87 is employed. Because of different life spans of the pumps, the annual worth method is employed for more simplicity as follows:

Pump A:

$$\frac{B}{C} = \frac{10{,}000 - 4000}{20{,}000\,(A/P, 10\%, 5) - 500\,(A/F, 10\%, 5)} = \frac{6000}{20{,}000 \times 0.263 - 500 \times 0.163} = 1.15.$$

Pump B:

$$\frac{B}{C} = \frac{15{,}000 - 6000}{70{,}000\,(A/P, 10\%, 10) - 1000\,(A/F, 10\%, 10)} = \frac{9000}{70{,}000 \times 0.162 - 1000 \times 0.0627} = 0.79.$$

As can be seen, the benefit–cost ratio for pump B is <1. Therefore it is rejected and pump A is selected.

8.5.3 Economic Models

Economic models relate to the demands of the consumer on the available resources for demand supply to satisfy the consumers. The demand and production functions for goods and services, and supply functions for resources, are included in this method. Production functions model the production of useful products by utilizing resources and certain technologies. This trade-off between economic activities for goods and services production and the environmental quality is called the production possibility curve (Figure 8.25). Depending on the technological capabilities and limitations in a

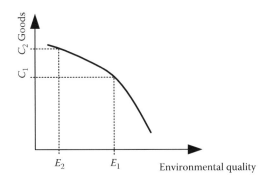

FIGURE 8.25 Production possibility curve. (Adapted from Karamouz, M., Szidarovszky, F., and Zahraie, B. 2003. *Water Resources Systems Analysis*. Lewis Publishers, CRC Press, USA.)

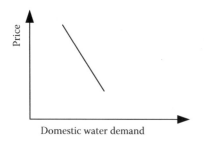
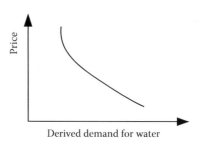

FIGURE 8.26 Examples of demand function. (Adapted from Karamouz, M., Szidarovszky, F., and Zahraie, B. 2003. *Water Resources Systems Analysis*. Lewis Publishers, CRC Press, Boca Raton, FL.)

specific time, each society chooses its locations on this curve. This is called social choice, which depends on the value that people in that society place on the environment (E_i) and the economic productions (c_i) (Field, 1997).

In resource-supply function, the quantity of resources made available per unit time via the price per unit of resources is considered. In water resources development projects, the development of well fields, interbasin water transfer projects, and other related projects should be considered in an accurate estimation of water price. In order to make a correct decision about the development alternatives and water allocation schemes, the distribution of water resources development costs should be considered. In water scarcity situations, more water might be allocated to the users, which could pay high prices for water, such as industries. However, the political and legal issues that are important in developing water allocation schemes affect this process (Viessman and Welty, 1985).

Demand functions show the demand of goods and services at any particular price. Figure 8.26 shows two typical demand functions. As illustrated in Figure 8.26, the price–quantity relation for domestic water use is very steep. This means that price increase does not affect the water demand significantly, because people would like to supply their needs for cooking, washing, and maintaining health standards. But the demand for products derived from water is significantly sensitive to price changes.

The demand functions that are considered in the engineering and planning process consider the variety of factors that affect the demand and are more complex. For example, the domestic demand functions are formulated as follows (Mays, 1996):

$$Q_W = Q_W(P_W, P_a, P, Y, Z), \tag{8.88}$$

where Q_W is the consumers' water demands in a specific period of time, P_W is the price of water, P_a is the price of an alternative water resource, P is an average price index representing all goods and services for water supply, Y is the consumer income, and Z is a vector representing other factors such as climate and consumer preferences (Mays, 1996).

PROBLEMS

1. Consider the following problem with two objective functions:

   Maximize $f_1 = x_1 + x_2$, $f_2 = x_1 - x_2$,
   Subject to $x_1, x_2 \geq 0$,
   $x_1 + 2x_2 \leq 4$,
   $2x_1 - x_2 \geq 2$.

   i. Represent graphically the decision space.
   ii. Display graphically the criteria space.

2. Consider the following linear programming problem:

$$\text{Maximize } x_1 + x_2,$$
$$\text{Subject to } x_1, x_2 \geq 0,$$
$$2x_1 + x_2 \leq \xi,$$
$$x_1 + 2x_2 \geq 3,$$

where ξ is randomly distributed with $\mu = 3$ and $\sigma^2 = 1$. Give the chance constraint formulation of the problem, and find the optimal solution.

3. Consider the two mutually exclusive alternatives A and B for a water diversion project as described in Table 8.19. Assume that the study period is 15 years and repeatability is applicable. Using the annual worth method, which project do you recommend?

TABLE 8.19
Characteristics of Considered Projects in Problem 3

	Project A	Project B
Capital investment	$160,000	$240,000
Annual revenue	$80,000	$112,000
Annual expense	$35,200	$68,800
Useful life (years)	8	12

4. A river supplies water to an industrial complex and agricultural lands located downstream of the complex. The average monthly industrial and irrigation demands and the monthly river flows in a dry year are given in
Table 8.20 (numbers are in million cubic meters). The prices of water for industrial and irrigation uses are $80,000 and $9000 per million cubic meters.

   i. Formulate the problem for optimizing the water allocation from this river.
   ii. Solve the problem using linear programming.

TABLE 8.20
Average Monthly Industrial and Irrigation Demands and the Monthly River Flows in a Dry Year of Problem 4

	Jan	Feb	Mar	Apr	May	Jun	Jul	Aug	Sep	Oct	Nov	Dec
Industrial demand	8	8.5	8.5	9	9	9.5	9.5	10	9.5	9	8.5	8.5
Irrigation demand	0	0	20	50	55	70	70	65	20	10	0	0
Flow	22.5	24.3	36	43.2	54	37.8	22.5	15.3	12.6	16.2	18	22.5

5. The variation of annual water demand of a city over a 30-year planning time horizon is exhibited in Table 8.21. The current demand of this city is about 90 million cubic meters. Three projects are studied in order to supply the demands of this city. Considering the cost of each project given in Table 8.22 and 3% rate of return, find the optimal sequence of implementing projects using linear programming.

6. Assume that you want to model the water demand of a city in order to forecast the municipal demands in the coming years. The historical data of water use, population, price, and precipitation in a 10-year time horizon are presented in Table 8.23. Formulate a multiple regression model for estimating the water use data and comment on selecting the independent variables.

TABLE 8.21
Variations of Annual Water Demand of the Considered City in Problem 5

Year	2008	2016	2025	2035	2040
Water demand (MCM)	110	125	135	150	170

TABLE 8.22
Cost of Considered Projects for Water Supply of the Considered City in Problem 5

Project	Capital Investment (10^6 $)	Capacity (MCM)
Dam A	4	30
Interbasin water transfer tunnel B	7	17
Dam C	6	30

TABLE 8.23
Historical Data of Water Use, Population, Price, and Precipitation in a 10-Year Time Horizon of Problem 6

Year	Population	Price (Monetary Unit/m^3)	Annual Precipitation (mm)	Water Use (m^3/year)
1991	21,603	0.25	812	2,032,410
1992	22,004	0.27	648	2,203,921
1993	23,017	0.29	670.4	2,178,329
1994	23,701	0.3	772	2,265,816
1995	24,430	0.31	670.4	2,323,782
1996	25,186	0.32	629.6	2,403,752
1997	26,001	0.33	576	2,647,942
1998	26,825	0.35	556	2,706,106
1999	27,781	0.31	656	2,700,314
2000	28,576	0.34	640	2,725,007

7. Consider a reservoir that supplies water to a city. The monthly water demand of this complex is 20 million cubic meters. The total capacity of the reservoir is 40 million cubic meters. Let S_t (reservoir storage at the beginning of month t) take the discrete values 0, 10, 20, 30, and 40. The cost of operation (Loss) can be estimated as a function of the difference between release (R_t) and water demand as follows:

$$\text{Loss}_t = \begin{cases} 0, & R_t \geq 10, \\ (20 - R_t)^2, & R_t > 10. \end{cases}$$

  i. Formulate a forward moving deterministic dynamic programming model for finding the optimal release in the next three months.
  ii. Formulate a backward moving deterministic dynamic programming model for finding the optimal release in the next three months.
  iii. Solve the DP model developed in part (a), assuming that the inflows to

the reservoir in the next three months ($t = 1, 2, 3$) are forecasted to be 10, 50, and 20, respectively. The reservoir storage in the current month is 20 million cubic meters.
8. Solve Example 8.1 considering an increase of 1.3% in monthly water consumption given in Table 8.2 (due to population increase), and increase of 2°C in temperature data presented in Table 8.4.
9. Find the minimum point of the function $f(x) = 2x^2 - 3x + 2$ in the $[-2, 2]$ interval using the GA.
10. Fuzzy sets A and B are defined in $X = [-\infty, \infty]$, by the membership functions as follows:

$$\mu_A(x) = \begin{cases} 2x & \text{if } 0 \leq x \leq 5 \\ 15 - x & \text{if } 5 < x \leq 10 \\ 0 & \text{otherwise.} \end{cases}$$

$$\mu_B(x) = \begin{cases} 9 - x & \text{if } 1 \leq x \leq 5 \\ 4 & \text{if } 5 \leq x \leq 15 \\ x - 11 & \text{if } 15 \leq x \leq 20 \\ 0 & \text{otherwise.} \end{cases}$$

Determine the membership functions of $A \cap B$, $A \cup B$, AB, and $A + B$ fuzzy sets.
11. Assume a project with a construction cost of $10,000,000. Consider its benefits follow the cash flow diagram shown in Figure 8.23. If the external rate of return is considered as 4% and MARR is equal to 2%, is this project economically acceptable?
12. Solve Example 8.2 with the changes assumed in Problem 8.

REFERENCES

Ahlfeld, D. P., Mulvey, J. M., Pinder, G. F., and Wood, E. F. 1988. Contaminated groundwater remediation design using simulation, optimization and sensitivity theory. 1. Model development. *Water Resources Research* 24(5): 431–441.

Atiya, A. and Parlos, A. 1992. Nonlinear system identification using spatiotemporal neural networks. In: Proceedings of the *International Joint Conference on Neural Networks*, Vol. 2, Baltimore, MD, pp. 504–509. The IEEE, Inc., Piscataway, NJ.

Au, T. and Au, T. P. 1983. *Engineering Economics for Capital Investment Analysis*. Allyn and Bacon, Boston.

Bose, N. K. and Liang, P. 1996. *Neural Network Fundamentals with Graphs, Algorithms, and Application*. McGraw-Hill, New York.

Carroll, D. L. 1996. Genetic algorithms and optimizing chemical oxygen-iodine lasers. In: H. B. Wilson, R. C. Batra, C. W. Bert, A. M. J. Davis, R. A. Schapery, D. S. Stewart, and F. F. Swinson, Eds, *Developments in Theoretical and Applied Mechanics*, Vol. XVIII. School of Engineering, The University of Alabama.

Chang, L. C., Shoemaker, C. A., and Liu, P. L.-F. 1992. Optimal time-varying pumping rates for groundwater remediation: Application of a constrained optimal control theory. *Water Resources Research* 28(12): 3157–3174.

Clouse, D. S., Giles, C. L., Horne, B. G., and Cottrell, G. W. 1997. Time-delay neural networks: Representation and induction of finite state machines. *IEEE Transitions on Neural Networks* 8(5): 1065–1070.

Coulibaly, P., Anctil, F., and Bobe, E. B. 2001. Multivariate reservoir inflow forecasting using temporal neural networks. *Journal of Hydrologic Engineering, ASCE* 6(5): 367–376.

Culver, T. B. and Shoemaker, C. A. 1992. Dynamic optimal control for groundwater remediation with flexible management periods. *Water Resources Research* 28(3): 629–641.

DeGarmo, E. P., Sullivan, W. G., Bontadelli, J. A., and Wicks, E. M. 1997. *Engineering Economy*. Prentice Hall, Upper Saddle River, NJ.

Demuth, H. and Beale, M. 2002. *Neural Network Toolboox for MATLAB*. User's Guide. Available at http://www.mathworks.com/.

Dougherty, D. E. and Marryott, R. A. 1991. Optimal groundwater management. 1. Simulated annealing. *Water Resources Research* 27(10): 2493–2508.

Elman, J. L. 1990. Finding structure in time. *Cognitive Science*, 2: 179–211.

Egles, R. W. 1990. Simulated Annealing: A Tool for Operational Research. *European Journal of Operational Research* 46: 271–281.

Fauset, L. V. 1994. *Fundamentals of Neural Networks, Architecture, Algorithms, and Application*. Prentice Hall, Englewood Cliffs, NJ.

Field, B. C. 1997. *Environmental Economics: An Introduction*. McGraw-Hill, New York.

Fransconi, P., Gori, M., and Soda, G. 1992. Local feedback multilayered networks. *Neural Computation* 4: 120–130.

Fujiwara, O., Puangmaha, W., and Hanaki, K. 1988. River basin water quality management in stochastic environment. *ASCE Journal of Environmental Engineering*, 114(4): 864–877.

Goldberg, D. E. 1989. *Genetic Algorithms in Search, Optimization, and Machine Learning*. Addison-Wesley, Reading, MA.

Hadley, A. 1964. *Nonlinear and Dynamic Programming*. Addison-Wesley, Reading, MA.

Holland, J. 1975. *Adaptation in Natural and Artificial Systems*. The University of Michigan Press., Ann Arbour.

James, L. D. and Lee, R. R. 1971. *Economics of Water Resources Planning*. McGraw-Hill, New York.

Jordan, M. I. 1986. Attractor dynamics and parallelism in a connectionist sequential machine. *Proceedings of the 8th Annual Conference, Cognitive Science Society*. MIT Press, Amherst, MA, pp. 531–546.

Karamouz, M., Szidarovszky, F., and Zahraie, B. 2003. *Water Resources Systems Analysis*. Lewis Publishers, CRC Press, Boca Raton.

Karamouz, M., Razavi, S., and Araghinejad, Sh. 2007. Long-lead seasonal rainfall forecasting using time-delay recurrent neural networks: A case study. *Journal of Hydrological Processes* 22(2): 229–238.

Lefkoff, L. J. and Gorelick, S. M. 1987. AQMAN: Linear and quadratic programming matrix generator using two-dimensional groundwater flow simulation for aquifer management modeling. U.S. Geological Survey Water Resources Investigations Report, 87-4061.

Mays, L. W. 1996. *Water Resources Handbook*. McGraw-Hill, New York.

McKinney, D. C. and Lin, M. D. 1995. Approximate mixed-integer nonlinear programming methods for optimal aquifer remediation design. *Water Resources Research* 31(3): 731–740.

Metropolis, N., Rosenbluth, A., Rosenbluth, M., Teller, A., and Teller, E. 1953. Equation of state calculations by fast computing machines. *Journal of Chemical Physics* 21: 1087–1092.

Milton, J. S. and Arnold, J. C. 1995. *Introduction to Probability and Statistics*, Third Edition. McGraw-Hill, New York.

Price, W. L. 1965. A controlled random search procedure for global optimization. *Computer Journal* 7: 347–370.

Ross, Sh. M. 1987. *Introduction to Probability and Statistics for Engineers and Scientists*. Wiley, New York.

Rubinstein, R. Y. 1981. *Simulation and Monte Carlo Method*. Wiley, New York.

Solomatine, D. P. 1999. Two strategies of adaptive cluster covering with descent and their comparison to other algorithms. *Journal of Global Optimization*, 14(1): 55–78.

StatSoft. Electronic statistics textbook. URL. 2002. Available at http://www.statsoftinc.com/textbook/stathome.html.

Sun, M. and Zheng, C. 1999. Long-term groundwater management by a MODFLOW based dynamic groundwater optimization tool. *Journal of American Water Resources Association* 35(1): 99–111.

Viessman, W. and Welty, C. 1985. *Water Management–Technology and Institutions*. Harper and Row Publishers, New York.

Wagner, B. J. and Gorelick, S. M. 1987. Optimal groundwater quality management under parameter uncertainty. *Water Resources Research* 23(7): 1162–1174.

Waibel, A., Hanazawa, T., Hintin, G., Shikano, K., and Lang, K. J. 1989. Phoneme recognition using time delay neural networks. *IEEE Transition on ASSP* 37(3): 328–339.

Wan, E. A. 1993. Time series prediction using a connectionist network with internal delay lines. In: A. S. Weigend and N. A. Gershenfeld, Eds, *Time Series Prediction: Forecasting the Future and Understanding the Past*, pp. 195–217. Addison-Wesley, Reading, MA.

Wang, M. and Zheng, C. 1998. Application of genetic algorithms and simulated annealing in groundwater management: Formulation and comparison. *Journal of American Water Resources Association* 34(3): 519–530.

Willis, R. L. 1976. Optimal groundwater quality management, well injection of waste waters. *Water Resources Research* 12: 47–53.

Willis, R. L. 1979. A planning model for the management of groundwater quality. *Water Resources Research* 15: 1305–1313.

9 Urban Water Infrastructure

9.1 INTRODUCTION

For sustaining the central role of water in societies, it should be managed to increase the water productive impact while protecting aquatic ecosystems, which are crucial for the environment. This could be achieved by developing the needed infrastructure with legal and institutional frameworks for integrated urban water management. Development and management of urban water can be done by a stock of facilities and installations called "urban water infrastructure"; this includes supply, delivery, treatment, and distribution of water (water mains) to its end users and also collection, removal, treatment, and disposal of sewage and wastewater. The objective of this chapter is to discuss urban water infrastructure and how to maintain the structural integrity of its components, with special attention to water mains.

In this chapter, first an introduction to different parts of urban water infrastructure is given. The interactions between different components as well as life cycle assessment (LCA) and sustainable design of urban infrastructures, are discussed. Finally, application of standard integrity monitoring (SIM) to water mains is presented. Infrastructure planning is a multidimensional task that consists of different steps as generating master plans, defining projects, and identification of alternatives, and finally, comparison of alternatives. Examples are given based on the concepts and tools discussed in Chapter 8 to economically compare these alternatives. For further economic and financial analyses of infrastructure development, refer to Goodman and Hastak (2006).

9.2 URBAN WATER INFRASTRUCTURES

The functions of urban water infrastructures can be reiterated to include delivery, treatment, supply, and distribution of water to the users as well as collection, removal, treatment, and disposal of sewage and wastewater. A stock of facilities and installations are needed to develop and manage water resources.

Water has always played a central role in human societies, but in order to sustain that role, it needs to be harnessed and managed to increase its productive impact and to reduce the risk of destruction, while protecting aquatic ecosystems as a vital component of the environment. This can be achieved by developing an adequate hydraulic infrastructure in parallel with legal and institutional frameworks for water management.

Adequate water infrastructure (mostly dams and reservoirs) is required to ensure the sustainability of water resources and to overcome scarcity problems. Physical infrastructure is also required to provide water-related services, primarily water supply and sanitation, for the population, agriculture, and industry, as well as for treatment and disposal of wastewater. Hydraulic infrastructure provides other benefits, such as hydroelectric power and navigation. In addition, water infrastructure is intended to supplement the natural ability of aquatic ecosystems to cope with drought and floods as well as to accommodate a certain pollution load.

9.2.1 INFRASTRUCTURE FOR WATER SUPPLY (DAMS AND RESERVOIRS)

Dams and reservoirs provide storage of water, including flood water, which can then be supplied for households and irrigation, as well as generation of power, thus reducing fossil fuel depletion and the negative environmental effects of fossil fuel burning. They often emerge as the priority in strategic planning with respect to water and energy. However, similar to the other infrastructure projects, there are also adverse environmental and social impacts that must be minimized or mitigated.

The world has around 55,000 large dams, most of them registered by the International Commission on Large Dams (ICOLD). About half of them are used solely for supplying water for irrigation purposes and roughly one-third of them are used for multipurpose. No reliable data exist on the total number of "small" dams, that is, those not meeting the ICOLD criteria. A very indicative figure is 800,000, almost all of them used for irrigation and water supply.

In the Asian and Pacific region, most dams and reservoirs have been built since 1950. Construction peaked in the 1970s, when hundreds of large dams were put into service each year. At that time, Japan, a flood-prone country with 40% of its population and 60% of economic assets located in vulnerable river plains, invested some 2 trillion Yen (about 20 Milliard US$) in hydraulic infrastructure, mostly dams and embankments. With these facilities, annual flood losses that, before the 1950s, could reach 20% of GDP were reduced to less than 1% of GDP. Even with its large stock of about 2800 large dams and reservoirs, Japan still spends about US$9 billion of public funds annually on expanding and maintaining hydraulic infrastructure, and another 100 large dams are under construction (White, 2005).

A major indicator of water resources development is the ratio between the available storage reservoir capacity and the volume of the annual renewable water resources. The ratio in the Asian and Pacific region is 0.10, less than the global average of 0.14 and far behind North America (0.33) and Europe (0.16) (Asian Development Bank, 2001).

EXAMPLE 9.1

Three water supply projects are implemented in a region. Water demand of a big city, a small city, and a village will be supplied through these projects. Level of income and distribution of the costs and benefits for these projects are presented in Table 9.1. Implementation of these projects reduces the tension on other resources that are expressed as beneficial in Table 9.1. Determine the distribution of costs and benefits and find which projects are progressive and regressive based on the methods introduced in Chapter 8.

Solution: The percent of income for the village, the small city, and the big city is shown in Table 9.2.

As can be seen in Table 9.2, the first program is regressive because the profits allocated to the groups with higher level of income are higher. The second program is progressive because it has allocated less profit to the people with higher income. In the third program, the difference between benefits and costs

TABLE 9.1
Level of Income and Distribution of Costs and Benefits Related to Example 9.1

Benefits and Costs	Village	Small City	Big City
Average Income ($)	1000	20,000	70,000
Water supply program 1			
Benefit to other resources	30	1400	7700
Costs of project implementation	20	800	4200
Water supply program 2			
Benefit to other resources	140	2200	7700
Costs of project implementation	40	1000	4200
Water supply program 3			
Benefit to other resources	30	300	840
Costs of project implementation	20	100	140

is 1% of the income for the village, the small city, and the big city. Therefore, the distribution of costs and benefits has not been very different in this program.

TABLE 9.2
Distribution of Costs and Benefits in the Three Water Supply Projects of Example 9.1

Percentage of Income	Village	Small City	Big City
Income	1000	20,000	70,000
Water supply program 1			
Benefit to other resources	3	7	11
Costs of project implementation	2	4	6
Difference between benefit and costs	1	3	5
Water supply program 2			
Benefit to other resources	14	11	11
Costs of project implementation	4	5	6
Difference between benefit and costs	10	6	5
Water supply program 3			
Benefit to other resources	3	1.5	1.2
Costs of project implementation	2	0.5	0.2
Difference between benefit and costs	1	1	1

9.2.2 INFRASTRUCTURE FOR THE WATER DISTRIBUTION SYSTEM

Water distribution systems (WDS) include the entire infrastructure from the treatment plant outlet to the household tap. Fire-flow requirements usually control the design aspect of a WDS. The actual layout of water mains, arteries, and secondary distribution feeders should be designed to deliver the required fire flows in all buildings and structures in the municipality, above and beyond the maximum daily usage rate. In developing the WDS layout, the effect of a break, joint separation, or other main failure that could occur during the WDS operation should be considered. Also, most of the water-quantity standards related to the WDS should be observed on the basis of general practices in this field.

In evaluating a WDS, pumps should be considered at their effective capacities when discharging at standard operating pressures. The pumping capacity, in conjunction with storage, should be sufficient to maintain the maximum daily usage rate plus the maximum required fire-fighting flow.

Storage tanks are frequently used to equalize pumping rates in the distribution system as well as provide water for fire fighting. In determining the fire flow from storage, it is necessary to calculate the rate of delivery during the specified period (fire fighting). Even though the volume stored may be large, the flow to a hydrant cannot exceed the carrying capacity of the mains, and the residual pressure at the point of use should not be less than what is needed (e.g., 140 kPa), which varies with different standards.

Depending on specific practices, some standards for urban water distribution networks are noted here for demonstrating the scale and range of figures. As an example, the recommended water pressure in a distribution system is typically 450–520 kPa, which is considered adequate for buildings up to 10 stories high as well as for automatic sprinkler systems for fire protection in buildings of 4–5 stories. For a residential service connection, the minimum pressure in the water distribution main should be 280 kPa; pressure in excess of 700 kPa is not desirable, and the maximum allowable pressure is 1030 kPa. Fire hydrants are installed at a spacing of 90–240 m and in locations required for fire fighting.

Although a gravity system delivering water without the use of pumps is desirable from a fire protection standpoint, the reliability of well-designed and properly safeguarded pumping systems can be developed to such a high degree that no distinction is made between the reliability of gravity-fed and pump-fed systems. Electric power should be provided to all pumping stations and treatment facilities by two separate lines from different sources (Statewide Urban Design and Specifications, 2007). In the following, an introduction to the status of water supply infrastructures in the world based on international reports is presented.

The provision of safe drinking water and improved sanitation has a high priority in the MDG and the Johannesburg Plan of Implementation as a primary means of eradicating poverty. For Asia and the Pacific, meeting the MDG targets, by 2015, the proportion of people without adequate and sustainable access to safe drinking water and improved sanitation presents a particularly formidable challenge. Two-thirds of the world's population do not have access to safe water and more than three quarters of the world's population are not served by adequate sanitation. The drastic increase projected in population over the next century will place enormous pressure on urban water infrastructure. The construction, extension, and rehabilitation of water supply and sanitation facilities, especially those serving the poor nations, will need to be accelerated (Report of the World Summit on Sustainable Development, 2002).

The Asian and Pacific region's lack of sufficient infrastructure denies a large part of its population access to safe water and decent sanitation. In 2002, every sixth person in the region, or an estimated 691 million people, did not have access to safe, sustainable water supplies and almost half the population did not have access to decent sanitation. A huge number of people have gained access to water and sanitation services as a result of the expansion in infrastructure since 1990. However, due to population growth, the absolute number of people without access to such services remained almost the same. Indeed, coverage of the region's urban population actually decreased due to population growth rate that surpassed the rate of development of urban water supply and sanitation (Asian Development Bank, 2005).

9.2.2.1 Water Mains

Water mains are provided solely for the purpose of supplying potable water and fire protection. "Water mains" refers here to the following categories, but not service lines:

1. A raw water transmission main transports raw water from the source to the treatment plant.
2. A treated water transmission main conveys water from the plant to the storage or directly to an arterial main.
3. Arterial mains transport treated water to the distribution mains from the treatment plant or storage.
4. The distribution mains transport water to the service lines, which convey the water to the end user.

For a given system, the transmission mains are typically larger in diameter, straighter, and have fewer connections than the distribution mains (Smith et al., 2000).

9.2.2.2 Design and Construction of the Water Main System

The important design requirements of water main systems are that they supply each user with sufficient volume of water for a particular designated use plus required fire flows at adequate pressure and that the system maintains the quality of the potable water delivered by the treatment plant. It is important in the design of water main systems to address the maintenance considerations, constantly. The performance of a water main system for health and fire-flow purposes depends on the Jurisdiction's ability to maintain the system at an affordable cost. Certain planning considerations related to a new system development or system expansion require the designer to consider factors such as future growth, cost, and system layout. For system layout, all major demand areas should be serviced by an arterial-loop system. High-demand areas are served by distribution mains tied to an arterial-loop system to form a grid without dead-end mains. Areas where adequate water supply must be maintained at all times for health and fire control purposes should be tied to two arterial mains where possible. Minor distribution lines or mains that make up the secondary grid system are a major portion of the grid since they supply the fire hydrants and domestic and commercial consumers. In the following paragraphs, the considerations that should be given to the optimal design of water mains are presented.

In the present context, to optimize a system means to maximize its net benefits to all water users. In order to measure net benefits, one has to evaluate the benefits and costs associated with the system. The benefits of a WDS are vast since modern urban centers can not exist without them; thus, the fundamental need for a WDS has been taken for granted. Consequently, the benefits are measured by the performance and quality of service that the system provides. The performance of a WDS may be measured by the degree to which the following objectives are accomplished:

1. To provide safe drinking water.
2. To provide water that is acceptable to the consumer in terms of aesthetics, odor, and taste.
3. To have an acceptable level of reliability.
4. To be capable of providing emergency flows, for example, for fire fighting at an acceptable pressure.
5. To provide the demand for water at an acceptable residual pressure during an acceptable portion of the time which are often subject to regulations and may vary with locale. For example, provide demand flow at a minimum 30-m pressure head for 99% of the time, are often subject to each country regulations.
6. To be economically efficient.

The cost of a WDS comprises all direct, indirect, and social costs that are associated with

1. Capital investment in system design, installation, and upgrading
2. System operation—energy cost, materials, labor, monitoring, and so on
3. System maintenance—inspection, breakage repair, rehabilitation, and so on

Thus, a WDS has multiple objectives, of which all but one (economic efficiency) are either difficult to quantify (e.g., reliability, water quality, and esthetics) or difficult to evaluate in monetary terms (e.g., level of service) or both. The situation is somewhat simpler concerning the cost aspects, where direct and indirect costs can be evaluated with relative ease, and social costs can (with some effort) be assessed, albeit with less certainty.

There are various methods to handle multiple objectives and criteria, which are generally noncommensurable and are expressed in different units. However, some of these techniques involve explicit (or implicit) assignment of monetary values to all objectives, for example, the linear scoring method and goal programming, which may introduce biases, and trying to assign a monetary value to reliability or to water quality. On the other hand, multiobjective evaluation techniques are suitable for only a moderate number of alternatives (e.g., the surrogate worth trade-off method, and utility matrices), which makes them unsuitable for handling the vast number of rehabilitation measures and scheduling alternatives involved in a typical WDS. An alternative approach is to formulate the problem as a "traditional" optimization problem in which the optimization criterion is minimum cost, while all the components that cannot be assigned monetary values are taken into consideration as constraints. A general formulation for the full scope of the problem can then be expressed as follows:

Minimize: {capital investment + operation costs + maintenance costs + rehabilitation costs}
Subject to
1. Physical/hydraulic constraints such as mass conservation, continuity equations, and so on
2. Network topology constraints such as layout of water sources, streets, tanks, and so on
3. Supply pressure head boundaries such as minimum and maximum residual pressure heads
4. Minimal level of reliability constraints
5. Minimal level of water-quality constraints
6. Available equipment constraints such as treatment plants, pumps, pipes, appurtenances and so on
7. Available supply water constraints such as quality, quantity, source location

All costs and constraints are considered for the entire life of the system. (The "life of a system" is a somewhat misleading idea, which is further discussed in Section 9.5.) It appears at present that any attempt to handle the full scope of the problem would be overly ambitious in light of the knowledge and computational tools available currently.

EXAMPLE 9.2

Consider two mutually exclusive alternatives for a water main distribution project, given in Table 9.3. Assume that a study period of 15 years is applicable. Using the annual worth and present worth methods, determine the preferred project.

Solution: Using the annual worth method, it can be written that

$$\text{AW of project A} = -\$200{,}000 \times \left(\frac{A}{P}, i\%, 15\right) - \$200{,}000 \times \left(\frac{P}{F}, i\%, 10\right) \times \left(\frac{A}{P}, i\%, 15\right)$$
$$+ \$100{,}000 - \$44{,}000,$$

$$= -\$200{,}000 \times \left(\frac{i \times (1+i)^{15}}{(1+i)^{15} - 1} - \frac{1}{(1+i)^{10}} \times \frac{i \cdot (1+i)^{15}}{(1+i)^{15} - 1}\right) + \$56{,}000.$$

$$\text{AW of project B} = -\$300{,}000 \times \left(\frac{A}{P}, i\%, 15\right) + \$140{,}000 - \$86{,}000,$$

$$= -\$300{,}000 \times \left(\frac{i \times (1+i)^{15}}{(1+i)^{15} - 1}\right) + \$54{,}000.$$

Figure 9.1 shows the estimated annual worth of projects A and B using a range of values for the interest rate (i). As can be seen in this figure, the annual worth of project A has always been greater than that of project B

TABLE 9.3
Alternatives for a Water Distribution Project (Example 9.2)

	Project A	Project B
Capital investment ($)	200,000	300,000
Annual revenues ($)	100,000	140,000
Annual expenses ($)	44,000	86,000
Useful life (years)	10	15

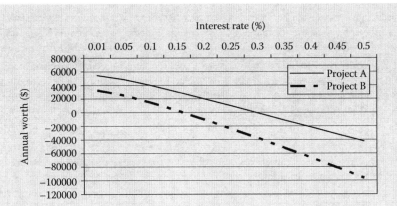

FIGURE 9.1 Variations of the annual worth of the projects in Example 9.2 with respect to interest rate.

regardless of the interest rate. Also, it can be seen that the annual worth of projects A and B becomes negative for the interest rates greater than 30% and 16%, respectively.

Using the present worth method, it can be written that

$$\text{PW of project A} = -\$200{,}000 - \$200{,}000 \times \left(\frac{P}{F}, i\%, 10\right) + \$100{,}000 \times \left(\frac{P}{A}, i\%, 15\right)$$

$$- \$44{,}000 \times \left(\frac{P}{A}, i\%, 15\right),$$

$$= -\$200{,}000 \times \left(1 + \frac{1}{(1+i)^{10}}\right) + \$56{,}000 \times \left(\frac{i \times (1+i)^{15}}{(1+i)^{15} - 1}\right).$$

$$\text{PW of project B} = -\$300{,}000 + \$140{,}000 \times \left(\frac{P}{A}, i\%, 15\right) - \$86{,}000 \times \left(\frac{P}{A}, i\%, 15\right),$$

$$= -\$300{,}000 + \$54{,}000 \times \left(\frac{i \times (1+i)^{15}}{(1+i)^{15} - 1}\right).$$

Figure 9.2 shows the estimated present worth of projects A and B using a range of values for the interest rate (i). As can be seen in this figure, for the interest rates greater than 7%, the present worth of project A has been greater than that of project B. So project A is the recommended project.

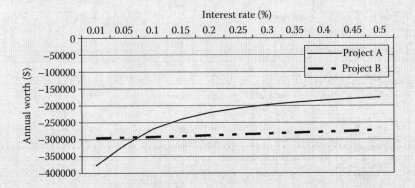

FIGURE 9.2 Variations of the annual worth of the projects in Problem 4 with respect to interest rate.

9.2.3 INFRASTRUCTURE FOR WASTEWATER COLLECTION AND TREATMENT

Urban populations require access to adequate sanitation and disposal of generated wastewater. Such objectives are achieved by wastewater management and sanitation. Water quality in receiving waters is affected mainly by wastewater discharges, solid wastes, and wet-weather flow pollution. Impacts of such discharges depend on the type of pollutant, the relative magnitude of the discharge versus the receiving waters, and the self-purification capacity of the receiving waters. In developed countries, even though the wastewater is treated at a secondary level, treated effluents still are polluted. Sewerage and wastewater treatment facilities are much less developed than water supply infrastructure. Moreover, even where such facilities exist, they may not operate properly because of financial and technical problems.

Adequately treated sewage is a valuable source of water for many applications. Sewage is reused mostly for agriculture and industrial applications; by treatment it can be brought up to acceptable drinking water standards. As stand-alone wastewater infrastructure projects are less attractive for investment, especially when drinking water is already available, developing and managing water supply and wastewater infrastructure in an integrated way will have a number of advantages.

The wastewater treatment plant (WWTP) design and operation philosophy depends very much on characterization of incoming wastewater (influent) concerning both quality and quantity. It is often overlooked the role of transformation processes occurring in the wastewater during retention in the sewerage system. Depending on the type and the design of the sewerage system, it can be expected that the wastewater composition at the moment of entering the sewers will change until the moment it enters the works. Factors like hydraulic retention time of the wastewater in the sewerage, natural aeration of sewer within the system, extent of the period of anaerobic conditions, slope/inclination of pipes, presence and type of auxiliary equipment, temperature, contribution of industrial discharges (trade effluents), and so on have a direct impact on the change in qualitative characteristics of the wastewater.

Consequently, the load of some contaminants can change dramatically from the beginning to the end of sewer collection system when the sewage enters the plant. For example, presence of a high load of ammonia (NH_4) and a low load of nitrate (NO_3) and/or nitrite (NO_2) in sanitary sewers is considered as a sign of "fresh" pollution as it corresponds to the characteristics of urine; however, the opposite situation implies that the wastewater has likely undergone several transformation processes (partial nitrification and sometimes denitrification) as a result of longer retention in the system, naturally intensified aeration, and creation of anoxic sections. In fact, a sewer system should be considered as a large bioreactor where many of the processes that are taking place at treatment plants also occur in the sewer pipes, but at a slower pace.

Qualitative and quantitative characterization of wastewater are the governing factors in the design and operation of the treatment plants. Furthermore, WWTPs are directly affected by the design and operation of the wastewater collection network that they serve. Wastewater quantity and its variation are usually properly taken into account by design engineers; however, this does not make less prominent the fact that sewage works are sensitive to hydraulic shocks and that these should be reduced as much as the local conditions allow.

It is not only this effect that can disturb the operation of the WWTP, although this is the main reason why full-scale plants are never in "steady state," but also variation in flow means also variation in velocities in sewer conduits and pipes. The special importance here is a minimum design velocity in the sewer, which is often set as 0.3 m/s and in many cases during dry-weather flow (DWF) conditions cannot be met.

Even at such a minimum velocity, depending on the local situation, infiltration rate, and type of the sewer system, a large amount of inorganic and organic materials settle and are deposited in the system. At higher flows/velocities, these deposits can be resuspended and carried out to the works causing again higher loads which, in extreme cases, can cause complete immobilization of the treatment units, especially those located at the influent entrance of the plant. As anaerobic conditions are likely to be formed (in wastewater as well as in deposits) during such DWF periods, settling within sewer systems is not desirable even from a qualitative perspective.

9.2.3.1 Costs of Wastewater Infrastructures

The general form of capital cost estimating equations for conveyance systems, pump stations, storage facilities, water treatment, and WWTPs is

$$C = aX^b, \qquad (9.1)$$

where C is the present worth of the infrastructure construction cost and X is the infrastructure size. The parameter C is the total cost including capital cost plus operation, maintenance, and replacement cost. The two parameters, a and b, are determined by fitting a power function to the available data. The traditional way to estimate a and b was by plotting the data on log–log paper and finding the parameters of the resulting straight line from approximation of the data in log–log space. Now, it is simple to find a and b from a least squares regression that is built into any spreadsheets.

As expressed in Equation 9.1, the economies of scale factor are represented by the exponent b. If b is less than 1.0, then unit costs decrease as size increases. All of these equations illustrated the economies of scale for the output measures of either flow or volume. Pipe flow exhibits very strong economies of scale with $b < 0.5$. The economy of scale factor for treatment plants is about 0.7. A generic economy of scale factor that has been used for years is $b = 0.6$ (Peters and Timmerhaus, 1980).

9.2.3.2 Cost of Stormwater Infrastructures

The cost of treating stormwater varies widely depending on the local runoff patterns and the nature of the treatment. Cost estimates for combined sewer systems are presented in U.S. EPA (1993) for swirl concentrators, screens, sedimentation, and disinfection. Typically, treatment will be combined with storage in order to moderate peak flows and allow bleeding water from storage to the treatment plant. This treatment–storage approach can be evaluated using continuous simulation and optimization to find the optimal mix of storage and treatment (Nix and Heaney, 1988). Ambiguities in such an analysis include the important fact that treatment occurs in storage and storage occurs during treatment for some controls (e.g., sedimentation systems).

Owing to the amount of lawn to be watered and the necessity for irrigation water, the cost per dwelling unit (DU) for urban water supply will be mainly variable. In more arid parts of the United States, most of the water entering cities is used for meadow watering. The major factor affecting the variability in wastewater treatment costs is the amount of wastewater. The required lengths of pipe for water supply and wastewater systems can be approximated based on DU and the ratios of the off-site pipe lengths to the on-site pipe lengths. Piping lengths per DU increase if central systems are used because of the longer collection system distances.

The costs of stormwater systems per DU vary widely as a function of the impervious area per DU and the precipitation in the area. The required stormwater pipe length per DU is almost equal to sanitary sewer lengths for higher-density areas in wetter climates. At the other extreme, very little use is made of storm sewers in arid areas, and runoff is routed down the streets to local outlets. Also, trade-offs exist between the pipe size and the amount of storage provided. Consequently, generalizing the expected total cost of stormwater systems is difficult.

> **EXAMPLE 9.3**
>
> The three alternatives described in Table 9.4 are available for wastewater treatment of an urban area for the next 25 years of their life span. Using 5% discount rate, compare the three projects with the present worth method.
>
> *continued*

Solution:

$$\text{PW of costs of project A} = -\$2{,}000{,}000 - \$7000\left(\frac{P}{A}, 5\%, 10\right) - \$8000\left(\frac{P}{A}, 5\%, 10\right) \times \left(\frac{P}{F}, 5\%, 10\right)$$

$$- \$9000\left(\frac{P}{A}, 5\%, 5\right) \times \left(\frac{P}{F}, 5\%, 20\right),$$

$$= -\$2{,}000{,}000 - \$7000 \times 7.7216 - \$8000 \times 7.7216 \times 0.6139$$

$$- \$9000 \times 4.3294 \times 0.3769 = -\$2{,}106{,}659.$$

$$\text{PW of project B} = -\$1{,}000{,}000 \times \left(1 + \left(\frac{P}{F}, 5\%, 10\right) + \left(\frac{P}{F}, 5\%, 20\right)\right) - \$4000\left(\frac{P}{A}, 5\%, 10\right)$$

$$- \$7000\left(\frac{P}{A}, 5\%, 10\right) \times \left(\frac{P}{F}, 5\%, 10\right) - \$9000\left(\frac{P}{A}, 5\%, 5\right) \times \left(\frac{P}{F}, 5\%, 20\right),$$

$$= -\$1{,}000{,}000 \times (1 + 0.6139 + 0.3769) - \$7000 \times 7.7216 - \$8000$$

$$\times 7.7216 \times 0.6139 - \$9000 \times 4.3294 \times 0.3769 = -\$1{,}884{,}141.$$

$$\text{PW of project C} = -\$1{,}000{,}000 \times \left(1 + \left(\frac{P}{F}, 5\%, 10\right) + \left(\frac{P}{F}, 5\%, 20\right)\right) - \$4000\left(\frac{P}{A}, 5\%, 10\right)$$

$$- \$7000\left(\frac{P}{A}, 5\%, 10\right) \times \left(\frac{P}{F}, 5\%, 10\right) - \$9000\left(\frac{P}{A}, 5\%, 5\right) \times \left(\frac{P}{F}, 5\%, 20\right),$$

$$= -\$1{,}500{,}000 - \$1{,}200{,}000 \times 0.6139 - \$6000 \times 7.7216 - \$8000$$

$$\times 7.7216 \times 0.6139 - \$9000 \times 4.3294 \times 0.3769 = -\$2{,}335{,}618.$$

The PW of project B is less than that of the others; therefore it is selected.

TABLE 9.4
Three Alternatives for Supplying a Community Water Supply

Costs	Timing	Project A ($)	Project B ($)	Project C ($)
Initial investment	Year 0	2,000,000	1,000,000	1,500,000
	Year 10	0	1,000,000	1,200,000
	Year 20	0	1,000,000	0
Operation and maintenance costs	Years 1–10	7000	4000	6000
	Years 10–20	8000	7000	8000
	Years 20–25	9000	9000	9000

9.3 INTERACTIONS BETWEEN THE URBAN WATER CYCLE AND URBAN INFRASTRUCTURE COMPONENTS

There are many interactions between the urban water cycle (UWC) and urban infrastructure components. Wastewater treatment, being somewhere at the end of such a cycle, is regularly "suffering" from the "upstream" components, while on other hand, it has itself a prominent impact on several "downstream" components. This section provides an overview of the interactions between wastewater treatment using water treatment, water transport and distribution, drainage, sewerage, solid waste, and transport. The overview of interactions concerning (1) drinking water quality and quantity;

(2) disposal of exhausted adsorption materials from treatment processes; (3) discharge of filter backwash water; (4) disposal of sludge generated during water purification into drains and sewerage; (5) design, construction, and operation of septic tanks and possible groundwater pollution; (6) quality and quantity of treated effluents and their discharge into ground or surface waters; (7) unaccounted for water losses such as leakage, linked with infiltration and exfiltration in the sewage systems; (8) type of sewerage; (9) sewage characteristics; (10) sewerage as a bioreactor—transformation processes in the sewer system; (11) sewer design; (12) the role of CSOs (combined sewer outflow); (13) the effect of the "first flush"; (14) separation at source; (15) dry sanitation; (16) impact of sewage works on the hydraulics of sewage collectors; (17) solid waste deposition on streets, drains, and collectors; (18) leachate from dumps and landfills; (19) impacts of fecal sludge on sewerage and works; (20) codigestion of sludge with solid waste; (21) design and maintenance of roads; (22) the "road as a drain" approach; and so on. All of the above mentioned factors and issues urge for a truly integrated approach toward urban water systems management by which all major interactions will be taken properly into account. This should also include latest thinking on the development of cites that calls for an integrated approach that encompasses all aspects of urban planning.

9.3.1 Interactions with the Wastewater Treatment System

Wastewater treatment is an important component of both UWC and urban infrastructure (Figure 9.3). Looking at it from a rather conventional perspective, it can be said that it is located

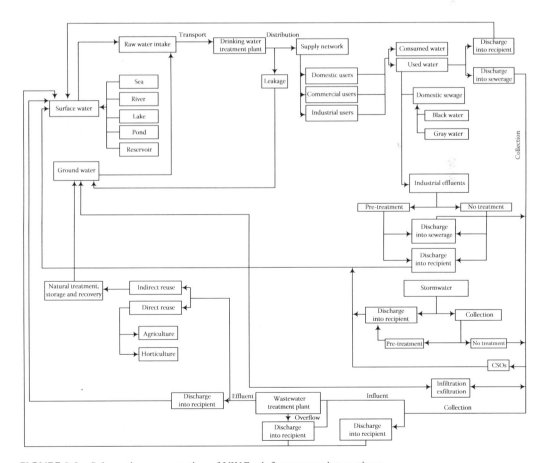

FIGURE 9.3 Schematic representation of UWC—infrastructure interactions.

somewhere at the end of this cycle, although purely by the definition of "cycle," such a statement is incorrect.

Nevertheless, it is important to understand that wastewater treatment is influenced by both upstream and downstream components of the cycle and that it may have an impact on various components of urban infrastructure. In general, wastewater treatment is subject to interactions among water treatment, water distribution, drainage and sewerage, solid waste, and transport.

9.3.2 Interactions between Water and Wastewater Treatment Systems

Water treatment may affect wastewater treatment by several means, such as (1) drinking water quality and quantity, (2) disposal of exhausted adsorption materials from treatment processes, (3) discharge of filter backwash water, and (4) disposal of sludge generated during water purification into drains and sewerage. On the other hand, wastewater treatment practices can have a significant impact on water treatment via (a) the design, construction, and operation of septic tanks and possible groundwater pollution and (b) the quality and quantity of treated effluents and their discharge into ground or surface waters.

From the perspective of conservation of water resources, wastewater effluent reuse is increasingly considered as a feasible option, especially in cases where water is scarce and reuse option is one of the few possibilities left. There are numerous examples where the effluent of the sewage works makes up a large part of the rivers, creeks, or channels. At many locations worldwide, it is not uncommon to have the intake of raw water for drinking water treatment below the discharge point of the WWTP effluent. These, among other factors, recommend that wastewater effluent be recognized as a resource and not waste. Because in the (urban) water cycle every drop of water becomes wastewater 2 or 3 times before it enters the global hydrological cycle as clean water, the effluent from WWTPs has a role in water supply from both the quality and the quantity perspective.

9.3.3 Interactions between Water Supply and Wastewater Collection Systems

The most obvious interaction between water supply and wastewater treatment is resulting from the fact that, without exceptions, water supply networks are not fully efficient systems, where unaccounted for water losses such as leakage present the major reason for it. Even in new networks, physical water losses could exceed 10%, while in regular practice, especially in developing countries, it is between 30 and 50%, and sometimes even higher (Kingdom et al., 2006).

Leakage management has often an important impact on the groundwater level which may, depending on local hydrogeological conditions, increase infiltration into sewer collection systems. It is a generally known fact that infiltration is to be minimized in the sewers as it further dilutes the wastewater, increases hydraulic load to sewers and consequently sewage works, helps earlier activation of combined sewer overflows, and induces additional problems to plant operators.

Similarly to water supply networks, sewerage systems are prone to exfiltration in cases where the groundwater table is below sewers, which is undoubtedly a problem by itself, but under certain situation can also cause contamination in the water supply network in case of loss of system pressure for whatever reason. The issue becomes more exaggerated in case of natural disasters (such as hurricanes or floods).

9.3.4 Interactions between Urban Drainage Systems and Wastewater Treatment Systems

Interaction between stormwater drainage, sewerage, and wastewater treatment is unavoidable as these components of the urban water cycle should be generally considered as a single integrated system. Obviously, the primary role of sewerage is to transport pollution from one point to another and ultimately to the WWTP (Wastewater Treatment Plant). Interactions between sewerage and sewage works depend very much on the type of the sewer system (separate or combined). In some countries, especially in newly developed and modern cities, one finds even three separate systems in place (sanitary sewers, industrial sewers, and stormwater drainage) (Brdjanovic et al., 2004).

The role of CSOs in stormwater and wastewater management is very important, because a decision on their position and number in the system, their design and moment of activation/spilling, and the pollution load allowed to be released into the environment through them has to be taken into account. Another stormwater-related issue that is closely related to CSOs and has direct consequences for plant operation is the effect of the "first flush." Furthermore, the impact of the first flush on a treatment plant depends very much on other components of urban infrastructure, such as transportation and solid waste management. For example, pollution accumulation and washoff collected by the first flush contain dry deposition of atmospheric pollutants on roofs; deposition of pollutants from traffic on streets, highways, and parking places; accumulation of dust, dirt, and larger residues on streets; solids generation by roadway deterioration; highway surface contamination by vehicle and tire wear, fluid spills, and leaks; road surface contamination by salting and sanding in winter conditions; washoff by the energy of storm-generated runoff (washoff and erosion); washoff by dissolution and elutriation due to acidity of rainfall; and so on.

In general, small sewer systems are more sensitive to the first flush relative to large systems where dilution effect is more prominent. The first flush, usually considered to be 15–20 min duration, removes the major part of pollution from the sources described above. It is also indicated that the first 25 mm of rainfall is responsible for 90% of such pollution. Furthermore, it is surprising that the effects of street cleaning using water are very low (contributes <10% to the reduction of pollution entering rainwater drainage system) (EPA, 1993).

Based on the literature (EPA, 1993), sewer networks are still the most efficient transport system, and the despite the fact that separate sewers are 50% more expensive than combined sewers (data for the Netherlands), there is an increasing trend toward connection of separate sewers at the expense of combined sewers, at least in the developed parts of the world (RIONED, 2005). Associate costs per capita served are still low in comparison with individual wastewater treatment systems in remote locations where sewer connections are not feasible (Wilsenach, 2006). Regardless of alternatives currently available, it is expected that sewer networks will remain a sanitation backbone for urban areas, especially to those densely populated. However, contrary to the opinion of those who are constantly working on improving conventional wastewater systems, there is an emerging movement supporting the philosophy of alternative sanitation and urban water drainage, which promote pollution prevention rather than control, separation at source rather than end of the pipe treatment, and reuse of valuable resources rather than wasting by discharge into the environment. It is believed that source separation of rainwater and urine has the best prospect of improving urban water management (Wilsenach, 2006).

9.3.5 Interactions between Urban Drainage Systems and Solid Waste Management

Inappropriate solid waste management is a large and emerging problem especially in developing countries, sometimes being of more concern than wastewater-related problems. Interactions of solid waste with urban water infrastructure components are numerous and only the most prominent are briefly discussed below. Uncontrolled deposition of solid waste within urban areas, especially on streets, presents not only a major problem for public health, but also has a direct impact on the functioning of the urban drainage system, as part of the waste can enter the sewerage system, be deposited/retained in the system, or eventually be discharged into the environment or reach the entrance to a sewage works.

As sewer networks and treatment plants are in general not designed to transport solid waste, such a practice has a negative impact on the efficiency, operation, and maintenance of urban wastewater infrastructure. There are numerous examples of dumping solid waste into open drains, which is accumulated at the banks of the channels, is deposited at the bottom, and is often accumulated at the entry of pumping stations, gates, weirs, and other infrastructure facilities within the system. Such a practice presents a huge obstacle to maintaining the sustainability of investments and is a continuous problem for system operators.

Another large problem caused by uncontrolled disposal of solid waste into rainwater drains is that with time these drains become loaded with waste, and their efficiency becomes drastically reduced. This is, in many cases, the major reason for the flooding during heavy rainfall and malfunctioning of the drainage system, which often has a consequence of flooding the urban areas.

In some cases of organized deposition of solid waste at designated locations, very often, and especially in developing countries, most of the sites are in fact nothing else than dumpsites rather than proper sanitary landfills. This consecutively means that most of the sites are characterized by uncontrolled leachate generation and its release into the environment. Such a practice has usually tremendous negative effects on the quality of receiving waters, especially groundwater. Taking into account that as a result of rapid population growth and often uncontrolled urbanization, urban and most frequently preurban agglomerations are getting more and more close to dumping sites, this issue becomes a very serious threat to public health affecting most of the world's poorest.

Fecal sludge is usually considered as a solid waste and its management can have a direct impact on water supply and sanitation infrastructure. Link with water supply is reflected by the risk of pollution of groundwater, which is often used as a source of on-site water supply where no piped water supply is available. In the case of leaking or overflowing septic tanks, there is a direct threat of bacteriological contamination of those humans who either get in contact with a septic tank content or drink contaminated water. Fecal sludge management practices have different impacts on sanitation infrastructure. In some cases, collected sludge is dumped to the nearest sewage manhole and transported with wastewater, or is discharged at the entrance to the treatment plant—both practices are unwanted by treatment plant operators. Because fecal sludge represents a high load of organic matter and nutrients, such discharge practices often have a negative impact on sewage works (shock loads). In order to cope with such variations in load, some treatment plants have specially designed receiving tanks for fecal sludge, which is mixed with incoming wastewater and introduced to the treatment plant during periods of low load.

Another interaction within the urban context is joint treatment of organic fraction of solid waste and of the sludge generated from wastewater treatment at sewage works. This process is called codigestion and has distinctive advantages over separate handling of solid waste and sludge concerning both financial and environmental aspects.

In addition, due to its high organic content, sewage sludge is sometimes used as a supplement fuel in solid waste incineration plants to improve combustion process. It is important here to obtain sludge with as little as possible water content which is directly linked with the efficiency of its pretreatment (dewatering), usually taking place at sewage works.

9.3.6 Interactions between Urban Water Infrastructure and Urban Transportation Infrastructures

Transportation is a very important segment of infrastructure whose elements, such as roads, often host the other urban infrastructure systems (water supply pipes, sewerage pipes, urban drainage, telecommunication cabling, electrical cabling, etc.). There is a strong link between the design and maintenance of roads, solid waste management, and sewer systems (especially in the case of combined sewers). In addition, in very difficult cases where urban planning failed (often the situation in developing countries), due to unavailability of space, roads can be retrofitted to serve as a part of the urban drainage system (the "road as a drain" approach).

Another important factor is that road renovation/construction takes place more frequently in comparison with replacement of the old sewers beneath them. Because of the fact that old (brick) sewers were not designed to sustain traffic load of modern and heavy vehicles, there are numerous cases of sewer pipes collapsing under the impact of heavy traffic load. Such unwanted interactions have negative effects on both the traffic and the integrity and functioning of the sewer pipes and drains.

9.4 FINANCING METHODS FOR INFRASTRUCTURE DEVELOPMENT

An essential ingredient of developing and maintaining viable urban water organizations is funding. Integrated urban water management offered the promise of improved economic efficiency and other benefits from the merging of multiple purposes and stakeholders. However, the benefits from integrated watershed management exacerbate problems of financing these more complex organizations because ways must be found to assess a "fair share" of the cost of this operation to each stakeholder (Heaney, 1997). Nelson (1995) provides a current overview of utility financing in the water, wastewater, and stormwater areas.

The main financing methods for urban infrastructures are (Debo and Reese, 1995)

1. Tax-funded systems
2. Service charge-funded systems
3. Exactions and impact fee-funded systems
4. Special assessment districts

9.4.1 Tax-Funded System

Usually, the public works department of a city (or similar organizations related to urban infrastructure development) is charged with maintaining and improving urban water infrastructure systems. Projects are funded through the budget of the department, whose source is mainly property tax revenue. However, if property taxes are used, then the infrastructure systems must compete for funds directly with public safety, schools, and other popular programs.

9.4.2 Service Charge-Funded System

In the service charge-funded system, an algorithm that divides the budget for the urban water infrastructure systems by some weighting of the demand for service should be used. This new funding method is being implemented because it has the advantage of separating the funding necessities according to the function on a user pays basis.

9.4.3 Exactions and Impact Fee-Funded Systems

System development charges have emerged as the way to compute the charges to be levied against new developments. This system charges the developer or builder an up-front fee that represents his equity buy-in to the urban water infrastructure systems. Usually this fee is calculated as a gauge of the depreciated value of the system, plus system-wide funding needs minus the existing users' share. The fee must be reasonable to avoid court challenges. Nelson (1995) defines the rational nexus test of reasonableness of system development charges. These test requirements are as follows:

- Establishment of a connection between new development and new or expanded facilities required to accommodate such development. This establishes the rational basis of public policy.
- Identification of the cost of those new or expanded facilities needed to accommodate new development.
- Appropriate apportionment of that cost to new development in relation to benefits it reasonably receives.

An important feature of this method is the ownership, or equity issue, of existing users. Usually the existing users are grouped into one class for ease of calculation; however, in actuality, different groups joined at different points in time. At the time of joining, some contractual agreement (written or unwritten) is initiated. A key weakness of the impact fee system is keeping track of these agreements over time and space when setting impact fees is extremely difficult. Because of this added database need and the wide variation in cost allocation techniques for apportioning costs, there can be wide fluctuations in impact fee computations. These shortcomings can be overcome, however, with better accounting and tracking of information.

9.4.4 Special Assessment Districts

This system funds the needs within a designated geographical area by dividing the funds, usually equally, among the parcels within the area. The calculation methods are inherently simple and, usually, the benefits and costs are roughly equally distributed.

In Example 9.4, the optimal planning for capacity building of urban water infrastructure due to finance limitations is presented.

Example 9.4

The variation of annual wastewater generation of a city over a 30-year planning time horizon is tabulated in Table 9.5. Assume that the wastewater generation of this city in year 2003 was about 90 MCM. Three projects have been proposed to treat wastewater projected generation until 2040. As presented in Table 9.6, three projects are studied for wastewater treatment of this city. The capital investments presented in Table 9.6 are based on prices of the year 2003. Considering the cost of each project and 3% rate of return, find the optimal sequence of implementing projects using linear integer programming.

TABLE 9.5
Variation of Annual Amount of Wastewater of a City (Example 9.4)

Year	2010	2020	2025	2035	2040
Wastewater (MCM)	100	120	130	145	170

TABLE 9.6
Description of Proposed Projects (Example 9.4)

Project	Capital investment (10^6 \$)	Capacity (MCM)
Treatment plant A	5	35
Treatment plant B	8	20
Treatment plant C	7	35

Solution: The formulation of the optimization model based on the proposed methodology for capacity building based on the previous chapter is as follows:

$$\text{Minimize} \quad C = \sum_{i=1}^{3} \sum_{j=1}^{5} X_{i,j} \cdot \text{Cost}_{i,j}.$$

Subject to

$$\text{Cap}_i \times X_{i,j} \geq D_j,$$

$$X_{ij} = \begin{cases} 1 & \text{if project } i \text{ is constructed in time period } j, \\ 0 & \text{otherwise,} \end{cases}$$

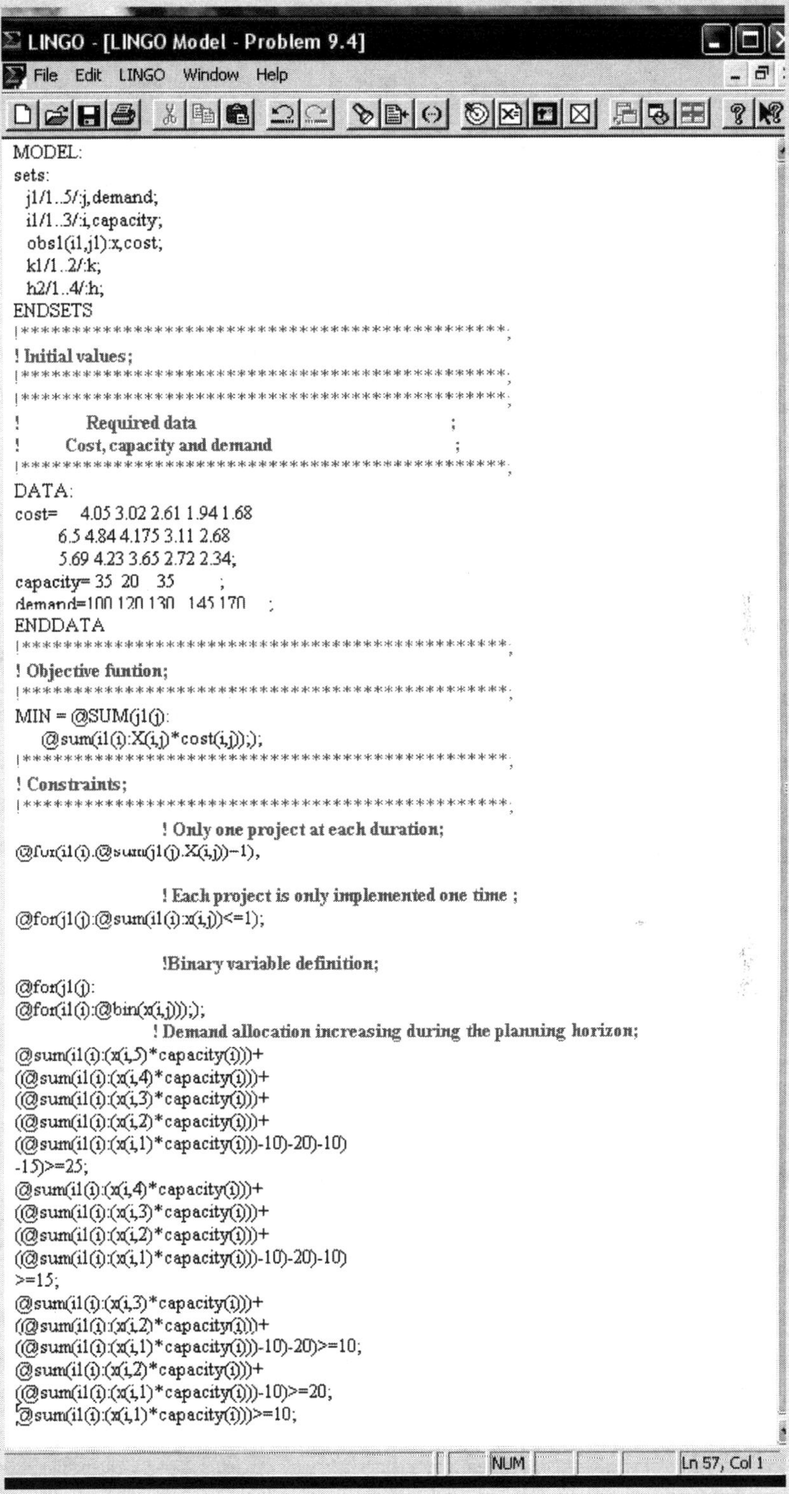

```
MODEL:
sets:
 j1/1..5/:j,demand;
 i1/1..3/:i,capacity;
 obs1(i1,j1):x,cost;
 k1/1..2/:k;
 h2/1..4/:h;
ENDSETS
!***************************************************;
! Initial values;
!***************************************************;
!***************************************************;
!      Required data                        ;
!      Cost, capacity and demand             ;
!***************************************************;
DATA:
cost=    4.05 3.02 2.61 1.94 1.68
         6.5 4.84 4.175 3.11 2.68
         5.69 4.23 3.65 2.72 2.34;
capacity= 35 20 35       ;
demand=100 120 130 145 170  ;
ENDDATA
!***************************************************;
! Objective funtion;
!***************************************************;
MIN = @SUM(j1(j):
   @sum(i1(i):X(i,j)*cost(i,j)););
!***************************************************;
! Constraints;
!***************************************************;
              ! Only one project at each duration;
@for(i1(i):@sum(j1(j):X(i,j))=1);

              ! Each project is only implemented one time ;
@for(j1(j):@sum(i1(i):x(i,j))<=1);

              !Binary variable definition;
@for(j1(j):
@for(i1(i):@bin(x(i,j))););
              ! Demand allocation increasing during the planning horizon;
@sum(i1(i):(x(i,5)*capacity(i)))+
(@sum(i1(i):(x(i,4)*capacity(i)))+
(@sum(i1(i):(x(i,3)*capacity(i)))+
(@sum(i1(i):(x(i,2)*capacity(i)))+
(@sum(i1(i):(x(i,1)*capacity(i)))-10)-20)-10)
-15)>=25;
@sum(i1(i):(x(i,4)*capacity(i)))+
(@sum(i1(i):(x(i,3)*capacity(i)))+
(@sum(i1(i):(x(i,2)*capacity(i)))+
(@sum(i1(i):(x(i,1)*capacity(i)))-10)-20)-10)
>=15;
@sum(i1(i):(x(i,3)*capacity(i)))+
(@sum(i1(i):(x(i,2)*capacity(i)))+
(@sum(i1(i):(x(i,1)*capacity(i)))-10)-20)>=10;
@sum(i1(i):(x(i,2)*capacity(i)))+
(@sum(i1(i):(x(i,1)*capacity(i)))-10)>=20;
@sum(i1(i):(x(i,1)*capacity(i)))>=10;
```

FIGURE 9.4 Formulation of Example 9.4 in LINGO software.

continued

TABLE 9.7
Present Value of the Initial Investment in the Projects

Year	Project A	Project B	Project C
2003	5	8	7
2010	4.065	6.5	5.69
2020	3.02	4.84	4.23
2025	2.61	4.175	3.65
2035	1.94	3.11	2.72
2040	1.68	2.68	2.34

where $Cost_{i,j}$ is the construction cost of project i at the jth time period, $X_{i,j}$ is the decision variable for construction of project i at the jth time period, Cap_i is the capacity of project i as mentioned in the third column of Table 9.6, and D_j is the water demand of the city at time period j (Table 9.5).

In order to estimate the present value of investment in year t of the planning time horizon, the following equation can be used:

$$P = \frac{F}{(1+i)^t}.$$

Table 9.7 shows the present value of the initial investment in the projects, which has been estimated using Equation 9.5 for the years in which water demand has increased. Then a linear programming routine is written to determine the best timing for projects construction. According to the results of linear programming that is formulated in LINGO software (Figure 9.4), the optimal projects implementation pattern is project A in 2003, project C in 2020, and project B in 2040.

9.5 SUSTAINABLE DEVELOPMENT OF URBAN WATER INFRASTRUCTURES

Several attempts have been made to define the sustainable development (SD) of urban water infrastructures including water resources and the wastewater system. The following definition has been proposed by ASCE (1998):

> A sustainable urban water infrastructure should provide required services over a long time perspective while protecting human health and the environment, with a minimum use of scarce resources.

One approach to the SD of urban water management would be to start from an analysis of required services and to identify the needs satisfied by the water system, rather than a stepwise improvement of existing technology (Larsen and Gujer, 1997). The services or the needs that the urban system should meet include

1. Reliable supply of safe water to all residents for drinking, hygiene, and household purposes
2. Safe transport and treatment of wastewater
3. Drainage of urban areas
4. Recovery of resources for reuse or recycling

Several publications present criteria for the SD of an urban water system, whose summary of the literature is divided into technical, environmental, economic, and social (including health) criteria.

Technical performance can be considered in terms of two aspects, effectiveness and efficiency. Effectiveness is the extent to which the objectives are achieved, such as how well the organization fulfills the needs identified above. Efficiency is the extent to which resources are utilized optimally to fulfill these needs. Suggested indicators of effectiveness are coverage of service, drinking water quality, and percent of wastewater that enters the treatment plant and percentage removal of selected compounds. Efficiency can be measured through the use of chemicals, energy, space, and the loss of water in the form of leakage.

Reliability, flexibility, and adaptability: Reliable systems are those that can provide their service even when unexpected events occur such as an electricity delivery stop or a sudden temperature drop. Systems may fail but must be capable of recovering without undue effort or cost. Seasonal variations in loading and climate should be considered. Technical systems should also be designed and operated in ways that enable them to cope with changes in the ecosystems (natural or man-made), in the technical components due to aging or innovations, or in the demands and desires in society. The potential to change (flexibility) and the ability to change (adaptability) need to be promoted (Jeffrey et al., 1997).

A basic concept of the SD of technology is the separation and selection of subsequent technologies. Separation of waste streams creates flexibility and provides a possibility for a combination of technologies. Municipal WWTPs are constructed to treat the wastewater from households and from industries with similar components, but not stormwater. In older urban areas, combined sewers are still a problem, causing sewer overflows and contamination of sludge. Flexibility and reliability of wastewater systems can hence be measured through the degree of separation, such as percentage of combined or separate sewer pipes. Similarly, further separation of different wastewater streams enables different treatment options.

Durability refers to the long life of the technical system. The lives of the components water systems are long, 20–40 years and even longer. A balance between endurance and flexibility is required since the components having long lifetime or high cost might decrease the flexibility.

Scale and degree of centralization need to be considered. A local water supply or wastewater treatment might be advantageous where natural conditions are appropriate and population density is low. In large cities where space is limited but the economic situation and potential for sophisticated technology are good, large-scale systems may be more appropriate and contribution to other technical systems such as the energy system relevant.

Environmental protection: Environmentally acceptable systems are those that fulfill the objectives of prevention of contamination and a sustainable use of resources. Quantity and quality of water supply and receiving water should be restored and maintained. Emissions of nutrients and oxygen-demanding substances should be as low as is required to maintain the quality and diversity of ecosystems. Harmful substance, such as cleansing agents, chlorine and its compounds, and substances from traffic should be avoided in order to protect aquatic life and meet quality directives for sewage sludge and prevent negative effects on agriculture and soil. Withdrawal, raw water quality, and protection are indicators that can be used for the assessment of water sources. Emissions to water from point sources and nonpoint sources are important indicators as well as emissions to air of some components.

Recycling and reuse of resources are especially important for an essential and limited resource, such as phosphorus, but should also include other plant nutrients, water, organic compounds, energy, and chemicals. A high degree of recycling or reuse also ensures lower emissions to water and air. Quality of sludge and recycling of nutrients are indicators that can be used for this purpose.

Noise, odor, and traffic are also important factors that need to be considered, for example, for the collection of sewage and transportation of sludge.

Cost-effectiveness affordability: Economic dimensions can be looked upon from different perspectives: household, organization, and society. Households require an affordable service, an organization requires a cost-effective system, and society requires a stable but flexible system. The price of water is a key determinant of both the economic efficiency and the environmental effectiveness of water services.

Personnel requirements: The time required for the operation and maintenance of the urban water system should be considered as well as the level of personnel skills. There is an increasing need, catalyzed by the ideas of SD, for civil engineers to embrace development in microbiology, chemistry, ecology, and IT. A diversity of humans in terms of education and gender are required to ensure a flexible and adaptable organization.

Demographics: Population, population growth and density, specific water use, and coverage of population supplied are examples of traditional indicators that are used to assess future and existing needs.

Social dimensions include the users as well as the staff. All humans have a need for healthy water supply and access to sanitation. Hygiene is a major consideration and the urban water system should be safe for those who use and operate the system or take care of the final products. It should be convenient as well as socially and culturally acceptable. Some individuals are willing to take a large responsibility and spend considerable time and effort to contribute to a reduction of the environmental impact. Others are not willing or may not be able to do so. The systems should be able to function well under both circumstances. Freedom of choice is therefore important.

Awareness and promotion of sustainable behavior: User behavior affects the function of an urban water system. The direct use of water and chemicals is the most obvious factor but lifestyle also affects resource use and emissions. For example, if more people would eat a less protein-based diet, less nitrogen would be emitted through the wastewater system. Nonpoint source pollution depends to a high degree on traffic travel habits since pollution from car exhausts reaches rivers and lakes through contaminated stormwater.

9.5.1 SELECTION OF TECHNOLOGIES

The development of the present system toward a more sustainable system may occur by improvements to existing systems and by developing new technology. A number of strategies can be identified, which are described in the following section.

9.5.1.1 Further Development of Large-Scale Centralized Systems

Large-scale, centralized water supply can become more sustainable by increasing the efficiency of treatment processes and decreasing the use of harmful chemicals by optimization of dosages or the use of alternative disinfection processes such as ozonation, nanofiltration, or biological processes. Water use can be decreased by the installation of water-saving appliances, repair of leakage, and by saving, recycling, and reuse. In regions suffering from by water shortage, reuse is an alternative to distant water reservoirs. Wastewater can be reclaimed for nonpotable purposes, such as toilet flushing and irrigation, and stormwater can be used to recharge groundwater and provide water areas in cities. When wastewater is used for irrigation, water and nutrients are recovered.

The centralized, large-scale wastewater system can be further developed toward higher effectiveness and a higher degree of removal through mechanical, biological, and chemical unit processes. Sewage sludge can be treated to extract and separate phosphorus, precipitation chemicals, and heavy metals. Heat may be recovered from the wastewater. Aerobic methods can be substituted by anaerobic methods to save electricity and produce biogas. Organic waste can be added to further increase biogas production. These strategies are by no means new, but seem to be the most common way to improve the environmental performance of urban water systems in large cities.

9.5.1.2 Separation for Recycling and Reuse

The separation of flows to improve the opportunities for recycling and reuse has been emphasized by several authors (Niemczynowicz, 1996; Butler and Parkinson, 1997) and may be feasible in both large-scale and small-scale systems. Separation of urine is a new technology that, apart from enabling the recycling of nutrients to agriculture, also has implications for treatment performance in wastewater systems. Human urine is the urban waste flow that contains the greater proportion of nutrients.

Separation of blackwater is also possible, enabling new technologies such as aerobic liquid composting or anaerobic digestion to function well, since blackwater is not diluted with gray water. The possibility of treating different waste streams makes the system flexible, but is not yet a generally accepted approach. Whereas solid household waste is source separated into plastics, glass, metals and so on, the separation of liquid waste seems to be more problematic. The technical and economic difficulties and people's willingness to accept the idea should be tested. But before a new technology is introduced, it is important to ensure that it does not lead to new human or environmental health problems.

9.5.1.3 Natural Treatment Systems

Local, small-scale systems often use the same type of treatment as large-scale systems. Several studies have shown that conventional small-scale systems are often less efficient in terms of energy; they need relatively more maintenance and are therefore more expensive than large-scale systems. A positive aspect of small-scale wastewater systems is the better quality of the recycled product and proximity to suitable land for the treatment and disposal of nutrients, which minimizes transport costs.

The possibility of using natural wastewater treatment systems is usually larger for smaller communities where the population density is low. Natural (or ecological) wastewater treatment systems refer to constructed wetlands, reed beds, or aquaculture, all of which use microbiological and macrophyte communities to assimilate pollutants. The usual purpose of wetlands and reed beds is to remove nutrients, not to recycle them. Aquaculture utilizes a series of ponds of algae, crop plants, and fish to recover nutrients into biomass. The energy requirements of constructed wetlands are low but if a significant reuse of nutrients is included (aquaculture) the energy requirements increase significantly (Brix, 1998). Therefore, it is necessary to recognize the importance of recycling of nutrients, not just removal, also in natural systems. Recent research in this direction has been reported (Farahbakhshazad and Morrison, 1999).

9.5.1.4 Combining Treatment Systems

The distinction between conventional and natural technologies is not clear and combinations of these technologies are common, such as combining chemical pretreatment to remove phosphorus and BOD and wetlands to remove nitrogen and remaining BOD. Wetlands are often used as a polishing step after conventional treatment. A combination of these technologies may be one possible way to achieve several of the standard requirements. Whereas conventional systems are reliable and effective for the removal of certain compounds, natural systems may contribute positively to other aspects such as biodiversity or landscape esthetics. Urine separation in combination with conventional treatment facilities also provides an environmentally improved option.

9.5.1.5 Changing Public Perspectives

Over and above the choice of technical system, the behavior of the users and the operators will affect functionality of urban water infrastructure. Changes of social and institutional aspects are challenging because they involve changing the way individuals (within and outside the organization) think and act. Users of the systems as well as the personnel who design and operate the urban water may need to be informed of the function of systems, the value of water, and how to use the system properly.

9.5.2 Assessing the Environmental Performance of Urban Water Infrastructure

Sustainable development index (SDI) should be used in the international, regional, and local scales for assessing the performance of urban water infrastructure. At a corporate level, the concept of SD is often limited to an improvement of environmental performance. An agency or an organization can turn to different methods to get information, which allows an understanding and improvement of its environmental performance. Examples of tools used for SDI assessment in urban infrastructure are LCA and environmental performance evaluation [by using special performance indicators (PIs)]. While LCA is a technique for assessing the environmental aspects associated with a product and/or a service system, environmental performance indicators (EPIs) focus on those areas that are under direct control by the agency/organization. Upstream and downstream activities are seldom included. The system boundaries for LCA and EPI are therefore not usually comparable. This section briefly describes how these two methods can be used to assess the environmental performance in general and an urban water system in particular and how this affects decisions on which upstream and downstream processes to include in the assessment of the system.

Within the urban water sector, PIs are being developed and used in several countries. A major objective of water utilities is to improve the quality of the service while keeping down the costs. Most of the PIs have been developed as a tool for monitoring these objectives. Environmental aspects have also been recognized as important recently. An international workshop was held in 1997 by the International Water Services Association in order to find a common reference set of PIs for water supply (IWSA, 1997).

The relative importance of environmental indicators depends on local and regional factors. If water is scarce, then the specific water use and leakage are highly relevant in order to save water. Where water is in abundance, other indicators are more important. The efficiency indicators have been ranked as having moderate importance, since these are not directly related to the objectives of the water sector but still apply to the definition of a sustainable urban water system.

An indicator should be clearly understandable, without long explanations. Percentages or dimensionless quotas are fairly easy to relate to, whereas an indicator such as drinking water quality is more difficult, at least for the public, and should therefore be related to a water-quality standard. Ease of understanding of indicators should be determined by the users, and this aspect still needs to be investigated.

The use of monetary indicators is a common procedure for companies wishing to monitor their performance or utility efficiency in comparison with others. In recent years, the use of EPIs has become increasingly popular within industry to describe agency pressure on the environment. In order to harmonize efforts, an international standard for environmental performance evaluation has been proposed and will be included in an ISO standard (ISO 14031, 1998). The objective of the ISO standard is to provide guidelines on the choice, monitoring, and control of EPIs. EPIs are categorized into three types: operational PIs, environmental condition indicators, and management PIs. In the ISO standard, it is emphasized that the indicators should be selected based on their ability to measure performance against the environmental targets set by the company. It is also stated that a life cycle approach should, if possible, be used to select indicators.

Another concept that has been used is the term "ecoefficiency," that is, producing more value with less environmental impact (Schaltegger, 1996). The prefix "eco" refers to both ecological and economic efficiency. Keffer et al. (1999) presented a set of defined general ecoefficiency indicators, as well as specifying guidelines for the selection of sector-specific indicators.

9.5.3 Life Cycle Assessment

LCA is a method that aims to analyze and evaluate the environmental impacts of projects, products, or services. The whole chain of activities required for the production of a certain product or service is taken into consideration. Both emissions of potentially harmful substances from these activities and their consumption of natural resources are analyzed. In this way, different technical systems producing the same utility (product or service) can be followed from cradle to grave (meaning that all activities are included, covering the mining of minerals, transport, production and disposal, and reuse or recycling) and compared with regard to their impacts on the environment.

LCA can briefly be described as a three-step procedure (ISO 14040, 1997). These steps include

- Goal and scope definition, where the purpose of the study is presented and the system boundaries are defined. The functional unit, that is, the basis for comparison in the study, is also defined in this step. A flow chart, which shows the relations between the different activities in the life cycle, is also produced.
- Life cycle inventory, the phase in which all information on the emissions and the resource consumption of the activities in the system under study is collected from various sources. Where reliable data are unavailable, assumptions may have to be made.
- Life cycle impact assessment (LCIA), where the environmental consequences of the inventory data are assessed. Often some kind of sensitivity analysis or discussion of uncertainties is included in this step.

In reality LCA is often an iterative process where the analyst will have to go through the different phases more than once. Often a fourth phase, interpretation, is added to the first three. LCA is included in the ISO 14000 series and further guidance on LCA can be found there (ISO 14040, 1997; ISO 14041-43, 1998).

9.5.4 Developing SDI Using a Life Cycle Approach

LCA, EPI, as well as SDI have several linkages concerning data collection, normalization, aggregation, and valuation. For these reasons, it seems appropriate that the knowledge gained in one field can be transferred to other fields. The question of data collection and data quality is crucial for LCA and is equally important for EPI. It is even likely that the same data can be used as a basis both for LCA and EPI, at least for the indicators categorized as pressure (in the PSR model). The characterization step of LCA where different environmental burdens are added into environmental themes can be useful for the aggregation of EPI, and the valuation step where different environmental issues are weighted may be useful in the selection of different indicators. Conversely, knowledge from the field of EPI can be useful for the interpretation and presentation of LCA results, and in fact LCIA is often described as an indicator system (Owens, 1999).

The life cycle perspective is important for SD but there is a need for simplified methods. One approach is a screening process to find which parameters should be included or omitted in the study. If a number of LCAs have been carried out on a sector or product group, it should be possible to select the most environmentally important parameters or categories. Another way to simplify LCA is at the system level where upstream or downstream processes are excluded, for example, by focusing on the steps in the life cycle that can specifically be controlled by the organization performing the LCA.

A life cycle perspective can guide the selection of indicators by revealing the most disturbing environmental loads. The LCIA step is useful since it aims to identify and quantify the most important resource uses and emissions. The related processes causing that load can thus be identified. The main idea is to avoid performing complete LCAs each time SDI is to be implemented in an organization, but instead to identify the most important processes and to find indicators related to the pressure indicator in the PSR framework.

In summary, the approach for using the LCA method to find the SDI is as follows:

The first step is to identify possible problematic flows (emissions and resource use) from different LCA case studies and through the literature studies. The result will be a long list of flows that should be reduced according to predefined criteria. Two different criteria are suggested, one related to socially and scientifically recognized environmental problems with possible help through the LCIA step, using classification or different weighting methods. The other criterion is to relate the flows to biogeochemical cycles. The large number of parameters can thus be reduced to a few compounds that are important because they are abundant and a few compounds that are important because they have a large impact on the environment or on human health.

The second step is to identify flows that occur in more than one LCA activity, since the point with this approach is to identify flows that should be managed at a systems level. Life cycle components that have a little potential adverse impact on the environment can hence be eliminated.

Further, the indicators that relate to important aspects of urban water systems but are not relevant to or cannot be considered in LCA, such as the quality of drinking water and sewage sludge, should be taken into account.

The resulting and suggested indicators might then be discussed with the end users to provide their input regarding important aspects of an indicator set. Finally, the indicators should be tested in one or more case studies and evaluated.

9.6 SIM OF WATER MAINS

In this section, the improvement of water mains' SIM capability as an approach for reducing high-risk drinking water mains breaks and inefficient maintenance scheduling is explored. Structural integrity* of water networks refers to the soundness of the pipe walls and joints for conveying water to its intended locations and preventing egress of water, loss of pressure, and entry of contaminants.

SIM is the systematic detection, location, and quantification of pipe wall and pipe joint damage and deterioration (e.g., wall thinning, cracking, bending, crushing, misalignment, or joint separation) of installed drinking water mains. Determination of the present condition and the deterioration rate of the pipe will be made possible by an effective SIM including (Royer, 2005) the following:

- The present condition of a particular pipe is determined by a single measurement of a structural parameter such as slope, soil type, and so on.
- Deterioration rate for a particular pipe emplacement is specified by periodic measurement of a structural parameter. If the measured parameter reaches a predetermined unacceptable level, then actions can be taken such as (a) small repairs to stop accelerated deterioration, and much larger repairs later or (b) repair, rehabilitation, or replacement or what is referred to as "R^3 (Repair, Rehabilitation, or Replacement)" to inhibit failures and associated damages. Conversely, if a measured parameter is at an acceptable level, then R^3 can be safely deferred. If suitable failure models exist, SIM data on the current condition and deterioration rate may also be useful for estimating future structural condition and remaining service life, and for selecting inspection frequencies.
- R^3 can be scheduled either too early or too late by the lack of cost-effective SIM capability for water mains. Either error can cause adverse effects.
- Late scheduling of R^3 can occur when sparse or inaccurate structural integrity data cause underestimation of pipe deterioration and/or loading. Potentially preventable main breaks will occur with permission of late scheduling of R^3. Particularly undesirable are high-risk main breaks, which can cause (1) sudden and significant losses of water and pressure, (2) serious health or drinking water quality effects on customers, (3) other adverse effects on critical customers, or (4) a major damage to the surroundings. A key target of this section is that high-risk mains breaks can be reduced by monitoring pipe deterioration more frequently, comprehensively, and/or accurately, and then using the collected data to generate more accurate and timely scheduling of prefailure R^3.

---

*Not including tuberculation and scale formation, leaching of pipe or liner constituents into the water, and permeation of contaminants through the pipe wall, coatings, linings, or gaskets. Also the structural integrity of pumps or valves is not included.

- Scheduling of R^3 can occur when inadequate quantity and quality of structural integrity data causes decision makers to overestimate the aerial extent and/or the severity of pipe deterioration. If the actual condition of the pipe can be accurately determined by examination, then selective R^3 on a small fraction of the pipe becomes an option when it is the more economical way to address the problem.
- Both the cost of high-risk main failures and the inefficiency of premature replacement exacerbate the funding gap between current and required spending for capital, operating, and maintenance expenditures. The size of the funding gap has been estimated by EPA at US$45–US$263 billion (i.e., ~$2.3–$13 billion per year) for the 20-year period from 2000 to 2019 [U.S. EPA, 2002a,b,c; Congressional Budget Office (CBO), 2002].

Numerous SIM options already exist for drinking water mains, but these options have many shortcomings. Various combinations of pipe materials, configurations, and failure modes are not adequately or economically characterized by existing SIM technology. A substantial amount of research, development, testing, and verification (R,D,T,&V) is under way or planned for improving SIM, but most of the effort is directed toward high-risk, nondrinking water applications. Example applications include oil and natural gas pipelines, nuclear power plants, large buildings, bridges, and aircraft. A significant amount of this investigation is government sponsored. Many of these SIM improvement efforts involve applying recent advances in technology to the creation of better, cheaper, and faster ways of acquiring and analyzing structural integrity data. The products, data, and procedures from some of this R,D,T,&V are a relatively untapped source of technology transfer opportunities for water mains' SIM improvements, but no systematic effort to identify, prioritize, and capitalize on these chances is made by the governments and urban water managers. Acceleration of the SIM improvement process is important, since

1. SIM capability improvement is a difficult, uncertain, slow, tedious, and expensive process.
2. A substantial portion of the transmission and distribution system is projected to be approaching or reaching the end of its service life. The need for decisions to implement or delay replacement will be extremely increasing between now and approximately 2035, when the annual replacement rate is projected to peak at about 2% (i.e., 26,000–32,000 km of pipe replaced/year), which is more than 4 times the current replacement rate in the U.S (U.S. EPA, 2002c).

It is important for the drinking water community, including the urban water managers, to define and prioritize drinking water mains' SIM capability improvement needs and/or to cooperate and collaborate in completing the required R,D,T,&V activities, due to resources limitation.

Prevention of mains breaks is emphasized over avoidance of main leaks. The adverse economic effects of major breaks (e.g., damages, disruption of business and traffic, and emergency response costs) and potential for health effects (e.g., potential contaminant intrusion or backflow due to pressure loss) appear to be more immediate and severe, and more likely to be linked to a specific mains break incident than for mains leaks. Although more water is probably lost from leaks than from mains breaks, leaks are often tolerated for a variety of reasons. As long as mains leaks can serve as a reliable sign of the location and timing of future main breaks, the detection, location, and quantification of mains leaks are relevant to prefailure detection, location, and prevention of main breaks. Structural condition assessment (SCA) is the determination of the present and future fitness-for-service of the pipe. It will be assumed that adequate SCA for at least the imminent failure case is, or will become, feasible for some indicators and situations (e.g., corrosion pitting, severe wall thinning, cross section deflection, misalignment, or bending). The value of the inspection to identification of failures that are either existing or obviously imminent will be limited by inspection without effective condition assessment. Condition assessment requires inspection data and is affected by their accuracy.

TABLE 9.8
Functional and Program Benefits of Effective and Affordable Inspection

	Program Benefits			
	Drinking Water-Quality Protection		Infrastructure Funding Gap Reduction	Water Conservation
Inspection of Functional Benefits	Short Term	Long Term		
Optimize asset management		•	•	•
Optimize inspection frequencies		•	•	
Optimize repair, rehab, and replacement scheduling		•	•	•
Optimize service life		•	•	
Reduce mains breaks	•	•	•	•
Reduce contaminant backflow via cross connections	•	•		
Reduce intrusion from break-induced pressure transients	•	•		
Reduce contaminant entry at/near mains break locations	•	•		
Reduce water loss			•	•
Reduce damage costs		•	•	
Reduce response costs		•	•	
Reduce leaks	•	•	•	•
Reduce water loss				•
Reduce failure-inducing conditions at pipe exterior	•	•	•	•

Source: U.S. Environmental Protection Agency (U.S. EPA). 2005. White paper on Improvement of Structural Integrity Monitoring for Drinking Water Mains. EPA/600/R-05/038.

Note: •, functional benefit causes program benefits.

A desirable objective is the integration of SIM with hydraulic and water quality monitoring in the distribution system.

9.7 PUBLIC BENEFITS FROM SIM IMPROVEMENTS

Reducing mains breaks supports the Safe Drinking Water Act's goals of protecting public health and drinking water quality. Reducing main breaks, optimizing maintenance planning, extending infrastructure service lives, and reducing water leakage are all in line with the EPA goals of improving utilities' infrastructure management capability. Reducing the infrastructure funding gap is a major challenge. The benefits of SIM capability improvement to the associated EPA goal areas are listed in Table 9.8.

Improved SIM can also reduce premature R^3. If the general condition of U.S. drinking water mains deteriorates due to the pipe replacement rate lagging behind the deterioration rate, there will be an increased need for efficient R^3 decision making. For more efficient decision making on R^3, accurate and economical pipe condition data are needed. For instance, premature abandonment of relatively new prestressed concrete cylinder pipelines (PCCP) has been prevented by the use of new SIM technologies that enable localized, accelerated deterioration of prestressing wires to be detected, located, and repaired during scheduled maintenance. Without this capability, main breaks would have caused random, frequent, and serious outages that would have made the pipeline too unreliable, hazardous, and costly to operate.

9.8 OPPORTUNITIES FOR SIM CAPABILITY IMPROVEMENT

Because of numerous significant trends and circumstances, there is an opportunity-filled time to accelerate the improvement and verification of SIM technologies and procedures for drinking water conveyance and storage systems. The feasibility of improving SIM capability for drinking water mains by adapting new technologies has already been demonstrated for a number of situations. Advances in relevant science and technology areas, such as sensors, communications, computing, and materials science, are occurring that can significantly improve the quantity, quality, timeliness, and cost of structural parameter data for determining structural strength, deterioration rates, loading, and approach to failure conditions for various structures. Improvement of SIM technology is typically a lengthy, difficult, costly, and uncertain process. Nonetheless, for various high-risk, impure water applications, multiple SIM improvement attempts are in progress. Opportunities to economically accelerate the improvement of SIM capability for water mains can be provided by products, such as concepts, prototypes, data, and demonstrations, from these attempts. Improving SIM technology for underground pipelines in spite of its difficulty is possible. For example, Table 9.9 lists six nondestructive evaluation (NDE) technologies (i.e., acoustic emission, electromagnetic, impact echo, remote field eddy current (RFEC), seismic, and ultrasonic) and describes the pipe materials and defect types to which they are applicable. Another example of SIM technology improvement is the development of instrumented cathodic protection.

Cathodic protection for electrically continuous metallic pipes is not a new technique and is required for gas and petroleum pipelines (Lawrence, 2001). It basically involves sensing and maintaining an electrical potential balance in the system that inhibits corrosion. Instrumented cathodic protection is an improvement that enables the cathodic protection system itself to be monitored, which precludes failures due to malfunctioning cathodic protection systems (e.g., Hock et al., 1994; Van Blaricum et al., 1998).

At least four SIM approaches for water mains seems to have the potential for significant improvement through the incorporation of better and more economical technologies for sensor positioning, sensing, and data storage, transmission, and analysis. These four SIM approaches are: mobile in-line inspection (MILI) systems, mobile nonintrusive inspection systems, continuous inspection devices, and intelligent systems.

9.8.1 MOBILE IN-LINE INSPECTION SYSTEMS

The measuring device for these systems requires to be physically inserted into, moved through, and removed from the pipe. The MILI sensors are usually in close contact with the pipe wall and measure structural parameters over short distances. MILI systems collect structural data for a given area or volume of the pipe for the short interval during each inspection cycle that the sensor is in measuring range of the parameter(s) of interest. Examples of MILI device output are (1) visual images of the inner surface indexed to pipe location, (2) continuous or discrete wall thickness profiles indexed to pipe location, and (3) void spaces outside the pipe indexed to pipe location. The pipe may have to be drained and/or cleaned to enable the MILI device scan to be employed. For small-diameter pipes, MILI devices may be operated either automatically or remotely. For large-diameter pipes, there is the additional option of direct inspection of the MILI device operation by a person.

9.8.2 MOBILE NONINTRUSIVE INSPECTION (MNII) SYSTEMS

These systems differ from MILI systems because the detector is not placed inside the pipe. MNII systems that can examine a substantial length of pipe from a single location, preferably without excavation, offer the promise of substantially reducing the time, cost, and disruption involved in pinpointing the pipe that should receive detailed inspection. Samples include (1) Lamb wave devices that transmit and receive ultrasonic waves for moderate distances (e.g., 30 m) along the pipe wall, which enables 60 m of the pipe to be inspected from one location; (2) electric field monitoring devices,

TABLE 9.9
Summary of NDE Method Issues That Affect Technique Selection for Various Water Pipe Materials

Inspection Method	Pipe Material	Defect Types	Notes
Acoustic emission	Pre-tensioned or pre-stressed concrete pipe	Breaks in reinforcing steel Slippage of broken reinforcement Concrete cracking	Pipe not removed from service Hydrophones left in place for several days to weeks
Electromagnetic	All-metallic pipe	Cracks	Commercial, off-the-shelf availability Detect environmental conditions that are likely to weaken the pipe Does not directly inspect the pipe Totally noninvasive
Impact echo	Concrete pipe containing steel	Delaminations and cracks at various concrete/mortar/steel interfaces	Requires dewatering and human access to the interior of the pipe Can be done externally if exterior access is available
RFEC	All-metallic pipe	Changes in metal mass graphitization Wall thinning gouges large cracks	Commercial, off-the-shelf availability Pig travels through the pipe via water hydrants May require cleaning before inspection Pig may dislodge material from the pipe wall, requiring flushing
Seismic	All-concrete pipe	Reductions in concrete modulus because of aging Reductions in concrete compression as a result of breakage or slippage of reinforcing steel	Requires dewatering and human access to the interior of the pipe
Ultrasonic	All-metallic pipe	Detection of wall thinning	Not commercially available for water pipes Does not require dewatering of pipes Developed for inspecting oil or gas pipelines systems are long, inflexible, and expensive

Source: Dingus, M., Haven, J., and Austin, R. 2002. *Nondestructive, Noninvasive Assessment of Underground Pipelines.* AWWA Research Foundation, Denver, CO, 116pp. With permission.

which temporarily electrify the pipe, then detect electric field changes at the ground surface that are indicative of pipe wall thinning and indicate problem locations for detailed investigation; and (3) aerial or satellite systems for remote monitoring of surface conditions symbol of pipe deterioration or failure. Like the MILI systems above, MNII systems provide a "snapshot" of the measured pipe structural parameter(s) during each mobilization/inspection/demobilization cycle.

9.8.3 CONTINUOUS INSPECTION DEVICES

These devices frequently collect structural data, which improves the capability for detecting and tracking transient deterioration indicators (e.g., acoustic emissions from cracking, leakage, or wire breaks in prestressed concrete cylinder pipes). These devices can be intrusive (i.e., require placement inside the pipe) or nonintrusive. These devices may be operated on a moderate to long-term basis (i.e., data collection for the inspected area may occur for a day, week, month, or years). The defect-to-sensor distance may be moderate to long. Cable-mounted hydrophones for acoustic emission monitoring of wire breaks in prestressed concrete cylinder pipes represent an example of them.

9.8.4 INTELLIGENT SYSTEMS

These are permanent, comprehensive, and automated SIM systems. The sensing and data storage/transmission/analysis capabilities are built-in or retrofitted to the monitored portion of mains. The sensing capability is selected and installed for the desired spatial, temporal, and failure mode coverage. Examples include instrumental cathodic protection (ICP) (e.g., EUPECRMS, 2003), which monitors coating integrity over long distances for cathodically protected pipelines or an electrically conductive composite pipe (ECCP), which is a prototype pipe with an embedded sensing layer (Meegoda and Juliano, 2002). Intelligent systems offer the potential for convenient, flexible, rapid, comprehensive, nondisruptive inspection, if they can provide the quality of data required at an affordable price.

9.9 CONCLUDING REMARKS

Urban water infrastructures include delivery, treatment, supply, and distribution of water by water mains to its end users and also for the collection, removal, treatment, and disposal of sewage and wastewater. Urban infrastructure systems are critically important for maintaining public health as well as for controlling the quality of the waters into which urban runoff and wastewater are discharged. In most urban areas in developed regions, government regulations require designers and operators, for good design and operation of urban water infrastructures, to meet the standards. Pressures must be adequate for fire protection, water quality must be adequate to protect public health, and urban drainage of wastewater and stormwater must meet effluent and receiving water body quality standards.

In this chapter, an introduction to different components of urban water infrastructures has been presented and their interactions among UWC have also been discussed. Financing and planning of the SD of urban water infrastructures have also been discussed in this chapter. SIM of water mains in order to reduce the repair, rehabilitation, and replacement cost of water mains and increase the reliability of supplying the demands are discussed. This requires monitoring as well as the use of various methods for detecting leaks and for predicting the impacts of alternative urban water systems (treatment, distribution, and collection) on operation, maintenance, and repair policies of these systems. Finally, nondestructive evaluation tests in the form of opportunities for SIM are presented.

The examples in this chapter are mostly related to economical comparison of alternatives, which are the key factors in infrastructure development activities. Urban infrastructure plays an important role in the daily life of millions of people around the world. The interactions of its components are not well integrated in the design, construction, rehabilitation, and maintenance of these systems. Economic and financial analysis plays an important role in infrastructure development. Its principle and deciding factors/parameters are not well understood and agreed upon by the analysts and decision makers, especially in the developing countries. These are the main challenges of current urban infrastructure development in addition to the size of capital investment needed to maintain and expand structures to meet the growing demand for urban water.

PROBLEMS

1. The net cash flows are shown in Table 9.10 for three alternatives for urban water supply projects. The MARR is 12% per year and the study period is 10 years. Which alternative is economically preferred based on the external rate of return method (external rate of return = MARR = 12% per year)?

TABLE 9.10
Cash Flows of Problem 1

End of Year	Project A ($)	Project B ($)	Project C ($)
0	171,200	126,400	143,600
1	14,800	24,200	20,000
.	.	.	.
.	.	.	.
10	14,800	24,200	20,000

2. The three alternatives described in Table 9.11 are available for supplying a community water supply for the next 25 years when the economic lives and the period of analysis terminate. Using the costs presented in the following table and 5% discount rate, compare the three projects by the present worth method.

TABLE 9.11
Costs of the Three Alternatives of Problem 2

Costs	Timing	Project A ($)	Project B ($)	Project C ($)
Initial investment	Year 0	2,000,000	1,000,000	1,500,000
	Year 10	0	1,000,000	1,200,000
	Year 20	0	1,000,000	0
Operation and maintenance costs	Years 1–10	7000	4000	6000
	Years 10–20	8000	7000	8000
	Years 20–25	9000	9000	9000

3. Compare the projects in Problem 2 by the benefit–cost ratio method.
4. Compare the projects in Problem 2 by the rate of return method.
5. Three water supply projects have been proposed for a city. Each of the projects has the useful life of 25 years and the interest rate is 5% per year. The benefits and costs of the projects are listed in Table 9.12. Compare the alternatives by using the benefit–cost ratio and present worth methods.
6. Suppose you deposit $6000 today, $3000 2 years from now, and $4000 5 years from now in a bank account paying 5% interest per year. How much money you will have after 10 years in your account?

TABLE 9.12
Benefits and Costs of the Three Water Supply Alternatives of Problem 6 (Million Dollars)

Benefits and Costs	Project A	Project B	Project C
Capital investment	8	9.5	11
Annual operation and maintenance costs	0.7	0.65	0.63
Annual expected reduction in leakage (benefits)	1.5	1.6	1.75

7. Consider that you have invested $1000 in a project. Based on the contract, you will receive $500 and $1500 benefits 3 and 5 years later, respectively. How much is the IRR of the investment?
8. Consider the following cash flow diagram shown in Table 9.13. If the external rate of return is considered 15% and MARR is equal to 20%, is the project rejected?

TABLE 9.13
Cash Flows of Problem 9

End of Year	Cost ($)	Benefit ($)
0	1000	0
1	500	0
2	100	600
.	.	.
6	100	600

9. Solve Example 9.4 by using dynamic programming.
10. A dam supplies water to a city and downstream agricultural lands. The average monthly domestic and irrigation demands and the monthly reservoir release in a dry year are as shown in Table 9.14. The prices of water for domestic and irrigation uses are $90,000 and $10,000 per million cubic meters, respectively.

   - Formulate the problem for optimizing the water allocation from this reservoir.
   - Solve the problem using linear programming.

TABLE 9.14
The Average Monthly Domestic and Irrigation Demands and the Monthly Reservoir Release of Problem 10 (Numbers are in Million Cubic Meters)

	Jan.	Feb.	Mar.	Apr.	May.	Jun.	Jul.	Aug.	Sep.	Oct.	Nov.	Dec.
Domestic	8	8.5	8.5	9	9	9.5	9.5	10	9.5	9	8.5	8.5
Irrigation	0	0	20	50	55	70	70	65	20	10	0	0
Release	25	27	40	48	60	42	25	17	14	18	20	25

11. In Problem 10, the shortages should be met by supplying from groundwater resources. The cost of meeting the demand by supplying from groundwater is a function of the volume of water that should be extracted. The cost of water extraction for domestic demands is estimated as $C_{Dom} = 1000 \cdot x_{Dom}^2$, where x_{Dom} is the volume of water

extracted for domestic uses. The cost of water extraction for irrigation purposes is estimated as $C_{irr} = 250 \cdot x_{irr}^2$, where x_{irr} is the volume of water extracted for irrigation.

- Formulate the problem for finding the optimal monthly volumes of water allocation.
- Solve the problem using linear programming.

12. In Problem 10, the city discharges its wastewater to the river upstream of the agricultural lands. The monthly wastewater discharge rate is about 20% of the water use of the city in each month. In order to keep the quality of the river flow in an acceptable range for irrigation, the pollution load of the city wastewater should be reduced by primary treatment, which has an additional cost equal to $12,000 per million cubic meters.

- Formulate the problem for optimizing the water allocation from this river.
- Solve the problem using linear programming.

REFERENCES

ASCE Task Committee on Sustainability Criteria. 1998. *Sustainability Criteria for Water Resource Systems*. ASCE, Reston, VA.

Asian Development Bank. 2001. *Water for All: The Water Policy of the Asian Development Bank*. Asian Development Bank, Manila.

Asian Development Bank. 2005. *Asia Water Watch 2015: Are Countries in Asia on Track to Meet Target 10 of the Millennium Development Goals?* (Summary) Asian Development Bank, Manila, November.

Brdjanovic, D., Park, W. C., and Figueres, C. 2004. Sustainable wastewater management for new communities: Lessons learned from Sihwa City in South Korea. In *Proceedings: Sustainable Communities Conference*, Burlington, VT.

Brix, H. 1998. How 'Green' are constructed wetland treatment systems? *Proceedings of 6th International Conference on Wetlands Systems for Water Pollution Control*. Aguas de Sao Pedro, Brazil, September 27–October 2.

Butler, D. and Parkinson, J. 1997. Towards sustainable drainage. *Water Science and Technology* 35(9): 53–63.

Congressional Budget Office. 2002. *Summary Report-Future Investment in Drinking Water and Wastewater Infrastructure*. Washington, DC, p. 16.

Debo, T. N. and Reese, A. J. 1995. *Municipal Storm Water Management*. CRC Press/Lewis Publishers, Boca Raton, FL.

Dingus, M., Haven, J., and Austin, R. 2002. *Nondestructive, Noninvasive Assessment of Underground Pipelines*. AWWA Research Foundation, Denver, CO, 116pp.

EUPEC Risk Management Systems. 2003. ATIS (All Terrain Information System) Cathodic Protection Remote Monitoring System. 1 pp. http://www.eupecrms.com/Products/ATIS, October 15, 2003.

Farahbakhshazad, N. and Morrison, G. M. 2000. A constructed vertical macrophyte system for the return of nutrients to agriculture. *Environmental Technology*, 21: 217–224.

Goodman, A. S. and Hastak, M. 2006. *Infrastructure Planning Handbook (Planning, Engineering and Economics)*. American Society of Civil Engineers and McGraw-Hill, New York.

Heaney, J. P. 1997. Cost allocation in water resources. In: C. Revelle and A. E. McGarity, Eds, *Design and Operation of Civil and Environmental Engineering Systems*, Chapter 13. Wiley, New York.

Hock, V. F., Lewis F., Setliff, W. A., Houtz, M. N., Lewis, B. J., and McLeod, M. E. 1994. Off-potential measurement systems for impressed current cathodic protection. U.S. Army Corps of Engineers Construction Engineering Laboratory, USACERL Technical Report FM-94/16, Champaign, IL.

ISO/DIS 14031. 1998. *Environmental Management—Environmental Performance Evaluation, Guidelines*. American National Standards Institute, New York.

ISO/DIS 14040. 1997. *Environmental Management—Life Cycle Assessment. Principles and Framework*. International Organization of Standardization, Geneva, Switzerland.

ISO/DIS 14041. 1998. *Environmental Management—Life Cycle Assessment, Goal and Scope Definition and Inventory Analysis*. International Organization of Standardization, Geneva, Switzerland.

IWSA. 1997. *Performance Indicators for Transmission and Distribution Systems.* Workshop Postprint, Lisbon, May 5–7.
Jeffrey, P., Seaton, R., Parsons, S., and Stephenson, T. 1997. Evaluation methods for the design of adaptive water supply systems in urban environments. *Water Science and Technology* 35(9): 45–51.
Keffer, C., Shimp, R., and Lehni, M. 1999. Eco-efficiency indicators & reporting. Report on the Status of the Project's Work in Progress and Guideline for Pilot Application, WBCSD Working Group on Eco-efficiency Metrics and Reporting, WBCSD, Geneva.
Kingdom, B., Liemberger, R., and Marin, R. 2006. The challenge of reducing non-revenue water (NRW) in developing countries, how the private sector can help: A look at performance-based service contracting. *Water Supply and Sanitation Sector Board Discussion Paper Series, World Bank*, paper no 8.
Larsen, T. A. and Gujer, W. 1997. The concept of sustainable urban water management. *Water Science and Technology* 35(9): 3–10.
Lawrence, C. F. 2001. To CP or not to CP? That is the question. In *Infrastructure Conference Proceedings*, March 11–14, 2001. American Water Works Association Research Foundation. Denver, CO, 14pp.
Meegoda, J. N. and Juliano, T. M. 2002. Research, Development, Demonstration and Validation of Intelligent Systems for Conveyance and Storage Infrastructure (Draft Final Report). National Risk Management Research Laboratory, U.S. EPA and U.S. Army Industrial Ecology Center, 94pp.
Nelson, A. C. 1995. *System Development Charges for Water, Wastewater, and Storm Water Facilities.* CRC Press/Lewis Publishers, Boca Raton, FL.
Niemczynowicz, J. 1996. Mismanagement of water resources. *Vatten* 52: 299–304.
Nix, S. J. and Heaney, J. P. 1988. Optimization of storage-release strategies. *Water Resources Research* 24(11): 1831–1838.
Owens, J. W. 1999. Why life cycle impact assessment is now described as an indicator system. *International Journal of LCA* 4(2): 81–86.
Peters, M. and Timmerhaus, K. 1980. *Plant Design for Chemical Engineers.* McGraw-Hill, New York.
Report of the World Summit on Sustainable Development. 2002. Johannesburg, South Africa, August 26–September 4 (United Nations Publication, Sales No. E.03.II.A.1 and corrigendum), Chapter I, Resolution 2, Annex.
RIONED. 2005. Urban Drainage Statistics 2005–2006. www.rioned.org. RIONED Foundation, Ede, the Netherlands.
Royer, M. D. 2005. White paper on improvement of structural integrity monitoring for drinking water mains. EPA/600/R-05/038, Cincinnati, OH, 45268.
Schaltegger, S. 1996. *Corporate Environmental Accounting.* Wiley, Chichester, U.K.
Smith, L. A., Fields, K. A., Chen, A. S. C., and Tafuri, A. N. 2000. *Options for Leak and Break Detection and Repair for Drinking Water Systems*, Battelle Press, Columbus, OH, 163pp.
Statewide Urban Design and Specifications. 2007. *Urban Design Standards Manual.*
U.S. Environmental Protection Agency (U.S. EPA). 1993. *Combined Sewer Control Manual.* EPA/625/R-93-007.
U.S. Environmental Protection Agency (U.S. EPA). 2002a. Potential contamination due to cross-connections and backflow and the associated health risks—An issues paper. Office of Ground Water and Drinking Water, August 13, 42pp, http://www.epa.gov/safewater/tcr/pdf/ccrwhite.pdf.
U.S. Environmental Protection Agency (U.S. EPA) 2002b. The clean water and drinking water infrastructure gap analysis. EPA-816-R-02-020, Office of Water, Washington, DC, September, 50pp.
U.S. Environmental Protection Agency (U.S. EPA). 2002c. The clean water and drinking water infrastructure gap analysis. EPA 816-F-02-017, EPA Office of Water, September, 2pp.
U.S. Environmental Protection Agency (U.S. EPA). 2005. White paper on Improvement of Structural Integrity Monitoring for Drinking Water Mains. EPA/600/R-05/038.
Van Blaricum, V. L., Flood Jack, T., Szeliga Michael, J., and Bushman, J. B. 1998. Demonstration of remote monitoring technology for cathodic protection systems: Phase II. FEAP TR 98/82, U.S. Army Construction and Civil Engineering Research Laboratories, Champaign, IL, May, 43pp.
White, W. R. 2005. World water storage in man-made reservoirs (FR/R0012). Review of Current Knowledge Series, April (Bucks, United Kingdom, Foundation for Water Research, available online at www.fwr.org.
Wilsenach, 2006. *Treatment of Source Separated Urine and its Effect on Wastewater Systems.* PhD thesis, Delft University of Technology, the Netherlands.

10 Urban Water System Dynamics and Conflict Resolution

10.1 INTRODUCTION

For considering integration and sustainability in water resources management, it is necessary to think over the social, economic, and environmental impacts of decisions. This integration in planning and management, especially in urban areas, needs a systematic approach, considering all the interactions among the elements of the system and with the outside world. Simulation of urban water dynamics will give the collective impacts of all possible water-related urban processes on issues such as human health, environmental protection, quality of receiving waters, and urban water demand. Individual processes are then planned and managed in a way that the collective impact, with due consideration of the interaction among processes, is improved as much as possible.

The art and science of systems analysis has evolved through developments in the separate disciplines of engineering, economics, and mathematics. Rapid developments have taken place and the availability of high-speed computers has contributed to its widespread applications. As the science of systems analysis has advanced over the last several decades and as the scale of modern water resource projects has grown, systems analysis has found widespread applications in water resources planning. The origin of the activity may be said to be in the 1950s in the United States, and the pioneering work has been done by a group of engineers, economists, and political scientists at Harvard University. Since then, the importance of systems planning has been increasingly recognized and continuous advances are being made.

A physical water system is a collection of various elements, which interact in a logical manner and are designed in response to various social needs, in the development and improvement of existing water resources for the benefit of human use. Simonovic (2000) described water resources systems analysis as an approach by which the components of such a system and their interactions are described by means of mathematical or logical functions. In an urban setting, the components of a water system is continuously affected by other components as well as considerable human interactions, which often places more burden on functionality of them.

In general, systems analysis is the study of all the interactions of the components. Very often systems analysis is concerned with finding that combination of components that generates an optimum solution, which consists of the best possible combination of elements for satisfying the desired objective. Thus, it involves defining and evaluating numerous water development and management alternatives. This can be done in a very detailed manner representing various possible compromises among conflicting groups, values, and management objectives. The dynamics of urban water supply and demand components, and how an optimum scheme can be developed for urban water management, are the first challenge. The second challenge is to operate and maintain the urban water infrastructure in an effective fashion.

The systems approach is especially useful when a project becomes so large that it cannot be considered as a unit, necessitating its decomposition. However, systems analysis is not an approach that can be used as a routine exercise and without thinking. Usually, the greatest effort of the analyst is to reduce the system to a manageable representation without losing its essential features and relationships. The analyst may overlook important relationships because he/she may lack access to all necessary data, and usually time is not sufficient in an actual planning environment to develop the ideal model and test it to its fullest extent or subject it to the scrutiny of several experts. Urban water systems are designed in different zones with known elements of dependency of each zone to others and the components within each zone. Therefore the analysis can be done in separate zones.

A prerequisite for a systems analysis is that all the elements of the system can be modeled either analytically or conceptually. It is important to distinguish the difference between a system and a model. A model is the mathematical and/or physical representation of the system and of the relations between the elements of the system. It is an abstraction of the real world, and, in any particular application, the quality of the model and thus of systems analysis depends on how well the model builder perceives the actual relationship and how well he/she is able to describe their functional form. Many advances in making analytical tools such as network analysis, pressure and leakage management, and simulation and optimization models have been made. Conceptual models in the form of laboratory experiments and/or building small-scale prototypes are used to physically model the real applications and in layers that can be focused piece by piece or as a whole in harmony.

Since models are abstractions of reality, they do not usually describe all the features that are encompassed by a real-world situation. A prerequisite for water resources systems analysis is the description of the system in terms of model components, which permits solutions to be obtained at reasonable costs and within a prescribed time frame. Therefore, the model builder should not attempt to examine the reality and accuracy of individual components, but only to meet the overall accuracy requirements for that system. Sensitivity analysis should be performed since urban water models are subject to high levels of uncertainty.

10.2 SYSTEM DYNAMICS

A system dynamics approach relies on understanding the complex interrelationships among different elements within a system. This is achieved by developing a model that can simulate and quantify the behavior of the system. Simulation of the model over time is considered essential to understand the dynamics of the system. Understanding the system and its boundaries, identifying the key variables, representing the physical processes or variables through mathematical relationships, mapping the structure of the model, and simulating the model for understanding its behavior are some of the major steps that are carried out in the development of a system dynamics simulation model. It is interesting to note that the central building blocks of system dynamics are well suited for modeling physical systems and for representing nonphysical models that have been developed using a systems thinking paradigm.

In a short period, the users of system dynamics simulation tools can experience the main advantages of this approach. The power of simulation is the ease of constructing "what-if" scenarios and tackling the physics of big, messy, real-world problems accurately. In addition, general principles of the system dynamics simulation tools also developed apply to social, economic, and ecological and biological systems.

System dynamics was introduced in the 1960s by investigators at the Massachusetts Institute of Technology. It was originally rooted in the management and engineering sciences but has gradually developed in other fields. In the field of system dynamics, a system is defined as a collection of elements that continually interact over time to form a unified body. The underlying pattern of interactions among the elements of a system is called the structure of the system. A good example of water resources engineering of a system is a storage tank. The structure of a tank is defined by the interactions among inflow, storage, outflow, and other site-specific issues such as release values. The structure of a storage tank includes the importance of variables influencing the system. The term "dynamics" refers to change over time. If something is dynamic, it is constantly changing in response to the stimuli influencing it. In an urban setting, storage tanks (Figure 10.1) and the stock of available

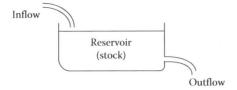

FIGURE 10.1 Schematic of an urban storage tank.

water as it is affected by input and output presents a unique challenge for the application of system dynamics in urban water.

In system dynamics, the way in which the elements or variables vary over time is referred to as the behavior of the system. In the storage tank example, the behavior is described by the dynamics of the storage tank filling and draining. This behavior is due to the influences of inflow, outflow, losses, and other factors affecting the system characteristics. By defining the structure of a storage tank, it is possible to use system dynamics analysis to trace out its behavior over time. System dynamics can also be used to analyze how structural changes in one part of a system might affect the behavior of the system as a whole. The visual effects of system dynamics mapping help the decision makers to clearly see the impact of different scenarios.

Perturbing a system allows one to test how the system will respond under varying sets of conditions. In the case of a storage tank, one can test the impact of storage fluctuations due to supply system shortages or transfer system failure that can cause the elimination of a particular user. In the different pressure zones in the water distribution system, system dynamics can display the variations in pressure and consumption in real time.

10.2.1 Modeling Dynamics of a System

There are many tools used for implementing the object-oriented modeling approach. Computer software tools such as STELLA developed by High Performance Systems Inc., VENSIM developed by Ventana System Inc., and POWERSIM developed by Powersim Software AS help the execution of these processes.

The power and simplicity of the system dynamics that lie on object-oriented programming are quite different from those used in functional algorithmic languages. The objects are

- Stocks
- Flows
- Converters
- Connectors

The power of object-oriented simulation is the ease of constructing "what-if" scenarios and tackling large-scale, messy, real-world problems. In Section 10.4, in order to simulate the state of the system under different water transfer capacities of a channel and conflict resolution in each demand zone, two simulation models are developed using object-oriented software.

The system dynamics simulation tool used in this example contains the above-mentioned specific objects that are used in representing the system structure. Stocks manifest as accumulations. They represent "conditions" (i.e., how things are) within a system such as water storage in a tank or a reservoir. Flow, representing "actions" (i.e., how things are going), is used to represent variables whose values are measured as rates of flow in and out of reservoirs (surface and underground tanks, treatment facilities and distribution systems, are examples of this type of object). A converter is used to transform inputs in the form of algebraic relationships and graphs into outputs such as mass balance equations, SCS, and rational method expressions, relating rainfall to runoff in urban catchments. A connector, representing the relationship between other objects, conveys information from one variable to the others. These diagrams also contain "clouds," which represent the boundaries of the system.

The steps in modeling the dynamics of a system are discussed below. The modeling process is *not* linear, and model developers usually find themselves cycling through the steps several times as they work on the model. It is sometimes necessary to cycle back through the entire sequence, as the model is refined based on the challenges that are made to its structure.

10.2.1.1 Define the Issue/System

For defining the system, the following steps should be followed:

- Clearly state the purpose of the model.
- Develop a base behavior pattern.
- Develop a system diagram.

In Step 1 of the modeling process, it is necessary to develop a clear purpose that can achieve the end results. For developing a clear statement of the purpose, additional tools will help make the purpose more operational. The first of these is a reference behavior pattern (RBP). The RBP is a graph over time for one or more variables that best capture the dynamic phenomenon we are seeking to understand and will help focus efforts on the behavioral dimensions. A system diagram is composed of the essential actors or sectors. The model will help focus our effort on the structural dimension.

In Step 2 of the modeling process, the hypotheses responsible for generating the behavior pattern(s) that are identified in Step 1 are represented. A "dynamic organizing principle" is a stock/flow-based, or feedback loop-based, framework that will reside at the core of the model. Think of it as providing the theme for your "story." Once the modeler has laid out some stocks and flows have been laid out, the next step in mapping a hypothesis is to characterize the flows. We seek to capture the nature of each flow as it works in reality and we strive to achieve an operational specification by using one of the generic flow templates. Once the flows associated with hypotheses have been characterized, the next step in mapping is to close the loops.

10.2.1.2 Test of Hypotheses

For testing the hypotheses responsible for generating the system dynamics, the following steps are taken:

- Mechanical mistake tests
- Robustness tests
- Reference behavior tests

Simulating the hypotheses in Step 3 on a computer is designed to increase the modeler's confidence that the developed model is useful for the purposes it is intended. These tests are also designed to make the modeler aware of the limitations of the model's utility. Thus, the modeler should emerge from this step both confident in what the model can do and knowledgeable about what it cannot do.

In order to ensure learning from each test, before each simulation, sketch out a best guess as to how the sector/model will behave. Then, the behavior should be explained and after the simulation is complete, either work to resolve any discrepancies in actual versus guessed behavior or resolve it in modeler rationale. If the test results are not acceptable, do not hesitate to cycle back through Steps 1 and 2 of the modeling process.

Policy tests are aimed at discovering ways to alter (often to improve) the performance of the system. Frequently, model behavior patterns are "insensitive" to even reasonably large changes in many model parameters. Few model parameters are "sensitive," meaning that a change in their value will result in a change in the characteristics, behavior pattern being exhibited by the model. Such parameters are known as "leverage points," because a small change in their value will result in a shift in the behavior mode.

10.2.1.3 Design and Test Policies

For designing and testing the considered policies, the following steps should be done:

- Policy tests
- Sensitivity tests
- Scenario tests

Once some effective policies/strategies have been identified, it is important to determine how sensitive their effectiveness is to both the behavioral assumptions included in the model and the assumptions about the external environment. Testing the former sensitivity is known as "sensitivity analysis." Testing the latter sensitivity is known as "scenario analysis."

10.2.2 Time Paths of a Dynamic System

The first step in modeling the system dynamics is to accurately model the dynamic behavior of systems. Dynamic behavior is considered as the behavior of systems over time, a which is also called a time path. The modeler should identify the patterns of system behavior using the most effective and important variables of the system. Then the appropriate relationships and functions are defined to mimic the observed behavior of the system. This step is considered the model calibration process and after that the model can be employed for testing policies aimed at altering a system's behavior in desired ways.

An example of a time path is shown in Figure 10.2. It is a straight line with a positive slope. A variety of time paths (often simultaneously) are exhibited in actual systems, but there are limited groups of distinct time paths that are used in system dynamics modeling. Time paths are grouped into five distinct families, which will be introduced briefly in the next sections. Other, more complicated, observable time paths can be modeled as combinations of these five groups.

10.2.2.1 Linear Family

The linear family of time paths is the first identifiable family, which is presented in Figure 10.3. This group of time paths includes equilibrium, linear growth, and linear decline. Regarding the simplicity and intuitive appeal of these paths, some important facts should be considered for appropriate application in system dynamics modeling.

The first fact is that the average person, who is not trained in dynamic modeling, considers the actual system's growth or decline in a linear fashion but commonly the actual growth and decline

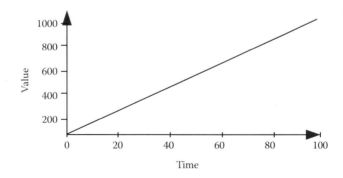

FIGURE 10.2 Example of time path.

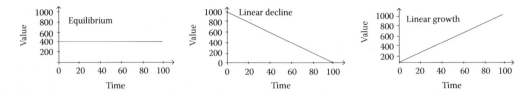

FIGURE 10.3 Linear family of time paths.

of systems follow exponential time paths rather than linear paths. Pure linear time paths are usually generated by systems devoid of feedback—a crucial building block of both actual systems and system dynamics models.

A second thing that should be considered is that the equilibrium time paths, such as that presented in Figure 10.3, are not commonly followed by actual dynamic systems. This kind of time path indicates a perfect balance with no external forces for change. In fact, from a system dynamics point of view, equilibrium implies that all of a system's state variables reach their desired values simultaneously, which cannot happen in reality. Curiously, much of modern economics and management science, especially in managing urban water systems, uses models based on the concept of equilibrium for considering complicated interactions among different components of the system. System dynamists, on the other hand, believe that the most interesting system behaviors (i.e., problems) are disequilibrium behaviors and that the most effective system dynamics models exhibit disequilibrium time paths. This is not to say, however, that the concept of equilibrium is useless. On the contrary, system dynamics models are often placed in an initial state of equilibrium to study their "pure" response behavior to policy shocks.

10.2.2.2 Exponential Family

The second distinct family of time paths is the exponential family, shown in Figure 10.4. The exponential family consists of exponential growth and exponential decay. As previously mentioned, real systems tend to grow and decline along exponential time paths, as opposed to linear time paths. For example, population growth and the water demand growth are presented as exponential time paths.

10.2.2.3 Goal-Seeking Family

The goal-seeking family is the third distinct family of time paths (Figure 10.5). All living systems (and many nonliving systems) exhibit goal-seeking behavior. Goal-seeking behavior is comparable with exponential decay. The only difference between these two time paths is that the exponential decay time path seeks a goal of zero, whereas the goal-seeking time path can seek a nonzero goal. For example, the water supply growth in an urban area follows a goal-seeking time path that will eventually converge to a maximum urban water supply potential. An important point that should be noted is that in a fixed situation the goal of this time path is fixed, but with some changes in the

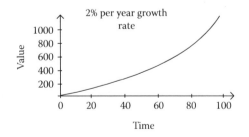
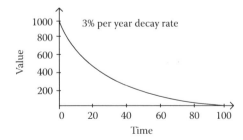

FIGURE 10.4 Exponential family of time paths.

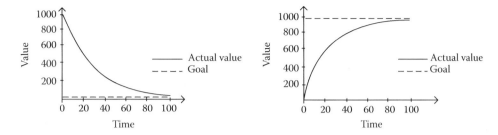

FIGURE 10.5 Goal-seeking family of time paths.

system, this goal can be moved. For example, by transferring water from adjacent watersheds to an urban area, the goal of the water resources development time path can be moved upward. This change can also result in some movements of other urban system time paths, such as water demand and population growth.

10.2.2.4 Oscillation Family

The fourth distinct family of time paths is the oscillation family. Oscillation is one of the most common dynamic behaviors in the world and is characterized by many distinct patterns. Four of these distinct patterns (sustained, damped, exploding, and chaotic) are shown in Figure 10.6.

Sustained oscillations that show periodical events can have periodicities (the number of peaks that occur before the cycle repeats) of any number. Damped oscillations are exhibited by systems that utilize dissipation or relaxation processes. An example of the dissipation process is pressure variations in a water distribution system after closing a valve. Exploding oscillations either grow until they settle down into a sustained pattern or grow until the system is torn apart. As a result, exploding oscillations usually do not occur very frequently in the real world and last only for a very short time period. Chaotic behavior is an oscillatory time path that unfolds irregularly and never repeats (i.e., its period is essentially infinite). Chaos is a unique type of time path, because it is an essentially random pattern

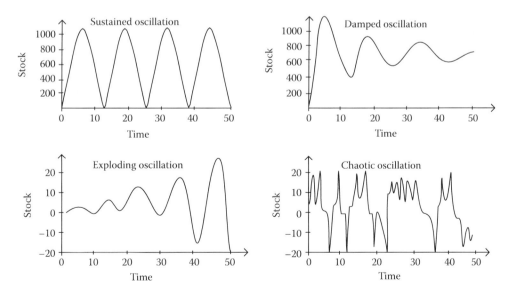

FIGURE 10.6 Oscillating family of time paths.

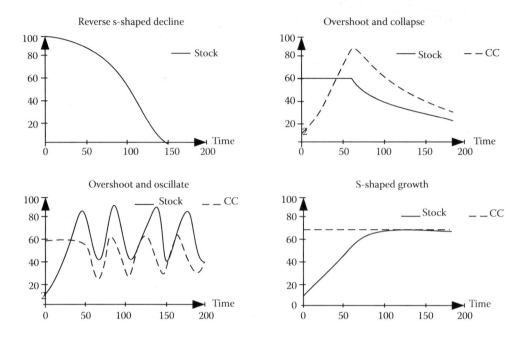

FIGURE 10.7 S-shaped family of time paths.

that is generated by a system devoid of randomness. Most of the hydroclimatological events in urban areas that affect water resources follow this kind of time path.

10.2.2.5 S-Shaped Family

The fifth distinct family of time paths is the s-shaped family, shown in Figure 10.7. A close inspection of the s-shaped growth time path pattern reveals that it is really a combination of two time paths: exponential growth and goal-seeking behavior. More precisely, in the case of s-shaped growth, exponential growth gives way to goal-seeking behavior, as the system approaches its limit or carrying capacity.

S-shaped growth is the characteristic behavior of a system in which positive and negative feedback structures fight for dominance but result in long-run equilibrium. As illustrated in Figure 10.8, the positive feedback exponential growth loop (on the right-hand side) initially dominates the system, causing the system variable to increase at an increasing rate. However, as the system approaches its limit or "carrying capacity," the goal-seeking negative feedback loop (on the left-hand side) becomes the dominant loop. In this new mode, the system approaches stasis asymptotically. This "shifting dominance" event from one loop to another is characteristic of nonlinear behavior and is found in a variety of physical, social, and economic systems.

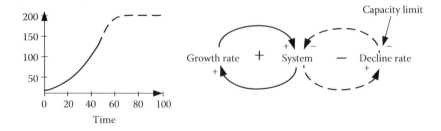

FIGURE 10.8 S-shaped structure and characteristic time path.

Sometimes, however, a system can overshoot its carrying capacity. If this occurs and the system's carrying capacity is not completely destroyed, the system tends to oscillate around its carrying capacity. On the other hand, if the system overshoots and its carrying capacity is damaged, the system will eventually collapse. This is referred to as an "overshoot and collapse" system response. A third possible outcome from an overshoot event is that a system will simply reverse direction and approach the system capacity in a reverse s-shaped pattern. As with a "normal" s-shaped pattern, a reverse s-shaped pattern is a combination of two time paths: exponential decay and a self-reinforcing spiral of decline. This behavior can be observed when the urban water demand exceeds the water supply capacity. In this situation, some strategies such as demand management and water conservation programs are employed to decrease the demand on the water supply capacity. The demand reduction process at first is most effective because unnecessary water consumptions are reduced and apparent losses are avoided. But by reaching the system capacity, decreasing the water demand becomes difficult, because people need to use more stringent water conservation plans.

EXAMPLE 10.1

A water and wastewater organization (W&WWO) is considering a law aimed at reducing the rate of groundwater and river discharge by untreated wastewater in the city. W&WWO will be given the authority to set an initial charge price for consumed water. The water consumers are initially charged \$0.1 per m^3, which is subtracted from their wastewater treatment bill based on the quantity of produced wastewater. A group of researchers collected and analyzed data about the monthly effects that resulted from this regulation. For this period, the following relationships are developed:

$$A_t = 6 + 7.2\,e^{-0.5P_{s,t}}, \tag{1}$$

$$B_t = 0.95(0.4 + 1.5P_{r,t})A_t, \tag{2}$$

$$C_t = 0.41(A_t - B_t), \tag{3}$$

$$D_t = A_t - B_t - C_t, \tag{4}$$

$$E_t = A_t - 0.5B_t, \tag{5}$$

$$P_{s,t+1} = 0.83 - 0.005B_t/A_t + P_{r,t}, \tag{6}$$

$$P_{n,t+1} = 0.35 - 0.02E_t, \tag{7}$$

where A_t is the quantity of consumed water per month t (1000 m^3/month), B_t is the quantity of treated wastewater per month t (1000 m^3/month), C_t is the quantity of water loss without any usage per month t (1000 m^3/month), D_t is the quantity of wastewater discharged to groundwater through absorption wells and septic tanks per month t (1000 m^3/month), E_t is the quantity of supplied water per month t (1000 m^3/month), $P_{s,t}$ is the sale price of 1 m^3 of water supplied by the W&WWO, $P_{n,t}$ is the cost of supplying 1 m^3 of water by W&WWO, and $P_{r,t}$ is the initial charge for 1 m^3 water usage.

Based on the above information, explain the expected changes in water usage and wastewater production and the corresponding economic impacts by answering the following questions.

1. Provide the meanings of Equations 1 through 7 in one-sentence statements. State very briefly whether or not and then why these equations are reasonable.
2. Draw a system graph that is primarily designed to support statements about the water flow in the city. Represent the water supply system of the city within a system boundary, SB, using three components called: (1) water treatment plants, (2) the water distribution system, and (3) the wastewater treatment system. Relate these three components of the system to three systems in its environment including (4) water loss, (5) groundwater, and (6) surface water resources. Then show what components are associated with each of the above described prices. Finally, show in your graph two cash flows, namely, NB_t = monthly water supply budget (10^3 \$/month) and R_t = monthly water supply revenue (10^3 \$/month).

continued

3. Solve the problem with initial values of $A_0 = 9.00$, $B_0 = 3.57$, $C_0 = 0.88$, $D_0 = 4.55$, $E_0 = 8.12$, and $P_{r,t} = 0.1$. Repeat the calculation with $P_{r,t} = 0.05, 0.15$, and 0.25.
4. Draw the variation of treated water and discharged water to groundwater as a function of $P_{r,t}$.
5. Plot the diagram of the effects of different $P_{r,t}$ on the monthly W&WWO revenue and water supplying budget.

Solution:

1. (a) $A_t = 6 + 7.2\,e^{-0.5 P_{s,t}}$: The consumption has a minimum and a maximum threshold equal to 6 (for $P_{s,t} = 0$) and 13.2 (for $P_{s,t} \to \infty$), respectively. As the sale price of water increases, the quantity of water consumption decreases proportionally from some consumption limitation and employing water conservation strategies. This follows the theory of the demand curve, which was discussed in Chapter 5.
   (b) $B_t = 0.95(0.4 + 1.5 P_{r,t}) A_t$: The treated wastewater is a portion of consumed water, and its quantity increases with an increase in the initial charge of consumers. The interesting point is that without considering any initial charge only 38% of consumed water is returned as wastewater for treatment and the maximum percentage of water treatment is 0.95, which is reached by considering an initial charge equal to 0.4.
   (c) $C_t = 0.41(A_t - B_t)$: 41% the nontreated water is lost because of problems in the water supply and distribution system.
   (d) $D_t = A_t - B_t - C_t$: The quantity of discharged wastewater to groundwater and rivers is obtained by the mass continuity equation knowing the quantity of consumed, treated, and lost water quantities in the water supply system.
   (e) $E_t = A_t - 0.5 B_t$: The required water supply of the month is obtained by using the continuity equation and considering the water consumption and the treated wastewater. It should be noted that because of the low quality of treated water it cannot be used for all water demands, so that only 50% of it can be used in urban demands.
   (f) $P_{s,t+1} = 0.83 - 0.005 B_t / A_t + P_{r,t}$: The water sale cost will decrease with the ratio of treated wastewater and water consumption at the month before, and will increase with the initial charge.
   (g) $P_{n,t+1} = 0.35 - 0.02 E_t$: The price of water production in a month will decrease with an increase of the quantity of water supply in the preceding month.
2. The system graph and associated costs are illustrated in Figure 10.9. Also, $NB_t = P_{n,t} E_t$ and $R_t = P_{s,t} A_t - P_{r,t} B_t$.
3. The results are tabulated in Tables 10.1 through 10.4.

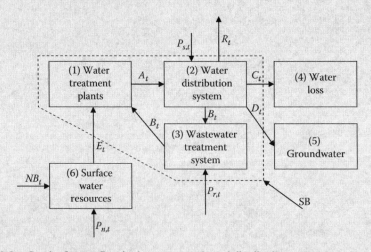

FIGURE 10.9 Graph of water flow in the water supply and distribution system.

TABLE 10.1
Variation of Water Quantity and Price for $P_{r,t} = 0.1$

t	A_t	B_t	C_t	D_t	E_t	$P_{r,t}$	$P_{s,t}$	$P_{n,t}$	NB	R
Month	$\dfrac{10^3 \text{ m}^3}{\text{Month}}$	$\dfrac{10^3 \text{ m}^3}{\text{Month}}$	$\dfrac{10^3 \text{ m}^3}{\text{Month}}$	$\dfrac{10^3 \text{ m}^3}{\text{Month}}$	$\dfrac{10^3 \text{ m}^3}{\text{Month}}$	$\dfrac{\$}{\text{m}^3}$	$\dfrac{\$}{\text{m}^3}$	$\dfrac{\$}{\text{m}^3}$	$\dfrac{10^3 \$}{\text{Month}}$	$\dfrac{10^3 \$}{\text{Month}}$
0	9.0000	3.5700	0.8800	4.5500	8.1200	0.1	—	—	—	—
1	10.5271	5.5004	2.0609	2.9657	7.7769	0.1	0.9280	0.1876	1.4589	9.2192
2	10.5285	5.5011	2.0612	2.9661	7.7779	0.1	0.9274	0.1945	1.5125	9.2139
3	10.5285	5.5011	2.0612	2.9661	7.7779	0.1	0.9274	0.1944	1.5124	9.2139
4	10.5285	5.5011	2.0612	2.9661	7.7779	0.1	0.9274	0.1944	1.5124	9.2139
5	10.5285	5.5011	2.0612	2.9661	7.7779	0.1	0.9274	0.1944	1.5124	9.2139

TABLE 10.2
Variation of Water Quantity and Price for $P_{r,t} = 0.05$

t	A_t	B_t	C_t	D_t	E_t	$P_{r,t}$	$P_{s,t}$	$P_{n,t}$	NB	R
Month	$\dfrac{10^3 \text{ m}^3}{\text{Month}}$	$\dfrac{10^3 \text{ m}^3}{\text{Month}}$	$\dfrac{10^3 \text{ m}^3}{\text{Month}}$	$\dfrac{10^3 \text{ m}^3}{\text{Month}}$	$\dfrac{10^3 \text{ m}^3}{\text{Month}}$	$\dfrac{\$}{\text{m}^3}$	$\dfrac{\$}{\text{m}^3}$	$\dfrac{\$}{\text{m}^3}$	$\dfrac{10^3 \$}{\text{Month}}$	$\dfrac{10^3 \$}{\text{Month}}$
0	9.0000	3.5700	0.8800	4.5500	8.1200	0.05	—	—	—	—
1	10.6417	4.8021	2.3942	3.4454	8.2406	0.05	0.8780	0.1876	1.5459	9.1035
2	10.6423	4.8023	2.3944	3.4456	8.2411	0.05	0.8777	0.1852	1.5262	9.1011
3	10.6423	4.8023	2.3944	3.4456	8.2411	0.05	0.8777	0.1852	1.5261	9.1011
4	10.6423	4.8023	2.3944	3.4456	8.2411	0.05	0.8777	0.1852	1.5261	9.1011
5	10.6423	4.8023	2.3944	3.4456	8.2411	0.05	0.8777	0.1852	1.5261	9.1011

TABLE 10.3
Variation of Water Quantity and Price for $P_{r,t} = 0.15$

t	A_t	B_t	C_t	D_t	E_t	$P_{r,t}$	$P_{s,t}$	$P_{n,t}$	NB	R
Month	$\dfrac{10^3 \text{ m}^3}{\text{Month}}$	$\dfrac{10^3 \text{ m}^3}{\text{Month}}$	$\dfrac{10^3 \text{ m}^3}{\text{Month}}$	$\dfrac{10^3 \text{ m}^3}{\text{Month}}$	$\dfrac{10^3 \text{ m}^3}{\text{Month}}$	$\dfrac{\$}{\text{m}^3}$	$\dfrac{\$}{\text{m}^3}$	$\dfrac{\$}{\text{m}^3}$	$\dfrac{10^3 \$}{\text{Month}}$	$\dfrac{10^3 \$}{\text{Month}}$
0	9.0000	3.5700	0.8800	4.5500	8.1200	0.15	—	—	—	—
1	10.4153	6.1841	1.7348	2.4964	7.3232	0.15	0.9780	0.1876	1.3738	9.2587
2	10.4175	6.1854	1.7352	2.4969	7.3248	0.15	0.9770	0.2035	1.4908	9.2504
3	10.4175	6.1854	1.7352	2.4969	7.3248	0.15	0.9770	0.2035	1.4906	9.2504
4	10.4175	6.1854	1.7352	2.4969	7.3248	0.15	0.9770	0.2035	1.4906	9.2504
5	10.4175	6.1854	1.7352	2.4969	7.3248	0.15	0.9770	0.2035	1.4906	9.2504

TABLE 10.4
Variation of Water Quantity and Price for $P_{r,t} = 0.25$

t	A_t	B_t	C_t	D_t	E_t	$P_{r,t}$	$P_{s,t}$	$P_{n,t}$	NB	R
Month	$\dfrac{10^3 \text{ m}^3}{\text{Month}}$	$\dfrac{10^3 \text{ m}^3}{\text{Month}}$	$\dfrac{10^3 \text{ m}^3}{\text{Month}}$	$\dfrac{10^3 \text{ m}^3}{\text{Month}}$	$\dfrac{10^3 \text{ m}^3}{\text{Month}}$	$\dfrac{\$}{\text{m}^3}$	$\dfrac{\$}{\text{m}^3}$	$\dfrac{\$}{\text{m}^3}$	$\dfrac{10^3 \$}{\text{Month}}$	$\dfrac{10^3 \$}{\text{Month}}$
0	9.0000	3.5700	0.8800	4.5500	8.1200	0.25	—	—	—	—
1	10.2000	7.5097	1.1030	1.5872	6.4451	0.25	1.0780	0.1876	1.2091	9.1183
2	10.2035	7.5123	1.1034	1.5878	6.4473	0.25	1.0763	0.2211	1.4255	9.1042
3	10.2035	7.5123	1.1034	1.5878	6.4473	0.25	1.0763	0.2211	1.4252	9.1042
4	10.2035	7.5123	1.1034	1.5878	6.4473	0.25	1.0763	0.2211	1.4252	9.1042
5	10.2035	7.5123	1.1034	1.5878	6.4473	0.25	1.0763	0.2211	1.4252	9.1042

continued

4. The variations of treated water and recharged wastewater to groundwater with $P_{r,t}$ are plotted in Figure 10.10. By increasing the $P_{r,t}$, the treated water quantity considerably increases and the recharged waste water to groundwater decreases. It should be noted that the rates of variation of these parameters are not the same.
5. The diagram of the effects of different $P_{r,t}$ on the monthly W&WWO revenue and water supplying budget is illustrated in Figure 10.11. The maximum water supplying revenue is reached when $P_{r,t}$ is 0.15, and the minimum water supply is obtained when $P_{r,t}$ is 0.25. The maximum water supplying revenue is more sensitive to $P_{r,t}$ variations in comparison with the water supply budget.

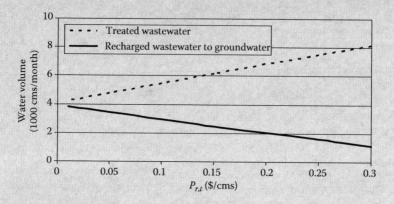

FIGURE 10.10 Variation of treated water and recharged wastewater to groundwater with $P_{r,t}$.

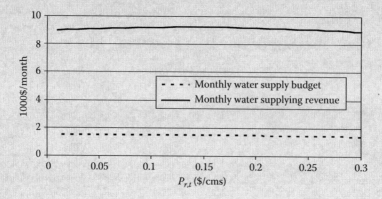

FIGURE 10.11 Diagram of the effects of different $P_{r,t}$ on the monthly W&WWO revenue and water supplying budget.

Example 10.2

Consider a city with limited resources and without any possibility for water transfer from adjacent basins. The quantity of surface and groundwater resources and their supply potential for domestic usage are deteriorating because of the urban development and water quality problems caused by wastewater, solid wastes, and industrial activities. The studies carried out on variations of population, water resources, and their supply potential for domestic usage have resulted in the following three relationships:

$$P_{t+1} = P_t + 0.9\left(\frac{W_t}{P_t} - 0.5\right) \times 10^9,$$

$$W_t = \frac{R_t P_t}{2 \times 10^8},$$

$$R_t = 10^8 + 0.1 P_t - 0.01 P_t^{1.143},$$

where P_t is the city population at year t, W_t is the water supply potential for domestic usage (MCM) at year t, and R_t is the available water resources (MCM) at year t. Assuming that the city population in the first year is 0.9×10^6, determine the maximum possible population of the city.

Solution: For determining the maximum possible population of the city, the three equations are solved based on the population factor. At first R_t (water resources relation) is replaced at W_t (water resources productivity) formulation:

$$W_t = \frac{(10^8 + 0.1 P_t - 0.01 P_t^{1.143}) P_t}{2 \times 10^8}.$$

Finally, the obtained relation is substituted into the population relation:

$$P_{t+1} = P_t + 0.9\left(\frac{(10^8 + 0.1 P_t - 0.01 P_t^{1.143}) P_t / 2 \times 10^8}{P_t} - 0.5\right) \times 10^9.$$

The population of each year is more than the previous year, so that the expression in brackets should be more than zero:

$$0.9\left(\frac{(10^8 + 0.1 P_t - 0.01 P_t^{1.143}) P_t / 2 \times 10^8}{P_t} - 0.5\right) > 0 \Rightarrow \frac{(10^8 + 0.1 P_t - 0.01 P_t^{1.143})}{2 \times 10^8} > 0.5,$$

$$10^8 + 0.1 P_t - 0.01 P_t^{1.143} > 10^8 \Rightarrow 0.1 P_t - 0.01 P_t^{1.143} > 0 \Rightarrow P_t > 0.1 P_t^{1.143}$$

$$10 > P_t^{0.143} \Rightarrow P_t < 10^7.$$

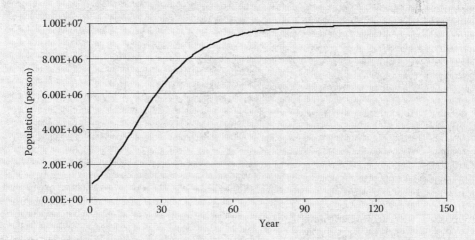

FIGURE 10.12 Variations of city population over time.

continued

Therefore the population cannot exceed 10^7 persons because of the limitations of water resources. Once the population reaches this value, it will fluctuate around it. The variations of P_t, R_t, and W_t are shown in Figures 10.12 through 10.14, respectively. In Figure 10.13, the graph is rescaled in order to have a better fit when it is drawn. The vertical axis shows the water volume in excess of 10^8 MCM. Therefore, when the graph converges to zero, it is actually holding a volume of 10^8 MCM. As is shown in these figures, after about 130 years, the curves converge to straight lines, reaching their limits.

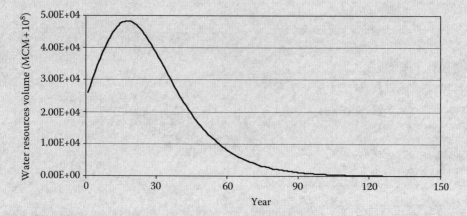

FIGURE 10.13 Variations of available water resources over time.

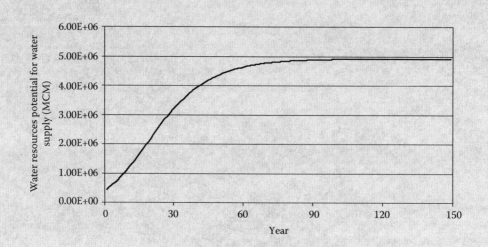

FIGURE 10.14 Variations of water supply potential for domestic usage over time.

10.3 CONFLICT RESOLUTION

Water management comprises of sharing water and resolving conflicts among users and stakeholders. A stakeholder is either directly affected by the decision or has the power to influence or block the decision. White et al. (1992) examined three sources of provoking conflicts. The first is human intervention in the environment when one or more of the stakeholders sees the activity as a disturbance to physical, biological, and social processes. The second source is disagreement over the management of the water supply at one location as it affects the use of it elsewhere. The third source is climatic variability independent of any human activity, which places more stress on the water resources. Urban areas are more subject to climate change due to added heat flux and the fate of contaminants in the urban area.

The use of water always involves an interaction between human users and soil, water, and air resources. The key indicators of the water conflicts are related to a number of issues, including water quantity, water quality, management of multiple uses, and level of national development. The political differences, geopolitical settings, and hydropolitical issues that are at stake and institutional control of water resources are the governing factors in water conflicts (Wolf, 1998).

10.3.1 Conflict Resolution Process

The conflict resolution process has been approached by many disciplines such as law, economics, engineering, political economy, geography, and systems theory. An excellent source of selected disciplinary approaches is available in Wolf (2002).

This section differentiates between traditional versus system approaches to water conflicts. Traditional conflict resolution approaches such as the judicial systems, state legislatures, commissions, and similar governmental systems provide resolutions in which one party gains at the expense of the other. This is referred to as the "zero-sum" or "distributive" solution. In water and environmental conflict resolution, a negotiation process referred to as alternative dispute resolution (ADR) is involved. ADR seeks a "mutually acceptable settlement." ADR generally move parties to what is referred to as "positive-sum" or "integrative" solutions (Bingham et al., 1994). Negotiation, collaboration, and consensus building are the key issues that facilitate ADR.

A systems approach to conflict resolution among stakeholders helped by exploring the underlying structural causes of conflict. It can transform problems into opportunities for all parties involved. Cobble and Huffman (1999) have explored the systems approach to conflict resolution in management science.

Some elements of the systemic approach have also been present in the work of Simonovic and Bender (1996), which proposes collaboration and a collaborative process with active involvement of stakeholders who agree to work together to identify problems and develop mutually acceptable solutions. Consensus building processes constitute a form of collaboration that explicitly includes the goal of reaching a consensus agreement on water conflicts. Wolf (2000) initiated an indigenous approach to water conflict reduction. Such methods include (a) prioritizing different demand sectors, (b) protecting downstream and minority rights, and (c) practicing forgiveness and compromise.

There are four steps that can be implemented in this process:

Step 1. Create the space and the intention among stakeholders to address a conflict. This can be achieved by encouraging participants to explore the source of the conflict. Stakeholders should identify critical issues, actions, and the thinking that led to the conflict situation.

Step 2. Build a shared understanding of the conflict through inquiry and the creation of a systems map. Causal diagrams like the one in Figure 10.15 may be of help. It is important to look for places of disagreement in the diagram.

Step 3. Build dialogues so that participants can directly address sources of conflict and understand their own role in it. During this process modeling "hot spots" should be considered rather than general problems.

Step 4. Create an action plan to develop and implement alternative solutions—new ways to work and interact. Participants are expected to reach agreements for trying out new solutions and behaviors.

10.3.2 A Systematic Approach to Conflict Resolution

A systematic approach plays three major roles in water conflict resolution. First, scientific investigation determines the relationship among the various components. Second, it helps us to describe the

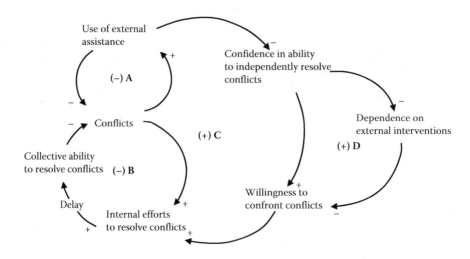

Legend: (−) Negative (balancing) feedback; (+) positive (reinforcing) feedback; +, a casual link between two variables where a change in one causes a change in the other in the same direction (one adds to another); −, a casual link between two variable where a change in one causes a change in the other in the opposite direction (one subtracts from another)

FIGURE 10.15 Advantages of a systematic approach to conflict resolution. (From Cobble, K. and Huffman, D. 1999. *The Systems Thinker* 10(2): 8–16. With permission.)

characteristics of the various components, including the physical systems, the ecosystems, affected social groups, and organizations with their preferences and modes of action. Third, it offers a means of estimating the significance of impacts not only in terms of physical quantities but also in terms of perceived impacts by the people and organizations affected.

A systematic approach's advantages are identified in Figure 10.15 using the systems language of causal diagrams. Negative (balancing) feedback loop A shows that relying on outside assistance, like hiring an outside mediator for example, to respond to conflict may serve parties involved in the short term. However, over the long term, it reduces the stakeholders' confidence in their own ability to resolve problems and willingness to confront conflicting situations as shown by the positive feedback C. Another unintended consequence is a rise in the stakeholders' dependence on external intervention, further decreasing their comfort with handling conflicting situations as indicated by the positive feedback D. The proposed approach offers a solution through building the stakeholders' conflict resolution skills (negative feedback B).

10.3.3 Conflict Resolution Models

As mentioned earlier, the presence of multiple decision makers makes the decision making process much more difficult, since they usually have different priorities and goals. Therefore reaching a compromise between the conflicting objectives is sometimes very difficult.

There are several alternative ways to solve conflict situations. One might consider the problem as a multiobjective optimization problem with the objectives of the different decision makers. Conflict situations can also be modeled as social choice problems in which the rankings of the decision makers are taken into account in the final decision. A third way of resolving conflicts was offered by Nash, who considered a certain set of conditions the solution has to satisfy, and proved that exactly one solution satisfies his "fairness" requirements. In this section, the Nash bargaining solution will be outlined (Karamouz et al., 2003).

Assume that there are I decision makers. Let X be the decision space and $f_i: X \mapsto R$ be the objective function of decision maker i. The criteria space is defined as

$$H = \left\{ \underline{u} \,\middle|\, \underline{u} \in R^I, \underline{u} = (u_i), u_i = f_i(x) \text{ with some } x \in X \right\}. \tag{10.1}$$

It is also assumed that in the case when the decision makers are unable to reach an agreement, all decision makers will get low objective function values. Let d_i denote this value for decision maker i, and let $\underline{d} = (d_1, d_2, \ldots, d_I)$. Therefore, the conflict is completely defined by the pair $(H, \underline{d})$, where H shows the set of all possible outcomes and $\underline{d}$ shows the outcomes if no agreement is reached. Therefore any solution of the conflict depends on both H and $\underline{d}$. Let the solution be therefore denoted as a function of H and $\underline{d}$: $\varphi(H, \underline{d})$. It is assumed that the solution function satisfies the following conditions:

1. The solution has to be feasible: $\varphi(H, \underline{d}) \in H$.
2. The solution has to provide at least the disagreement outcome to all decision makers: $\varphi(H, \underline{d}) \geq \underline{d}$.
3. The solution has to be nondominated. That is, there is no $\underline{f} \in H$ such that $\underline{f} \neq \varphi(H, \underline{d})$ and $\underline{f} \geq \varphi(H, \underline{d})$.
4. The solution must not depend on unfavorable alternatives. That is, if $H_1 \subset H$ is a subset of H such that $\varphi(H, \underline{d}) \in H_1$, then $\varphi(H, \underline{d}) = \varphi(H_1, \underline{d})$.
5. Increasing linear transformation should not alter the solution. Let T be a linear transformation on H such that $T(\underline{f}) = (\alpha_1 f_1 + \beta_1, \ldots, \alpha_I f_I + \beta_I)$ with $\alpha_i > 0$ for all i, then $\varphi(T(H), T(\underline{d})) = T(\varphi(H, \underline{d}))$.
6. If two decision makers have equal positions in the definition of the conflict, then they must get equal objective values at the solution. Decision makers i and j have equal position if $d_i = d_j$ and any vector $\underline{f} = (f_1, \ldots, f_I) \in H$ if and only if $(\bar{f}_1, \ldots, \bar{f}_I) \in H$ with $\bar{f}_i = f_j, \bar{f}_j = f_i, \bar{f}_l = f_l$ for $l \neq i, j$. Then we require that $\varphi_i(H, \underline{d}) = \varphi_j(H, \underline{d})$.

Before describing the method to find the solution satisfying these properties, some remarks are in order. The feasibility condition requires that the decision makers cannot get more than the amount being available. No decision maker would agree to an outcome that is worse than the amount he/she would get anyway without the agreement. This property is given in condition 2. The requirement that the solution is nondominated shows that there is no better possibility available for all. The fourth requirement says that if certain possibilities become infeasible but the solution remains feasible, then the solution must not change. If any one of the decision makers changes the unit of his/her objective, then a linear transformation is performed on H. The fifth property requires that the solution must remain the same. The last requirement shows a certain kind of fairness stating that if two decision makers have the same outcome possibilities and the same disagreement outcome, then there is no reason to distinguish among them in the final solution.

If H is convex, closed, and bounded, and there is at least one $\underline{f} \in H$ such that $\underline{f} > \underline{d}$, then there is a unique solution $\underline{f}^* = \varphi(H, \underline{d})$, which can be obtained as the unique solution of the following optimization problem:

$$\text{Maximize}: (f_1 - d_1)(f_2 - d_2) \cdots (f_I - d_I), \tag{10.2}$$

$$\text{Subject to}: f_i \geq d_i \quad (i = 1, 2, \ldots, I),$$

$$\underline{f} = (f_1, \ldots, f_I) \in H.$$

The objective function is called the *Nash product*. Note that this method can be considered as a special distance-based method when the geometric distance is maximized from the disagreement point $\underline{d}$.

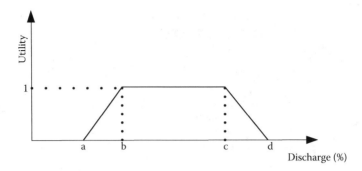

FIGURE 10.16 Utility function of the allocated water.

In many practical cases, conflict resolution is made when the decision makers have different powers. In such cases, Equation 10.2 is modified as

$$\text{Maximize: } (f_1 - d_1)^{c_1} (f_2 - d_2)^{c_2} \cdots (f_I - d_I)^{c_I}, \tag{10.3}$$

$$\text{Subject to: } f_i \geq d_i \quad (i = 1, 2, \ldots, I),$$

$$(f_1, \ldots, f_I) \in H.$$

Here $c_1, c_2, \ldots, c_I$ shows the relative powers of the decision makers. The solution to this problem is usually called the nonsymmetric Nash bargaining solution. The general format of the utility function for different water users/stakeholders is considered in Figure 10.16. In this figure, the utility function of different sectors varies between 0 and 1 when the allocated water to each sector is in the range of b–c.

Conflict resolution is a special section in game theory. See Forgo et al. (1999) for more details.

EXAMPLE 10.3

Consider a multipurpose reservoir upstream of an urban area with conflicting objectives. Discuss the general considerations in resolving conflicts over a multipurpose reservoir operation.

Solution: The objectives of the operation of the reservoir have been categorized as follows:

1. *Downstream water supply*: There are several users downstream of the reservoir with urban, agricultural, and environmental demands. The major conflicts in reservoir operation occur mostly in dry seasons when there is not enough water and the reservoir is not capable of supplying the demands.
2. *Flood damage control*: The reservoir is also intended for flood control, which is necessary for protection of cities, infrastructures, and irrigation projects from floods downstream of the reservoir. This objective is in conflict with water conservation strategies for storing water in high-flow seasons.
3. *Environmental water demand*: Supplying downstream river instream requirements is in conflict with reservoir water conservation and supplying other downstream demands.
4. *Power generation*: Power generation is another objective that has conflict with the other objectives. This is because the trends in supplying demands and power generation do not follow the same pattern. Therefore, supplying one of these demands in a specific period of time might be in conflict with supplying the others.

For resolving conflicts, first the utility functions are defined based on the priorities and favorable ranges of water supply for the agencies associated with each demand. Different stakeholders and water users and their objectives are described in the next example.

Each of these sectors has its own set of priorities for allocating water to different demands within their own line of operation and responsibilities considering the relative priority of each water use. For example, household demands may have the highest priority for the city department of water supply.

EXAMPLE 10.4

In Example 10.3, consider the following data that show the demand and the utility function for different sectors that are obtained from compromising sessions. If the relative weights of agriculture, domestic water use, industrial water use, environmental protection, and power generation are considered as 0.135, 0.33, 0.2, 0.2, and 0.135, respectively, find the optimal solution and the most appropriate water allocation scheme for a year with an average flow of 15 MCM and an initial water storage of 25 MCM.

The results of the compromising sessions are as follows:

Environmental sector: The environmental water quantity and quality in the downstream river is the main concern of this sector. Assume that the discharge of 8 MCM per month is needed for the river ecosystem; the environmental utility function for river flow (f_{env}) is formulated as

$$f_{env}(Q_{env}) = \begin{cases} 1 & \text{if } Q_{env} \geq 8 \text{ MCM}, \\ 1 - 0.33(8 - Q_{env}) & \text{if } 5 \leq Q_{env} < 8 \text{ MCM}, \\ 0 & \text{if } Q_{env} < 5 \text{ MCM}, \end{cases} \quad (10.4)$$

where Q_{env} is the instream flow downstream.

Agricultural sector: The main objective of this sector is water supply with acceptable quality to fulfill agricultural demands and reduction of the return flow removal cost. The utility of this sector related to the water supply is based on the water supply reliability. The considered agricultural water demand in this example is about 8 MCM. Therefore, the utility function for agricultural demand is assumed to be

$$f_{agr}(Q_{agr}) = \begin{cases} 1 & \text{if } Q_{agr} > 8, \\ 1 - 0.17(8 - Q_{agr}) & \text{if } 2 < Q_{agr} \leq 8, \\ 0 & \text{if } 0 < Q_{agr} \leq 2, \end{cases} \quad (10.5)$$

where f_{agr} is the utility function related to the water supply reliability and Q_{agr} is the water allocated to agricultural water.

Industrial sector: The main objective of this sector is water supply to fulfill industrial demands. The utility function of the decision makers in this sector for assessing the reliability of the industrial water supply is as follows:

$$f_{ind}(Q_{ind}) = \begin{cases} 1 & \text{if } Q_{ind} > 8, \\ 1 - 0.18(8 - Q_{ind}) & \text{if } 3 < Q_{ind} \leq 8, \\ 0 & \text{if } 0 < Q_{ind} \leq 3, \end{cases} \quad (10.6)$$

where f_{ind} is the utility function related to the water allocated to the industrial complex (Q_{ind}). *Water and wastewater sector*: The main objective of these companies is water supply with acceptable quality to domestic demands and wastewater collection and disposal. The utility function of the decision makers in this sector for water allocated to domestic water users is assumed to be

$$f_d = \begin{cases} 1 & \text{if } Q_d > 10, \\ 1 - 0.14(10 - S_d) & \text{if } 3 < Q_d < 10, \\ 0 & \text{if } 0 < Q_d \leq 3, \end{cases} \quad (10.7)$$

continued

where f_d is the utility function related to domestic water supply reliability and Q_d is the percentage of the supplied domestic water demand.

Water supply and energy production sector: The main objectives of this sector are electrical power generation and water storage for future demands. The reservoir storage utility is developed considering the minimum and maximum allowable water storage and water level over the hydropower intake in each month. The utility function of the decision makers in this sector for reservoir water storage is

$$f_{s,m}(S_{m+1}) = \begin{cases} 0 & \text{if } S_{t+1} \leq 10 \text{ million m}^3, \\ 0.0066 \times (S_{t+1} - 30) & \text{if } 10 < S_{t+1} < 25 \text{ million m}^3, \\ 1 & \text{if } 25 \leq S_{t+1} \leq 30 \text{ million m}^3, \\ 0 & \text{if } S_{t+1} > 30 \text{ million m}^3, \end{cases} \qquad (10.8)$$

where $f_{s,m}$ is the utility function related to the reservoir storage and S_{m+1} is the reservoir storage at the end of month m.

Solution: The nonsymmetric Nash solution of the example is the unique optimal solution of the following problem:

$$\text{Maximize: } Z = \prod_{i=1}^{5} (f_i - d_i)^{w_i},$$

$$\text{Subject to: } S_i + Q_i - R_i = S_{i+1},$$

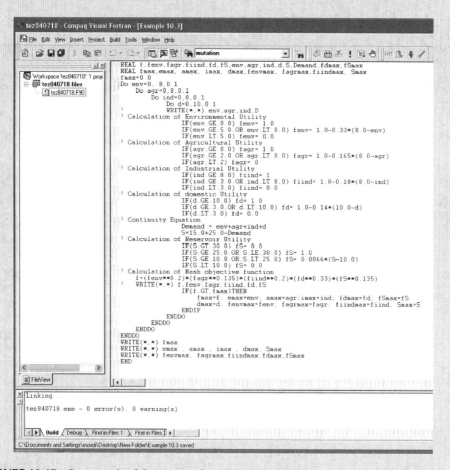

FIGURE 10.17 Source code of the program for solving Example 10.3.

where w_i is the relative weight, f_i is the utility function of each sector, d_i is the disagreement point, S_i is the reservoir storage at each stage, I_i is the inflow to the reservoir at each stage, and R_i is the reservoir release at each stage which is equal to the summation of water allocation to each community at each stage.

In order to find the optimal solution of the Nash product in this example, a Fortran code is generated for searching all feasible environments and this is shown in Figure 10.17. After running the shown program, the optimal solution of the problem is as follows:

The water allocated for environmental flow: 7.5 MCM.
The water allocated to the agricultural demand: 4.4 MCM.
The water allocated to the industrial demand: 6.1 MCM.
The water allocated to the domestic demand: 9 MCM.
The maximum value of the Nash objective function based on the above optimum values is about 0.445.

10.4 CASE STUDIES

In this section, are present the application of methodologies discussed earlier to some practical case studies, which were done by Karamouz et al. (2004) and Karamouz et al. (2005). The first case includes the application of conflict resolution in water pollution control in an urban areas. The second case includes the application of object-oriented programming models to decision making in urban areas.

10.4.1 Case 1: Conflict Resolution in Water Pollution Control in Urban Areas

In this case, a conflict resolution approach to water pollution control in the Tehran's metropolitan area, with its complex system of water supply, demand, and water pollution, is discussed. Tehran's annual domestic consumption is close to 1 billion m³. Water resources in this region include water storage in three reservoirs, Tehran Aquifer, as well as local rivers and channels that are mainly supplied by urban runoff and wastewater. The sewer system is mainly composed of the traditional absorption wells. Therefore, the return flow from the domestic consumption has been one of the main sources of groundwater recharge and pollution. Some parts of this sewage are drained into local rivers and drainage channels and partially contaminate the surface runoff and local flows. These polluted surface waters are used in conjunction with groundwater for irrigation purposes in the southern part of Tehran. Different decision makers and stakeholders are involved in water pollution control in the study area, which usually has conflicting interests. In this study, the Nash bargaining theory (NBT) is used to resolve the existing conflicts and provide the surface and groundwater pollution control policies considering different scenarios for the development of projects. Results show how significant an integrated approach is for water pollution control of surface and groundwater resources in the Tehran region.

10.4.1.1 Water Resources Characteristics in the Study Area

The Tehran plain lies between 35° and 36° 35′ Northern latitude and 50° 20′ and 51° 51′ Eastern longitude, in the south of the Alborz mountain ranges. About 800 million m³ of water per year are provided for domestic consumption by over 8 million inhabitants. More than 60% of water consumption in Tehran returns to the Tehran Aquifer via traditional absorption wells. Some of the sewage is also drained into local rivers and is used for irrigation in the southern part of Tehran (Karamouz et al., 2001).

The groundwater in the Tehran Aquifer, which is used as drinking water, is polluted due to return flow from domestic absorption wells. In order to overcome the current problems, several development plans are being investigated and implemented. These projects will change the current balance of recharge and water use in the region (Karamouz et al., 2004).

TABLE 10.5
Forecasted Water Demand, Wastewater, and Aquifer Pollution Load in Case Study 1

Year	Population	Water Demand (MCM)	Wastewater (MCM)	Recharge to the Groundwater (MCM)	Aquifer Pollution Load (Nitrate) (kg/year)
2006	8,120,402	889.184	693.564	522.744	40,251,288
2011	9,042,420	990.145	772.313	601.493	46,314,961
2016	9,956,493	1090.236	850.384	679.528	52,323,656
2021	10,786,922	1181.168	921.311	750.491	57,787,807

TABLE 10.6
Main Characteristics of the Tehran Wastewater Collection Project

Project Phase	Population Covered by TWCP	Construction Cost (Cumulative) (US Dollars)	Nitrate Load Reduction (kg/year)	Reduction in Discharged Wastewater (m³/year)
1	2,100,000	13,650,000	13,810,797	179,361,000
2	4,200,000	27,300,000	27,621,594	358,722,000
3	6,300,000	40,950,000	41,432,391	538,083,000
4	8,400,000	54,600,000	55,243,188	717,444,000
Total	10,500,000	68,250,000	69,053,985	896,805,000

Table 10.5 presents the forecasted Tehran Aquifer recharge due to return flow from domestic water use (without considering the effects of Tehran Wastewater Collection Project (TWCP)) and the corresponding nitrate load as a water quality indicator. TWCP is the most important ongoing project for solving the current quantity and quality problems of sewage disposal in the study area. The initial study of the TWCP was performed with the aid of the World Health Organization (WHO) and the United Nations (UN). Table 10.6 and Figure 10.18 present the main characteristics of the Tehran wastewater collection system.

Another completed project is a network of drainage wells to lower the groundwater table in the southern part of the city. In this project, more than 100 drainage wells have been constructed. The

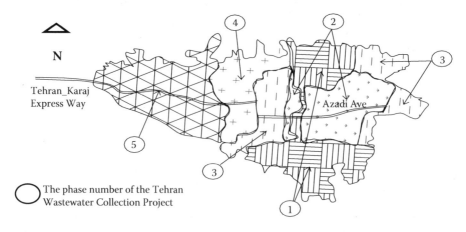

FIGURE 10.18 Area covered by different phases of the TWCP.

Urban Water System Dynamics and Conflict Resolution

pumped water is discharged to the local streams and channels and contributes to the surface water in the southern part of the city. For more detailed information about surface water resources in the study area, see Karamouz et al. (2004).

In this study, optimal operating policies for drainage wells as well as the optimal coverage of the TWCP are developed for different scenarios in the development stages, considering the objectives and utility functions of the decision makers and the stakeholders of the system.

10.4.1.2 Conflict Resolution Model

Based on the available data and the results of some brain-storming sessions, the main decision makers and stakeholders of the system and their utilities are as follows:

Department of environment: The water quality in the Tehran Aquifer is the main concern of this department. The available data show that the concentration of some water quality variables such as nitrate, total dissolved solids (TDS), and coliform bacteria deviate from groundwater and drinking water quality standards. Nitrate is considered as the representative water quality variable in this study. As the suggested nitrate concentration for groundwater is usually between 25 and 50 mg/L, the utility function of this department (f_{env}) is formulated as

$$f_{env} = \begin{cases} 1 & \text{if } c < 25\,\text{mg/L}, \\ 1 - 0.04(c - 25) & \text{if } 25 \leq c < 50\,\text{mg/L}, \\ 0 & \text{if } c \geq 50\,\text{mg/L}, \end{cases} \tag{10.9}$$

where c is the average nitrate concentration in the Tehran Aquifer.

Tehran Water and Wastewater Company: The main objectives of this company are water supply to domestic demands and wastewater collection and disposal. Considering the importance of the water supply reliability in urban areas, its most favorite range in the study area is more than 94%. Therefore, the utility function of the decision makers in this company for reliability of water supply is assumed to be

$$f_{wws} = \begin{cases} 1 & \text{if } s > 94, \\ 1 - 0.0156(s - 30) & \text{if } 30 < s < 94, \\ 0 & \text{if } 0 < s \leq 30, \end{cases} \tag{10.10}$$

where f_{wws} is the utility function related to the water supply reliability and s is the percentage of the supplied domestic water demand. As the construction costs of different phases of TWCP are different, the utility function of the Tehran Water and Wastewater Company is assumed as

$$f_{wwx} = \begin{cases} 1 & \text{if } x \leq 13.65, \\ 0.95 - 0.00366(x - 13.65) & \text{if } 13.65 < x \leq 27.3, \\ 0.85 - 0.00366(x - 27.3) & \text{if } 27.3 < x \leq 40.95, \\ 0.75 - 0.00366(x - 40.95) & \text{if } 40.95 < x \leq 54.6, \\ 0.65 - 0.00366(x - 54.6) & \text{if } 54.6 < x \leq 68.25, \end{cases} \tag{10.11}$$

where f_{wwx} is the construction cost of TWCP and x is the total allocated budget to TWCP (million $). The utility function of the construction cost of TWCP is evaluated based on the estimated cost of the different phases of the TWCP.

Tehran Health Department: The main objective of this department is to make sure that the water supplied to the domestic sector meets the drinking water quality standards. As the recommended nitrate concentration in the allocated water is less than 10 mg/L, the utility function of the decision makers of this department (f_{hd}) is assumed to be

$$f_{hd} = \begin{cases} 1 & \text{if } c \leq 10\,\text{mg/L}, \\ 1 - 0.02(c - 10) & \text{if } 10 < c \leq 25\,\text{mg/L}, \\ 0.7 - 0.028(c - 25) & \text{if } 25 < c \leq 50\,\text{mg/L}, \\ 0 & \text{if } c > 50\,\text{mg/L}, \end{cases} \quad (10.12)$$

where c is the average nitrate concentration in the Tehran Aquifer.

Water Supply Authority (Water and Wastewater Company): The main objective of this authority is to supply water in order to meet different water demands in the study area. To control the variations of the groundwater table elevation in the Tehran Aquifer and reduce the pumping costs of drainage wells, which are used to draw down the rising groundwater table, are the main utilities of this company. The most favorite range of the average groundwater table level variation in the Tehran Aquifer is about 2 cm due to its impacts on the subsidence of the buildings' foundation, especially in the southern part of the city that has experienced considerable groundwater table variations. The utility function of the Tehran Water Supply Authority for groundwater table variations is assumed to be

$$f_{wsl} = \begin{cases} 1 & \text{if } 0 \leq l < 2\,\text{cm}, \\ 1 - 0.0556(l - 2) & \text{if } 2 \leq l < 20\,\text{cm}, \\ 0 & \text{if } l \geq 20\,\text{cm}, \end{cases} \quad (10.13)$$

where f_{wsl} is the utility function related to the variation of the groundwater table level and l is the average annual variation of the groundwater table in the Tehran Aquifer. The existing pumping capacity of the drainage wells located in the southern part of Tehran is about 100 million m^3. The utility function of Tehran Water Supply Authority for the total pumping capacity of drainage wells is

$$f_{wsd} = \begin{cases} 1 & \text{if } 0 \leq d < 100\,\text{MCM}, \\ 1 - 0.033(d - 100) & \text{if } 100 \leq d < 400\,\text{MCM}, \\ 0 & \text{if } d \geq 400\,\text{MCM}, \end{cases} \quad (10.14)$$

where f_{wsd} is the utility function related to the discharge volume from the drainage wells and d is the total annual discharge from drainage wells located in the southern part of Tehran (million m^3). Considering the Nash product as the objective function, the following model is formulated for water pollution control in the study area:

$$\text{Maximize: } Z = (f_{env} - d_{env})^{w_1} (f_{wws} - d_{wws})^{w_2} (f_{wwx} - d_{wwx})^{w_3}$$
$$(f_{hd} - d_{hd})^{w_4} (f_{wsl} - d_{wsl})^{w_5} (f_{wsd} - d_{wsd})^{w_6}, \quad (10.15)$$

$$\text{Subject to: } \Delta V_A = (R + I) - (s + d), \quad (10.16)$$

$$L = \frac{S_A}{A_A} \Delta V_A, \quad (10.17)$$

$$c = \frac{c_R R + c_0 V_A}{V_A + R + I}, \quad (10.18)$$

$$x = f(A), \quad (10.19)$$

$$R = f(A, P), \quad (10.20)$$

where

$d_{env}/d_{wws}/d_{wwx}/d_{hd}/d_{wsl}/d_{wsd}$: Disagreement points of each decision maker/agency in the study area.

- S_A: Storage coefficient.
- A_A: Aquifer area.
- ΔV_A: Variation in aquifer water storage.
- I: Aquifer recharge due to infiltration from precipitation and infiltration from streams, canals, and underground inflow at the boundaries.
- R: Infiltration from absorption wells.
- c_R: Concentration of water quality variable in discharging flow.
- c_0: Initial average concentration of water-quality variable in the Tehran Aquifer.
- A: The area covered by TWCP.
- L: Variation in water table elevation.
- d: Discharge from drainage wells
- s: Allocated groundwater to domestic demand.
- P: Population.
- x: TWCP construction cost.
- w_i: The relative weight/authority of decision maker/stakeholder i.
- d_j: Disagreement point of decision makers/stakeholders corresponding to the utility function j.

The results of applying this model for water pollution control in the Tehran Aquifer are presented in the next section.

10.4.1.3 Results and Discussion

The conflict resolution model presented in this study is applied for determining the optimal area that should be covered by TWCP as well as the pumping discharge of drainage wells in the development stages of the project. In the water balance of the Tehran Aquifer, it is assumed that only the domestic water supply, the return flows from absorption wells, and the water discharge from drainage wells are variables, and the other terms in the water balance equation are constant. The discharges from the drainage wells usually control the annual groundwater table variation.

As mentioned earlier, the Aquifer is polluted mainly due to wastewater discharge through absorption wells. Based on the available data, nitrate can be considered as the water quality indicator. The average concentration of nitrate and the total nitrogen in raw domestic sewage are 90 and 5 mg/L, respectively. Close to 86% of the total nitrogen can be converted to nitrate through nitrification process. Therefore, it is assumed that the ultimate nitrate concentration in wastewater discharged to the Tehran Aquifer is equal to 77 mg/L. As presented in Table 10.6, the per capita water use and the corresponding wastewater have been estimated. Therefore, the aquifer pollution load can be estimated considering different scenarios for Tehran population and the subregions covered by TWCP.

The relative authority of utility functions $w_1, w_2, \ldots, w_6$ is assumed to be 0.18, 0.27, 0.13, 0.2, 0.13, and 0.09, respectively. Based on the form of utility function, the disagreement points of all decision makers are set to zero. The results of the proposed model for different planning horizons are presented in Table 10.7. As can be seen in Table 10.3, the population covered by TWCP will reach 7.4 million people in 2006. In this year, the nitrate concentration will be less than the drinking water standard (10 mg/L). The results show that the discharge capacity of the drainage wells should be maintained at 50 million m^3/year. Based on the derived operating policies, the quality of groundwater and also the groundwater table variation can be effectively controlled during the planning horizon.

TABLE 10.7
Results of the Conflict Resolution Model for Water and Wastewater Management—Case Study 1

	Year			
Decision Variables	2006	2011	2016	2021
Average nitrate concentration (mg/L)	9.65	9.68	12.21	9.5
Reliability of domestic water supply	94.7	95.9	94.6	95.3
Average rise in groundwater table level (m)	0.077	0.072	0.0696	0.0652
Average discharge from drainage wells (MCM)	50	50	50	50
Total cost of TWCP (million $)	48.1	53.3	54.6	62.4
Population covered by TWCP (million person)	7.4	8.2	8.4	9.6
Population covered by TWCP based on the timetable of the project (million persons)	2.5	4.6	6.7	8.8

10.4.2 CASE STUDY 2: DEVELOPMENT OF AN OBJECT-ORIENTED PROGRAMMING MODEL FOR WATER TRANSFER

In this case, a systematic approach to water transfer from southeast to southwest of Tehran, the capital of Iran, is discussed. By the construction of this channel, up to 7 m^3/s of most urban drainage water can be transferred from the east to west in the region to overcome the negative impacts of water rise in the east and to utilize the urban water drainage in the cultivated lands and for other municipal applications in the west. The channel intercepts several local rivers and drainage channels and can partially collect water from these outlets (Karamouz et al., 2001). In order to simulate the state of the system under different water transfer scenarios and water allocation schemes, an object-oriented simulation model has been developed. Using different objects, the user can change the allocation schemes for any given demand points. An economic model based on conflict resolution methods is developed for resource planning in the study area.

10.4.2.1 Area Characteristics

Distribution of lands with different potential for agricultural development may significantly affect the water supply planning in the southern part of Tehran. Figure 10.19 shows a system of local rivers, which includes Sorkheh-Hesar River, Absiah and Firouz-abad drainage channels in the east, and Beheshti and Yakhchi-abad drainage channels and Kan River in the west (Karamouz, 1998). This collection of rivers and channels drains sanitary and industrial wastewater as well as urban surface water of different parts of the region and supplies water to irrigation zones in the southern part of Tehran. As shown in this figure, there are five agricultural water supply zones, which include Fashfooyeh, Kahrizak, Varamin, Eslamshahr & Khalazir, and Ghaleno, and are named as Zones 1 through 5, respectively.

Because of shortage of water and having more suitable and cultivated lands in the western part compared with the eastern part, a water transfer channel from the east to west is proposed by Karamouz et al. (2005). This channel crosses all the local rivers and wastewater channels from the east to west in the study area before reaching the irrigation lands.

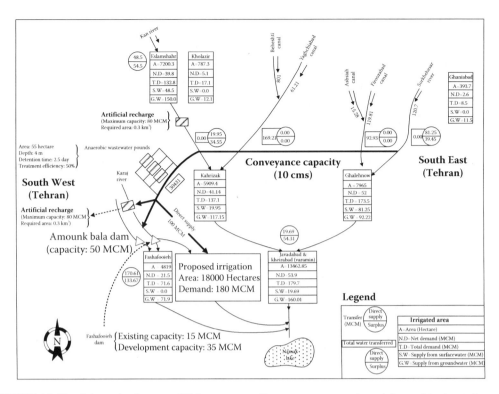

FIGURE 10.19 Schematic diagram of Tehran metropolitan area water supply and demand in the study area. (From Karamouz, M. 1998. Integrated Water Resources Management, Tehran, Final Report, Tehran Water Authority Board, Tehran, Iran. With permission.)

10.4.2.2 Conflict Resolution Model for Land Resources Allocation in Each Zone

For each irrigated area, a system dynamics model of the water sharing conflict has been developed. The dynamic hypothesis of this model is shown in Figure 10.20. In this model, it is assumed that each irrigated area has only three different types of crops including wheat, barley, and tomato. By knowing the area and water demand of each crop, the first estimate of its water allocation has been made.

The weight of each crop is defined based on their usage and price. According to the demand and price of each crop, the weights of wheat, barley, and tomato are considered to be 3, 1, and 2, respectively. Figure 10.21 shows a schematic diagram for finding land resources allocations in each zone, based on the dynamic hypothesis of Figure 10.20.

Finally, when water allocation to all crops is determined, the area in which each crop can be planted by regarding its water demand determines in converters named area crops 1, 2, and 3 at Figure 10.21. The converter available land refers to total lands that can be planted in a specific plain. Ultimate estimate of each crop area is made regarding the total available lands and crop area that has been determined before. The converter benefit represents the total benefit of the plain due to its land resources allocations.

The formulation of the above descriptions is as follows:

$$Ds_i = As_i - Al_i, \qquad (10.21)$$

$$F_i = \sqrt{Ds_i} \times w_i, \qquad (10.22)$$

$$Al_i = f\left(\frac{\sum_t F}{\sum_i \sum_t F}\right), \qquad (10.23)$$

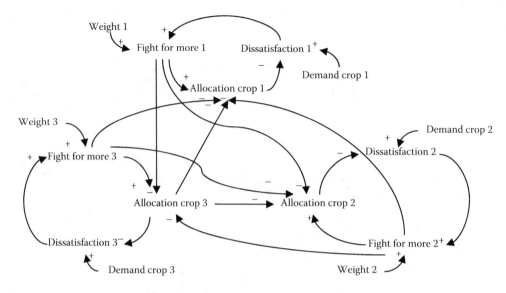

FIGURE 10.20 Dynamic hypothesis of land resources allocations in each zone.

where Ds_i is the dissatisfaction of stakeholder i, As_i is the aspiration of stakeholder i, Al_i is the water allocated to stakeholder i, F_i is the fight for more water of stakeholder i, and w_i is the weight of stakeholder i.

10.4.2.3 Results of the Conflict Resolution Model

As groundwater withdrawal is expensive in the study area, all residents prefer to use surface water for water supply. The surface water allocated to each zone and the land associated with each zone are

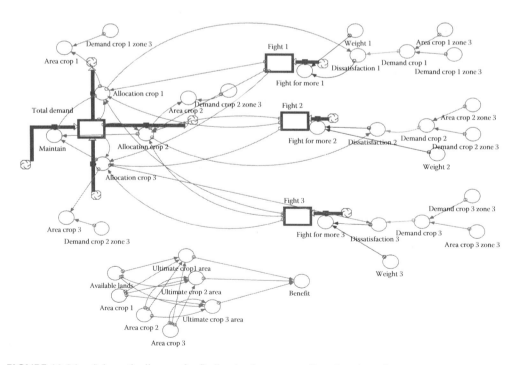

FIGURE 10.21 Schematic diagram for finding land resources allocations in each zone.

TABLE 10.8
Resource Allocation to Each Zone

Zone	Water Demand (MCM)	Allocated Water (MCM)	Area Allocated (ha)	Income (10^6 $)
Zone 1	53.8	40.3	1754.1	1.6
			2113.3	
			951.6	
Zone 2	102.9	75.8	2200.9	2.6
			1297.8	
			2410.7	
Zone 3	135.0	102.6	6299.8	5.4
			3198.8	
			3964.3	
Zone 4	112.3	79.3	2745.5	2.6
			3624.9	
			1619.6	
Zone 5	130.0	101.7	6296.4	2.7
			1280.6	
			379.0	

presented in Table 10.8. The allocated water is calculated based on each zone's demand. The relative weight of each zone is also related to the condition of the groundwater table in that zone. The land resources allocation in each zone is calculated from its weight and water demand. The relative weight of each crop is considered based on its price.

10.4.2.4 Optimal Groundwater Withdrawal in Each Zone

Owing to the shortage of surface water in the study area, the remaining water demand of each zone is allocated from groundwater storage. Figure 10.22 shows the variation of groundwater withdrawal

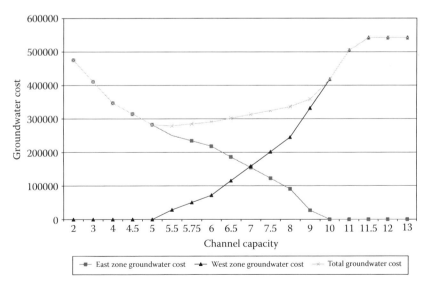

FIGURE 10.22 Variation of groundwater withdrawal cost at the western and eastern plains due to different water transfer channel capacities.

TABLE 10.9
Groundwater Depth in Different Zones

Zone	Zone 1	Zone 2	Zone 3	Zone 4	Zone 5
Groundwater depth (m)	50	14.5	21	20	8.6

cost in the eastern and western plains. The cost of groundwater withdrawal is calculated by using Equation 10.19:

$$\text{Cost} = \frac{h \times R}{\eta \times E} \quad (10.24)$$

where R is the cost of each kW/h of the power used for water purposes and is equal to $0.004, h is the depth of groundwater, E is the pumping efficiency and is assumed to be 90%, and $\eta = 0.102$.

As the groundwater table is down in the western plains, it is economical that more surface water is allocated to the western plains because of their high cost of groundwater withdrawal. The groundwater depth in different zones is shown in Table 10.9.

10.4.2.5 Sizing Channel Capacity

As discussed earlier, groundwater withdrawal is economical in the eastern plains, because the higher groundwater table in these zones is higher than that in the western plains. Hence, construction of a transfer channel, 15 km long from east to west of the study area, will be economical. For finding the optimal channel capacity, different scenarios are considered and each scenario's net benefit of agricultural lands is calculated.

The net benefit of each zone is equal to the benefit resulting from selling agricultural products (due to optimal land resources allocation) less the groundwater withdrawal costs. Figure 10.23 shows the variation of agricultural income and channel cost in the study area. As presented in Figure 10.23, the channel should be designed to carry 7 m^3 of water/s.

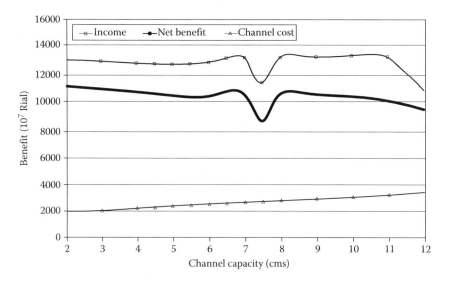

FIGURE 10.23 Variation of agricultural income and conveying capacity of the channel.

10.5 CONCLUSION: MAKING TECHNOLOGIES WORK

Many communities around the world are searching for solutions for water problems. Their ability to solve is limited by scarce resources and data, and in some instances by undefined/unsettled water governance, and by their planning schemes affected by politics and subjective decisions.

As a result, regions and municipalities are looking for quick results and planning schemes for the period of a decision maker's term in his/her office. In developing countries, there is a misrepresentation that developed countries have access to better techniques and hardware/software and therefore utilizing technologies from those countries will bring the tools that would work miracles. They ended up with black boxes with little or no adaptation capability and without the environment and experts that could utilize them. So the reliance will be shifted to in-house models.

Again, some of the in-house models soon become obsolete because of many assumptions and simplifications that had to be made, in order to build the models with scarce data and also with substandard software support systems. So the new challenge in planning, more than the development of models and tool boxes, is geared toward making transparent data and algorithms that are adaptable to regional and native characteristics and can be used to bring different decision makers and stakeholders together and create consensus. A shared vision planning is needed to make the selected techniques and allocation schemes useable, adaptable, and expandable. Participatory planning is the key to sound water management. System dynamics and conflict resolution techniques can be used to formulate the problems with the intension of bringing stakeholders into the decision making process.

PROBLEMS

1. In an unconfined aquifer system, the following agencies are affected by the decisions made to discharge water from the aquifer to fulfill water demands:

   Agency 1: Department of Water Supply
   Agency 2: Department of Agriculture
   Agency 3: Industries
   Agency 4: Department of Environmental Protection

TABLE 10.10
Utility Function Parameters for Different Sectors

Sector	a	b	c	d
Department of Agriculture	50	90	200	500
Department of Domestic Water	20	80	100	150
Department of Water Supply	80	95	100	150
Industries	60	95	100	150
Department of Environmental Protection	40	60	120	132

TABLE 10.11
Relative Weights or Relative Authority of Agencies

Agencies	Relative Weight (Case 1)	Relative Weight (Case 2)
Department of Agriculture	0.133	0.17
Department of Domestic Water	0.33	0.2
Department of Water Supply	0.2	0.23
Industries	0.133	0.1
Department of Environmental Protection	0.2	0.3

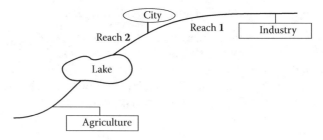

FIGURE 10.24 Components of the river system.

Department of Water Supply has a twofold role, namely, to allocate water to different purposes and to control the groundwater table variations. The decision makers in different agencies are asked to set their utility functions as given in Table 10.10.

The analyst should set the weights shown in Table 10.11 on the role and authority of different agencies in the political climate of that region:

The initial surface, volume, and TDS concentration of the water content of the aquifer are 600 km^2, 1440 million m^3, and 1250 mg/L, respectively. The net underground inflow is 1000 million m^3/year, with a TDS concentration of 1250 mg/L. Assume that 60% of allocated water returns to the aquifer as the return flow and that the average TDS concentration of the return flow is 2000 mg/L and the average storage coefficient of the aquifer is 0.06. Find the most appropriate water allocation scheme for this year using the NBT.

2. Determine the monthly water allocation to domestic, industrial, agricultural, and recreation demands in a river system shown in Figure 10.24. The average monthly river discharges upstream of the system, in a 2-year time horizon, are presented in Table 10.12.

The return flow of domestic and industrial sectors is assumed to be 20% of allocated water and the initial volume of the lake is 30 million m^3. The utility functions of different sectors are presented in Table 10.12 and Figure 10.25. The values of utilities have been normalized between 0 and 1, and the higher utility shows the higher priority of a decision

TABLE 10.12
Monthly River Discharge and Utility Functions of Agricultural Sector

Agriculture Demand (MCM/Month)	River Discharge Upstream of the System (MCM)	Month
(0,10,15,25)[a]	31	1
(0,8,12,20)	31	2
(0,10,15,25)	31	3
(0,10,15,25)	40	4
(0,10,15,25)	45	5
(0,5,10,15)	60	6
(5,10,25,50)	75	7
(15,30,60,90)	100	8
(20,35,55,90)	90	9
(25,35,60,90)	51	10
(30,40,60,90)	31	11
(25,35,60,10)	31	12

[a] The entries in the parentheses are a, b, c, and d in the utility function, respectively.

Urban Water System Dynamics and Conflict Resolution

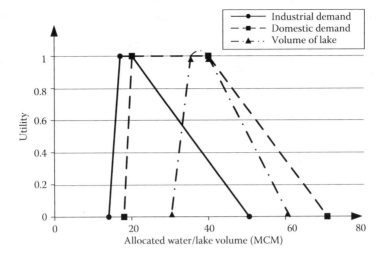

FIGURE 10.25 Utility function of different sectors for water allocated to different sectors and volume of the lake.

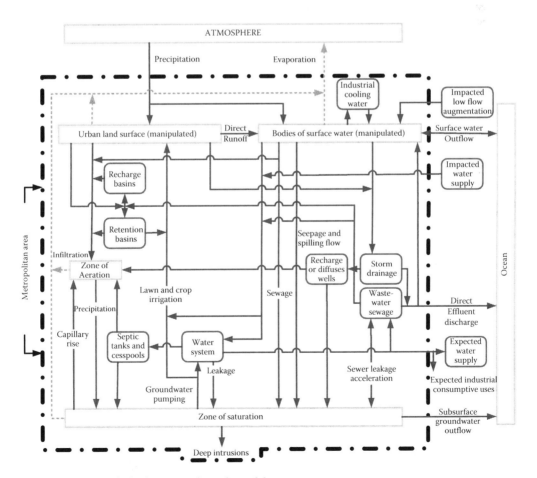

FIGURE 10.26 Typical urban water dynamic model.

maker or a sector. The shape of utilities is considered to be trapezoidal, and the array ($a, b, c,$ and d) in Table 10.12 shows the values of water allocated to the agricultural sector, corresponding to utilities of 0, 1, 1, and 0, respectively.

3. In Figure 10.26, a typical dynamic model of urban water management has been shown. Using your own city data, create this model with system dynamic simulation software. Explain the relations you have considered between different parts of the urban water system. Discuss the future of water if water demand could be decreased by 5%?

4. Considering the urban water dynamic model in Figure 10.26, what are the main parts of urban water system? How are different parts of the urban water supply affected by each other?

5. Formulate a population growth model in an urban area using an objective oriented model/software based on the following data:

    The initial value of "Population" is 50,000 people. "Population" increases at the rates of "Births" and "In Migration," and decreases at the rates of "Deaths" and "Out Migration."

    The rate of "Births" is approximately 1.5% per year. The rate of "Deaths" is also a constant percentage of the "Population." The "Average Lifetime" of a person is assumed as 67 years. The rate of "Out Migration" is equal to 8% of the initial "Population."

6. Develop an urban growth model by considering the dynamics of population and business growth within a fixed area of a city based on the following assumptions. The model should contain two sectors: the "Business Structures" and the "Population" sectors.

    A city's life cycle is characterized by a period of economic growth followed by a period of transition toward equilibrium. During the growth period, the city appears to be economically healthy: business activity is expanding and unemployment is low. The "Population" and the number of "Business Structures" grow quickly. During the transition period, conditions become less desirable and pressures arise which impede further growth. The "Construction" rate of "Business Structures" becomes smaller, and so does the rate of "In Migration" into the city. In equilibrium, when the "Population" and the number of "Business Structures" stop growing, the city suffers from problems such as high unemployment.

    The availability of "Jobs" and promise of higher incomes are the prime motivations for "In Migration" into an urban area. People tend to move to areas where employment opportunities are favorable. People tend to migrate to cities that, at a given time, are perceived to offer more job opportunities.

    When people move into or out of an area, they add to or subtract from the number of people in the area's "Labor Force." Businesses cannot ignore the "Labor Availability" in their location and expansion decisions. Readily available labor allows businesses greater flexibility in choosing employees and shortens the time necessary to fill open positions. Moreover, high "Labor Availability" tends to decrease wage competition for labor among businesses.

7. Based on Problems 5 and 6, formulate a simple model to explain growth and stagnation in an urban area using an objective oriented model/software based on the following data assumed for the urban area:

    In spite of dynamics considered in Problem 5, the rate of "In Migration" is also a product of "Population" and several other factors.

    The "In Migration Normal" of 8% is the fraction of the "Population" that migrates into the city each year in normal conditions. The "Job Attractiveness Multiplier" depends on "Labor Availability." When there are many available jobs ("Labor Availability" is less than 1 because "Labor Force" is smaller than "Jobs"), the "Job Attractiveness Multiplier" increases to values greater than 1. Then people are inclined to move to the city. When available jobs are scarce ("Labor Availability" is greater than 1 because "Labor Force" is greater than "Jobs"), the "Job Attractiveness Multiplier" decreases to values between 0 and 1. Then people will not migrate to the city.

8. Using the data given in Problem 7, map the two interacting sectors of the "Business Structures" and the "Population" in an objective oriented model/software.

9. In Problem 7, what causes the growth of "Population" and "Business Structures" of an urban area during the early years of its development? Use the structure of model developed in the Problem 7 to answer this question. How does this answer compare to the answer in Problem 8?
10. In Problem 7, what is the behavior of "Labor Availability" during the time horizon? What causes this behavior?
11. In Problem 7, how does the finite "Land Area" limits the population growth? Will all the available "Land Area" be occupied in equilibrium condition ($20\,km^2$)? Use the model to show the effects of annexing surrounding land to increase the "Land Area" to $30\,km^2$. Initialize the model in equilibrium. Use the STEP function to increase the "Land Area" after 10 years to $40\,km^2$. How does the behavior change? Will all the "Land Area" be occupied in equilibrium condition?
12. In Problem 7, does the assumption of a fixed "Land Area" invalidate the model? Most cities can and have expanded from their original areas. How does such expansion influence the results given by the model? For example, what would be the likely consequences of expanding the "Land Area" within which the city is allowed to grow? Simulate the model with a sequence of such expansions, say every 20 years.

REFERENCES

Bingham, G., Wolf, A., and Wohlgenant, T. 1994. *Resolving Water Disputes: Conflict and Cooperation in the U.S., Asia, and the Near East*. Washington DC: U.S. Agency for International Development.

Cobble, K. and Huffman, D. 1999. Learning from everyday conflict. *The Systems Thinker* 10(2): 8–16.

Forgo, F., Szep, J., and Szidarovszky, F. 1999. *Introduction to the Theory of Games*. Kluwer Academic Publishers, Dordrecht, the Netherlands.

Karamouz, M. 1998. Integrated Water Resources Management, Tehran, Final Report, Tehran Water Authority Board, Tehran, Iran.

Karamouz, M., Zahraie, B., Araghi-Nejhad, Sh., Shahsavari, M., and Torabi, S. 2001. An integrated approach to water resources development of Tehran region in Iran. *Journal of the American Water Resources Association* 37(5): 1301–1311.

Karamouz, M., Szidarovszky, F., and Zahraie, B. 2003. *Water Resources Systems Analysis*. Lewis Publishers, CRC Press, Boca Raton, FL.

Karamouz, M., Kerachian, R., and Moridi, M. 2004. Conflict resolution in water pollution control in urban areas: A case study. *Proceeding of the 4th International Conference on Decision Making in Urban and Civil Engineering*, Porto, Portugal, October 28–30.

Karamouz, M., Moridi, A., and Nazif, S. 2005. Development of an object oriented programming model for water transfer in Tehran metropolitan area. *Proceedings of the 10th International Conference on Urban Drainage*, Copenhagen, Denmark, June 10–13.

Simonovic, S. P. and Bender, M. J. 1996. Collaborative planning-support system: An approach for determining evaluation criteria. *Journal of Hydrology* 177: 237–251.

Simonovic, S. P. 2000. Tools for water management: One view of the future. *Water International* 25(1): 76–88.

White, I. D., Mottershead, D. N., and Harrison, S. J. 1992. *Environmental Systems: An Introductory Text*, 2nd Edition. Chapman & Hall, New York.

Wolf, A. 1998. Conflict and cooperation along international waterways. *Water Policy* 1(2): 251–265.

Wolf, A. 2000. Indigenous approaches to water conflict negotiations and implications for international waters. *International Negotiation* 5(2): 357–373.

Wolf, A. 2002. *Conflict Prevention and Resolution in Water Systems*. Edward Elgar, Cheltenham, U.K.

11 Urban Water Disaster Management

11.1 INTRODUCTION

Water shortages, water pollution, flood disasters, aging water infrastructures as well as the increasing demand for water, and other problems associated with water resources are becoming more devastating and the need for disaster planning and management is growing. The rapid osculation of these problems has the gravest effects on developing countries. Therefore, water disaster of a higher magnitude can affect everyone with rapid expansion of water sectors and inadequate institutional and infrastructural setups. The developed societies are also becoming more vulnerable due to their high dependence on water.

Let us put water disaster in some perspective. Water is vital for the sustenance of life and there is no substitute for this essential resource. Water is the most abundant substance on the Earth. Moreover, it is a renewable natural resource, cleansing and redistributing itself through natural cycles. However, the global quantity of freshwater is finite. Increasing demands for water from various sectors and decreasing water availability due to overuse, pollution and inefficient water management, aging infrastructure, and accidental and intentional widespread contamination lead to local and regional water disasters.

During the 1980s, improvements in water supply and sanitation facilities made some progress; however, water-related problems grew more serious due to rapid urban and industrial growth. In the 1990s, various water-related problems were recognized as urgent global issues. It was hoped that these problems can be solved through continued efforts by the governments and international organizations to put into practice the declarations and action plans, in facing the water disaster. Last year, some of the most devastating water disasters hit Southeast Asia, North America, and societies suffering from global and regional conflicts.

Natural causes are the source of many disasters. Natural disasters cause suffering to people around the world. This can be particularly devastating for those living in developing countries, where the social infrastructure is not fully in place. This further exacerbates poverty. In recent years, devastating floods, droughts, and hurricanes have caused regional disasters. At certain moments, it seems we have become more vulnerable to water disaster due to the complexity of our societies and urban settings as well as our means of transportation and communication. Furthermore, the world is getting more hostile and regional conflicts with political and economic motives have devastated the daily life of people in some regions. Climate change has also intensified flood and drought impacts all over the world with an adverse effect on public water services and utilities.

To demonstrate the global water disasters, some examples of recent water-related disasters could be of some value:

- Serious shortages of safe drinking water continue to exist in Iraq, especially outside of Baghdad and Basra despite progress in recent months to get water treatment and pumping stations back into service. In these and other cities, water pressure and chlorination levels have been inadequate for essential needs.

When allied bombing destroyed the national electrical grid, water treatment plants ceased to function throughout most of Iraq. Iraqis in most cities turned to highly polluted rivers as the principal source of water for drinking, bathing, and for the disposal of untreated sewage, leaving many Iraqis vulnerable to lethal waterborne diseases (Iraq Health, 2006).

> Hurricane Katrina smashed the entire neighborhoods in New Orleans, Louisiana, in the United States and killed at least 185 people. This event shows that even developed countries are subject to water disaster that has not been experienced before.

In this chapter, the nature of a water disaster and the factors contributing to the formation and extent of changes caused by a disaster will be discussed. The notion of water demand as a "load" and the supply resources as a "resistance" will be discussed in the context of reliability and risk-based design. The elements of uncertainty and how risk management can be coupled with disaster management are presented. Then, dealing with a disaster that requires strategic decisions involving many agencies and the public as a whole is discussed. The concept of regional vulnerability and water allocation based on minimizing the impacts on vulnerable areas is also presented. Finally, disasters and technology and utilizing different tools as well as a list of some actions to curb the consequences of disaster are discussed. This chapter is divided into five sections: First, an introduction to UWDM is presented. Then the planning process for UWDM is presented, followed by situation analysis and disaster indices. Finally, guidelines for UWDM and the conclusion are given.

11.2 SOURCES AND KINDS OF DISASTERS

Many regional problems are the direct result of the regional climates, geography, and hydrology. Others are as a result of human activities and mismanagement. Traditionally, water disaster has been attributed to the quantity of water that causes droughts and floods. A drought is a shortage of water beyond a level that a water supply/distribution system can deliver to meet the basic demands. Expansion of urban, agricultural, and industrial sectors has introduced new concerns about water distribution and supply system safety and their structural integrity.

Floods are the result of excess runoff, which can increase depending on various factors such as intensity of rainfall, snow melt, soil type, soil moisture conditions, and land use and land cover. Runoff from rural and urban areas is generally a response of excess water after the processes of infiltration and evapotranspiration have taken place. Obviously, urban regions have more impervious surfaces and the surface runoffs are more intense. So they are more prone to flooding.

Drought and flood severity and their impacts on urban areas are often more significant in undeveloped parts of the cities. Inadequate recharge of groundwater also contributes to the shortage of water at the time of low flows/precipitation. Water supplies in urban areas are more fragile during drought events because of limited storage facilities and high water demands variability.

In addition to the quantity of water, its quality might also be the cause of a water disaster. In many developed countries, issues such as widespread contamination of water mains and water supply reservoirs by accidental or intentional acts have been the focus of many activities that created a new subject of water security in public and private agencies dealing with water. Developing countries are also subject to water disasters of increasing magnitude due to their complex infrastructure as well as their rapid industrial and urban growth.

11.2.1 Drought

When we think of water disasters, we tend to think of cyclones, hurricanes, storms, and floods. Less attention is given to droughts, which is also classified as a disaster or even the most long-term disaster with intense social and environmental predicaments.

What is a drought? It is viewed as a sustained and regionally extensive occurrence of below average or any low-flow thresholds on natural water availability (Figure 11.1). The more precise definitions for specific areas of concern that are most commonly used are

Urban Water Disaster Management

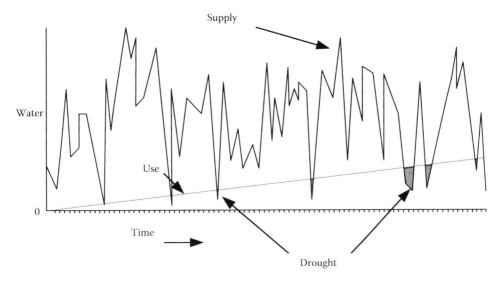

FIGURE 11.1 Graphical definition of drought.

- Meteorological or climatic drought is a period of well-below-average or normal rainfall that spans from a few months to several years.
- Agricultural drought is a period when soil moisture is inadequate to meet the demands for crops to initiate and sustain plant growth.
- Hydrological drought is a period of below-average or normal stream flow where rivers dry up completely and remain dry for a very long period or where there is significant depletion of water in aquifers.
- Socioeconomic drought refers to the situation that occurs when physical water shortage begins to affect people due to inability of the transfer and distribution systems to perform.
- Among different types of droughts, hydrological drought has more severe effects on urban areas because of the high rate of water use in these areas and significant social and economic issues related to water shortages in urban areas.

Long-term demand measurement in the spirit of disaster management involves land use and conservative programs that can promote the efficient use of water in normal water condition as well as during extreme events such as floods and droughts. The effect of droughts on public water supplies necessitates cooperation among water users and local, regional, and national public officials. Little efforts have been invested in long-term planning of drought in many semiarid and arid countries. Developing a national or regional drought policy and an emergency response program is essential for reducing societal vulnerability and disasters and hence reducing the water disaster impacts. Recently, resiliency in the context of drought management has received attention (Karamouz et al., 2004).

Strategies for drought management focus on reducing the impact of droughts by regional analysis to determine the most vulnerable part of a region. Figure 11.2 shows the vulnerable areas of Lorestan Province for a drought with 10 years return period. The area with dark shadows requires more attention if the region is facing a 10-year drought. Drought planning and management must take into account not only the risk of potential economic and social (psychological) damages resulting from droughts, but also the ecological and economic costs and benefits of exercising alternative options such as aquifer recharge and capacity (physical) building (filling the reservoir) schemes, and helps us into determining how these options can reduce potential damages and provides a way of facing future droughts (Karamouz, 2002).

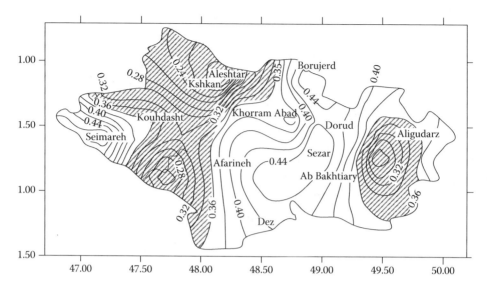

FIGURE 11.2 Vulnerable areas of Lorestan Province in Iran in a drought with 10 years return period.

11.2.2 Floods

Floods continue to be the most destructive natural hazard in terms of short-term damages and economic losses to a region. According to the Federal Emergency Management Agency (FEMA) of the USA, floods are the second most common and widespread of all natural disasters (drought is first, globally). Within the USA, an average of 225 people are killed and more than $3.5 billion in property damage is caused by heavy rains and flooding each year (FEMA, 2008).

In addition to water supply and drainage needs, urbanization also creates new demands in flood management. Increasing urbanization decreases infiltration capacities and water retention characteristics of the natural land, which makes flood runoff accumulate in rivers in a short time. The increasing quantity as well as the short duration of drainage produces high flood peaks that cannot be contained within the urban streams. As flood frequencies increase, the total economic and social damages in the urban areas tend to make significant impacts on the national economies (Herath, 2008).

Floods convey both dangers and benefits to people in the developing nations. Floods account for about 40% of all deaths caused by natural disasters, most of which are in the developing nations. For example, 3.7 million people were killed in a 1931 flood on China's Yangtze River. In 2000, four of the world's five largest natural disasters were floods.

A common scenario in an uncontrolled urbanization is that floodplain occupation by the population takes place, in a sequence of years with small flood levels. When higher flood levels return, damage increases and the public administrations have to invest in population relief. Structural solutions have higher costs and it is feasible only when damages are greater than their development or due to intangible social aspects. Nonstructural measures have lower costs, but there are some difficulties in their implementation because they are not politically attractive.

Flooding is caused primarily by hydrometeorological mechanisms, acting either by a single factor or by a combination of different factors. Most of the urban areas are subject to some kind of flooding after spring rains, heavy thunderstorms, or winter snow thaws.

Floods can be slow or fast rising but generally develop over a period of hours and days. Flash floods move at very fast pace and can roll boulders, tear out trees, destroy buildings, and obliterate bridges. Walls of water can reach the heights of 3–6 m and, generally, are accompanied by a deadly cargo of debris. Flood damage causes both direct and indirect costs. Direct costs reflect immediate losses and repair costs as well as short-term costs such as flood fighting, temporary housing, and administrative assistance. By contrast, indirect costs are incurred in an extended period following a flood and include loss of business and personal income (including permanent loss of employment), reduction in property values, increased insurance costs and loss of tax revenue, psychological traumas, and disturbance to ecosystems.

Flood-related property losses can be minimized by making flood insurance available on reasonable terms and encouraging its purchase by people who need flood insurance protection, particularly those living in flood-prone areas. The program identifies flood-prone areas in the urban area and makes flood insurance available to property owners in communities that participate in the program and meet floodplain building standards to mitigate flood hazards. Flood hazards are mitigated through a variety of grant programs.

Measures can be taken, such as the repair of river embankments to make rivers capable of disposing of rainfall up to the conveyance capacity of the river and the improvement and expansion of reservoirs, diversion channels, and sewer systems to reduce/eliminate the danger of floods. Moreover, in order to promptly and correctly counter local changes in precipitation and the danger of high tides, the effective use of the information transmission system of a comprehensive flood prevention program is needed.

For example, concurrently with the expansion and improvement of rivers and sewerages, the Tokyo Metropolitan Government is constructing facilities to store rainwater and make it permeate the ground, particularly in public spaces such as roads and parks and large-scale private facilities, for the purpose of reducing the flood prevention load on urban systems. There is also a subsidy system to help individuals improve the drainage of their home lots (Tokyo Metropolitan Area, 2006).

11.2.2.1 Principles of Urban Flood Control Management

The reliable experience in flood control of many countries has now led to some main principles in urban drainage and flood control management as follows:

- Flood control evaluation should be done in the whole basin and not only in specific flow sections.
- Urban drainage control scenarios should take into account future city developments.
- Flood control measures should not transfer the flood impact to downstream reaches, giving priority to source control measures.
- The impact caused by urban surface washoff and others related to urban drainage water quality should be reduced.
- More emphasis should be given to nonstructural measures for floodplain control such as flood zoning, insurance, and real-time flood forecasting.
- Management of the control starts with the implementation of Urban Drainage Master Plan in the municipality.
- Public participation in the urban drainage management should be increased.
- The development of the urban drainage should be based on the cost recovery investments.

These principles have been applied in most of the developed countries. The urban drainage practices in most of the developing countries do not fulfill these principles. The main causes are the following:

- Urban development in the developing countries' cities occurs too fast and unpredictably. Usually, the tendency of this development is from downstream to upstream, which increases the damage impacts.
- Urbanization in preurban areas is usually developed without taking into account the city regulations. This form of urbanization is as follows:
  - Unregulated developments: In preurban areas of big cities, the real estate is low priced. Regulation of this area requires investments that almost equal the price of the land.

As a consequence, private land owners develop urbanization without the infrastructure, selling it to the low-income population.
- Invasion of public areas (such as public green areas) that were planned in Urban Master Plan for future parks, public construction, and even streets; owing to low-income conditions (the homeless) and slow decision making by public administration, these developments are consolidated receiving water and electricity.
- Preurban and risk areas (floodplains and hillside slope areas) are occupied by low-income population without any infrastructure. Spontaneous housing developments in risk areas of the Humid Tropics' cities are prone to flooding: Bangkok, Bombay, Guayaquil, Lagos, Monrovia, Port Moresby, and Recife; hillsides prone to landslides: Caracas, Guatemala City, La Paz, Rio de Janeiro, and Salvador.
- Municipality and population usually do not have sufficient funds to supply the basic needs of water, sanitation, and drainage needs.
- Lack of appropriate garbage collection and disposal decreases the water quality and the capacity of the urban drainage network due to filling. Desbordes and Seravat (1988) stated that in some African countries there is no urban drainage and when system drainage exists it is filled with garbage and sediments.
- There is no prevention program for risk area occupation and when the flood occurs nonreturnable funds are given to the local administration to cope with the problem, without any requirement of future prevention programs.
- Population, States and Counties administrations do not have enough knowledge on how to deal with floods using the above principles.
- Lack of institutional organization in urban drainage at a municipal level such as regulation, capacity building, and administration. In Asian cities, there is a lack of comprehensive project organization and clear allocation of responsibilities; adequate urban land use planning and enforcement; and capability to cover all phases and aspects of technical and nonstructural planning (Ruiter, 1990).
- In cases where a separate system is applied to solid waste and is dumped into storm sewers. This is a situation that also occurs in some developed countries.
- When there are unrealistic regulations for urban occupation related to social and economic conditions, the owner uses different procedures in order not to comply.

11.2.3 Widespread Contamination

Traditionally, water disasters have been attributed to droughts or floods. Expansion of urban, agricultural, and industrial sectors has introduced new concerns about urban water systems' safety and about their structural integrity. In many developed countries, issues such as widespread contamination of water mains and water supply reservoirs by accidental or intentional acts have been the focus of many activities that created a new subject; water security in public and private agencies dealing with water. Recent, a truck carrying MTBE (a highly toxic petrochemical substance) rolled over near a water supply reservoir of a major city in the western part of Iran and contaminated about 100 MCM of water. Water security, due to this terrorist act, has the highest priority in water resources planning in North America and some European countries.

Pollution is one of the major contributors to the urban water problems. It affects rivers, lakes, and groundwater and much of it comes from poor sanitation and uncontrolled solid waste disposal. Industrial effluent is another major source of pollution. Pollution leads to a variety of health problems. Diarrhea, for example, resulting from the lack of water and poor sanitation, causes 3.3 million deaths a year, mainly of children under the age of five. This is equivalent to a child dying every 10 s, or, as someone has said, a jumbo jet crashing every hour. Pollution from all sources, large- and small-scale industry, informal sector or domestic, must be reduced to protect surface water and groundwater.

Urban stormwater runoff is often a major source of pollution affecting the quality of receiving waters. Pollutants accumulate on the land surface of an urban area during dry periods. Typical pollutants include gasoline and oil from motor vehicles, sediment from construction activities, chemicals from lawns, litter and solid waste from people, and fecal droppings from animals. These pollutants are washed off the land surface by rain, discharged quickly into the drainage system, and ultimately end up in the receiving waters.

Stormwater flows from impervious surfaces (e.g., roofs, car parks, and roads) are removed from urban area instreams, pipes, and channels to prevent flooding. At times, stormwater infiltrates into wastewater flows, resulting in CSOs and pollution of the environment as well as extra loads at WWTP (Akan and Houghtalen, 2003).

Many cities lack effective storm drainage systems, and ill-planned construction closes off natural water courses. In Algiers, massive flooding in November 2001 caused 800 deaths (700 in densely populated neighborhoods), where a natural stormwater drainage channel in the city had been converted to a paved road. Overflowing of clogged storm drains and sewers during high rainfall is projected to become a greater source of disasters in major cities than river flooding. The key institutional issue is that drainage has no clear constituency until major problems occur. Local governments may become motivated to act on drainage issues when flooding affects the business district, as in Cabanatuan in the Philippines, where the local business community put pressure on the mayor to invest in drainage infrastructure. For years, in Kampala, the local authorities had neglected to protect past investments in the Nakivubo channel from settlement encroachment and obstruction with solid waste. There, and in communities in Ethiopia, recent reforms expanding local democracy have raised the profile of drainage as a priority for public expenditure.

11.2.4 SYSTEM FAILURE

System failures are uncertain events due to a combination of factors. The failure can lead to supply interruption, physical damage, and/or unacceptable water quality. The reaction to a failure is repair, replacement, and/or some other types of protective actions. In order to quantify the uncertainty of failure, there is a wide range of failure models that can be used. The output of such models is the distribution function and/or probabilities of outcomes (e.g., failures). The state of the network can be assumed either through some linear models (sometimes nonlinear) or through monitoring conditions. However, due to high costs, this is primarily considered only if the consequences of failure are very high. The question is how risk assessment concerns the costs and consequences and other sorts of outcomes. In doing so, there is a choice to include social costs, which are third-party costs. A line needs to be drawn regarding cost/consequences far into the future.

Urban water systems are the natural, modified, and built water systems that exist in small communities, towns, and cities. The natural system includes the network of streams, rivers, groundwater, wetlands and estuaries, and coastal and marine areas. The built system includes the network of water supply reservoirs, water supply plants, pipes, concrete channels, drains, WWTP, and outfalls. The modified systems are a combination of the other two. The functions provided by the built system of water supply, wastewater, and stormwater infrastructure are commonly referred to as water services.

Urban water systems includes the following:

- Natural water systems
  - Streams
  - Rivers
  - Wetlands
  - Estuaries
- Built water systems
  - Dams
  - Pipes
  - Treatment plants
  - Outfalls

The natural and built systems are interconnected and interact in both positive and negative ways as part of a much larger urban ecosystem. For example, the natural water system replenishes water supply reservoirs and aquifers, and streams and wetlands receive and can process stormwater from urban areas.

However, water taken in times of drought, or discharges from wastewater systems, intensifies the adverse environmental impacts on the natural water systems and the communities. Urban water services, particularly sewers and stormwater services, transport and redirect nutrients and persistent pollutants such as heavy metals and organic chemicals. These infrastructural functions can impact significantly the natural material flow processes of urban ecosystems and adjacent rural ecosystems.

11.2.5 Earthquakes

Somehow an urban water disaster (UWD) is a consequence of a more destructive disaster such as earthquakes or tsunamis. All the water systems are subject to seismic activities and are highly vulnerable due to nonelastic design of components. Drawing lessons from past earthquakes and applying them to regions with a similar setup helps in the planning for earthquakes and tsunami-type events, which result in flooding and/or water supply and distribution collapses.

11.3 WHAT IS UWDM?

UWDM is a systematic process for the reduction of water disaster impacts on social, economic, and environmental systems. Here are a few reasons why many decision makers argue that an UWDM is necessary to be implemented:

- Water disasters are increasingly under pressure from population growth, economic activity, and intensifying competition for the water among users.
- Water withdrawals have increased more than twice as fast as population growth and currently one-third of the world's population live in countries that experience medium to high water stress. The higher the water stress, the higher the vulnerability to disaster-scale water shortages.
- Pollution is further enhancing water scarcity by reducing water usability.
- Shortcomings in the management of water, a focus on developing new sources rather than managing existing ones better, and top-down sector approaches to water management result in uncoordinated development and management of the disaster.
- More and more development means greater impacts on the environment.
- Current concerns about climate variability and climate change require improved management of water disasters to cope with more intense floods and droughts.

11.3.1 Policy, Legal, and Institutional Framework

Attitudes are changing as officials are becoming more aware of the need to manage disasters efficiently. Officials also see that the construction of new infrastructure has to take into account environmental and social impacts and the fundamental need for systems to be economically viable for maintenance purposes against disaster. However, they may still be inhibited by the political implications of such a change. The process of revising water policy is therefore a key step, requiring extensive consultation and demanding political commitment.

Water legislation converts policy into law and should

- Clarify the entitlement and responsibilities of users and water providers at the time of disaster as well as the chain of command.
- Clarify the roles of the municipalities to notify commend and illegal actions at the time of disaster.

- Formalize the transfer and transport of water from adjacent areas at the time of water disaster by different means.
- Provide legal status for water management institutions of government and water user groups.
- Ensure equitable use of water at the time of disaster.

Bringing some of the principles of UWDM into a water sector policy and achieving political support may be challenging, as hard decisions have to be made. It is therefore not surprising that often major legal and institutional reforms are only stimulated when serious water management problems have been experienced.

For many reasons, the governments of developing countries consider water disaster planning and management to be a central government responsibility. However, public and voluntary participation is the key issue in disaster management. This view is consistent with the international consensus that promotes the concept of government as a facilitator and regulator, rather than an implementer of projects even at the time of disaster. The challenge is to reach mutual agreement about the level at which, in the same specific instance, government responsibility should cease, or be partnered by autonomous water services management bodies and/or community-based organizations. The role of government and national guards should be converted from legislations into laws in order to allow them to take over basic operation and management of facilities and equipment at the time of disasters.

In order to bring UWDM into effect, institutional arrangements are needed to enable

- The functioning of a consortium of stakeholders and water users to get involved in decision making, with representation of all sections of society in planning for disaster management
- Water disasters management based on levels of disaster and how widespread it is
- Organizational structures at provincial and local levels to enable decision making at the lowest appropriate level
- Government to coordinate the national management of water disasters across water use sectors

11.4 SOCIETAL RESPONSIBILITIES

Protecting people's lives and property from disasters and keeping social assets safe are the basic issues for the development of cities and their activities. Thus, an important problem is to make the cities more resistant to different types of disasters. In order to create a city for the twenty-first century where inhabitants can feel secure, all parties need to come together to promote disaster-resistant urban development through the mutual integration of livability. Water-related disasters are of significant importance because they include every facet of our urban life with high social and economic drawbacks.

Urban developers are instructed to take safety measures and tighten regulations on urban and structural standards to prevent the danger of the failure of concrete tunnels, tanks, conduits, and other water-related substructures due to seismic loads and aging infrastructures. Pipe connections as well as the impact of water mains breaks in a district should be assessed and their potential threat to the gas and electricity lines should be evaluated. In particular, tighten regulations and guidelines on the urban planning and structure of buildings, including those on the design of the building and the minimum open spaces for passage of water mains, tributaries, pumping stations, and service connections as well as shafts and conduits for distributing water through neighborhoods and buildings.

11.5 PLANNING PROCESS FOR UWDM

11.5.1 Taking a Strategic Approach

The recognition that water disaster in most parts is a result of failure of water management systems leads to long-term planning. The impact is more obvious at the time of water disasters. What we expect to achieve is a UWDM plan, endorsed and implemented by government to overcome the additional stress imposed on systems that normally fail to meet normal levels of water demand and services in many municipalities. In the process, the stakeholders and politicians will become more informed about water issues, nature of disasters, and the importance and benefits of addressing disaster management. The plan may be more or less detailed depending on the present situation in the country but will identify longer-term steps that will be required to continue along a path to security against system failure and natural events such as droughts and floods.

Being strategic means seeking the solutions that attack the causes of the water problems rather than their symptoms. It takes a long-term view. Understanding the underlying forces that cause water-related problems helps in building up a shared water vision and commitment to make that vision come true. In that sense, a strategy sets the long-term framework for incremental action that moves toward a basis for disaster prevention and formatting of the UWDM principles.

11.5.2 Scope of the Strategy Decisions

A policy (or a problem statement) is often the starting point being a statement of intent. The essential difference is to translate the policies into a strategy. A strategy seeks to meet certain goals through specific actions and investments. In a strategy, the possible disasters and options to reach the goals of minimizing their unsuitable effects have to be assessed and a plan is devised for managing the probable disasters with expected consequences.

The vision is water security across the board. It involves educating the public about disasters and how ready the system is to face disasters and what is expected from them. Calm, cooperative, and careful handling of disasters brings people together rather than separate them with anguish, sense of not belonging and fear of isolation and disparity at the time of water disaster.

The strategic goals describe how the vision might be achieved. Each goal should cover a given issue (problems or certain opportunities may arise from a disaster), address the main changes required to make the transition to minimize unsuitable effects in a way that is broad enough to encompass all aspects of the issue and ensure all relevant stakeholders. The strategy should cover sufficient goals to address the main economic, social, and environmental concerns of urban water disaster management.

Targets for each goal describe specific and measurable activities, accomplishments, or thresholds to be achieved within a given duration after a disaster's occurring. These form the core of any action plan, and serve to focus disasters and guide the selection of options for action. Reaching such goals will often require a legal and institutional reform supported by specific management skills and instruments.

Institutional roles cover the roles, partnerships, and systems required to implement the strategy. This may include linkage between the UWDM plan and other strategic plans and between plans at different spatial levels: national, subnational, local, or for different sectors or geographical regions. It would identify which institutions are responsible for which parts of the strategy action plan. It might also signal a rationale for streamlining institutions (especially where responsibilities overlap or conflict).

The action plan is developed from the outcome of the strategy. There is an inseparable link that refers the action plan back to strategy as further assessment and adjustment takes place.

11.5.3 UWDM as a Component of a Comprehensive DM

In a long-term strategic plan, a UWDM is a component of a comprehensive plan for lifeline disasters, which turns into a comprehensive regional and national plan for disaster management. Activities and guidelines of UWDM should be synchronized with the plans provided to compensate for the impacts of large-scale disasters such as an earthquake.

A disaster-resistant city (DRC) with targeted districts covers areas that are close to earthquake centers and seismic activities and their outskirts in the adjacent cities. The plan for DRC should aim at completion over a medium time horizon (say 10 years), but in districts in urgent need of development, completion within a shorter time period is required.

Designated in this plan are closely redesigned/reinforced housing districts, particularly within high-risk districts designated as key development areas where development of infrastructure and restoration in those areas must be pursued to create disaster-resistant urban areas through multilayered, intensive implementation of earthquake design standards. This includes all activities, especially water mains above the ground and underground tanks. Work in key districts should be actively promoted through multilayered, concentrated work according to a development plan.

The minimum size of a disaster-prevention base should be determined so that adequate protection can be provided at the time of disaster. Disaster-prevention bases should be located in districts where there is much land space that can be purchased, such as factory sites and large factories that can be moved, as well as public lands.

Swift and systematic urban restoration following a major earthquake is an extremely important issue that will decide the future course of the city and the lives of the people who live in the area. Therefore, it is necessary to study plans and procedures to rebuild the city before another major disaster strikes.

A guideline for the procedures and the planning of the administrative process must follow to ensure the swift and systematic revival of urban areas. Urban revival simulation training should be carried out with a carefully designed manual as a guide, with the aim of strengthening the coordination and deploying new methods (Tokyo Metropolitan Area, 2006).

11.5.4 Planning Cycle

Planning is a logical process, which is most effective when viewed as a continuous cycle. This cycle entails a vision and work plan for UWRM surrounded by situation analysis, strategy choice through implementation, and evaluation process (CIDA, 2005).

The triggers to start a planning process may be internal or external or a combination of both. However, once it is agreed that improved management and development of a water disasters plan is important, the question immediately arises as to how we get a plan in place to achieve it. That is the purpose of this section.

UWDM planning requires a team to organize and coordinate efforts and facilitate a series of actions before, during, and after the water disasters. An important starting point for government commitment is an understanding of UWDM principles and ensuring the allocation of water at least at a minimal level at the time of disaster.

Commitment from the public is necessary as they are the ones who strongly influence handling the disaster, through voluntary participation, and change in water use habits. Thus planning requires recognizing and mobilizing relevant interest groups, stakeholders, volunteers, and the decision makers, despite their multiple and often conflicting goals in the use of water at the time of disaster. Politicians are a special group of stakeholders as they are both responsible for approving a plan and are also held accountable for its success or failure. Thus a UWDM contains

- Management of the process
- Maintaining political commitment
- Ensuring effective public participation
- Creating awareness of UWDM guidelines to be followed by different players in disaster management
- Possible involvement of National Guard or Army at the time of disaster

11.6 WATER DISASTER MANAGEMENT STRATEGIES

Possible solutions could be sought at the same time or immediately after defining the problems. Such solutions need to be analyzed, considering the requirements, the advantages and disadvantages involved, and their feasibility. Establishing the goals for the UWDM plan is important at this stage now that the extent of the problem and the hurdles to be faced are known. For each goal, the most appropriate strategy is selected and assessed for feasibility as well as its conformity to the overall goal of disaster management. The scope for technical and managerial action is considerable given the complexity of the water sector.

A national vision that water bridges between many cultural, social, and economic ties and incentives combined with a national drive for resource management and protection on conversion captures many common grounds about the use and management of water in a country. A vision provides direction to the actions on utilizing water and protecting it with high or low flow consequences and, in particular, guides the planning process. The vision may or may not be translated into water policies but would be expected to provide a guideline toward sustainable supply of the demands. The vision is particularly important in facing water disasters. It goes beyond quantitative and qualitative aspects of water and captures the heart of a nation and city and a community faced with water disaster. The question to be answered is whether a vision for disaster management is tied to with a water vision on integral water resources management. The answer could be found in the complexity of water-related activities of social vulnerability of society on water. When a city of 5 million habitants receives just enough water to meet the domestic demand in a normal year, and this city has exercised an ad hoc planning for water supply and waste disposal to the stream flows, urban flood control, and distribution system without enough integration, then the water vision shifts from opportunity seeking and preventive maintenance to solving and overcoming problems.

A change of paradigm is also in shifting from supply management to demand and shortfall managements, and it has made a drastic shift in our vision and actions on disaster management. In order to define the actions needed to reach such a change in vision, it is important to know the existing situation and to consult with stakeholders and various government entities to understand the needs for contingency plans for water disasters.

On the basis of the vision for disaster prevention and management, the situation analysis and the water disasters strategy can be prepared. Several drafts may be required not only to achieve feasible and realistic activities and budgets, but also to get politicians and stakeholders to agree on the various trade-offs and the level of risk that can be taken. Approval by government is essential for disaster mobilization and implementation.

These are not dealt with in this training material. To obtain the UWDM plan is a milestone but not the end product. Too often, plans are not implemented and the main reasons are important to know and avoid:

- Lack of political commitment to the process, usually due to the drive coming from external sources or a lack of engagement of key decision makers in initiating the process.
- Unrealistic planning with disaster requirements beyond the reach of local government.
- Unacceptable plans. Plans rejected by one or more influential groups due to inadequate consultation or unrealistic expectations of compromise. With water disasters, where economic losses are incurring, adequate consultation is vital.

To achieve any sustainability in management for water disasters management, a long-term commitment is required and therefore the plan should be seen as a revolving plan with features of performance evaluation and reformulation at periodic intervals, studying and simulating past and similar disasters.

Mobilizing a team, development of a work plan, drawing in the public, and stakeholders, and ensuring political commitment are all components of the startup of the planning process. The actions of the team and other process management structures take place throughout the planning cycle and are the means to keep the cycle turning.

The mobilization of efforts should also seek to build trust in the process. Trust is one nontangible social capital that gives cohesion by reducing uncertainty. Trust reduces conflict and eases the consultation process and improves acceptance of the result. Loss of trust has the opposite effects.

11.6.1 Experience on Disaster Management

According to UNEP (2005), major observations and lessons learnt from the way governments handle disaster management are as follows:

- *Enhance the communication system*: Due to a failure of the telecommunication system, direct information on damages and intensity of the disaster will not be properly transmitted to other municipalities. Therefore, there is a strong need to improve the communication system, particularly to share the information before and during the disasters.
- *Incorporate environment issues in disaster management*: Current disaster management plans of the municipality do not concretely incorporate environmental issues. In contrast, environmental issues are handled by a separate department that has few links to disaster issues.
- *Relief versus long-term rehabilitation and reconstruction*: In most cases, the emergency committee, after the disaster, focuses on the emergency response and interagency coordination only. Reconstruction efforts are handled by individual departments as part of their everyday duties. Thus, there is a lack of integration of different issues and disciplines, and environment aspects are no exception. Long-term impacts of environment are often overlooked as a result.
- *A proactive link between research and practice*: It is often found that the practitioners at municipality and city governments are not aware of research results conducted in universities and research institutions. Thus, opportunities for information exchange in the form of study groups and/or participatory training programs should be created, where practitioners can interact with experts and resource persons on a continual basis.
- *Training of community leaders*: Similar to the above interaction, opportunities for training of community leaders should also be created, and interactions among the practitioners and the community members promoted.
- *Dissemination of experiences to other municipalities*: It is important and relevant to share experiences of good practices with other municipalities and regions. Incentives to the municipality to undertake certain innovative approaches are also important, for example, in the form of awards or prizes. The municipality should take pride in the incentive and should promote its dissemination to other municipalities, thereby motivating its neighbors (UNEP, 2005).
- *Interlinkages between disaster preparedness and environmental management*: The interlinkages between disaster preparedness and environmental management in policies and plans at the local and national levels are the most important point in urban disaster management. Disaster management plans should incorporate environmental dimensions and should anticipate the impact of disasters on the environment as well as the impact of environmental practices (e.g., forest and river management) on the impacts of disasters.
- *Plans and Programs*: While disaster and environmental management should be linked to development planning, broader river basin management should also be considered in disaster programs and should be linked to the overall city management strategy.
- *Dissemination, adaptation, and implementation*: Many of the lessons from the past disasters in the same country are relevant to other cities/areas, but are often not disseminated properly. Cooperation of cities and municipalities in this regard is essential for adaptation and implementation of the lessons learnt.
- *Training and human resource development*: Proper training and human resource development programs should be undertaken not only for disaster managers, but also for high-level decision makers such as city mayors. Training in decision-making systems is essential in this regard to integrate disaster preparedness and response into the larger development and management of cities in general (UNEP, 2005).

11.6.2 Initiation

The initiation of a UWDM planning process may arise from several sources. Internationally, governments have agreed at the global summit on SD to put in place plans for sustainable management and development of water disasters by 2005. This is being followed up with support from the international community and donors. As a result, the drive for UWDM plans may appear to come from outside by donors and international agencies offering assistance to achieve the goal.

At the national level, many governments are aware of the problems that their own water sector is facing from issues, such as pollution, scarcity, emergencies, competition for use, and have identified action as a priority. Many have also taken the first step of developing a water policy or water vision or have contributed to the development of such visions in their region. The focus on specific water problems or problem areas is also adequate stimulus for the government to act, and even if this results in a more focused plan of action to solve a specific issue, it may lead to the gradual development of a fully integrated approach to water disaster management. The process for the development of the UWDM plan requires a different process than is usually taken in governmental planning. Key differences include

- A multisectoral approach: To manage water in an integrated way means developing linkages and structures for management across sectors. For such strategies to be successful, the main water use sectors should be involved in planning and strategy development from the outset.
- A dynamic process: The development of a sustainable management system for water disasters and the integrated approach will be a long process. This will require regular review, adaptation, and possibly reformulation of plans to remain effective.
- Stakeholder participation: Because most problems with UWDM are felt at the lowest levels and therefore changes in water management are required down to the individual participation, the strategy development process requires extensive consultation with the stakeholders.

There are several key activities regardless of the underlying reasons; they include obtaining government commitment, raising awareness on principles for management of water disasters (UWDM), and establishing a management team.

11.6.2.1 Political and Governmental Commitment

Political support and commitment are essential for the success of any part of the disaster management process and the changes across ministries or legal and institutional structures require the highest level of political commitment (cabinet, head of the state). Some of the reasons for strong political support are as follows:

- Multiple levels will be involved at the time of disaster, and only through the right political process can a coordinated emergency be handled properly.
- Ensure that the water disasters vision and objectives incorporate political goals consistent with other national goals especially for allocation of immense financial resources needed at the time of widespread water disaster.
- Conversely, ensure that the water vision and objectives are reflected in political aspirations.
- Ensure that the policy implications of the strategy are followed and considered throughout the process and not merely at some formal end point (to allow a continuous improvement approach to the work).
- Make decisions on recommended policy plans with legal and institutional supports.
- Ensure that the plan is adopted and followed through.
- Commit government funds (and, if necessary, mobilize donor assistance).

Political commitment needs to be of long term and therefore across political parties so that it is not rejected when a new government takes office. For this reason, a compelling vision is that all parties can aspire to provide a good foundation for action.

11.6.2.2 Policy Implications for Disaster Preparedness

Urban disasters (natural and man-made) can cause large loss of life and have enormous economic and financial costs. They are especially devastating to poor people, who often live and work in tough conditions. The drive for mitigation increases as the effects of disasters, and the costs of failures, become more immediate and widely spread.

Institution establishments are needed that can motivate action in advance of disaster and can share the costs and benefits of preventive measures among affected people in a fair manner. Hazard mitigation requires improving knowledge, building constituencies for risk reduction, and strengthening institutions and partnerships across levels of government and the private sector.

Stakeholder analysis essentially involves identification of key players in disaster management. In identifying the key players, one can ask the following:

- Who are adversely impacted and who are the potential losers?
- Have vulnerable groups that may be impacted by the plan been identified?
- What are the functional relationships among the stakeholders at the time of disasters?

Certain important policy developments and actions in countries facing different disasters include (UNEP 2005)

- Raising awareness about the integration of environment and disaster issues.

Proposed actions: Arrange meetings, seminars, and forums at different levels.

- Documentating and disseminating information.

Proposed actions: Undertake field surveys after different disaster events to document concrete examples and disseminate it through the Internet, and in training and study programs.

- Bridging the gaps between knowledge and practice.

Proposed actions: Undertake and facilitate training programs involving experts, resource persons, practitioners, and community leaders through forums.

- Implementing practical exercises and techniques.

Proposed actions: Undertake small projects at different locations having different socioeconomic contexts. These will be learning experiences, which can be also disseminated to other areas.

- Developing guidelines and tools on environment and disaster management.

Proposed actions: Guidelines and tools should be developed based on the field experiences, and incorporating the comments and expertise of practitioners and professionals from different parts of the world.

- Establishing a continuous monitoring system.

Proposed actions: A network or partnership should be developed, which will continuously monitor different activities, identify best practices, and disseminate through training and capacity-building programs (UNEP, 2005).

Building knowledge about the physics of hazards may be inadequate or absent, even among residents at risk, yet community awareness of physical hazards is fundamental for mitigation efforts. Comprehensive vulnerability assessments using remote sensing, satellite imagery, and risk and loss estimation modeling can help document and reduce physical, social, and economic vulnerability. Changing physical infrastructure and innovative techniques for retrofitting buildings can improve disaster prevention. Soft nonstructural methods can increase hazard information, create new knowledge, build physical and human resources capacities, and train and raise the awareness of decision makers and the communities at risk.

Estimating losses can justify the financial investment for preparedness. Memphis, Tennessee, calculated a $0.5 million cost for retrofitting water pumping stations to be disaster resistant compared with $17 million to replace each pump and $1.4 million for each day the system is out of service. But developing countries rarely have well-documented, location-specific, and hazard-specific costing of hard and soft mitigation measures. Even rarer is the cost of public education systematically. Without an educated constituency, collective decisions on disaster policies are usually dominated by better-off members of the community. Their priorities can differ greatly from those of poor people, who risk a larger share of their assets in a disaster.

11.6.2.3 Public Participation

The provision of a foundation and strategy for involving the stakeholders in the various stages of preparing and implementing the UWDM plan is needed, so that the stakeholder engagement strategy runs efficiently through the planning and political processes as well as planning and supplementation processes.

11.6.2.3.1 Benefits of Public Involvement

- It can lead to informed decision making as stakeholders often possess a wealth of information that can benefit the project.
- Stakeholders are the most affected by lack of or poor water disasters management.
- Consensus at early stages of the project can reduce the likelihood of conflicts at the time of water disaster and public participation in successful handling of the disaster.
- Stakeholder involvement contributes to the transparency of public and private actions, as these actions are monitored by the different stakeholders that are involved.
- The involvement of stakeholders can build trust between the government and civil society, which can possibly lead to long-term collaborative relationships.

11.6.2.3.2 Methods for Public Participation

Stakeholders should be engaged at all critical steps in the process of developing the plan. These stages should be planned and the work plan should identify the timing, the purpose, the target stakeholders, the method, and the expected outcome. The scale and strategy of stakeholder participation must be carefully determined and be coordinated with the emergency response team and perhaps national guards.

Methods may include

- Stakeholder workshops, in which selected stakeholders are invited to discuss water issues
- Representation in the management structure for the planning process
- Local consultations "on the ground"
- Surveys
- Consultations with collaborating organizations (such as NGOs, academic institutions, etc.)

Using multiple sources of information not only has the advantage that the information obtained is more likely to be accurate, but especially the participatory methods of information gathering can also contribute to creating a sense of local responsibilities of the process and consensus about the actions to be taken. Stakeholder participation techniques range from a low level of involvement to a high level of involvement depending on the state of disaster.

11.6.2.4 Lessons on Community Activities

- *Perception versus action*: There is a difference between perception and the action taken. Although people are quite aware of the risks involved and of risk mitigation measures, it is not reflected in the action that they take.
- *Preparedness for evacuation*: The experiences in different cities also demonstrate the need for the local government and community to work together in organizing evacuation simulation exercises, and designating "disaster evacuation areas" for residents to assemble during a disaster.
- *Self-reliability versus dependence*: A new system of emergency radios that the city had introduced made people more reliant on the system. A balanced approach that introduces new ideas without destroying local existing systems is necessary.
- *Public awareness*: It is important to raise awareness among people and communities about the link between environmental management and disaster preparedness. The link between upstream and downstream issues should not only be reflected in river basin management, but should also be incorporated in community awareness raising and implementing projects at the local level (UNEP, 2005).

11.6.3 Case Study 1: Drought Disaster Management

The ability of a water supply system to face a disaster caused by drought can be improved through long-term management such as construction of infrastructures development and rehabilitation, reduction of water distribution and water use losses, dual WDS, and reuse of wastewater.

In short-term water disaster management, the severity of a hydrological drought can be reduced by supply/demand management through delivery restrictions, augmentation of water availability by a lower quality or by a greater cost than ordinary sources, and making better use of available water resources by optimal-operation-of-reservoir rules.

Common UWDM during a drought consists of the following steps

1. Monitoring drought indices/forecasting of water resources and demands
2. Consideration of drought management options
3. Establishment of levels of indicators that trigger the various options of a water disaster program
4. Adaptation of a management plan at the levels indicated by the drought indices

Water supply managers prefer a number of smaller shortages to a few very large shortages (droughts), suggesting that damages are convex by the amount of shortfalls. In a general situation of actual drought, realistic water management rules would suggest that during periods of incipient drought, reductions should be made in demand even if it can be fully delivered from available water. This reduction prevents larger shortages in later periods. To minimize the impact of current and consequent droughts, water management rules are coupled with the water disaster management to balance the shortage during droughts and to ensure that a sufficient amount of water remains for subsequent water supply. Decision makers in water resources systems must evaluate trade-offs among immediate and future uses of water before the volume of the future supply becomes known. In the face of this uncertainty, forecast of future climate can be helpful in determining efficient operating decisions. When anticipating a drought and during a drought, climatic and hydrologic forecasts would be more valuable.

A water manager needs quantitative drought triggers to activate the onset and extent of the restrictions that should be utilized. A drought trigger value is typically a single number, more useful than raw data for decision making. Although none of the major drought indices is inherently superior to the rest in all circumstances, some indices are better suited than others for certain uses. Anticipated water and the available volume of water storage are suitable drought indices for urban water supply systems. These indices are dependent on both hydrological and climatologic features of a region as well as the water supply system. Indices such as the standard precipitation index (SPI) and the surface water severity index, which deal with the climatic and hydrological droughts, respectively, would be suitable candidates for a drought monitoring index in urban areas.

11.6.4 Case Study 2: Drought Management in Georgia, USA

Periods of drought have naturally come and gone throughout history. In recent years, however, some regions of the United States such as Georgia have become more vulnerable to the effects of drought as pressures increase on natural resources. Rain deficits are rising, crops are dying on the vine, and lake levels are plummeting. Meanwhile, concern is growing among water managers and others. Already, 31 states have implemented drought management plans, five of which attack the situation proactively, rather than emphasizing an emergency response. Georgia will soon be among the proactive.

Georgia's plan links indicators, which characterize stages of drought severity, with responses. For instance, when an indicator shows that a drought is developing, a response could be to curtail nonessential water uses. Then during a drought, the plan also provides guidance, such as when and how to make water use restrictions more stringent, or when and how to implement water use surcharges to manage demand. The idea is that it is more effective—and less painful—to prepare for drought, and take timely actions, than to wait until a full-blown disaster has developed.

Officials hope that the new statewide plan will provide guidance for local and regional entities to tailor their plans to the type of drought, the users affected, and the effects of various mitigation strategies on local users, as well as those elsewhere in the state. An example of such strategies at an individual level is an alternative for chemical applications to lawns, which is one of the major sources of nonpoint pollution. This alternative, called xeriscaping, uses more native vegetation to reduce the need for water, pesticides, and fertilizers. Xeriscaped lawns also cost less to maintain while lowering public health risks. It is important to think about the impacts of each individual decision.

Businesses and industries can make a difference by lessening their dependence on water. At the local government level, it is hoped that officials will understand the need for drought planning, particularly as population growth occurs, because drought occurs when demands exceed supplies. Demand for water has skyrocketed in Georgia, while new supplies are limited.

At the state level, officials are beginning to take a risk-management approach to drought management, which is a slow process to change the historic way of doing things. However, identifying the likely impacts of drought, and the factors behind that vulnerability, allows the state to address the risks associated with drought in advance. In time, these actions will lessen risks and therefore impact the level of government intervention required in the form of emergency assistance.

11.6.5 Case Study 3: Management of Northern California Storms and Floods of January 1995

An unusual series of storms from January 5 through 26, 1995, caused heavy, prolonged, and, in some cases, unprecedented precipitation across California. This series of storms resulted in widespread minor to record-breaking floods from Santa Barbara to the Oregon border. Several stream-gaging stations used to measure the water levels in streams and rivers recorded the largest peaks in the history of their operation.

11.6.5.1 Flood Characteristics

Before the January storms, rainfall was near normal across northern California. The most intense storms occurred during the week of January 8 and produced an average 33 cm of precipitation over most of northern California. Precipitation amounts of as much as 61 cm were recorded for the week. Maximum 1-day rainfall data of these storms were compared with the theoretical 100-year, 24-h, precipitation, and rainfall amounts were greatest in Humboldt, Lake, Mendocino, Napa, Sacramento, Shasta, Sonoma, and Trinity Counties.

Flooding was significant throughout northern California from January 9 through 14. The highest peak flows in the region occurred from January 8 through 10 with a peak height of 8 m. Flooding in small basins was unusually rapid because of the high-intensity, short-duration microbursts of rainfall. Small streams rather than large rivers caused most of the damage in the Central Valley. Flooding along large rivers in the Central Valley was controlled by diversions and flood-control reservoirs that had large storage capacities available. Small streams and rivers in the coastal areas also caused widespread flooding. More than 10 years of drought and low stream flows resulted in the accumulation of dense riparian vegetation in most stream channels. Flooding along the Russian River was due, in part, to the accumulation of vegetation and debris that had reduced the capacity of the stream channel. This reduced capacity resulted in higher water levels with less stream flow.

11.6.5.2 Response

Personnel of the USGS provided the crucial information needed to determine peak discharges during the January 1995 floods. Accurate recording of peak discharges at long-term stream-gaging stations is essential for computations of flood frequency and magnitude. This information is used by federal, state, and local officials to prepare for and minimize damages from future floods. Most measurements of water levels in streams are automated. In recent years, information compilation has been improved by the use of electronic data collection platforms (DCPs). These platforms use automated earth–satellite telemetry for the immediate transmission of data from remote sites. Cellular telephones and modems also are being used to obtain timely information, but not all stream-gaging stations are equipped with this instrumentation. This on-time information is used for early warning of flood. When on-time information shows increasing of water levels at upstream rain gauges, a warning is given that a flood is about to occur. According to the amount of increase in water levels, suitable decisions are needed to mitigate or decrease impacts of flood in habitable areas downstream. During the recent floods, the USGS was able to rapidly compile and disseminate near-real-time information for many of its gaging stations by using the telephone and computer networks. The USGS has also made data available on the Internet to provide immediate access to flood data and other hydrological data. During January 1995, most of data were made available within hours of a flood peak.

11.7 SITUATION ANALYSIS

Output from the situation analysis is a report elaborating the progress with implementing improved management of water disasters, the outstanding issues, the problems, and some of the solutions. The purpose of this step is to help characterize the present situation and to use the information to predict possible future UWDs for developing a UWDM approach.

The situation analysis examines the key factors of influence in a given situation. It is especially important to view the situation first from the perspective of those directly affected. Awareness of the problems caused by UWDs and the motivation to seek solutions are a function of the condition experienced by the stakeholders.

For the purposes of a UWDM plan, the situation analysis is assessed against the principles of those embodied in the UWDM approach. Analysis and interpretation made against predefined goals and the national disaster management vision or policy can be focused and targeted to address the main constraints and causes rather than the symptoms.

The analysis should adequately reflect the concerns and impacts of the water disasters on citizens, the environment, and the society as a whole. The analysis report should be shared widely, and this means summarized as appropriate. The situation analysis should serve as an important indicator of the transparency of the process. The sharing of the report with politicians and other senior members of the government helps to maintain political commitment, enlist their support for the solutions and action emerging, and create awareness of the implementation implications of the forthcoming plan.

An essential aspect in the management of UWDs is the anticipation of change including the changes in the natural system due to geomorphological processes, the changes in the engineered components due to aging, the changes in the demands, and even changes in the supply potential of water, possibly due to changing climate and weather. System dynamics is a new approach that helps managers to see all the impacts of a disaster. System dynamics is the essence of the limit to growth and how understanding that limit could help us to prevent disaster and/or manage it intelligently. We are not elaborating on this but we are making a reference to the ongoing study on the future water allocation from Karkheh Reservoir in Iran (Karamouz, 2004).

11.7.1 Steps in the Development of Situation Analysis

11.7.1.1 Approach

There is a role for specialist expertise in conducting the analysis when high-tech skills are required, large baseline surveys need to be done, or there is particular need for an independent viewpoint. There are several related principles for coordinating the collection of knowledge:

- Multistakeholder groups themselves should design the information gathering, analysis, and research process to ensure ownership of the strategy and its results.
- All the "analysis" tasks are best implemented by bringing together, and supporting, existing centers of technical expertise, learning, and research.
- Since analysis is central to strategy development, it should be commissioned, agreed, and endorsed at the highest level (i.e., by key government ministries or by the planning steering committee). This will increase the chance that analysis will be well focused and timely in relation to the plan's evolution and timetable and that it will be implemented.
- In the same way, analysis needs good coordination. It is logical for the disaster management team to coordinate the analysis but it should not undertake all the analyses itself.

11.7.1.2 Objectives

The objectives of the analysis must be clear. The field of UWDM is large and covers a huge number of issues. The purpose of the situation analysis is to study the UWDs in terms of the UWDM principles. Weaknesses, problems, and issues identified in UWDM may arise from the following areas:

- Policies affecting UWDM
- Legislation on preventive measures as well as mobilization of people/equipment during disaster and relief effort after that
- Charter of water disaster management institutions and commissions
- Practicing UWDM for different types of disaster

An analysis of the present water disasters management situation in the country should therefore identify gaps in the management framework and allow a prioritization of action.

11.7.1.3 Data Collection

Data for the situation analysis comes from a variety of sources. For reasons of efficiency and effectiveness, the planning process should build on and explore earlier knowledge and experience and draw on lessons learned. Part of such useful knowledge is not readily available or well documented. It often exists in an ad hoc form among professionals and practitioners as well as among government and nongovernment staff within disaster-management and water-relevant sectors. The political level holds important knowledge on the various processes involved in achieving overall endorsement of the goals of the plan and rallying support to its implementation.

The knowledge to be compiled and made available includes the following areas:

- UWDM experience at the country level, where elements of UWDM frameworks may have been completed in part or in full. National disaster management policies, management organizations, and water disasters assessment tools are constantly changing in many countries around the world and constitute important requisites for UWDM.
- International UWDM experience, which can mean both experiences collected from several countries or groups of countries and experiences where regional aspects of disasters are dominating.
- Experience from past and present national planning processes within other sectors and in particular those cutting across several sectors. Examples of such processes are development of damage reduction strategies, strategies and plans for sustainable level of water demand, development of water rationing, and conservation strategies.

The analysis of the current situation can be started by reviewing the potential of possible UWDs and their indicators and examining the role of water disasters management in relation to the achievement goal of minimizing undesired effects. The social/cultural/institutional context and macroeconomic policies that condition current policies and practices for the water disasters management should also be taken into account.

11.7.2 Urban Disasters Situation Analysis

The situation analysis should examine all types of UWDs with respect to different quantity and quality problems. It should identify the pertinent parameters of the hydrological cycle and urban morphological condition and evaluate the cost of damages caused by different types of disasters. The analysis should pinpoint the major water disasters issues, their severity, and social implications, as well as the risks and hazards of them. For the purposes of UWDM planning, care should be taken not to embrace an approach that is too technical but to emphasize the implementation process and the enabling environment for efficient, and effective handling of the disaster.

Socioeconomic aspects are important when looking at the impacts of the water disasters on water users (including environment) and society as a whole. In UWDs assessment for UWDM, social and economic management issues such as urban-affected population, land use patterns, and environmental issues are as important as hydrological and hydraulics structural integrities of water systems. The above approach is of central importance to regional cooperation in UWDM. The relevant unit of analysis is the affected region as a whole, irrespective of whether it crosses national boundaries.

Articulating priority goals can focus more attention on the future management situation and less attention on the means to arrive there, similarly providing an initial perspective and basis for discussion of the priority goals ahead of the strategy and plan development. The involvement of and feedback to the political level is important at this point to maintain political commitment to the process and ensure that they embrace the situation analysis and are aware of the likely solutions and actions to emerge.

11.8 DISASTER INDICES

Perception of disaster depends on the degree to which a system feels vulnerable. One way to measure the relative level of water disaster is to look at the indicators of water supply security. A weighted combination of reliability of supply and any reserve resources; vulnerability of the system to a high impact on the available safe water supply; and resiliency of the system to return to the satisfactory state of operation shows the state of the systems' readiness to face a disaster.

Disaster indices provide ways by which we can quantify relative levels of disaster. They can be defined in a number of ways. One way is to express relative levels of disaster as separate or weighted combinations of reliability, resilience, and vulnerability, the measures of various criteria that contribute to human welfare in time and space. These criteria can be economic, environmental, ecological, and social. To do this, one must first identify the overall set of criteria, and then for each one decide which ranges of values are satisfactory and which ranges are not. These decisions are subjective. They are generally based on human judgment or social goals, not scientific theories. In some cases, they may be based on well-defined health and safety standards. Most criteria will not have predefined or published standards or threshold values separating what is considered satisfactory and what is not. For many criteria, the duration as well as the extent of individual and cumulative failures may be important.

A set of criteria should be selected and for each criterion a range of values that are satisfactory should be defined. The duration as well as the extent of individual and cumulative failures of water resources systems should be determined for each criterion (Karamouz and Moridi, 2005).

- Development of a shared vision for dealing with the disaster and identifying ways by which all parties involved can make the shared vision come true is the key strategy in handling a disaster.
- Developing coordinated approaches among all concerned and responsible agencies.
- Establish baselines for system functioning against drastic changes in supply and delivery.
- Monitoring and evaluating disaster progression and recovery.

11.8.1 Reliability

System performance indices are essential for the evaluation of system performance to identify optimal situations. These indices show the system's ability to work without any problem. System performance can be desired or undesired. Undesired conditions are named as failure. A system's weaknesses in delivering water with the desired pressure and the requested quantity and quality are different aspects of system failure. Mean and standard deviation of the system's output are applicable for the evaluation of system performance but are is not sufficient. Figure 11.3 shows the weakness of mean and standard deviation indices for definition of severity and frequency of system failure. Diagrams of Figures 11.3a and 11.3b are symmetrical to the axis x, so their mean and standard deviation are the same. There are two failure events in Figure 11.3b but there is no failure situation in the other figure. Furthermore, it cannot be identified how an increase or decrease in the system's means can affect the system performance. Because of these weaknesses in using mean and standard deviation indices for the evaluation of probability of failure and system performance, reliability is used.

The random variable of X_t is the system's output classified into two groups of failure and success outputs. The system reliability is the probability that X_t belongs to the success outputs group (Hashimoto et al., 1982). The reliability of a system can be calculated by

$$\alpha = \text{Prob}\,[X_t \in S], \quad \forall t, \tag{11.1}$$

where S is the set of all satisfactory outputs. Based on this definition, reliability is the opposite of risk, in which the probability of system failure could be expressed upon.

All systems, whether natural or man-made, fail, and they do so for any number of reasons, including structural inadequacies, natural causes exceeding the design parameters of the system (e.g., droughts and floods), and human causes such as population growth that raises the system's demands above the supply capacity. Thus, reliability is conceptually related to the probability of system failure, and the

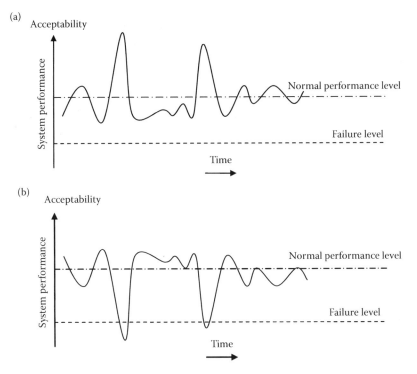

FIGURE 11.3 Comparison of two different functions with the same mean and standard deviation. (After Hashimoto, T., Stedinger, J. R., and Loucks, D. P. 1982. *Journal of Water Resources Research* 18(1): 14–20.)

rate, and consequences of failure. It can be measured in several different but related ways, depending on the needs and relevance of a particular situation. For example, it may be useful to express and characterize the expected length of time between successive failures (i.e., time to failure), similar to the notion of a 100-year flood event, an event that is expected to occur on average once in 100 years.

While reliability is frequently used in the planning and operations documents, trade, and water, it is

- Often not formally defined by agencies or institutions
- Concerned with different water management entities' operating and planning processes with substantial variation
- Often considered in qualitative rather than quantitative terms

Thus, there is no existing and widely accepted framework of definitions and measurement approaches that are applicable in all aspects of influence and applications.

Reliability at key points within the sphere of influence ultimately depends on the complexity of interactions of a web of rights, regulations, laws, and rules that affect all institutions that manage water in a municipality. It will be necessary to understand these interactions in order to develop meaningful definitions of water supply reliability and, based on these definitions, related indicators could be defined. These interactions play out across numerous dimensions that include, primarily

- System scale and complexity
- Specific management function (i.e., operations, planning)
- Purpose of the service, that is, the intended use of water

A major determinant of how reliability is defined as the scale or slice of the water stock. Discussions with water professionals and a review of the literature indicate that differences in the system scale (i.e., the boundaries of the system) will have a significant influence on how reliability is defined. Scale can be system-wide, regional, watershed-specific, city-wide, or local. The main categories of scale, from smaller to larger, are

- WDS
- Urban water and irrigation districts
- Watershed
- State and federal water projects
- Statewide

While some of these states are much larger than urban scale, it is important to emphasize that many urban water problems go far beyond the city limit and may include other basins (i.e., water transfer) and regions. These categories of scale combine geographical, hydrological, and institutional dimensions, and their respective boundaries do not always coincide. Scale is also a function of complexity, in terms of both the number and interconnections of components (e.g., a single component such as a reservoir versus a system such as the combined CVP/SWP network) and the mix of engineered infrastructure and natural elements. As complexity increases, many of the more straightforward engineering concepts and tools for measuring reliability become inadequate. At present, the largest scale at which water supply reliability is utilized to any extent is in the projects' scale, which considers reliability in both daily operations and planning for system expansion. At any given scale, the relevant management function also influences how reliability is defined.

Two critical functions in the water management arena are planning and operations. In public utilities, the planning and operations functions tend to proceed in parallel and are often not fully integrated. In operations, the focus is on achieving specific goals within short time frames (hourly, weekly, and monthly), given the existing infrastructure and sources of supply. In planning, in contrast, planners determine how to match supply and demand in the longer term through a combination of measures that restrict demand and extend the supply resources and/or needed infrastructure. While reliability is typically a key goal for both operations and planning, their different time frames lead to somewhat different definitions and measures of reliability.

Water users have historically been divided into two main categories: consumptive users and instream flows or ecological users. Consumptive uses have historically been the primary focus of agencies charged with managing water supply. The building and operation of reservoir, water transfer tunnels, and distribution systems were undertaken for the sole purpose of collecting and delivering water supplies. The application of reliability concepts to environmental uses appears to lag behind the application of reliability concepts to consumptive use. Despite that, water supply reliability indicators and metrics must therefore be developed to monitor the reliability of both types of service. There are two different methods that are used for reliability estimation: (1) statistical analysis of recorded data of failures of similar systems and (2) indirect analysis by studying combination of significant parameter impacts on reliability.

In reliability theory, the Weibull distribution is assumed to model lifetime (the time between two consecutive system failures) most appropriately. So the cumulative probability function of the system failures could be formulated as

$$F(t) = 1 - e^{-\alpha t^\beta} \quad (t > 0), \tag{11.2}$$

where $\alpha, \beta > 0$ are parameters to be estimated. Another definition of reliability is the probability that no failure will occur within the planning horizon (Karamouz et al., 2003), which could be formulated as

$$R(t) = 1 - F(t) = e^{-\alpha t^\beta}. \tag{11.3}$$

Urban Water Disaster Management

The density function of the time-of-failure occurrences is obtained by simple differentiation of Equation 11.2:

$$f(t) = \alpha\beta t^{\beta-1} e^{-\alpha t^\beta} \quad (t > 0). \tag{11.4}$$

Note that in the special case of $\beta = 1$, $f(t) = \alpha e^{-\alpha t}$. It is the density function of the exponential distribution. The exponential distribution is only rarely used in reliability studies, since it has the so-called "forgetfulness" property. If X is an exponential variable indicating the time when the system fails, then for all $t, \tau > 0$.

$$P(X > t + \tau | X > \tau) = P(X > t). \tag{11.5}$$

This relation shows that the probability of the working condition in any time length t is independent of how long the system was working before. Equation 11.5 can be shown as

$$P(X > t + \tau | X > \tau) = \frac{P(X > t + \tau)}{P(X > \tau)} = \frac{R(t+\tau)}{R(\tau)} = \frac{e^{-\alpha(t+\tau)}}{e^{-\alpha\tau}} = e^{-\alpha t} = R(t) = P(X > t). \tag{11.6}$$

In most practical cases the breakdown probabilities are increasing in time as the system becomes older, so the exponential variable is inappropriate in such cases.

The hazard rate is given as

$$\rho(t) = \lim_{\Delta t \to 0} \frac{P(t \leq X \leq t + \Delta t \,|\, t \leq X)}{\Delta t} \tag{11.7}$$

and shows how often failures will occur after time period t. It is clear that

$$\rho(t) = \lim_{\Delta t \to 0} \frac{P(t \leq X \leq t + \Delta t)}{P(t \leq X)\Delta t} = \lim_{\Delta t \to 0} \frac{F(t+\Delta t) - F(t)}{\Delta t} \cdot \frac{1}{R(t)} = \frac{f(t)}{R(t)}. \tag{11.8}$$

Therefore the hazard rate can be computed as the ratio of the density function of time between failure occurrences and the reliability function (Karamouz et al., 2003).

EXAMPLE 11.1

Assume that the time between flood occurrences in urban area follows the Weibull distribution. Formulate the reliability and hazard rate of this region in dealing with floods.

Solution: The region reliability is calculated as

$$R(t) = 1 - F(t) = e^{-\alpha t^\beta}$$

and therefore from relation 11.8 the hazard rate is calculated as

$$\rho(t) = \frac{\alpha\beta t^{\beta-1} e^{-\alpha t^\beta}}{e^{-\alpha t^\beta}} = \alpha\beta t^{\beta-1},$$

which is an increasing polynomial of t showing that failure will occur more frequently for larger values of t. In the special case of an exponential distribution $\beta = 1$; so $\rho(t) = \alpha$ is a constant.

continued

It is an important problem in reliability engineering to reconstruct $F(t)$ or the reliability function if the hazard rate is given (Karamouz et al., 2003). Note first that

$$\rho(t) = \frac{f(t)}{1-F(t)} = \frac{(F(t))'}{1-F(t)} = -\frac{(1-F(t))'}{1-F(t)}.$$

By integration of both the sides in the interval $[0, t]$,

$$\int_0^t \rho(\tau)\,d\tau = [-\ln(1-F(\tau))]_{\tau=0}^t = -\ln(1-F(t)) + \ln(1-F(0)).$$

Since $F(0) = 0$, the second term equals zero; so

$$\ln(1-F(t)) = -\int_0^t \rho(\tau)\,d\tau,$$

implying that

$$1 - F(t) = \exp\left(-\int_0^t \rho(\tau)\,d\tau\right),$$

and finally,

$$F(t) = 1 - \exp\left(-\int_0^t \rho(\tau)\,d\tau\right). \tag{11.9}$$

Example 11.2

Construct the cumulative distribution function (CDF) of failure occurrences in two given situations: (1) the hazard rate is constant, $\rho(t) = \alpha$; (2) $\rho(t) = \alpha\beta t^{\beta-1}$.

Solution: Using Equation 11.9, for $\rho(t) = \alpha$, it is obtained that

$$F(t) = 1 - \exp\left(-\int_0^t \rho(\tau)\,d\tau\right) = 1 - \exp\left(-\int_0^t \alpha\,d\tau\right) = 1 - e^{-\alpha t}.$$

Similarly, when $\rho(t) = \alpha\beta t^{\beta-1}$, then

$$F(t) = 1 - \exp\left(-\int_0^t \alpha\beta\tau^{\beta-1}\,d\tau\right) = 1 - \exp\left(-[\alpha\tau^\beta]_0^t\right) = 1 - e^{-\alpha t^\beta},$$

showing that the distribution is Weibull.

For $F(t)$, which is increasing, $F(0) = 0$ and $\lim_{t\to\infty} F(t) = 1$, whereas $R(t)$, which is decreasing, $R(0) = 1$ and $\lim_{t\to\infty} R(t) = 0$. In the case of $\beta > 2$ and the Weibull distribution, $\rho(0) = 0$, $\lim_{t\to\infty} \rho(t) = \infty$, where $\rho(t)$ is a strictly increasing and strictly convex function. If $\beta = 2$, then $\rho(t)$ is linear, and if $1 < \beta < 2$, then $\rho(t)$ is strictly increasing and strictly concave. If $\beta = 1$, then $\rho(t)$ is a constant. In the case of $\beta < 1$, the hazard rate is decreasing in t, in which case defective systems tend to fail early. So hazard rate decreases for a well-made system. Note that

$$\rho'(t) = \frac{f'(t)(1-F(t)) + f(t)^2}{(1-F(t))^2},$$

which is positive if and only if the numerator is positive:

$$f'(t) > -\frac{f(t)^2}{1-F(t)} = -f(t)\rho(t). \tag{11.10}$$

If inequity of Equation 11.10 is satisfied, $\rho(t)$ is locally increasing; otherwise $\rho(t)$ decreases.

The above discussions about reliability are applicable to a component of a system. The reliability of a system with several components depends on what kind of configuration of the components is the entire system. Systems could be arranged in "series," "parallel," or combined combinations. The failure of any component in a serial system results in system failure, but in a parallel system, only when all components fail simultaneously, failure occurs.

A typical serial combination is shown in Figure 11.4. If $R_i(t)$ denotes the reliability function of component i ($i = 1, 2, \ldots, n$), then the reliability function of the system could be estimated as

$$R(t) = P(X > t) = P((X_1 > t) \cap (X_2 > t) \cap \cdots \cap (X_n > t)),$$

where X is the time when the system fails, and $X_1, \ldots, X_n$ are the same for the components. It can be assumed that the failures of the different system components occur independently of each other; then

$$R(t) = P(X_1 > t)P(X_2 > t) \cdots P(X_n > t) = \prod_{i=1}^{n} R_i(t). \tag{11.11}$$

Note that the inclusion of a new component into the system results in a smaller reliability function, since it is multiplied by the new factor $R_{n+1}(t)$, which is below one.

In a parallel system as shown in Figure 11.5, the system fails if all components fail, so the system failure probability is calculated as

$$P(X \leq t) = P((X_1 \leq t) \cap (X_2 \leq t) \cap \cdots \cap (X_n \leq t)).$$

If the components fail independently of each other, then

$$R(t) = 1 - F(t) = 1 - P(X \leq t) = 1 - \prod_{i=1}^{n} P(X_i \leq t) = 1 - \prod_{i=1}^{n} F_i(t) = 1 - \prod_{i=1}^{n} (1 - R_i(t)), \tag{11.12}$$

where F and F_i are the cumulative distributions until the first failure of the system and component i, respectively. In this case, including a new component in the system will increase the system reliability, since the second term is multiplied by $1 - R_{n+1}(t)$, which is less than one.

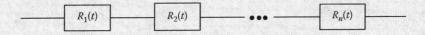

FIGURE 11.4 Serial combinations of components.

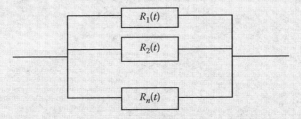

FIGURE 11.5 Parallel combinations of system components.

continued

In practical cases, commonly a mixture of series and parallel connections between system components is used. In these cases, relations 11.11 and 11.12 should be combined appropriately as it is shown in the following example.

EXAMPLE 11.3

The system illustrated in Figure 11.6 includes five components, where components 2, 3, and 5 are parallel. Calculate the system reliability at $t = 0.1$ assuming that $R_1(t) = R_4(t) = e^{-2t}$ and $R_2^i(t) = R_3^j(t) = R_5^i(t) = e^{-t}$ ($1 \leq i \leq 2, 1 \leq j \leq 3$).

Solution: At first the reliability of parallel components is calculated using Equation 11.12.

$$R_2(t) = 1 - \left(1 - R_2^1(t)\right)\left(1 - R_2^2(t)\right),$$

$$R_3(t) = 1 - \prod_{i=1}^{3}\left(1 - R_3^i(t)\right),$$

$$R_5(t) = 1 - \left(1 - R_5^1(t)\right)\left(1 - R_5^2(t)\right).$$

The reliability function of the system using Equation 11.11 is estimated as

$$R(t) = R_1(t)\left[1 - \prod_{i=1}^{2}\left(1 - R_2^i(t)\right)\right] \cdot \left[1 - \prod_{i=1}^{3}\left(1 - R_3^i(t)\right)\right] \cdot R_4(t) \cdot \left[1 - \prod_{i=1}^{2}\left(1 - R_5^i(t)\right)\right].$$

At $t = 0.1$,

$$R_1(0.1) = R_4(0.1) = e^{-0.2} = 0.8187,$$

$$R_2^i(0.1) = R_3^j(0.1) = R_5^i(0.1) = e^{-0.1} = 0.9048.$$

Then,

$$R_2(0.1) = R_5(0.1) = 1 - (1 - 0.0952)^2 = 0.9909,$$

$$R_3(0.1) = 1 - 0.0952^3 = 0.9991.$$

Hence

$$R(0.1) = 0.8187^2(0.9909)^2(0.9991) = 0.6575 = 65.75\%.$$

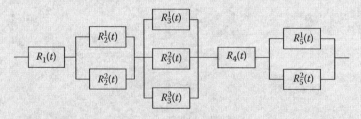

FIGURE 11.6 Combined connections in a system.

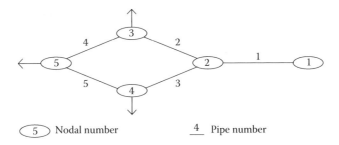

FIGURE 11.7 Example water distribution network.

11.8.1.1 Reliability Assessment

11.8.1.1.1 State Enumeration Method

In this method, all possible states of the system components that define the state of the entire system are listed. For a system with M components each of which has N operating states, there will be N^M possible states for the entire system. For example, if the state of each component is classified into failed and operating states, there will be 2^M possible states for the entire system.

After the enumeration of all possible system states, the probability of the occurrence of the identified successful states is computed. Finally, all of the successful state probabilities are summed to obtain the system reliability.

Consider the simple water distribution network in Figure 11.7. The tree diagram for a five-pipe system as shown in Figure 11.8 is called an event tree and the analysis involving the construction

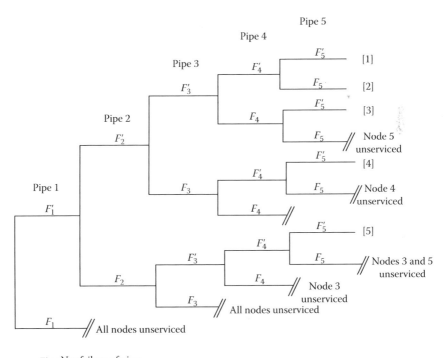

F_1' = Nonfailure of pipe
F_1 = Failure of pipe
// = Branch associated with unserviceability of one or more nodes

FIGURE 11.8 Event tree for the reliability of example water distribution network.

of an event tree is referred to as an event-tree analysis. An event tree simulates the topology of a system as well as the sequential or chronological operation of the system, which is highly important (Mays, 2004).

EXAMPLE 11.4

Calculate the system reliability of the water distribution network shown in Figure 11.7 considering that node 1 is the source node and nodes 3, 4, and 5 are demand nodes. All of the pipes have the same failure probability equal to 8%. System reliability is defined as the ability of a system to supply water to all of the demand nodes. The pipes performances are independent of each other.

Solution: The event tree of the subjected WDS is shown in Figure 11.8. The complete event tree of this WDS has $2^5 = 32$ branches because of two possible states for each of five pipes. However, the event tree is summarized based on the role of each pipe component in the network connectivity. For example, if pipe 1 fails, all demand nodes regardless of the state of other pipes cannot receive water and this situation causes a system failure. So, branches in the event tree beyond this point are not shown. As for Figure 11.8, there are only five branches corresponding to system success.

Considering that $p(B_i)$ indicates the probability that the branch B_i of the event tree provides full service to all users, the probability associated with each branch resulting in satisfactory delivery of water to all users is calculated, owing to independence of serviceability of individual pipes, as

$$p(B_1) = p(F'_1)p(F'_2)p(F'_3)p(F'_4)p(F'_5) = (0.92)(0.92)(0.92)(0.92)(0.92) = 0.659,$$

$$p(B_2) = p(F'_1)p(F'_2)p(F'_3)p(F'_4)p(F_5) = (0.92)(0.92)(0.92)(0.92)(0.08) = 0.057,$$

$$p(B_3) = p(F'_1)p(F'_2)p(F'_3)p(F_4)p(F'_5) = (0.92)(0.92)(0.92)(0.08)(0.92) = 0.057,$$

$$p(B_4) = p(F'_1)p(F'_2)p(F'_3)p(F'_4)p(F_5) = (0.92)(0.92)(0.92)(0.92)(0.08) = 0.057,$$

$$p(B_5) = p(F'_1)p(F'_2)p(F'_3)p(F'_4)p(F_5) = (0.92)(0.92)(0.92)(0.92)(0.08) = 0.057.$$

Therefore, the system reliability, which is the sum of all the above probabilities associated with the operating state of the system, is equal to 0.887.

11.8.1.1.2 Path Enumeration Method

In this method, a path is considered as a set of components or modes of operation, which result in a specified outcome from the system. The desired system outcomes in system reliability analysis include failed state or operational state. Another useful definition in applying this method is "minimum path," which is a path that no component has traversed more than once in going along it. Tie-set analysis and cut-set analysis are two techniques that are frequently used in this methodological category.

Cut-set analysis. A set of system components or modes of operation whose failure results in system failure is referred as a cut set. For instance, in a WDS, a set of system components including pipes, pumps, storage facilities, valves, and so on, which, when they fail jointly, would disrupt the service to certain users, is considered as a cut set. Since cut sets are directly related to the system failure modes, they are used to identify the distinct and discrete ways that cause a system failure.

For calculating the system failure probability, the minimum cut-set concept is utilized. A minimum cut set is a set of system components, in which the failure of all components at the same time results in the failure of the system, but when any one component of the set does not fail, the system does not fail. Therefore, it can be concluded that in a minimum cut set, the components or modes of operation are effectively connected in parallel and the minimum cut sets perform in series. Therefore, the failure

Urban Water Disaster Management

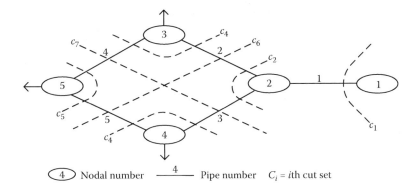

FIGURE 11.9 Minimum cut sets for the water distribution network of Figure 11.7.

probability of a system is formulated as

$$p_{f,\text{sys}} = p\left[\bigcup_{i=1}^{I} C_i\right] = p\left[\bigcup_{i=1}^{I} \left(\bigcap_{j=1}^{J_i} F_{ij}\right)\right], \quad (11.13)$$

where C_i is the h minimum cut set; J_i is the total number of components or modes of operation in the ith minimum cut set; F_{ij} represents the failure event associated with the jth component or mode of operation in the ith minimum cut set; and I is the total number of minimum cut sets in the system. For a large number of minimum cut sets, the bounds for probability of a union can be computed. For achieving adequate precision, closeness of the bounds on failure probability of the system should be examined (Mays, 2004). The cut sets for the simple WDS shown in Figure 11.7 are illustrated in Figure 11.9.

Example 11.5

Evaluate the reliability of the water distribution network of Example 11.4 using the minimum cut-set method.

Solution: Based on the definition of the minimum cut set, seven minimum cut sets are characterized in the example water distribution network as follows (Figure 11.9):

$$C_1: F_1 \quad C_2: F_2 \cap F_3 \quad C_3: F_2 \cap F_4 \quad C_4: F_3 \cap F_5 \quad C_5: F_4 \cap F_5 \quad C_6: F_2 \cap F_5 \quad C_7: F_4 \cap F_3,$$

where F_k is the failure state of pipe k. The system failure probability, $p_{f,\text{sys}}$, is equal to the probability of occurrence of the union of the cut sets and the system reliability, $p_{s,\text{sys}}$, is obtained by subtracting it from 1 as follows.

$$p_{s,\text{sys}} = 1 - p\left[\bigcup_{i=1}^{7} C_i\right].$$

Based on the probability theory, the above equation is reformed as

$$p_{s,\text{sys}} = 1 - p\left[\bigcup_{i=1}^{7} C_i\right] = p\left[\bigcap_{i=1}^{7} C_i'\right].$$

continued

Since all cut sets behave independently, all their complements also behave independently. The probability of the intersection of a number of independent events is

$$p_{s,\text{sys}} = p\left[\bigcap_{i=1}^{7} C_i'\right] = \prod_{i=1}^{7} p(C_i')$$

and based on the definition of minimum cut sets the probability of each of them is calculated as

$$p(C_1') = 0.92, \quad p(C_2') = p(C_3') = \cdots = p(C_7') = 0.92 \times 0.92 = 0.85.$$

Thus the total system reliability of the example water distribution network is 0.34.

Tie-set analysis. A tie set is a minimal path of the system in which system components or modes of operation are arranged in series. Therefore, failure of any component or mode in a tie set results in tie set failure. Since the tie sets are connected in parallel, the system serviceability continues if any of its tie sets act successfully. In this method, the system reliability is formulated as

$$p_{s,\text{sys}} = p\left[\bigcup_{i=1}^{I} T_i\right] = p\left[\bigcup_{i=1}^{I}\left(\bigcap_{j=1}^{J_i} F_{ij}'\right)\right], \quad (11.14)$$

where T_i is the ith tie set; F_{ij}' refers to the nonfailure state of the jth component in the ith tie set; J_i is the number of components or modes of operation in the ith tie set; and I is the number of tie sets in the system. In the cases with a large number of tie sets, bounds for system reliability could be computed.

The main shortcoming of this method is that it is not directly related to failure states. The minimum tie sets of WDS of Figure 11.7 are depicted in Figure 11.10.

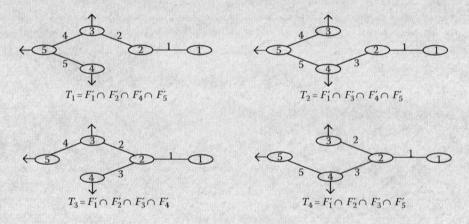

FIGURE 11.10 Minimum tie sets of the illustrated water distribution network in Figure 11.7.

EXAMPLE 11.6

Calculate the reliability of the water distribution network of Example 11.4 using the tie-set method.

Solution: The four distinguished minimum tie sets of the system are illustrated in Figure 11.10 and can be expressed as

$$T_1 : F_1' \cap F_2' \cap F_4' \cap F_5';$$
$$T_2 : F_1' \cap F_3' \cap F_4' \cap F_5';$$
$$T_3 : F_1' \cap F_2' \cap F_3' \cap F_4';$$
$$T_4 : F_1' \cap F_2' \cap F_3' \cap F_5'.$$

The system reliability is equal to the union of these paths and is formulated as follows based on the probability theory:

$$\begin{aligned} p_{s,sys} &= p(T_1 \cup T_2 \cup T_3 \cup T_4) \\ &= [p(T_1) + p(T_2) + p(T_3) + p(T_4)] \\ &\quad - [p(T_1, T_2) + p(T_1, T_3) + p(T_1, T_4) + p(T_2, T_3) + p(T_2, T_4) + p(T_3, T_4)] \\ &\quad + [p(T_1, T_2, T_3) + p(T_1, T_2, T_4) + p(T_1, T_3, T_4) + p(T_2, T_3, T_4)] \\ &\quad - p(T_1, T_2, T_3, T_4). \end{aligned}$$

Because of independency of pipes, the joint probability of tie-set occurrence is equal to the product of the probabilities of individual events. This results in

$$p(T_1) = p(F_1')p(F_2')p(F_4')p(F_5') = (0.92)^4 = 0.7164,$$
$$P(T_2) = P(T_3) = P(T_4) = 0.7164.$$

It should be noted that in this example, the union of two and more tie sets is equal to the intersections of the nonfailure state of all five pipes. As an example,

$$T_1 \cup T_2 = (F_1' \cap F_2' \cap F_4' \cap F_5') \cup (F_1' \cap F_3' \cap F_4' \cap F_5') = (F_1' \cap F_2' \cap F_3' \cap F_4' \cap F_5').$$

Therefore, the system reliability could be summarized as

$$\begin{aligned} p_{s,sys} &= [p(T_1) + p(T_2) + p(T_3) + p(T_4)] - 6p(F_1' \cap F_2' \cap F_3' \cap F_4' \cap F_5') \\ &\quad + 4p(F_1' \cap F_2' \cap F_3' \cap F_4' \cap F_5') - p(F_1' \cap F_2' \cap F_3' \cap F_4' \cap F_5') \\ &= 4(0.7164) - 3(0.92)^5 = 0.8884. \end{aligned}$$

In summary, the path enumeration method involves the following steps (Henley and Gandhi, 1975):

1. Distinguish all minimum paths.
2. Calculate the required unions of the minimum paths.
3. Give each path union a reliability expression in terms of module reliability.
4. Compute the total system reliability based on module reliabilities.

11.8.1.2 Reliability Analysis: Load-Resistance Concept

External loads in combination with uncertainties in analysis, design, build, and operation with their random nature lead to system failure once the system resistance fails to withstand. An urban system fails when the external loads, L (e.g., demand), exceeds the system resistance, R_e (e.g., capacity and water supply limitations in the water resources system). Reliability (R) of any component or the entire system is equal to its safety probability:

$$R = p[L \leq R_e]. \tag{11.15}$$

In hydrology and hydraulics, load and resistance are functions of some random variables:

$$R_e = h(X_{R_e}), \tag{11.16}$$

$$L = g(X_L). \tag{11.17}$$

So reliability is a function of random variables:

$$R = p\big[g(X_L) \leq h(X_{R_e})\big]. \tag{11.18}$$

Reliability variations with time are not considered in the above equation and X_{R_e} and X_L are the stationary random variables, and the resulting model is called the static reliability model. The word "static," from the reliability computation point of view, represents the worst single stress, or load, applied. Actually, the loading applied to many hydraulic systems is a random variable. Also, the number of times a loading is imposed is random. The static reliability model is used for the evaluation of system performance in a special situation (the most critical loading).

11.8.1.3 Reliability Indices

For reliability analysis, a performance function $W(X)$ of X_L and X_{R_e}, which are system load and resistance, respectively, proposed by Mays (2001) is defined.

$$W(X) = W(X_L, X_{R_e}), \tag{11.19}$$

$$R = p\big[W(X_L, X_{R_e}) \geq 0\big] = p[W(X) \geq 0]. \tag{11.20}$$

$W(X) = 0$ is called failure surface or limit. $W(X) \geq 0$ is the safety region and $W(X) < 0$ is the failure region. The performance function $W(X)$ could be defined in different forms such as

$$W_1(X) = R_e - L = h(X_{R_e}) - g(X_L), \tag{11.21}$$

$$W_2(X) = \left(\frac{R_e}{L}\right) - 1 = \left[\frac{h(X_{R_e})}{g(X_L)}\right] - 1, \tag{11.22}$$

$$W_3(X) = \ln\left(\frac{R_e}{L}\right) = \ln\big[h(X_{R_e})\big] - \ln\big[g(X_L)\big]. \tag{11.23}$$

Another reliability index is the β reliability index, which is equal to the inverse of the coefficient of variation of the performance function, $W(X)$, and is calculated as

$$\beta = \frac{\mu_w}{\sigma_w}, \tag{11.24}$$

where μ_w and σ_w are mean and standard deviation of the performance function, respectively.

By assuming an appropriate probability density function (PDF) for random performance functions, $W(X)$, using Equation 11.20 the reliability is estimated as

$$R = 1 - F_W(0) = 1 - F_{W'}(-\beta), \tag{11.25}$$

where F_W is the cumulative distribution function of variable W, and W' is the standardized variable:

$$W' = \frac{W - \mu_W}{\sigma_W}. \tag{11.26}$$

11.8.1.4 Direct Integration Method

Following the reliability definition, the reliability is expressed by Mays (2001) as

$$R = \int_0^\infty f_{R_e}(r) \left[\int_0^r f_L(l)\, dl \right] dr = \int_0^\infty f_{R_e}(r) F_L(l)\, dr, \tag{11.27}$$

where $f_{R_e}()$ and $f_L()$ are the PDFs of resistance and load functions, respectively. If load and resistance are independent, then reliability is formulated as:

$$R = E_{R_e}[F_L(r)] = 1 - E_L\left[F_{R_e}(l)\right], \tag{11.28}$$

where $E_{R_e}[F_L(r)]$ is the expected value of the cumulative distribution function of load in the probable resistance limitation. The reliability computations require the knowledge of the probability distributions of loading and resistance. A schematic diagram of the reliability analysis by Equation 11.27 is shown in Figure 11.11.

To illustrate the computation procedure involved, the exponentially distribution has been considered for loading L and the resistance R_e:

$$f_L(l) = \lambda_L e^{-\lambda_L l}, \quad l \geq 0, \tag{11.29}$$

$$f_{R_e}(r) = \lambda_{R_e} e^{-\lambda_{R_e} r}, \quad r \geq 0. \tag{11.30}$$

The static reliability can then be derived as

$$R = \int_0^\infty \lambda_{R_e} e^{-\lambda_{R_e} r} \left[\int_0^r \lambda_L e^{-\lambda_L l}\, dl \right] dr = \int_0^\infty \lambda_{R_e} e^{-\lambda_{R_e} r} \left[1 - e^{-\lambda_L r}\right] dr = \frac{\lambda_L}{\lambda_{R_e} + \lambda_L}. \tag{11.31}$$

For some special combinations of load and resistance distributions, the static reliability can be derived analytically in the closed form. Kapur and Lamberson (1977) considered the loading L and resistance R_e to be log normally distributed and computed reliability as

$$R = \int_0^\infty \phi(z)\, dz = \Phi(z), \tag{11.32}$$

where $\phi(z)$ and $\Phi(z)$ are the PDF and the cumulative distribution function, respectively, for the standard normal variant z:

$$z = \frac{\mu_{\ln R_e} - \mu_{\ln L}}{\sqrt{\sigma_{\ln R_e}^2 - \sigma_{\ln L}^2}}. \tag{11.33}$$

The values of the cumulative distribution function $\Phi(z)$ for the standard normal variant are given in normal distribution tables (Mays, 2001).

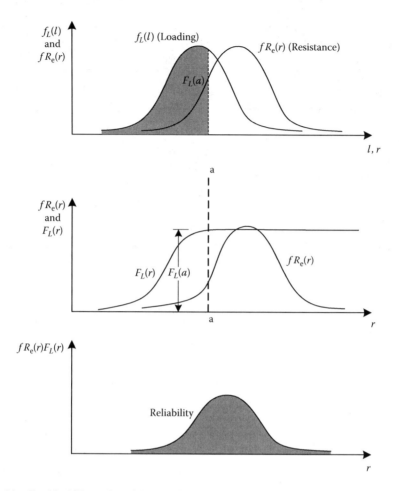

FIGURE 11.11 Graphical illustration of the steps involved in reliability computation. (From Mays, L.W. 2001. *Water Resources Engineering*. Wiley, New York. With permission.)

By considering exponential distribution for the load and the normal distribution for the resistance, the reliability was expressed by Kapur and Lamberson (1977) as

$$R = 1 - \Phi\left(\frac{\mu_{R_c}}{\sigma_{R_c}}\right) - \exp\left[-\frac{1}{2}\left(2\mu_{R_c}\lambda_L - \lambda_L^2 \sigma_{R_c}^2\right)\right] \times \left[1 - \Phi\left(-\frac{\mu_{R_c} - \lambda_L \sigma_{R_c}^2}{\sigma_{R_c}}\right)\right]. \quad (11.34)$$

EXAMPLE 11.7

A storage tank serves a WDS through a 60 cm diameter cast iron pipe 1.5 km long. The head elevation of the storage tank is maintained at a constant height of 30 m above the elevation at the user end. The required pressure head at the user end is fixed at 150 kPa with variable demand on flow rate. Assume that the demand is a random variable with a log-normal distribution with the mean and the standard deviation of 0.027 and 0.008 m³/s, respectively. The pipe has been installed for about 5 years so that the pipe roughness is not certain. The studies show that the Hazen–Williams C coefficient has a log-normal distribution with a mean of 130 and a standard deviation of 20. Estimate the reliability of justifying the water demand in this WDS.

Solution: In this example, the resistance of the system is the water supply from the source, and the loading is the water demand by the user. Both supply and demand are random variables. By the Hazen–Williams equation, the supply is calculated as

$$\text{Resistance} = Q_s = \frac{C}{231.6}\left(\frac{\Delta h}{L}\right)^{0.54} D^{2.63},$$

where Δh is the head difference (m) between the source and the user, D is the pipe diameter (m), and L is the pipe length (m). The roughness coefficient C is a random variable, so is the supply. Due to the multiplicative form of the Hazen–Williams equation, the logarithmic transformation leads to a linear relation among variables, that is,

$$\ln R_e = \ln C - \ln(231.6) + 0.54 \ln\left[\frac{30 - 150 \times 10 \times 0.01}{1500}\right] + 2.63 \ln(0.6) = \ln C - 9.28.$$

Since the roughness coefficient C is log normally distributed, $\ln C$ is normally distributed, as is the resistance. From the moment relations given for log-normal distribution, the mean and the standard deviation of $\ln C$ are determined as

$$\sigma_{\ln C} = \sqrt{\ln(1 + \Omega_C^2)} = \sqrt{\ln\left(1 + \left(\frac{20}{130}\right)^2\right)} = 0.153$$

and

$$\mu_{\ln C} = \ln \mu_C - \frac{\sigma_{\ln C}^2}{2} = \ln(130) - \frac{0.153^2}{2} = 4.856.$$

From these results, the mean and the standard deviation of $\ln R_e$ are -4.419 and 0.153, respectively.

Because the water demand (loading) has a log-normal distribution, the mean and the standard deviation of its log-transformed scale can be calculated in the same manner as for the roughness coefficient C. That is,

$$\sigma_{\ln L} = \sqrt{\ln\left(1 + \left(\frac{0.008}{0.027}\right)^2\right)} = 0.290 \quad \text{and}$$

$$\mu_{\ln L} = \ln(0.027) - \left(\frac{0.29^2}{2}\right) = -3.654.$$

Knowing the distributions and statistical properties of the load (water demand) and resistance (water supply), the system reliability can be calculated as

$$R = \Phi\left[z = \frac{-4.419 - (-3.654)}{\sqrt{0.153^2 + 0.290^2}}\right] = \Phi(z = -2.333) = 1 - \Phi(2.333) = 0.010.$$

This means that the water demand by the user is met with a probability of 1.0% of the time.

11.8.1.5 Margin of Safety

The margin of safety (MS) is defined as the difference between the project capacity (resistance) and the value calculated for the design loading, MS $= R_e - L$. The reliability is equal to the probability that $R_e > L$, or equivalently,

$$R = p(R_e - L > 0) = p(MS > 0). \tag{11.35}$$

If R_e and L are independent random variables, then the mean value of MS is given by

$$\mu_{MS} = \mu_{R_e} - \mu_L \tag{11.36}$$

and its variance by

$$\sigma_{MS}^2 = \sigma_{R_e}^2 + \sigma_L^2. \tag{11.37}$$

If the MS is normally distributed, then

$$z = \frac{MS - \mu_{MS}}{\sigma_{MS}}, \tag{11.38}$$

where z is a standard normal variant. By subtracting μ_{MS} from both sides of the inequality in Equation 11.35 and dividing both sides by σ_{MS}, it can be seen that (Mays, 2001)

$$R = p\left(z \leq \frac{\mu_{MS}}{\sigma_{MS}}\right) = \Phi\left(\frac{\mu_{MS}}{\sigma_{MS}}\right). \tag{11.39}$$

The key assumption of this analysis is that it considers that the MS is normally distributed but does not specify the distributions of loading and capacity. Ang (1973) indicated that, provided $R > 0.001$, R is not greatly influenced by the choice of distribution for R_e and L and the assumption of a normal distribution for MS is satisfactory. For lower risk than this (e.g., $R = 0.00001$), the shape of the tails of the distributions for R_e and L becomes critical, in which case accurate assessment of the distribution of the MS of the direct integration procedure should be used to evaluate the risk or probability of failure.

EXAMPLE 11.8

Apply the MS approach to evaluate the reliability of the simple WDS described in Example 11.7.

Solution: Calculate the mean and the standard deviation of the resistance (i.e., water supply) as

$$\mu_{R_e} = \exp\left(\mu_{\ln R_e} + \frac{1}{2}\sigma_{\ln R_e}^2\right) = \exp\left(-4.419 + \frac{1}{2}(0.153)^2\right) = 0.012 \text{ cms}$$

and

$$\sigma_{R_e} = \sqrt{\mu_{R_e}^2 \left(\exp(\sigma_{\ln R_e}^2) - 1\right)} = \sqrt{0.012^2 \left(\exp(0.153)^2 - 1\right)} = 0.007 \text{ cms}.$$

From the problem statement, the mean and the standard deviation of the load (water demand) are $\mu_L = 0.027$ cms and $\sigma_L = 0.008$ cms, respectively. Therefore, the mean and the variance of the MS can be calculated as

$$\mu_{MS} = \mu_{R_e} - \mu_L = 0.012 - 0.027 = -0.015 \text{ cms},$$

$$\sigma_{MS}^2 = \sigma_{R_e}^2 + \sigma_L^2 = (0.007)^2 + (0.008)^2 = 0.0001 \text{ (cms)}^2.$$

Now, the reliability of the system can be assessed, by the MS approach using the table of normal distribution, as

$$R = \Phi\left[\frac{-0.015}{\sqrt{0.0001}}\right] = \Phi(-1.5) = 1 - \Phi(1.5) = 1 - 0.933 = 0.067.$$

> The reliability computed by the MS method is not identical to that of the direct integration method, although the difference is practically negligible. It should be noted that it is assumed that the distribution of the MS in this example is normal. This is not always true; therefore, the reliability obtained should be regarded as an approximation of true reliability.

11.8.1.5.1 Factor of Safety

The factor of safety (FS) is given by the ratio R_e/L and the reliability can be specified by $p(\text{FS} > 1)$. Several factors of safety measures and their usefulness in hydraulic engineering are discussed by Yen (1978) and Mays and Tung (1992). By taking the logarithm of both sides of this inequality, it yields

$$R = p(\text{FS} > 1) = p(\ln(\text{FS}) > 0) = p\left(\ln\left(\frac{R_e}{L}\right) > 0\right),$$

$$R = p\left(z \leq \frac{\mu_{\ln \text{FS}}}{\sigma_{\ln \text{FS}}}\right) = \Phi\left(\frac{\mu_{\ln \text{FS}}}{\sigma_{\ln \text{FS}}}\right), \quad (11.40)$$

where $z = (\ln(\text{FS}) - \mu_{\ln \text{FS}})/\sigma_{\ln \text{FS}}$ and Φ is the standard cumulative normal probability distribution. It is assumed that the FS is normally distributed.

If the resistance and loading are independent and log normally distributed, then the risk can be expressed as (Mays, 2001)

$$\overline{R} = \Phi\left\{\frac{\ln\left[\left(\overline{Q}_{R_c}/\overline{Q}_L\right)\sqrt{\left(1+\Omega_L^2\right)/\left(1+\Omega_{R_c}^2\right)}\right]}{\ln\left(\sqrt{\left(1+\Omega_L^2\right)\left(1+\Omega_{R_c}^2\right)}\right)}\right\}. \quad (11.41)$$

11.8.1.6 Mean Value First-Order Second Moment (MFOSM) Method

Mean and standard deviation of the performance function, $W(X)$, are estimated using the first-order variance method in uncertainty analysis (a detailed discussion on uncertainty analysis is given in Section 11.9.3). These values are used for estimation of the β reliability index as

$$\beta = \frac{W(\mu)}{\sqrt{S^T C(X) S}}, \quad (11.42)$$

where μ is the mean vector, $C(X)$ is the covariance matrix of random variable X, and S is the N-dimensional matrix equal to the partial differential of W to X (Mays, 2001).

Studies have shown that in a confidence boundary of <0.99, the reliability value is not sensitive to the distribution function of W and assumption of a normal distribution for W would be sufficient. But for confidence levels that are more restrict than 0.99, such as 0.995, the extreme edges of the distribution function of W will be critical.

The application of this model is easy; however, it must be noted that there are some weaknesses because of the nonlinearity and noninvariability of the performance function in critical situations. Therefore the application of this method in the following situations is not recommended:

- Estimation of risk and reliability with high precision
- Highly nonlinear performance function
- Existence of random variables with high skewness in performance function

In order to decrease the errors of this method, another method called the advanced first-order second-moment (AFOSM) method has also been developed by Tung (1996).

11.8.1.7 AFOSM Method

The main thrust of the AFOSM method is to reduce the error of the MFOSM method, while keeping the advantages and simplicity of the first-order approximation. The expansion point in the AFOSM method lies on the failure surface defined by $W(X) = 0$. Among all the possible values of X that fall on the limit state surface, one is more concerned with the combination of stochastic variables that would yield the lowest reliability or highest risk. The point on the failure surface with the lowest reliability is the one having the shortest distance to the point where the means of stochastic variables are located. This point is called the design point or the most probable failure point. With the mean and standard deviation of the performance function computed at the design point, the AFOSM reliability index can be determined.

Owing to the nature of nonlinear optimization, the algorithm AFOSM does not necessarily converge to the true design point associated with the minimum reliability index. Therefore, different initial trial points are used and the smallest reliability index is selected to measure the reliability (Mays, 2001).

11.8.2 Time-to-Failure Analysis

Any system will fail eventually; it is just a matter of time. Due to the presence of many uncertainties that affect the operation of a physical system, the time when the system fails and its performance is unsatisfactory, are intended as random variables. Instead of considering detailed interactions of resistance and loading over time, a system or its components can be treated as a black box or a lumped parameter system and their performances are observed over time. This reduces the reliability analysis to a one-dimensional problem involving time as the only random variable.

11.8.2.1 Failure and Repair Characteristics

For a repairable system or component, its service life can be extended indefinitely if repair work can restore the system as if it was new. Intuitively, the probability of a repairable system available for service is greater than that of a nonrepairable system.

For repairable urban water systems, such as pipe networks, pump stations, and storm runoff drainage structures, failed components within the system can be repaired or replaced so that the system can be put back into service. The time required to have the failed system repaired is uncertain and, consequently, the total time required to restore the system from its failure state to its operational state is a random variable. Like the time to failure (TTF), the random time to repair (TTR) has the repair density function describing the random characteristics of the time required to repair a failed system when failure occurs at time zero. The repair probability is the probability that the failed system can be restored within a given time period and it is sometimes used for measuring the maintainability or repairability. They are usually expressed in the form of resiliency discussed in Section 11.8.3.

11.8.2.2 Availability and Unavailability

The term "availability" is generally used for repairable systems to indicate the probability that the system is in operating condition at any given time t. On the other hand, reliability is appropriate for nonrepairable systems indicating the probability that the system has been continuously in its operating state starting from time zero up to time t.

Availability can also be interpreted as the percentage of time that the system is in operating condition within a specified time period. On the other hand, unavailability is the percentage of time that the system is not available for the intended use during a time period, given that it is operational at time zero.

11.8.3 Resiliency

Resiliency describes how quickly a system recovers from failure once failure has occurred. In other words, resiliency is basically a measure of the duration of an unsatisfactory condition. Resiliency could be considered as the most important indicator of system performance in disaster recovery and how successful the management of disaster strategies has been. In general form, the resiliency of a system can be calculated as (Karamouz et al., 2003)

$$\beta = \text{Prob}[X_{t+1} \in S \mid X_t \in F], \qquad (11.43)$$

where S and F are the set of all satisfactory and unsatisfactory outputs, respectively. The application of this definition in practical cases has been discussed in Section 11.11.1.

Developing resilience systems and living with disaster are the major challenges of the recent decades. Alternative solutions have been explored and the concept of resilience has been introduced into risk management of all the water planning and management fields. The principle of resilience is from ecology. Holling (1973) used the following definition for resiliency: "a measure of the persistence of systems and of their ability to absorb changes and disturbances and still maintain the same relationships between populations and state variables." Pimm (1991) has defined resiliency as follows: "the speed with which a system disturbed from equilibrium recovers some proportion of its equilibrium." More generally, "resiliency" could be considered as a tendency to stability and as resistance to perturbation. There is no general and overall accepted definition for resiliency, although everybody agrees that it is desirable.

Resilience has also been defined as the ability of a system to persist if exposed to a perturbation by recovering after the response. Using this definition of resiliency, it is the opposite of resistance, which is the ability of a system to persist if disturbed, without showing any reaction at all.

The resiliency concept definition and formulation is specialized regarding the field of application. For example, De Bruijn and Klijn (2001) defined resilience in the context of flood risk management. Strategies for flood risk management in which resilience is used focus on reducing the impact of floods by "living with floods" instead of "fighting floods," as in the traditional strategy. Therefore, resilient flood risk management is the flood risk management that aims at giving room to the floods, but with concurrent impact minimization. This also implies that the consequences of floods have to be taken into account and that safety standards must be differentiated on the basis of land-use and spatial planning. The area as a whole is more resilient if the less valuable parts are flooded prior to the more valuable parts. A resilient flood risk management strategy also considers measures to reduce the impacts of flooding, such as the design of warning systems and evacuation plans and the application of spatial planning and building regulations. Resilience strategies may also include measures to accelerate the recovery after a disaster, for example, damage compensation and insurances.

11.8.4 Vulnerability

Vulnerability is the distance of failed character state from the desired condition in historical time series. There are three ways of vulnerability estimation. Some researchers consider the maximum distance from the desired situation, other groups consider the expected value of system failure cases, and the last group considers the probability of failure occurrence. Among these different ideas, the second idea is more common. Considering vulnerability as the expected value of system failure, the vulnerability of a system is defined as

$$\text{vulnerability} = \frac{\text{no. of situations with positive } (X_T - X_t)}{\text{no. of failure situations}},$$

where X_T is the desired situation and X_t is the system situation.

Using the probability of failure occurrence, the vulnerability of a system can be calculated by the following formula:

$$v = \sum_{j \in F} s_j e_j, \qquad (11.44)$$

where e_j is the probability that x_j corresponds to s_j.

Vulnerability analysis is a worthwhile tool in evaluating the entire event chain causing water disaster. Recovery analysis means a systematic investigation of how a system returns to the state of normal operation (Karamouz et al., 2003). Further discussion on using this index for the evaluation of system performance in practical cases is given in Section 11.11.1.

11.9 UNCERTAINTIES IN URBAN WATER ENGINEERING

Urban water engineering design and analysis deal with the occurrence of water in various parts of a hydrosystem and its effects on environmental, ecological, and socioeconomical settings. Due to the extremely complex nature of the physical, chemical, biological, and socioeconomical processes involved, major efforts have been put by different investigations to have a better understanding of the processes.

In general, uncertainty due to the inherent randomness of physical processes cannot be eliminated. On the other hand, uncertainties such as those associated with lack of complete knowledge about the process, data, models, parameters, and so on could be reduced through data collection, research, and careful design and implementation/manufacturing. In urban water engineering, uncertainties involved can be divided into four basic categories: hydrological, hydraulic, structural, and economic. More specifically, in urban water engineering analyses and designs, uncertainties could arise from the various sources including natural or intrinsic uncertainties, model uncertainties, parameter uncertainties, data uncertainties, and operational uncertainties.

Natural uncertainty is associated with the inherent randomness of natural processes such as the occurrence of precipitation and flood events. The occurrence of hydrological events often displays variations in time and space. Their occurrences and intensities could not be predicted precisely in advance. Because a model is only an abstraction of the reality that generally involves certain simplifications, model uncertainty reflects the inability of a model or design technique to represent precisely the system's true physical behavior. Parameter uncertainties result from the inability to quantify accurately the model inputs and parameters. Parameter uncertainty could also be caused by a change in operational conditions of hydraulic structures, inherent variability of inputs and parameters in time and space, and a lack of sufficient data.

Data uncertainties include (1) measurement errors, (2) inconsistency and nonhomogeneity of data, (3) data handling and transcription errors, and (4) inadequate representation of the data sample due to time and space limitations. Operational uncertainties include those associated with construction, manufacturing, deterioration process, maintenance impact, and human factors. The magnitude of this type of uncertainty is largely dependent on the workmanship and quality control during the construction and manufacturing.

11.9.1 IMPLICATIONS AND ANALYSIS OF UNCERTAINTY

In urban water engineering design and modeling, the design quantity and system output are functions of several system parameters; not all of them can be quantified with absolute accuracy. The task of uncertainty analysis is to quantify the uncertainty features of the system outputs as a function of model characteristics and the stochastic variables involved. It provides a formal and a system framework to quantify the uncertainty associated with the system's output. Furthermore, it offers the designer useful insights regarding the contribution of each stochastic variable to the overall uncertainty of the system outputs. Such knowledge is essential to identify the "important" parameters to which more attention should be given to have a better assessment of their values and, accordingly, to reduce the overall uncertainty of the system outputs.

11.9.2 Measures of Uncertainty

Several expressions have been used to describe the level of uncertainty of a parameter, a function, a model, or a system. In general, the uncertainty associated with the latter three is a result of the combined effect of the uncertainties of the contributing parameters. The most complete and ideal description of uncertainty is the PDF of the quantity subject to uncertainty. However, in most practical problems, such a probability function cannot be derived or obtained precisely.

Another measure of the uncertainty of a quantity is to express it in terms of a reliability domain such as the confidence interval. A confidence interval is a numerical interval that would capture the quantity subject to uncertainty with a specified probabilistic confidence. Nevertheless, the use of confidence intervals has a few drawbacks: (1) The parameter population may not be normally distributed as assumed in the conventional procedures, and this problem is particularly important when the sample size is small. (2) No means is available to directly combine the confidence intervals of individual contributing random components to give the overall confidence interval of the system. A useful alternative to quantify the level of uncertainty is to use the statistical moments associated with a quantity subject to uncertainty. In particular, the variance and standard deviation that measure the dispersion of a stochastic variable are commonly used.

11.9.3 Analysis of Uncertainties

In the design and analysis of hydrosystems, many quantities of interest are functionally related to a number of variables, some of which are subject to uncertainty. A rather straightforward and useful technique for the approximation of uncertainties is the first-order analysis of uncertainties, sometimes called the delta method.

The use of the first-order analysis of uncertainties is quite popular in many fields of engineering because of its relative ease in application to a wide array of problems. First-order analysis is used to estimate the uncertainty in a deterministic model formulation involving parameters that are uncertain (not known with certainty). More specifically, first-order analysis enables one to estimate the mean and variance of a random variable that is functionally related to several other variables, some of which are random. By using first-order analysis, the combined effect of uncertainty in a model formulation, as well as the use of uncertain parameters, can be assessed.

Consider a random variable y that is a function of k random variables (e.g., the pressure of a specific point in WDS is a function of demands in different points that could be considered as random variables):

$$y = g(x_1, x_2, \ldots, x_k). \tag{11.45}$$

This can be a deterministic equation such as the rational formula or the Darcy–Veisbakh equation or this function can be a complex model that must be solved on a computer. The objective is to treat a deterministic model that has uncertain inputs in order to determine the effect of the uncertain parameters $x_1, x_2, \ldots, x_k$ on the model output y.

Equation 11.45 can be expressed as $y = g(X)$, where $X = x_1, x_2, \ldots, x_k$. Through a Taylor series expansion about k random variables, ignoring the second- and higher-order terms, we obtain

$$y \approx g(\overline{X}) + \sum_{i=1}^{k} \left[\frac{\partial g}{\partial x_i}\right]_{\overline{X}} (x_i - \overline{X}). \tag{11.46}$$

The derivations $\left[\partial g/\partial x_i\right]_{\overline{X}}$ are the sensitivity coefficients that represent the rate of change of the function value $g(\overline{X})$ at $x = \overline{X}$ (Mays, 2001).

Assuming that the k random variables are independent, the variance of y is approximated as

$$\sigma_y^2 = \text{Var}[y] = \sum a_i^2 \sigma_{x_i}^2. \tag{11.47}$$

And the coefficient of variation, Ω_y, which is often used as the measure of uncertainty, is estimated as

$$\Omega_y = \left[\sum_{t=1}^{k} a_i^2 \left(\frac{\bar{x}_i}{\mu_y}\right)^2 \Omega_{x_i}^2\right]^{1/2}, \tag{11.48}$$

where $a_i = (\partial g/\partial x)_{\bar{X}}$. Refer to Mays and Tung (1992) for a detailed derivation of Equations 11.47 and 11.48.

Example 11.9

Appling the first-order analysis, formulate σ_Q and Ω_Q in Manning's equation $[Q = (0.311/n)S^{1/2}D^{8/3}]$, where diameter D is a deterministic parameter and n and S are considered to be uncertain.

Solution: Since n and S are uncertain, Manning's equation can be rewritten as

$$Q = Kn^{-1}S^{1/2},$$

where $K = 0.311D^{8/3}$. The first-order approximation of Q is determined using the above equation, so that

$$Q \approx \bar{Q} + \left[\frac{\partial Q}{\partial n}\right]_{\bar{n},\bar{S}} (n-\bar{n}) + \left[\frac{\partial Q}{\partial S}\right]_{\bar{n},\bar{S}} (S-\bar{S})$$

$$= \bar{Q} + \left[-K\bar{n}^{-2}\bar{S}^{1/2}\right](n-\bar{n}) + \left[0.5K\bar{n}^{-1}\bar{S}^{-1/2}\right](S-\bar{S}),$$

where $\bar{Q} = K\bar{n}^{-1}\bar{S}^{1/2}$.

The variance of the pipe capacity can be computed as

$$\sigma_Q^2 = \left[\frac{\partial Q}{\partial n}\right]_{\bar{n},\bar{S}}^2 \sigma_n^2 + \left[\frac{\partial Q}{\partial S}\right]_{\bar{n},\bar{S}}^2 \sigma_S^2,$$

$$\sigma_Q = \left\{\left[\frac{\partial Q}{\partial n}\right]_{\bar{n},\bar{S}}^2 \sigma_n^2 + \left[\frac{\partial Q}{\partial S}\right]_{\bar{n},\bar{S}}^2 \sigma_S^2\right\}^{1/2}.$$

The coefficient of variation of Q is determined as

$$\Omega_Q^2 = \sum_{i=1}^{2} \left[\frac{\partial Q}{\partial x_i}\right]^2 \left[\frac{\bar{x}_i}{\bar{Q}}\right]^2 \Omega_{x_i}^2,$$

$$\Omega_Q^2 = \left[\frac{\partial Q}{\partial n}\right]_{\bar{n},\bar{S}}^2 \left[\frac{\bar{n}}{\bar{Q}}\right]^2 \Omega_n^2 + \left[\frac{\partial Q}{\partial S}\right]_{\bar{n},\bar{S}}^2 \left[\frac{\bar{S}}{\bar{Q}}\right]^2 \Omega_S^2,$$

$$\Omega_Q^2 = \left[\frac{-K\bar{S}^{1/2}}{\bar{n}^2}\right]^2 \left[\frac{\bar{n}}{\bar{Q}}\right]^2 \Omega_n^2 + \left[\frac{0.5K}{\bar{n}\bar{S}^{1/2}}\right]^2 \left[\frac{\bar{S}}{\bar{Q}}\right]^2 \Omega_S^2,$$

$$\Omega_Q^2 = \left[\frac{-K\overline{S}^{1/2}}{\overline{Q}}\right]^2 \left[\frac{1}{\overline{n}^2}\right]\Omega_n^2 + (0.5)^2 \left[\frac{K}{\overline{n}\,\overline{S}^{1/2}}\right]^2 \left[\frac{\overline{S}}{\overline{Q}}\right]^2 \Omega_S^2,$$

$$\Omega_Q^2 = \left[\overline{n}^2\right]\left[\frac{1}{\overline{n}^2}\right]\Omega_n^2 + 0.25\left[\frac{1}{\overline{S}}\right]\left[\overline{S}\right]\Omega_S^2,$$

$$\Omega_Q^2 = \Omega_n^2 + 0.25\Omega_S^2,$$

$$\Omega_Q = \left[\Omega_n^2 + 0.25\Omega_S^2\right]^{1/2}.$$

EXAMPLE 11.10

Determine the mean capacity of a storm sewer pipe, the coefficient of variation of the pipe capacity, and the standard deviation of the pipe capacity using Manning's equation (refer to Example 11.9). The following parameter values can be considered as shown in Table 11.1.

Solution: Manning's equation for full pipe flow is $Q = (0.311/N)S^{1/2}D^{8/3}$, so for first-order analysis we have

$$\overline{Q} = \frac{0.311}{0.015}(0.001)^{1/2}(1.5)^{8/3} = 1.93 \text{ m}^3/\text{s}.$$

Using the results of Example 11.9, it is found that

$$\Omega_Q = \left[(0.01)^2 + 0.25(0.25)^2\right]^{1/2} = 0.027,$$

$$\sigma_Q = \overline{Q}\Omega_Q = 1.93(0.027) = 0.052 \text{ (m}^3/\text{s}).$$

TABLE 11.1

Parameters of the Considered Storm Sewer Pipe in Example 11.10

Parameter	Mean	Coefficient of Variation
n	0.015	0.01
d	1.5 m	0
S	0.001	0.05

11.10 RISK ANALYSIS: COMPOSITE HYDROLOGICAL AND HYDRAULIC RISK

The risk of a hydraulic component, subsystem, or system is defined as the probability of the loading exceeding the resistance, that is, the probability of failure. The relationship between reliability R and risk $\overline{R}$ is

$$R = 1 - \overline{R}. \tag{11.49}$$

Therefore the same formulations of reliability can be used for the analysis of system risk.

11.10.1 Risk Management and Vulnerability

Systems that are highly exposed, sensitive, and less able to adapt are vulnerable. This is illustrated in Figure 11.12. In order to develop adaptation strategies, at first the systems/elements that are vulnerable to change should be identified. Then the scope should be determined to increase the coping capacity of those systems—their resilience—to decrease the vulnerability.

As illustrated in Figure 11.12, the vulnerability of a system is a function of three instinct factors, namely, exposure, sensitivity, and adaptive capacity. This approach to vulnerability assessment is important because it highlights the key elements that are combined to alleviate the risks and costs that disasters can impose on a system. Taking into account these factors can help us identify the extent of disasters. Developing action plans in each of these areas can help us reduce or deal with that different types of disasters.

There are many changes that may adversely affect the management of urban water systems. These changes may involve risks that can be caused by climatic variation (causing droughts or floods), mechanical failures, poor management, and health risks from pollution and waterborne diseases. The consequences can be considered in terms of financial, environmental, health, cultural and professional ethic impacts, short-term, long-term, and cumulative risks. Managing risk involves understanding the factors that contribute to the cause of that risk.

By understanding the complex interrelationships in the urban system, one can actively work toward reducing the factors that contribute to the consequences of risk. These factors include communications across resources availability and allocation and/or ecosystem variability. In this way, better decision-making processes can be implemented to create improved solutions for UWDs throughout a region. An example of this process might be the comparison between major water schemes and disaggregated household or community schemes.

Once the risk attributes are better understood, the effect of implementing suitable control strategies to address a particular risk can be evaluated. Sound risk-management planning does reduce risk, but rarely eliminates it. Contingency plans should be developed to effectively manage risk issues. For example, adequate capability to monitor public health and respond to outbreaks of widespread disease contamination should be provided.

In the United States, disaster prevention started with coalitions of scientists, emergency-relief organizations, professional associations, and other civic groups who lobbied governments to fund research and hazard mitigation strategies. This movement received impetus when the FEMA, armed

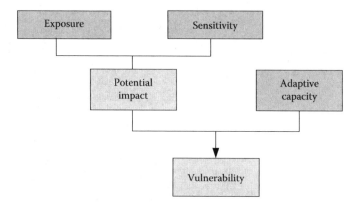

FIGURE 11.12 Vulnerability and its components. (From Allen Consulting Group. 2005. Climate Change Risk and Vulnerability Promoting an Efficient Adaptation Response. Department of Environment and Heritage Australian Green House Office. Available at: http://www.greenhouse.gov.au/impacts/publications/pubs/risk-vulnerability.pdf)

with a federal mandate and incentives, took the lead and promoted local and state initiatives (such as the regional Earthquake Preparedness Projects in California), but still worked through civic and professional partners.

The public needs to decide on acceptable levels of risk, comparing the immediate benefits of expenditures on other social priorities with the delayed benefits of reduced loss of life and asset replacement cost following a potential disaster. These trade-offs can be eased when well-designed incentives change private behavior to help prevent hazards. Examples include reducing insurance premiums on residential property when basic hazard-resistant steps are taken, offering disaster insurance with strict enforcement of building code provisions, or providing tax holidays or grants for mitigation. Poor residents, for whom insurance or fiscally based incentives may not be practical, would benefit from urban planning for slum prevention, enforceable environmental zoning in cities, and resettlement combined with community-based upgrading.

In Australia, there is a new disaster management authority to coordinate with all aspects of the response, working with NGOs, private sectors, universities, local communities, and external donors. The program includes predisaster preparedness and postdisaster response, reconstruction, and disaster prevention. Incentives are being introduced to build constituencies for disaster prevention by capitalizing on the population's awareness and willingness to change (Allen Consulting Group, 2005).

11.10.2 Risk-Based Design of Water Resources Systems

Reliability analysis can be applied to the design of various hydraulic structures with or without considering the risk costs that are associated with the failure of hydraulic structures or systems. The risk-based least cost design of hydraulic structures promises to be, potentially, the most significant application of reliability analysis. The risk-based design of water resources engineering structures integrates the procedures of economic, uncertainty, and reliability analyses in the design practice. Engineers using a risk-based design procedure consider trade-offs among various factors such as risk, economics, and other performance measures in hydraulic structure design. When risk-based design is embedded into an optimization framework, the combined procedure is called optimal risk-based design.

Because the cost associated with the failure of a hydraulic structure cannot be predicted from year to year, a practical way to quantify it is to use an expected value on the annual basis. The total annual expected cost is the sum of the annual installation cost and the annual expected damage cost.

In general, as the structural size increases, the annual installation cost increases, while the annual expected damage cost associated with failure decreases. The optimal risk-based design determines the optimal structural size, configuration, and operation such that the annual total expected cost is minimized.

In the optimal risk-based designs of hydraulic structures, the thrust of the exercise is to evaluate annual expected damage cost as a function of the PDFs of loading and resistance, damage function, and the types of uncertainty considered. The conventional risk-based hydraulic design considers only the inherent hydrological uncertainty associated with the random occurrence of loads. It does not consider hydraulic and economic uncertainties. Also, the probability distribution of the load to the water resources system is assumed to be known, which is generally not the case in reality. However, the evaluation of annual expected cost can be made by further incorporating the uncertainties in hydraulic and hydrological models and parameters.

To obtain an accurate estimation of annual expected damage associated with structural failure, it would require the consideration of all uncertainties, if they can be quantified. Otherwise, the annual expected damage would, in most cases, be underestimated, leading to an inaccurate optimal design.

11.10.3 Creating Incentives and Constituencies for Risk Reduction

Indispensable for mitigation strategies are strong disaster prevention proponents and the political will to lead regulatory changes and financial appropriations. With limited resources, developing countries must rely on partnerships of all actors. In the United States, disaster prevention started with coalitions of scientists, emergency-relief organizations, professional associations, and other civic groups who lobbied governments to fund research and hazard mitigation strategies. This movement received impetus when the FEMA, armed with a federal mandate and incentives, took the lead and promoted local and state initiatives (such as the regional Earthquake Preparedness Projects in California), but still worked through civic and professional partners.

The public needs to decide on acceptable levels of risk, comparing the immediate benefits of expenditures on other social priorities with the delayed benefits of reduced loss of life and asset replacement cost following a potential disaster. These trade-offs can be eased when well-designed incentives change private behavior to help prevent hazards. Examples include reducing insurance premiums on residential property when basic hazard-resistant steps are taken, offering disaster insurance with strict enforcement of building code provisions, or providing tax incentives or grants for mitigation. Poor residents, for whom insurance or fiscally based incentives may not be practical, would benefit from urban planning for slum prevention, enforceable environmental zoning in cities, and resettlement combined with community-based upgrading and tenure regularization schemes.

Recent disasters can motivate countries to undertake some of these measures and instill in them longer-term thinking to better enforce risk reduction. Gujarat state in India is trying to establish effective disaster management institutions following the January 2001 earthquake that killed 15,000 people. The state has a new disaster management authority to coordinate with all aspects of the response, working with NGOs, private sectors, universities, local communities, and external donors. The program includes predisaster preparedness and postdisaster response, reconstruction, disaster prevention, and risk reduction. Incentives are being introduced to build constituencies for disaster prevention by capitalizing on the population's heightened awareness and willingness to change.

Reducing risks caused by climate change may be more difficult since the risks mount gradually and less visibly but no less urgently. Coastal cities and other population centers (especially small island states) will need to invest in protective barriers and possibly to relocate residences and essential public facilities through managed retreat.

Priorities for such adaptation should be given to built areas and infrastructure that require urgent attention in any case, such as vulnerable informal settlements and outgrown sanitation and drainage systems. Adaptive expenditures will place a significant burden on the public sector, private utility companies, and, indirectly, on the urban economy. Low-income residents living with social and economic stress will need particular assistance.

11.11 SYSTEM READINESS

For increasing urban water systems readiness, the following steps can be taken:

- *Raising awareness on the integration of environment and disaster issues*: This is the first step and should be done at different levels, targeting (1) policy makers, to initiate policy dialogue at central, prefecture, and city levels; (2) professionals, to raise awareness among managers; and (3) individuals, to raise awareness at community levels.

Proposed actions: Arrange meetings, seminars, and forums at different levels.

- *Documenting and disseminating examples*: Documentation and its dissemination is one of the best ways to raise awareness among different stakeholders on the need for good cooperation for disaster preparedness.

Proposed actions: Undertake field surveys after different disaster events to document concrete examples and disseminate it through the Internet, and in training and study programs.

- Bridge gaps between knowledge and practice: To enhance the understanding of the need to integrate environment management and disaster management, gaps between knowledge and practice should be reduced. Experts, resource persons, and practitioners should have opportunities to interact and learn from each other's experiences, for example, in preparing disaster contingency plans.

Proposed actions: Undertake and facilitate training programs involving experts, resource persons, practitioners, and community leaders through forums.

- *Implementing practical examples*: While it is important to raise awareness, provide training, and initiate policy dialogue, it is also important to implement environment and disaster management practices at the local level.

Proposed actions: Undertake small projects at different locations having different socioeconomic contexts. These will be learning experiences, which can also be disseminated to other areas.

- *Developing guidelines and tools on environment and disaster management*: To provide a standard of environment and disaster management practices, guidelines and tools will have to be developed.

Proposed actions: Guidelines and tools should be developed based on the field experiences, and incorporating the comments and expertise of practitioners and professionals from different parts of the world.

- *Continuous monitoring of systems*: It is required to continuously monitor the progress in the practice of environment management and disaster preparedness, and to distribute the lessons to develop policies and strategies.

Proposed actions: A network or partnership should be developed that will continuously monitor different activities, identify best practices, and disseminate information through training and capacity building programs.

11.11.1 Evaluation of WDS Readiness

The objective of every urban WDS is to deliver enough water of acceptable quality with adequate pressure to different demand points. There are many interruptions and occasional disasters that impact the performance of the WDS; some could be quite devastating. The most common disasters are main breaks that may cause considerable water losses and bring up the system to partial or complete shutdown. Evaluation of the state of the system's readiness helps managers make better decisions to prevent disasters and respond better to emergencies. Nazif and Karamouz (2009) have quantified the state of the system's readiness in dealing with disasters using a hybrid index called the system readiness index (SRI). The three indices of a system's performance, namely reliability, resiliency, and vulnerability, are integrated through the demand-weighted average of pressure deficits at the critical nodes of WDS. Critical nodes are those with high demand and/or pressure variations. Available pressures at the demand nodes are the governing factor in assessing WDS performance. This is a new approach in system readiness evaluation.

The SRI in each week is calculated based on the demand-weighted average of pressure at the critical nodes:

$$\text{SRI}_t = -\sum_{j=1}^{k} \frac{\sum_{i=1}^{m} D_i \text{PD}_{ijt}}{\sum_{i=1}^{m} D_i} \times \frac{T_{jt}}{168}, \quad i \in S : \{\text{Set of critical nodes}\}, \qquad (11.50)$$

where PD_{ijt} is the pressure deficit from a normal situation at the critical node i in the jth water main-break event at time step t, D_i is the demand at the ith critical node, T_{jt} is the duration of the jth water main-break event during time step t (h), m is the number of selected critical nodes, and k is the total number of water main-break events that have happened in a given week. In the above equation, 168 is the total hours of operation per week and is used to incorporate the weekly fraction of failure in the equation. In this index, by increasing pressure deficit, the system's readiness decreases. The state of different nodes in the system's readiness has been considered in the calculation of demand-weighted average of pressure deficits.

It should be noted that two groups of nodes are selected as critical nodes. The first group consists of those with higher water dependencies (water demands of top 20 percentile) and the second group consists of nodes with highest head loss [i.e., away from the storage tank(s)]. This step plays an important role in achieving reliable results on the state of the system's readiness. It must be noted that the critical nodes should be revised during the simulation procedure if the hydraulic state of the system changes.

The negative sign in the SRI formulation is to show more severity of system failure as the SRI absolute value increases. Five main categories of SRI are defined in this study, namely, sound, good, fair, poor, and critical. These classes are determined based on a probability distribution fitted to the SRI values. The ranges of probabilities on SRI classifications are shown in Table 11.2. It is assumed that if there is more than 95% probability of failure, the system should be in full alert (red zone). The other zones are assumed arbitrarily to cover its range of variation. As the class number of SRI increases, the system could be considered at a lower state of readiness. According to Table 11.2, the system readiness is categorized as sound when the SRI has the probability of failure of <5% based on the probability distribution of observed/generated SRI values. As another sample of SRI interpretation, one could say for example if SRI is in the fourth class, the system could be subject to failure with 65–95% probability of occurrence, with 80% as a representative of that range. In this situation, the critical nodes have high pressure deficits and there is no water delivery in certain nodes. The characteristics of the other classes are given in Table 11.2.

The following steps are carried out for SRI determination using a probabilistic neural network (PNN):

1. Reliability, resiliency, and vulnerability indices of the system during the main breaks in the weekly time intervals are calculated.
2. Critical nodes in WDS are identified based on the hydraulic analysis of the WDS.

TABLE 11.2
System Readiness Based on SRI Classification

The State of System Readiness	SRI Class	Probability of Failure
Sound	1	<0.05
Good	2	0.05–0.35
Fair	3	0.35–0.65
Poor	4	0.65–0.95
Critical	5	>0.95

3. Deficit pressure from normal situations in the critical nodes and their demand-weighted average are calculated.
4. SRI values for main-break events in each week using demand-weighted average pressures at the critical nodes are determined using Equation 11.48.
5. A PNN is used to train a system for prediction of SRI, as the new information on the state of the system becomes available. The flowchart of the proposed approach for determining the SRI is shown in Figure 11.13.

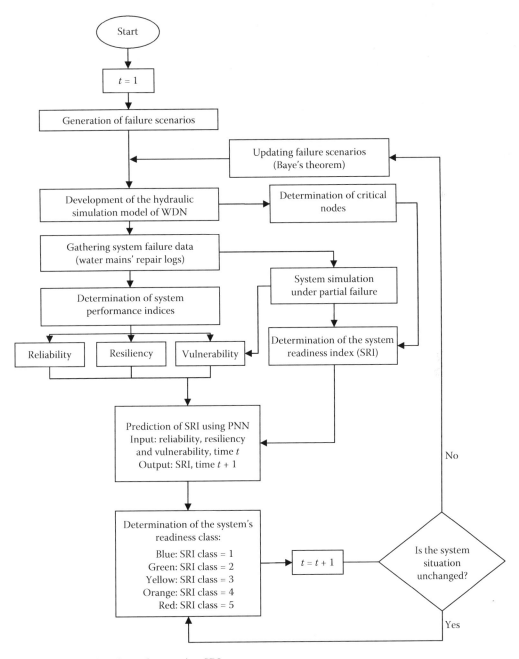

FIGURE 11.13 Flowchart of measuring SRI.

The reliability index in time t (Rel$_t$) (during each week) is estimated as

$$\text{Rel}_t = \left(1 - \frac{\sum_{j=1}^{k} T_{jt}}{168}\right) \times 100, \quad (11.51)$$

where T_{jt} is the duration of the jth water main-break event at time step t (h) and k is the total number of water main-break events that have happened in a given week. In the above equation, 168 is the total hours of operation per week.

The inverse of average break time has been considered as the resiliency index in time t (Res$_t$):

$$\text{Res}_t = \left(1 - \frac{k}{\sum_{j=1}^{k} T_{jt}}\right) \times 100, \quad (11.52)$$

where T_{jt} is the duration of the jth water main-break event at time step t (h) and k is the total number of water main-break events that have happened in a given week.

The vulnerability in time t (Vul$_t$) is assumed to be the total water shortage that is caused by water main-break events in time step t:

$$\text{Vul}_t = \frac{\Delta Q_t}{Q_{\text{req},t}} \times 100, \quad (11.53)$$

where ΔQ_t and $Q_{\text{req},t}$ are the water shortage and the total water demand in time step t, respectively.

As the flow and pressure in each node of WDS are related, the operation of the system should be simulated under pressure-deficit conditions, satisfying the relationship between pressure and flow. The head-flow relationship proposed by Wagner et al. (1998) is used here for the evaluation of water shortages in the water main-break situation. The head-flow relationship is a continuous function with preselected upper and lower bounds for pressure as follows:

$$Q_j^{\text{avl}} = 0 \qquad \text{if } P_j \leq P_j^{\min}, \quad (11.54)$$

$$Q_j^{\text{avl}} = Q_j^{\text{req}} \left(\frac{P_j - P_j^{\min}}{P_j^{\text{des}} - P_j^{\min}}\right)^{1/n_j} \qquad \text{if } P_j^{\min} < P_j < P_j^{\text{des}}, \quad (11.55)$$

$$Q_j^{\text{avl}} = Q_j^{\text{req}} \qquad \text{if } P_j \geq P_j^{\text{des}}, \quad (11.56)$$

where Q_j^{avl} is the available flow and Q_j^{req} is the required flow (i.e., the demand) at node j, P_j is the pressure at node j, and $P_j^{\min}$ is the pressure at which there is no outflow at node j and it is taken as the minimum. The desirable pressure P_j^{des} corresponds to the pressure above which the demand at node j is completely supplied. From the above relations, the water shortage in each week is calculated as

$$\Delta Q_t = \sum_{j=1}^{m} \sum_{i=1}^{n} Q_{jit}^{\text{req}} - Q_{jit}^{\text{avl}}, \quad (11.57)$$

where ΔQ_t is the water shortage in time step t. Q_{jit}^{req} and Q_{jit}^{avl} are the required and the available water at node i in the jth system failure in time step t, respectively. The n and m denote the number of nodes in the water distribution network and the number of failures in each time step, respectively.

Urban Water Disaster Management

EXAMPLE 11.11

Evaluate the readiness of the Tehran (capital of Iran) WDS. Because of the aging water distribution infrastructure of Tehran, the water pipe breaks are a common problem in this city. As there is not any rehabilitation program of aging pipes in this city, this problem is getting more severe over the time. The total length of pipes in the Tehran WDS is about 9323 km and an informal figure of 700 pipe breaks in a day is quoted. This results in a break rate of about 8 per 100 km of pipes per day, which causes considerable water losses during a day. Furthermore, a lack of emergency teams in the nearby posts (say within 10 min response time) has resulted in oscillating adverse factors of disengagements. So evaluation of system readiness in dealing with this problem and developing strategies for increasing the system readiness are essential for the Tehran WDS.

The considered part of Tehran water distribution supplies water for 120,000 residents in an area of about 450 ha (1 ha = 10,000 m^2). Maximum daily water consumption in this system is about 60,000 m^3. The simulated part of the WDS of Tehran has been shown in Figure 11.14. There are 35 main demand nodes with the total demand of about 0.51 cms and 78 water mains. The characteristics of the network are shown in Table 11.3. The minimum and desirable pressures for this water distribution network are 5 and 20 m, respectively.

Solution: The selected WDS has been simulated under failure scenarios for 60 weeks. The system performance indices of reliability, resiliency, and vulnerability have been calculated. The results have been shown in Table 11.4 and Figure 11.15.

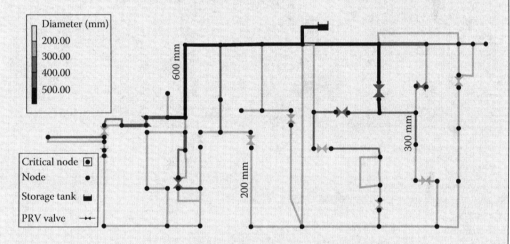

FIGURE 11.14 Schematic of the WDS of Tehran supported by storage tank no. 83.

TABLE 11.3
Characteristics of Water Mains in the Simulated WDS

Diameter (mm)	Number of Pipes	Total Length (m)
200	30	9458
250	18	4960
300	11	2931
350	5	690
400	5	1585
>400	9	3197

continued

TABLE 11.4
System Performance Indices

Week Number	Reliability	Resiliency	Vulnerability	SRI	Week Number	Reliability	Resiliency	Vulnerability	SRI
1	58.3	85.7	14.7	−3.7	32	73.7	93.2	5.7	−1.0
2	64.2	85.0	3.7	−0.6	33	56.2	91.9	4.3	−1.0
3	67.4	87.2	4.8	−0.5	34	77.7	86.7	4.3	−0.4
4	79.3	88.5	11.26	−0.3	35	61.8	89.1	3.4	−0.6
5	67.1	87.3	15.1	−1.4	36	73.7	86.4	15.0	−1.3
6	67.7	83.4	8.3	−1.4	37	63.5	86.9	16.5	−1.5
7	68.1	92.5	8.8	−0.5	38	63.2	88.7	5.7	−0.8
8	65.2	88.0	10.9	−0.9	39	60.1	91.0	15.0	−2.9
9	80.5	78.6	8.8	−0.4	40	90.6	87.4	3.0	−0.1
10	81.6	83.8	9.2	−0.3	41	69.1	88.4	8.3	−1.1
11	84.3	88.7	7.2	−0.2	42	57.7	84.5	6.3	−1.0
12	86.1	78.6	3.3	−0.3	43	57.3	84.7	11.2	−2.0
13	62.4	87.3	7.3	−0.7	44	30.3	90.6	5.6	−1.3
14	60.5	84.9	5.5	−0.8	45	56.7	90.4	6.8	−1.3
15	75.7	87.7	3.9	−0.5	46	84.4	88.6	4.4	−0.3
16	64.8	88.2	5.6	−1.1	47	57.2	87.5	3.6	−0.7
17	92.6	75.8	5.1	−0.3	48	86.1	82.9	12.6	−1.0
18	79.0	85.8	3.1	−0.3	49	48.2	88.5	9.3	−1.5
19	85.2	83.9	3.4	−0.2	50	67.1	85.5	9.4	−1.4
20	86.5	77.9	4.1	−0.2	51	80.3	87.9	2.9	−0.3
21	71.2	83.5	7.6	−0.7	52	44.9	89.2	5.1	−1.1
22	67.3	89.1	3.2	−0.5	53	79.2	79.9	10.0	−1.7
23	45.3	90.2	17.9	−3.3	54	65.4	88.0	5.3	−1.0
24	33.6	86.5	16.8	−5.0	55	73.5	84.3	3.2	−0.4
25	47.7	89.7	6.1	−2.7	56	40.0	91.1	11.2	−3.6
26	86.6	86.6	9.6	−1.0	57	84.5	84.6	7.2	−0.7
27	50.9	90.3	3.2	−0.8	58	77.9	83.8	4.1	−0.7
28	76.7	87.2	3.6	−0.5	59	79.3	85.6	9.0	−0.8
29	93.8	80.7	3.6	−0.1	60	60.0	85.1	10.1	−1.4
30	68.8	88.5	7.5	−1.5	min	30.3	75.8	2.9	**−0.1**
31	57.5	84.6	6.2	−1.4	max	**93.8**	**93.2**	**17.9**	−5.0

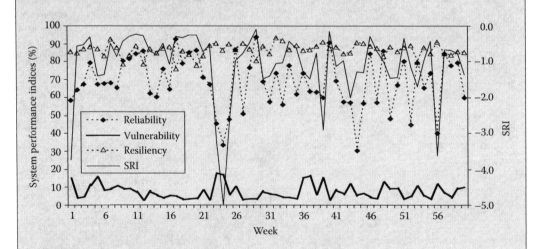

FIGURE 11.15 Variation of system performance indices in comparison with SRI.

As can be seen, the system reliability is relatively low and varies between 30.3% and 93.8% in a given week. Only in 35% of the time period the reliability is more than 75%, but in most of the time period (88%), the reliability is more than 50%. Although the low end of reliability figures is acceptable, the high end is not satisfactory. The general low reliability of the system could be because of the lack of well-equipped emergency response groups and the absence of the preventive maintenance program in the study area.

The resiliency of the system varies between 75.8% and 93.2%. There are 42 out of 60 weeks with a resiliency of more than 85%, which shows the system's ability to recover well after any system failure. Implementing some modification and rehabilitation programs as well as extending the current GIS-based monitoring system, the resiliency of the system could be even better than the current situation.

The minimum and maximum values of vulnerability (percent of water shortage) are 2.9% and 17.9%, respectively. In only 13% of the time, the water shortage is above 10% and, almost always, 80% of the demand is supplied, which shows the system's ability to continue its water delivery relatively well. The high values of the vulnerability index are due to some primary water main breaks that supply water to the relatively large section of the system. This is perhaps an inherent characteristic of WDSs with large water mains without enough redundancy.

The calculated SRI values based on the proposed algorithm are shown in Table 11.4. It should be noted that the SRI values are rescaled to be within 0 and -5 in order to bring it to a manageable range as is done for many indices. This has been done by dividing the SRI values by the minimum calculated SRI value, multiplied by -5. Figure 11.15 shows a close correlation between the system performance indices and SRI. The SRI has relatively high correlations with vulnerability ($R = 0.83$) and reliability ($R = 0.82$). Figure 11.15 also illustrates that when all of the system performance indices have low values, the value of SRI suddenly decreases and the system readiness situation goes to the red zone, which is expected. The worst-case scenarios are caused due to the failure of large-size mains (pipes) in the WDSs. In such cases, there is not any water delivery to some of the water demand nodes. Then, the flow to those sections is stopped and diverted to other nodes. These additional flow causes increase in pressure losses and pressure deficits in those nodes. This leads to high values of system vulnerability. Furthermore, to recover large failures, more time is needed and this will adversely affect the reliability and resiliency indices and lower the SRI. The failure of smaller pipes may result in less or zero pressure deficit because of the system redundancy. It also takes less time to recover, so the critical situations are not happening too often and the system readiness is classified as fair or better. As can be seen in Figure 11.15, only in few weeks during the operation horizon, the SRI takes values <-3.0. During those weeks, the situation could be devastating and could result in extensive structural and economic damages. By prediction of these situations, the expected damages could be reduced by declaring a state of emergency and implementing the corrective measures such as pipe replacement and rehabilitation program.

In order to determine the probability ranges of SRI, the log-normal probability distribution has been fitted to the observed SRI values. The corresponding values of the probability range of SRI classes based on the fitted log-normal distribution are shown in Table 11.5.

TABLE 11.5
Corresponding Values of the Probability Limits of SRI Classes

SRI Class	Probability of Occurrence	SRI Value
1	<0.05	$-0.2 < \text{SRI} < 0.0$
2	0.05–0.35	$-0.6 < \text{SRI} \leq -0.2$
3	0.35–0.65	$-1.1 < \text{SRI} \leq -0.6$
4	0.65–0.95	$-3.0 < \text{SRI} \leq -1.1$
5	>0.95	$\text{SRI} \leq -3.0$

11.11.2 Hybrid Drought Index

Selection of an integrated index for quantifying drought severity is a challenge for decision makers in developing water resources and operation management policies. Karamouz et al. (2009) developed a hybrid drought index (HDI) by combining the three important indices of meteorological, hydrological, and agricultural droughts, namely, SPI (McKee et al., 1993), surface water supply index (SWSI) (Shafer and Dezman, 1982), and palmer drought severity index (PDSI), by utilizing the drought damage data. The data obtained from drought damage are good indicators of combined drought severity and include all aspects of its gradual and longer-term impacts. Therefore, it can be used to identify the combined effects of drought and to analyze its variability in order to develop preventive schemes and reduce drought impacts on different water-user sectors.

The framework of the adjusted SWSI index developed by Garen (1993) has been adapted in the definition of HDI according to Equation 11.58:

$$\text{HDI}_t = f(\text{Damage}_t) = \frac{p_t(\text{Damage}) - 100}{24}, \quad (11.58)$$

where p_t is the cumulative probability of damage in month t. The characteristics of the damages for the three types of droughts are complex with a significant degree of nonlinearity and uncertainties. Therefore, for analyzing the nonlinear relations between them, the ANN models are applied.

The following steps are carried out for HDI determination and utilization for the drought severity prediction using the ANN:

1. Calculation of the time series of SPI, SWSI, and PDSI as indices for meteorological, hydrological, and agricultural aspects of droughts
2. Analysis and adjustment of historical drought damage data and estimating the monthly damage
3. Finding the most appropriate cumulative probability distribution that fits to the historical monthly damages
4. Determination of the HDI values from the cumulative probability of damages (Equation 11.56)
5. Classification of the HDI values into subcategories where class 1 and the last class denote the normal and extremely severe drought conditions, respectively
6. Estimation of HDI subcategories and consequently the range of drought damage through calculated indices of different drought types by training a system

The flowchart of the proposed approach for determining the HDI is shown in Figure 11.16. In comparison to the SWSI, HDI is utilized only for evaluation of dry spans (unlike SWSI, HDI only has negative values, so it can be used only for drought classification). The HDI values vary within $(-4, 0)$, which makes it comparable with other indices of drought such as SWSI.

According to the HDI, drought starts when the HDI goes below -2 and continues unless it reaches above -1. The reason for selecting -2 as a threshold of drought beginning is that when the HDI is above -2, there is no significant change in the amount of damages, especially for the HDI values > -1. It can also be mentioned that at the beginning of drought spells, observed impacts are not very high because of gradual occurrence of droughts.

There are four main categories of drought severity according to the HDI: negligible, low, moderate, and high severity. This classification of drought, is shown in Table 11.6 which is based on the cumulative probability of drought damage. For damages with the cumulative probability and the HDI value of $<25\%$ and -1, respectively, drought severity is considered to be negligible. This is because of the fact that damage exists but it is not significant enough to be attributed only to drought occurrence. When the value of HDI reaches -2, the damage caused by drought is moderate and it can be assumed that drought span starts. Finally, when the value of HDI reaches -3, different water users,

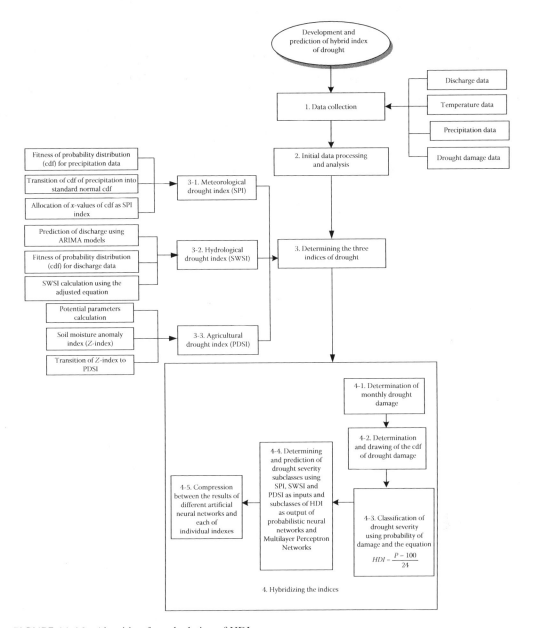

FIGURE 11.16 Algorithm for calculation of HDI.

especially agriculture and animal husbandry sectors, suffer from water shortages and it is considered the critical stage of drought.

In regions where generally drought damages are high, small climate changes may cause extensive variations in damages. Therefore, in these regions, in order to evaluate the variations of drought impact, the classification of drought severity in smaller intervals can be helpful. For a more accurate evaluation of drought, each of the four major categories of drought severity obtained through the HDI classification has been divided into four subcategories except for the fourth category, which has five subcategories considering the fact that values <-4 might be predicted by the neural networks (see Table 11.6). As a result, 17 subcategories have been defined according to the cumulative probability of drought damages.

TABLE 11.6
Classification of Drought Severity Based on the HDI

Drought	Number of Subclasses	HDI Interval	Average of HDI
Negligible severity	1	$-0.25 < \text{HDI} < 0$	-0.125
	2	$-0.5 < \text{HDI} < -0.25$	-0.375
	3	$-0.75 < \text{HDI} < -0.5$	-0.625
	4	$-1 < \text{HDI} < -0.75$	-0.875
Mild	5	$-1.25 < \text{HDI} < -1$	-1.125
	6	$-1.5 < \text{HDI} < -1.25$	-1.375
	7	$-1.75 < \text{HDI} < -1.5$	-1.625
	8	$-2 < \text{HDI} < -1.75$	-1.875
Moderate	9	$-2.25 < \text{HDI} < -2$	-2.125
	10	$-2.5 < \text{HDI} < -2.25$	-2.375
	11	$-2.75 < \text{HDI} < -2.5$	-2.625
	12	$-3 < \text{HDI} < -2.75$	-2.875
Extremely severe	13	$-3.25 < \text{HDI} < -3$	-3.125
	14	$-3.5 < \text{HDI} < -3.25$	-3.375
	15	$-3.75 < \text{HDI} < -3.5$	-3.625
	16	$-4 < \text{HDI} < -3.75$	-3.875
	17	$\text{HDI} < -4$	-4.125

EXAMPLE 11.12

Evaluate the HDI for the Gavkhooni/Zayandeh-rud basin in central Iran. This basin has five sub-basins with a total area of 41,347 km². The dominant climate in this region is arid and semiarid. The precipitation varies throughout the basin between 2300 mm in the west (where most of the precipitation is in snow form) and 130 mm in the central part of Iran (where Isfahan City is located). Annual average precipitation in this basin is about 1500 mm. The average of precipitation in the Zayandeh-rud basin has been used for the calculation of drought indices. The Zayandeh-rud River is the main surface resource for supplying the irrigation demands in this basin. As water and energy demands increase in Isfahan, water withdrawals from the river increase and it is important to incorporate climate variability into water resources decision making.

The Zayandeh-rud reservoir controls the stream flow with a volume of 1470 MCM. The location of the Zayandeh-rud reservoir is shown in Figure 11.17. The average annual inflow to the Zayandeh-rud reservoir is about 1600 MCM, of which an average flow of 600 MCM is transferred from the adjacent Karoon river basin. Drought trends in the basin have been studied between years 1971 and 2004.

The Gavkhooni/Zayandeh-rud basin has special effects on the development and economy of the region through agricultural, industrial, and tourism activities. Statistics show that the amount of precipitation, especially in the high altitudes from October 1999 to April 2000, has decreased ~35–45% compared to the long-term average and resulted in 250 MCM (million cubic meters) water shortages in the Isfahan region in the year 2000.

Solution: As the first step in the HDI calculations, the monthly time series of SPI, modified SWSI, and PDSI are calculated from 1971 to 2004. The time series of damage is converted to year 2006 values for comparison purposes. The interest rate is considered as 17%. This time series of monthly damages are shown in Figure 11.18. As mentioned in the steps of HDI calculation, the cumulative probability function of damage occurrences is developed using the Weibull probability distribution (Figure 11.19)

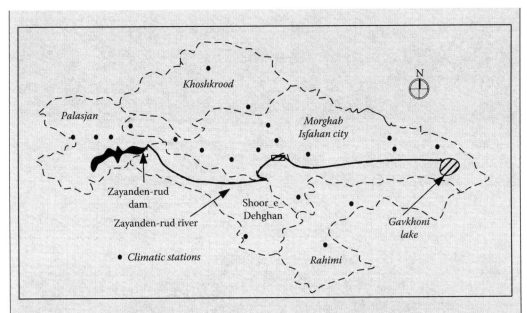

FIGURE 11.17 The Gavkhooni/Zayandeh-rud basin with sub-basins and climatic stations.

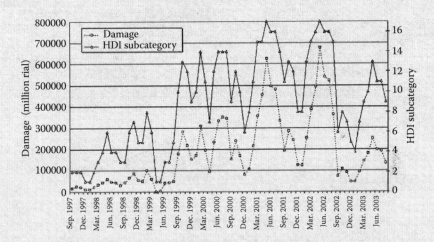

FIGURE 11.18 Monthly time series of damage (converted to values in the year 2006).

and used for determining the HDI values (Equation 11.56). Different thresholds of drought intensity according to the HDI, which are obtained from the probability–damage diagram, are also shown in Figure 11.19.

The values of three indices (which indicate different climatic, hydrological, and agricultural aspects of drought) including SPI, SWSI, and PDSI as well as the hybrid index, HDI, have been calculated for the study area. The time series of these indices, their associated number of drought spans and duration of dry periods are illustrated in Figure 11.20. This figure shows that during the drought periods determined by HDI, in most of the cases three types of agricultural, climatic, and hydrological droughts have occurred. According to the figure, agricultural drought spans, because of their gradual progress, have lower frequencies but greater durations than that of hydrological and particularly meteorological

continued

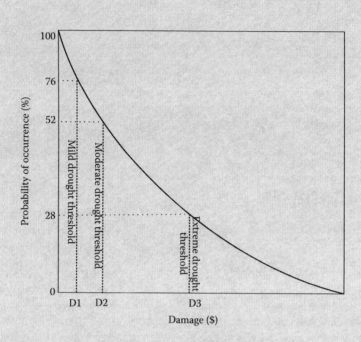

FIGURE 11.19 Diagram of cumulative probability of historical damages and thresholds of different drought severity.

droughts. Table 11.7 compares quantitatively the drought spans determined by the traditional indices and their overlapping periods with the HDI. This shows that HDI can cover different aspects of drought and also it can be concluded that the hydrological and agricultural sectors have the main role in the magnitude of drought damages due to the higher overlapping of drought periods according to the SWSI and PDSI as hydrological and agricultural drought indices with those reported according to the HDI.

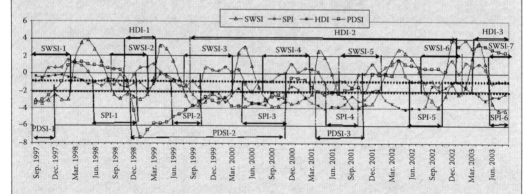

FIGURE 11.20 Time series of three indices of drought in comparison to the HDI.

TABLE 11.7
Durations of Drought Spans (month)

Drought Span's Number	SPI	SWSI	PDSI	HDI
1	7	6	4	5
2		6	24	
3	5	8		40
4	7	8		
5	6	7	8	
6	5	6		
7	3	3		6
Overlapping with HDI	22	34	27	51

11.11.3 Disaster and Scale

If we maintain too broad an interpretation of disaster management, it becomes difficult to determine the progress made in achieving it. In particular, concern only with the water disaster of a major city could overlook the unique attributes of particular local watershed economies, environments, ecosystems, resource substitution, and human health. On the other hand, not every hectare of land should be relieved. This highlights the need to consider the appropriate spatial scales when applying disaster management to a specific urban region.

We also need to consider the appropriate temporal scales when dealing with reliability of water disaster of a specific urban region. We cannot prevent disasters from happening; given the variations in natural water supplies and the fact that floods and droughts do occur, it is impossible, or at least very costly, to design and operate urban water systems that will never fail. During periods of failure, the economic benefits derived from such systems may decrease. However, ecological benefits may depend on these events. One of the challenges faced in measuring our success is to identify the appropriate temporal scales in which these measurements should be made.

An example of temporal scale and disaster variability is as follows: If a city of 2–3 million habitants faces water disaster as a drought over a period of 2–5 years, the tension could be reduced by many well-thought water conservation and rationing schemes as well as taping into other resources. If the same city faces a water tunnel collapse or a massive accidental or terrorist act of water contamination, then there will be a disaster with a potential for many losses of life and epidemic illnesses. The second scenario requires a different level of the systems' readiness at a much shorter time scale (Karamouz and Moridi, 2005).

11.11.4 Design by Reliability

What level of reliability are we expecting for an urban water system? How this reliability, defined as P (Supply > Demand) $\geq P$ (where P is the acceptable level of the systems reliability), could be increased by

a. Possibility of taping into other resources
b. Conservation and rationing programs with the objective of first providing the basic water needs of a region

c. Prioritizing water allocations according to the demand importance
d. Determining the possible allocation of water to other case at the time of disaster
e. Designing under a normal and acceptable level of risk reliability but making provisions and plans to meet the extreme events at a lower reliability level

In hydraulic engineering, a design can be based on hydrological and biological inputs, which encompass the notion of "load." The load $s(t)$ must be compared with the resistance $r(t)$, which could be the ability of a system to tolerate additional pressure. If the load exceeds that resistance, a failure will occur. In hydraulic structure, the load $r(t)$ is the force exerted by flow and other loads, and the resistance $r(t)$ is the force that the structure can accept without failing. In water distribution, water demand is a load that should be compensated for by a supply as a resistance.

Figure 11.21 shows the design procedure involving the above notion of load and resistance and how reliability and uncertainty could be integrated into the design process. In urban water systems, there are two types of basic input: the natural input such as rainfall, temperature, sediment yield, and so on, which are the random variables describing the natural environment, and inputs coming from human activities, such as water demand, sewage effluents, pollutants, and so on, which are random numbers describing the interaction of man with his environment. The hydrological, hydraulic, and structural uncertainties are identified and coupled with a loss function to determine the expected damage resulting from a disaster/failure.

11.11.5 WATER SUPPLY RELIABILITY INDICATORS AND METRICS

In order to use reliability concept in design, it must be quantified based on a number of indicators. No single indicator can adequately measure and communicate all the critical dimensions of reliability. Multiple indicators are needed, many of which will be strongly interrelated due

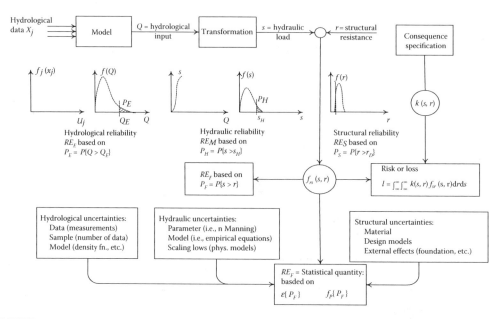

FIGURE 11.21 Generalized concept of risk and reliability analysis for structures. (From Plate, E. J. and Duckstain, L. 1988. *Water Resources Bulletin*, 24(2).)

to the network characteristics of the water supply system. Some indicators will be more useful for monitoring supply reliability for certain types of service (e.g., residential/municipal versus environmental) or for different functions (e.g., real-time operations versus policy development). The following paragraphs discuss a variety of features that could distinguish reliability indicators.

Some indicators will require greater breadth, reflecting the effects of many factors, while some will be more narrowly focused to isolate a particular factor or reflect a certain system scale. Operations functions are particularly concerned with having indicators that measure reliability in terms of the factors under their control, in addition to broader indicators dominated by actual inflow levels. Some of the dimensions of breadth or focus are as follows:

- *Type of service*: Clearer definitions of the services currently provided to different end users (e.g., consumptive, environmental, and inflow) would facilitate the development of indicators and metrics. In other industries, service is often described or defined by specifying requirements related to the purpose of use, quantity and temporal/geographical distribution of deliveries, and quality and delivery curtailment provisions.
- *Temporal scale and distribution*: For example, hourly, monthly, or annual.
- *Spatial (or geographical) scale and distribution*: The scale refers to, for example, storage data that are aggregated to the combined project level or are disaggregated at the individual reservoir level.
- *Climate normalization*: For the available supply indicators, both metrics that are normalized for climatic variability and those that are not will be valuable. When normalized, metrics will reflect how efficiently the existing storage and transportation system is being used. When not normalized, reliability of the system including the biggest failure in supply availability or inflow is tracked.
- *Condition and outcome indicators/metrics*: Factors that affect reliability can be translated into indicators of reliability, for which metrics can be crafted. Indicators are statements on what is expected to change if the program shows progress toward the objective or if problems begin to arise. Indicators may identify outcomes, or conditions and features that are believed to influence outcomes. Indicators of condition can provide insight into the mechanics of reliability and may be components of outcome indicators.
- *Physical and financial indicators/metrics*: Reliability indicators that reflect physical phenomena such as the supply delivered or allocated need to be paired with financial indicators that describe the costs associated with reliability as the level of those physically based indicators change. It is important to monitor cost in order to develop policy positions on the desired level of reliability, and to answer questions about the cost of different possible levels of reliability.
- *Forecast versus actual*: Some outcome indicators could compare actual to forecasted or expected/desired service dimensions. Some metrics might utilize forecast data, while others may reflect what has actually happened. For example, one measure of reliability at the project level may be the initial monthly allocation to contractors (developed in the first quarter of the water year) relative to the final allocation that is produced in the third quarter of the water year (summer). For some customers (e.g., agricultural), the closer these values are, the more reliable project supplies are to them. Furthermore, for indicators/metrics that embody forecast values, the dependency of the outcome on the nature of the forecast must be carefully considered. For example, for metrics that involve the allocation as a measure of supply, the degree of conservativeness in the runoff forecast will influence the values of the reliability metric. In other words, the less risky (more conservative) the projects are in the inflow forecasts that are the basis of allocations, the more likely they will be to actually meet that allocation.

- *Deterministic and probabilistic indicators*: Both deterministic metrics and probabilistic metrics (discrete probabilities or probability distributions) will be useful.
- *Quantitative and qualitative indicators/metrics*: Some metrics will be quantitative while others will be qualitative (e.g., those related to institutional failures).

11.11.6 Issues of Concern for the Public

The public plays an important role in the situation analysis being able to reach beyond numbers and statistics to the real impact of the UWDM on society. The aim of including them at this stage is to identify, prioritize, and formulate the problems in a clear way and with a common understanding. It is important to be aware of the conflicts, views, and interest of the public. Negotiation skills and conflict resolution techniques will be a useful skill.

This first step analyses the major disaster management actors, their interests and goals, and their interrelationships. It aims to shed light on the social reality and power relationships prevailing in the institutional setting of the disaster management. Major actors include potential losers. Furthermore, it is important to determine who volunteers to help.

11.12 GUIDELINES FOR DISASTER MANAGEMENT

A set of criteria should be selected and for each criterion a range of values that are satisfactory should be defined. The duration and the extent of individual and cumulative failures of water resources systems should be determined for each criterion.

11.12.1 Disaster and Technology

All stakeholders involved in or impacted by the UWD can be aided by the use of modern information-processing technology. This technology includes computer-based interactive optimization and simulation models and programs, all specifically developed to perform a more comprehensive multisector, multipurpose, multiobjective water systems assessment at a short period right before, during, and after a water disaster. Without such models, programs, and associated databases, it would be difficult to predict the expected impacts of any proposed plan and management policy.

Without the development and use of computer programs incorporating various models, programs, and databases within an interactive, menu-driven, graphics-based framework, it would be difficult for many to use these tools and databases to explore their individual ideas, to test various assumptions, and to understand the output of their analyses. Such programs allow the stakeholders to create their own models, rather than to be forced to use someone else. Models can help achieve a shared vision among all stakeholders as to how their system functions, if not how they would like it to function. The models that help us to predict the impacts of possible actions we take today are based on the current conditions of our urban water systems.

What we might do to improve or increase the derived benefits, however measured, of our UWDM is, to a large extent, dependent on the state of the urban water systems that exist today. There may well be trade-offs between what we would like to do today to plan for management of the next water disaster for our own benefit and what the managers and decision maker at the time of the next disaster wish we had done. Modeling can help us identify these possible trade-offs. While models cannot determine just what decisions to make, the trade-off information derived from such models can contribute to the decision-making process.

Urban water systems simulation during the disasters is an effective tool that involves many decision makers to participate in a real-time hypothetical disaster management rehearsal to identify where the weaknesses are and investigate ways and means to strengthen them at the time of a real disaster management.

On the use of science and technology:

- Integrating the best science available into the decision-making process, while continuing scientific research to improve knowledge and understanding
- Establishing baseline conditions for system functioning and sustainability against which change can be measured
- Monitoring and evaluating actions to determine if goals and objectives are being achieved

Finally, it should be stated that disaster of any magnitude could be converted to opportunities if we learn from them and utilize the unification of efforts that has been developed through the disaster for better practices and implementation of sound water resources, urban water systems, and demand management as well as disaster prevention.

11.12.2 Disaster and Training

A key to UWDM is the existence of sufficiently well-trained personnel in all of the disciplines needed in the planning, development, and management processes. In regions where such a capacity is needed but does not exist, it should be developed. Training and education are a key input, and requirement, of disaster management. While outside experts and aid organizations can provide temporary assistance, each urban region must inevitably depend primarily on its own professionals to provide the know-how and experience required for UWDM.

Capacity building is one of the most essential and important long-term conditions required for disaster management. Another important factor in disaster management is that the local people must not only be capable, but must also be willing to assume the responsibility for their urban water systems. One of the drawbacks of a centralized dominating government that takes the responsibility for local system design and operation is that the local people become accustomed to looking to the government for help, rather than looking to them. The ideal local urban water systems managers are well-trained persons who know the behavior of that system, have experience with its disasters such as floods and droughts, and know the concerns and customs of the people of the region, a group to which they belong.

11.12.3 Institutional Roles in Disaster Management

The government institutions, agencies, local authorities, the private sector, civil society organizations, and partnerships all constitute an institutional framework that ideally should be geared toward the implementation of the policy and the legal provisions. Whether building existing water management institutions or forming new ones, the challenge will be to make them effective and this requires capacity building. Awareness creation, participation, and consultations should serve to upgrade the skills and understanding of decision makers, water disaster and water resources managers, and professionals in all sectors. The key goals for the institutional framework are as follows:

- Separate water disasters management functions from service delivery functions (water supply, sanitation, and sewerage) and consolidate the government as the main manager of the water disasters—the enabler but not the provider of services. This will avoid conflicts of interest and encourage commercial autonomy.
- Manage surface water disasters within the boundaries of urban, not within administrative boundaries, decentralizing regulatory and service functions to the lowest appropriate level and promoting stakeholder involvement and public participation in planning and management decisions.

- To ensure balance between the extent and complexity of regulatory functions and the skills and humans required to deal with them. A continued capacity building program is required to develop and maintain the appropriate skills to facilitate, regulate, and encourage the private sector potential contributions to financing and delivery of services (water supply, sanitation, and sewerage).

11.13 DEVELOPING A PATTERN FOR THE ANALYSIS OF THE SYSTEM'S READINESS

The following sections contain a paradigm for a system's readiness during a UWD.

11.13.1 Assessment of a Disaster Caused by Water Shortage/Drought

- Identification of the system's components, data gathering, and analysis
- Assessment of the baseline estimates for the appropriate operation of a system from the hydrological and hydraulic viewpoints
- Developing an algorithm to convert the indicators of the existing and anticipated situations of an urban water system to the disaster triggers
- Evaluation of the severity of a disaster by analyzing the following criteria:
  - Reliability
  - Resiliency
  - Vulnerability
- Assessment of the alternatives for water supply during a disaster according to its scale and severity
- Assessment of the system's readiness from the hydrological viewpoint

11.13.2 Assessment of a Disaster Caused by System Failure

- Assessment of the potential factors that may cause an urban water pollution disaster
- Evaluation of the severity of a disaster by an analyzing the following criteria:
  - Reliability
  - Resiliency
  - Vulnerability
- Assessment of the system's readiness from the hydrological viewpoint.

11.13.3 Assessment of a Disaster Caused by Widespread Contamination

- Assessment of the baseline estimates for the appropriate operation of a system from the hydraulic and structural standpoints
- Evaluation of the severity of a disaster by an analysis of the reliability and vulnerability criteria
- Assessment of the system's readiness from the contamination viewpoint

11.13.4 Developing a Comprehensive Qualitative/Quantitative Monitoring System for the Water Supply, Transfer, and Distribution Network

- Identification of the characteristic variables, proper indicators, and disaster triggers in a system
- Providing necessary facilities and equipments for the system
- Using a telemetric system for either online or offline data/information transformation

Urban Water Disaster Management

- Design and application of a monitoring system on the water pressure and heavy metals concentration in pipelines
- Design and application of a nondestructive monitoring system
- Analyzing the life cycle data of different components of a system

11.13.5 Guidelines for the Mitigation and Compensatory Activities

- Assessment of the consequences of applying alternate policies during a disaster
- Managing water distribution during a disaster
- Identification of the most vulnerable areas and priorities to be considered in compensatory activities
- Isolation of the most vulnerable areas to prevent the spread of a disaster

11.13.6 Organization and Institutional Chart of Decision Makers in a Disaster Committee

As an example, Figure 11.22 shows the organization chart of the group of UWDM in the Tehran Metropolitan Water and Wastewater Company, Tehran, Iran. This chart, for a more general situation, has been shown in Figure 11.23.

The described steps for evaluation of system readiness are summarized in Figure 11.24. As can be seen from this figure, there are five phases in system readiness evaluation that finally result in a hybrid index that determines what is the system's readiness situation. According to the hybrid index, appropriate decisions are made for increasing system preparedness to deal with disasters. This hybrid index in this methodology is predicted based on system performance indices, which have been described before. This is very useful to avoid disaster occurrence. For this purpose, the system performance indices are calculated based on the available data in the management information system (MIS) and then they are accumulated and result in the hybrid index of system readiness.

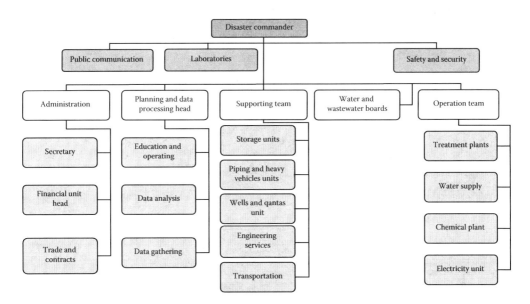

FIGURE 11.22 Network of responsibilities during a UWD.

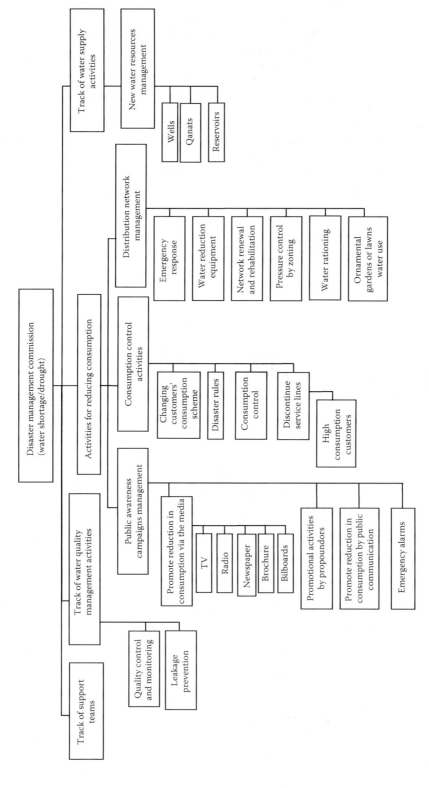

FIGURE 11.23 Organization chart for disaster management commission.

Urban Water Disaster Management

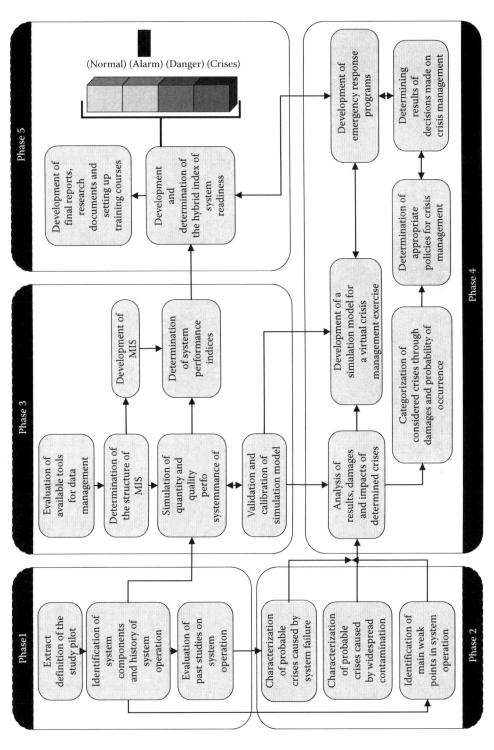

FIGURE 11.24 Diagram of system readiness evaluation methodology.

11.14 CONCLUSION

Today, realizing the severity of UWDs has finally infiltrated into public consciousness. Clearly, major financial resources will be required to augment urban water systems and to upgrade water systems quality for dealing with disasters, but these will be much less than the funds that will be needed to rebuild failed systems and rehabilitate contaminated surface water, and groundwater due to mismanaged environmental factors and/or intentional acts of war and terrorism.

Quite clearly, a considered strategy for management of water resources in urban areas needs to be put in place in order to avoid the disaster. Developing a framework for urban water management can draw inspiration and guidance from several global agreements and norms. Water conservation and protection is an integrated part of every water and environmental convention.

The consequences of the increasing global water scarcity will largely be felt in the arid and semiarid areas, in rapidly growing coastal regions, and in the mega cities of the developing world. Water scientists predict that many of these cities already are, or will be, unable to provide safe, clean water and adequate sanitation facilities for their citizens—two fundamental requirements for human well-being and dignity.

The problem will be magnified by rapid urban growth. In 1950, there were less than 100 cities with a population in excess of 1 million; by 2025, that number is expected to rise to 650. By the year 2000, some 23 cities—18 of them in the developing countries—have had populations exceeding 10 million. On a global scale, half of the world's people will live in urban areas.

Some of the world's largest cities, including Beijing, Buenos Aires, Dhaka, Lima, and Mexico City, depend heavily on groundwater for their water supply, but it is unlikely that dependence on aquifers, which take many years to recharge, will be sustainable. Groundwater from aquifers beneath or close to Mexico City, for example, provides it with more than 3.2 billion liters per day, but already water shortages occur in many parts of the capital. As urban populations grow, water use is shifting from agriculture to municipal and industrial uses, making decisions about allocating water between different sectors more difficult.

In this chapter, three levels of water system failure to the points of widespread disaster have been discussed. Water scarcity in many regions could easily be triggered to become a disaster when these regions are facing drought or any other impact that makes their limited resources scarce or unusable.

Water scarcity is intensified by six principal factors:

- Reluctance to treat water as an economic as well as a public good, resulting in inefficient water use practices by households, industries, and agriculture. Farmers pay too little to cover the whole cost of water resources development; very often in developing countries, households pay a lump sum tariff for their water use.
- Excessive reliance in many places on inefficient institutions for water and wastewater services. There is no incentive to improve their efficiency and reliability under the current organizations.
- Fragmented management of water between sectors and institutions, with little regard for conflicts between social, economic, and environmental objectives.
- Inadequate recognition of the health and environmental concerns associated with current practices. Lack of a board for licensing engineers and the shortage of trained engineers, in many developing countries data on water quality, and information dissemination systems further aggravate this problem. Experts still do not have comprehensive understanding of water-quality issues in developing countries as most of them are trained in developed, often temperate, countries.
- Environmental degradation of water sources, in particular, reduced water quality and quantity due to pollution from urban or land-based activities. Too little money and attention are paid to improve such basic infrastructures as water and wastewater systems, while more money is spent for economic growth, building dams, and massive infrastructures. Lack of

consensus on "who should pay for water and wastewater" very often makes it difficult to build sustainable water and wastewater systems. One of the examples is the sewage treatment systems in Thailand; municipalities often refuse to manage and operate sewage treatment plants because people do not want to pay for their wastewater.
- Inadequate use of alternative water sources. Alternative water sources other than groundwater and surface water are rarely explored. Desalination is too expensive, and rainwater harvesting is only good for small communities in remote areas. Wastewater reuse may be a future alternative but requires a better understanding of the risks and benefits of water reuse.

Instead, we must adopt a new approach to water resources management in the new millennium so as to overcome these failures, reduce poverty, and conserve the environment—all within the framework of SD (Parliamentary Commissioner for the Environment).

The solution to the water disaster is closely linked to improving the water governance of our cities. A paradigm shift is urgently needed in urban water governance. What is needed is a broad-based partnership of public, private, and community sectors. The private sector brings in efficiency gains in water management. Community participation facilitates transparency, equity, and sense of ownership and helps in cost recovery. The government's role is most important in policy-setting and as a regulatory agency. The new paradigm must build on the relative strengths of all actors, avoiding overlaps, and redundancies.

Managing the growing demand must also be addressed as a priority. Debt-ridden countries can buy precious time by postponing costly investments. Water saved through demand management is nothing but a new supply without additional investment. A $2 million investment in demand management could delay a new investment of $100 million for 10 years. That makes good economic sense. Demand management not only saves new investment in water projects but also significantly reduces the cost of wastewater treatment. Pricing policy and tariff structures are important instruments for demand management and should go beyond cost recovery. Profligate users usually change their water-use behavior, only when faced with a punitive tariff.

There are many good ideas around the world on how to conserve water. The City of Windhoek has clearly shown that effective demand management measures could contain excessive demand for water and postpone costly investments by several years. We should exchange information and share from these experiences. We should replicate the good practices, like the lifeline tariff in Durban in South Africa, where low-income communities are given a fixed volume of water a day for which they pay an affordable amount. What is needed is an effective information clearing house which could help sharing of information and experience by city managers and policy makers (Karamouz and Moridi, 2005). These are some examples how the world reacted to water scarcity and reducing the risk of water disaster in these regard.

Furthermore, in this chapter, many initiatives for UWDM have been presented. Action items at this stage involve raising questions and trying to create consensus among the decision makers and stakeholders. Without a share vision on UWDM, we are only relying on sacrifices, sporadic and noncoordinated actions, and outside helps to reduce the level of suffering and discomfort caused by a disaster. The accelerated development plans to build expand, and invest in the water infrastructures, prevent us from sitting back and thinking hard about disaster prevention and management and coming with solid contingency plans for different probable disasters to prevent public and water infrastructure facilities. Perhaps we can conclude this chapter by making suggestions and raising some more questions.

- What are the probable disasters, their magnitude, and duration, and what, who, and where will be effected?
- What is the state of systems readiness to face the disasters?

- What makes us more vulnerable when expending and implementing our development projects?
- How our management infrastructure system as well as monitoring systems and DSS equips us with tool, models, gadgets, and sirens to predict, handle, recover, and learn about disaster?

We can also conclude by initiating the following action items and situation scenarios:

On the disaster indices:

- A set of criteria should be selected and for each criterion a range of values that are satisfactory should be defined. The duration as well as the extent of individual and cumulative failures of water resources systems should be determined for each criterion.
- Development of a shared vision of dealing with the disaster and identifying ways for all parties involved to achieve the shared vision is the key strategy in handling a disaster.
- Developing coordinated approaches among all concerned, responsible, and affected agencies.
- Establish baselines for system functioning against drastic changes in supply and delivery.
- Monitoring and evaluating disaster progression and recovery.

On the risk-based design:

- Settling a risk level of say 5% per year of failure to meet 95% of a given demand target
- Determining the cost of investment to reduce it to 3% to meet a given target of 95% of the time
- Determining the trade-off between accepting a risk of 5% to meet 95% of the demand target and a risk of 3% to meet 80% of the demand

On the use of science and technology:

- Integrating the best science and engineering techniques available into the decision-making process, while continuing scientific research to improve knowledge and understanding
- Establishing baseline conditions for system functioning and sustainability against which change can be measured
- Monitoring and evaluating actions to determine if goals and objectives are being achieved

On the other issues:

- How much water we have that is considered secure water? (i.e., for a given city like Tehran with 30 cms daily water demand, how much water is considered secure?)
- How evenly secure water is distributed through the city?
- What is the national plan to meet the disaster of different magnitudes?
- What are the regional plans to meet droughts higher than the last or most sever historical drought?
- How secure the reservoirs and transfer systems are?
- When was the last time that key transfer tunnels were thoroughly inspected?
- Do we have a regional preventive maintenance programs?
- Do we have an emergency response program for water disaster that has been simulated in real time to see the state of system's readiness to face a water disaster?

Finally, it should be stated that disaster of any magnitude could be converted to opportunities if we learn from the past disasters and utilize the unification of efforts that has been resulted from disaster. Experience has shown that the chance of better practices and implementation of sound water resources and demand management as well as disaster prevention is higher when people

Urban Water Disaster Management

and societies have experienced or witnessed the suffering of societies/regions. Perhaps what has happen in 2005 around the world dealing with water disasters could be a strong motive for many governments/municipalities to place action plans for water disaster management, especially in urban areas on their national/regional agenda.

PROBLEMS

1. Construct the CDF of failure occurrences in two given situations. (1) $\rho(t) = \alpha t$, (2) $\rho(t) = e^t$.
2. Calculate the reliability of the system illustrated in Figure 11.25, at $t = 0.5$. The reliability function of each component is given in system diagram.

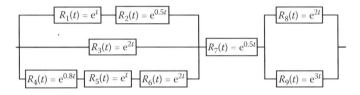

FIGURE 11.25 System of Problem 2.

3. Calculate the system reliability of the water distribution network of Figure 11.26 using the state enumeration method. Node 1 is the source node and other nodes are demand nodes. All of the pipes have the same failure probability equal to 5%. System reliability is defined as the ability of the system for supplying water to all of the demand nodes. The pipes' performances are independent of each other.

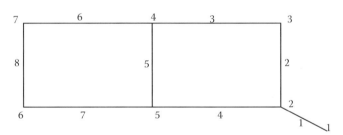

FIGURE 11.26 Layout of water distribution network of Problem 3.

4. Calculate the reliability of the network of Problem 3 using cut-set analysis.
5. Employ the tie-set analysis to calculate the reliability of water network described in Problem 3.
6. Determine the coefficient of variation of the loading and the capacity for a main pipe of a water distribution network with the parameters given in Table 11.8. Consider a uniform distribution for definition of the uncertainty of each parameter.

TABLE 11.8
Characteristics of Main Pipe of Problem 6

Parameter	Mode	Range
Water demand	10 m³/s	8–12 m³/s
C_{HW}	150	145–155
d	800 mm	600–1000 mm
S_0	0.001 m/m	0.0009–0.0011 m/m

7. Using the results of Problem 6, determine the risk of the loading exceeding the capacity of the main pipe. Consider that MS = $Q_C - Q_L$ and MS follow the normal distribution.
8. Determine the coefficient of variation of the loading and the capacity for a sewer pipe with the parameters given in Table 11.9. The uncertainty of each parameter is defied with a triangular distribution.

TABLE 11.9
Characteristics of Sewer Pipe of Problem 8

Parameter	Mode	Range
C	0.65	0.6–0.70
i	20 cm/h	18–22 cm/h
A	7 ha	6.9–7.1 ha
f	0.002	0.0019–0.0021
d	150 mm	145–155 mm
S_0	0.0005 m/m	0.0004–0.0006 m/m

9. Determine the risk of the loading exceeding the capacity of the sewer pipe in Problem 8. The MS is normally distributed and is calculated as MS = Q_C/Q_L.
10. Apply the first-order analysis of uncertainty to the Darcy–Weisbach equation. Consider that S_0, f, and d are uncertain.
11. Consider a 800-mm water transfer tunnel with mean loading equal to 2 m³/s and coefficient of variation equal to 0.20. Calculate the risk of failure of a pipe using safety margin approach when MS is normally distributed. The energy slope of tunnel is 0.0002 with a coefficient of variation of 0.10. The Darcy–Weisbach friction factor is 0.0002 and has a coefficient of variation of 0.15.
12. The monthly supplied water for an urban area in 3 years is presented in the Table 11.10. The monthly mean water demand of the urban area is equal to 100 MCM. Evaluate the reliability, resiliency, and vulnerability of this urban water supply system for supplying more than 90% of urban water demand. How would the system reliability change if the water demand in months June to August is decreased up to 15%.

TABLE 11.10
Monthly Supplied Water for Urban Area Described in Problem 12

Month	First Year	Second Year	Third Year
January	101	82	93
February	120	102	110
March	110	110	107
April	85	86	83
May	77	64	55
June	73	76	67
July	84	93	59
August	105	100	72
September	111	119	94
October	106	92	101
November	112	99	121
December	118	108	111

REFERENCES

Akan, A. O. and Houghtalen, R. J. 2003. *Urban Hydrology, Hydraulics and Stormwater Quality*. Wiley, Hoboken, NJ.

Allen Consulting Group. 2005. Climate Change Risk and Vulnerability Promoting an Efficient Adaptation Response. Department of Environment and Heritage Australian Green House Office. Available at: http://www.greenhouse.gov.au/impacts/publications/pubs/risk-vulnerability.pdf.

Ang, A. H. S. 1973. Structural risk analysis and reliability based design. *Journal of the Structural Engineering Division, ASCE* 99(ST9): 1891–1910.

Canadian International Development Agency (CIDA). 2005. Integrated Water Resources Management, Training Manual and Operation Guide. Available at: http://dsp-psd.pwgsc.gc.ca/Collection/BT31-4-27-1998E.pdf.

Desbordes, M. and Seravat, E. 1988. Towards a specific approach of urban hydrology in Africa. *International UNESCO Symposium on Hydrological Process and Water Management in Urban Areas*, Duisburg, the Netherlands.

De Bruijn, K. M. and Klijn, F. 2001. Resilient flood risk management strategies. In: L. Guifen and L. Wenxue, Eds, *Proceedings of the IAHR Congress*, September 16–21, Beijing China, ISBN 7-302-04676-X/TV Tsinghua University Press, Beijing, pp. 450–457.

FEMA. 2008. http://www.fema.gov/library.

Garen, D. 1993. Revised surface water supply index for Western United States. *Journal of Water Resource Planning and Management ASCE* 119(4): 437–454.

Hashimoto, T., Stedinger, J. R., and Loucks, D. P., 1982. Reliability, resiliency and vulnerability criteria for water resources performance evaluation. *Journal of Water Resources Research* 18(1): 14–20.

Herath, G. 2008. Soil erosion in developing countries: A socio-economic appraisal. *Journal of Environmental Management*, 68(4): 343–353.

Holling, C. S. 1973. Resilience and stability of ecological systems. *Annual Review of Ecology and Systematic* 4: 1–24.

Iraq Health. 2006. The Children of Iraq on the Brink of Disaster. Available at: http://www.phrusa.org/research/health_effects/humiraq.htm.

Kapur, K. C. and Lamberson, L. R. 1977. *Reliability in Engineering Design*. Wiley, New York.

Karamouz, M. 2002. Decision support system and drought management in Lorestan Province. Final Report to Lorestan Water Board, Iran.

Karamouz, M., Szidarovszky, F., and Zahraie, B. 2003. *Water Resources Systems Analysis*. Lewis Publishers, CRC Publishing, Boca Raton, FL.

Karamouz, M. 2004. Qualitative and quantitative planning and management of water allocation with emphasis on conflict resolution. Second Report to Water Research Council, Iranian Water Resources Management Organization, Ministry of Energy, Iran.

Karamouz, M., Torabi, S., and Araghi Nejad, Sh. 2004. Analysis of hydrologic and agricultural droughts in central part of Iran. *ASCE, Journal of Hydrologic Engineering* 9(5): 402–414.

Karamouz, M. and Moridi, A. 2005. Water crisis management. *Joint Workshop of Tehran–Madrid–Melbourne on Optimal Management of Water Demand and Consumption*. Tehran. Iran.

Karamouz, M., Rasouli, K., and Nazif, S. 2009. Development of a hybrid index for drought prediction: A case study. *ASCE Journal of Hydrologic Engineering*, 14(16): 617–627.

Mays, L. W. and Tung, Y. K. 1992. *Hydrosystems Engineering and Management*. McGraw-Hill, New York.

Mays, L. W. 2001. *Water Resources Engineering*. Wiley, New York.

Mays, L. W. 2004. *Water Supply Systems Security*. McGraw-Hill, New York.

Nazif, S. and Karamouz, M. 2009. An algorithm for assessment of water distribution system's readiness: Planning for disasters. *ASCE Journal of Water Resources Planning and Management*, 135(4): 244–252.

Pimm, S. L. 1991. *The Balance of Nature? Ecological Issues in the Conservation of Species and Communities*. University of Chicago Press, Chicago.

Plate, E. J. 1986. Trends in stochastic hydraulics: stochastic design for water quality of a river, Ed: M. L. Albertson and C. N. Papadakis, Megatrends in Hydraulic Engineering, Colorado State University, Fort Collins, CO.

Plate, E. J. and Duckstain, L. 1988. *Water Resources Bulletin*, 24(2).

Ruiter, W. 1990. Watershed: Flood protection and drainage in Asian cities. *Land & Water*, 168: 17–19.

Shafer, B. A. and Dezman, L. E. 1982. Development of a surface water supply index (SWSI) to assess the severity of drought conditions in snow pack runoff area. *Proceeding of the Western Snow Conference*, pp. 164–175, Colorado State University, Fort Collins, CO.

Tokyo Metropolitan Area. 2006. Urban Disaster Prevention project. Available at http://www.toshiseibi.metro.tokyo.jp/plan/pe-010.htm.

Tung, Y. K. 1996. Uncertainty and reliability analysis. In: L.W. Mays, Ed., *Water Resources Handbook*, Chapter 7, McGraw-Hill, New York.

United Nations Environment Program (UNEP). 2005. Environmental management and disaster preparedness. Available at www.unep.or.jp/ietc/wcdr/unep-tokage-report.pdf.

Wagner, J. M., Shamir, U., and Marks, D. H. 1998. Water distribution reliability: Simulation methods. *ASCE Journal of Water Resource Planning and Management* 114(3): 276–294.

Yen, B. C. 1978. Safety factor in hydrologic and hydraulic design. *Proceedings of the International Symposium on Risk Reliability in Water Resources*, University of Waterloo, Waterloo, Ontario, Canada.

12 Climate Change

12.1 INTRODUCTION

Climate is often described as the synthesis of weather recorded over a long period of time either in part or generally of the earth as a whole. It is defined in terms of long-term averages and other statistics of weather conditions, including the frequencies of extreme events. Climate is considered a dynamic phenomenon. A natural process that plays a major part in shaping of the earth's climate is the greenhouse effect. It produces a relatively hospitable environment near the earth's surface where humans and other life forms have been able to live. Through a range of activities such as accelerated use of fossil fuels and broad-scale deforestation and land use changes which are intensified in urban areas, humans have contributed to an enhancement of the natural greenhouse effect. This enhanced greenhouse effect results from an increase in the atmospheric concentrations of the so-called greenhouse gases (GHGs), such as carbon dioxide and methane. All these effects that change the natural variations of climate are considered as climate change phenomena. Typical human activities in cities, such as polluting industries, commercial businesses, and heavy traffic as well as dependency on cars, contribute to emission of GHGs and therefore climate change.

There are many aspects in climate change impacts that come together to produce a picture of potentially significant implications for urban water utilities. This can create an information overload that, coupled with uncertainties, presents a barrier to understanding and to developing responses. Climate change impacts in urban areas are commonly intensified and affect the urban water supply and management and also results in urban flash floods that impact the design of urban drainage systems. Planners should be aware of the specific effects of climate change on their region. Some areas will require planning specifically for the possible changes in that region.

The aim of this chapter is to provide an essential understanding of climate change and its impacts and then turn to consideration of the issues involved in developing suitable urban water planning and water management sector responses to climate change. A general description of climate change processes and effects follows the introduction. The impacts of these climatic changes on urban water suppliers are then identified and described, including regional differences. Responses to climate change are then discussed, both in terms of "adaptation strategies" to reduce or avoid impacts of climate change and in terms of "mitigation strategies" that utilities may adopt to reduce the contribution of water utility operations to the production of GHG emissions. Finally, the principals of climate change projection and downscaling are addressed and exemplified through some case studies.

12.2 CLIMATE CHANGE PROCESS

The understanding of climate change processes is supported by extensive scientific consensus led by the Intergovernmental Panel on Climate Change (IPCC). The role of the IPCC is to assess the science of human-induced climate change, mainly on the basis of peer-reviewed and published scientific/technical literature. In 2006, the National Center for Atmospheric Research (NCAR) produced a primer on climate change for the American Water Works Association Research Foundation (AWWARF) that provides a comprehensive overview of the current science supporting the understanding of climate change processes and effects. It builds on the consensus findings of the Third Assessment Report of the IPCC, published in 2001, and closely matches findings of the Fourth Assessment Report of the IPCC, published in 2007.

Based on the NCAR definition, any persistent change in the statistical distribution of climate variables is considered as climate change. Climate change is commonly known as a change in the global average temperature, which is more severe in urban areas because of some special characteristics such as existence of heat islands. A precise knowledge of this change is of little predictive value for water resources especially due to high uncertainties in climate variations in urban areas. On a global scale, the warming would result in increased evaporation and increased precipitation. What is more important is how the temperature will change regionally and seasonally, affecting precipitation and runoff, which is by no means clear. In general, the hydrological effects of climate change are likely to influence water storage patterns throughout the hydrologic cycle and to impact the exchange among aquifers, streams, rivers, and lakes. Chalecki and Gleick (1999) have provided a bibliography of the impacts of climate change on water resources in the United States. The effects of climate change on different components of the water cycle are described in the following subsections.

12.2.1 INCREASING TEMPERATURE

The fundamental climate change process is global warming. Since 1900, the global average temperature has increased by about 0.74°C. It has been demonstrated that the GHGs produced by human activities have resulted in increasing of the global temperature. This trend will accelerate, if the concentration of GHGs increases in the atmosphere. By 2100, the additional rise in global average temperature is projected to be in a range 1.1–6.6°C above the 1990 levels. This demonstrates the developmental state of the art of modeling global climate processes. But there are two main concerns in this process. First is the uncertainty about the relation between the amount of warming and the concentration of GHGs. For example, the IPCC concluded that there is a two-thirds probability that global temperatures will rise by 2–4.5°C with a doubling of CO_2 levels in the atmosphere. This means that there is a one-third probability that the warming could be lower or higher than that range. GHG emissions growth, which is affected by future development paths including population growth, economic growth, and technological changes, is another source of uncertainty. Due to different assumptions on GHG production, climate models produce varying estimates of both CO_2 concentrations and global mean temperature, but all of them show that the global temperature will increase in the following decades (Mays, 2004; Cromwell et al., 2007). The water shortage severity will be more sensible in urban areas due to limited water resources and increasing water demand as a result of increasing temperature and population growth.

12.2.2 VARIATION IN EVAPORATION AND PRECIPITATION

Climate change will basically accelerate the pace of the hydrological cycle because of global warming. As temperatures rise, the availability of energy for evaporation increases and the atmospheric demand for water in the hydrological cycle also increases. Therefore water evaporates more readily and causes the total amount of precipitation to increase at a global level. This accelerated hydrological cycle will result in an overall increased intensity of rainfall events. Consistent with the fact that global warming has already been occurring during the last century, stream flow records already document this increase in storm intensity. But it should be noted that the forecasts of precipitation in different regions are completely different from each other. Based on these forecasts, climate change will intensify the water shortage in arid and semiarid regions, although the climate model results are not the same in a regional scale (Mays, 2004, Cromwell et al., 2007). The water shortage severity will be more sensible in urban areas due to limited water resources and increasing water demand as a result of increasing temperature and population growth.

12.2.3 SOIL MOISTURE

Subsurface water flow is a major part of the hydrological cycle, which consists of infiltration of surface water into the soil to become soil moisture, subsurface (unsaturated) flow through the soil, and groundwater (saturated) flow through soil or rock strata. Since climate changes alter the precipitation

Climate Change

and evapotranspiration, it will affect soil moisture storage and groundwater recharge dynamics (see Rind et al., 1990; Vinnikov et al., 1996), which play an important role in urban water supply schemes especially in arid and semiarid regions. The increase of impervious areas in urban areas also intensifies this effect of climate change.

The impact of climate change on soil moisture varies in different regions due to changes in precipitation and evaporation. In regions where the amount of precipitation decreases, the soil moisture will decrease, but in regions with increasing precipitation, different situations might happen. In such regions, the soil moisture on the average or over certain periods may decrease because of the considerable increases in evaporation if temperature is higher or the timing of the precipitation or runoff changes.

12.2.4 Snowfall and Snowmelt

Rising temperature will lead to dramatic changes in snowfall and snowmelt dynamics, especially in basins with considerable snow (Glieck, 2000). Global warming will decrease the snow part of precipitation, and areas accustomed to snowfall will experience the shorter and warmer winters that start with delay. The shorter cold season means that the spring melt can arrive much earlier and have significant implications for stream flows available downstream in the late summer and early fall. This will affect the regime of surface resources which is much important in urban water supply scheme especially in regions with high dependency to surface resources. Both climate models and theoretical studies of snow dynamics have long projected that higher temperatures will lead to a decrease in the extent of snow cover in the North Hemisphere. More recently, field studies have corroborated these findings.

12.2.5 Storm Frequency and Intensity

The global climate change will result in a wetter world; however, the regional rainfall patterns change differently. Unfortunately, climate simulation models cannot produce detailed precipitation patterns at regional scales. Glieck (2000) concludes that since intense convective storms tend to occur in regional scales, the changes in rainfall intensity and variability cannot be modeled by global climate models. These difficulties are more evident in urban areas due to different microclimates that are affecting climate characteristics in these regions.

Based on Wigley and Jones (1985), it can be concluded that

- Runoff variations due to climate changes are intensified in arid and semiarid regions.
- Runoff variations are affected more by precipitation changes than by evaporation.
- Runoff changes are always greater than precipitation variations.
- Runoff changes usually do not exceed the evapotranspiration changes except in the regions where the ratio of long-term average runoff to long-term average precipitation is <0.5.
- The global average of runoff will highly increase in response to climate change effects and on the other hand, there is a large increase in land evapotranspiration that will balance the global water cycle.

For characterizing and predicting the effects of climate change on the hydrological cycle, first the stream flow variability in historical data should be investigated because stream flow is climatically determined but highly variable. These changes will highly affect the urban water resources. The urban water supply management strategies should be modified based on the probable changes.

12.2.6 Groundwater

Alley et al. (1999) have mentioned the following effects of climate change on groundwater sustainability:

- Significant changes in groundwater recharge because of changes in regional precipitation and temperature or time pattern of precipitation.

- More severe and longer-lasting droughts that intensify the withdrawal from groundwater.
- Changes in evapotranspiration resulting from changes in vegetation.
- Increased water supply from groundwater as a main or alternative source of water supply in urban areas especially in arid and semiarid areas.

Surficial aquifers that supply flow to streams, lakes, wetlands, and springs are much more sensitive to climate changes than other components of the groundwater system. Groundwater resources assessments and simulations are based on average conditions, such as annual recharge and/or average annual discharge to streams. This is due to the slow response of groundwater systems to variations of a climate system. But it should be mentioned that the use of average conditions may underestimate the importance of droughts (Alley et al., 1999). The special impacts of climate change on groundwater in different regions and their quantity and quality variations have received less attention in the literature compared with surface water. However, in arid and semiarid regions where water supply for domestic demand mostly relies on groundwater, a more precise evaluation of possible changes in water table is needed. Furthermore, some evaluations on spatial patterns of groundwater changes could help in developing longer-term urban water supply schemes.

12.2.7 Water Resource System Effects

Lattenmaier et al. (1996) evaluated the sensitivity of the six major water systems in the United States through analysis of system performance indices including reliability, resiliency, and vulnerability, whose definitions are discussed in Chapter 11. They finally concluded that the most important factors in system performance's sensitivity to climate changes are changes in runoff and water demands. They also mentioned that the system sensitivities depend on the water usage purposes and their priorities. For example, in urban water supply system because of demand for high-quality water with considerable temporal changes, and limited water resources, any small change in temporal variation of water resources will highly impact the system performance. In such cases, flexible water supply schemes are needed to provide enough resiliencies in the system and decrease possible water shortages.

12.3 IMPACTS OF CLIMATE CHANGE ON URBAN WATER SUPPLY

The general impacts of climate change on water cycle components were described in the previous section, but the impacts of climate change on urban water supply are different to some extent. The impacts on urban water supply may involve many additional cause-and-effect relationships and are more remote and uncertain in terms of both the chain of the events and timing.

For appropriate planning of urban water systems, a working knowledge of the cause-and-effect relationships that produce impacts is needed. Therefore, in this section, the climate change impacts on urban water supply are discussed under three distinct groups: direct, indirect, and compound impacts.

12.3.1 Direct Impacts

The effects of climate change on urban water utility functions and operations are discussed as direct impacts. These impacts and their possible effects on urban water supply can be summarized as follows (Cromwell et al., 2007):

- Warmer and shorter winter seasons
  - Increase glacial melting
  - Decreased seasonal snowpacks
  - More rain
  - Earlier spring snowmelt
    - Changes in recharge pattern of groundwater
    - Earlier runoff into surface waters

Climate Change

- Lower baseflows in surface waters, especially in the summer and fall
- Lower reservoir levels in the summer and fall
- Warmer and potentially drier summer seasons
  - Changes in vegetation of watershed and aquifer recharge areas
    - Changes in recharge pattern of groundwater
    - Changes in quality [e.g., total organic carbon (TOC), alkalinity] and quantity of runoff
  - Increased water temperature
    - Increased evaporation and eutrophication in surface sources
    - Water treatment and distribution challenges
  - Increased water demand
    - Increased irrigation water demand because of longer growing seasons
    - Increased domestic water demand because of longer dry spells
    - Increased discharge of groundwater resources for supplying water demand
- Increasing the frequency and intensity of rainfall events
  - Increased turbidity and sedimentation of runoff and surface water resources
    - Loss of reservoir storage due to increased evaporation
      - Shallower, warmer water, and eutrophication
      - Potential conflicts with flood control objectives
    - Water filtration or other treatment challenges
  - Increased risk of flood damage to water utility facilities

In considering the cause–effect outlines, distinct differences between urban areas such as their natural climatical, physical, and population characteristics should be considered although there are some similarities across regions. But it should be noted that these impacts do not happen at the same time and all together. Theses impacts happen over time because there are gradient functions, threshold effects, and conceivably many other influences involved in their happening. These impacts should be evaluated in the following decades in order to be considered in mid-to-long-term (20–50 year) planning. Although some of these impacts are evident in current situations, urban utility planners will have to struggle with many of them in the future rather than as phenomena that are already observed, except at the leading edge of the trend.

These cause–effect outlines make it possible to scan the full spectrum of potential direct impacts for each regional scenario that follows a simple logic and organizes the impacts into related groupings. Taken together, it is believed that these outlines provide complete coverage of direct impacts. Despite some common elements between them, it is also believed that these outlines constitute the minimum set needed to cover the direct impacts on the urban water supply system.

12.3.2 Indirect Impacts

Indirect impacts are defined as resulting from the effects of climate change on the baseline environment in which urban water utility functions and operations are carried out. But the baseline operating environment is also changing. The understanding of the impacts of climate change and responses must incorporate this critical dimension. There are at least three major categories, as listed below, of indirect impacts on urban water supply utilities that can result from the impacts of climate change on environmental and socioeconomic systems. Indirect impacts may be more significant than the direct impacts. Unlike the direct impacts, the main categories of indirect impacts apply uniformly throughout all regions.

1. *Baseline impacts on terrestrial and aquatic ecosystems.* Changes in basic climate variables such as temperature, rainfall, seasonal patterns, runoff characteristics, and recharge patterns of both ground water and surface water can produce significant baseline changes in both terrestrial and aquatic ecosystems. Changes in the vegetative composition of terrestrial ecosystems can change the

baseline geophysical and chemical character of watershed runoff and of waters recharging groundwater aquifers. The character of clay and silt particles comprising turbidity can get changed by changed runoff patterns.

2. *Baseline impacts on water pollution.* The expectation is that more severe storms resulting from increased frequency and intensity of rainfall will produce more severe flooding which is much intensified in urban areas due to high percentage of impervious areas. This will inevitably result in additional water pollution from a variety of sources that enter the stormwater affecting especially wastewater treatment, storage, and conveyance systems. Preliminary work by the EPA has confirmed that, for the most part, wastewater treatment plants and combined sewer overflow control programs have been designed on the basis of the historical hydrological record, taking no account of prospective changes in flow conditions due to climate change. As a result, it is conceivable that urban water suppliers will face a continually increased influent variation from sewage overflows.

3. *Baseline impacts on socioeconomic systems.* The fates of urban water and wastewater utilities are always closely tied together by the water bill because affordability is indifferent to organizational establishments. Threats to the cost effectiveness of wastewater programs raised by climate change can have consequences for the water utility because capital resources and ratepayer resources are constrained. If the wet-weather programs referred to above require expensive resizing to manage higher flows, it may reduce influent challenges for urban water utilities, but will still have an impact on the urban water supply side of the business through this financial connection. A different and equally worrisome set of financial threats to wastewater utilities may result from the dry-weather extremes of climate change since discharge permits and waste load allocations are quite often grounded in the low flows documented in the historical hydrological record.

Because the baseline operating environment is changing and will continue to change in a manner that may not be linear, it may not be sufficient to simply replace old arrangements with new ones; it may be necessary to develop and apply a completely different dynamics in conceiving and managing institutional relationships in urban areas.

At the highest level of abstraction, coping with climate change challenges the baseline concepts of environmental quality protection. There is a good understanding of the objectives of these efforts in a static environment. But the objectives in an environment where the baseline environment is changing should be well recognized. There are two distinct ways: dealing with the warming trends and preserve the status quo, or utilizing a more dynamic approach to managing environmental quality. In this way, devising new partnering relationships with stakeholders and regulators will be pivotal in solving the problems in future urban water supply under climate change effects (Cromwell et al., 2007).

12.3.3 Compound Impacts

The cumulative effect of direct and indirect impacts on urban water supply system can amount to much more than a summation. Because many of the individual direct and indirect impacts affect the same natural systems or utility systems, the total impact can get magnified. It has more of the character of a compounding process. There are many conceivable challenges relating to water treatment, for example, that may have to be met in order to cope with alternative water sources in response to climate change. Evaluating them individually may not reveal all the issues, but looking at all of them together can indicate the need for a more extreme treatment solution (such as membrane treatment). Unintended side effects also result from such complex optimizations. It may not be a simple matter, for example, to distribute membrane-treated water through an aged distribution system. Side effects can be manifested as health threats.

12.4 ADAPTATION WITH CLIMATE CHANGE

Urban water utilities should respond to climate change through "adaptation" to modify plans and operations to minimize impacts. These adaptation efforts are classified into two groups. The first

Climate Change

consists of vulnerability analyses that are intended to identify the most near-term priorities in dealing with impacts that happen sooner. The second is long-term planning, or more formally, integrated resource planning (IRP) that adopts the broadest possible strategic view of how a utility can plan to cope with such systemic changes over the longer term. In addition, utilities are responding through adoption of measures to help mitigate the onset of climate change by reducing energy consumption that contributes to the production of GHGs. All of these strategies are discussed in the following subsections.

12.4.1 Vulnerability Analysis

Goals of an vulnerability evaluation of urban water utilities to climate change processes include obtaining a better analytical assessment of the possibility of getting that current water resource development and facility plans disrupted by near-term (20–50 year) manifestations of climate change processes. This initial focus on vulnerability is a good means of identifying an urban water utility's priority issues relating to climate change. Two alternative approaches to vulnerability analysis have been articulated: "top-down" and "bottom-up."

Some of these efforts have employed climate models (referred to as GCMs—General Circulation Models) to attempt to build climate change estimation into the urban water supply planning. This has been labeled the "top-down" approach to vulnerability analysis. The major drawback of this approach lies in the current level of analytical resolution of the GCMs. The nearly two dozen most recognized GCMs are consistent in projecting an increasing global mean temperature, but across a range of variability that is also, in part, the product of various GHG "emission scenarios" that reflect alternative global assumptions about the future path of economic growth and efficacy of GHG controls. This variability is more evident in urban areas regarding to high uncertainties in GHG emission and the factors affecting the climate. Also, the "top-down" approach is challenged by one additional source of variability in that GCMs precipitation forecasts must be converted to changes in surface runoff and groundwater recharge in order to connect them with water resource planning models. There is reason to question whether climate modeling can catch up with climate change to a sufficient extend to provide useful precision, but in some cases, the research insights gained from such modeling may be as, or more, useful than the predictions.

The central idea of the "bottom-up" approach is that urban water utilities can work with their own water resources planning models to assess the vulnerability of their 20–50 year supply plans to climate change. Based on the general findings of climate change research, utilities can identify the likely cause–effect pathways that can be proven to be troublesome. A utility's own water resource modeling tools can then be applied to examine extreme scenarios, involving such features as decreased inflow from runoff, decreased recharge, increased evaporation losses, and seasonal shifts.

The "bottom-up" analysis enables a utility to test the robustness of current plans to deal with changes in key climate-related variables. Once the thresholds or tipping points of a utility's plans have been identified in this manner (using familiar models in which a utility has relatively good confidence), it is then possible to turn to the climate experts and ask how plausible such breaking point scenarios seem in light of the results of broader research with the various GCMs. This type of comparative and interpretive analysis of climate modeling across the range of GCMs may be an important research priority for water utilities, enabling continuing improvements in the "bottom-up" approach to vulnerability analysis.

12.4.2 Integrated Resource Planning

The long-term response that is most prevalent in discussions of climate change is IRP. A hallmark of this approach to long-term planning is the adoption of a very broad view of the problem that "integrates" all aspects of environmental, socioeconomic, and engineering issues as a basic strategy for keeping a wide range of options open and providing a maximum of flexibility in the operation of urban water supply systems. The IRP approach has been used extensively in urban water development and supply planning through Integrated Water Resources Planning. The continued improvement and refinement of best practices in IRP will be of significant value to urban water utilities in coping with climate change.

An essential part of maintaining a broadly "integrated" approach is the continuous involvement of all stakeholders. IRP is most appropriate to problems of sweeping proportions that involve complex trade-offs between multiple objectives and multiple constraints such as water supply in urban areas with different categories of water demand and different sources of water pollutants. The best solutions are made possible in such situations because IRP recognizes that stakeholders have the capacity to redefine some of the objectives and constraints when necessary to avoid an impasse.

Partnership between water and wastewater utilities within a region and regulatory agencies constitute important subgroups of stakeholders with whom it may be necessary to devise entirely new institutional structures and methods of collaborating to meet multiple objectives and constraints on a changing play field. Such restructuring is consistent with the IRP approach of redefining objectives and constraints to broaden the boundaries and admit a broader range of solutions for urban water supply.

Even before climate change emerged as such a dominant theme, it had been asserted that managing water resources within an urban area to a "triple bottom line" criteria (essentially containing all the same elements as "integrated" resource planning) can produce better environmental outcomes than managing a disjointed array of standards established at the national and state levels. The additional imperatives imposed by climate change may provide enough reasons to consider such innovative approaches. If programs such as those intended to control sewage overflows and waste load allocations are undermined by changes in hydrology, there may still be room to maneuver within the altered hydrological system and attain good outcomes, if regulatory constraints are flexible.

Hydrological changes will also change the balances between utilities within an urban regions. There are no guarantees in the existing resource allocations. The potential for broader regional collaboration—up to and including consolidation of smaller utilities into larger entities—has often run around because existing resource allocations create "have" versus "have not" relationships that are difficult to convert into win–win relationships. Climate change has the potential to create situations in which no utility has a sure advantage but in which there are clear advantages to operating at a larger regional scale to make more options available. Restructuring of regional and related institutional relationships may be an important theme in coping with climate change.

The bottom line in urban water supply planning has always been a matter of coping with variability. With the coming changes in climate, there will be a heightened need to respond to increased variability especially in urban areas. The net effect of the direct impacts of global warming will be to change the variability of key parameters affecting the quantity and quality of water that would normally be expected to be available at any specific time and place. In addition, the capability to store water in various forms and the demand for water will be changed. This variations may cause serious problems in urban areas due to limited water resources and high priority of supplying water with adequate quantity and quality for habitants life.

Given the widespread extent of these changes, the IRP approach of taking the broadest possible view of the problem is indeed an ideal adaptation strategy. Working within broadly established system boundaries, it is possible to optimize in a manner that first derives maximum advantages from flexible operating strategies to expand the number of ways in which supplies can be managed to meet demands. This has the benefit of deferring irreversible capital projects that may present more risk in a changing environment. Urban water utilities already employ sophisticated modeling tools to design such operating strategies, and it seems likely that applying such modeling tools to the design of adaptation strategies will be as important priority as the climate models discussed previously.

Significantly, the IRP approach also features a comprehensive assessment of strategies that can be applied to manage the urban water demand side. Warming processes will lead to altered demand patterns as a result of seasonal shifts in precipitation, more evaporation, more frequent heat waves, and

Climate Change

more extensive droughts. Conservation programs offer a bonus in reducing both water supply needs and energy use. Conservation incentives (and disincentives to outdoor water use) may become more essential if warming processes otherwise increase water demands, especially during peak demand periods when both water supply and electric power capacities are stretched to their limits (Cromwell et al., 2007).

12.4.3 Reducing GHG Emissions

As it has happened in many water-short areas, it is possible to conceive of many ways to enhance the reliability of urban water resource management by essentially utilizing more energy to produce more water. But evaluation of these options must acknowledge the reality that urban water utilities account for a significant share of total electric consumption and power plant emissions that account for a significant share of GHGs. The key for urban water utilities will be to fully incorporate the objective of reducing GHG emissions as an additional objective within the IRP optimization framework.

As urban areas have grown, and continue to grow in the suburbs, water resource managers have already devised elaborate portfolio strategies to tap into multiple sources of supply coupled with strategic investments in system expansion to move and store water. While this flexibility in transmission and storage will be a valuable asset in reoptimizing current schemes to meet the challenges ahead, some features of the systems designed under the current understanding of climatic variability may not be reversible, or easily adaptable, under different operating regimes that were never previously envisioned or constrained to limit the production of GHGs.

Urban areas are known to attain higher air temperatures than the surrounding rural areas, because of differences in thermal balances. This phenomenon may be somewhat mitigated by strong winds, and it may increase cloudiness and precipitation in the city, as a thermal circulation sets up in between the city and the surrounding region. Air pollution originates in urban and industrial areas from gaseous emissions, volatilization of toxic compounds from water and soil, and particles from land use activities and wind erosion. The transport of pollutants in the atmosphere, with respect to the direction and distance traveled and concentration levels, can be described as a pollution plume, with the highest concentrations near the emission source.

Air pollution contributes to pollution of urban precipitation in the form of acid or polluted rain. Acid rain causes a lot of damages and can be generated and deposited locally, but it can also be transported to hundreds or thousands of kilometers by wind and can affect large areas. Air in cities contains solid particles of diverse sizes and origins. Such particles are scavenged by rain and transported into soils and surface waters. The composition of particles is of great concern, because they frequently absorb metals, toxic organic compounds, and gases. Besides water quality effects (and human health effects), the presence of particulate in the atmosphere contributes to increased precipitation, because such particles serve as condensation nuclei in the formation of a rainfall droplet.

The rising cost of electric power has caused many water utilities to re-examine their operating strategies in transmission and distribution in search of ways to curtail electric usage during peak periods and generally conserve the use of electric power. Renewable energy supply strategies, such as solar or wind-powered pumping, or in-line hydropower generation, may be more expensive initially, but may provide fuel cost savings and avoid GHG emissions. The challenge is to integrate such strategies within the IRP process to produce the best operating outcomes for the system as a whole in terms of cost, reliability, and social/environmental consequences.

Many urban water suppliers in overconstrained settings have also turned to energy-intensive membrane treatment processes to enable desalination of saline water sources and reuse of highly treated wastewater effluent. These processes make it possible to overcome a deterioration in the reliability of normal sources of supply by making it possible to meet part of the demand from sources that will be abundant under most climate change scenarios (i.e., yields from water reuse and desalt supply options are drought resistant). Although the cost—and especially the energy cost—of these technologies are significant, they should also be evaluated in the context of the overall IRP portfolio of options. If these technologies can plug a gap, or shore up a vulnerability produced by climate change processes, in a way that enables a broader scope for optimization across the entire portfolio, then they could play a critical role in improving the overall system planning.

12.5 CLIMATE CHANGE PREDICTION

The first step in the adaptation of urban water systems to climate change impacts is to make a quantitative and qualitative evaluation of the climate changes in the future. Based on the predicted impacts on urban water resources, appropriate strategies should be addressed for operation of water systems. GCMs are commonly used for climate change predictions.

These models solve the mass and energy conservation equations (in discrete space and time) that describe the geophysical fluid dynamics of the atmosphere. GCMs have the same general structure as numerical weather prediction models that are used for weather prediction (Mays, 2004). Lattenmaier et al. (1996) discussed the various types of climate change scenarios based on GCM simulations that include scenarios based directly on GCM simulations, check and balance methods, scenarios based on GCM atmospheric circulation patterns, stochastic downscaling models, and regional models. They concluded that perhaps the most useful scenarios of future climatic conditions are those developed by using a combination of methods. In the following section, the procedure for developing climate change scenarios, GCMs structure, and the downscaling procedure are introduced.

12.5.1 CLIMATE CHANGE SCENARIOS

The choice of climate change scenarios is important because it can determine the outcome of a climate change impact analysis, based on which urban water resources planning and management strategies will be developed. Extreme scenarios can produce extreme results, and moderate scenarios can produce moderate results. The selection of scenarios is controversial because scenarios are often criticized for being too extreme, too moderate, too unreliable, and too dependent on important factors such as changes in variability. All of these inaccuracies in developing climate change scenarios may cause socioeconomic losses or impacts on the urban water system if it is not backed by an analysis of risk and uncertainties.

Therefore, more than one scenario should be used to show the extent of uncertainty about regional climate change. Using one scenario can be misinterpreted as a prediction. Using multiple scenarios, particularly if they reflect a wide range of conditions (e.g., wet and dry), indicates some of the uncertainty about regional climate change. Climate change scenarios have been typically developed for a particular point in the future. Many climate change scenarios examine the climate associated with a doubling of carbon dioxide levels in the atmosphere over preindustrial levels ($2 \times CO_2$). This will most likely happen in the last half of the twenty-first century. These can be considered static scenarios because they are based on a false presumption that a stable climate will be reached in the future. This assumption is made to simplify analysis, not because it is widely believed that climate will reach a static condition. In contrast, transient scenarios examine how climate may change over time. They typically start in the present day and cover a number of decades into the future.

The selected climate change scenarios for impact assessment should meet the following five conditions:

Condition 1: The scenarios should be consistent with the broad range of global warming projections based on increased atmospheric concentrations of GHGs.
Condition 2: The scenarios should be physically plausible; that is, they should not violate the basic laws of physics.
Condition 3: The scenarios should estimate a sufficient number of variables on a spatial and temporal scale that allows for impact assessments (Smith and Tirpak, 1989; Viner and Hulme, 1992).

Climate Change

Condition 4: The scenarios should, to a reasonable extent, reflect the potential range of future regional climate change.

Condition 5: The scenarios should be straightforward to obtain, interpret, and apply for impact assessment.

There are three generic types of climate change scenarios: scenarios based on outputs from GCMs, synthetic scenarios, and analogue scenarios. Since a majority of impact studies have used the scenarios based on GCMs, this method is described in the following section.

12.5.2 General Circulation Models

GCMs are mathematical representations of the atmosphere, ocean, ice cap, and land surface processes based on physical laws and physically based empirical relationships. GCMs are used to examine the impact of increased GHG concentrations on future climate. GCMs describe the atmosphere as a rectangular grid covering the earth with cubes of air above which currently are 2° to 5° grids in latitude and longitude, and 6° to 15° vertical levels which are assumed to be representative of volume elements in the atmosphere as presented in Figure 12.1. Therefore their resolution is quite coarse. GCMs estimate changes for dozens of meteorological variables for these grid boxes. These models solve a series of equations in the atmosphere describing the movement of energy and momentum and conservation of mass and water vapor (Equations 12.1 through 12.4).

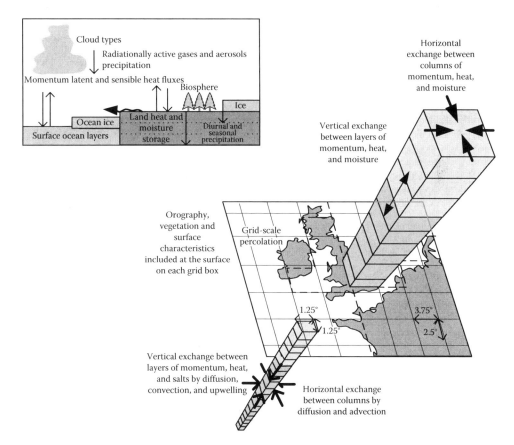

FIGURE 12.1 Schematic description of the atmosphere for utilizing GCMs. (After Viner, D. and Hulme, M. 1992. *Climate Change Scenarios for Impact Studies in the UK*. Climatic Research Unit, University of East Anglia, Norwich, U.K. With permission.)

1. Momentum equation:

$$\frac{D\bar{v}}{Dt} = -2\bar{\Omega} \wedge \bar{v} - (\rho^{-1})\nabla \cdot p + \bar{g} + \bar{F}. \quad (12.1)$$

2. Mass conservation equation:

$$\frac{Dr}{Dt} = -\rho(\nabla \cdot \bar{v}) + C - E. \quad (12.2)$$

3. Energy conservation equation:

$$\frac{DI}{Dt} = -p\frac{d\rho^{-1}}{dt} + Q. \quad (12.3)$$

4. Gas law:

$$p = \rho RT. \quad (12.4)$$

Here $(D/Dt) = (\partial/\partial t) + \bar{v} \cdot \nabla$, $\bar{v}$ is the velocity relative to the rotating earth, t is the time, $\bar{\Omega}$ is the angular velocity vector of the earth, ρ is the atmosphere density, g is the apparent gravitational acceleration, $\bar{F}$ is the force per unit mass, R is the density, C and E are the rate of creation and destruction of atmosphere constitutions, I is the internal energy per unit mass ($I = c_v \cdot T$), Q is the heating rate per unit mass, R is the gas constant, T is the temperature, c_v is the specific heat of air at constant volume, and p is the atmospheric pressure. These equations are solved with defined values of divergence terms at each grid points, and defined input and output for the upper and lower surfaces of the array for each defined element of the atmosphere.

The most advanced GCMs coupling the atmosphere and ocean models are referred to as coupled ocean–atmosphere GCMs (Gates et al., 1992). GCMs deal best with the large-scale dynamics and parameterize regional and local processes. This is partly because the hydrodynamics of the atmosphere is nonlinear and the energy that is fed into the system at the cyclonic scale is cascaded, through nonlinear interactions, to the smallest scales. Because of numerical modeling limitations, this cascade is parameterized. The coarse resolution in GCMs is another reason why many subgrid-scale processes, such as cloud formation, convective rainfall, infiltration, evaporation, runoff, and so on, are parameterized.

The major advantage of using GCMs as the basis for creating climate change scenarios is that they are the only tool that estimates changes in climate due to increased GHGs for a large number of climate variables in a physically consistent manner. The GCMs estimate changes in a host of meteorological variables, for example, temperature, precipitation, pressure, wind, humidity, and solar radiation (Schlesinger et al., 1997), which are consistent with each other within a region and around the world, and thus they fully meet Conditions 1 and 2, and partially satisfy Condition 3.

A major disadvantage of using GCMs is that, although they accurately represent global climate, their simulations of current regional climate are often inaccurate (Houghton et al., 1996). In many regions, GCMs may significantly underestimate or overestimate current temperatures and precipitation. Another disadvantage of GCMs is that they do not produce output on a geographical and temporal scale fine enough for many impact assessments (Condition 3) as it is needed in urban climate studies. GCMs estimate uniform climate changes in grid boxes several hundred kilometers across, and although they estimate climate on a daily or even twice daily basis, results are generally archived and reported only as monthly averages or monthly time series. An additional disadvantage of GCM-based scenarios is that a single GCM, or even several GCMs, may not represent the full

Climate Change

TABLE 12.1
Characteristics of the Available GCM Models for the Globe

	Acronym	Model	SRES Scenario Runs					
Max Planck Institute für Meteorologie, Germany	MPIfM	ECHAM4/OPYC3				A2		B2
Hadley Centre for Climate Prediction and Research, UK	HCCPR	HADCM3		A1FI		A2	B1	B2
						A2b		B2b
						A2c		
Australia's Commonwealth Scientific and Industrial Research Organization, Australia	CSIRO	CSIRO-Mk2	A1			A2	B1	B2
National Centre for Atmospheric Research, USA	NCAR	NCAR-CSM				A2		
		NCAR-PCM			A1B	A2		
Geophysical Fluid Dynamics Laboratory, USA	GFDL	R30				A2		B2
Canadian Center for Climate Modelling and Analysis, Canada	CCCma	CGCM2				A2		B2
						A2b		B2b
								B2c
						A2c		
Center for Climate System Research, National Institute for Environmental Studies, Japan	CCSR/NIES	CCSR/NIES AGCM + CCSR OGCM	A1	A1FI	A1T	A2	B1	B2

Source: IPCC, Fourth Assessment Report, Intergovernmental Panel on Climate Change (IPCC). 2007. Available at http://www.ipcc.ch.

range of potential climate changes in a region (Condition 4). Although GCMs have clear limitations for scenario construction, they do provide the best information on how global and regional climate may change as a result of increasing atmospheric concentrations of GHGs.

The name and characteristics of available GCMs all over the world are given in Table 12.1. In order to get the results of different GCMs for defined scenarios using the data available on the IPCC website, the following steps should be tracked:

1. First the IPCC data distribution center site should be visited: http://www.ipcc-data.org/. At this site, one gets interesting information about GCMs and climate change scenarios.
2. Then use the "environmental data and scenarios" module. In this page, one can get some information about CO_2 concentration, its changes in the future, and the possible impacts.
3. The next step is selecting the following link "scenario data for the atmospheric environment." In this link, the considered concentrations for air pollutants in different SRES scenarios are described.

TABLE 12.2
List of All Available GCM Models, Scenarios, and Variables

Models	Scenarios	Variables
BCC:CM1[anomalies[a]]	1PTO2X[anomalies]	Specific humidity[anomalies], precipitation flux[anomalies], air pressure at sea level[anomalies], net upward shortwave flux in air[anomalies], air temperature[anomalies], air temperature daily max[anomalies] air temperature daily min[anomalies], eastward wind[anomalies] northward
BCCR:BCM2[anomalies]	1PTO4X[anomalies]	
CCCMA:CGCM3_1-T47 [anomalies]	20C3M	
CCCMA:CGCM3_1-T63 [anomalies]	COMMIT	
CNRM:CM3[anomalies]	PICTL	
CONS:ECHO-G[anomalies]	SRA1B[anomalies]	
CSIRO:MK3[anomalies]	SRA2[anomalies]	
GFDL:CM2[anomalies]	SRB1[anomalies]	
GFDL:CM2_1[anomalies]		
INM:CM3[anomalies]		
IPSL:CM4[anomalies]		
LASG:FGOALS-G1_0 [anomalies]		
MPIM:ECHAM5[anomalies]		
MRI:CGCM2_3_2[anomalies]		
NASA:GISS-AOM[anomalies]		
NASA:GISS-EH[anomalies]		
NASA:GISS-ER[anomalies]		
NCAR:CCSM3[anomalies]		
NCAR:PCM[anomalies]		
NIES:MIROC3_2-HI[anomalies]		
NIES:MIROC3_2-MED[anomalies]		
UKMO:HADCM3[anomalies]		
UKMO:HADGEM1[anomalies]		

Source: IPCC, Fourth Assessment Report, Intergovernmental Panel on Climate Change (IPCC). 2007. Available at http://www.ipcc.ch.

[a] Anomaly is the difference from the average value.

4. Click the following link: "AR4 (2007): SRES Scenarios" on the left-hand side of the page; in a table similar to what is shown in Table 12.2, a list of all of the models, scenarios, and climatic variables simulated by GCMs is given. Each entry on the table links to a page that displays the data available for that model/scenario/variable. By choosing each of the models, scenarios, or variables, the related data will be described and one can choose the appropriate data as the mean of different time periods.

EXAMPLE 12.1

Using the downloaded monthly mean relative humidity data developed by the HADCM3 GCM for the period 1961–1999 based on climate change scenario B1a, draw the diagram of monthly variations of relative humidity at point (20, 41) of the model grid.

Solution: Following the described steps in Section 12.5.2, the related file to the desired data is downloaded, which is called "HADCM3_B1a_RHUM_1980.tar." The file is extracted and is shown in Figure 12.2.

Climate Change

There 96 × 73 points on the model grid where data are produced through the HADCM3 model. Data start from January 1961 and ends in December 1999. The data needed at point (20, 41) of the model grid are extracted from the "HADCM3_B1a_RHUM_1980.tar" file for different months as given in Table 12.3. The diagram of monthly variation of mean relative humidity in the period 1961–1999 is shown in Figure 12.3.

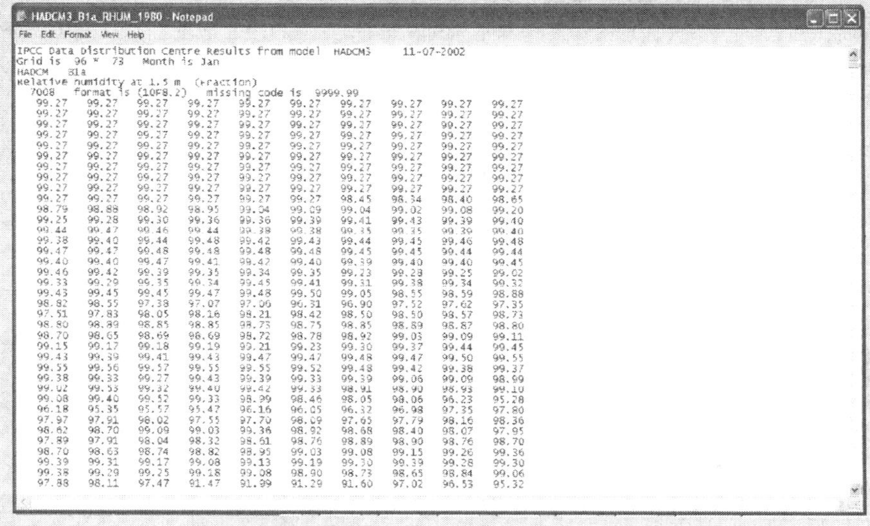

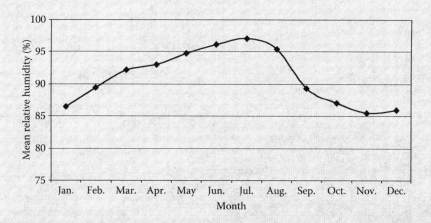

FIGURE 12.2 View of downloaded data of mean relative humidity from HADCM3 for the period 1961–1999.

TABLE 12.3
Monthly Variations of Mean Relative Humidity at Point (20, 41) of the HADCM3 Model Grid (1961–1999)

Month	January	February	March	April	May	June	July	August	September	October	November	December
Mean relative humidity (%)	86.47	89.41	92.23	92.98	94.78	96.12	97.08	95.38	89.32	87.00	85.46	85.90

FIGURE 12.3 Diagram of monthly variations of mean relative humidity in the period 1961–1999 at point (20, 41) of the HADCM3 model grid.

12.5.3 Spatial Variability

There are a number of options for manipulating the spatial variability of a climate change scenario. Some scenarios contain only a uniform change in climate over an area that may suggest, for example, the same land uses for urban and agricultural areas. This may be contrary to what is really expected considering responses to climate change impacts for these areas. The regional changes in climate are defined by GCM grid boxes, between 250 and 600 km depending on a specific GCM resolution listed in Table 12.1. Because of the lack of precision about regional climates in a GCM, Von Storch et al. (1993) advocate that in general the minimum effective spatial resolution should be defined by at least four GCM grid boxes. The accuracy of GCM simulations for an individual grid box will depend, however, on the spatial autocorrelation of a particular weather variable.

In the last few years, the hydrological schemes have also been considered in GCMs with higher complexity and integration. It is difficult to directly use deterministic estimation of precipitation from GCMs for hydrological modeling purposes due to nonlinearly in climate systems. Methods for transforming output from GCMs to local weather variables are needed. Numerical models of natural systems describe nonclosed systems, which are impossible to validate. There is, however, urgent needs to improve the quality of these models and the benefits in operational use. The spatial resolution of GCMs is increasing, but still they simulate climate at a very coarse spatial scale: in the order of 250 by 250 km, which is not enough for the evaluation of climate change impacts on urban areas. In the following sections, ways to improve the spatial precision of the GCM outputs are described.

12.5.3.1 Downscaling

The GCMs have trouble simulating climate variable variations correctly and one interesting method to overcome this is to use a stochastic model in space and time. The UK Department of Environments (DOE, 1996) stated several years ago, "Even if global climate models in the future are run at high resolution there will remain the need to 'downscale' the results from such models to individual sites or localities for impact studies. Downscaling methodologies are still under development and more work needs to be done in intercomparing these methodologies and quantifying the accuracy of such models." Downscaling is defined as the creation of a relationship between the large-scale circulation (predictors) and local weather variables (predictands). The term "downscaling" is a bit misleading since the methodology is actually increasing the resolution, therefore scaling up the picture. But downscaling is more correctly, referring to the process of moving from the large-scale predictor to the local predictand. Figure 12.4 shows a schematic of the process of moving from the large scale to the small scale.

Sophisticated downscaling techniques calculate subgrid box-scale changes in climate as a function of larger-scale climate or circulation statistics. The two main approaches to downscaling have used either regression relationships between a large area and site-specific climates (e.g., Wigley et al., 1990) or relationships between atmospheric circulation types and local weather (e.g., Von Storch et al., 1993). When applied to daily GCM data, these techniques offer the prospect of generating daily climate change scenarios for specific sites or catchments and therefore meet Condition 3 for a climate change scenario. The disadvantage of downscaling approaches is that they require large amounts of observed data to calibrate the statistical relationships and can be computationally very intensive. Such methods are also very time-consuming since unique relationships need to be derived for each site or region. Downscaling methods are also based on the fundamental assumption that the observed statistical relationships will continue to be valid in the future under conditions of climate change. This assumption may violate Condition 2 for a climate change scenario. There are a wide range of techniques used for downscaling. As statistical downscaling is commonly used for downscaling, it is further described in Section 12.5.4.

12.5.3.2 Regional Models

Downscaling techniques are statistical methods for generating greater spatial variability in a climate change scenario. An alternative approach involves the use of high-resolution regional climate models

Climate Change

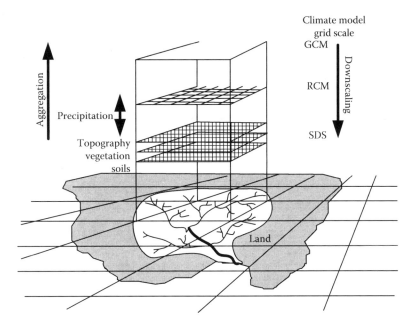

FIGURE 12.4 Schematic drawing of the basic principles of downscaling from the GCM output. (From Wilby, R. L. and Dawson, C. W. 2004. *Using SDSM Version 3.1—A Decision Support Tool for the Assessment of Regional Climate Change Impacts*, Climate Change Unit, Environment Agency of England and Wales, Nottingham, U.K., and Department of Computer Science, Loughborough University, Leics, U.K. With permission.)

(RCMs; also called limited area models, LAMs). RCMs are typically constructed at a much finer resolution than GCMs (often 50 km), but their domain is limited to continents or subcontinents. In such applications, detailed information at spatial scales down to 10–20 km may be established at temporal scales of hours or less. Although RCMs yield greater spatial details on climate, they are still constrained at their boundaries by the coarse-scale output from GCMs. Therefore, to an extent, the performance of an RCM can only be as good as that of the driving GCM. A number of RCM climate change experiments have now been performed all over the world, for example, Giorgi et al. (1994), Mearns et al. (1992), Jones et al. (1995), Walsh and McGregor (1997), and Bhaskaran et al. (1996), but their performance in relation to downscaling techniques has not yet been fully evaluated. The costs of establishing a RCM for a new region and running a climate change experiment are extremely high, both computationally and in terms of human resources. Therefore, they are not easily accessible. Furthermore, there are some difficulties in creating an interface between the GCM and the regional model (e.g., how to relate the coarse resolution grid cell data of the GCM, often below the GCM skill scale, to boundary conditions for the far finer-scale nested model). Nevertheless, in the long term, this technique can result in useful results and their development should be encouraged. To date, most of the regional models are employed on the Northern Hemisphere, whereas many of the areas that are most vulnerable to global change lie in the Southern Hemisphere. For the moment, it remains premature for the regional model output to be used extensively in climate impact assessments, at least for many regions in the globe.

There are many advantages to dynamic downscaling, particularly around the variety of factors (e.g., temperature, precipitation, soil moisture, wind direction and strength, etc.) generally available in both GCMs and the nested finer-scale grid. However, dynamic downscaling requires supercomputer systems and there are too few supercomputer systems available to perform such meso-scale simulations for all the areas around the world to more fully assess the local/regional impacts and consequences of climate change. As supercomputer capabilities and availability increase, dynamic downscaling will become more widely accessible.

The first step to demonstrate that the RCM has predictive skill is to integrate an atmospheric GCM with observed sea surface temperature (SST) for several seasons into the future. The GCM output would then be downscaled using the RCM. There is expected to be some regional skill and this needs to be quantified. This level of skill, however, will necessarily represent the maximum skill theoretically even possible with atmospheric-oceanic general circulation models (AOGCM) as applied for forecasts years and decades into the future since, for these periods, SSTs must also be predicted, and not specified. Such experiments have not been systematically completed. Indeed, does the concept of predictive skill even make sense when we cannot verify the models until decades into the future?

12.5.4 Statistical Downscaling

Statistical downscaling is developed based on the assumption that empirical relationships exist between atmospheric processes at different spatial and temporal scales (Wilby et al., 1998). These relationships can be related to predictor variables and broadscale atmospheric variables, such as mean sea-level pressure (MSLP) and geopotential height, through a transfer function. The choice of predictor variables to incorporate into the model has an important role in the statistic downscaling problem. The following criteria have to be fulfilled for a downscaling method to be successful (Hewitson and Crane, 1996):

- Large-scale predictors are adequately reproduced by GCMs.
- The relationship between the predictors and the predictands are valid for periods outside the calibration period.
- Climate change signals are sufficiently incorporated into the predictors.

Satisfying the second criterion is the most difficult and questionable issue. No predictor fulfills all three criteria; a restricted number of predictors that meet one or two criteria or a wide range of predictors to cover all should be selected. The argument for using many predictors is obvious: all criteria will be met. The disadvantage is that the degree of freedom in the predictor set is increased, and the results become more difficult to analyze. The major steps of downscaling is outlined by Hewitson and Crane (1996) and are presented here, with some modifications, as follows:

1. Reduction or processing of the predictor
2. Comparison of GCM circulation with observed circulation
3. Spatial analysis and selection of predictor and predictand domains
4. Temporal analysis of the predictor and the predictand
5. Temporal manipulation of the predictor, for example, time lagging
6. Calibration of the transfer function between the predictor and the predictand
7. Evaluating the relationship between the predictor and the predictand
8. Application of the developed model to GCM data

Items 3–7 are carried out in a recursive fashion until an optimum result is reached for the target objective. During this procedure, predictor variables and weighting functions are selected regarding the downscaling method used. All these steps may not always be taken in each method, but they should always be considered to achieve reliable results from downscaling. The statistical downscaling methods are divided into three main groups: analogue methods, regression methods, and conditional probability approaches. The last group has two main subgroups, weather patterns and weather states, which are closely related. This classification may not be perfect, since some hybrid methods are available (Xu, 1999), but it serves as a starting point for discussion. The analogue method is not as common in the literature as the others, but offers a simple technique that can be used as a benchmark (Zorita and von Storch, 1999). There are advantages and shortcomings of statistical downscaling in comparison with regional downscaling. The main features of statistical downscaling are

- Personal computers can be used for the calculations.
- There is no need for detailed knowledge about the physical processes.
- Long and homogeneous time series are needed for fitting and confirming the statistical relationship.

The second feature is modified by the fact that many statistical downscaling methods rely on known physical understanding and processes that are important for the predictand. However, the exact formulation of the processes is not necessarily considered for statistical downscaling.

The transfer function in statistical downscaling is created by translating anomalies of the large-scale flow into anomalies of some local climate variable (Zorita and von Storch, 1999). The main advantages of the statistical downscaling approach are easy implementation and cheap computation. Another feature of this procedure is that calibration to the local level is an integrated part of the downscaling procedure. Furthermore, downscaling methods require very few parameters, which make these methods attractive for many hydrological applications (Wilby et al., 2000). Precipitation is a strongly intermittent and nonlinear process, and precipitation fields are characterized by anomalous scaling laws. It is also shown that the spatial behavior of precipitation is dependent on the time scale; precipitation is more intermittent for shorter time periods. Using stochastic models for the downscaling of precipitation helps gain a better understanding of the probabilistic structure of precipitation. These models have long been used to model hydrologic events, such as runoff. If they are derived from the atmospheric processes that drive precipitation, they can predict what governs precipitation patterns.

Given the wide range of downscaling techniques (both regional and statistical), there are a number of model comparisons using generic data sets and model diagnostics. Until recently, these studies were restricted to statistical-versus-statistical or regional-versus-regional model comparisons. However, a growing number of studies are undertaking statistical-versus-regional model comparisons and Table 12.4 summarizes relative strengths and weaknesses that have emerged.

An important point in the application of statistical downscaling is considering the seasonal dependencies. For the regions with clearly defined seasons based on temperate variations, the driving forces for the precipitation differ over the year, and intraannual anomalies in the downscaling results. Precipitation during summer months is driven by convection to a greater extent than during winter months, which should be considered in developing classification schemes. A solution to the problem can be to divide the year into four seasons. The growth season of plants is an example that is crucial as there is a nonlinear relationship between the amount of rain and the water use efficiency of plants.

Furthermore, downscaling methods must be valid in a perturbed climate in order to be useful tools in studies of a future climate change. The methods are supposed to estimate the difference between climate situations and must therefore function under a situation different from that under which it was developed. For the validity of a downscaling scheme outside the calibrated and validated range, the climate should be stationary and no major shift in the physical processes governing the climate must occur. One approach to evaluate the ability of a model to cope with different climate situations is to calibrate the model on the data from the driest seasons and then validate it on the wettest seasons in the calibration period and vice versa. This is a simple test of the method's sensitivity to different climate situations. The model performance in extreme years is considered in order to test its stability. Long time series that are used in the downscaling reasonably contain many different situations, including those situations that will be more frequent in a perturbed future climate. The problem will be minimized if the method can model those situations and if the range of variability of the large-scale variable in a future climate is of the same order as today. There is some difficulty in dealing with stability and nonstationarity in downscaling and one should be careful when applying a model to the output of GCM simulations. If the climate is nonstationary, the underlying assumptions of the statistical link have to be considered to be unaffected by the climate change for model validity.

TABLE 12.4
Main Strengths and Weakness of Statistical Downscaling and Regional Model

	Statistical Downscaling	Regional Model
Strengths	• Station-scale climate information from GCM-scale output • Cheap, computationally undemanding and readily transferable • Ensembles of climate scenarios permit risk/uncertainty analyses • Applicable to "exotic" predictands such as air quality and wave heights	• 10–50 km resolution climate information from GCM–scale output • Respond in physically consistent ways to different external forcings • Resolve atmospheric processes such as orographic precipitation • Consistency with GCM
Weakness	• Dependent on the realism of GCM boundary forcing • Choice of domain size and location affects results • Requires high-quality data for model calibration • Predictor–predictand relationships are often non-stationary • Choice of predictor variables affects results • Choice of empirical transfer scheme affects results • Low-frequency climate variability problematic • Always applied off-line; therefore, results do not feedback into the host GCM	• Dependent on the realism of GCM boundary forcing • Choice of domain size and location affects results • Requires significant computing resources • Ensembles of climate scenarios seldom produced • Initial boundary conditions affect results • Choice of cloud/convection scheme affects (precipitation) results • Not readily transferred to new regions or domains • Typically, applied off-line; therefore results do not always feedback into the host GCM

Source: Wilby, R. L. and Dawson, C. W. 2004. *Using SDSM Version 3.1—A Decision Support Tool for the Assessment of Regional Climate Change Impacts*, Climate Change Unit, Environment Agency of England and Wales, Nottingham, U.K. and Department of Computer Science, Loughborough University, Leics, LE11 3TU, U.K. With permission.

The physical laws governing the climate are not likely to change, but the physical processes are parameterized for a specific climate, which may make the model results questionable for a changed climate. It can be noted that in order to overcome the problem with nonstationarity of climate, a multivariate approach is recommended. This is not possible in all methods, in which case sufficient data should be included in order to ensure statistical integrity of the results. A stochastic model conditioned on the observed interannual variability can also take into account the nonstationarity.

Regression-based methods are commonly used for downscaling. These methods are defined by Wilby and Wigley (1997) as follows: "Generally involve establishing linear or nonlinear relationships between sub grid-scale (e.g. single-site) parameters and coarser-resolution (grid-scale) predictor variables." The common regression-based methods in this field include ANN (Olsson et al., 2004; Hewitson and Crane, 1996) and Canonical Correlation Analysis (CCA) (von Storch et al., 1993). The models and techniques range from simpler multiple regression schemes to more sophisticated models, such as a method to link the covariances of atmospheric circulation and of local weather variables in a bilinear way (Burger, 1996). The regression methods can also be classified according to the parameters that are used in the downscaling procedure. One approach uses the same variable for the large-scale predictors as for the local-scale predictands, for example, temperature, but most studies use different parameters. The steps followed in regression methods are

1. Identifying a large-scale parameter G (predictor) that controls the local parameter L (predictand). If the intent is to calculate L for climate experiments, G should be simulated well by climate models.

Climate Change

2. Developing a statistical relationship between L and G.
3. Validation of the relationship with independent data.
4. If the relationship is confirmed, G is derived from GCM experiments to estimate L. The idea proposed in step 1, that the predictor should be well simulated by GCMs, is a key issue in this method. As mentioned before, the whole idea of using statistical downscaling of climate variables is to evaluate the effects of climate change. The GCM is still the best tool to predict the future climate. A weak point in this context is the assumption that the statistical relationships derived will remain fixed in the future.

12.6 DOWNSCALING MODELS

In this section, two downscaling models that are widely employed in climate change studies are introduced.

12.6.1 Statistical Downscaling Model

The models that are used in climatic studies to simulate the weather parameters with the same variability as observed values are called weather generators and the most used is WGEN proposed by Richardson (1981). The statistical downscaling model (SDSM) (Wilby et al., 1999) is a hybrid between a multilinear regression method and a stochastic weather generator.

Large-scale point predictors are used to condition local-scale weather generator parameters, both concerning precipitation occurrence and amount. The employed climate variables in the SDSM are listed below:

- Mean temperature
- MSLP
- Zonal velocity component near surface
- Zonal velocity component at 500 hPa height
- Zonal velocity component at 850 hPa height
- Meridional velocity component near the surface
- Meridional velocity component at 850 hPa height
- Vorticity
- Divergence near the surface
- Divergence at 500 hPa height
- Divergence at 850 hPa height
- 500 hPa geopotential height
- 850 hPa geopotential height
- Specific humidity at 500 hPa height
- Relative humidity at 850 hPa height
- Near-surface specific humidity
- Near-surface relative humidity

The parameter values are conditioned on a monthly basis. Identifying empirical relationships between gridded predictors (such as MSLP) and single site predictands (such as station precipitation) is a central issue in all statistical downscaling methods. This remains one of the most challenging stages in the development of any SDSM since the choice of predictors largely determines the character of the downscaled climate scenario. The decision process is also complicated by the fact that the explanatory power of individual predictor variables varies both spatially and temporally.

The model has been applied in many catchments in North America (Wilby and Dettinger, 2000) and Europe (Wilby et al., 2003). The structure and operation of SDSMs can be divided into five distinct tasks: (1) preliminary screening of potential downscaling predictor variables, (2) assembly

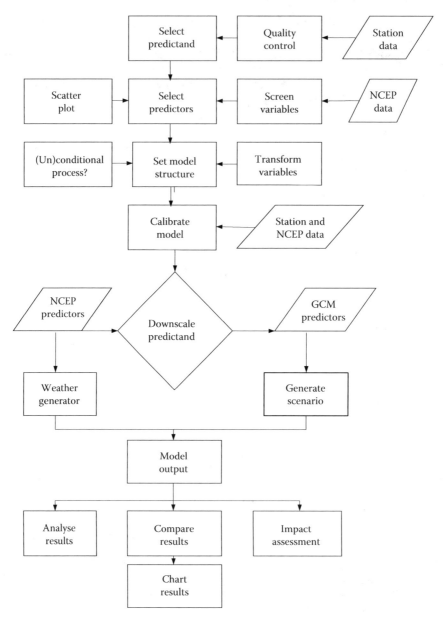

FIGURE 12.5 The SDSM climate scenario generation process. (From Wilby, R. L. and Dawson, C. W. 2004. *Using SDSM Version 3.1–A Decision Support Tool for the Assessment of Regional Climate Change Impacts*, Climate Change Unit, Environment Agency of England and Wales, Nottingham, U.K., and Department of Computer Science, Loughborough University, Leics, U.K. With permission.)

and calibration of SDSM, (3) synthesis of ensembles of current weather data using observed predictor variables, (4) generation of ensembles of future weather data using GCM-derived predictor variables, and (5) diagnostic testing/analysis of observed data and climate change scenarios (Figure 12.5).

In the application of SDSM, the first step is determining whether in each day rainfall occurs or not, as follows:

$$\omega_i = \alpha_0 + \sum_{j=1}^{n} \alpha_j \hat{u}_i^{(j)}, \tag{12.5}$$

where ω_i is the candidate in a random process generator that indicates a state of rainfall occurring or not occuring on day i, $\hat{u}_i$ is the normalized predictor on day i, and α_j is the estimated regression coefficient. Precipitation on day i occurs if $\omega_i \leq r_i$, where r_i is a stochastic output from a linear random-number generator.

The value of rainfall in each rainy day is estimated in the second step using z-score as follows:

$$Z_i = \beta_0 + \sum_{j=1}^{n} \beta_j \hat{u}_i(j) + \varepsilon, \tag{12.6}$$

where Z_i is the z-score for day i, β_j are the estimated regression coefficients for each month, ε is a normally distributed stochastic error term, and

$$y_i = F^{-1}[\phi(Z_i)], \tag{12.7}$$

where ϕ is the normal cumulative distribution function and F is the empirical function of y_i daily precipitation amounts. The standardized predictors used in this method are provided by subtracting the means and dividing them by the standard deviations over the calibration period. During downscaling with the SDSM, a multiple linear regression model is developed between few selected large-scale predictors and the local-scale predictands such as rainfall.

In the SDSM, the parameters of the regression model are estimated using the efficient dual simplex algorithm. Large-scale relevant predictors are selected using correlation, partial correlation, and P-value analysis, and also considering physical sensitivity between selected predictors and predictand for the site in question. The correlation statistics and P-values indicate the strength of the association between the predictor and predictand. Higher correlation values imply a higher degree of association and lower P-values indicate a greater probability for association between variables. It is important to know that the P-values of <0.05 as threshold does not necessarily mean that the result can be statistically significant. On the other hand, the high P-value would indicate that the predictor–predictand correlation could be probability related (Wilby and Dawson, 2004).

Example 12.2

A rain gauge of an urban area called Station A is located at 51°46′ longitude and 31°38′ latitude with 1200 m altitude. The precipitation data for 10 years from 1992 to 2001 are available in this rain gauge. The monthly data are given in Table 12.5. Produce the precipitation data of Station A under the HadCM3A2 scenario outputs from 2002 to 2010 using the SDSM.

TABLE 12.5
Monthly Observed Precipitation for Station A under HadCM3 A2 Scenario for the Period of 1992–2001

Year	January	February	March	April	May	June	July	August	September	October	November	December
1992	7.74	21.72	17.42	15.00	44.52	29.00	6.77	0.00	5.00	4.84	32.00	22.26
1993	—	—	—	—	—	—	—	—	—	—	—	—
1994	8.71	6.43	—	19.00	—	20.00	11.61	0.00	0.00	15.48	10.00	—
1995	11.61	17.14	3.87	38.00	33.87	26.00	5.81	0.00	2.00	22.26	10.00	2.90
1996	18.39	26.90	28.06	53.00	41.61	27.00	0.00	15.48	3.00	6.77	27.00	30.97
1997	15.48	23.57	36.77	29.00	15.48	63.00	70.65	9.68	0.00	12.58	14.00	7.74
1998	23.23	34.29	49.35	16.00	42.58	44.00	20.32	2.90	7.00	2.90	29.00	0.00
1999	16.45	10.71	—	—	—	—	—	—	—	—	—	—
2000	—	9.31	51.29	30.00	44.52	24.00	17.42	0.00	0.00	0.00	32.00	70.65
2001	33.87	7.50	45.48	58.00	21.29	13.00	4.84	2.90	16.00	12.58	73.00	7.74

continued

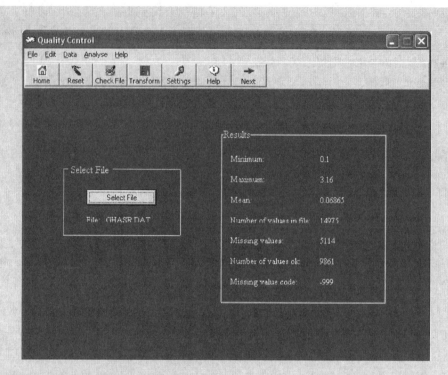

FIGURE 12.6 Quality control results for Station A.

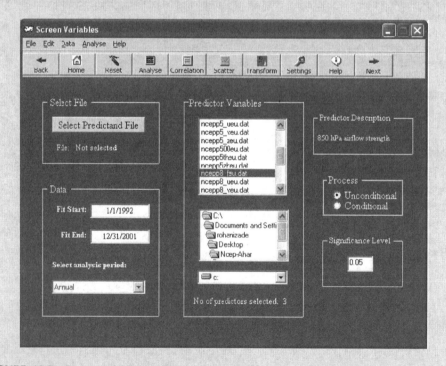

FIGURE 12.7 Screen variables page for selecting predictors for Station A (1992–2001).

Climate Change

Solution: The downscaling procedure at Station A is described step by step in the following:

Step 1: The input file is prepared; the daily precipitation data are sorted based on the time in two columns; first the date and second the precipitation amount (the monthly data are interpolated to develop the daily data needed as input to the SDSM).

Step 2: Then the process of quality control is run to test the format of the prepared input file as shown in Figure 12.6 (from the SDSM program).

Step 3: In this step, the 16 available weather variables are screened to find the variables that could be used as precipitation predictand. For Station A, data start from 1/1/1992 and end at 12/31/2002. The analysis is done in an annual period under an unconditional process, as shown in Figure 12.7.

At the end of this analysis, the results of the explained variance and correlation matrix between precipitation and selected weather variables are given (Figures 12.8 and 12.9). Finally, three variables are selected for precipitation simulation as follows:

1. Surface divergence
2. MSLP
3. 500 hPa vorticity

The related correlation analysis of these variables is given in Figures 12.8 and 12.9. The *P*-value is low and partial correlations are high, which show the accuracy of the downscaling process.

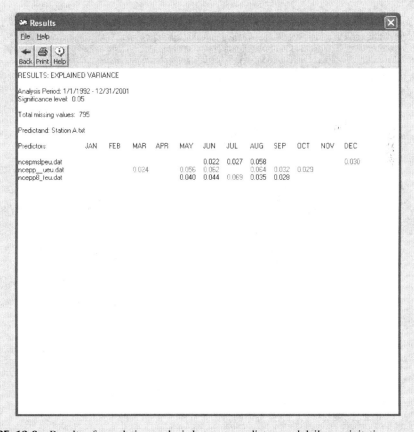

FIGURE 12.8 Results of correlation analysis between predictors and daily precipitation.

continued

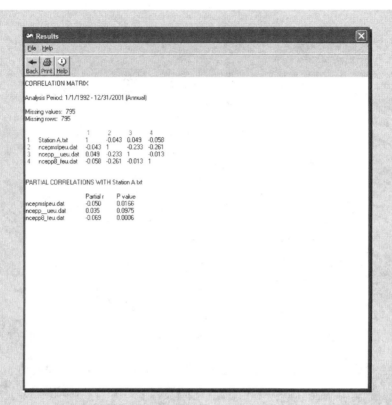

FIGURE 12.9 Matrix of correlation between predictors and predictands.

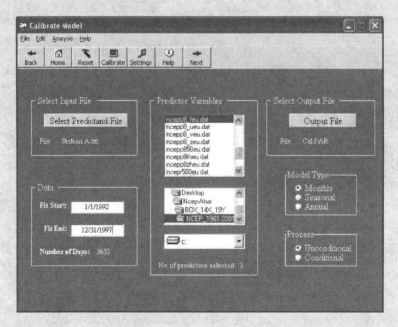

FIGURE 12.10 Model calibration page.

Step 4: The model is calibrated using the observed data and the selected signals. At the end of this step, the *R*-squared and standard error are reported (Figures 12.10 and 12.11).

Step 5: Weather generator: In this step, data for the considered period and selected climate change scenario (here HadCM3 A2) are developed. The start date of data generating and the number of data to generation according to considered years will be entered. In this step, you can choose production of one or more ensembles; here one ensemble is selected (Figure 12.12).

Step 6: In this step, the statistical analyses of the observed and developed data through tables and graphs are given (Figures 12.13 through 12.15).

In Table 12.6, the monthly downscaled precipitation data of Station A are given. As can be seen, although there are some fluctuations, the monthly precipitation, especially in January, February, and June, is decreasing. This may result in considerable water shortages for domestic water supply, since the main source of urban water supply is provided in January and February.

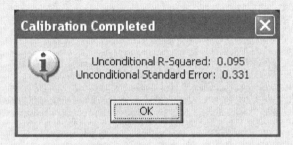

FIGURE 12.11 Model calibration results.

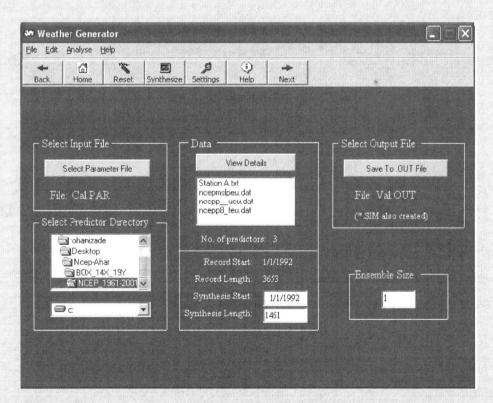

FIGURE 12.12 Weather generator application for Station A rain gauge.

continued

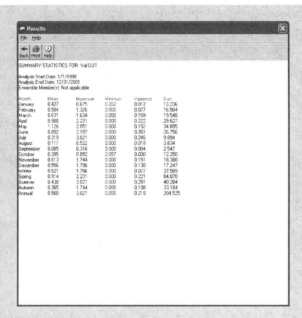

FIGURE 12.13 Statistics of the simulated precipitation data of Station A rain gauge for the period 2002–2010 under the HadCM3 A2 climate change scenario.

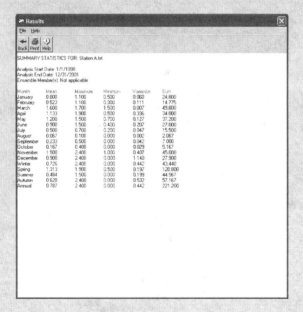

FIGURE 12.14 Statistics of the observed precipitation data of Station A rain gauge for the period 1992–2001 under the HadCM3 A2 climate change scenario.

Climate Change

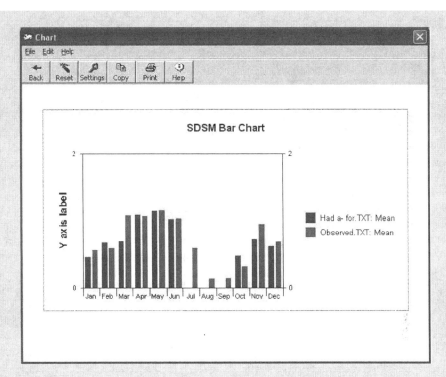

FIGURE 12.15 Comparison of the HadCM3 A2 scenario results and observed precipitation.

TABLE 12.6
Monthly Simulated Precipitation (mm) for Station A under the HadCM3 A2 Scenario for the Period of 2002–2010

Year	January	February	March	April	May	June	July	August	September	October	November	December
2002	15.57	19.57	19.36	19.92	26.64	18.99	23.04	1.23	2.46	13.26	21.01	14.26
2003	15.04	19.56	21.48	18.92	29.06	21.93	12.44	5.07	2.26	11.11	20.04	17.61
2004	14.78	21.15	21.24	20.74	30.36	17.82	14.68	2.22	2.58	9.75	18.51	17.44
2005	14.78	20.25	22.30	26.93	33.80	18.80	10.31	2.18	2.58	10.78	19.33	14.84
2006	14.14	21.03	22.48	22.10	31.65	23.54	10.60	0.90	2.32	9.21	18.00	17.49
2007	14.19	18.03	14.60	23.31	34.73	10.55	11.11	1.10	2.56	9.09	15.97	17.55
2008	13.25	19.78	21.25	21.17	29.84	15.48	12.13	2.08	2.63	9.23	18.58	15.25
2009	13.69	17.11	21.81	17.36	31.36	19.17	23.84	1.16	2.55	8.30	18.81	15.36
2010	13.77	18.64	17.10	18.21	32.56	25.97	15.97	2.40	2.02	6.86	18.80	14.32

12.6.2 LARS-WG Model

The Long Ashton Research Station Weather Generator (LARS-WG) is a stochastic weather generator that can be used for the simulation of weather data at a single site (Racsko et al., 1991; Semenov et al., 1998; Semenov and Brooks, 1999), under both current and future climate conditions. These data are in the form of daily time series for a group of climate variables, namely, precipitation, maximum and minimum temperatures, and solar radiation.

Stochastic weather generators were originally developed for two main purposes:

1. To provide a means of simulating synthetic weather time series with statistical characteristics corresponding to the observed statistics at a site, but which were long enough to be used in an assessment of risk in hydrological or agricultural applications
2. To provide a means of extending the simulation of weather time series to unobserved locations, through the interpolation of the weather generator parameters obtained from running the models at neighboring sites

It is worth noting that a stochastic weather generator is not a predictive tool that can be used in weather forecasting, but is simply a means of generating time series of synthetic weather statistically "identical" to the observations.

LARS-WG is based on the series weather generator described in Racsko et al. (1991). It utilizes semiempirical distributions for the lengths of wet and dry day series, daily precipitation, and daily solar radiation. The semiempirical distribution Emp = $\{a_0, a_i; h_i, i = 1, \ldots, 10\}$ is a histogram with 10 intervals, $[a_{i-1}, a_i)$, where $a_{i-1} < a_i$, and h_i denotes the number of events from the observed data in the ith interval. Random values from the semiempirical distributions are chosen by first selecting one of the intervals (using the proportion of events in each interval as the selection probability), and then selecting a value within that interval from the uniform distribution. Such a distribution is flexible and can approximate a wide variety of shapes by adjusting the intervals $[a_{i-1}, a_i)$. The cost of this flexibility, however, is that the distribution requires 21 parameters (11 values denoting the interval bounds and 10 values indicating the number of events within each interval) to be specified compared with, for example, three parameters for the mixed-exponential distribution used in an earlier version of the model to define the dry and wet series (Racsko et al., 1991).

The intervals $[a_{i-1}, a_i)$ are chosen based on the expected properties of the weather variables. For solar radiation, the intervals $[a_{i-1}, a_i)$ are equally spaced between the minimum and maximum values of the observed data for the month, whereas for the lengths of dry and wet series and for precipitation, the interval size gradually increases as i increases. In the latter two cases, there are typically many small values but also a few very large ones, and this choice of interval structure prevents a very coarse resolution being used for the small values.

The simulation of precipitation occurrence is modeled as alternate wet and dry series, where a wet day is defined to be a day with precipitation >0.0 mm. The length of each series is chosen randomly from the wet or dry semiempirical distribution for the month in which the series starts. In determining the distributions, observed series are also allocated to the month in which they start. For a wet day, the precipitation value is generated from the semiempirical precipitation distribution for the particular month independent of the length of the wet series or the amount of precipitation in previous days.

Daily minimum and maximum temperatures are considered as stochastic processes with daily means and daily standard deviations conditioned on the wet or dry status of the day. The technique used to simulate the process is very similar to that presented in Racsko et al. (1991). The seasonal cycles of means and standard deviations are modeled by finite Fourier series of order 3, and the residuals are approximated by a normal distribution. The Fourier series for the mean is fitted to the observed mean values for each month. Before fitting the standard deviation Fourier series, the observed standard deviations for each month are adjusted to give an estimated average daily standard deviation by removing the estimated effect of the changes in the mean within the month. The adjustment is calculated using the fitted Fourier series already obtained for the mean.

The observed residuals, obtained by removing the fitted mean value from the observed data, are used to analyze a time autocorrelation for minimum and maximum temperatures. For simplicity, both of these are assumed to be constant throughout the whole year for both dry and wet days with the average value from the observed data being used. Minimum and maximum temperature residuals have a preset cross-correlation of 0.6. Occasionally, simulated minimum temperature is greater than simulated maximum temperature, in which case the program replaces the minimum temperature

Climate Change

by the maximum less 0.1. The analysis of daily solar radiation over many locations showed that the normal distribution for daily solar radiation, commonly used in other weather generators, is unsuitable for certain climates (Chia and Hutchinson, 1991). The distribution of solar radiation also varies significantly on wet and dry days.

Therefore, separate semiempirical distributions were used to describe solar radiation on wet and dry days. An autocorrelation coefficient was also calculated for solar radiation and assumed to be constant throughout the year. Solar radiation is modeled independently of temperature. LARS-WG accepts sunshine hours as an alternative to solar radiation data. If solar radiation data are unavailable, then sunshine hours may be used; these are automatically converted to solar radiation using the approach described in Rietveld (1978).

The process of generating synthetic weather data can be divided into three distinct steps:

1. Model calibration—SITE ANALYSIS—observed weather data are analyzed to determine their statistical characteristics. This information is stored in two parameter files.
2. Model validation—QTEST—the statistical characteristics of the observed and synthetic weather data are analyzed to determine whether there are any statistically significant differences.
3. Generation of synthetic weather data—GENERATOR—the parameter files derived from observed weather data during the model calibration process are used to generate synthetic weather data having the same statistical characteristics as the original observed data, but differing on a day-to-day basis. Synthetic data corresponding to a particular climate change scenario may also be generated by applying global climate model-derived changes in precipitation, temperature, and solar radiation to the LARS-WG parameter files. The application and input/output requirements of LARS-WG are described in detail in its user manual prepared by Semenov and Barrow (2002).

EXAMPLE 12.3

Employ the LARS model to downscale Station A data given in the previous example.

Solution:

Step 1: For applying the LARS model, two input files should be developed as follows:

1. Station A.st file: This file includes the site name, latitude, longitude, and altitude of the Station A rain gauge, the directory path location and name of the file containing the observed weather data for Station A and the format of the observed weather data in the file, which here is the year (YEAR), Julian day (JDAY, i.e., from 1 to 365), minimum temperature (MIN; °C), maximum temperature (MAX; °C), precipitation (RAIN; mm), and solar radiation (RAD; MJm-2day-1) (Figure 12.16).
2. Station A.sr file: Contains the daily data of rainfall, maximum temperature, minimum temperature, and solar radiation at Station A in the order mentioned in Station A.st file (Figure 12.17).

Step 2: After preparing the mentioned files, for calibrating the model the SITE ANALYSIS option is selected and Station A.st is addressed to the model. Two files called Station A.sta and Station A.wg are created. These two files provide the statistical characteristics of the data and parameter information which will be used by LARS-WG to synthesize artificial weather data in the GENERATOR process.

In Station A.sta file (Figure 12.18), the first few lines give the site name and location information. This is followed by the statistical characteristics of the observed weather data.

continued

```
[SITE]
Station A
[LAT, LON and ALT]
       38.5    46.85    1200
[WEATHER FILES]
C:\Program Files\LARS-WG 3.0\debr\Station A.sr
[FORMAT]
YEAR JDAY MIN MAX RAIN RAD
```

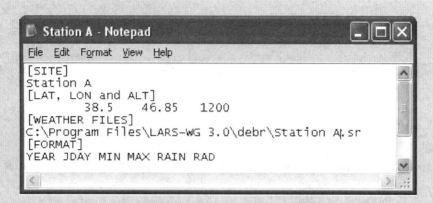

FIGURE 12.16 Layout of basic characteristics of Station A (Station A.st file).

First the empirical distribution characteristics for the length of wet and dry series of days in the observed data are obtained. This information is given in four lines by seasons [i.e., winter (DJF), spring (MAM), summer (JJA), and autumn (SON)] where first two lines of each section correspond to the wet series, and the last two lines represent the dry series. The wet and dry series are modeled based on histograms constructed from the observed data. The histograms in this case consist of 10 intervals and the cutoff points for each interval are given in the first line, and the number of events in the observed data falling into each interval is given in the second line.

So, using Station A.sta, it can be seen that the wet series intervals (hi) are $0 \leq h1 < 1$, $1 \leq h2 < 2$, $2 \leq h3 < 3$, $3 \leq h4 < 4$, $4 \leq h5 < 5$, $5 \leq h6 < 6$, $6 \leq h7 < 7$, $7 \leq h8 < 8$, $8 \leq h9 < 9$, and $9 \leq h10 < 10$, with corresponding frequencies of occurrence of 236, 161, 44, 16, 8, 3, 0, 2, 0, and 0, respectively (Figure 12.18). The same data are given for precipitation. The summary statistics (min, max, mean, and standard deviation) of wet and

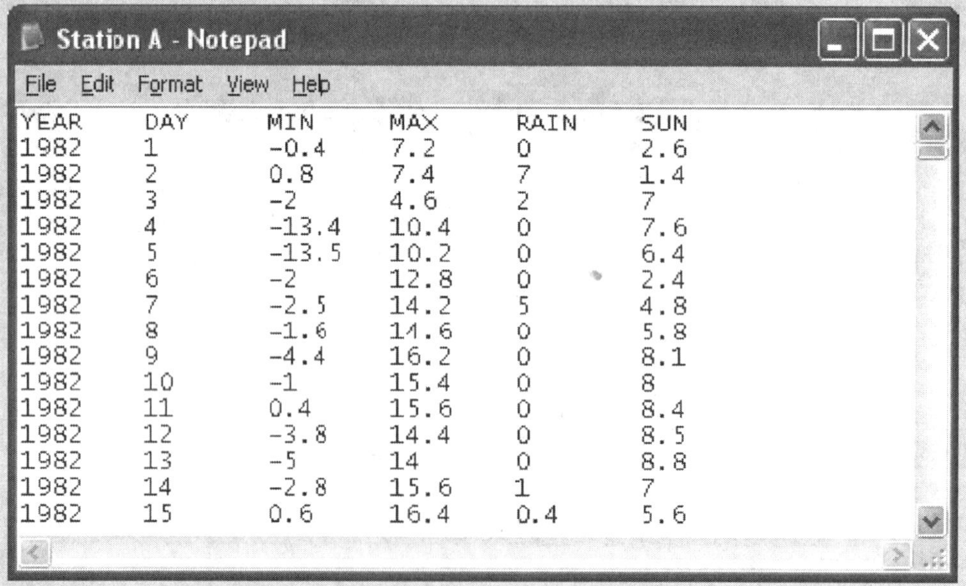

YEAR	DAY	MIN	MAX	RAIN	SUN
1982	1	-0.4	7.2	0	2.6
1982	2	0.8	7.4	7	1.4
1982	3	-2	4.6	2	7
1982	4	-13.4	10.4	0	7.6
1982	5	-13.5	10.2	0	6.4
1982	6	-2	12.8	0	2.4
1982	7	-2.5	14.2	5	4.8
1982	8	-1.6	14.6	0	5.8
1982	9	-4.4	16.2	0	8.1
1982	10	-1	15.4	0	8
1982	11	0.4	15.6	0	8.4
1982	12	-3.8	14.4	0	8.5
1982	13	-5	14	0	8.8
1982	14	-2.8	15.6	1	7
1982	15	0.6	16.4	0.4	5.6

FIGURE 12.17 Input data of Station A for the LARS model (Station A.sr).

Climate Change

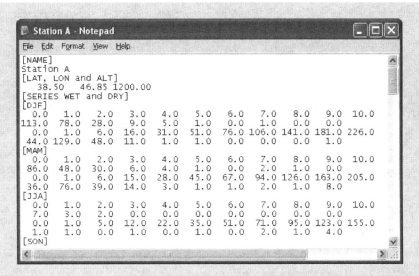

FIGURE 12.18 Summary statistics produced through LARS for Station A at a seasonal scale (Station A.sta file).

dry days, precipitation, minimum and maximum temperatures, and solar radiation in monthly and daily scales are also given in this file.

File Station A.wg contains the monthly histogram intervals and frequency of events in these intervals for wet and dry rainfall series in four lines similar to Station A.sta, four Fourier coefficients, $a[i]$ and $b[i]$ $i = 1, \ldots, 4$, for the means and standard deviations of different combinations of minimum and maximum temperatures, and solar radiation on wet and dry days (Figure 12.19).

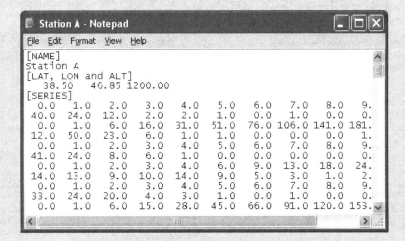

FIGURE 12.19 Summary statistics produced through LARS for Station A at a monthly scale (Station A.wg file).

continued

Step 3: For model validation, the Test option is selected from the QTest module in the LARS model. The QTest option carries out a statistical comparison of synthetic weather data generated using LARS with the parameters derived from observed weather data in the calibrated model. For selecting the Test option, first, name of Station A is selected; then, the simulation period (20 years) and random segment related to a stochastic predicting model are selected (Figure 12.20). The performance of the model in simulating the observed precipitation pattern in the calibration period is evaluated in Figure 12.21. The model output matches closely the observed values.

Step 4: The Generator option in the LARS model is used to simulate synthetic weather data. This option is used to generate synthetic data that have the same statistical characteristics as the observed weather data, or to generate synthetic weather data corresponding to a scenario of climate change.

First the scenario related to the considered area would be determined. Here, scenarios obtained according to the GCM climatic model, namely ECHO-G A1 for grid 17–31, is used. Initial inputs of this stage are the name and the address of the GCM scenario.

After selecting the site and setting up the appropriate scenario files, there are three further options to be completed. The first is the scaling factor. This factor can be used if a climate change scenario is implemented for a particular future time period, say 2100, to obtain data for an earlier time period without having to create a scenario file for this earlier time period. The scaling factor allows doing this by assuming that the changes in climate over time are linear (Figure 12.22). Here the scaling factor has been considered equal to one and the model is simulated in 20 years with a random seed of 877. The simulated monthly data for the period 2002–2010 are given in Table 12.7.

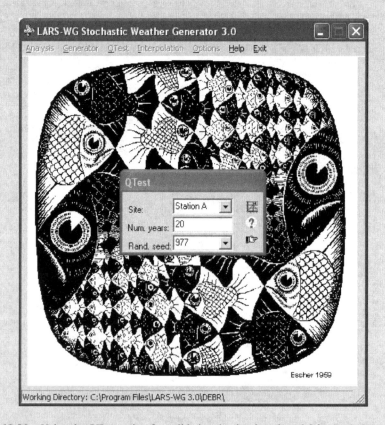

FIGURE 12.20 Using the QTest option for validating the developed model for Station A.

Climate Change

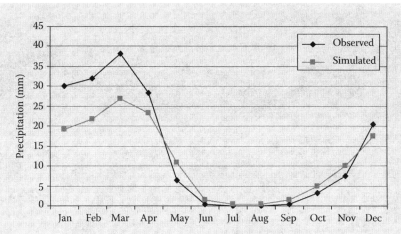

FIGURE 12.21 Simulated and observed precipitation at Station A.

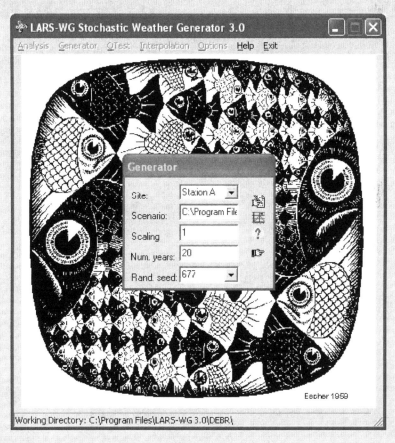

FIGURE 12.22 Using the generator option to produce simulation of future climate at Station A.

continued

TABLE 12.7
Average Monthly Simulated Precipitation (mm) at Station A for the Period of 2002–2010

Year	January	February	March	April	May	June	July	August	September	October	November	December
2002	32.1	31.9	24.5	72.4	89.6	98.5	198.1	112.4	43.8	77.9	60.2	69.4
2003	30.9	91.2	65	96.6	104.9	93.5	137.1	122	67.8	37.1	40.7	45.7
2004	39.6	19.8	25.4	56.4	72	112.5	142.5	61	73.2	30.6	37.9	86.6
2005	47.9	56.3	35.3	110.4	69.2	155.1	127.8	129.6	52.1	36.4	65.3	38.5
2006	44.1	36.3	47.8	74	130.4	77.1	175	98.1	65.9	26.1	14.6	47.2
2007	27.8	108.2	46.2	74.7	92	137.1	109.4	115.2	86.8	38.4	48.2	81
2008	15.2	65.2	156.4	94.4	82.6	106.1	159.8	77.3	66.9	29.8	66.6	97.9
2009	34.3	36.4	55	76.3	161	163.7	174.2	116.1	61.4	53.8	39.8	42.2
2010	23.9	64	35.7	64.9	132.5	149.8	154.2	109.8	63	88.9	67.9	98.3

12.7 CASE STUDIES

In this section, two studies on the impacts of climate change on different components of the urban water cycle are discussed. Theses case studies include the impacts of climate change on urban floods and the temperature and rainfall in a watershed.

12.7.1 ASSESSMENT OF THE PRECIPITATION AND TEMPERATURE VARIATIONS IN THE AHARCHAI RIVER BASIN UNDER CLIMATE CHANGE SCENARIOS

Karamouz et al. (2009) have used two climate scenarios namely A2(a) and B2(a), according to forecasted values from GCM daily data for HADCM3 predictors for evaluating possible climate changes in the Aharchai river watershed upstream of the Satarkhan reservoir in north-western Iran as a case study (Figure 12.23). The A2(a) scenario assesses reinforcing forces that support the increasing population based on family values and traditions without relying on a lifestyle that depends on

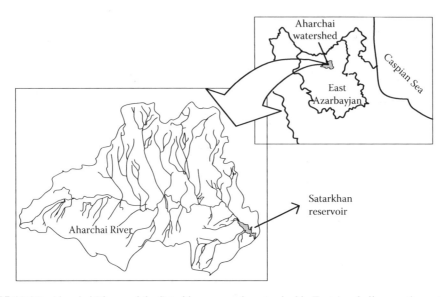

FIGURE 12.23 Aharchai River and the Satarkhan reservoir watershed in East Azerbaijan, north-western Iran.

economical advancement. However, in the B2(a) scenario, more emphasis is given to reinforcing economical, social, and environmental issues. Both of them are regional scenarios of climate change.

The SDSM was calibrated for 1971–1990 with a correlation factor (R^2) of 0.81 and 0.79 for A2(a) and B2(a) scenarios, respectively. The validation results for 1991–2000 give a correlation factor (R^2) of 0.64 and 0.58 for A2(a) and B2(a), respectively. The forecasted monthly average precipitation and temperature of two selected scenarios [A2(a) and B2(a)] and the observed data are shown in Figures 12.24 and 12.25. The 50-year forecasted average monthly data are from 2006 to 2055 and the observed data are based on the monthly average from 1961 to 2005.

The results show that the average of temperature during the simulation period will increase by 0.83°C and 0.67°C for the two scenarios in comparison with the observed data. Moreover, the precipitation will increase by 3.47% and 8.67% in comparison with the observed data. As shown in Figure 12.24, the peak of precipitation happens in May in both scenarios and the minimum precipitation is in July. According to the observed data in the study region, these results seem to be correctly estimated. Comparing the results of these two scenarios with the observed data shows that

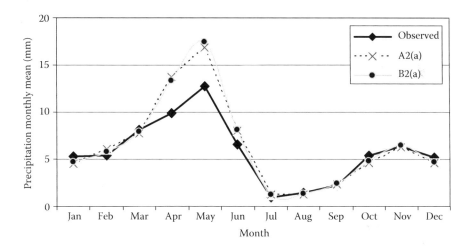

FIGURE 12.24 Comparison of precipitation output of A2(a) and B2(a) scenarios from SDSM, and observed monthly precipitation.

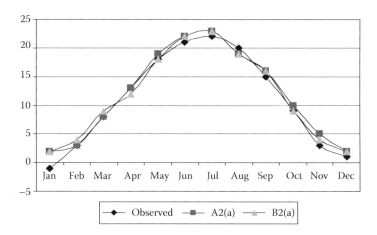

FIGURE 12.25 Comparison of temperature output of A2(a) and B2(a) scenarios from SDSM, and observed monthly temperature.

there is little difference between the two climatic scenarios for precipitation, but in comparison with the observed data, a 25% increase occurs between March and June.

12.7.2 Evaluation of Urban Floods Considering Climate Change Impacts

Climate change impacts all of the components of the world's water cycle. In urban areas, due to some special characteristics such as population concentration and limitations on natural water systems, the effects of climate change are intensified. One of the most important components of the urban water cycle is urban runoff, which is highly affected by climate change and urbanization. Karamouz et al. (2008) have studied the urban rainfall and runoff variations under climate change impacts in Tehran metropolitan area in Iran. In recent years, the rapid development of Tehran, the capital of Iran, without considering the impacts on the urban water cycle, has made considerable quality and quantity challenges. There has not been a systematic approach to runoff management, which has led to overflow of channels and some health hazard problems. They have considered the north-eastern part of Tehran, between 51° 22′ and 51° 30′ longitudes and between 35° 42′ and 35° 53′ latitudes as the case study. The total area of the study region is about 110 km^2. The percentage of the impervious area is about 85%, and the average slope of the region is about 21%. In recent years, some river training projects have been performed on these channels, and these have changed the characteristics of the surface water collection system. For the evaluation of the effects of these projects on the runoff collection system of Tehran, three scenarios have been considered:

- Scenario 1: It is developed for the evaluation of the surface water collection system about 10 years ago. The model for this scenario is developed to evaluate the effectiveness of development projects done in recent years for improving the past situation.
- Scenario 2: In this scenario, the present surface water collection system is modeled.
- Scenario 3: The future plans for improvement and the development of the case study drainage system have been modeled in this scenario.

The effects of climate change on the runoff volume have been considered in these scenarios. The simulated drainage systems in these scenarios and their characteristics are shown in Figure 12.26. The circled areas in Figure 12.26 show the differences between different scenarios.

For using the SDSM, a combination of three predictors is selected: (1) relative humidity at 850 hPa height, (2) near-surface specific humidity, and (3) near-surface relative humidity. This combination has been selected because of its maximum correlation with the daily rainfall of the north-eastern

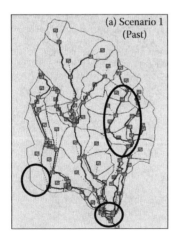
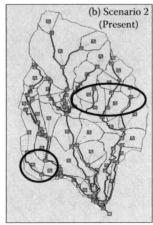
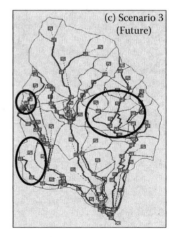

FIGURE 12.26 Different drainage system scenarios.

Climate Change

part of Tehran. The model has been calibrated with rainfall data of 1977–1984 and validated for the remaining available data (1985–2000). Rainfall is evaluated in each calendar season to determine the climate change effects. The characteristics of identified extreme rainfalls of the future in specified seasons are presented in Table 12.8 for every 10 years from 2007 until 2097. It seems that the magnitude and frequency of the extreme rainfalls are slightly affected by the climate change, and

TABLE 12.8
Characteristics of the Extreme Rainfalls in the Future under Climate Change Effects

Year	Number of Extreme Rainfalls				Average of Extreme Rainfalls			
	Spring	Summer	Fall	Winter	Spring	Summer	Fall	Winter
2007	12	1	10	13	29.03	17.8	27.06	46.56
2017	11	2	8	17	49.7	33.46	48.42	42.45
2027	23	1	12	17	22.18	76.99	49.45	38.61
2037	20	1	9	11	26.25	15.4	32.87	62.65
2047	15	0	8	8	26.12	0	48.6	50.77
2057	18	0	9	16	31.24	0	37.94	40.27
2067	18	0	9	9	37.39	0	35.69	38.66
2077	18	0	14	12	32.85	0	56	45.15
2087	19	1	16	9	40.1	71	55.9	46.44
2097	15	0	8	14	42.85	0	47.41	59.44

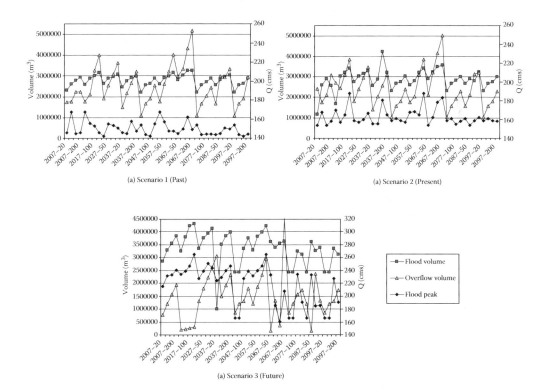

FIGURE 12.27 Results of the drainage system simulation under extreme rainfalls with different return periods for the evaluation of climate change effects.

also the severity of wet and dry periods is increasing in the study area due to the effects of climate change.

The identified extreme rainfalls are applied in the developed hydraulic model in three scenarios defined for considering the developments on the system. The results are illustrated in Figure 12.27. As can be seen, there are no obvious trends in the volume and peak of floods due to climate change effects, but the man-made changes in the drainage system of the case study have changed the volume and peak of floods considerably. In response to the training projects, the flood peaks and volumes have been increased considerably; however, as the capacity of the system increases, the overflow volumes have been decreased. This may lead to considerable damage and it is necessary to revise the training projects of the drainage channels in the study area. There are useful applications of some new training programs for dredging the channels and construction of some detention and retention ponds regarding urban developments.

12.8 CONCLUSION

The results of many recent studies around the world have demonstrated that climate change is happening. Although some of its impacts are depending on our actions, the situation in the future could be worse than the present condition. At the moment, the scientific consensus supports the view that (1) global warming is already happening, (2) it is likely to accelerate over the next several decades, (3) it is possible to meaningfully mitigate it through GHG reductions, and (4) it will get worse until we can stabilize the situation through mitigation. The impacts of climate change in urban areas are intensified because of the special systems and processes in urban areas. At first, different types of climate change impacts on water resources should be identified and then appropriate responses should be adopted to decrease the possible damages especially those related to water resources. The considered responses include (1) vulnerability analysis to identify near-term priorities for adaptation of capital and operating plans, (2) IRP to provide a comprehensive framework to further study the change processes and devise a broad array of adaptive measures that can sustain water supplies despite the ongoing environmental changes, and (3) GHG reductions to help mitigate the global warming process. The planning element of these response measures is especially significant. It is worth noting that climate change is not something that the present generation of utility managers can solve. However, the present generation of managers can establish the right planning and related research processes needed to enable future success in coping with climate change.

For evaluating the climate change impacts, different socioenvironmental scenarios for emission of GHGs are defined and then simulated through GCMs. With improvements in GCMs and new techniques such as downscaling and RCMs, more sophisticated scenarios can be created. Nonetheless, there is fundamental uncertainty about regional climate change. The magnitudes and even direction of change of many important meteorological variables are uncertain. There is even greater uncertainty about changes in variability and extreme events—changes that may be critical for climate impacts assessment. Readers should always remember that climate change scenarios do not yield predictions of the future; they only help us to understand the potential implications of climate change and the vulnerability of human and natural systems to this change.

PROBLEMS

1. Download the data of mean maximum, mean, and mean minimum temperature at grid 50×26 from the HADCM3 model for scenario A2 in the period 1961–1999 and 2000–2049. (a) Evaluate the effects of climate change in the next 50 years (2000–2049) and discuss the probable reasons for the assumptions in developing scenario A2. (b) The water per capita consumption and annual mean temperature data for a city located in this grid are given in Table 12.9. Determine the annual water demand variations of this city in the future based on the results of part (a).

TABLE 12.9
Water per Capita Consumption and Annual Mean Temperature Data for Problem 1

Year	1961	1962	1963	1964	1965	1966	1967	1968	1969	1970
Mean temperature (°C)	15.1	18	13.8	19	18.5	16.4	15.9	17.3	16.6	12
Water per capita consumption (l/day)	160	200	140	200	180	175	170	177	180	160
Year	1971	1972	1973	1974	1975	1976	1977	1978	1979	1980
Mean temperature (°C)	17.5	18.7	14.9	20	19.5	20.5	16.7	17.6	21	19.3
Water per capita consumption (l/day)	180	185	165	210	210	205	194	200	220	210
Year	1981	1982	1983	1984	1985	1986	1987	1988	1989	1990
Mean temperature (°C)	18.4	19.2	20	17.5	16	20.6	19.5	17	18.2	19
Water per capita consumption (l/day)	190	195	200	188	185	195	195	185	190	198
Year	1991	1992	1993	1994	1995	1996	1997	1998	1999	
Mean temperature (°C)	18.5	19.3	19.9	20.5	18.8	19.1	19.8	21	21.1	
Water per capita consumption (l/day)	195	205	207	214	206	208	212	218	222	

2. Based on the available GCM data of relative humidity, temperature, and rainfall compare the results of scenarios B1 and B2 from the HADCM3 model for grid 46 × 13 in the period 2000–2049. Discuss the possible reasons for the observed differences. Based on these changes, what points should be considered for planning a water resources development project in an urban area located in this grid (discuss on the basis of direct and indirect impacts).

3. Using the precipitation data of a station in your own city, evaluate the changes in mean, maximum, and minimum monthly rainfall and wet and dry spells under scenarios HadCM3-A2. Discuss the changes that have happened in each scenario and the intensity of the climate change effects. How would water allocation be planned in response to these changes?

4. Utilize the rainfall, minimum temperature, and maximum temperature data of a station in a city in the LARS model and downscale it. Define different scenarios and evaluate the effect of climate change on the statistics of the considered variables.

5. Which parameter in the SCS, Horton and Green and Ampt models are affected by climate change? Discuss about parameters affected directly or indirectly.

6. As increasing the rainfall intensity is one of the important impact of climate change, which models listed in Problem 5 are more sensitive to this change? Also, determine the losses due to infiltration in Example 3.3, when reducing the storm duration by half and doubling the amount of precipitation intensity.

7. Repeat the assumption of Problem 6 and solve Examples 3.5, 3.6, 3.7, and 3.9. Compare the results and check the following impacts of climate change:
   - Changes in recharge pattern of groundwater
   - Increased risk of flood damage to facilities of water utilities
   - Other impact presented in Section 12.3

8. For estimation of climate change effects on demand variation, develop an IDNN model to simulate the monthly water consumption of the considered city in Problem 1. Compare the results and discuss the impact on water shortage caused by climate change.

9. Develop an MLP model to simulate the monthly water consumption of the city in Problem 4 as a function of rainfall, minimum, maximum temperature.

10. Discuss different elements of a typical urban water dynamic model presented in Figure 10.26, affected by climate change directly or indirectly.
11. Based on the materials in Sections 12.2 and 12.3, and also Example 8.1, draw the causal loops of climate change impacts on urban water consumption.
12. The data of monthly precipitation that is shown in Table 8.3, are related to a rain gauge of an urban area located at 51 longitude and 35 latitude with 900 m altitude. Generate the precipitation data of this gauging station under the GCM, HadCM3-A2 scenario outputs from 2004 to 2014 using the SDSM.
13. Solve Example 8.1 considering the results of Problem 12. Compare the demand variations and discuss the climate change impacts on demand variation. How much the urban water consumption will increase by increasing the temperature by 1° centigrade?
14. Formulate a simple model based on dynamic hypotheses of climate change impacts on water balance in a region; in this model consider the following causal loops and their interactions. For simplicity, separate primary and secondary impacts considering their interactions and feedbacks.

    - Evaporation
    - Soil moisture
    - Vegetation
    - Runoff coefficient
    - Precipitation characteristics
    - Reservoir dependable capacity

15. Downscale the data measured at the rain gauge discussed in Problem 12 and Example 8.2 using the LARS model.
16. Solve Example 8.2 considering the results of Problem 15. Compare the demand variations and discuss the climate change impacts on demand variation.
17. Solve Example 6.3 by inflow hydrograph obtained from Example 3.3 and Problem 6. Compare the peak discharges and the time delay in both models.
18. Solve Example 6.1, by reducing the storm durations to half and doubling the amount of precipitation (rate rainfall excess rate of 30 mm/h for two rainfall durations of 7.5 and 2.5 min). Compare peak discharges and their time delays.
19. Solve Example 6.2, by reducing the soil saturation percent to half and doubling the rainfall duration to half and by doubling the precipitation intensity.
20. One of secondary impact of Climate change is variation of hydrological parameters which are the basis for design of urban infrastructures. Keeping this in mind, solve Examples 6.4, 6.6 and 6.7 by doubling the precipitation intensity and the design discharge. Compare the infrastructure dimensions and discuss the impact of climate change on the design criteria.

REFERENCES

Alley, W. M., Reilly, T. E., and Franke, O. L. 1999. Sustainability of groundwater resources. U.S. Geological Survey Circular 1186. U.S. Geological Survey, Denver, CO.

Bhaskaran, B., Jones, R. G., Murphy, J. M., and Noguer, M. 1996. Simulations of the Indian summer monsoon using a nested regional climate model: Domain size experiments. *Climate Dynamics* 12: 573–588.

Burger, G. 1996. Expanded downscaling for generating local weather scenarios. *Climate Research* 7: 111–128.

Chalecki, L. H. and Glieck, P. H. 1999. A comprehensive bibliography of the impacts of the climate change and variability on water resources of the United States. *Journal of American Water Resources Association* 35: 1657–1665.

Chia, E. and Hutchinson, M. F. 1991. The beta distribution as a probability model for daily cloud duration. *Agricultural and Forest Meteorology* 56: 195–208.

Cromwell, J. E., Smith, J. B., and Raucher, R. S. 2007. *Implications of Climate Change for Urban Water Utilities*. Stratus Consulting Inc. Washington, DC and Boulder, CO.

Gates, W. L., Mitchell, J. F. B., Boer, G. J., Cubasch, U., and Meleshko, V. P. 1992. Climate modeling, climate prediction and model validation. In: J. T. Houghton, B. A. Callander, and S. K. Varney, Eds, *Climate*

Change 1992: The Supplementary Report to the IPCC Scientific Assessment. Cambridge University Press, Cambridge, U.K.

Giorgi, F., Shields, B. C., and Bates, G. T. 1994. Regional climate change scenarios over the US produced with a nested regional climate model. *Journal of Climate* 7: 375–399.

Glieck, P. H. 2000. Water the potential consequences of climate variability and change for the water resources of the United States. A report of the national water assessment group for the U.S. Global Change Research Program. Pacific Institute for Studies in Development, Environment and Security, Oakland, CA.

Hewitson, B. C. and Crane, R. G. 1996. Climate downscaling: Techniques and application. *Climate Research* 7: 85–95.

Houghton, J. T., Meira, Filho, L. G., Callander, B. A., Harris, N., Kattenberg, A., and Maskell K. Eds. 1996. Climate change 1995: The science of climate change. Contribution of Working Group I to the Second Assessment Report of the Intergovernmental Panel on Climate Change. Cambridge University Press, Cambridge, U.K.

IPCC, Third Assessment Report, Intergovernmental Panel on Climate Change (IPCC). 2001. Available at http://www.ipcc.ch/.

IPCC, Fourth Assessment Report, Intergovernmental Panel on Climate Change (IPCC). 2007. Available at http://www.ipcc.ch/.

Jones, R. G., Murphy, J. M., and Noguer, M. 1995. Simulation of climate change over Europe using a nested regional climate model. Part 1: Assessment of control climate, including sensitivity to location of lateral boundaries. *Quarterly Journal of the Royal Meteorological Society* 121: 1413–1449.

Karamouz, M., Hosseinpoor, A., and Nazif, S. 2008. Evaluation of urban floods considering climate change impacts. *Proceedings of the 4th International Conference on Flood Defense*. Toronto, Canada.

Karamouz, M., Rouhanizadeh, B., Taheriyoun, M., and Emami, F. 2009. Impact of climate change on sediment loads transport in a watershed scale: A case study. *Proceedings of the ICWR Conference*, Malaysia.

Lattenmaier, D. P., McCabe, G., and Stakhiv, E. Z. 1996. Global climate change: Effect on hydrologic cycle. In: L. W. Mays Ed., *Water Resources Handbook*. McGraw-Hill, New York.

Mays, L. W. 2004. *Urban Water Supply Management Tools*. McGraw-Hill, New York.

Mearns, L. O., Rosenzweig, C., and Goldberg, R. 1992. Effect of changes in interannual climatic variability of CERES-Wheat yields: Sensitivity and $2HCO_2$ general circulation model studies. *Agricultural and Forest Meteorology* 62: 159–189.

Olsson, J., Uvo, C. B., Jinno, K., Kawamura, A., Nishiyama, K., Koreeda, N., Nakashima, T., and Morita, O. 2004. Neural networks for rainfall forecasting by atmospheric downscaling. *Journal of Hydrologic Engineering* 9: 1–12.

Racsko, P., Szeidl, L., and Semenov, M. 1991. A serial approach to local stochastic weather models. *Ecological Modeling* 57: 27–41 (Agriculture, Washington, DC).

Richardson, C. W. 1981. Stochastic simulation of daily precipitation, temperature and solar radiation. *Water Resources Research* 17: 182–190.

Rietveld, M. R. 1978. A new method for the estimating the regression coefficients in the formula relating solar radiation to sunshine. *Agricultural and Forest Meteorology* 19: 243–252.

Rind, D. R., Goldberg, R., Hansen, J., Rosensweig, C., and Ruedy, R. 1990. Potential evapotranspiration and the likelihood of future drought. *Journal of Geophysical Review* 95: 9983–10004.

Schlesinger, M. E., Andronova, N., Ghanem, A., Malyshev, S., Reichler, T., Rozanov, E., Wang, W., and Wang, F. 1997. *Geographical Scenarios of Greenhouse-Gas and Anthropogenic-Sulfate-Aerosol Induced Climate Changes*. Climate Research Group, Department of Atmospheric Sciences. University of Illinois. Urbana-Champaign, USA. Available at http://crga.atmos.uiuic.edu/public/CRG_Publications/Geographical.pdf.

Semenov, M. A. and Barrow, E. M. 2002. *LARS-WG, A Stochastic Weather Generator for Use in Climate Impact Studies*. User Manual, Version 3.0. Available at http://www.iacr.bbsrc.ac.uk/mas-models/larswg.html.

Semenov, M. A., Brooks, R. J., Barrow, E. M., and Richardson, C. W. 1998. Comparison of the WGEN and LARS-WG stochastic weather generators in diverse climates. *Climate Research* 10: 95–107.

Semenov, M. A. and Brooks, R. J. 1999. Spatial interpolation of the LARS-WG stochastic weather generator in Great Britain. *Climate Research* 11: 137–148.

Smith, J. B. and Tirpak, D. Eds. 1989. The potential effects of global climate change on the United States. EPA-230-05-89-050. U.S. Environmental Protection Agency, Washington, DC.

UK Department of the Environment. 1996. Review of the potential effects of climate change in the United Kingdom. Second Report of the UK Climate Change Impacts Review Group, 248pp.

Viner, D. and Hulme, M. 1992. *Climate Change Scenarios for Impact Studies in the UK.* Climatic Research Unit, University of East Anglia, Norwich, U.K.

Vinnikov. K. Y., Robock, A., Speranskaya, N. A., and Srhlosser, C. A. 1996. Scales of temporal and spatial variability of midlatitude soil moisture. *Journal of Geophysical Research* 101: 7163–7174.

Von Storch, H., Zorita, E., and Cubasch, U. 1993. Downscaling of global climate change estimates to regional scales: An application to Iberian rainfall in wintertime. *Journal of Climate* 6: 1161–1171.

Walsh, K. and McGregor, J. 1997. An assessment of simulations of climate variability over Australia with a limited area model. *International Journal of Climatology* 17: 201–224.

Wigley, T. M. L., Jones, P. D., Briffa, K. R., and Smith, G. 1990. Obtaining sub-grid-scale information from coarse-resolution general circulation model output. *Journal of Geophysical Research* 95: 1943–1954.

Wilby, R. L. and Wigley, T. M. L. 1997. Downscaling general circulation model output: A review of methods and limitations. *Progress in Physical Geography* 21: 530–548.

Wilby, R. L., Hassan, H., and Hanaki, K. 1998. Statistical downscaling of hydrometeorological variables using general circulation model output. *Journal of Hydrology* 205: 1–19.

Wilby, R. L., Hay, L. E., and Leavesley, G. H. 1999. A comparison of downscaled and raw GCM output: Implications for climate change scenarios in the San Juan River Basin, Colorado. *Journal of Hydrology* 225: 67–91.

Wilby, R. L., Hay, L. E., Gutowski, W. J., Arritt, R. W., Takle, E. S., Pan, Z., Leavesley, G. H., and Clark, M. P. 2000. Hydrological responses to dynamically and statistically downscaled climate model output. *Geophysical Research Letters* 27: 1199–1202.

Wilby, R. L. and Dettinger, M. D. 2000. Streamflow changes in the Sierra Nevada, CA simulated using a statistically downscaled General Circulation Model scenario of climate change. In: S. J. McLaren and D. R. Kniveton, Eds, *Linking Climate Change to Land Surface Change*, pp. 99–121. Kluwer Academic Publishers, Dordrecht, the Netherlands.

Wilby, R. L., Tomlinson, O. J., and Dawson, C. W. 2003. Multi-site simulation of precipitation by conditional resampling. *Climate Research* 23: 183–194.

Wilby, R. L. and Dawson, C. W. 2004. *Using SDSM Version 3.1–A Decision Support Tool for the Assessment of Regional Climate Change Impacts,* Climate Change Unit, Environment Agency of England and Wales, Nottingham, NG2 5FA, UK and Department of Computer Science, Loughborough University, Leics, LE11 3TU, U.K.

Wigley. T. M. L. and Jones, P. D. 1985. Influence of precipitation changes and direct CO_2 effects on streamtlow. *Nature* 314: 149–151.

Xu, C. Y. 1999. From GCMs to river flow: a review of downscaling methods and hydrologic modelling approaches. *Progress in Physical Geography* 23: 229–249.

Zorita, E. and von Storch, H. 1999. The analog method as a simple statistical downscaling technique: Comparison with more complicated methods. *Journal of Climate* 12: 2474–2489.

Index

A

ABR. *See* Anaerobic baffled reactor (ABR)
Accidental spill, 311, 334
Acidification, 330
Action plan, 232–233, 472
Activated sludge process, 297
ADALINE. *See* Adaptive linear (ADALINE) network
Adaptability, 411
Adaptation, 508, 510, 539, 544, 548
Adaptation with climate change, 544
Adaptive, 29, 33, 35–36
Adaptive linear (ADALINE) network, 352–353. *See also* Neural network; Temporal neural network
Adaptive random sampling (ARS), 359
ADR. *See* Alternative dispute resolution (ADR)
Advanced, 50, 297, 300, 359, 502, 550
Advanced first-order second-moment (AFOSM) method, 502
AES. *See* Adaptive random sampling (AES)
Affected area, 10
AFOSM. *See* Advanced first-order second-moment (AFOSM) method
Aging, 17, 54, 411, 420, 463, 471, 482, 515
Agricultural, 213–214, 445, 456
Agricultural drought, 465
Agricultural income, 456
Agricultural sector, 445
Aharchai river basin, 574–576
AIC. *See* Average incremental cost (AIC)
Air pollution, 547
Algebraic product, 378
Algebraic sum, 160, 378
Algebraic sum of fuzzy sets, 378
Allocation, 21, 38–39, 47–48, 206–207, 453–454, 455
Alternative, 9–10, 41, 112, 133, 192, 226, 234, 384–386, 397, 398, 402
Alternative dispute resolution (ADR), 441
AMC. *See* Average moisture conditions (AMC)
AMC I (AMC in dry watershed soil), 73
AMC II (AMC before the beginning of a rainfall), 73
AMC III. (AMC in wet watershed soil), 73
American Water Works Association (AWWA), 143
American Water Works Association Research Foundation (AWWARF), 539
Ammonia, 318, 321, 400
Anaerobic baffled reactor (ABR), 297, 298
Analysis of uncertainty, 504, 536
Analytical hierarchy process (AHP), 374
ANN. *See* Artificial neural network (ANN)
Annual real losses, 201–203
Annual Worth, 384, 386, 387, 389, 398, 399
Annual Worth method, 385
Anomaly, 519, 552

Apparent losses, 200
Aquaculture, 413
Aquatic biota, 311, 315, 318, 327, 329
Aquatic ecosystem, 543–544
 and stormwater discharge, relationship, 327
Aquatic habitat, 311–313, 315, 327, 334
Aquatic habitat, degradation of, 327
Aquatic life, 4, 293, 311, 316, 329, 411
Aquifer storage and recovery (ASR), 246, 247, 332–333
Aquifer storage recovery, 332
Artificial neural network (ANN), 346–358, 519. *See also* Neural network
 multilayer perceptron (MLP) network, 348–352
 temporal neural network, 352–358
Artificial recharge, 453
Asian and Pacific region, dams and reservoirs in, 394
ASR. *See* Aquifer storage recovery system (ASR)
Atmospheric-oceanic general circulation models, 556
Attribute, 29, 38, 42–44
Authorized consumption, 200
Availability, 44, 45–46, 128–129, 233, 502
Average incremental cost (AIC), 225
Average moisture conditions (AMC), 73
AWWA. *See* American Water Works Association (AWWA)
AWWARF. *See* American Water Works Association Research Foundation (AWWARF)

B

Back propagation algorithm (BPA), 347
Back propagation network, 347
Backward, 188, 368, 369, 390
Bacteria, 318–319, 321, 334
Base flow, 323, 331
Baseline impacts, 543, 544
Basic solution, 364
Basic variable, 364
Basin Development Factor (BDF), 85, 105, 106, 122
BASINS. *See* Better Assessment Science Integrating Point and Non-Point Sources (BASINS)
BDF. *See* Basin Development Factor (BDF)
Benefit, 48, 383, 397, 418, 478, 523, 526
Benefit-cost ratio method, 386–387
Bernoulli, 340, 344
Bernoulli distribution, 340
Best management practice (BMP), 1, 19, 241, 242
 for stormwater drainage system, 242–246
Best planning practices (BPPs), 18, 19
 applications of, 19
Best practice standard water balance, 201
Better Assessment Science Integrating Point and Non-Point Sources (BASINS), 325
BHP. *See* Brake horsepower (BHP)

583

Binomial, 160, 340, 344
Binomial distribution, 340
Bioretention basins, 244
Bioretention swales, 243–244
Black water, 14, 403
Block rate dummy variable, 196
BMP. *See* Best management practice (BMP)
BPA. *See* Back propagation algorithm
BPPs. *See* Best planning practices (BPPs)
Brake horsepower (BHP), 154
Buffer strips, 244
Built water system, 469, 470
Bulk decay, 165

C

Calibration, 326, 564, 565, 569
Capacity expansion, 134, 364
Capital recovery, 383
Carbon dioxide, 8, 539, 548
Caribbean wastewater treatment, 298–299. *See also* Wastewater—management, in urban drainage system
CARL. *See* Current annual real losses (CARL)
Case study, 447–456, 479–481, 574–578
Cash flow, 383, 384, 385
Catchment basin, 5
Causal diagram, 441, 442
Centralization, scale and degree of, 411
Channel flow, 246–251, 256–259, 270
Channel modification, 102, 312
Channel routing, 256, 290
Chaos, 433–434
Chezy–Manning formula, 170
Chi, square, 385, 344
Chloride, 7, 143, 172, 173, 174, 317, 340
Chlorine decay, 164, 165, 169, 173
Clark method, 98–101. *See also* Time–area UHs
Clark's UH theory, 98
Climate, 5
  arid and semiarid, 6
  categories of, 5
  cool coastal, 6
  effects of, 5–6
  rain shadow, 6
  subtropical, 6
  tropical, 6
Climate change, 292, 463, 510, 519
  adaptation with, 544–545
    GHG emissions, reducing, 547–548
    integrated resource planning, 545–547
    vulnerability analysis, 545
  case studies
    Aharchai river basin, 574–576
    urban floods evaluation, 576–578
  downscaling models
    LARS-WG model, 567–574
    statistical downscaling model, 559–567
  evaporation and precipitation, variation in, 540
  groundwater, 541–542
  prediction, 548
    climate change scenario, 548–549
    downscaling, 554
    general circulation models, 549–553
    regional models, 554–556
    spatial variability, 554
    statistical downscaling, 556–559
  snowfall and snowmelt, 541
  soil moisture, 540–541
  storm frequency and intensity, 541
  temperature increase, 540
  in urban water supply
    compound impacts, 544
    direct impacts, 542–543
    indirect impacts, 543–544
  water resource system effects, 542
Climate change process, 539–542, 545
Climate change scenarios, 548–549, 554, 574–576
Climatic drought, 465
CN. *See* Curve Number (CN)
Codigestion, 406
Coefficient, 84, 90, 100, 112, 148, 149, 151, 152, 165, 170, 198, 216, 250, 251, 270, 271, 272, 280, 281, 313, 321, 322, 346, 364, 499, 569
Coefficient of roughness, 148
Combined sewer overflows (CSOs), 299, 300, 317, 405, 416, 544
Combined sewers, 317, 405, 406, 411
Commitment, 25, 190, 211, 326, 470, 472, 473, 474, 476, 477, 482, 483
Community activities, lessons on, 479
Complement, 47, 49, 378, 494
Composite CN, 73
  with connected impervious area, 75
  with total impervious area, 76
  with unconnected impervious area, 76
Compound pipeline, 154, 155
Computer-based models, 325–326
Computer software tools, 429
Concave, 378, 488
Confidence interval, 505
Confined aquifer, 457
Conflict issue, 227
Conflict management, 47
Conflict resolution, 47–48, 440–447, 449–451, 453–455
  models, 442–447
  process, 441
  systematic approach, 441–442
Connectors, 429–430
Consortium of stakeholders, 471
Constraint, 41, 44, 117, 360, 361, 362, 363, 367, 369, 376
Constructed wetlands, 245
Consumptive use, 135, 183, 184, 188, 212, 486
Consumptive user, 486
Contaminant, 163, 164, 169, 317, 329, 332, 400
Contamination, 13, 317, 332, 411, 468–469, 528
  disaster assessment caused by, 528
Continuity equation, 100, 256, 262, 398, 436
Continuity principle, 77
Continuous, 171–172, 207, 288, 340, 341, 421, 511
Control, 14, 30, 48, 117, 199, 203, 204, 277–280, 283, 295, 447, 450, 451, 458, 466–468, 474–482
Conventional risk-based hydraulic design, 509
Conversion constant, 95
Converters, 429, 453
Convex, 363, 368, 378, 443, 479, 488
Conveyance loss, 135, 184

Index

Conveyance (channel) storage, 290–291
Cost, 330, 401
Cost-effectiveness affordability, 411
Coupled ocean–atmosphere GCMs, 550
Covariance, 501, 558
Critical depth, 277, 278, 282
Critical Values for Univariate Analysis, 114
Cross correlation, 568
Crossover, 369, 370, 371
Cross section, 269, 270
C-series standards, 143
CSOs. *See* Combined sewer overflows (CSOs)
Cultural, 9
Culvert, 275–284
    inlet control flow, 277–279, 280
    outlet control flow, 279, 283
        full-flow conditions, 281
        partly full-flow conditions, 282
    protection downstream of, 283–284
    sizing of, 282–283
Culvert inlet control flow coefficients, 280
Cumulative distribution function, 340, 488, 497, 561
Current annual real losses (CARL), 202
Curse of dimensionality, 368
Curve Number (CN), 73
    modified for different soil moisture conditions, 75
    for urban land uses, 74
Cut-set analysis, 492, 535

D

DAF. *See* Dissolved-air flotation (DAF)
Damage, 292, 294, 466, 469, 509, 518–522
Dams, 131, 393–394
Dams and reservoirs, infrastructure for, 393–394
Darcy–Weisbach equation, 145–148, 151, 159, 536. *See also* Hazen–Williams equation
    application of, 147
Data uncertainty, 504
Decision maker, 25, 38, 50, 442–447, 449, 479, 529
Decision-making, 25–26, 190
Decision Space, 363, 364, 366, 368, 371, 379, 388, 443
Decision support systems (DSS), 13, 28, 42
    conceptual system structure, 44
Decision variable, 359–360
Deficit, 116, 231, 237, 360, 480, 511–514, 517
Definition, 118–119, 135, 183–184, 200, 248, 410, 415, 464, 465
Degree of freedom, 556
Degree of saturation, 60, 255
Delta method, 505
Demand fluctuation, 135
Demand forecasting, 139, 192, 193, 195
Demand function, 388
Demand management, 183–237, 533. *See also* Demand-side management (DSM)
Demand-side management (DSM), 185, 186, 225
    alternatives, 228–232
    awareness creation as, 204–205
    demographic consideration in, 224
    downsizing a capital facility as, 227
    drought demand management plan as, 227
    factors affected by, 225
    free water dilemma in, 205
    incentives as, 205
    issues in, 187, 227
    long-term conservation, 225
        investment in conservation program, 226
    pricing as, 205–207
    resource combination as, 227
    short-term conservation as, 226
Density function, 342, 343, 487, 496, 502
Department of Environment, 449, 457, 508, 554
Depression storage, 54, 55, 58, 66, 67, 71–73, 322
Desalination, 533, 547
Desegregation, 197, 292, 329
Design of channel, 269, 270, 272
Design Problems, 151
Detention, 290, 291
Detention basin, 241, 288, 289, 292
Detention time, 133, 244, 245, 290
Deterministic, 33, 35, 284, 339, 368, 369, 505, 526
Diagram of system readiness, 531
Differential, 33, 163, 250, 251, 360, 368, 501
Differential dynamic programming, 360, 368
Direct impact, 38, 315, 542–544, 546
Direct integration method, 497–499, 501
Direct reuse, 213, 403
Direct runoff (P_d), 87, 88, 100, 102
Direct runoff hydrograph, 87, 88, 103, 124
Disagreement point, 229, 231, 443, 447, 451
Disaster committee, 529
Disaster indices, of water management, 464, 483, 484, 534
    reliability, 484
        AFOSM method, 502
        analysis, 495–496
        assessment, 491–495
        direct integration method, 497–499
        estimation, 486
        indices, 496–497
        margin of safety, 499–501
        mean value first-order second moment (MFOSM) method, 501–502
        water users, 486
    resiliency, 503
    time-to-failure (TTF) analysis, 502
        availability and unavailability, 502
        failure and repair characteristics, 502
    vulnerability, 503–504
Disaster management
    guidelines for, 526
        disaster and technology, 526–527
        disaster and training, 527
        institutional roles in, 527–528
    handling, 475
Disaster management commission
    organization chart, 530
Disaster preparedness, policy implications for, 477–478
Disaster prevention, 473, 474, 508, 510
Disaster-resistant city (DRC) plan, 472–473
Disasters, sources and kinds of, 464
    drought, 464–466
    earthquakes, 470
    floods, 466–468
    system failure, 469–470
    widespread contamination, 468–469
Discharge of chemicals to receiving waters, 292, 293

Discount gradient, 383
Discount rate, 224, 383, 385, 401, 422
Discrete, 340, 368, 375, 383
Discrete differential dynamic programming, 368
Discrete dynamic programming, 368
Discrete volume method (DVM), 169
Dissemination, 27, 28, 47, 475, 510, 532
Dissolved-air flotation (DAF), 297
Dissolved oxygen (DO), 4, 292, 329
    depletion, in streams, 292, 293
Distance, based method, 443
Distribution, 5, 21, 138, 139, 141–144, 162–163, 171, 172–175, 286, 289, 340–345, 395–396, 397, 398, 491–494, 497–501, 535, 568–569, 570
Distribution function, 340–345, 469, 488, 497, 501, 561
Distributive solution, 441
Diversion, 116, 135, 185, 187, 228, 291, 302, 339, 389, 467, 481
Diversion channel, 467
DO. *See* Dissolved oxygen (DO)
Domestic consumption, 138, 447
Dominated, 9, 81, 374, 376, 377, 443, 478, 525
Downscaling, 539, 548, 554–559, 561
    LARS-WG model, 567–574
    statistical downscaling model, 559–567
DP. *See* Dynamic programming (DP)
DPSIR model. *See* Driving force, Pressure, State, Impact, and Response (DPSIR) model
Drainage area, 53–54, 55, 57, 73, 79, 82–83, 84, 305
Drainage system, 85–86, 241–307, 404–406, 469, 576, 577
Drainage system performance
    channel improvements, 85
    channel linings, 86
    curb and gutter streets, 86
    storm drains or storm sewers, 86
DRC plan. *See* Disaster-resistant city (DRC) plan
Dredging, 578
Driving force, 38–44, 127, 247, 251, 557
Driving force, Pressure, State, Impact, and Response (DPSIR) model, 38
Driving force, State, and Response (DSR) model, 38
Drought, 5, 225–227, 464–466, 479–480, 518–523, 528
    disaster management, case study, 479–480
    in Georgia, 480
    graphical definition of, 465
Drought indices, 479, 480, 520, 522
Drought management, 465, 479, 480
Drought planning, 465, 480
DSM. *See* Demand-side management (DSM)
DSR model. *See* Driving force, State, and Response (DSR) model
DSS. *See* Decision support systems (DSS)
Durability, 411
Duration, 79–81, 95, 98, 100, 140, 285, 289, 484, 523
DVM. *See* Discrete volume method (DVM)
Dwelling unit, 136, 401
Dynamic behavior, 353, 431, 433
Dynamic hypothesis, 453, 454
Dynamic interactions, 38, 39, 42
Dynamic organizing principle, 430
Dynamic programming (DP), 360, 366–369, 390, 423
Dynamic relationship, 38, 43
Dynamic strategy planning, 38, 39, 40, 41, 42, 43, 44
    DSR dynamic strategy planning procedure, 39–43

dynamic relationships identification, 38–39
Sustainable Urban Land Use Management Systems identification, 38

E

Earthquakes, 470
EAs. *See* Evolution algorithms (EAs)
Ecoefficiency, 414
Ecological impacts, 224, 326, 327
Ecological Sanitation (EcoSan) approach, 246, 296–297. *See also* Sanitation
Economic analysis, of urban water system, 202, 206, 224, 381, 383, 384, 385, 386
    Annual Worth method, 385
    benefit–cost ratio method, 386–387
    demand function, 388
    external rate of return method, 385
    Future Worth method, 385
    internal rate of return method, 385
    Present Worth method, 385
    production functions model, 387–388
    resource-supply function, 388
Economic consideration, 381
Economic evaluation, 224, 386
Economic incentive, 199, 205
Economic Leakage Index (ELI), 202
    factors in leakage control, 202
    levels in, 202
    water cost vs. NRV control cost, 202
Economic life, 11
Economic model, 339, 387, 452
Economic tool, 49
EcoSan. *See* Ecological Sanitation (EcoSan) approach
EDM. *See* Event-driven method (EDM)
Education and training, 199, 204, 205, 229, 232
Effective porosity, 61
Effective rainfall, 57, 63, 68, 99, 103, 121
Effect of water storage reservoir location on pressure distribution, 140
Efficiency, 453
Effluent standards, 163, 193, 299
EIA. *See* Environmental Impact Assessment (EIA)
Electrical conductivity, 329
Electrically conductive composite pipe, 421
Electronic leak detector, 203
ELI. *See* Economic Leakage Index (ELI)
Elman RNN, 353, 354, 358
Embankment, 275, 281, 287, 291, 292, 394, 467
Emerging paradigm, 37
End-use analysis, 197
Energy conservation equation, 548, 550
Energy equation, 144, 148, 151, 153, 154, 155, 157, 281
Energy grade line, 144, 248, 249, 251
Energy head, 248
Engineering economy, 383
Environmental effects, 7, 333, 393
Environmental Impact Assessment (EIA), 27, 45
Environmentally acceptable systems, 411
Environmental management, 19, 475, 479
Environmental performance, assessment, 413–414
Environmental performance indicator, 414
Environmental protection, 411
Environmental sector, 445

Index

EPANET, 164, 169, 170, 171
Equality, 361, 362
Equilibrium time, 55, 83, 251, 252, 254, 432
Equipment, 173
ERR. *See* External rate of return (ERR)
Espey 10-min UH, 88, 89, 91, 92, 123
Eutrophication, 313, 330, 543
Evaporation and precipitation, variation in, 540
Event-driven method (EDM), 165, 171
Event tree, 491, 492
Event-tree analysis, 492
Evolution algorithms (EAs), 369–373
    genetic algorithms, 360, 369–372
    simulation annealing, 372–373
Exactions and impact fee-funded systems, 407–408
Excess rainfall, 57–79, 88, 246–247
Exponential, 62, 340–342, 432
Exponential distribution, 341–342, 487, 498, 568
Exponential family, of time paths, 432
Exponential random variable, 340
External, 206, 225, 382–383, 495
External Rate of Return (ERR) method, 385, 391, 422, 423

F

Factor of safety (FS), 137, 501
Failure, 287, 288–289, 469–470, 484–489, 492–493, 502, 503–504, 517, 528
Falling limb, 89, 99, 253, 254
FDM. *See* Finite difference method (FDM)
Feasible, 363, 364, 366, 374, 376
Federal Emergency Management Agency (FEMA), 466
Feedback loop, 36, 430, 434, 442
Feedback structure, 434
Feed forward network. *See* Multilayer perceptron (MLP) network
FEMA. *See* Federal Emergency Management Agency (FEMA)
F_f (constant rate of percolation), 69, 71. *See also* Φ-index
Finite, difference, 169
Finite difference method (FDM), 169
Fire demand, 136
Fire-fighting flow, 395
Fire-flow requirements, 395
Fire storage calculation, 140
Firm, 33, 194, 382, 544
First flush, 7, 132, 293, 403, 405
First flush effect, 7
First-order analysis, of uncertainties, 505
Fitness, 369, 370, 417, 519
Fixed, 143, 197, 348, 381–382
Flexibility, 411, 547, 568
Flood control, 295, 467–468
Flood disasters, 463
Flooding, 285, 292, 293, 466, 469, 481
Flooding, of urban drainage systems, 292, 293, 294–295
    control principles, 295
Floodplain, 294, 313, 314
Floodplain, urbanization effects on, 313, 314
Flood record
    adjusting, 107, 108
        procedures of, 107

homogeneity Testing, 112
peak adjustment factors
    for urbanizing watersheds, 107
revisiting, 104
Floods, 5, 464, 466–468, 481, 503
    direct costs, 466–467
    indirect costs, 467
    measures, 467
    routing method, 256–259
    urban, 294
        control management, principles of, 295, 467–468
Flow(s), 158, 162, 245, 326, 331, 415, 416, 429, 469
    area, 249, 251, 262, 264
    depth, 248–249, 251, 262, 264, 274
    equation, 282
    measurement, 175
    rate, 145, 174, 175
    splitter, 268, 269
Forecast, 139, 192–199, 448, 525, 575
Forecasting, 139, 192–199
Forward, 348, 368
Frequency analysis, 110–111, 285–287
Freshwater, 130
Friction
    factor, 145, 146, 148, 249, 250, 536
    loss, 139, 152, 281
    slope, 148, 149, 156, 249, 250
Froude number, 249
FS. *See* Factor of safety (FS)
Future
    value of money, 383
    worth, 384–386
    Worth method, 385
Fuzzy, 11, 339, 377, 378–381, 391
    decisions, 377
    modeling, 339
    sets, 11, 377–381, 391

G

Gamma, 96, 97, 273, 342, 344
    distribution, 342
    function, 7, 96, 97, 342
    function UH, 96, 97
    variable, 342
GAs. *See* Genetic algorithms (GAs)
Gas law, 550
GCMs. *See* General circulation models (GCMs)
GEF. *See* Global—Environment Facility (GEF)
GEF IW. *See* Global—Environmental Facility International Waters (GEF IW)
General circulation models (GCMs), 545, 549–553, 556
    bottom-up approach, 545
    characteristics, 551
    top-down approach, 545
Generation, 41, 42, 43, 369, 370, 560
Generator, 569
Genetic algorithms (GAs), 11, 360, 369–372
Geographical information system (GIS), 5
Geometric, 94, 248, 264, 265, 269, 344, 443
Geomorphology, 311
Geophone, 203
GHGs. *See* Greenhouse gases (GHGs)

GI. *See* Growth index (GI)
GIS. *See* Geographical information system (GIS)
Glacial melting, 542
Global
    Environment Facility (GEF)
        public involvement processes, 28
    Environmental Facility International Waters
        (GEF IW), 28
    warming, 8, 540, 546, 548, 578
    water disasters, 463
Goal-seeking
    behavior, 432, 434
    family, 432, 433
Goodness of fit, 285
Goods, 48, 207, 381, 387, 388
Governance model, 22–23, 26
    characteristics of, 26–27
    good governance processes, 23
        coherency, 23
        creating objectives, 23
        creating policies, 23–24
        prioritization, 23
    principles, 23
    responsiveness, 25
    sound information, 25
    stakeholders participation, 26
        importance of participation, 26
    transparency, 25
        recommendations, 25
Government institution, 527
Grass, lined channel, 272, 273
Gravity, 54, 56, 138, 246, 270, 396
Gray water, 14, 18, 211, 403, 413
Green and Ampt Model, 60, 62, 579
Greenhouse
    effect, 8
    gases (GHGs), 5, 8, 539
        emission reduction, 547–548
Grid, 397, 553
Gross, 11, 193, 236
Gross national product, 193
Groundwater, 5, 131, 214, 226, 311, 404, 447, 450, 454,
        541–542
    intentional discharges into aquifers, 332–333
    recharge, 18, 214, 330
    unintentional discharges into aquifers, 331–332
    urbanization effects on, 330–333, 330
    withdrawal, 454–456
Growth index (GI), 69
    estimation of, 69
Guideline, 48, 473, 477, 511, 526, 529
Gumbel distribution, 285–286, 286
Gutter hydraulic capacity, 265
Gutters, 261–265
    with composite cross slopes, 262, 263–264
    hydraulic capacity, 265
    triangular section, 261–262
    V-shaped sections, 262, 265

H

HADCM3 model, 552, 553
HADCM3 predictors, 574
Hard-limit transfer function, 347

Hardy Cross method, 159
    procedure, 160–162
Hazard rate, 487, 488
Hazen–Williams equation, 148, 150, 151, 158–160, 499.
        See also Darcy–Weisbach equation
    head loss calculation, 149
    Nomogram, 150
    sources of error in, 148
HDI. *See* Hybrid drought index (HDI)
HDPE. *See* High-density polyethylene (HDPE) pipes
Head, 143, 144, 145–151, 152, 275, 276, 282, 514
    losses, 149, 154, 159, 176
Heat, 8–9
HEC, 79
HEC-HMS. *See* Hydrologic Modeling System (HEC-HMS)
Heterogeneity, 5, 224, 331
Hidden
    layer, 346, 347, 353, 354, 355, 358
    variables, 346
High-density polyethylene (HDPE) pipes, 275
High Performance Systems Inc., 429
Holistic Paradigm, 32, 33, 35, 37, 50, 185, 68
Holtan method, 68
Homogenous, 110
Horton Method, 62, 67
HSPF. *See* Hydrologic—Simulation Program
        FORTRAN model
Hybrid drought index (HDI), 518–523
    algorithm, 519
    drought severity classification, 520
Hydraulic, 53, 144–151, 169, 170, 248, 249–251, 256, 265,
        272, 275, 302, 393, 496, 507–510, 524
    conductivity, 60, 61, 255
    grade line, 81, 139, 140, 144, 153
    head, 249, 304
    radius, 148, 166, 249, 261, 270, 272, 281
Hydraulics in distribution system
    energy equation, 144
    energy grade line, 144
    EPANET, 170
        gradient algorithm, 170
    flow rate calculation, 145
    flow velocity calculation, 145
    head loss calculation, 145, 148, 149
        EPANET, 170
    hydraulic grade line, 144
    minor head loss, 149
        coefficients of, 151
        EPANET, 170
    pump head calculation, 148
    system loss calculation, 145
    total head calculation, 145
    turbulent flow, 145
        friction faction factor, 145
        Moody diagram, 146
Hydrocarbons, 318
Hydrograph, 2, 4, 5, 7, 79, 87–104, 115, 118, 119,
        253, 256, 306
Hydrologic, 1, 6–7, 73, 82, 102, 115–119, 256, 284–285,
        291, 311–315, 507–510, 524, 540, 546
    Cycle, 1–5, 9, 11, 205, 311, 540
    different elements of, 3
    evaporation, 2
    infiltration, 2

Index

MODELING System (HEC-HMS), 115
  analysis of meteorological data, 117
  application of, 115
  modeling basin components, 116
  parameters estimation, 117
  rainfall–runoff simulation, 117
  starting the program, 118
precipitation, 2
runoff, 2
Simulation Program FORTRAN (HSPF) model, 325, 326
temperature, 2
transpiration, 2
urbanization effects on, 311–315
Hydrological
  drought, 465
  risk, 284, 285, 289, 306
  soil groups, 73
Hydrological and hydraulic risk, 507–510
Hydrologic MODELING System (HEC-HMS), 115
  analysis of meteorological data, 117
    precipitation methods, 117
  application of, 115
  modeling basin components, 116
    basin model screen, 116
    loss rate methods, 116
    transform methods, 116
  parameters estimation, 117
  rainfall–runoff simulation, 117
    control specification screen, 117
  starting the program, 118
    data input, 119
    data output, 119
    project definition screen, 118
Hydropower, 183, 185, 383, 446, 547
Hyetograph, 63, 64, 66, 77, 79, 103, 117, 122, 124, 292
Hypolimnion, 329

I

ICP. *See* Instrumental cathodic protection (ICP)
ICT. *See* Information and communication technologies (ICT)
ICWE. International Conference on Water and the Environment (ICWE)
IDF. *See* Intensity–duration–frequency (IDF)
IDNN. *See* Input-delayed neural network (IDNN)
ILI. *See* Infrastructure leakage index (ILI)
Impervious
  area, 53, 54, 72–76
    depression storage intensity vs. time for, 72
    directly connected, 73
    hydraulically connected, 53
  cover (%), 54
  surfaces, 72, 251–255, 311–315, 469. *See also* Imperviousness
    overland flow in, 251–254
Imperviousness. *See also* Impervious—surfaces
  effects on runoff and infiltration, 314
  and stream quality, relationship between, 327–328
Improvement, 85–86, 416, 417, 418–421, 463
Incremental, 31, 78, 186, 225, 368, 482
Indirect
  impacts, 315, 543, 544, 579
  reuse, 213, 403
Industrial sector, 445–446
Inequality, 277, 278, 361, 362
Infiltration, 1–5, 7, 12, 13, 14, 18, 54, 57–61, 245, 246, 291
  capacity, 59–71, 73, 119, 122
  imperviousness effects on, 314
  rate, 2, 4, 58–61, 63, 66, 68, 70–73, 400
  trenches, 2
Infiltration capacity estimation, 59, 63
  Green and Ampt model, 60
    average soil parameters for, 62
    ponding time (tp), 61, 62
    subsurface moisture, 60
  Holtan method, 68
    advantage of, 69
    factor A estimate, 69
    rainfall losses estimation using, 70
  Horton method, 62
    disadvantage of, 67
    f_f, 69
    Horton infiltration capacity, 63
  Modified Horton method, 67
    infiltration loss determination using, 68
  Richard equation, 59
  simple infiltration models. *See* Φ-index method
    Kostiakov equation, 71
    Phillip equation, 71
    two-parameter equation for, 71
Information and communication technologies (ICT), 28
Information technology (IT), 27
Infrastructure
  development, 12, 134, 381, 393, 407, 421, 422
  facilities, 405, 533
  interaction, 403
  leakage index (ILI), 202
    current annual real losses (CARL), 202
    unavoidable annual (UARL), 201
Infrastructure, of urban water, 393
  development, financing methods for, 407
    exactions and impact fee-funded systems, 407–408
    service charge-funded system, 407
    special assessment districts, 408–410
    tax-funded system, 407
  SIM capability improvement, opportunities for, 419
    continuous inspection devices, 421
    intelligent systems, 421
    mobile in-line inspection systems, 419
    mobile nonintrusive inspection (MNII) systems, 419–421
  SIM improvements, public benefits from, 418
  sustainable development, 410–412
    environmental performance assessment, 413–414
    life cycle assessment, 414–415
    SDI development, using life cycle approach, 415–416
    technologies selection, 412–413
  urban drainage systems
    and solid waste management, interactions between, 405–406
    and wastewater treatment systems, interactions between, 404–405
  urban water cycle
    and urban infrastructure components, interactions between, 402–403

Infrastructure, of urban water (*continued*)
  waste water treatment system, interactions with, 403–404
  urban water infrastructure
    and urban transportation infrastructures, interactions between, 406
    waste water collection and treatment, 400–402
  water
    and wastewater treatment systems, interactions between, 404
  water distribution system, 355–359
  water mains, SIM of, 416–418
  water supply, 393–395
    and wastewater collection systems, interactions between, 404
Initial abstraction, 76, 77, 119, 121
Inlets, 265–267, 276
  control flow in culverts, 277–279, 280
  location of, 267
  time, 83
Input-delayed neural network (IDNN), 354–357. *See also* Neural network; Temporal neural network
Input-delayed recurrent neural network, 358
Instantaneous unit hydrograph (IUH), 98
Institutional
  organization, 241, 295, 468
  role, 472, 527
Instrumental cathodic protection (ICP), 421
Intangible, 294, 383, 466
Integrated
  Resource Planning (IRP), 225, 545–547
  Urban Water Management (IUWM), 1, 45, 393, 407
  Water Resource Management (IWRM), 183
Integrative solutions, 441
Intelligent, 419, 421, 482
Intensity–duration–frequency (IDF), 79–81, 83, 288, 289, 292
Interbasin water transfer, 20, 390
Interception, 57–58, 59
  storage (IS), 54, 57, 58
Interest rate, 383, 384, 398, 399, 520
Intergovernmental Panel on Climate Change (IPCC), 539
Internal rate of return (IRR) method, 385
International Commission on Large Dams, 394
International Conference on Water and the Environment (ICWE), 205
  Dublin principles, 205
International Water Association. *See* IWA
International Waters Learning Exchange and Resource Network (IW: LEARN), 27
Intersection, 146, 265, 267, 282, 378, 380, 494, 495
  of fuzzy sets, 378
IPCC. *See* Intergovernmental Panel on Climate Change (IPCC)
IRP. *See* Integrated Resource Planning (IRP)
IRR. *See* Internal rate of return (IRR)
Irrigation efficiency, 229, 232
IS. *See* Interception—storage (IS)
IT. *See* Information technology (IT)
IUH. *See* Instantaneous unit hydrograph (IUH)
IUH to UH Conversion, 100
IUWM. *See* Integrated—Urban Water Management (IUWM)

IW: LEARN. *See* International Waters Learning Exchange and Resource Network (IW: LEARN)
IWA (International Water Association), 200
IWRM. *See* Integrated Water Resource Management (IWRM)

J

Japan, hydraulic infrastructure in, 394
Jump, 139, 195, 277
Junction chambers, 269

K

Kinematic wave, 55, 56, 83, 102, 116, 302
Knowledge, 20, 45, 478, 482, 483

L

Lag time, 4, 57, 95–97, 306, 313, 134
Lake(s), 245
  urbanization impacts on, 329
Laminar flow, 56, 146, 147, 249, 250, 251, 252
Land
  resources allocation, conflict resolution model for, 453–454
  use, 2, 18, 38–44, 48, 50, 54, 69, 73, 74, 82, 83, 316, 317, 319, 320
  use management, 19, 38, 39, 40, 41, 42, 44
  use planning, 2, 18, 19, 38, 48, 50, 468
Large-scale centralized systems, development of, 412
LARS-WG model. *See* Long Ashton Research StationWeather Generator (LARS-WG) model
LCA. *See* Life cycle assessment (LCA)
LCP. *See* Least cost planning (LCP)
Leak, 199–204, 332, 404
  control, 203
  detector, 203
  noise correlator, 203
Leakage
  control, 189, 201–204, 228, 229, 232
  detection, 199, 203
Least cost planning (LCP), 224
Least mean squares, 352
Length, 149, 401, 419, 568
Levees, 312
Levelized unit cost (L), 225
Leveragepoints, 431
Life cycle assessment (LCA), 10, 393, 414
  application of, 11
  defined as, 10
  functions of, 10
  phases in, 10–11
  of urban infrastructures, 414–415
Life cycle impact assessment, 415
Life span, 20, 285, 384, 385, 387, 401
Lifetime, 342, 384, 411, 460, 486
Limited area models, 555
Linear
  decline, 431, 432
  family, 431, 432
  growth, 431, 432
  programming (LP), 360–365, 389, 410
    simplex method, 364–365
  transfer function, 347

Index

Linearization, 366
Load-resistance concept, 495–496
Local
   authorities, 469, 527
   head losses, 149, 469, 527
   microclimate, 2, 5
Lodz sewerage system, 299, 300, 301. *See also* Wastewater—management, in urban drainage system
   problems in, 300
Log, normal, 108, 498, 499, 517
Log-sigmoid transfer function, 347
Long Ashton Research Station Weather Generator (LARS-WG) model, 567–574
Looping pipes, 157
Loss
   coefficient, 149, 151, 156, 157, 158, 177, 281
   function, 524
LP. *See* Linear—programming (LP)
Lumped, 5, 81, 502

M

Management strategy, 42, 237, 475, 503
Manholes, 268, 269
Manning's
   coefficient, 250, 251, 252
   roughness coefficient, 56, 89, 90, 123, 251, 262, 271, 272
Marginal, 20, 197, 198, 208, 225, 241
Margin of safety (MS), 499
   factor of safety (FS), 501
Mass conservation equation, 169, 550
Master Plan, 31–32, 290
Maximum contaminant levels (MCL), 171
MCL. *See* Maximum contaminant levels (MCL)
MDGs. *See* Millennium development goals (MDGs)
Mean sea-level pressure, 556
Mean value first-order second moment (MFOSM) method, 501–502
Measure of uncertainty, 319, 506
Mechanical mistake test, 430
Membership function, 377–381, 391
Metals, 317
Meteorological drought, 465
Methane, 8, 173, 539
MFOSM. *See* Mean value first-order second moment (MFOSM) method
Microbial pollution, 318–319
MILI system. *See* Mobile in-line inspection (MILI) systems
Millennium development goals (MDGs), 127, 183
   in water, 127
Minimum
   attractive rate of return, 385
   cut set, 492
      for water distribution network, 493
Minor head losses, 149, 151, 154
Mitigation, 477, 510, 529, 539
Mix, 4, 7, 9, 141, 172
Mixed integer linear programming, 360
MLP. *See* Multilayer perceptron network (MLP)
MNII systems. *See* Mobile nonintrusive inspection (MNII) systems
Mobile in-line inspection (MILI), 419, 421
Mobile nonintrusive inspection (MNII) systems, 419, 421
Model, 1, 5, 17, 18, 21, 22, 26–27, 38, 60–72, 99, 119, 139, 163–165, 168–169, 170–172, 175, 227, 288, 325, 326, 428–430, 449, 450–454, 504, 557, 558, 559–576
Modeling, 72, 115–119, 162–175, 300–304
Modified
   Horton method, 67
   water system, 469
Momentum equation, 55, 249, 550
Monitoring, 29, 45, 171–172, 174, 511, 528–529
   network, 204
Monte Carlo simulation, 359. *See also* Simulation—techniques
Moody diagram, 146, 147, 148, 153
MS. *See* Margin of safety (MS)
Multicriteria decision making (MCDM), 373
Multicriteria optimization, 373–377. *See also* Optimization—techniques
Multilayer perceptron (MLP) network, 348–352, 519, 579. *See also* Neural network
Multipurpose reservoir operation, resolving conflicts over, 444–445
Multivariate, 358, 558
Muskingum–Cunge method, 256–258, 259, 305
Mutation, 369–371
Mutually exclusive, 345, 389, 398

N

Nash
   bargaining solution, 442, 444
   bargaining theory (NBT), 447
   product, 231, 441, 443, 447, 450
   solution, 446
   theorem, 228, 229
National Center for Atmospheric Research, 539
National Fire Protection Association (NFPA), 140
Natural disasters, 404, 463, 466
Natural/intrinsic uncertainty, 504
Natural wastewater treatment systems, 413
Natural water system, 469, 470
NBT. *See* Nash—bargaining theory (NBT)
NDE technologies. *See* Nondestructive evaluation (NDE) technologies
Negative effect, 9, 311, 406, 411
Network of responsibilities, 529
Neural network, 339, 343, 346, 347, 352, 519
   artificial, 346–358
   input-delayed, 354–357
   recurrent, 353–354
   temporal, 352–358
   time delay, 357–358
   time delay recurrent, 358
Newtonian Paradigm, 18, 32–37, 50, 184
NFPA. *See* National Fire Protection Association (NFPA)
Nitrate, 317, 448–452
Nitrite, 318, 400
Node, 143
Nondestructive evaluation (NDE) technologies, 419, 420
Nondominated, 374, 376, 443
Nonlinear programming (NLP), 360, 366
Non point sources, 325

Nonrevenue water (NRW), 189, 191, 200–203
    to reduce, 203–204
Nonstructural, 17, 18, 19, 212, 294, 295, 466–468, 478
Normal, 251, 343, 344, 377, 378, 435, 525, 561
Northern California storms and floods of January 1995, case study, 480
    flood characteristics, 481
    response, 481
NRW. *See* Nonrevenue water (NRW)
Numerical method, 169
Nutrients, 296, 317, 329, 330, 406, 413

O

Objective function, 41, 117, 360, 361, 364, 366, 369, 373, 374, 443
Object oriented, 429, 447, 452
    program, 339, 429, 447, 452
    programming, 429, 452–456
    simulation, 429, 452
OLS. *See* Ordinary least squares (OLS)
Onset, 480, 545
Open channel, 54, 57, 90, 246–249, 269–270, 332
Open-channel flow, 247–251
    channel section, elements of, 248
    classification of, 249
    hydraulic analysis of, 249–251
Open drainage channel design, 269–275
    grass-lined, 272–275
    unlined channels, 270–271
Operational uncertainty, 504
Operation and maintenance, 205, 387, 412
Opportunity
    cost, 206
    seeking, 474
Optimal risk-based hydraulic design, 509
Optimization
    model, 40–42, 237, 339, 374, 408, 428
    techniques, 339, 359–377
        dynamic programming, 366–369
        evolution algorithms, 369–373
        linear programming, 360–365
        multicriteria optimization, 373–377
        nonlinear programming, 366
Ordinary least squares (OLS), 197
Organic matter, 7, 293, 294, 297, 317, 318, 329, 406
Organization chart, 529, 530
Oscillation family, of time paths, 433–434
Outlet control flow, in culverts, 279
    full-flow conditions, 281, 281
    partly full-flow conditions, 282
Overflow, 294, 469
Overland flow, 55, 251–256, 302
    on impervious surfaces, 251–254
    on pervious surfaces, 255–256
Overshoot and collapse system response, 435

P

Palmer drought severity index (PDSI), 518
Paradigm shift, 17, 32, 37, 184, 188, 233, 533
Parallel connections, 490
parallel pipe system, 157–159. *See also* Looping pipes
Parameters, 56–57, 71, 72, 81, 84, 87–89, 91, 97, 100, 117, 257, 269, 270, 342, 377–381, 401, 430–431, 457, 504, 558, 561
Participation, 20, 26–27, 478–479
Partnership, 478, 511, 546
Path enumeration method, 492, 495
Pathogen, 133, 316, 318, 330
Pavement drainage inlets, 265–267
    location of, 267
P_d. *See* Direct runoff (P_d)
PD. *See* Population—density (PD)
PDF. *See* Probability density function (PDF)
PDSI. *See* Palmer drought severity index (PDSI)
Peak discharge
    adjustment Factors for Urbanization, 105
    formulae, 84
    ratio of urban to rural, 106
    urban effects on, 104
Peaking factor, 95. *See also* Conversion constant
Pending time (t_p), 61
Performance indicators, 414
Pervious surfaces
    nonhydraulically connected, 53
    overland flow in, 255–256
    unconnected, 73
Φ-index method, 70
    factors affect, 58
    representation of, 71
Phosphate, 225, 317
Phosphorus, 317–319, 321, 330, 332, 411, 412, 413
Physical impact, 4, 315, 326, 328
Physical water system, 427
Pipes, 7, 148, 158, 168, 170, 303
Pipes in series, 154–157. *See also* Compound pipeline
Pipeline analysis and design, 151
    parallel pipe system, 157–159
    pipe networks, 159
        EPANET, 170
        Hardy Cross method, 159
        linear theory method, 159
        Newton–Raphson method, 159
    pipes in series, 154–157
    pipe size determination, 155–157
Pipe network, 159–162, 170
Plain, 362, 363, 447, 453, 455, 456
PNEC. *See* Predicted No Effect Concept (PNEC)
Point sources, 5, 325, 330, 411
Poisson, 340, 344, 345
    distribution, 340–341, 345
    process, 345
Policy
    optimization, 41
    tests, 430, 431
Political commitment, 470, 473, 474, 476, 477, 482
Pollutant export, 322, 323
Pollution, 468
    control, 163, 166, 211, 447–452
    diffuse, 7
    load, 448, 451
    source, 7, 311, 316, 319, 331, 333
    stormwater, 7
        sources contributing in, 7
Ponding time (t_p), 61

Index

Ponds, 245
Population
    density (PD), 53–55, 304, 311, 323, 324, 334, 411, 413
        vs. impervious cover (%), 55
        in imperviousness estimation, 54
    estimation, 129
    projection, 128, 193
Porous pavement, 246, 247
Positive-sum, 441
Powersim Software AS, 429
Predicted No Effect Concept (PNEC), 294
Prediction, 193, 518, 548–559
Preliminary, 292, 318, 328, 544, 557
Preparation, 172, 299
Preparedness, 475, 477–478, 479
Present value, 11, 224–227, 365, 383–386, 410
Present worth, 383, 386, 398, 399, 401, 422
Present Worth method, 385
Pressure-regulating valve (PRV), 143
    functions of, 143
Pressure, State, and Response (PSR) model, 38
Prestressed concrete cylinder pipelines, 418
Price–restrictions interaction, 196
Primary, 171, 192, 241
Private, 21, 382–383, 533
Probability density function (PDF), 496
Probability distribution, 108, 285, 286, 509
Production, 2, 296, 370, 387–388
    functions model, 387–388
Project lifetime, 384
Proofing, 232
PRV. *See* Pressure-regulating valve (PRV)
PSR model. *See* Pressure, State, and Response (PSR) model
Public, 8, 20, 22, 27–28, 29, 334, 407, 413, 418, 478–479, 526
    involvement benefits, 478
Public participation
    absence of, 27
    methods for, 478–479
    overcome
        approaches to, 27, 29
        decision support systems (DSS) to, 28
        information technology approaches to, 27
        international approaches to, 28
    in UWDM planning process, 478–479
        community activities, lessons on, 479
        public involvement, benefits of, 478
        public participation, methods for, 478–479

Q

QTest, 569
Qualitative, 212, 400, 528
    aspects, 7, 128, 474
Quality, 4, 8, 38–39, 132, 133, 162–175, 241–246, 294, 303, 315–326, 400, 449
    management, 137, 530
Quantitative, 400, 474, 480, 528

R

Rainfall, 284–289
    abstractions, 57
    design runoff, 287–288
    design-storm-duration and depth, 289
    frequency analysis, 285–287
    hydrological risk, 285
    intensities (i), 79
    losses, 54
    measurement, 79
    return period, 284–285
        design, 288–289
    spatial and temporal distribution of, 289
Rainfall abstractions
    combined loss models, 73
    depression storage
        empirical estimates of, 72
        factors affecting, 72
        intensity vs. time for impervious area, 72
    infiltration, 58
        capacity or rate, 59
        curve, 65
        factors affecting, 58
    interception storage, 57
        estimation of, 58
    main sources of, 57
Rainfall, excess
    calculation, 57
    computation of, 66
    depth of, 77
Rainfall-harvesting system
    collecting surfaces in, 131
    storage tank volume in, 132
    water quality in, 132
    water usage possibilities, 132
Rainfall measurement, 79
    intensity–duration–frequency curves, 79
        attributes of, 80
        disadvantages, 80
Rainwater collection system, 9, 132, 222, 235
Rainwater harvesting, 18, 50, 131, 132, 214–223, 533
Rainwater roof catchment systems (RRCS), 214
    advantages of, 214–215
    designing
        dry season demand versus supply, 216
        mass curve analysis, 216–220
        mass curve with dimensionless constant analysis, 220–221
    disadvantages of, 215
    economic feasibility of, 216, 221–223
    support services needed for user, 223
Random time to repair (TTR), 502
Random variable, 340, 341, 343, 484, 505
Rate constant, 166, 170
Rational method, 80–83, 104, 261, 267, 291, 429
RCM. *See* Regional—climate models (RCM)
RCP. *See* Reinforced concrete pipes (RCP)
Real losses, 201
    classification of, 201
    managing, 203
    unavoidable annual (UARL), 201
Real, time operation, 525
Recession limb time, 94, 95
Reclaimed wastewater, 135, 184, 225
Recovery analysis, 504
Recreation, 214
Recurrent neural network (RNN), 353–354, 358. *See also* Neural network; Temporal neural network

Recursive dynamic programming, 11, 340, 360, 366, 367, 368, 390, 423, 556
Recursive function, 367, 368
Recycling and reuse, separation for, 412–413
Reference behavior
  pattern, 430
  test, 430
Regional
  analysis, 465
  climate models (RCM), 554–555
  plans, 41, 534
Regressive programs, 394
Regulation, 48, 225, 295, 299, 467
Rehabilitation, 130, 475
Reinforced concrete pipes (RCP), 275
Relative precision, 395
Reliability, 142, 192, 411, 479, 484–502, 509, 510, 514, 517, 523–526
  AFOSM method, 502
  analysis, 495–496
  assessment
    path enumeration method, 492–495
    state enumeration method, 491–492
  based design, 9, 463, 464, 509, 534
  design by, 523–524
  direct integration method, 497–499
  estimation, 486
  indices, 496–497
  margin of safety (MS), 499–501
  mean value first-order second moment (MFOSM) method, 501–502
  water users, 486
Reliable
  supply, 410
  systems, 411
Relief, 143, 294, 466, 475, 482, 508, 510
Remote
  field eddy current, 419
  sensing, 478
Removal, 1, 191, 242, 244, 245, 298, 393, 411–413, 421, 445
Repair, 191, 203, 204, 502
  probability, 502
Repairable urban water systems, 502
Repeatability assumption, 385
Replacement, 190, 381, 382, 417
Requirement models, 195
Reservoir(s)
  operation, 23, 444
  storage, 368, 390, 391, 446, 447, 543
  urbanization impacts on, 330
Resiliency, 503, 517
Resilient flood risk management strategy, 503
Resolution, 47–48, 427, 440, 443–447, 449, 452, 453, 454, 455
Resources, recycling and reuse of, 411
Resource-supply function, 388
Retention, 290, 291
Return period, 80–87, 260, 284–285, 288–289
Revenue, 20, 200, 201, 384, 407, 438
Reynolds number, 146, 166, 249
Risk
  analysis, 507
    and reliability analysis concept, 524
  risk management and vulnerability, 508–509
  risk reduction, incentives and constituencies for, 510
  water resources systems, risk-based design of, 509
  assessment, 46
  based design, 509, 534
  management, 294, 469
  management and vulnerability, 508–509
  reduction, incentives and constituencies for, 510
River, 29, 39–41, 228, 329–330, 452, 458, 467, 574–576
RNN. See Recurrent neural network (RNN)
Robustness test, 430
Root mean squared error (rmse), 348
Roughness, 90, 146, 148, 270–272, 499
RRCS. See Rainwater roof catchment systems (RRCS)
R_S. See Spearman correlation coefficient (R_S)
Rules, 34, 162, 302, 479, 485, 530
Runoff, 1–7, 53–58, 74, 76, 216
  impact on receiving waters
    categories of, 315
    pollutant sources and constituents, 316–319
  imperviousness effects on, 314
  coefficients, 81–84, 216, 313, 321, 322
  for rational formula vs. hydrologic soil group, 82
  for rational formula vs. slope range, 82
  pollution, 315, 316, 318

S

Safe yield, 130
Salts, 549
Salvage value, 386, 387
Sampling, 172, 174, 319, 359
Sand filter, 244
Sanitation, 188, 295, 296–297, 396, 406. See also Ecological Sanitation (EcoSan) approach
Satellite, 297, 421, 478, 481
SBR. See Sequential batch reactors
SCADA. See Supervisory Control And Data Acquisition (SCADA)
Scenario tests, 431
SCS method. See Soil Conservation Service (SCS) method
SCS UH, 93, 96, 97
SD. See Sustainable development (SD)
SDI. See Sustainable development index (SDI)
Sea surface temperature, 556
Security, 17, 23, 163, 175, 206, 234, 464, 468, 472, 483
Sediment basins, 242, 243
Sensitivity tests, 431
Sequential batch reactors (SBR), 297
Service charge-funded system, 407
Severity, 5, 331, 464, 518, 519, 520, 522, 540
  index, 480, 518
Sewer networks, 405–406
Shape, 46, 261, 265, 275, 276, 280, 434, 435
  factor, 97
SIM. See Standard—integrity monitoring (SIM)
Simple
  empirical method, 320–323
  Infiltration Models, 70
  method, 159, 166, 319–323
Simplex method, 364–365, 366
Simulated annealing, 360, 372–373

Index

Simulation, 117, 170, 171, 288, 292, 339–359, 372–373, 428, 429
  annealing, 372
  model, 171, 292, 339, 541
  techniques, 339–359
    artificial neural network, 346–358
    Monte Carlo simulation, 359
    stochastic process, 344–345
    stochastic simulation, 339–344
Sinclair–Dyes Inlet TMDL Project, 326
Single payment
  compound amount, 383
  present worth, 383
Sinking fund, 383
Siphones, 269
Site analysis, 569
Situation analysis, 463, 473, 474, 481, 482, 483, 526
  urban disasters, 483
  of urban water disaster management, 481
    approach, 482
    data collection, 482–483
    objectives, 482
Slack variables, 364
SLAMM. *See* Source Loading and Management Model (SLAMM)
Slope, 139, 243, 244, 249, 250, 261–266, 269–270
Small-scale systems, 413
Snow, 1, 6, 8, 213, 302, 464, 466, 520, 541, 542
Snowfall, 541
Snowmelt, 1, 541, 542
Social, 12, 45, 134, 382, 388, 412, 413, 415, 427, 463–466, 468–472, 483, 484
  choice, 388, 442
Socioeconomic
  drought, 465
  systems, climate change impacts in, 544
Soft paths, 187
Soil
  erosion, 325
  moisture, 60, 75, 288, 540–541
  permeability, 73
Soil Conservation Service (SCS) Method, 73
  hydrological soil groups, 73
    rate of rainfall loss estimation, 79
  runoff curve number, 73
  runoff equation, 76
  variables in, 76
Solid waste, 242, 402–406
Source Loading and Management Model (SLAMM), 325, 326
SPARROW. *See* Spatially Referenced Regression on Watershed Attributes (SPARROW)
Spatial, 5, 87, 289, 302, 352, 525, 554, 555–556, 557
  variability, 554
Spatially Referenced Regression on Watershed Attributes (SPARROW), 325
Spearman–Conley test, 111–115, 120
Spearman correlation coefficient (R_S), 111
  critical values for, 112
Special assessment districts, 408–410
Specific storage, 4
SPI. *See* Standard—precipitation index (SPI)

SRBLUM-DSRD-DSS. *See* Sustainable urban land use management DSR dynamic decision support system (SRBLUM-DSRD-DSS)
SRI. *See* System—readiness index (SRI)
S-shaped
  family, 434–435
  growth, 434
Stakeholders, 20, 21, 26–27, 46, 440–442, 477–479, 526, 546
  participation, 24, 28, 46, 476, 477, 479
Standard, 41, 48, 137–138, 163, 192, 395, 414, 568
  integrity monitoring (SIM), 393, 416
    capability improvement, opportunities for, 419–421
    public benefits from, 418
    of water mains, 416–418
  normal distribution, 343–344
  precipitation index (SPI), 480, 518, 521
State
  enumeration method, 491–492
  incremental, 368
  variable, 360
Static network, 348, 352, 354
Stationarity, 557, 558
Statistical, 540
Statistical downscaling, 556–559
Statistical downscaling model (SDSM), 559, 560, 561, 575, 576
  regression-based methods, 558
  strengths and weaknesses, 558
STELLA, 43, 429
Stethoscope, 203
Stochastic, 339, 344–345, 372, 502, 504, 524, 558, 568
  process, 344, 345, 568
  simulation, 339–344. *See also* Simulation—techniques
Stocks, 429, 430
Storage
  coefficient, 100, 101, 123, 451, 458
  facilities, in urban drainage system, 289–292
  tank, structure of, 428, 429
  Treatment, and Overflow Runoff Model (STORM), 325
  volumes, sizing of, 291–292
Storm
  frequency and intensity, 541
  peak flow estimation, 83
STORM. *See* Storage—Treatment, and Overflow Runoff Model (STORM)
Stormwater, 7, 14, 53, 259–275, 289–292, 300–302, 311–313, 315–320, 334, 401, 405, 469
  drainage system, 241–246
  Infrastructure, 225, 317, 401, 463
  infrastructures cost, 401
  Management Model (SWMM), 325
Strategic goals, 473
Strategy, 30–31, 33, 38–44, 190, 292, 359, 472
  choice, 473
Stratification, 7, 329
Stream
  bank erosion, 312, 326
    vs. healthy stream bank, 328–329
  flow, 311, 312, 314, 315, 327, 481, 486, 541
  slope
    urbanization effects, 313, 314
Street gutters, 241, 259, 261, 289

StromNET, 300–304
　applications of, 303
　detention pond modeling capabilities, 301–302
　hydraulic modeling capabilities, 301
　hydrology modeling capabilities, 301
　program output, 303–304
　water-quality modeling capabilities, 302
Structural condition assessment, 417
Structure, 11, 17, 18, 19, 288, 294, 416–419, 421, 441, 466, 509
St. Venant continuity and momentum equations, 249
Subsurface flow, 3, 99, 288
Subsurface moisture, 60
Sum squared error (SSE), 348, 353
Supervisory Control And Data Acquisition (SCADA), 175
Supply
　cost, 206
　curve, 193, 217
　Management, 1, 10, 21, 23, 32, 50, 184, 233, 474, 541
Support of a fuzzy set, 378
Surface
　sewer systems, 268–269
　water severity index, 480
　water supply index (SWSI), 518
　Water Treatment Rule (SWTR), 163
Surplus, 207, 364, 453
Sustainable design, 393
Sustainable development (SD), 11, 17, 31, 38, 410–412
　environmental performance assessment, 413–414
　life cycle assessment, 414–415
　SDI development, using life cycle approach, 415–416
　technologies selection, 412–413
Sustainable development index (SDI), 413, 414
　development, using life cycle approach, 415–416
Sustainable urban land use management DSR dynamic decision support system (SRBLUM-DSRD-DSS), 42
Sustainable water resources management, 50
Swales, 244
SWMM. *See* Stormwater—Management Model (SWMM)
SWSI. *See* Surface—water supply index (SWSI)
SWTR. *See* Surface—Water Treatment Rule (SWTR)
Symptoms, 13, 472, 481
System
　analysis, 427
　definition, 428
　dynamics, 41, 43, 185, 339, 428–440, 453, 457, 482
　failure, 287, 469–470, 472, 484, 489, 492, 503, 528, 532
　　built system, 469
　　disaster assessment caused by, 528
　　modified system, 469
　　natural system, 469
　input volume, 200
　readiness, 510–515, 517, 529, 531
　readiness index (SRI), 511, 513
　　determination, 512–513
　　flowchart, 513
　thinking, 38, 428
　thinking paradigm, 428
System dynamics, 41, 43, 185, 339, 428–440, 453, 457, 482
　case studies, 447
　　model for water transfer, 452–456
　　water pollution control conflict resolution, 447–452
　and conflict resolution, 440–441
　　models, 442–447
　　process, 441
　　systematic approach, 441–442
　examples, 435–440
　modeling, 429–430
　　design and test policies, 431
　　hypotheses, testing, 430–431
　　issue/system, defining, 430
　technologies work, making, 457
System readiness analysis, 528–531
　disaster assessment
　　on system failure, 528
　　on water shortage/drought, 528
　　on widespread contamination, 528
　mitigation guidelines, for, 529
　monitoring system, 528–529
　organization chart, 529–531

T

Tabu search, 360
Tapped delay line (TDL), 352. *See also* Neural network
Tariff, 49, 191, 207, 210, 211, 533
Tax-funded system, 407
t_c. *See* Time of concentration (t_c)
TCR. *See* Total—Coliform Rule (TCR)
TDL. *See* Tapped delay line (TDL)
TDM. *See* Time-driven method (TDM)
TDNN. *See* Time delay neural network (TDNN)
TDRNN. *See* Time delay recurrent neural network (TDRNN)
Technical performance, 411
Tehran
　aquifer, 447, 448, 449, 450, 451
　Health Department, 449–450
　Waste water Collection Project (TWCP), 448, 449, 451
　Water and Waste water Company, 449
　water pollution control, 447–452
TEIA. *See* Transboundary Environmental Impact Assessment (TEIA)
Temperature increase, 540
Temporal neural network, 352–358. *See also* Neural network
　adaptive linear filter, 352–353
　input-delayed neural network, 354–357
　recurrent neural network, 353–354
　tapped delay line, 352
　time delay neural network, 357–358
　time delay recurrent neural network, 358
Termination, 373
Terrestrial and aquatic ecosystems
　climate change impacts in, 543–544
Texture, 61, 73, 243
Theory, 486, 493, 495
TIA. *See* Total—impervious area (TIA)
Tie-set analysis, 492, 494, 535
Time–area UHs, 98
Time delay neural network (TDNN), 354, 357–358. *See also* Neural network; Temporal neural network
Time delay recurrent neural network (TDRNN), 358. *See also* Neural network; Temporal neural network
Time-driven method (TDM), 169
Time of concentration (t_c), 55–57, 83, 90, 97, 100, 251–253, 267

Time paths, of dynamic system, 431
  exponential family, 432
  goal-seeking family, 432–433
  linear family, 431–432
  oscillation family, 433–434
  S-shaped family, 434–435
Time series, 139, 195, 520–522, 550, 557, 568
Time-to-failure (TTF) analysis, 485, 502
  availability, 502
  failure and repair characteristics, 502
  unavailability, 502
Time to repair, 502
TOPSIS, 374
Total
  Coliform Rule (TCR), 163
  impervious area (TIA), 54
  maximum daily load, 319
  resource cost (TRC), 224
Toxic metals, 313
t_p. *See* Ponding time (t_p)
Tracer study, 172, 173, 174, 175
Training, 204–205, 346–347, 348, 473, 474, 475, 477, 527, 576, 578
  process, 346, 347, 348
Transboundary Environmental Impact Assessment (TEIA), 27
Transformation, 2, 32, 171, 325, 340, 400, 443
Translation hydrograph, 99
Transpiration, 2, 57, 313, 314, 540, 541, 542
Transportation infrastructures, 406
TRC test. *See* Total—resource cost (TRC) test
Treatment
  plant, 131, 172–173, 297, 298–299, 364, 400, 401, 406
  train, 17
Trend, 18, 33, 107, 540, 543, 544, 578
Triangular gutter, 261, 262, 267, 306
Triple bottom line criteria, 546
Tropical areas, 6, 297
TTF. *See* Time-to-failure (TTF) analysis
Turbine, 143, 144
Turbulent flow, 56, 146, 155, 250, 251, 252
Turnover, 245
TWCP. *See* Tehran—Waste water Collection Project (TWCP)

U

UARL. *See* Unavoidable annual real losses (UARL)
UASB. *See* Up-flow anaerobic sludge blanket (UASB)
UH. *See* Unit hydrographs (UH)
UHI. *See* Urban heat islands (UHI)
UH theory, 102
Unaccounted water, 188, 200, 202, 203, 228, 229
Unavoidable annual real losses (UARL), 201, 203
UNEP. *See* United Nations Environment Program (UNEP)
Uniform, 71, 144, 248, 251, 256, 261, 340, 341, 343, 344
  distribution, 340, 341, 535, 568
  series compound amount, 383
  series gradient, 383
Union, 26, 128, 223, 378, 380, 493, 495
Unit-area loading, 319, 320, 334
United Nations Environment Program (UNEP), 475
Unit hydrographs (UH), 2, 4, 5, 57, 87, 88, 89, 90, 98, 123
  25-min UH development, 97
  35 UH development, 96
  application of, 102, 103
  basis of UH theory, 102
  curvilinear, 94
  development
    basin features useful in, 89
    storm selection criteria in, 88
  espey 10-min
    calculation for an urban watershed, 91
    elements of the, 89
    estimated parameters for describing, 88
    parameters calculated as, 89
    values of ϕ for, 90
  gamma function, 96
    relationship of parameters, 97
  mass curve ratios, and, 93
  SCS UH
    and mass curve (S hydrograph), 93
  time–area, 98
    15-min UH development, 101
    advantages of, 98–99
    components of, 99
    parameters for, 100
    practical limitations of, 99
Universal Soil Loss Equation (USLE), 325
Unlined channel, 270, 272
Up-flow anaerobic sludge blanket (UASB), 297, 298
UQ_i. *See* Urban peak discharge (UQ_i)
Urban area
  drainage system
    characters of, 53
    main part of, 54
    time of concentration (t_c), 55–57
  impervious
    depression storage intensity vs. time for, 72
    directly connected, 73
    hydraulically connected, 53
  imperviousness estimation, 53
  peak runoff estimation
    rational method for, 81
  pervious surfaces
    nonhydraulically connected, 53
    unconnected, 73
  rainfall–runoff analysis in, 53
    flow rate Estimation, 55
  runoff estimation in, 53, 78, 81
Urban disasters situation analysis, 483
Urban drainage system, 54, 241, 242, 246, 265, 292–294, 300
  and solid waste management, interactions between, 405–406
  and wastewater treatment systems, interactions between, 404–405
Urban effect
  test of significance, 110
    Spearman–Conley test, 113–115
    Spearman test, 111, 113
Urban floods
  control management
    causes, 467–468
    principles of, 467
  evaluation of, 576–578

Urban heat islands (UHI), 8
   issues in
      global warming, 8
      high energy use, 8
      poor air quality, 8
      risks to public health, 8
   phenomenon, 8
   reduction strategies, 9
Urban infrastructure, 241, 300, 402–406, 407, 421, 422
Urbanization, 466
   effect on peak discharge, 104
   environmental impacts of, 311–334
      on groundwater, 330–333
      on hydrologic cycle, 311–315
      on lakes, 329
      physical and ecological impacts, 326–329
      on reservoirs, 329
      on rivers, 329
      on water quality, 315–326
   impact on water cycle, 4
   index of, 85
Urban peak discharge (UQ_i), 85
Urban planning, 13, 17, 29–32, 50, 241–246
   land and water strategy
      objectives considered in, 30
      targets in, 30
   master plans, 31
      statements covered in, 31
      uses of, 31
   regional and local strategies, 30
Urban storm water runoff, 468
Urban water assessment, 44, 45
   communication system, 46–47
   conflict resolution in, 47
   demand management in, 45
   environmental impact assessment, 45
   gauging importance in, 45
   importance, 45
   knowledge base for, 45
   monitoring importance in, 45
   objectives in, 45
   regulatory instruments in, 48
      direct controls, 48
      economic instruments, 49
      technological instruments, 50
   risk management, 46
   water allocation in, 47
Urban water cycle (UWC), 1, 2, 12, 14, 17, 402, 404, 576
   community-based enabling systems, 13
   facts about urban water, 13–14
   management, 12–14
   urban infrastructure components, 402–403
Urban water demand management (UWDM), 183–237, 464, 470–474, 476, 477, 479, 481, 527, 533
   as component of comprehensive DM, 472–473
   demand forecasting, 193
      parametric models in, 195
      regression analysis in, 194–195
   demand-side management in, 185, 186
   demand variation effect in, 192
   Dublin principles, 205
   end-use analysis in, 197
      survey of, 197
   estimation and forecasting, 193
   health factor in, 192
   integrated management in, 211
   planning process
      disaster preparedness policy, 477–478
      dynamic process, 476
      multisectoral approach, 476
      planning cycle, 473
      political and governmental commitment, 476–477
      public participation, 478–479
      stakeholder participation, 476
      strategic approach, 471–472
      strategy decisions, scope of, 472
   policy, legal, and institutional framework, 470–471
   poverty reduction and, 185
   resource combinations in, 227
   soft path approach, 188
      backcasting, 188
      bulk meter management system, 190
      ecological water requirements, 188
      increased end-use efficiency, 188
      supply reduce points, 188
      technology efficiency, 188
      water conservation, 188
      water saving technologies, 189
   strategic objectives of, 190–191
   supply management in, 184–185
      resource alternatives, 226
   sustainable water development in, 232
      decreasing urbanization effects for, 232
      demand management for, 232, 233
      determining resources availability for, 233
      efficiency task force for, 233
      metering for, 232
      recycling for, 232
      research and development for, 232
      stakeholders in decision-making for, 233
      system for data analysis for, 233
   water demand screening in, 192
   water reuse in, 211
      criteria in, 212–213
      direct, 213
      indirect, 213
      possibilities of, 213–214
Urban water disaster (UWD) management, 463, 470
   analysis, 528–531
   disaster indices, 483
      reliability, 484–502
      resiliency, 503
      time-to-failure (TTF) analysis, 502
      vulnerability, 503–504
   disaster management, guidelines for, 526
      disaster and technology, 526–527
      disaster and training, 527
      institutional roles in, 527–528
   disasters, sources and kinds of, 464
      drought, 464–466
      earthquakes, 470
      floods, 466–468
      system failure, 469–470
      widespread contamination, 468–469
   risk analysis, 507
      risk management, 508–509
      risk reduction, 510
      water resources systems, 509

situation analysis, 481
  developmental steps, 482–483
  urban disasters situation analysis, 483
societal responsibilities, 471
system readiness, 510
  disaster and scale, 523
  hybrid drought index, 518–523
  public concern, 526
  reliability, design by, 523–524
  supply reliability indicators, 524–526
  WDS readiness, evaluation of, 511–517
urban water engineering uncertainties, 504
  analysis of, 505–507
  implications and analysis of, 504–505
  measures of, 505
UWDM, 470
  component of comprehensive DM, 472–473
  planning cycle, 473
  policy, legal, and institutional framework, 470–471
  strategic approach, 471–472
  strategy decisions, scope of, 472
water disaster management strategies, 473
  disaster management experience, 475
  drought disaster management case study, 479–480
  initiation, 476–479
  management of storms and floods case study, 480–481
Urban water drainage system, 241
  Best Management Practices for, 242–247
  channel flow, 256–259
  components of, 259–275
    design considerations, 260–261
    flow in gutters, 261–265
    open drainage channel design, 269–275
    pavement drainage inlets, 265–267
    surface sewer systems, 268–269
  culverts, 275–284
    inlet control flow, 277–280
    outlet control flow, 279–283
    protection downstream of, 283–284
    sizing of, 282–283
  flooding in, 292, 293, 294–295
    control procedures, 295
  open-channel flow, 247–251
  overland flow, 251–256
  risk issues in
    chemicals discharge, 292, 293
    DO depletion, 292, 293
    flooding, 292, 293
  sanitation, 295
  storage facilities, 289–292
  surface drainage design, 284–289
  wastewater management, 295–300
Urbanwater engineering uncertainties, 504
  analysis of, 505–507
  features, 504
  implications and analysis of, 504–505
  measures of, 505
Urban water infrastructure, 9, 11, 14, 127, 381, 393–424
  life of, 11
  management, 9–10
    challenges, 9
    economic aspect in, 12
    environmental aspect in, 12
    life cycle assessment (LCA), 10
    social aspect in, 12
Urban water management, 17, 45, 184, 185, 187, 233, 296, 393, 407, 410, 427, 460, 532
  demand side management, 37
  supply side management, 37
  traditional challenges, 17
Urban water planning
  linear systems, 33
    closed entity system, 34
    Newtonian paradigm, 32–34
    reductionism, 34
  nonlinear systems, 34
    communicative planning, 37
    Holistic–Ecological paradigm in, 34
    modernist planning, 37
Urban water pricing, ix
Urban water supply, 128, 210, 297, 396, 401, 427, 432, 486, 539, 541, 545, 546, 565
  compound impacts, 544
  direct impacts, 542–543
  indirect impacts, 543–544
  reduction of degraded, 127
Urban water system, 10, 11, 12, 134, 186, 224, 225, 339, 403, 410, 412, 416, 421, 428, 432, 460, 468, 469, 502, 510, 523, 524, 526, 527, 528, 542, 548
  components of, 127
  drainage and effluents in, 127
    driving forces controlling, 127
  fire fighting, for, 138
  history of, 128
  objectives of, 127
  power supply guideline, 138
  pressure requirement, 138
  pumping efficiency, 138
  quantitative guidelines for, 138
  storage capacity, 138, 139
    calculation, 141
    factors in, 139
    fire, 140
    operating, 139
    standpipes, 139
  water demand, 135
    forecasting, 139
    variation in, 136
Urban water system readiness, 510
  disaster and scale, 523
  hybrid drought index, 518–523
  public concern, 526
  reliability, design by, 523–524
  water supply reliability, 524–526
  WDS readiness, evaluation of, 511–517
US-EPA method, 323–325, 324
U.S. Geological Survey (USGS), 84
USGS. *See* U.S. Geological Survey (USGS)
USLE. *See* Universal Soil Loss Equation (USLE)
Utility function, 230, 444, 445, 446, 447, 449, 450, 451, 457, 458, 542
UWC. *See* Urban water cycle (UWC)
UWD. *See* Urban water disaster (UWD)
UWDM. *See* Urban water demand management (UWDM)
UWDM strategies, 183, 189, 192, 232, 234

V

Valve, 143, 151, 152, 177, 203, 433, 515
Vapor, 8, 549
Variable, 55, 540, 541, 543, 545, 548, 549, 552, 554, 556, 557, 558, 559, 560, 561, 562, 563, 568, 578, 579
Variable-speed pumps, 139
Variance, 343, 344, 359, 500, 501, 506
Vector, 348, 352, 353, 354, 355, 358, 372, 388, 443, 501
VENSIM, 43, 429
Ventana System Inc., 429
Verification, 319, 326, 417, 419
Virus, 318, 330
Vision, 17, 29, 30, 50, 296, 457, 472, 473, 474, 477, 533, 534
VOCs. *See* Volatile organic compounds (VOCs)
Volatile organic compounds (VOCs), 8
Vulnerability, 292, 294, 464, 470, 478, 484, 503–504, 508, 514, 517, 545
    analysis, xv 504, 545, 578
    components, 508
    definition, 503
    evaluation, 545
    and risk management, 508–509
    ways of estimation, 503
Vulnerable area, 464, 465, 466, 529

W

W&WWO. *See* Water and wastewater organization (W&WWO)
Wall decay, 165, 166
Walls, 144, 173, 243, 275, 276, 281, 416, 466
Washoff, 7, 295, 303, 325, 326, 405, 467
Wastewater, 4–5, 211, 212, 214, 225, 292–301, 316–317, 332, 393, 400–408, 435, 436, 438, 445, 447–453, 459, 533, 544
    collection and treatment, 5, 400–402
    collection system, 5, 9, 404, 428
    Infrastructure, 20, 400, 401, 405
    management, in urban drainage system, 295–300
    reuse, 533
    treatment, 193, 296, 298–299, 317, 362, 400, 401–405, 411, 413, 544
    treatment plant (WWTP), 297, 298, 299, 317, 400, 403, 404, 544
Water
    consumption
        calculation, 137
        global, 128
        loss calculation, 130
        urban population effect on, 130
    consumptive use of, 135
    conveyance loss of, 135
    delivery and release of, 135
    demand, 135
        forecasting, 139
        variation in, 136
    instream use of, 135
    offstream use of, 135
    per capita consumption
        domestic sector, in, 138
        limiting methods, 138
        measuring methods, 138
    pollution, 163, 544
    public supply of, 135
    quality
        criterion definition, 138
        monitoring, 171
        standard definition, 137
    reclaimed, 135
    resources for supply, 131
    return flow of, 135
    safety measures
        in distribution system, 133
    self-supplied, 135
    supply
        challenges, 134
        in different continents, 130
        at fire hydrant, 138
        pressure requirement, 138
    sustainable resource development, 130
    treatment
        common treatment, 133
        contact tank capacity calculation, 134
        point of consumption treatment, 133
        required chlorine calculation, 134
        unit design calculation, 133
    use
        in fire flow, 136–137, 138
        fluctuation, 136
        homogeneous areas, in, 135
    withdrawal, 135
Water allocation, 47–48, 360, 388, 445, 447, 452, 453, 524
Water and wastewater organization (W&WWO), 435
Water and wastewater treatment systems interactions, 404
Water conservation, 224
    least cost planning (LCP) in, 224
    mandatory restrictions as, 224
    price change aids, 224
    restriction sensitivity aids, 224
    supply conservation as, 224
Water demand, 134–138, 139, 183–237, 389, 390, 435, 448, 450, 453, 455
    definition, 183, 198
    domestic
        factors in, 199
        in house hold, 195–196
    models
        demand functions curve, 198
        parametric models, 195
        regression analysis, 194–195
Water demand and supply, 189, 207
    demand elasticity in, 207
    price elasticity in, 207, 209–211
Water disaster management strategies, 473
    disaster management experience, 475
    drought disaster management case study, 479–480
    initiation, 476
        disaster preparedness policy, 477–478
        political and governmental commitment, 476–477
        public participation, 478–479
    management of storms and floods case study, 480
        flood characteristics, 481
        response, 481
Water distribution network, 17, 128, 139, 164, 194, 201, 204, 205, 395

Index

Water distribution system (WDS), 5, 136, 164, 167, 195, 200, 429, 433, 435
  components of, 141
  function of, 141
  hydraulics of, 144–171
  infrastructure for, 395–399
  levels in, 141
  metering devices in, 143
  pumps in, 143
  readiness, evaluation of, 511–517
  storage tanks in, 143
  typical branch in, 142
  typical looped network in, 142
Water governance, 14, 17, 19–20, 29, 50, 457, 533
  effects of poor, 20
  hybrid models, 21
  models, 21
    for local public utility services, 22
  public transparency in, 20
  at regional level, 20–21
Water infrastructure, 9–12, 17, 20, 393–424, 533
Water losses, 200
Water loss reduction, 199, 200
  unaccounted water, 200
    reasons, 200
Water mains, 143, 311, 334, 393, 396, 397, 416, 417, 418, 419, 421
  design and construction, 396–399
Water management, 1, 17, 18, 21, 23–25, 27–34, 37, 50, 184–188, 295–297, 452, 457
  critical functions in
    operations, 486
    planning, 486
Water per capita, 138, 578, 579
Water pollution, 38–41, 163, 316, 447, 544
  control conflict resolution, 447–452
Water price, 49, 193, 207, 209, 235, 388
Water pricing, 206
  factors in, 207–211
  opportunity cost in, 206
  price and value in, 206
  supply cost in, 206
Water quality, 38–41, 44, 128, 131, 132, 134, 162–175, 243–245, 294, 303, 330, 333, 334, 449
  management, 530
  monitoring, 171, 418
    continuous monitoring, 171
    factors in, 172
    grab samples, 171
  urbanization impacts on, 315–326
    pollutant resources, 316–319
    stormwater, modeling of, 319–326
Water quality modelling
  advection–dispersion transport, 163
    conservation of mass in, 164, 169
  chlorine decay
    booster chlorination stations, 169
    bulk decay rate, 165
    overall decay rate, 166–168
    wall decay rate, 165
  contaminant propagation equations
    discrete volume method (DVM), 169
    EPANET, 170
    event-driven method (EDM), 165, 169, 171
    finite difference method (FDM), 169
    time-driven method (TDM), 169
  event-driven method (EDM), 165, 169, 171
    advantages of, 171
    principle of, 171
  spatial and temporal distribution, 170
    QUALNET, 170–171
  statistic model, 168–169
  steady-state model, 166–168
  tracer study, 172–175
    application of, 172–173
    conservative chemicals used in, 172, 173–174
    injection procedures in, 174
    objective of, 172
    phases of, 173
    principle of, 172
    tracer concentrations monitoring, 174
    tracer selection in, 173–174
    water system operation and, 175
Water quantity, 137–138, 437
Water quantity standards, 137, 138
Water-related disasters, 463, 471
Water resources assessment, 18, 44, 45, 51, 542
  climatic impact on, 5
  cultural aspects in, 9
  greenhouse effect impact on, 8
  hydrologic effects in, 6–7
  qualitative aspects in, 7
Water resources systems
  risk-based design of, 509
  effects, 542
Water reuses, 127, 211–214, 533
Water Sensitive Urban Design (WSUD), 17, 18–19, 131
  benefits of, 17
  BPPs in, 19
  main goal of, 17
Watershed, 1
  area, 84, 104
  effect on peak discharge, 87
  espey 10-min UH calculation for, 91
  espey UH for, 92
  geomorphological impacts, 326–327
  inlet flow determination, 83
  runoff estimation in
    coefficient method for, 84
    regression method for, 84
Water shortage/drought
  disaster assessment on, 528
Water shortages, 37, 465, 467, 514, 520, 528, 540
Water supply, 20–22, 25–27, 127–179, 183, 205, 215–217, 225–227, 230, 233, 297, 390, 393–396, 401, 404, 406, 423, 427, 435, 436, 438, 449–450, 452, 479, 480, 483, 484, 486, 524–530, 539, 541–546, 565
  and energy production sector, 446
  management, 184–185
    resource alternatives for, 226
  reliability indicators, 524–526
  and wastewater collection systems, 404
Water Supply Authority, 191, 450
Water use, 9, 11, 130–131, 135–136, 163, 186, 187, 192–195, 204, 205, 233, 236, 388–390, 397, 471, 480, 518, 530, 532
  classifications, 185, 194

consumptive, 183
efficiency by return flow, 184
efficiency task force, 233
elastic ties of demand in domestic, 210
forecasting, 194
instream, 183
nonconsumptive, 183
offstream, 184
per-capita, 192
urban, 192
Water users, 486
consumptive uses, 486
instream flows/ecological users, 486
WDS. *See* Water distribution system
Weibul, 342, 344, 486, 487, 488, 520
distribution, 342, 486, 487, 488
Weighting
function, 556
method, 374, 415

Wetlands, 298
Wetting front, 60, 61
Widespread contamination, 463, 464, 468, 528, 531
Willingness to pay, 206, 381
Withdrawal, 130, 184, 187, 233, 312, 454, 455, 456, 470, 520
Work plan, 42, 173, 473, 474, 478
WSUD. *See* Water Sensitive Urban Design (WSUD)
WWTP. *See* Wastewater treatment plant (WWTP)

X

Xeriscaping, 480

Z

Zero-sum, 441